Semiconductor Physics

Karlheinz Seeger

Ludwig Boltzmann-Institut für Festkörperphysik, Wien
and
Institut für Angewandte Physik
der Universität Wien

Springer-Verlag
Wien New York

Sponsored by the Ludwig Boltzmann-Gesellschaft
zur Förderung der wissenschaftlichen Forschung in Österreich

Library of Congress Catalog Card Number 73-10413
Printed in Austria by Bors & Müller, A-1010 Wien

Logo design: Peter Klemke, Berlin

With 372 Figures

ISBN 3-211-81186-9 Springer-Verlag Wien-New York
ISBN 0-387-81186-9 Springer-Verlag New York-Wien

Preface

This book has been designed primarily as a text book for a three-semester, three-hour per week senior or graduate course in semiconductor physics for students in electrical engineering and physics. It may be supplemented by a solid state physics course. Prerequisites are courses in electrodynamics and - for some of the chapters - basic quantum mechanics. Emphasis has been laid on physical rather than technological aspects. Semiconductor physics is in fact an excellent and demanding training ground for a future physicist or electrical engineer giving him an opportunity to practice a large variety of physical laws he was introduced to in the more fundamental courses.

A detailed treatment of the transport and optical properties of semiconductors is given. It was decided to omit the usual description of the material properties of certain semiconductors and instead to include the "in-between" equations in mathematical derivations which I hope will make life simpler for a non-theoretician. In view of the many thousands of papers which appear every year in the field of semiconductor physics and which are distributed among more than 30 journals, it would have been impossible for a single person to write a comprehensive book unless there had not been some excellent review articles on special topics published in the series "Solid State Physics", "Festkörper-Probleme / Advances in Solid State Physics", "Semiconductors and Semimetals", and "Progress in Semiconductors", and I have leaned heavily on such review articles.

All equations are written in MKS rationalized units, although numerical examples are given in units of cm instead of m and in units of g instead of kg as it is normal practice in semiconductor physics. It is certainly very important for a student to be able to occasionally check the theoretical treatment of a problem by performing a dimensional analysis which is much simpler in the MKS system than in the cgs system. I hope the theoreticians among my readers will forgive me for breaking with their tradition.

The book has been set on an IBM 72 Composer which proved to be quite satisfactory. However, I would like to draw the attention of the reader to a point which arises from the design of the founts: the letter l very closely resembles to figure 1. Even so, I hope there will be no confusion since in cases of possible doubt the meaning of a symbol is always explained and a list of important symbols is given at the beginning. This will also be helpful where letters have several meanings or - as for "frequency" - different symbols are well-established in electricity ("1/f noise") and optics (greek letter ν).

Problems to be solved by the student have been omitted in view of the existence of excellent "problems-and-solutions" books such as "Problems in Solid State Physics" (H. J. Goldsmid, ed., New York: Academic Press. 1969) which contains three chapters on semiconductor physics, and "Aufgabensammlung zur Halbleiterphysik" by W. L. Bontsch-Brujewitsch et al. (Basel, Switzerland: C. F. Winter, and Braunschweig, Germany: Vieweg. 1970). The student is strongly urged to try his newly acquired skills on those problems. For further studies the Proceedings of the biannual International Conference on the Physics of Semiconductors are recommended.

The book will, of course, also be useful to scientists and engineers involved in research and development; over 650 references have been included in the text.

I am indebted to my colleague Dr. M. W. Valenta, formerly with Prof. Dr. J. Bardeen and with Los Alamos Scientific Laboratories, for reading the manuscript and giving valuable advice for the English presentation. I am particularly grateful to Dr. M. Kriechbaum, University of Graz, Austria, for helpful criticism. Further valuable comments and suggestions came from Prof. Dr. K. Baumann and Dr. P. Kocevar, both also at the University of Graz, from Prof. Dr. H. W. Pötzl, Technical University, Vienna, and from the colleagues at my institute. I am happy to thank Frau Jitka Fucik for her invaluable services in setting up the book in type, to Frau Viktoria Köver for drawing the 372 figures, and to Dr. H. Kahlert for computer and photographic services. Last, not least, I would like to take the opportunity to express my gratitude to my wife Lotte for her patience during the preparation of the manuscript.

Vienna, June 1973 K. S.

Contents

List of Important Symbols

Subscripts : n = electrons; p = holes; i = intrinsic conduction; L light, H heavy hole; l = longitudinal; t = transverse; – right-hand, + left-hand polarized light

Superscripts : x = neutral; ± singly ionized; * in compressed coordinate system

$\vec{a}\vec{b}$ and $(\vec{a}\vec{b})$ are scalar products, $[\vec{a}\vec{b}]$ is the vector product ("cross product")

---

a	Lattice constant (p. 11)
$\vec{a}$	Primitive translation vector of elementary cell (p. 22); polarization of radiation (unit vector) (p. 340)
ac	Alternating current
a_B	Bohr radius
A	Acceptor (p. 7)
$\vec{A}$	Vector potential; primitive translation vector of reciprocal lattice (p. 22)
A_o	Amplitude of vector potential (p. 340)
A_s	Sound amplitude (p. 175)
Å	Angström unit (equal to 10^{-8} cm)
ac	Alternating current
b	Mobility ratio μ_n/μ_p (p. 60); width of a sample
$\vec{b}$	Primitive translation vector of elementary cell (p. 22)
$\vec{B}$	Magnetic induction; primitive translation vector of reciprocal lattice (p. 22)
B_{12}, B_{21}	Einstein coefficients (p. 437)
c	Velocity of light in free space
$\vec{c}$	Primitive translation vector of elementary cell (p. 22)
c_l	Average longitudinal elastic constant (p. 177)
c_v	Lattice specific heat at constant volume (p. 90)
cw	Continuous wave
C	Unit cell; capacity (p. 139); intrinsic carrier density constant (p. 5)
$\vec{C}$	Primitive translation vector of reciprocal lattice (p. 22)
d	p-n junction width (p. 137); emitter collector distance (p. 135); sample thickness

dc	Direct current
D	Donor (p. 6); diffusion coefficient (p. 130, 161); spring constant (p. 193); optical deformation potential constant (p. 209)
$\vec{D}$	Dielectric displacement vector (p. 190)
D_ρ	Matrix of rotation (p. 239)
e	Electronic charge (e^- electron; e^+ hole) (p. 4)
$\vec{e}$	Unit vector in direction of $\vec{E}$ (p. 253); deformation vector or tensor (p. 114)
e_C	Callen effective charge (p. 201)
e_{pz}	Piezoelectric constant (p. 190)
e_S	Szigeti effective charge (p. 201)
$\vec{E}$	Electric field intensity
E_o	Effective field strength for polar optical scattering (p. 218)
E_s	Electric field intensity at the surface barrier (p. 152)
f	Distribution function (p. 40); van-der-Pauw factor (p. 485); rf frequency (p. 126)
f_o	Zero-field distribution function (p. 54); spherically symmetrical part of distribution (p. 204)
f_1	Drift term of distribution in the diffusive approximation (p. 202)
f_t	Number of filled traps or acceptors relative to total number (p. 142)
f_{vc}	Oscillator strength for interband transition (p. 371)
F	Free energy (p. 39)
$\vec{F}$	Electrothermal field intensity (p. 64); force (p. 15)
g	Spin degeneracy of impurity level (p. 37); density of states (p. 41)
g_j	Degree of degeneracy of level (p. 36)
G	Generation rate (p. 131)
h	Planck's constant; $\hbar = h/2\pi$
$\vec{h}_\rho$	Unit vector to ρ-th valley (p. 253)
H	Hamilton operator
$\vec{H}$	Magnetic field strength
H_{mn}	Matrix element of the perturbing potential H_1 between the states m and n (p. 168)
I	Current; light intensity (p. 421)
Im	Imaginary part
$\vec{j}$	Current density (p. 52)
j_s	Saturation current density of reverse biased p-n junction (p. 142)
j_t	Laser threshold current density (p. 440)
J	Acoustic energy flux (p. 280)
k	Extinction coefficient (p. 332)
$\vec{k}$	Wave vector of a carrier (p. 12); $\vec{k}'$: after scattering (p. 168); k_E in the electric field direction (p. 206)
k_B	Boltzmann's constant

K	Electromechanical coupling coefficient (p. 191); anisotropy coefficient K_m/K_τ (p. 248)
K_m	Effective mass ratio m_l/m_t (p. 248)
K_τ	Ratio of momentum relaxation times τ_l/τ_t (p. 248)
l	Length of a sample; effective Franz-Keldysh length (p. 352); phonon mean free path (p. 99)
l_{ac}	Mean free path of carrier for acoustic scattering under equipartition (p. 53)
L	Diffusion length (p. 132); Lorenz number (p. 86)
L_D	Debye length (p. 132)
m	carrier effective mass (p. 9); elasto-resistance tensor (p. 114)
m_d	Density-of-states effective mass (p. 289)
m_o	Free electron mass
m_r	Reduced effective mass (p. 344)
m_H	Hall effective mass (p. 243)
m_M	Magnetoresistance effective mass (p. 406)
m_σ	Conductivity effective mass (p. 289)
M	Atomic mass (p. 193)
n	Electron concentration (p. 5); real refractive index (p. 332); magnetic quantum number (p. 307)
n_i	Intrinsic carrier concentration (p. 5)
n_s	Electron concentration at the surface barrier (p. 152)
N	Number of atoms in a crystal (p. 35); density of scattering centers (p. 164); number of equivalent valleys (p. 252); complex refractive index (p. 489)
N_A	Acceptor concentration (p. 7)
N_c	Density-of-states factor at the conduction band edge (p. 42)
N_D	Donor concentration (p. 6)
N_I	Concentration of ionized impurities (p. 172)
N_q	Number of phonons with wave vector $\vec{q}$ at thermal equilibrium (p. 177)
N_u	Number of unit cells in a crystal (p. 198)
p	Hole concentration (p. 5); impact parameter (p. 170)
$\vec{p}$	Crystal momentum operator (p. 286)
P	Kane parameter (p. 16)
$\vec{P}$	Dielectric polarization (p. 198)
P_E	Ettingshausen coefficient (p. 106)
P_H	Planar Hall coefficient (p. 84)
q	Ratio of ionized impurity scattering and acoustic deformation potential scattering (p. 184)
$\vec{q}$	Wave vector of phonon (p. 176)
Q	Heat produced per unit time per unit volume (p. 101)
Q_N	Nernst coefficient (p. 105)

Q_{Th}	Thomas heat (p. 102)
r	Reflection coefficient (r_∞: for vertical incidence) (p. 489); exponent in $\tau_m \propto \epsilon^r$ (p. 53)
$\vec{r}$	Position vector
r_c	Collector impedance (p. 148)
r_e	Emitter impedance (p. 148)
r_H	Hall factor (p. 69, 243)
R	Resistance (p. 79); reflectivity of a plate (p. 490); absorption rate of phonon flux (p. 339)
Re	Real part
R_B	Resistance in a magnetic field (p. 79)
R_H	Hall coefficient (p. 62)
Ry	Rydberg energy
s	Surface recombination velocity (p. 156)
s_{RL}	Righi-Leduc factor (p. 111)
S	Scattering rate (p. 169) (S_+ for emission, S_- for absorption of a phonon); entropy (p. 39); entropy transport parameter (p. 92); strain (p. 191)
$\vec{S}$	Poynting vector (p. 340)
S(E)	Conductivity ratio $\sigma(E)/\sigma(0)$ in a symmetry direction of the many-valley model (p. 253)
S_{RL}	Righi-Leduc coefficient (p. 111)
t	Time
T	Lattice temperature; stress, tension (p. 191); transmittivity of a plate (p. 490);
T_e	Carrier temperature (p. 118)
T_M	Magnetoresistance scattering coefficient (p. 75)
$\vec{u}_s$	Sound velocity (p. 280)
U	Internal energy (p. 39)
$\vec{v}$	Carrier velocity (p. 54)
$\vec{v}_d$	Drift velocity (p. 52); v_{ds}: saturation value (p. 215)
V	Voltage or potential; crystal volume
V_D	Diffusion voltage in a p-n junction (p. 137)
w	Transistor base width (p. 146)
$\vec{w}$	Heat flow density (p. 86, 100)
w_1	Acoustic rate constant (p. 250)
w_2	Intervalley rate constant (p. 250)
W	Thermodynamic probability (p. 39)
x_c	Collector base width (p. 146)
x_e	Emitter base width (p. 146)
X	Hydrostatic pressure (p. 113)
X_{kl}	Stress tensor (p. 114)
Z	Ionic charge in units of e (p. 170); thermoelectric figure of merit (p. 105)

α	Tensor of inverse effective mass in units of m_σ (p. 238); current amplification factor (p. 147); absorption coefficient (p. 490); polar constant (p. 218)
β	$(\mu - \mu_0) / \mu_0 E^2$ (p. 118); laser gain factor (p. 440)
γ	Anisotropic part of $(\mu - \mu_0) / \mu_0 E^2$ (p. 257)
Δ	Kane parameter (p. 16)
ϵ	Carrier energy (p. 9)
ϵ_{ac}	Acoustic deformation potential constant (p. 176)
ϵ_A	Acceptor energy $\epsilon_v + \Delta\epsilon_A$ (p. 7)
ϵ_c	Conduction band edge (p. 4)
ϵ_D	Donor energy $\epsilon_c - \Delta\epsilon_D$ (p. 6)
ϵ_{exc}	Exciton energy (p. 344)
ϵ_F	Faraday ellipticity (p. 402)
ϵ_G	Energy gap, $\epsilon_c - \epsilon_v$, (p. 4)
ϵ_{ik}	Deformation tensor (p. 234)
ϵ_n	Energy eigenvalues (p. 307)
ϵ_t	Trap energy (p. 434)
ϵ_v	Valence band edge (p. 4)
ζ	Fermi energy (p. 39); ζ_n: relative to the conduction band (p. 91); ζ_p: relative to the valence band (p. 91)
ζ_n^*, ζ_p^*	Quasi Fermi levels (p. 144)
η	Luminescence efficiency (p. 434)
θ	Scattering angle $\measuredangle(\vec{k}, \vec{k}')$ (p. 170); $\measuredangle(\langle 001 \rangle, \vec{E} \text{ or } \vec{B})$
θ'	Scattering angle $\measuredangle(\vec{a}, \vec{q})$ (p. 377)
Θ	Absolute thermoelectric force (p. 93); optical phonon (Debye) temperature (p. 196)
θ_F	Faraday angle (p. 401)
Θ_i	Intervalley phonon temperature (p. 250)
θ_V	Voigt angle (p. 406)
κ	Static relative dielectric constant (p. 198); thermal conductivity (p. 86); compressibility (p. 117)
$\overline{\kappa}$	Complex relative dielectric constant (κ_r real, $-\kappa_i$ imaginary part) (p. 491)
κ_0	Permittivity of free space
κ_{opt}	Optical relative dielectric constant (p. 198)
λ	T/T_e (p. 184); wavelength $= 2\pi\lambda\!\!\!^-$ (p. 489)
Λ_ρ	Matrix for transformation in $\vec{k}$-space (p. 241)
μ	Mobility (p. 52)
μ_0	ohmic (p. 118)
μ_{ac}	acoustic (p. 184)
μ_H	Hall (p. 71)
μ_{Ion}	ionized impurity (p. 184)
μ_M	magnetoresistance (p. 244)

μ_{FE}	field effect (p. 455)
μ_B	Bohr magneton (p. 310)
μ_o	Permeability of free space = $1/\kappa_o c^2$
μ_{Th}	Thomson coefficient (p. 102)
ν	Frequency
Ξ_d, Ξ_u	Deformation potential constants for ellipsoidal constant energy surfaces (p. 235)
π_{ik}	Piezoresistance (p. 113)
Π	Peltier coefficient (p. 102)
ρ	Resistivity (p. 79); mass density (p. 90)
ρ_B	Resistivity in a magnetic field (p. 79)
σ	Electrical conductivity (p. 52); scattering cross section (p. 164)
σ_w	Conductivity tensor (p. 68)
τ	Lifetime (p. 143)
τ_d	Dielectric relaxation time (p. 132)
τ_i	Intervalley scattering time (p. 250)
τ_m	Momentum relaxation time (p. 52)
τ_r	Radiative lifetime in the upper level of a laser (p. 437)
τ_ϵ	Energy relaxation time (p. 118)
φ	Work function in units of e (p. 448); azimuth
ϕ	Acoustic potential (p. 279)
Φ	Electrostatic potential (p. 137)
ψ	Electron wave function (p. 9); angle $\sphericalangle (\vec{j}, \langle 001 \rangle$ direction) (p. 254)
ω	Angular frequency
ω_c	of cyclotron resonance (p. 292)
ω_e	eigenfrequency of oscillator (p. 363)
ω_l	of longitudinal phonon (p. 198)
ω_o	of optical phonon (p. 208)
ω_p	of plasma (p. 363)
ω_s	of phonon (p. 175)
ω_t	of transverse phonon (p. 198)
Ω	Solid angle

1. Elementary Properties of Semiconductors

1a. Insulator-Semiconductor-Semimetal-Metal

A consequence of the discovery of electricity was the observation that metals are good conductors while non-metals are poor conductors. The latter were called insulators. Metallic conductivity is typically between 10^6 and $10^4\,(\text{ohm-cm})^{-1}$ while typical insulators have conductivities of less than $10^{-10}\,(\text{ohm-cm})^{-1}$. Some solids with conductivities between 10^4 and $10^{-10}\,(\text{ohm-cm})^{-1}$ are classified as "semiconductors". However, substances such as alkali-halides whose conductivity is due to electrolytic decomposition shall be excluded. Also we restrict our discussion to chemically uniform, "homogeneous" substances and prefer those which can be obtained in monocrystalline form. Even then we have to distinguish between semiconductors and semimetals. This distinction is possible only as a result of thorough investigation of optical and electrical properties and how they are influenced by temperature, magnetic field, etc. Without giving further explanations at this time the statement is made that semiconductors have an "energy gap" while semimetals and metals have no such gap. However, very impure semiconductors show a more or less metallic behavior and with many substances the art of purification by e.g. zone refining [1] is not so far advanced that a distinction can easily be made. The transition between semiconductors and insulators is even more gradual and depends on the ratio of the energy gap to the temperature of investigation. Very pure semiconductors become insulators when the temperature approaches the absolute zero.

Typical element semiconductors are germanium and silicon. An inspection of the periodic table of elements reveals that these materials belong to the fourth group while typical metals such as the alkalis are in the first group and typical non-metals such as the halides and the noble gases which crystallize at low tem-

[1] H. Schildknecht: Zonenschmelzen. Weinheim/Germany: Chemie. 1964; A. Holden and P. Singer: Crystals and Crystal Growing. Watertown, Mass.: Educational Services, Inc. 1960. Growth of Crystals (A. V. Shubnikov and N. N. Sheftal', eds.), several volumes. New York: Consultants Bureau. 1962 and following years; M. Tanenbaum: Semiconductors (N. B. Hannay, ed.), p. 87-144. New York: Reinhold Publ. Co. 1959.

perature are in the seventh and eighth group, respectively. Other semiconducting elements in the fourth group are diamond which is a modification of carbon, and "grey" tin (α-Sn) which is stable only at low temperatures. All fourth-group semiconductors crystallize in a structure known as the diamond structure in which neighboring atoms are arranged in tetrahedral symmetry. In the third group the lightest element, boron, and in the sixth group the heavy elements selenium and tellurium are semiconductors. A typical semimetal is the heaviest fifth group element, bismuth, and also the lighter elements of this group, arsenic and antimony, may be classified as such although they are at present less thoroughly investigated.

Typical compound semiconductors are the III-V compounds such as gallium arsenide, GaAs, and indium antimonide, InSb, and the II-VI compounds such as zinc sulfide, ZnS, ("zinc blende"). They cristallize in the zinc blende structure which can be obtained from the diamond structure by replacing the carbon atoms alternatively by e.g. zinc and sulfur atoms. These compounds have a stoichiometric composition, just as e.g. the semiconductor silicon carbide, SiC, while germanium silicon alloys may be obtained as semiconducting mixed crystals for any arbitrary composition. Many metal oxides and sulfides are semiconductors, often with non-stoichiometric composition. Some of them are of technical importance such as cuprous oxide, Cu_2O, (formerly used for rectifiers), lead sulfide, PbS, (for infrared detectors), and the ferrites (iron oxides) for their magnetic properties. Today silicon is mainly used for the fabrication of transistors which serves for amplification of electric signals. This is the most important technical application of semiconductors nowadays.

Semiconduction is specified by the following properties:

1) In a "pure" semiconductor conductivity rises exponentially with temperature ("thermistor" action). At lower temperatures a smaller concentration of impurities is required in order to ensure this behavior.
2) In an impure semiconductor the conductivity depends strongly on the impurity concentration. E.g. nickel oxide NiO in a pure condition is an insulator. By doping (which means intentionally adding impurities) with 1 % lithium oxide, Li_2O, the conductivity is raised by a factor of 10^{13}. In the doped material, however, the conductivity changes only slightly with temperature, just as in a metal.
3) The conductivity is changed (in general: raised) by irradiation with light or high-energy electrons or by "injection" of carriers from a suitable metallic contact (injection will be explained in Chap.5a).
4) Depending on the kind of doping the charge transport may be either by electrons or by so-called "positive holes". The electric behavior of positive holes is the same as that of positrons but otherwise there is no similarity. It is possible to dope a single crystal non-uniformly such that in some parts charge transport is by (negative) electrons and at the same time in others by positive holes. Semiconductor diodes and transistors are single crystals of that kind.

Semiconducting behavior is not restricted to solids. There are liquid semiconductors. However, because of atomic diffusion, regions with different dopings will mix rapidly and a stable device with an inhomogeneous structure is not possible. Recently attention has been paid to glassy and amorphous semiconductors which may possibly find technical application as inexpensive fast switches if their reproducibility can be improved.

As mentioned before, semiconductors become metallic when heavily doped. Superconductivity, known for some metals at very low temperatures, has also been observed with some heavily doped semiconductors. Transition temperatures are below 0.5 K.

Some aromatic hydrocarbons were found to be semiconductors. No practical application has as yet been found for organic semiconductors. Raising the conductivity in these compounds by heating is limited by thermal decomposition.

1b. The Positive Hole

As mentioned in the above section charge transport may be due to "positive holes". In this chapter we shall explain this idea qualitatively by considering a lattice of carbon atoms which for simplicity is assumed 2-dimensional. In Fig.1.1 the bonds between neighboring atoms are covalent. Each C-atom contributes 4 valence electrons and receives from its 4 neighbors 1 electron each so that a "noble gas configuration" with 8 electrons in the outer shell is obtained. This is similar in a 3-dimensional lattice.

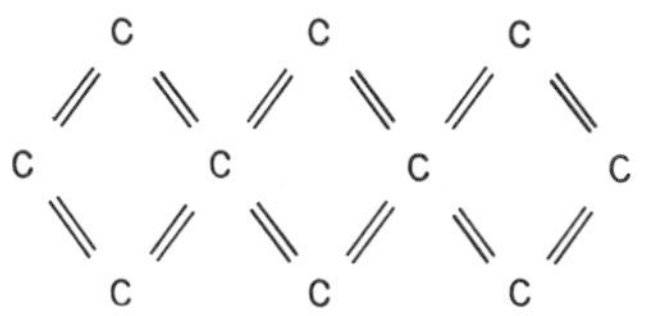

Fig.1.1 Schematic two-dimensional representation of perfect diamond lattice. Each covalent bond (=) represents two valence electrons of opposite spin.

Now imagine an extra electron being transferred somehow into the otherwise perfect diamond crystal. (By "perfect" we also mean that the surface with its free bonds is far enough away from the electron, ideally an infinitely large crystal). No free bonds will be available for this electron where it could attach itself. The electron will move randomly with a velocity depending on the lattice temperature. However, if we apply an external electric field to the crystal, a drift motion will be superimposed on the random motion which in the simplest case will have a direction opposite to the field because of the negative charge of the electron. The extra electron which we call a "conduction electron" makes the crystal "n-type" which means that a negative charge is transported. In practice the extra electron will come from an impurity atom in the crystal.

We can also take a valence electron away from an electrically neutral diamond. The crystal as a whole is now positively charged. It does not matter which of the many C-atoms looses the electron. The main point is that this atom will now replace its lost electron by taking one from one of its neighbors. The neighbor in turn will react similarly with one of its neighbors. This process is repeated over and over again with the result that the hole produced by taking away an electron from the crystal moves in a random motion throughout the crystal just as the extra electron did in the n-type crystal. What happens if now we apply an external electric field? Wherever the hole is, a valence electron will fill it by moving in a direction opposite to the electric field with the effect that the hole is drifting in the direction of the field. This is exactly what one would expect from a positive charge. Since the crystal as a whole is charged positively we may think of this charge as being localized at the position of the hole. In semiconductor physics positive holes are treated as if they were positively charged electrons. Conductivity by positive holes is called "p-type".

For a comparison consider the carbon dioxide bubbles in mineral water. Instead of an electric field there is the gravitational field and instead of an electric charge there is the mass of the water molecules. Since the bubbles drift in a direction opposite to the field direction they can formally be treated like negative mass particles as long as they are in the bulk of the liquid although, of course, carbon dioxide has a positive mass and is subjected to a lift only. Similarly the assumption of positively charged particles called "holes" in semiconductors is a very simple formal description of an otherwise quite involved process; but one should keep in mind that the hole is actually a lacking valence electron, and in case of doubt one has to show that both ways of looking at the problem give the same result.

The picture of conduction processes we have developed so far is oversimplified. It does not show that electrons and holes have different "effective masses". Only by wave mechanical methods shall we obtain a more realistic view of the problem.

1c. Conduction Processes, Compensation, Law of Mass Action

Before becoming involved in wave mechanics we will continue with the classical model investigating thermal pair generation and annihilation. Let us call ϵ_G the binding energy of a valence electron to an atom in the crystal (G stands for "gap" which will be explained in Chap.2). If the energy ϵ_G is supplied thermally a conduction electron may be generated which leaves a hole where the electron has been. The electron and the hole independently from each other move through the crystal. Since we consider the hole as a particle similar to the electron except for the sign of its charge, we have created an electron hole pair. Occasionally a conduction electron will recombine with a hole which actually means that it finds a free bond and "decides" to stay there. The binding energy ϵ_G is transformed either into electro-magnetic radiation ("recombination radiation") or atomic vibrations ("phonons"). From the particle point-of-view, the annihilation of the electron hole pair is usually called "recombination". Denoting electrons by the symbol e^- and holes by e^+, a chemical reaction equation of the form

$$e^- + e^+ \rightleftharpoons \epsilon_G \tag{1c.1}$$

will be an adequate description of the process. Assuming that no radiation is incident, the generation energy ϵ_G is taken from the lattice vibrations. Therefore with increasing temperature the equilibrium is shifted towards the left side of the equation, the number of carriers and therefore the conductivity is increased which is so characteristic of semiconductors. Of course, radiation is incident on the crystal even if it is in thermal equilibrium with its environment. This is the "black-body radiation". It compensates exactly the recombination radiation of energy ϵ_G (see the right-hand side of Eq.(1c.1)).

It is shown in statistical mechanics that a "small system" which is in thermal contact with a "large system" can acquire an energy ϵ_G at a rate proportional to $\exp(-\epsilon_G/k_B T)$, where k_B is Boltzmann's constant and T the absolute temperature (at room temperature $k_B T = 25.9$ meV). In the present case the "small system" is a valence electron and the "large system" the crystal. The exponential is multiplied by a power of T; however, the temperature dependence is determined essentially by the exponential as is well-known from e.g. the law of thermionic emission from metals. For Eq.(1c.1) the power function is T^3 if we apply the law of mass action. Denoting the concentrations of conduction electrons and holes by n and p, respectively, it is

$$np = C.T^3.\exp(-\epsilon_G/k_B T) \tag{1c.2}$$

The value of the constant C depends on the semiconductor material. The form of Eq.(1c.2) is similar to the one describing the concentrations of H^+ and OH^- in water where these concentrations are always small compared with the concentration of the neutral water molecules. In a semiconductor the electron and hole concentrations will also be small relative to the concentrations of atoms because otherwise the conductor would have to be classified as a metal rather. A rigorous derivation of Eq.(1c.2) including a calculation of the constant of proportionality C will be given in Chap.3, Eq.(3a35).

In a pure semiconductor for every conduction electron a hole is also produced, and n = p. We call this "intrinsic conduction" and add a subscript i to n. Eq.(1c.2) yields

$$n_i = C^{1/2}\, T^{3/2} \exp(-\epsilon_G/2k_B T) \tag{1c.3}$$

In Figs.1.2a and 1.2b n_i is plotted vs temperature for silicon ($\epsilon_G = 1.12$ eV at 300 K) and germanium ($\epsilon_G = 0.665$ eV at 300 K). At temperatures above 250 K (for Si) and 200 K (for Ge) ϵ_G varies linearly with temperature

$$\epsilon_G(T) = \epsilon_G(0) - aT \tag{1c.4}$$

the coefficient a being 2.84×10^{-4}/K for Si and 3.90×10^{-4}/ K for Ge. However, this does not change the exponential law, Eq.(1c.3),except for a change in the factor C, since

$$\exp(-\epsilon_G + aT)/2k_B T = \exp(a/2k_B)\exp(-\epsilon_G/2k_B T)$$

Therefore in Eq.(1c.3) we use for ϵ_G its value obtained by extrapolation from the range linear in T to absolute zero. At low temperatures a T^2 term with the same sign as the aT term is found in ϵ_G. One obtains these additional terms from optical investigations.

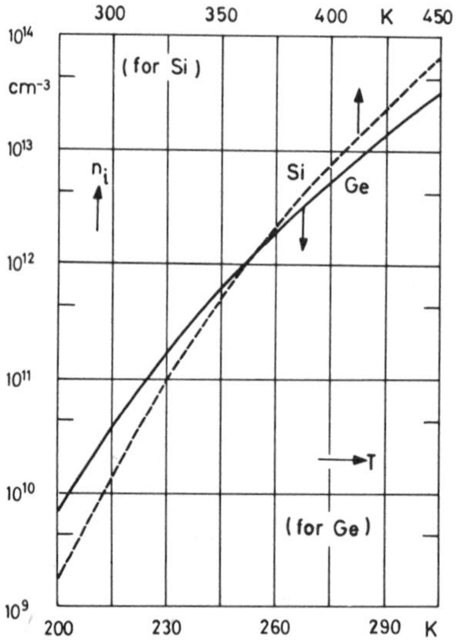

Fig.1.2a Intrinsic carrier concentration as a function of temperature for silicon and germanium: 200 to 305 K (Ge); 275 to 450 K (Si), after R. B. Adler, A. C. Smith, and R. L. Longini: Introduction to Semiconductor Physics. New York: J. Wiley and Sons. 1964.

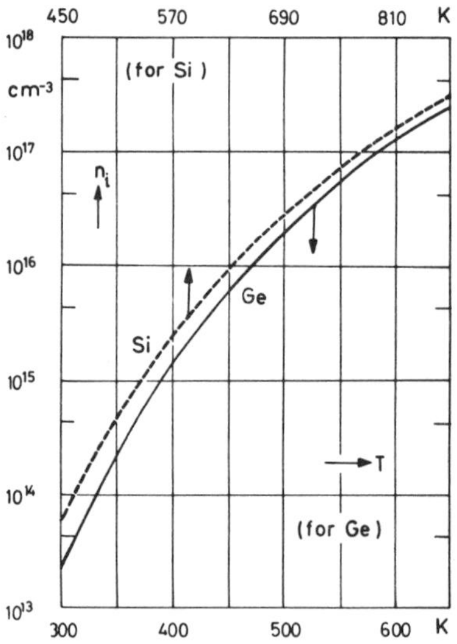

Fig.1.2b Same as Fig.1.2a except for temperature range: 300 to 650 K (Ge); 450 to 870 K (Si).

Now we consider a doped semiconductor. Assume one atom to be replaced by an impurity atom, e.g. a 4-valent C atom by a 5-valent phosphorous atom. Only 4 of the 5 valence electrons of the phosphorus are required to bind the 4 neighboring C atoms. The fifth electron is bound very losely. The binding energy $\Delta\epsilon_D$ of the fifth electron is considerably lower than the binding energy ϵ_G of a valence electron to a C atom. An impurity which releases one or several electrons in this way is called a "donor" D. If we denote a neutral donor by D^x and a singly-ionized donor by D^+, the equation of reaction is

$$e^- + D^+ \rightleftharpoons D^x + \Delta\epsilon_D \tag{1c.5}$$

At high temperatures all donors are thermally ionized and the concentration n of conduction electrons is equal to that of donors, N_D. The concentration p of holes is given by

$$np = n_i^2 \tag{1c.6}$$

where n_i is given by Eq.(1c.3). Charge carriers which are in the minority are called "minority carriers". At still higher temperatures n_i will become larger than N_D and the semiconductor will then be intrinsic. The temperature range where

$n = N_D$ independent of temperature, is called "extrinsic" *). Fig.1.3 gives a schematic diagram of carrier concentration vs the inverse temperature. At the right end of the curve where temperature is low, the balance of Eq.(1c.5) is shifted towards the right-hand side. There is carrier "freeze-out" at the donors. The application of the law of mass action again yields an essentially exponential temperature dependence with $\Delta\epsilon_D$ in the exponent. Because $\Delta\epsilon_D$ is smaller than ϵ_G the slope of the curve is less steep than in the intrinsic range.

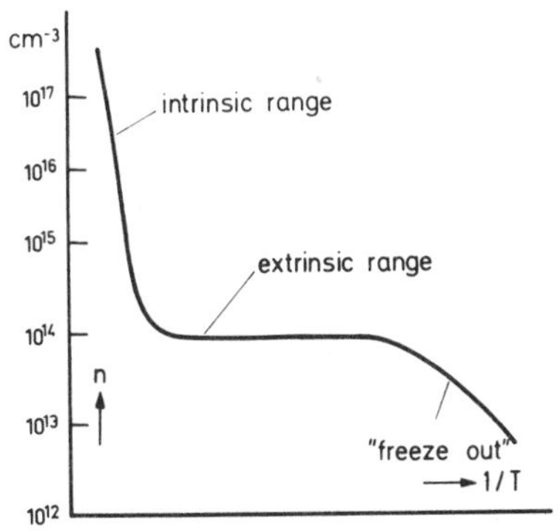

Fig.1.3 Schematic diagram of the carrier concentration as a function of the reciprocal of the temperature.

If in diamond a 4-valent C atom is replaced by a 3-valent boron atom a valence electron is lacking. Supplying an energy $\Delta\epsilon_A$, an electron from a C atom is transferred to the B atom and a mobile hole is thus created. The B atom becomes negatively charged. Boron in diamond is an "acceptor" A. Denoting the neutral acceptor by A^x and the singly-ionized acceptor by A^- the chemical reaction equation can be written in the form

$$e^+ + A^- \rightleftharpoons A^x + \Delta\epsilon_A \qquad (1c.7)$$

The temperature dependence of the hole concentration p is similar to that of n in n-type semiconductors.

If both donors and acceptors are distributed at random in the semiconductor we have a "compensated" semiconductor. At ideal compensation there would be equal numbers of donors and acceptors. Quite often a high-resistivity material supposed to be very pure is compensated. A convenient method of compensation of acceptors is the lithium drift process. The positive lithium ion Li^+ has the helium configuration of the electron shell and diffuses nearly as easily through some solids as helium gas. Diffusion takes place via the interstitial mechanism. The negatively charged acceptors attract the positively charged lithium ions forming ion pairs with the result of ideal compensation. In a p-n junction during diffusion an external electric field is applied such that ion drift produces a p-i-n junction with a large perfectly compensated intrinsic region between the p and the n regions (e.g. a type of γ-ray counters are produced by this method [1]).

The normal case is partial compensation of impurities. Whether $N_D > N_A$ or vice versa the semiconductor is n- or p-type, respectively.

[1] See e.g. G.Bertolini and A.Coche: Semiconductor Detectors. Amsterdam: North-Holland Publ.Co. 1968; G.Dearnally and D.C.Northrop: Semiconductor Counters for Nuclear Radiations. London: Spon. 1966.

*) For a "partly compensated" semiconductor (which is actually the normal case, see below) N_D should by replaced by $N_D - N_A$.

Compound semiconductors with non-stoichiometric composition are n- or p-type depending on which component of the compound is excessive. Assume that in e.g. CdS one S^{--} ion is lacking somewhere in the crystal. There is then an excessive Cd^{++} ion which will be neutralized by two localized electrons. Therefore the lattice vacancy acts as an acceptor and makes the semiconductor p-type.

In small-band-gap polar semiconductors like PbTe and HgTe the free carriers are mainly produced by deviations from stoichiometric composition (while in large-gap ionic solids, such as sodium chloride, the gap energy is larger than the energy to generate a vacancy which compensates the ionized impurities and therefore makes the crystal insulating; this is called "self-compensation").

2. Energy Band Structure

The "energy band structure" is the relation between energy and momentum of a carrier in a solid. For an electron in free space the energy is proportional to the square of the momentum. The factor of proportionality is $1/2m_o$ where m_o is the free electron mass. In the "simple model of band structure" the same relation between energy and momentum is assumed except that m_o is replaced by an "effective mass" m. This may be larger or smaller than m_o. Why this is so will be seen later in this chapter. Quite often the band structure is more complex and can be calculated even with electronic computers only semiempirically. A short description of some typical band structures will be given in Chap.2d and used for the calculation of charge transport in Chap. 7 and 8 while in Chap. 4 and 5 the transport properties will be calculated assuming the simple model of band structure (which is quite a good approximation for most purposes).

2a. Single and Periodically Repeated Potential Well

A charge carrier in a crystal passing an atom first is subject to an acceleration and then, when leaving the atom, to a deceleration until it gets into the field of the next atom and these acceleration processes are repeated. The crystal field can be approximated by a periodic array of potential wells. The calculation is simplified if each minimum is assumed to be a square well. This is the one-dimensional "Kronig-Penney model" of the crystal [1].

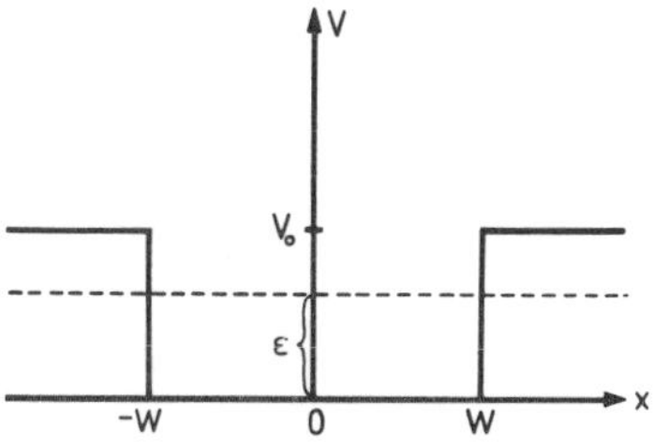

Fig.2.1 Rectangular potential well.

As a preliminary study we will calculate the energy levels of a particle in a single potential well [2] (Fig.1). The depth of the well is V_o and its width b. The following notation is introduced

$$a^2 = 2m\hbar^{-2}\epsilon; \; \beta^2 = 2m\hbar^{-2}(V_o - \epsilon) \quad (2a.1)$$

Inside the well where $V(x) = 0$, the Schrödinger equation takes the form

$$d^2\psi/dx^2 + a^2\psi = 0; \; -b^2/2 \leqslant x \leqslant b/2 \quad (2a.2)$$

[1] R.de L.Kronig and W.J.Penney, Proc.Roy.Soc.(London) A130 (1930) 499.
[2] This has been adapted from A.Nussbaum: Semiconductor Device Physics. sec.1.9 and 1.13. Englewood Cliffs, N.J.:Prentice-Hall Inc. 1962.

while outside the well where $V(x) = V_o$ it is

$$d^2\psi/dx^2 - \beta^2\psi = 0 \;; \qquad x < -b/2 \quad \text{and} \quad x > b/2 \tag{2a.3}$$

The boundary condition is that the position probability density of the particle $|\psi|^2$ vanishes at infinity. The solutions are

$$\begin{aligned} \psi &= Ce^{i\alpha x} + De^{-i\alpha x} && \text{for } -b/2 \leqslant x \leqslant b/2 \\ \psi &= Ae^{-\beta x} && \text{for } x > b/2 \\ \psi &= Ae^{\beta x} && \text{for } x < -b/2 \end{aligned} \tag{2a.4}$$

where A, C and D are integration constants.
The continuity condition for ψ and $d\psi/dx$ at $x = b/2$ yields

$$\begin{aligned} Ae^{-\beta b/2} &= Ce^{i\alpha b/2} + De^{-i\alpha b/2} \\ -\beta Ae^{-\beta b/2} &= i\alpha Ce^{i\alpha b/2} - i\alpha De^{-i\alpha b/2} \end{aligned} \tag{2a.5}$$

Eliminating $e^{-\beta b/2}$ we have

$$-\beta Ce^{i\alpha b/2} - \beta De^{-i\alpha b/2} = i\alpha Ce^{i\alpha b/2} - i\alpha De^{-i\alpha b/2} \tag{2a.6}$$

Hence the ratio D/C is obtained

$$D/C = (\beta+i\alpha)\, e^{i\alpha b}/(-\beta+i\alpha) \tag{2a.7}$$

Similarly the continuity condition at $x = -b/2$ yields

$$D/C = (-\beta+i\alpha)\, e^{-i\alpha b}/(\beta+i\alpha) \tag{2a.8}$$

Eliminating D/C from these 2 equations we have

$$(\beta+i\alpha)^2\, e^{i2\alpha b} = (-\beta+i\alpha)^2 \tag{2a.9}$$

and

$$(\beta+i\alpha)\, e^{i\alpha b} = \pm(-\beta+i\alpha) \tag{2a.10}$$

For D/C from Eq.(2a.7) is obtained

$$D/C = \pm 1 \tag{2a.11}$$

Assuming for the moment the plus sign, we obtain from Eq.(2a.10) for $-\cot(\alpha b/2)$:

$$-\cot(\alpha b/2) = i\,\frac{e^{i\alpha b}+1}{e^{i\alpha b}-1} = i\,\frac{-\beta+i\alpha+\beta+i\alpha}{-\beta+i\alpha-\beta-i\alpha} = \frac{\alpha}{\beta} \tag{2a.12}$$

while for the minus sign in Eq.(2a.11)

$$\tan(\alpha b/2) = i\,\frac{e^{i\alpha b}-1}{e^{i\alpha b}+1} = i\,\frac{-(-\beta+i\alpha)-(\beta+i\alpha)}{-(-\beta+i\alpha)+\beta+i\alpha} = \frac{\alpha}{\beta} \tag{2a.13}$$

is found. Solving for β and putting a factor -b/2 on both sides yields

$$-\beta b/2 = \begin{cases} (\alpha b/2)\tan(\alpha b/2) & \text{for + sign} \\ (-\alpha b/2)\cot(\alpha b/2) & \text{for - sign} \end{cases} \tag{2a.14}$$

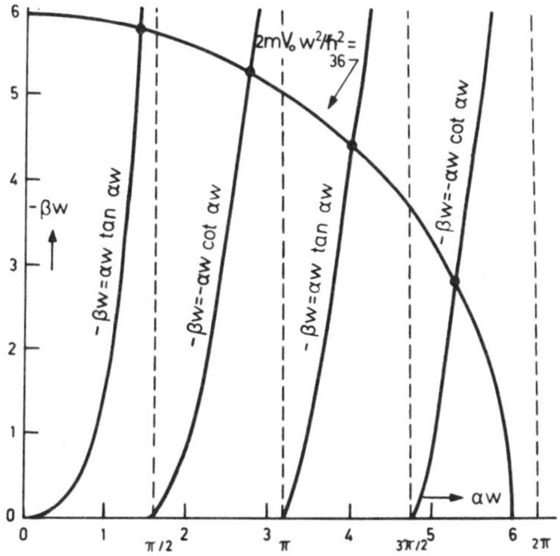

Fig.2.2 Graphical determination of the energy levels for the potential well shown in Fig.2.1 (after L.I. Schiff: Quantum Mechanics. New York. McGraw-Hill. 1949).

These functions are plotted in Fig.2.2 as well as the circle which results from Eq.(2a.1) given in the form

$$(\alpha b/2)^2 + (-\beta b/2)^2 = 2m\hbar^{-2}\, V_0 (b/2)^2 \tag{2a.15}$$

The intersections of this circle and the curves in Fig.2.2 yield the energy eigenvalues (in units of V_0) of the Schrödinger equation.

$$\frac{\epsilon}{V_o} = \frac{(\alpha b/2)^2}{2m\hbar^{-2}\, V_0 (b/2)^2} \tag{2a.16}$$

If the denominator is large compared to $(\pi/2)^2$ the first eigenvalue is near $\alpha b/2 = \pi/2$ as shown in Fig.2.2; the following eigenvalues are not equidistant. Taking e.g. 36 for the denominator on the right-hand side of Eq.(2a.16) we obtain $\epsilon/V_0 = 0.05,\ 0.20,\ 0.43$, and 0.75. These are discrete energy levels. Electron transitions between these energy levels caused by the emission or absorption of light yield a line spectrum.

Next we consider the Kronig-Penney model [1] which is shown in Fig.2.3. The lattice constant is a = b+c. The potential is periodic in a :

$$V(x) = V(x+a) = V(x+2a) = \cdots \tag{2a.17}$$

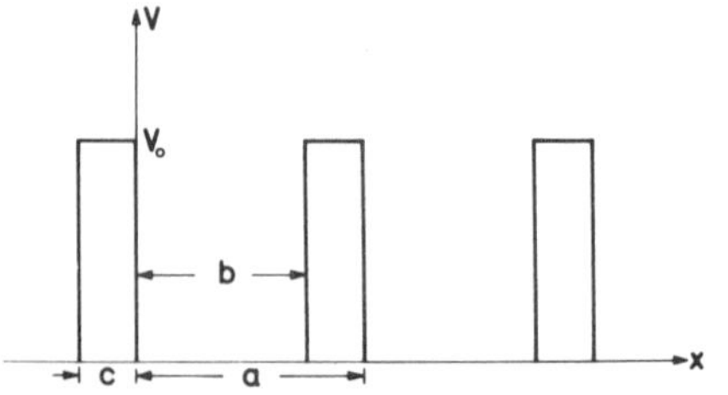

Fig. 2.3 Periodic potential wells (Kronig-Penney model).

It is reasonable to try for $\psi(x)$ a solution of the form

$$\psi(x) = u(x)e^{ikx} \text{ where} \qquad u(x) = u(x+a) = u(x+2a) = \cdots \tag{2a.18}$$

This is called a „Bloch function". Taking Eq.(2b.2) for the range $0 \leqslant x \leqslant b$ (inside the well) we have

$$\frac{d^2u}{dx^2} + 2ik\frac{du}{dx} + (\alpha^2 - k^2)\,u = 0\,; \quad 0 \leqslant x \leqslant b \tag{2a.19}$$

and accepting Eq.(2b.3) for the range $-c < x < 0$ (outside the well) we have

$$\frac{d^2u}{dx^2} + 2ik\frac{du}{dx} - (\beta^2 + k^2)\,u = 0: \quad -c < x < 0 \tag{2a.20}$$

The solutions are with constants A, B, C, and D :

$$u = Ae^{i(\alpha-k)x} + Be^{-i(\alpha+k)x}\,; \quad 0 \leqslant x \leqslant b \tag{2a.21}$$

and

$$u = Ce^{(\beta-ik)x} + De^{-(\beta+ik)x}\,; \quad -c < x < 0 \tag{2a.22}$$

The condition of continuity for u and du/dx at x = 0 yields

$$\begin{aligned} A + B &= C + D \\ i(\alpha-k)A - i(\alpha+k)B &= (\beta-ik)C - (\beta+ik)D \end{aligned} \tag{2a.23}$$

The condition of periodicity for u and du/dx means that these quantities in Eq. (2b.21) at x = b shall be equal to those in Eq.(2b.22) at x = -c :

$$\begin{aligned} &Ae^{i(\alpha-k)b} + Be^{-i(\alpha+k)b} = Ce^{-(\beta-ik)c} + De^{(\beta+ik)c} \\ &i(\alpha-k)Ae^{i(\alpha-k)b} - i(\alpha+k)Be^{-i(\alpha+k)b} = (\beta-ik)Ce^{-(\beta-ik)c} - (\beta+ik)De^{(\beta+ik)c} \end{aligned} \tag{2a.24}$$

Eqs.(2a.23) and (2a.24) are a homogeneous set of equations for A, B, C, and D which can be solved if the determinant Δ of these coefficients vanishes. With b+c=a the determinant becomes

$$\Delta = i8\alpha\beta e^{ik(c-b)}\left\{\cos(ka) - \frac{\beta^2 - \alpha^2}{2\alpha\beta}\sinh(\beta c)\sin(\alpha b) - \cosh(\beta c)\cos(\alpha b)\right\} \tag{2a.25}$$

where α and β are defined by Eq.(2b.1). Introducing a function $L = L(\epsilon/V_0)$ by

$$L = \frac{1-2\epsilon/V_0}{2\sqrt{(\epsilon/V_0)-(\epsilon/V_0)^2}} \sinh\left\{\sqrt{\frac{2mV_0}{\hbar^2}(1-\frac{\epsilon}{V_0})c}\right\} \sin\left\{\sqrt{\frac{2mV_0}{\hbar^2}\cdot\frac{\epsilon}{V_0}}\cdot b\right\}$$
$$+ \cosh\left\{\sqrt{\frac{2mV_0}{\hbar^2}(1-\frac{\epsilon}{V_0})c}\right\} \cos\left\{\sqrt{\frac{2mV_0}{\hbar^2}\cdot\frac{\epsilon}{V_0}}\cdot b\right\} \tag{2a.26}$$

the condition $\Delta = 0$ can be written in the form

$$k = k(\epsilon) = \frac{1}{a} \arccos L(\epsilon/V_0) \tag{2a.27}$$

In this way the function $k(\epsilon)$ is calculated rather than $\epsilon(k)$. Let us compare these results with those obtained for a single well. Assuming again a value of 36 for $2m\hbar^{-2}V_0(b/2)^2$ and, in addition, for the ratio $c/b = 0.1$ the function $L(\epsilon/V_0)$ as plotted in Fig.2.4 is obtained. In certain ranges of ϵ/V_0 it is larger than +1 or

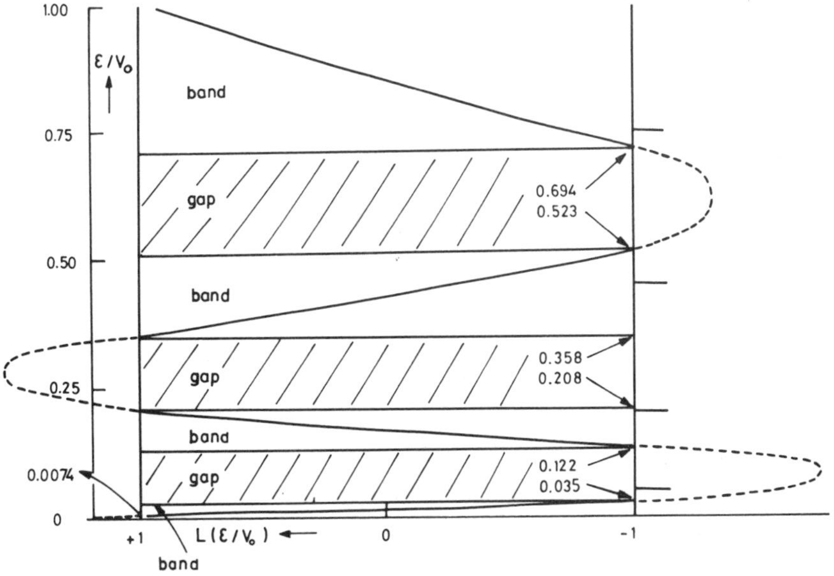

Fig.2.4 Band structure of a one-dimensional lattice. Marks at the right ordinate scale indicate discrete levels of a single potential well.

less than -1. On the other hand the cosine can have real values only between -1 and +1. This means there are ranges of ϵ/V_0 where k is not real. These ranges are called "forbidden energy bands" or "gaps". Between the gaps there are "allowed energy bands". These bands arise from the discrete N-fold degenerate levels of N atoms very far apart from each other while they are brought together. The right ordinate scale of Fig.2.4 shows these discrete levels. In the limiting case b = 0 (or c >> b) the bands become discrete levels, of course.

For the assumed c/b ratio 4 gaps are obtained. The lowest gap starts at the bottom of the well ($\epsilon = 0$) and goes up to $\epsilon/V_o = 0.0074$. This gap is too narrow to be shown in Fig.2.4.For smaller c/b ratios the gaps are smaller. For the case of e.g. c/b = 1/240 only two very narrow gaps at $\epsilon/V_o = 0.07$ and 0.63 are found. In the limiting case c = 0, a = b, Eq.(2a.27) yields $\epsilon = \hbar^2 k^2/2m$ as in free space.

For a plot of $\epsilon(k)$ it is useful to remark that in Fig.2.4 the function $L(\epsilon/V_o)$ is nearly linear within a band. Denoting the band edges ϵ_1 (at L = +1) and $\epsilon_2 > \epsilon_1$ (at L = -1) Eq.(2a.27) yields

$$\epsilon = \frac{\epsilon_2 + \epsilon_1}{2} - \frac{\epsilon_2 - \epsilon_1}{2} \cos(ka)\ ; \quad \epsilon_2 > \epsilon_1 \tag{2a.28}$$

The following band is passed by $L(\epsilon/V_o)$ in the opposite direction. This changes the sign of the cosine term :

$$\epsilon = \frac{\epsilon_4 + \epsilon_3}{2} - \frac{\epsilon_4 - \epsilon_3}{2} \cos(ka)\ ; \quad \epsilon_4 > \epsilon_3 \tag{2a.29}$$

etc. These functions are plotted in Fig.2.5. The dash-dotted parabola represents

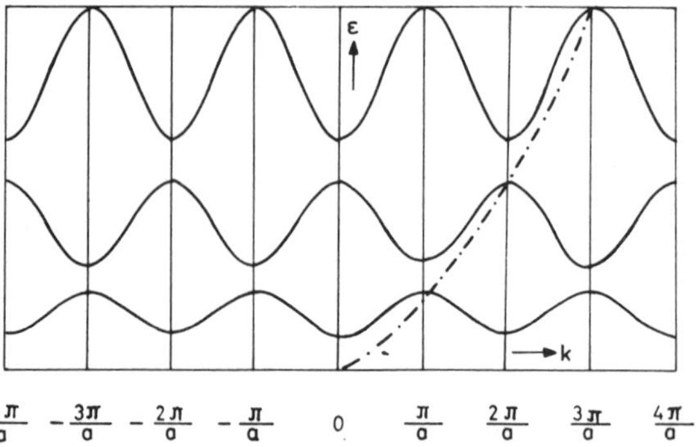

Fig.2.5 $\epsilon(\vec{k})$ diagram for an electron in a one-dimensional lattice.

the function $\epsilon = \hbar^2 k^2/2m$ valid for a free electron. For an electron inside the crystal the parabola is replaced by S-shaped parts of a sine curve which are separated from each other by discontinuities in energy at $k = n\pi/a$ where n = 1, 2, 3...; these are the energy "gaps" discussed before. At the lower edges of the S-shaped parts $d^2\epsilon/dk^2$ is positive and we shall see that it is this quantity which leads to a positive "effective mass" m defined by

$$m = \frac{\hbar^2}{d^2\epsilon/dk^2} \tag{2a.30}$$

It is shown in wave mechanics that a particle known to exist at a position somewhere between x and x+Δx cannot be represented by a wave with a discrete k value but rather by a "wave packet" composed of waves with k values essentially between k and k+Δk where the product $\Delta x.\Delta k$ is of the order of 1. After introducing the momentum $\hbar k$ this relation illustrates the Heisenberg Uncertainty Principle which states that the position of a particle and its momentum can be observed simultaneously only with precision given by this relation. It is the group velocity $v = d\omega/dk = \hbar^{-1} d\epsilon/dk$ of the wave packet which can be identified with the particle velocity.

Assume that a force F is acting upon the particle in the crystal. The energy gain $d\epsilon$ of the particle in a time dt will be Fvdt. On the other hand $d\epsilon = (d\epsilon/dk)dk = vd(\hbar k)$. Hence the force F is equal to $d(\hbar k)/dt$ which is analogous to Newton's law if we associate $\hbar k$ with a momentum as mentiond before. $\hbar k$ is called the "crystal momentum".

The mass m of a free particle is defined by Newton's law: The derivative of v with respect to time is the acceleration of the particle by the force F which, if no other forces were acting upon the particle, would be equal to F/m. For a carrier in a solid it is possible to retain this relation by neglecting the crystal forces and introducing the "effective mass" m. Hence $Fm^{-1} = dv/dt = \hbar^{-1}(d^2\epsilon/dk^2)\cdot dk/dt$ and on the other hand $Fm^{-1} = d(\hbar k)/dt\, m^{-1}$ which yields Eq.(2a.30). In the special case $\epsilon = \hbar^2 k^2/2m$ valid for a free particle of mass m, Eq.(2a.30) can be verified easily.

Now let us consider the upper edge of a band displayed in Fig.2.6. Here the effective mass of an electron would be negative. Fortunately, it is the ratio of the charge to the mass rather than the mass itself that is important in transport theory. Consequently one can talk either of negatively charged electrons with a negative effective mass or positively charged particles called "holes" with a positive effective mass. The convention is to adopt to talk of holes with positive effective mass.

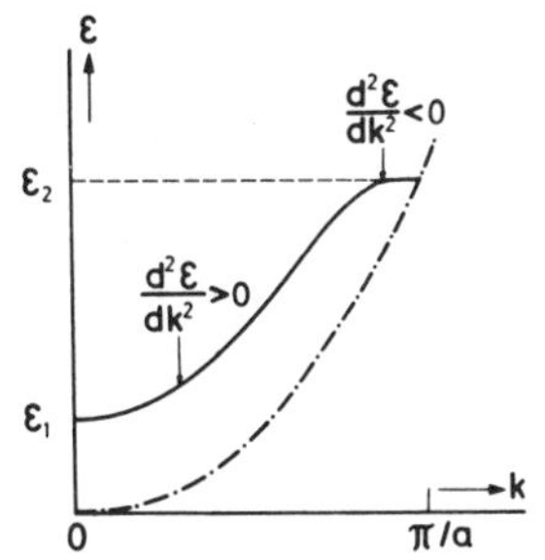

Fig.2.6 Section of Fig.2.5 for $\vec{k}$ between 0 and π/a.

For a quantitative calculation of the effective mass Eq.(2a.28) turns out to be too crude an approximation. However, even for the one-dimensional case considered here the calculation is quite involved. For the special case $|V(x)| \ll \epsilon$ it yields [3]

$$m = m_o \cdot \frac{\epsilon_G}{4(\epsilon_c - \frac{1}{2}\epsilon_G)} \tag{2a.31}$$

[3] See e.g. R.A.Smith: Wave Mechanics of Crystalline Solids, 2nd ed., Chap.6.2. London: Chapman and Hall Ltd. 1969.

where $\epsilon_c - \frac{1}{2}\epsilon_G$ is the middle and ϵ_G the width of the nearest gap. The effective mass is smaller for a smaller gap. In e.g. n-InSb at room temperature an effective mass of 0.013 m_o and a gap of 0.18 eV are observed. A material with a larger gap, 1.26 eV, is n-InP where the effective mass, 0.05 m_o, is also larger. For more data see Table 1.

Table 1: Band gaps and electron effective masses of compounds*) with lowest conduction band minimum at $\vec{k} = 0$.

Compound	Optical Gap (300 K) eV	m/m_o (optical)	El.Gap (extrap. to 0 K) eV	m/m_o (electrical)
InSb	0.180	0.0116	0.27	0.013
InAs	0.36	0.023-0.027	0.47	0.025
InP	1.26	0.077	1.34	0.05
GaSb	0.67-0.725	0.047	0.77-0.82	0.05
GaAs	1.43	0.043-0.071	1.4	0.072

*) From O.Madelung: Physics of III-V Compounds. New York: J. Wiley. 1964.

These examples suggest that also in 3 dimensions the relation $m \propto \epsilon_G$ holds to a good approximation for a simple band structure.

For a more detailed calculation of the band structure such as for InSb by Kane [4], the atomic wave functions as well as the electron spin have to be taken into account. With two parameters P and Δ which will not be considered in detail, Kane obtains for InSb the following $\epsilon(\vec{k})$ relation with zero energy at the upper edge of the valence band and 3 kinds of holes which we will discuss in Chap.8a :

Conduction band : $$\epsilon = \epsilon_G + \frac{\hbar^2 k^2}{2m} + \frac{P^2 k^2}{3} \cdot \left(\frac{2}{\epsilon_G} + \frac{1}{\epsilon_G + \Delta}\right)$$

Valence band :

$$\begin{cases} \text{heavy holes:} & \epsilon = -\dfrac{\hbar^2 k^2}{2m} \\ \text{light holes:} & \epsilon = -\dfrac{\hbar^2 k^2}{2m} - \dfrac{2P^2 k^2}{3\epsilon_G} \\ \text{split-off band:} & \epsilon = -\Delta - \dfrac{\hbar^2 k^2}{2m} - \dfrac{P^2 k^2}{3(\epsilon_G + \Delta)} \end{cases} \qquad (2a.32)$$

[4] E.O.Kane, J.Phys.Chem.Solids 1 (1957) 249.

In these relations the anisotropy of the valence band is neglected (see Chap. 2d). Interactions which determine the heavy hole effective mass are not included in Kane's model which itself is still very simplified. The heavy hole mass m may be assumed infinite in this model [5]. The valence band consists of 3 subbands two of which are degenerate at $\vec{k} = 0$. The third band is split-off by spin orbit coupling of the electrons. The split-off energy at $\vec{k} = 0$ is denoted by Δ. In the case of $\Delta \gg \epsilon_G$ and Pk, we find for an approximation

Conduction band : $$\epsilon - \epsilon_G = \hbar^2 k^2/2m + \frac{1}{2}\left(\sqrt{\epsilon_G^2 + 8P^2k^2/3} - \epsilon_G\right)$$

Valence band :
$$\begin{cases} \text{heavy holes:} & \epsilon = -\hbar^2 k^2/2m \\ \text{light holes:} & \epsilon = -\hbar^2 k^2/2m - \frac{1}{2}\left(\sqrt{\epsilon_G^2 + 8P^2k^2/3} - \epsilon_G\right) \\ \text{split-off band:} & \epsilon = -\Delta - \hbar^2 k^2/2m - P^2k^2/(3\epsilon_G + 3\Delta) \end{cases} \qquad (2a.33)$$

The conduction band and the light-hole valence band are "nonparabolic" which means that ϵ is not simply proportional to k^2. Introducing for simplification the quantities

$$\epsilon' = \frac{1}{2}\left(\sqrt{\epsilon_G^2 + 8P^2k^2/3} - \epsilon_G\right) \qquad (2a.34)$$

and

$$m_n = 3\hbar^2 \epsilon_G/4P^2 \qquad (2a.35)$$

we can solve Eq. (2a.34) for $\hbar^2 k^2/2m_n$:

$$\hbar^2 k^2/2m_n = \epsilon'(1 + \epsilon'/\epsilon_G) \qquad (2a.36)$$

If in Eq.(2a.33) we put $m = \infty$ we get for the conduction band $\epsilon' = \epsilon - \epsilon_G$ where one notices that the energy zero is shifted to the lower edge of the conduction band. For a finite value of m Eq.(2a.36) is essentially retained except that ϵ_G is replaced by an "energy parameter" depending on m. For n-InSb and n-InAs (gap 0.18 eV and 0.36 eV at 300 K, respectively; effective mass $m_n = 0.013\ m_0$ and $0.025\ m_0$) the energy parameter has the values of 0.24 eV and 0.56 eV, respectively [6].

---

[5] D.Long: Energy Bands in Semiconductors, p.49. New York: J.Wiley and Sons. 1968.

[6] D.Matz, J.Phys.Chem.Solids 28 (1967) 373.

Data like gap and effective masses still have to be obtained experimentally. Even in the computer age crude approximations have to be made in the calculation of band structures. It is a normal procedure to include adjustable parameters which are determined from experimental data like the gap and the effective masses. In Chap.11a we shall consider fundamental optical reflection where the band structure is of vital importance. For a clear picture of the origin of gaps we now discuss the tight-binding approximation.

2b. Energy Bands by Tight Binding of Electrons to Atoms

In the preceding treatment the effect of the crystal lattice on electrons was approximated by that of a periodic potential. Another way of looking at the problem is to construct a crystal from single atoms which are infinitely far away from one another, and to see how the discrete atomic energy levels are changed during this process. When the atoms are far away from each other the electrons are "tightly bound" to each atom. At what atomic distance will an exchange of electrons between neighboring atoms begin to become important ?

If the wave functions ψ of atom A and B overlap, linear combinations $\psi_A \pm \psi_B$ characterize the behavior of the compound AB. In Fig.2.7 the energy as a function of the distance between the two atoms is plotted. The sum of the ψ-functions yields the bound state while the difference represents the situation where the atoms repel each other. A calculation is presented in monographs on quantum mechanics [1] and will not be attempted here. If N rather than 2 atoms come together the atomic level is split not only in 2 but in N levels. In a crystal of $N \approx 10^{23}$ atoms the levels are so close to each other that this can be called a band. Let us consider again diamond. In a free carbon atom the two 2s levels and two of the six 2p levels which have a higher energy than the 2s levels, are occupied by the four valence electrons. These four filled levels per atom form the valence band if many atoms are put together to form a diamond crystal while the other four levels form the empty conduction band. At a lattice constant of 3.6 Ångström-units (10^{-8} cm) there is a gap of about 5.3 eV between these two bands. At absolute zero temperature pure diamond is an insulator

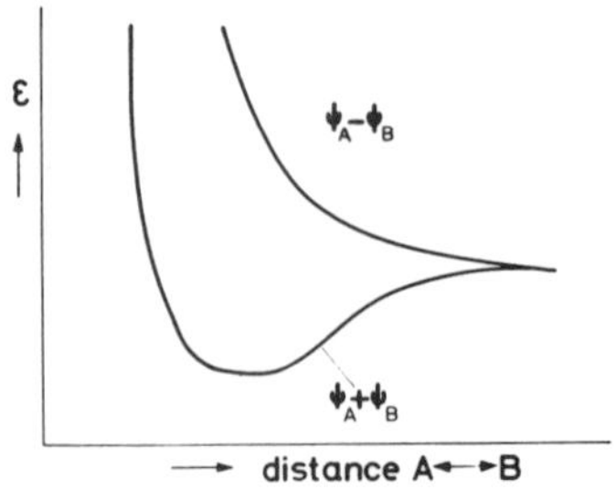

Fig.2.7 Energy of a two-atomic molecule AB as a function of the atomic distance.

[1] See e.g. S.Flügge: Rechenmethoden der Quantentheorie, 3rd ed., Problem Nr.59. (Heidelberger Taschenbücher, Vol.6.) Berlin - Heidelberg - New York: Springer. 1966.

since it will be shown in Chap.3 that all electrons are in the valence band completely filling it and for a drift motion an electron would have to increase its energy by a small amount which is considerably less than the gap, but there is no free place in the valence band. Therefore no drift is possible and diamond is an insulator then. However, at high temperatures a few electrons are lifted thermally into the conduction band enabling both electrons and holes to drift : diamond is semiconducting then.

In metals there are either partly-filled bands or overlapping free and filled bands. Why is there a gap in diamond and not in e.g. metallic tin ? In fact both the crystal structure and the number of valence electrons decide upon this question : Crystals of metallic Sn are tetragonal, of Pb (always metallic) face-centered cubic while the semiconducting group-IV elements crystallize in the diamond lattice. Hund and Mrowka [2] investigated this problem very carefully with different crystal structures. The one-dimensional chain, the 2-dimensional hexagonal lattice (coordination number 3), the diamond lattice, and the element-wurtzite lattice (Fig.2.17) require only s-functions (even functions of space coordinate) and p-functions (odd functions, e.g. xf(r)) for the calculation of the crystal wave function while the 2-dimensional square lattice (coordination number 4) and the body-centered cubic lattice require also d-functions (proportional $(x^2-y^2)g(r)$ for the square lattice). The overlap of bands calculated for the cubic lattice yielding a metallic behavior is shown in Fig.2.8.

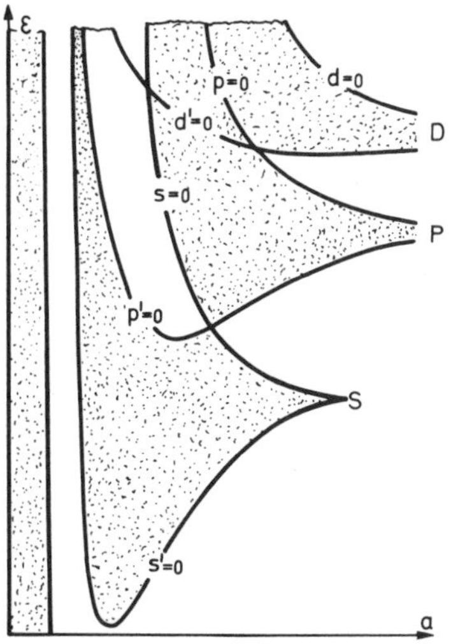

Fig.2.8 Energy bands of a cubic lattice as a function of the lattice constant (after F.Hund and B.Mrowka, ref.2).

However, also the band type composed of s- and p-functions may lead to a metal. A typical example is graphite. The bonds in the hexagonal planes are filled by 3 valence electrons per atom. The fourth valence electron of each carbon atom is located in the conduction band and produces a rather metallic conductivity.

Let us briefly go through the "cell-method" calculation of the diamond lattice. In two lattice points which are apart by the distance of neighboring atoms the potential has the same symmetry but a different space orientation relative to the coordinate axes. The two atoms are therefore centers of cells a and b. While the ψ-function in cell a is given by

$$\psi_a = A_a s(r) + \sqrt{3}\,(B_a x + C_a y + D_a z)\, p(r)/r \qquad (2b.1)$$

[2] F.Hund und B.Mrowka, Ber.Sächs.Akad.Wiss.87 (1935) 185; 325.

the ψ-function in cell b centered at $\vec{e} = (a/4; a/4; a/4)$ is given by

$$\psi_b = \exp(2\pi i\vec{k}\vec{e})[A_b s(r') + \sqrt{3}\{B_b(x-a/4)+C_b(y-a/4)+D_b(z-a/4)\}]p(r')/r' \qquad (2b.2)$$

where r' is the distance from the center of cell b. The continuity condition of ψ and $d\psi/dr$ at the halfway intermediate points between the atom and its 4 nearest neighbors yields 8 equations for the 8 unknown coefficients $A_a \ldots D_b$. A solution of the homogeneous system of linear equations exists if the determinant vanishes yielding $p^2=0$; $p'^2=(dp/dr)^2=0$ and

$$\frac{sp'+s'p}{sp'-s'p} = \frac{1}{2}\{1+\cos(\pi ak_x)\cos(\pi ak_y)+\cos(\pi ak_y)\cos(\pi ak_z)+\cos(\pi ak_z)\cos(\pi ak_x)\}^{1/2} \qquad (2b.3)$$

The right-hand side of Eq.(2b.3) can have values between +1 and -1 only. This function as well as s, p, s', and p' are plotted vs r in Fig.2.9. The Schrödinger equation was solved by a "perturbation method" assuming an arbitrary not too unrealistic potential [1]. In the lower part of Fig.2.9 the energy is plotted vs r, the distance between nearest neighbors. The numerical calculation was made for 8 values of the energy. The curve representing p'=0 is two-fold degenerate due to the fact $p'^2=0$ as mentioned above. This is the upper edge of the valence band. This together with the two-fold degeneracy of Eq.(2b.3) (positive and negative values) makes room for 4 valence electrons per atom which is the case in diamond. The upper edge of the conduction band is also two-fold degenerate since it is formed by the solution $p^2=0$. If the energy comes close to the atomic d-level the simplified calculation becomes invalid, however [3].

It is interesting to note that for the lattice constant of diamond at normal atmospheric pressure the upper edge of the valence band increases with r while the lower edge of the conduction band decreases with r. Decreasing the interatomic distance by external hydrostatic pressure thus increases the energy gap which yields a decrease in the number of carriers and therefore an increase in resistivity, in agreement with observations (see Chap.41).

[3] F.Hund: Theorie des Aufbaues der Materie, chap.VII. Stuttgart: Teubner. 1961.

Fig.2.9a Radial wave functions and their derivatives (multiplied by 10) for $\epsilon/2Ry = -0.635$, where Ry is the Rydberg energy.

Fig.2.9b The function (sp'+s'p)/(sp'-s'p) for $\epsilon/2Ry = -0.635$.

Fig.2.9c Calculated energy bands of diamond (a_B is the Bohr radius; after ref.2).

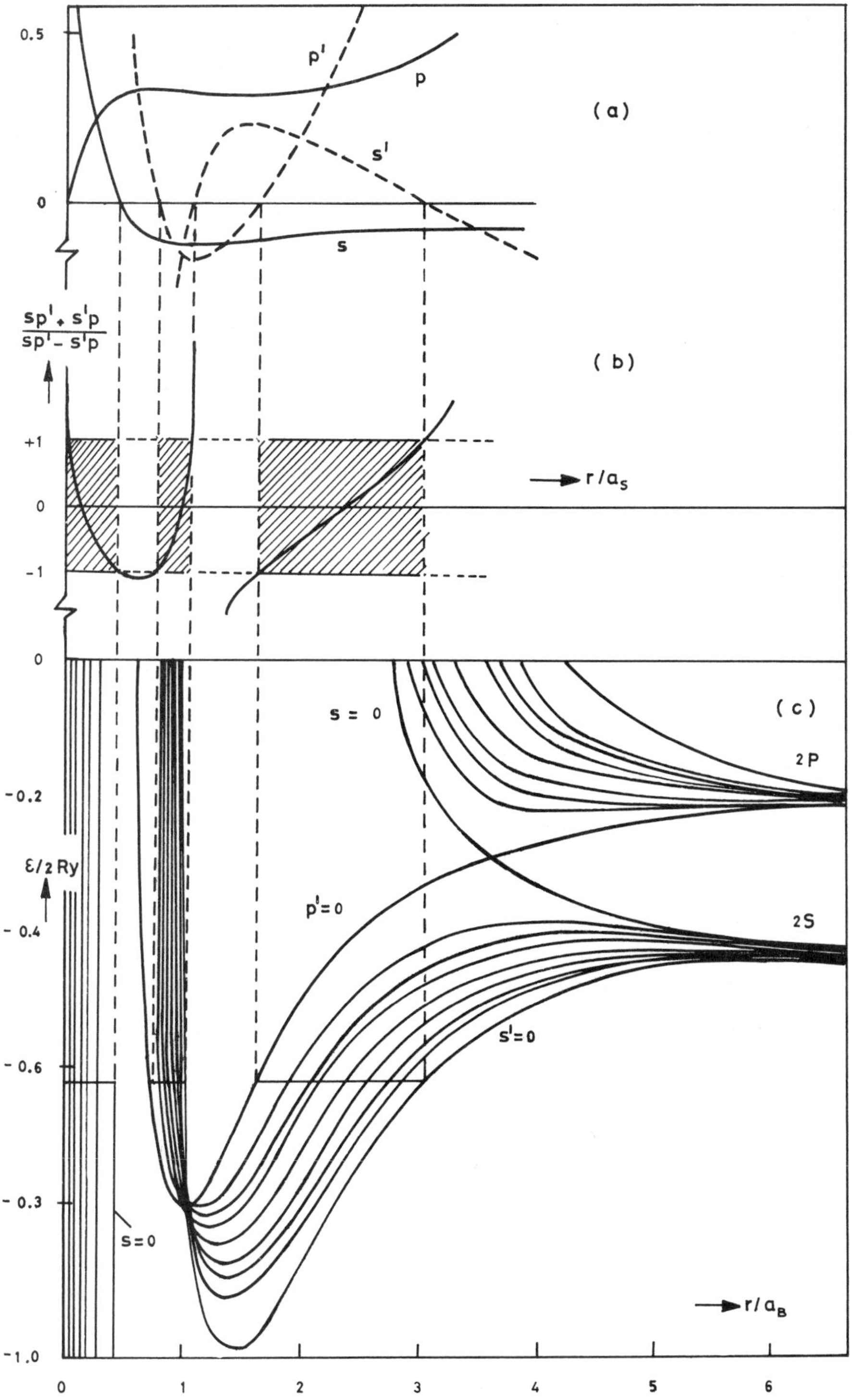

0.5
0
p'
p
(a)
s'
s
sp' + s'p
sp' − s'p
(b)
+1
0
−1
r/a_s
0
s = 0
(c)
2P
−0.2
ε/2 Ry
p' = 0
−0.4
2S
s' = 0
−0.6
−0.3
s = 0
r/a_B
−1.0
0
1
2
3
4
5
6

2c. The Brillouin Zone

As demonstrated by Fig.2.5 the band structure given for the range $-\pi/a \leqslant k \leqslant \pi a$ is repeated indefinitely. It is therefore sufficient to consider only this range which is called the "first Brillouin zone". The neighboring regions from $-2\pi/a$ to $-\pi/a$ and from π/a to $2\pi/a$ form the "second Brillouin zone" etc. Since for most purposes it is sufficient to consider the first Brillouin zone this is sometimes just called Brillouin zone.

What is the Brillouin zone of a 3-dimensional lattice ?

The history of Brillouin zones started with scattering processes of X-rays by a multidimensional lattice. The scattering corresponds to that of electron waves. An electron moving in a crystal is represented by an electron wave packet continuously scattered at the crystal lattice. The scattering process obeys the Laue equation

$\exp\{i(\vec{p}\vec{G})\} = 1$ where the vector

$$\vec{p} = l\vec{a} + m\vec{b} + n\vec{c}\,; \quad (l, m, n = 0, 1, 2, 3, \cdots) \tag{2c.1}$$

characterizes an atom of the crystal lattice which has a primitive unit cell defined by the primitive base vectors $\vec{a}$, $\vec{b}$, and $\vec{c}$ (in two dimensions: Fig.2.10). The vector $\vec{G} = \vec{k}' - \vec{k}$ represents the change of wave vector due to the scattering process.

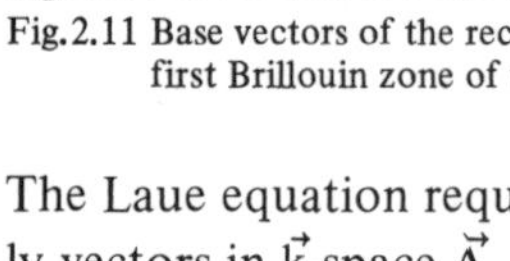

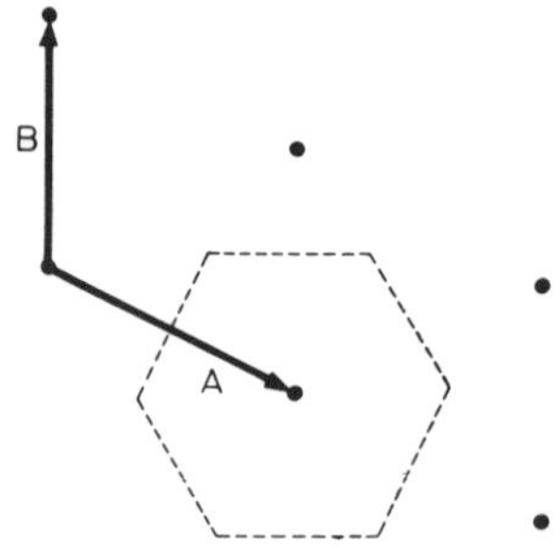

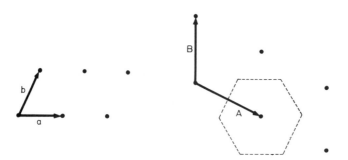

Fig.2.10 Base vectors of a two-dimensional lattice.

Fig.2.11 Base vectors of the reciprocal lattice. The area of the hexagon is the first Brillouin zone of the real lattice shown in Fig.2.10.

The Laue equation requires $(\vec{p}\vec{G}) = 2\pi$ or an integral multiple thereof. Accordingly vectors in $\vec{k}$-space $\vec{A}$, $\vec{B}$, and $\vec{C}$ are defined such that

$$\vec{G} = g\vec{A} + h\vec{B} + k\vec{C}\,; \quad (g, h, k = 0, 1, 2, 3, \cdots) \tag{2c.2}$$

where $(\vec{a}\vec{A}) = (\vec{b}\vec{B}) = (\vec{c}\vec{C}) = 2\pi$ and other products such as $(\vec{a}\vec{B}) = 0$; $(\vec{b}\vec{C}) = 0$ etc. $\vec{A}$, $\vec{B}$, and $\vec{C}$ are denoted as the base vectors of the reciprocal lattice. They are perpendicular to the corresponding base vectors of the real lattice. In Fig.2.11 the base vectors of the reciprocal lattice are shown in comparison to the real lattice of Fig.2.10. If $\vec{x}_e$ and $\vec{y}_e$ are unit vectors in a 2-dimensional Cartesian coordinate system the base vectors of Fig.2.10 and 2.11 may be represented by $\vec{a} = 2\vec{x}_e$; $\vec{b} = \vec{x}_e + 2\vec{y}_e$; $\vec{A} = \frac{\pi}{2}(2\vec{x}_e + 2\vec{y}_e)$ and $\vec{B} = \pi\vec{y}_e$. Of course from the Laue equation the well-known Bragg equation is easily derived.

The "primitive cell" contains all points which cannot be generated from each other by a translation (Eq.2c.1). The "Wigner-Seitz cell" contains all points which are nearer to one considered lattice point than to any other one. In the reciprocal lattice the Wigner-Seitz cell is denoted as the "first Brillouin zone". For a construction of the Brillouin zone, draw arrows from a lattice point to its nearest neighbors and cut these in half. The planes through these intermediate points perpendicular to the arrows form the surface of the Brillouin zone. The hexagon shown in Fig.2.11 is the Brillouin zone of the lattice of Fig.2.10.

It is easy to see that the reciprocal lattice of the face-centered cubic lattice is the body-centered cubic lattice and vice versa. The important semiconductors of the fourth group of the periodic table crystallize in the diamond lattice which is shown in Fig.2.12. Each atom is tetrahedrically bonded to four nearest neighbors,

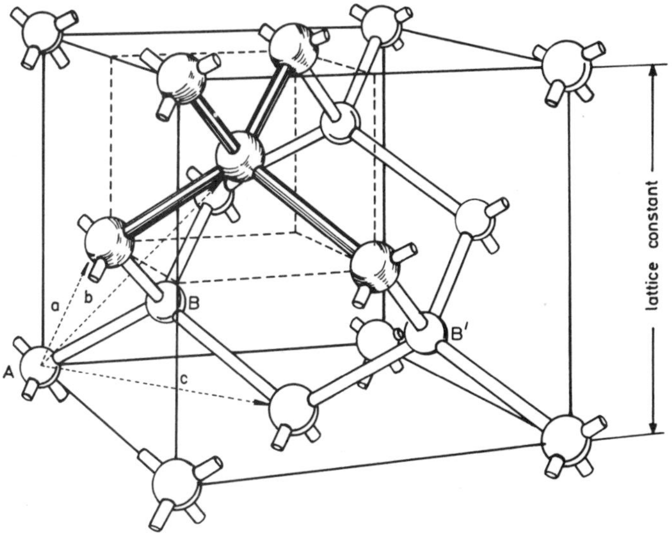

Fig.2.12 Diamond crystal structure. The lattice constant for C, Si, and Ge is 3.56, 5.43, and 5.66 Å, respectively (after W.Shockley: Electrons and Holes in Semiconductors. New York: D.van Nostrand. 1950.).

similar to the bonds in the CH_4 molecule. Let us take the atom A as the corner of a face-centered cubic lattice. In going from A along the cube diagonal halfway towards the cube center one encounters atom B which is the corner of a second face-centered cubic lattice of which besides B only the atoms B', C, and C' are shown in Fig.2.12. The vectors $\vec{a}$, $\vec{b}$, and $\vec{c}$ from B to the face centers are the base vectors. Perpendicular to these are the base vectors $\vec{a}'$, $\vec{b}'$, and $\vec{c}'$ of the reciprocal lattice, which is a body-centered cubic lattice shown in Fig.2.13.

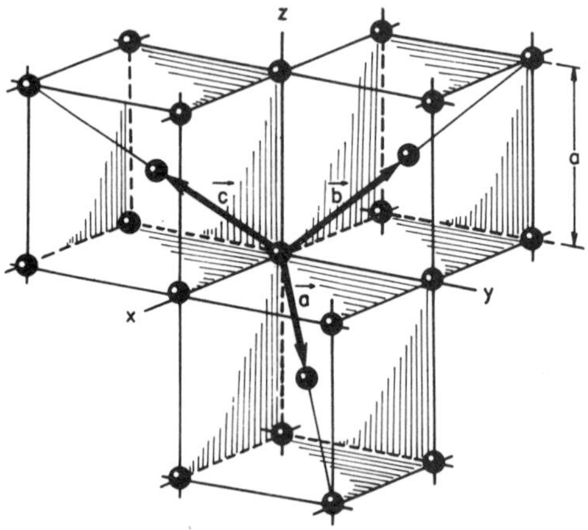

Fig.2.13 Primitive translations in the body centered cubic lattice (after C.Kittel: Einführung in die Festkörperphysik. München: R.Oldenbourg. 1968).

By calculating the scalar product of e.g. < 1, 1, 1> with < 1, -1, 0> it is obvious that each cube diagonal is perpendicular to a face diagonal. The construction of the Brillouin zone is then not difficult. (Fig.2.14). The faces normal to the cube diagonals are the hexagons while the faces normal to the lines from one cube center to the next are the hatched squares. At one of the hexagon faces a second "truncated octahedron" is attached in order to demonstrate how $\vec{k}$-space is

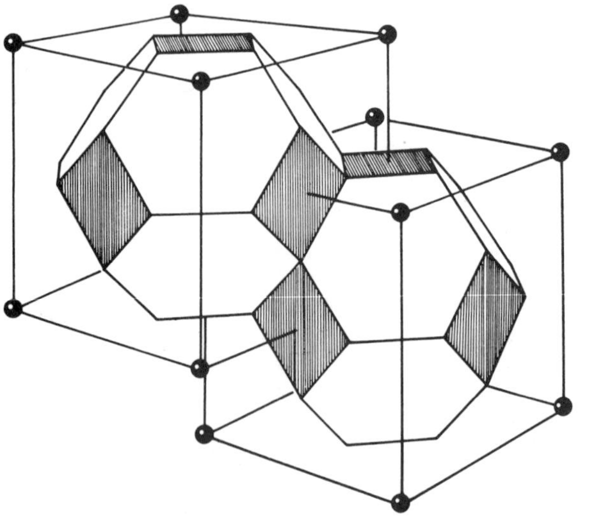

Fig.2.14 First Brillouin zone for the face centered cubic lattice (after C.Kittel: Einführung in die Festkörperphysik. München: R.Oldenbourg. 1968).

filled. After suitably cutting the second truncated octahedron and attaching the parts to the first Brillouin zone this would be the second Brillouin zone [1]. A 2-dimensional example is shown in Fig.2.15. Similarly the 3rd Brillouin zone is constructed etc. All Brillouin zones of a lattice have equal volume. In Fig.2.16 the most important symmetry points and lines in the first Brillouin zone of the face-centered cubic lattice are given. There are six equivalent X-points, eight L-points etc. [2].

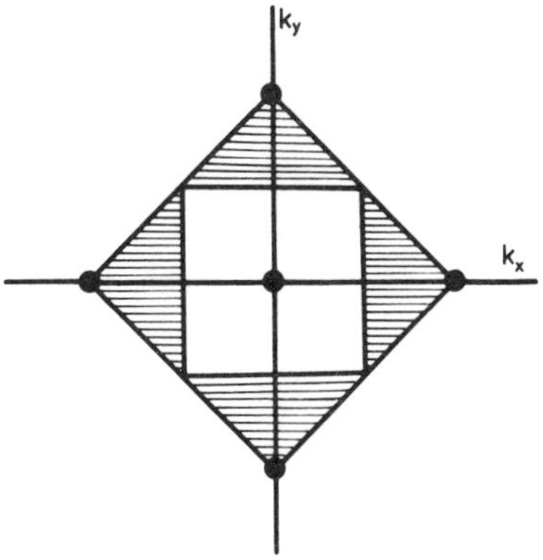

Fig.2.15 First two Brillouin zones for the single square lattice (after R.A.Smith: Wave Mechanics of Crystalline Solids. London: Chapman and Hall Ltd. 1961).

The zinc blende lattice so typical for III-V compounds can be generated from a diamond lattice by replacing half of the atoms by group-III atoms and half by group-V atoms. Each bond connects a group-III atom with a group-V atom. If e.g. an InSb plate cut with <111> broad faces is etched, one side consists of In atoms and the opposite side of Sb atoms.

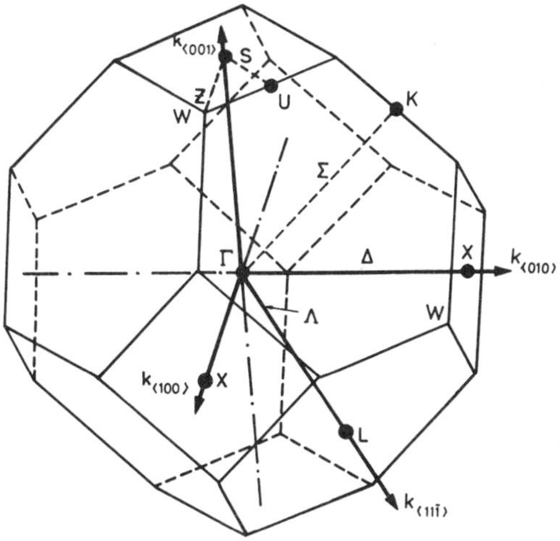

Fig.2.16 First Brillouin zone for the diamond and zinc blende lattice, including important symmetry points and lines (after D.Long, ref.4).

---

[1] See e.g. H.Jones: The Theory of Brillouin Zones and Electronic States in Crystals. Amsterdam: North Holland Publ.Co. 1960; F.Seitz: Modern Theory of Solids. New York: McGraw-Hill. 1940,

[2] L.P.Bouckaert, R.Smoluchowski, and E.P.Wigner, Phys.Rev.<u>50</u> (1936) 58.

The inversion symmetry of the diamond lattice therefore is not found in the zinc blende lattice. (For further details see e.g.ref.3).

The wurtzite structure is closely related to the zinc blende structure. In Fig. 2.17 the orientations of adjacent atomic tetrahedra in both structure types are

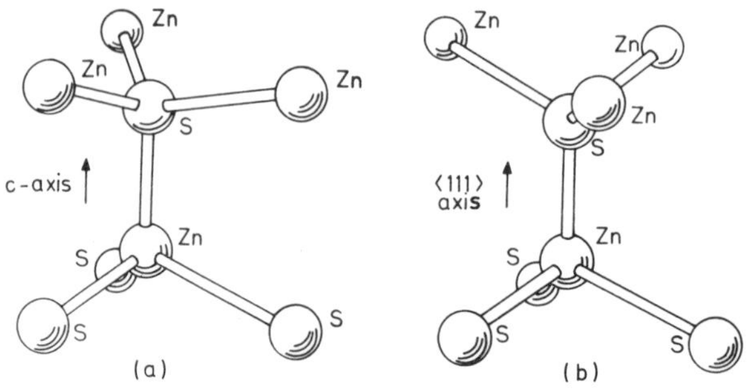

Fig.2.17 Atomic arrangements in wurtzite (a) and zinc blende (b) (after D.Long, ref.4).

compared [4].The hexagonal wurtzite crystal has a unique crystallographic axis of symmetry, designated the c-axis, which corresponds to the <111> axis in the cubic zinc blende structure. There are 4 atoms per unit cell, while in the zinc blende structure there are only two. The band structures for both types of structures are very similar. In CdS the band gap has been determined in both modifications from the fundamental optical reflectivity. The gap of the cubic structure is only a few hundredth of an eV lower than that of the wurtzite structure which is 2.582 eV (at 0 K). CdS ordinarily crystallizes in the wurtzite structure but can be forced to accept the zinc blende structure by depositing it epitaxially from the vapor phase onto cubic GaAs. The Brillouin zone is the same as that of Te to be treated below except that it has a hexagonal symmetry.

The crystal structure of Te is shown in Fig.2.18. It consists of spiral chains arranged in a trigonal structure. The chains may either be right-handed or left-handed; Fig.2.18 shows only the right-handed version. The chains are interconnected by van der Waals and weak electronic forces. There is no center of inversion symmetry. The unit cell contains 3 atoms. The Brillouin zone which is a hexagonal prism is shown in Fig.2.19 including the group-theoretical notation of zone points [2]. The symmetry of the crystal about the c-axis is not sixfold but threefold ("trigonal"). Optical experiments (exciton absorption and inter-

[3] H.W.Streitwolf: Gruppentheorie in der Festkörperphysik. Leipzig: Akad.Verlagsges. Geest und Portig. 1967.

[4] D.Long: Energy Bands in Semiconductors. New York: J.Wiley and Sons.1968.

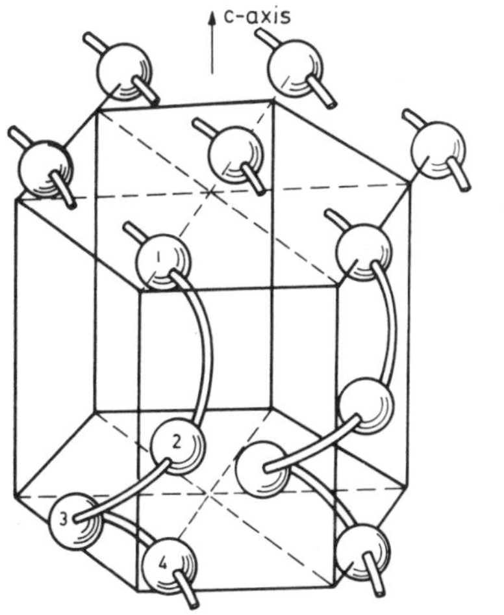

Fig.2.18 Crystal lattice of tellurium: Atomic spiral chains in trigonal arrangement (after D.Long, ref.4).

band magnetoabsorption at 10 K, polarization parallel to c-axis) have shown that it is a "direct semiconductor" (see Chap.11b) with a band gap of 0.3347 (±0.0002) eV which increases with decreasing temperature and increasing pressure [5].

A semiconductor of some interest for thermoelectric applications is Bi_2Te_3. Its rhombohedral crystal structure [6] consists of Te and Bi layers in the sequence Bi-Te-Te-Bi-Te-Bi-Te-Te-Bi. The unit cell contains one Bi_2Te_3 molecule. The Brillouin zone is shown in Fig.2.20. Conduction and valence bands seem to consist of 6 ellipsoids each, which are oriented at angles of 76° and 56°, respectively, with respect to the c-axis. The optical absorption

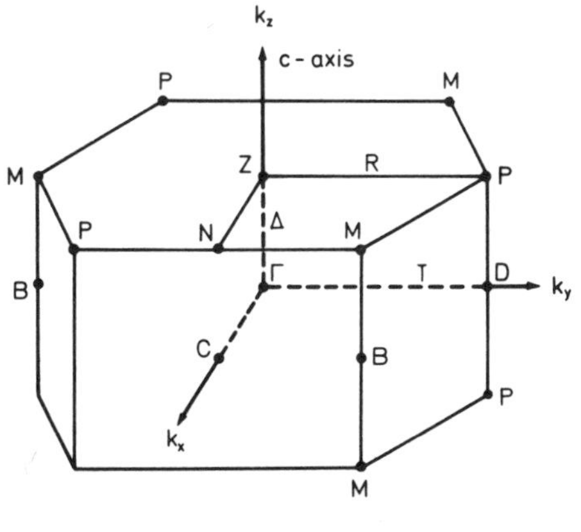

Fig.2.19 First Brillouin zone for tellurium, including important symmetry points and lines. The trigonal symmetry about the c-axis is indicated by the distinction between the M and P points (after D.Long, ref.4).

[5] C.Rigaux and G.Drilhorn, Proc.Intern.Conf.Phys.Semic. Kyoto. J.Phys.Soc. Japan 21, Suppl. (1966) 193.

[6] J.R.Drabble: Progr. in Semic. (A.F.Gibson and R.E.Burgess,eds.). Vol.7, p.45. London: Temple Press. 1963.

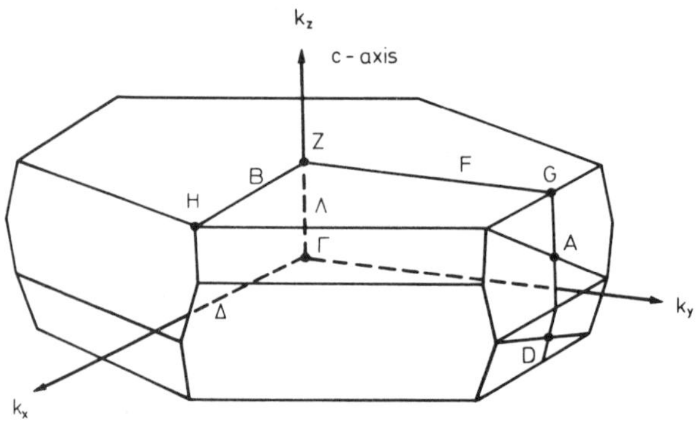

Fig.2.20 First Brillouin zone for Bi_2Te_3, including important symmetry points and lines (after D.Long, ref.4).

yields values for the band gap of [7] 0.145 eV at 300 K and 0.175 eV at 85 K.

The lead salts PbS, PbSe, and PbTe crystallize in the cubic NaCl lattice and thus have 2 atoms per unit cell. The optically determined band gaps at absolute zero temperature are 0.29, 0.15, and 0.19 eV, respectively, and increase with temperature in contrast to most other semiconductors (e.g. in PbTe from 0.19 eV at 0 K to 0.32 eV at 300 K). With pressure the gaps decrease.

A very interesting behavior is exhibited by PbTe-SnTe mixed crystals which have a smaller band gap than each of the components. Technological interest in narrow gap materials arises from their possible applications as long-wavelength infrared detectors and lasers. Similar systems are PbSe-SnSe, CdTe-HgTe, Bi-Sb, and – under uniaxial stress – gray tin (α-Sn). Fig.2.21 shows [8] the band gap of $Pb_{1-x}Sn_xTe$ alloys as a function of composition for temperatures of 12 and 77 K. E.g. at 77 K the gap vanishes for x = 0.4. At this composition the valence band edge (of PbTe) changes its role and becomes the conduction band edge [9] (of SnTe). ("Band inversion"). The somewhat similar situation in

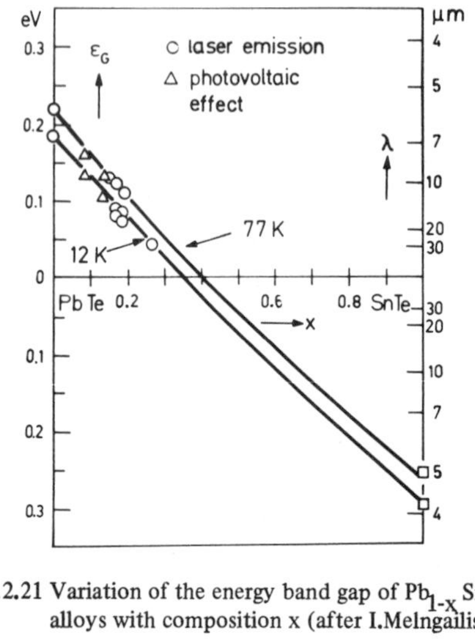

Fig.2.21 Variation of the energy band gap of $Pb_{1-x}Sn_xTe$ alloys with composition x (after I.Melngailis, ref.8).

---

[7] D.L.Greenaway and G.Harbeke, J.Phys.Chem.Solids 26 (1965) 1585.
[8] I.Melngailis, Journal de Physique (suppl.No.11-12) 29 (1968) C 4-84.
[9] J.O.Dimmock, I.Melngailis, and A.J.Strauss, Phys.Rev.Lett. 16 (1966) 1193.

$Cd_xHg_{1-x}Te$ is shown in Fig.2.22 where the band gap is plotted vs x. A second horizontal scale indicates the dependence on pressure for x = 0.11. The hatched

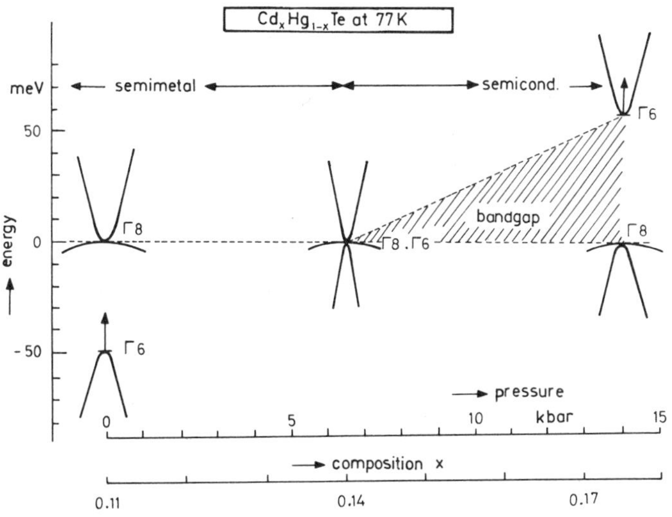

Fig.2.22 Band structure of $Cd_xHg_{1-x}Te$ alloys near the semimetal-semiconductor transition. The vertical arrows indicate the change in band gap with increasing temperature or pressure (after Vérié, ref.10).

area shows how the band gap widens with increasing values of x. The alloys with $x < 0.14$ are semimetals. The graph may suggest a negative band gap in the semimetal. In Fig.2.23, where the band gap is plotted vs the mean atomic number $\bar{Z}$ for five II-VI compounds with zinc blende structure it is shown that with increasing values of $\bar{Z}$ one may arrive at a negative band gap. The semimetal HgTe is in fact indicated at a negative value of the band gap. Finally band structure calculations for HgTe by the "k.p method" (see Chap.8a) yield a gap of value [10] ±0.15 eV; the negative sign seems to be more appropriate in this case.

The band structure of gray tin (α-Sn) shown [11] in Fig.2.24 is similar to that of HgTe. The "negative band gap" ($\Gamma_7^- - \Gamma_8^+$) is 0.3 eV. The top of the valence band is identical with the conduction band minimum at $\vec{k} = 0$. However, in electrical measurements a band gap of 0.09 eV is observed which is the gap between the top of the valence band and the L_6^+ minima in the <111> direction in $\vec{k}$-space. Most conduction electrons are in these minima because of their higher density of

[10] C.Vérié: Festkörper-Probleme/Advances in Solid State Physics (O.Madelung, ed.), Vol.X. Oxford: Pergamon. Braunschweig: Vieweg. 1970.

[11] S.Groves and W.Paul, Phys.Rev.Lett. 11 (1963) 194.

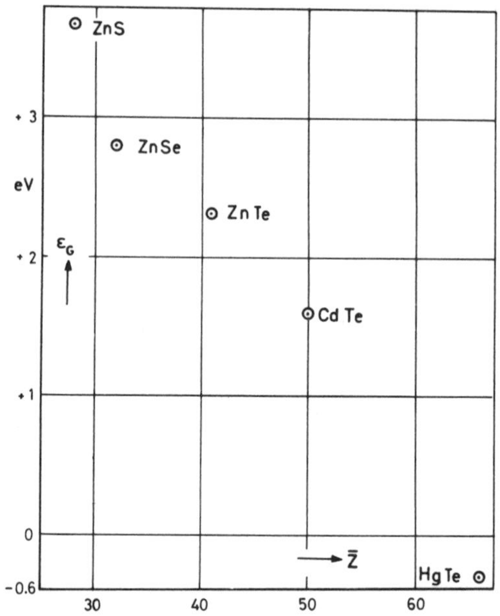

Fig. 2.23 Energy gaps of various II-VI compounds with zinc blende structure, as a function of their mean atomic number (after C. Vérié, ref. 10).

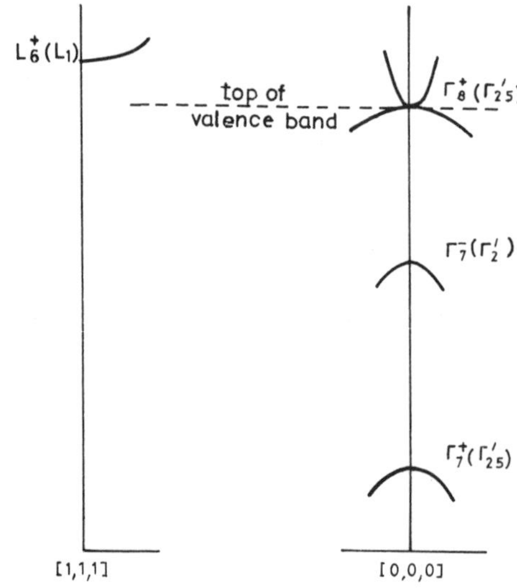

Fig. 2.24 Energy band structure of gray tin (after Groves and Paul, ref. 11).

states (see Chap. 3a) which increases with the number of valleys (4 in <111> direction) and with the effective mass (which is inversely proportional to the curvature of the valley). Therefore gray tin although actually being a semimetal in electrical measurements acts like a semiconductor having a gap of 0.09 eV.

The fact that PbTe-SnTe mixed crystals are semiconducting on both sides of the L_6^+ - L_6^- crossing point can be explained by the fact that these states have only a twofold spin degeneracy. Otherwise, as in the CdTe-HgTe crystal system, one side would be semimetallic. A review on narrow-band-gap semiconductors has been given in ref.10.

2d. Constant Energy Surfaces

The $\epsilon(\vec{k})$ diagrams which have been calculated for silicon [1] and germanium [2] are shown in Figs.2.25 and 2.26. If in Fig.2.25 a cut is made parallel to the $\vec{k}$-axis not too far above the edge of the conduction band, 2 points are found which for silicon are on the <100> axis, for germanium on the <111> axis.

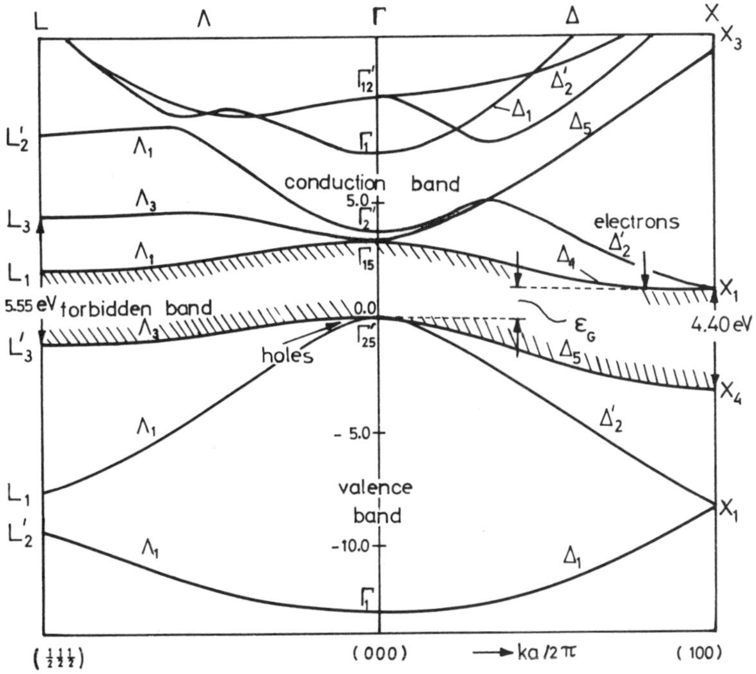

Fig.2.25 Energy band structure of silicon (after Cardona and Pollak, ref.1).

---

[1] M.Cardona and F.H.Pollak, Phys.Rev. 142 (1966) 530.

[2] J.C.Phillips, D.Brust, and F.Bassani: Proc.Int.Conf.Phys.Semicond. Exeter. 1962,p.564. London: The Institute of Physics and The Physical Society.

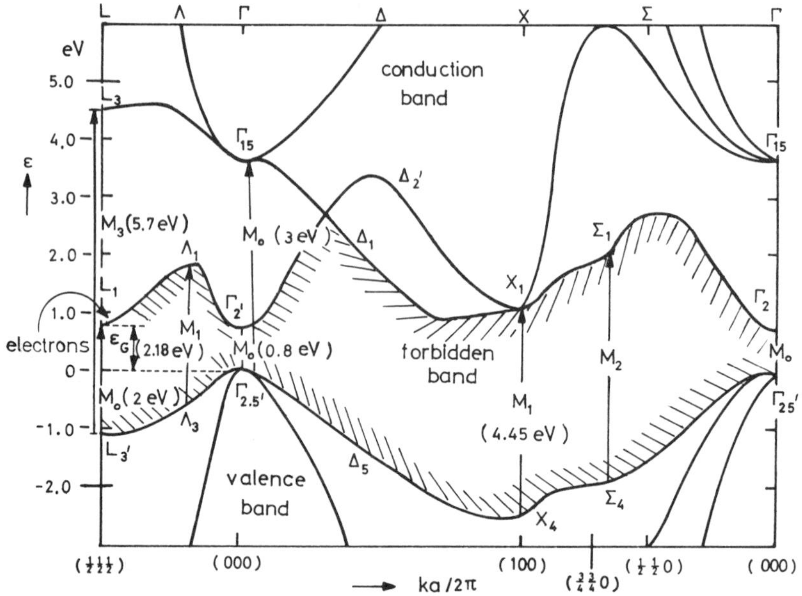

Fig.2.26 Energy band structure of germanium. The indicated transitions will be discussed in Chap.11i (after Phillips et al., ref.2).

In 3-dimensional $\vec{k}$-space these points become surfaces of constant energy. These surfaces turn out to be ellipsoids of revolution with their long axis on <100> and equivalent axes in silicon and on <111> and equivalent axes in germanium. In silicon there are six equivalent energy minima (or "valleys") and for a given value of energy there are six ellipsoids in $\vec{k}$-space (Fig.2.27). In germanium the valleys are at the intersections of <111> directions with the surface of the Brillouin zone. So there are eight half-ellipsoids which form four complete valleys (Fig.2.28).

Similar cuts near the valence band edge yield pairs of constant-energy points which in 3-dimensional $\vec{k}$-space become warped spheres around $\vec{k} = 0$ (Figs.2.29 and 2.30; the warping has been exaggerated in these figures). These warped spheres are concentric, one of them representing heavy holes and the other one light holes (in Eq.(2a32) the anisotropy was neglected). For comparison it may be interesting to consider constant energy surfaces in metals [3] which are often multiply-connected at the L point. These are denoted as "open Fermi surfaces".

---

[3] A.B.Pippard, Phil.Trans.A250 (1957) 325; see also W.Brauer: Einführung in die Elektronentheorie der Metalle. Leipzig: Geest und Portig K.G. 1966; J.M.Ziman, Contemporary Physics 3 (1962) 241.

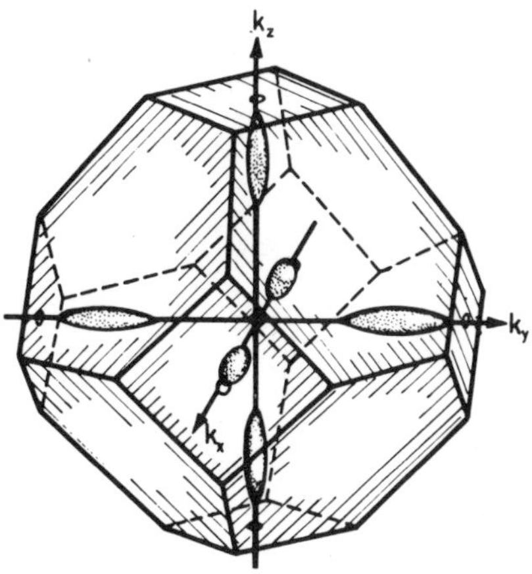

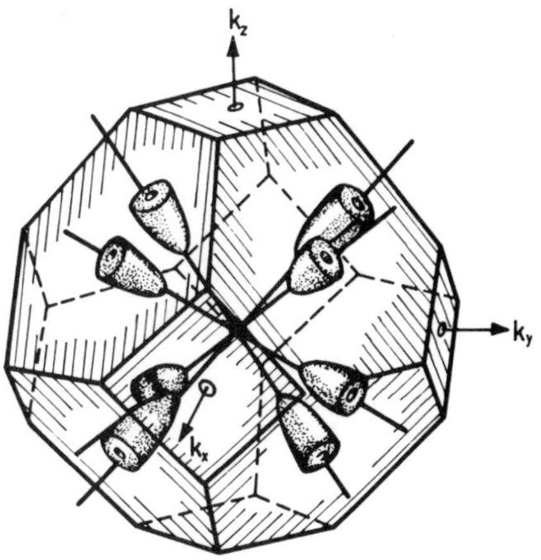

Fig. 2.27 Surfaces of constant energy in $\vec{k}$-space for the conduction band edge of silicon. The first Brillouin zone is the same as shown in Fig. 2.16 (after J.P. McKelvey: Solid-State and Semiconductor Physics. New York: Harper and Row. 1966).

Fig. 2.28 Surfaces of constant energy in $\vec{k}$-space for the conduction band edge of germanium: 8 half-ellipsoids of revolution centered at L points on the zone boundary (after J.P. McKelvey: Solid-State and Semiconductor Physics. New York: Harper and Row. 1966).

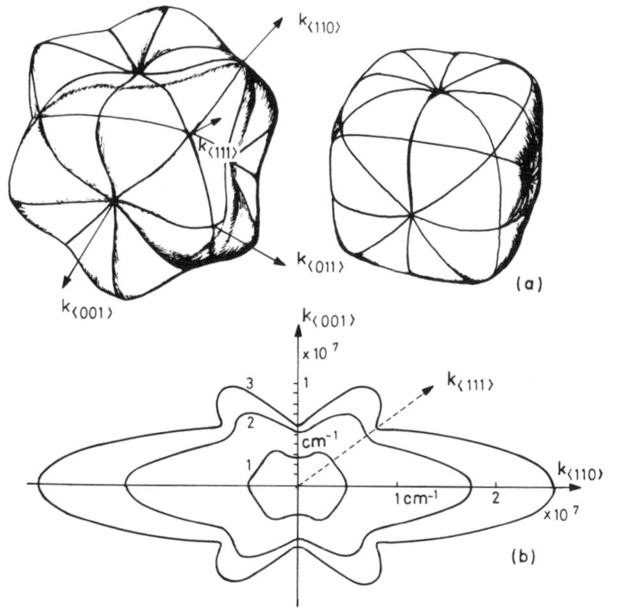

Fig. 2.29a Surfaces of constant energy in $\vec{k}$-space for the heavy-hole valence band edge of silicon and germanium. Right-hand side: Low energy. Left-hand side: High energy.

Fig. 2.29b Cross sections of the constant energy surfaces in the [110] plane.

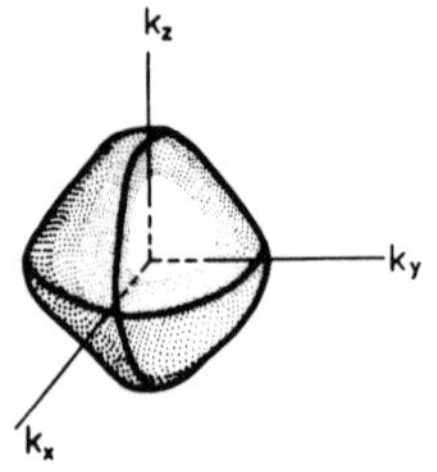

Fig. 2.30 Surface of constant energy in $\vec{k}$-space for the light-hole valence band edge of silicon and germanium.

3. Semiconductor Statistics

The periodic potential distribution of an electron in a crystal shown in Fig.2.5 involves N discrete levels if the crystal contains N atoms as we have seen in Fig. 2.9. A discussion of these levels can be confined to the first Brillouin zone. We have seen in the last chapter that due to the crystal periodicity the electron wave functions, which in one dimension are $\psi(x) = u(x) \exp(ikx)$, have also to be periodic ("Bloch functions"). Hence from

$$u(x + Na) = u(x) \tag{3.1}$$

and

$$\exp(ikx + ikNa)\, u(x + Na) = \exp(ikx)\, u(x) \tag{3.2}$$

we obtain

$$\exp(ikNa) = 1 \tag{3.3}$$

or

$$k = n2\pi/Na\,; \quad n = 0, \pm 1, \pm 2, \cdots \pm N/2 \tag{3.4}$$

where a is the lattice constant. We notice that Eq.(3.1) is actually valid for a ring-shaped chain which means that we neglect surface states (see Chap. 14a). Since for the first Brillouin zone k has values between $-\pi/a$ and $+\pi/a$, we find that the integer n is limited to the range between $-N/2$ and $+N/2$. In Fig.3.1 the discrete levels are given for a "crystal" consisting of N = 8 atoms.

How are the N electrons distributed among these levels including impurity levels if there are any ? This problem can be treated with statistical methods since a real crystal will consist not only of N = 8 atoms but of $N \approx 10^{23}$ atoms, a number which is $\gg 1$.

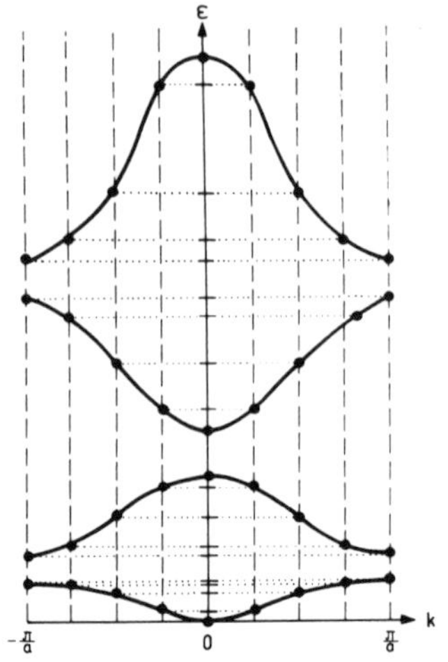

Fig.3.1 Discrete energy levels for a "crystal" consisting of eight atoms.

3a. Fermi Statistics

The following assumptions are made :

1.Electrons cannot be distinguished from one another. 2.Each level of a band can be occupied by not more than two electrons with opposite spin. This is due to the Pauli exclusion principle originally formulated for electrons in an atom. 3.Considering only singly ionized impurities for simplicity we also postulate that each impurity level can be occupied by just one electron. We shall see later on that these considerations can similarly be applied to holes.

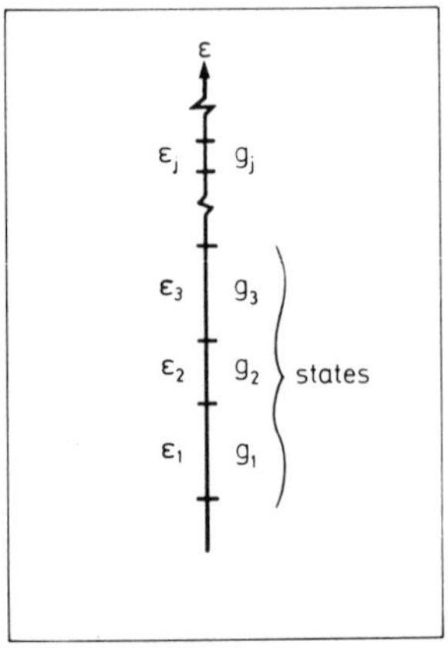

Fig.3.2 Energy levels $\epsilon_1, \epsilon_2, \cdots$ in energy bands, with $g_1, g_2, \cdots$ states, respectively.

Bands containing n electrons may consist of N levels $\epsilon_1, \epsilon_2, \cdots \epsilon_N$ with $g_1, g_2, \cdots g_N$ states, respectively, as shown schematically in Fig.3.2. $n_j < g_j$ of these states are assumed to be occupied by one electron each and hence g_j-n_j be empty. Now we will calculate Boltzmann's "thermodynamic probability" W [1].The number of times we can exchange g_j states with one another is $g_j!$

[1] This calculation has been adapted from E.Spenke: Electronic Semiconductors, chap.VIII. New York: McGraw-Hill. 1958.

which is often denoted by $\Gamma(g_j + 1)$. However, since electrons were assumed to be indistinguishable, an exchange of states should count only if an occupied state is exchanged with an unoccupied state. Hence by $g_j!$ we counted too many exchanges, namely $n_j!$ exchanges of n_j occupied states among one another and $(g_j - n_j)!$ exchanges of $g_j - n_j$ unoccupied states among one another, leaving only

$$W_j = \frac{g_j!}{n_j!\,(g_j - n_j)!} \tag{3a.1}$$

distinguishable distributions. Considering all N sets of states we obtain for the total number of distinguishable distributions

$$W = \prod_j W_j = \prod_j \frac{g_j!}{n_j!\,(g_j - n_j)!} \tag{3a.2}$$

Now we assume that the gap of a group IV semiconductor contains N_D donor levels of which N_{Dx} are occupied by one electron each and thus are neutral. At first sight one might expect the number of distinguishable distributions of the N_{Dx} electrons among N_D levels to be $N_D!/N_{Dx}!\,(N_D - N_{Dx})!$. In fact,the number is larger. A vacancy at an empty donor level may be occupied by an electron of either spin but once an electron is on the donor another one will not be taken up. The two possibilities for the spin of the electron adds a factor of $g_D = 2$ to the distribution number. For N_{Dx} electrons this factor is $g_D^{N_{Dx}}$ with the result that W is

$$W = g_D^{N_{Dx}} \frac{N_D!}{N_{Dx}!\,(N_D - N_{Dx})!} \tag{3a.3}$$

In this case of a donor it is the <u>end</u> product of the occupation (see Eq.(1c.5)) namely the neutral donor that has an unpaired electron. In the case of an acceptor it is <u>before</u> the reaction

$$A^x + e^- \rightleftharpoons A^- \tag{3a.4}$$

begins that the impurity contains an unpaired electron (one of 3 valence electrons of e.g. indium). Since $N_{Ax} = N_A - N_{A^-}$ we obtain for W in this case

$$W = g_A^{N_A - N_{A^-}} \frac{N_A\,!}{N_{A^-}!\,(N_A - N_{A^-})!} \tag{3a.5}$$

where $g_A = 2$. If there are both donors and acceptors present the total number of distinguishable distributions is given by

$$W = g_D^{N_{Dx}} \frac{N_D!}{N_{Dx}!\,(N_D - N_{Dx})!}\, g_A^{N_A - N_{A^-}} \frac{N_A!}{N_{A^-}!\,(N_A - N_{A^-})!} \cdot \prod_j \frac{g_j!}{n_j!\,(g_j - n_j)!} \tag{3a.6}$$

The most probable distribution of the electrons over all the states is obtained from the equation [2]

$$dW/dn_j = 0 \tag{3a.7}$$

which is subject to the accessory conditions that both the total number of electrons, n, and the total energy, U, remain constant. For the differentiation of $n_j!$ we can use Stirling's formula in the form

$$\ln n_j! = n_j \ln n_j - n_j \;;\quad d \ln n_j!/d\,n_j = \ln n_j \tag{3a.8}$$

valid for $n_j \gg 1$. Obviously it is more convenient to differentiate ln W than W. By adopting the method of Lagrange undetermined multipliers, α und β, and introducing the donor and acceptor energy levels, ϵ_D and ϵ_A, respectively, the sum

$$\Sigma = \ln W + \alpha(n - N_{Dx} - N_{A^-} - \sum_j n_j) + \beta(U - N_{Dx}\epsilon_D - N_{A^-}\epsilon_A - \sum_j n_j\epsilon_j) \tag{3a.9}$$

subject to the following differentiations has to vanish:

$$0 = \frac{d\Sigma}{dN_{Dx}} = \ln g_D - \ln N_{Dx} + \ln(N_D - N_{Dx}) - \alpha - \beta\epsilon_D \tag{3a.10}$$

$$0 = \frac{d\Sigma}{dN_{A^-}} = -\ln g_A - \ln N_{A^-} + \ln(N_A - N_{A^-}) - \alpha - \beta\epsilon_A \tag{3a.11}$$

$$0 = \frac{d\Sigma}{dn_j} = -\ln n_j + \ln(g_j - n_j) - \alpha - \beta\epsilon_j \tag{3a.12}$$

These equations are solved for N_{Dx}, N_{A^-}, and n_j :

---

[2] See e.g. E.Schrödinger: Statistical Thermodynamics. Cambridge: Univ.Press. 1948.

$$N_{Dx} = N_D / \{ \frac{1}{g_D} \exp(\alpha + \beta\epsilon_D) + 1 \} \tag{3a.13}$$

$$N_{A^-} = N_A / \{ g_A \exp(\alpha + \beta\epsilon_A) + 1 \} \tag{3a.14}$$

$$n_j = g_j / \{ \exp(\alpha + \beta\epsilon_j) + 1 \} \tag{3a.15}$$

The maximum value W_{max} of W, denoted as "thermodynamic probability", is obtained from Eq.(3a.5) by substituting for N_{Dx}, N_{A^-}, and n_j the values just obtained

$$\ln W_{max} = \alpha(N_{Dx} - N_{Ax} + \sum_j n_j) + \beta(N_{Dx}\epsilon_D - N_{Ax}\epsilon_A + \sum_j n_j\epsilon_j) + $$
$$+ N_D \ln\{1 + g_D \exp(-\alpha - \beta\epsilon_D)\} + N_A \ln\{1 + g_A \exp(\alpha + \beta\epsilon_A)\} + \tag{3a.16}$$
$$+ \sum_j g_j \,.\, \ln\{1 + \exp(-\alpha - \beta\epsilon_j)\}$$

According to Boltzmann, the entropy S is given by

$$S = k_B \ln W_{max} \tag{3a.17}$$

where k_B is Boltzmann's constant. The "free energy" F is given by

$$F = U - TS \tag{3a.18}$$

where the temperature T is defined by the equation

$$\frac{1}{k_B T} = \left(\frac{\partial \ln W_{max}}{\partial U}\right)_{n = const} \tag{3a.19}$$

and n and U are obtained by equating the expressions in parenthesis in Eq.(3a.9) to zero.

The Fermi energy [1] ζ (also denoted as "chemical potential" or "Gibbs' potential") is defined by the equation

$$\zeta = (\partial F / \partial n)_{T = const} \tag{3a.20}$$

These definitions applied to the previous equations yield for the Lagrange multipliers α and β the following physical interpretations to be derived below :

$$1/k_B T = \beta \tag{3a.21}$$

and

$$\zeta/k_BT = -a \tag{3a.22}$$

Let us first prove Eq.(3a.21). From Eqs.(3a.16) and (3a.19) we obtain

$$\frac{1}{k_BT} = \frac{\partial a}{\partial U}(n-N_A)+\beta+\frac{\partial \beta}{\partial U}(U-N_A\epsilon_A)-N_D\frac{g_D\exp(-a-\beta\epsilon_D)}{1+g_D\exp(-a-\beta\epsilon_D)}\left(\frac{\partial a}{\partial U}+\frac{\partial \beta}{\partial U}\epsilon_D\right)+$$

$$+N_A\frac{g_A\exp(a+\beta\epsilon_A)}{1+g_A\exp(a+\beta\epsilon_A)}\left(\frac{\partial a}{\partial U}+\frac{\partial \beta}{\partial U}\epsilon_A\right)-\sum_j g_j\frac{\exp(-a-\beta\epsilon_j)}{1+\exp(-a-\beta\epsilon_j)}\cdot \tag{3a.23}$$

$$\cdot\left(\frac{\partial a}{\partial U}+\frac{\partial \beta}{\partial U}\epsilon_j\right)$$

Using Eqs.(3a.13-15) this can be simplified to give

$$\frac{1}{k_BT} = \frac{\partial a}{\partial U}n+\beta+\frac{\partial \beta}{\partial U}U-(N_{Dx}+N_{A^-}+\sum_j n_j)\frac{\partial a}{\partial U}-$$

$$-(\epsilon_D N_{Dx}+\epsilon_A N_{A^-}+\sum_j \epsilon_j n_j)\frac{\partial a}{\partial U} \tag{3a.24}$$

Now the factors of the last two terms are n and U, respectively, and we obtain Eq.(3a.21). Eq.(3a.22) is derived quite similarly from Eqs.(3a.20) and (3a.16).

The ratio n_j/g_j is denoted as the "thermal equilibrium probability of occupancy" of a state of energy ϵ_j. From Eq.(3a.15) this is found to be

$$n_j/g_j = [\exp\{(\epsilon_j-\zeta)/k_BT\}+1]^{-1} \tag{3a.25}$$

which is called the "Fermi-Dirac distribution function" $f(\epsilon_j)$. It is plotted in Fig.3.3. For a finite temperature T the step at $\epsilon = \zeta$ has a width of the order of

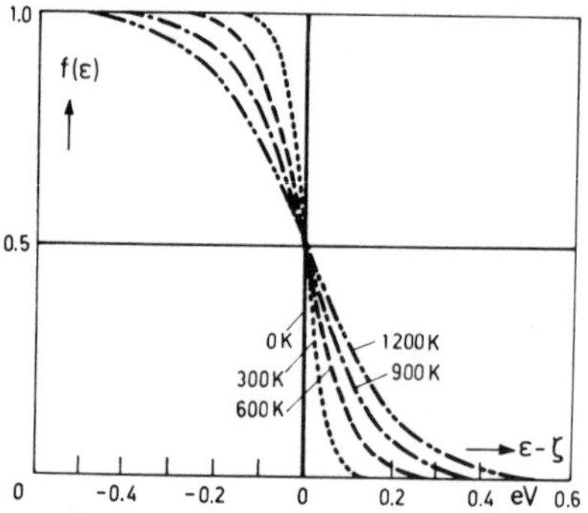

Fig.3.3 Fermi-Dirac distribution function for various values of the temperature. For the temperature dependence of ζ which is not indicated here see Eq.(3a.30) for extrinsic conduction and Eq.(3a.37) for intrinsic conduction.

k_BT and the high-energy tail of the function is well approximated by an exponential :

$$n_j/g_j \approx \exp\{-(\epsilon_j - \zeta)/k_BT\} \propto \exp(-\epsilon_j/k_BT) \qquad (3a.26)$$

This is the "Maxwell-Boltzmann distribution function". If the Fermi level is in the gap and separated by more than 4 k_BT from either band edge, the semiconductor is called "non-degenerate" and the distribution function (3a.26) may be applied to the "gas" of carriers.

The states of energy ϵ_j shall now be assumed to form an energy band. The sum over all states has to be replaced by an integral. Let us take for simplicity a parabolic band where

$$\epsilon = \epsilon_c + \hbar^2 k^2/2m_n \qquad (3a.27)$$

and the electron effective mass, m_n, is a scalar quantity. The surfaces of constant energy in $\vec{k}$-space are concentric spheres. Since the crystal momentum is given by $\hbar\vec{k}$, the volume element in phase space is $dxdydzd(\hbar k_x)d(\hbar k_y)d(\hbar k_z)$. According to Quantum Statistics phase space can be thought to consist of cells of volume $h^3 = (2\pi\hbar)^3$ with up to 2 electrons of opposite spins per cell. Integration over coordinate space yields the crystal volume V. Hence the number of states in an energy range $d\epsilon$ is given by

$$g(\epsilon)d\epsilon = 2\,\frac{V\hbar^3 d^3k}{(2\pi\hbar)^3} = 2\,\frac{V}{4\pi^2}\cdot\left(\frac{2m_n}{\hbar^2}\right)^{3/2}(\epsilon-\epsilon_c)^{1/2}\,d\epsilon \qquad (3a.28)$$

The total concentration of carriers in the band is then given by

$$n = \frac{1}{V}\int_{\epsilon_c}^{\infty} f(\epsilon)g(\epsilon)d\epsilon \qquad (3a.29)$$

where $f(\epsilon)$ is the Fermi-Dirac distribution given by the right-hand side of Eq. (3a.25). Since this distribution decreases exponentially at large energies, the upper limit of integration may be taken as infinity without introducing appreciable error. With $g(\epsilon)$ given by Eq.(3a.28) we find for the integral

$$n = 2\,\frac{1}{4\pi^2}\left(\frac{2m_n}{\hbar^2}\right)^{3/2}\int_{\epsilon_c}^{\infty}\frac{(\epsilon-\epsilon_c)^{1/2}d\epsilon}{\exp\{(\epsilon-\zeta)/k_BT\}+1} = N_c F_{1/2}(\zeta_n/k_BT) \qquad (3a.30)$$

where we have introduced the Fermi energy relative to the band edge, $\zeta_n = \zeta - \epsilon_c$, the "effective density of states"

$$N_c = 2\frac{1}{4\pi^2}\left(\frac{2m_n k_B T}{\hbar^2}\right)^{3/2} \cdot (1/2)! = 2\left(\frac{m_n k_B T}{2\pi\hbar^2}\right)^{3/2} \quad (3a.31)$$

and the Fermi integral

$$F_j(\eta) = \frac{1}{j!}\int_0^\infty \frac{x^j dx}{\exp(x-\eta)+1} \quad (3a.32)$$

for $j = 1/2$. This integral has been tabulated [3]. For $\eta > 1.25$ it is approximated by

$$F_{1/2}(\eta) = (4\eta^{3/2}/3\pi^{1/2}) + \pi^{3/2}/6\eta^{1/2} \quad (3a.33)$$

with an error of $< 1.5\%$. For $\eta < -4$ which is the case of non-degeneracy (Maxwell-Boltzmann statistics), $F_j(\eta)$ is approximately

$$F_j(\eta) = \exp(\eta) \quad (3a.34)$$

$\eta_n = \zeta_n/k_B T$ is denoted as the "reduced Fermi energy". A corresponding result is obtained for holes in the valence band (subscript v). For T = 300 K and $m_n = m_p = m_0$, $N_c = N_v = 1.27 \times 10^{19}/cm^3$. Hence the constant C of Eq.(1c.2) is obtained from the product np for non-degeneracy

$$C = N_c N_v / T^3 = 4\left(\frac{\sqrt{m_n m_p}\, k_B}{2\pi\hbar^2}\right)^3 = \left(\frac{m_n}{m_0}\frac{m_p}{m_0}\right)^{3/2} \frac{9.28 \times 10^{31}}{cm^6 K^3} \quad (3a.35)$$

In an intrinsic semiconductor n equals p and from

$$n = N_c \exp\{(\zeta-\epsilon_c)/k_B T\}; \quad p = N_v \exp\{(\epsilon_v-\zeta)k_B T\} \quad (3a.36)$$

the Fermi energy

$$\zeta = \frac{1}{2}(\epsilon_c+\epsilon_v) + \frac{3}{4} k_B T \ln(m_p/m_n) \quad (3a.37)$$

is obtained. For equal effective masses of electrons and holes the Fermi level is independent of the temperature and located in the middle of the gap. For $m_n \ll m_p$ the Fermi level approaches the conduction band edge with increasing temperature.

[3] J.S.Blakemore: Semiconductor Statistics, appendix B. Oxford: Pergamon. 1962.

If two bands, Γ and L, with band edges at ϵ_c and $\epsilon_c + \Delta_L$, respectively, are considered the carrier concentration

$$n = n_\Gamma + n_L = N_{c\Gamma} F_{1/2}(\zeta_n/k_BT) + N_{cL} F_{1/2}(\{\zeta_n - \Delta_L\}/k_BT) \qquad (3a.38)$$

where

$$N_{c\Gamma} = 2\left(\frac{m_\Gamma k_BT}{2\pi\hbar^2}\right)^{3/2} \quad \text{and} \quad N_{cL} = 2\left(\frac{m_L k_BT}{2\pi\hbar^2}\right)^{3/2} \qquad (3a.39)$$

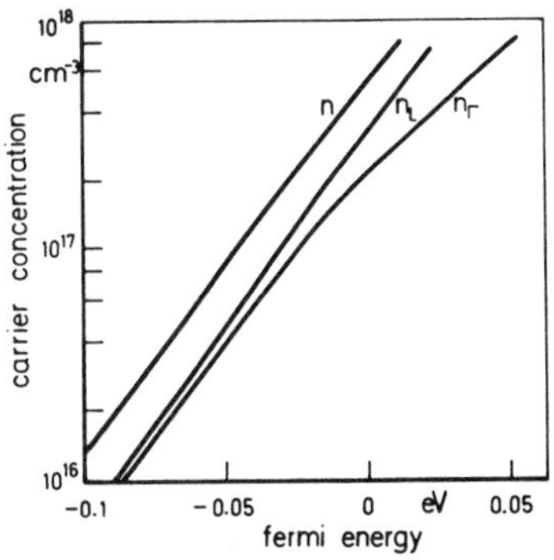

Fig.3.4 Calculated light-(Γ) and heavy-(L) electron concentrations in GaSb at 300 K, as a function of the Fermi energy (after W.Jantsch, Thesis, Univ.Wien, Austria, 1971).

Fig.3.4 shows n, n_Γ, and n_L as a function of ζ_n for n-GaSb at 300 K where $m_\Gamma/m_o = 0.05$; $m_L/m_o = 0.74$; $\Delta_L = 101.7$ meV.

Let us finally calculate the entropy density S given by Eqs.(3a.17) and (3a.16) where we replace the sums by the corresponding integrals :

$$\left.\begin{aligned} \sum_j n_j &= N_c F_{1/2}(\eta_n) \;; \\ \sum_j n_j \epsilon_j &= \frac{3}{2} k_BT\, N_c F_{3/2}(\eta_n) \end{aligned}\right\} \qquad (3a.40)$$

and

$$\sum g_j \ln\left[1+\exp\{(\zeta-\epsilon_j)/k_BT\}\right] = \frac{N_c}{(1/2)!}\int_0^\infty dx.x^{1/2}\ln(1+e^{\eta_n - x}) = N_c F_{3/2}(\eta_n) \qquad (3a.41)$$

The integral is solved by partial integration. The total electron concentration is then given by

$$n = N_{Dx} + N_{A^-} + N_c F_{1/2}(\eta_n) \qquad (3a.42)$$

and the internal energy density U_n (subscript n for electrons) by

$$U_n = N_{Dx}\epsilon_D + N_{A^-}\epsilon_A + \frac{3}{2} k_BTN_c F_{3/2}(\eta_n) \qquad (3a.43)$$

For the entropy density a rather lengthy expression is obtained [4] which is simplified by neglecting all donors and acceptors :

---

[4] See e.g. ref.1 Chap.VIIIh.

$$S_n = -(\zeta/T)n + U_n/T + k_B N_c F_{3/2}(\eta_n) = (\frac{5}{3} U_n - n\zeta)/T \quad (3a.44)$$

From Eq.(3a.18) we now find for the free energy

$$F_n = n\zeta - \frac{2}{3} U_n \quad (3a.45)$$

Eq.(3a.19) in the form

$$1/T = (\partial S_n/\partial U_n)_{n = \text{const}} \quad (3a.46)$$

and Eq.(3a.20) are easily verified by taking into account that $\zeta=\zeta(n)$, $U_n=U_n(n)$, and $\partial U_n/\partial \zeta = \frac{3}{2}$ n.

For the consideration of holes we introduce the concentration p_j of unoccupied states in a band

$$p_j = g_j - n_j = g_j/[1 + \exp\{(\zeta-\epsilon_j)/k_B T\}] \quad (3a.47)$$

which is similar to Eq.(3a.25) except that energies including the Fermi energy have the opposite sign. The free hole concentration is then given by

$$p = N_v F_{1/2}(\eta_p) \quad (3a.48)$$

and the internal energy density U_p of free holes by

$$U_p = -\frac{3}{2} k_B T N_v F_{3/2}(\eta_p) \quad (3a.49)$$

The entropy density is similarly

$$S_p = -(\frac{5}{3} U_p - p\zeta)/T \quad (3a.50)$$

For mixed conduction the total entropy density S is then given by

$$S = S_n + S_p \quad (3a.51)$$

In deriving these equations we have assumed the density of states, $g(\epsilon)$, given by Eq.(3a.28). In Chap.9b we will consider quantum effects in a strong magnetic field which yield a different density of states.

3b. Occupancy Factors for Impurity Levels

For the ratio of neutral and total donor densities we obtain from Eqs.(3a.13), (3a.21), and (3a.22)

$$N_{Dx}/N_D = [\frac{1}{g_D} \exp\{(\epsilon_D - \zeta)/k_BT\} + 1]^{-1} \tag{3b.1}$$

Since N_D equals $N_{Dx} + N_{D+}$ we have for the ratio N_{Dx}/N_{D+} :

$$N_{Dx}/N_{D+} = g_D .\exp\{(\zeta - \epsilon_D)/k_BT\} \tag{3b.2}$$

According to Eqs.(3a.30) and (3a.34) for nondegeneracy the concentration of free electrons is given by

$$n = N_c .\exp\{(\zeta - \epsilon_c)/k_BT\} \tag{3b.3}$$

where N_c is the effective "density of states" in the conduction band. Introducing the donor ionization energy $\Delta\epsilon_D$ by

$$\Delta\epsilon_D = \epsilon_c - \epsilon_D \tag{3b.4}$$

we obtain from Eq.(3b.2) the law of mass action

$$N_{Dx}/(N_{D+}n) = (g_D/N_c).\exp(\Delta\epsilon_D/k_BT) \tag{3b.5}$$

The energy $\Delta\epsilon_D$ is the heat produced by the reaction

$$D^+ + e^- \rightleftharpoons D^x + \Delta\epsilon_D \tag{3b.6}$$

If in Eq.(3b.5) N_{Dx} is replaced by $N_D - N_{D+}$ and the equation is solved for N_{D+} this yields

$$N_{D+} = \frac{N_D}{\frac{g_D n}{N_c} \cdot \exp(\Delta\epsilon_D/k_BT) + 1} \tag{3b.7}$$

The charge neutrality requires the number of positive charges to be equal to the number of negative charges :

$$p + N_{D+} = n + N_{A-} = n + N_A - N_{Ax} \tag{3b.8}$$

For an n-type semiconductor the holes and the neutral acceptors may be neglected. Hence

$$N_{D^+} \approx n+N_A \tag{3b.9}$$

which with Eq.(3b.7) yields

$$\frac{n(n+N_A)}{N_D-N_A-n} = \frac{N_c}{g_D} \exp(-\Delta\epsilon_D/k_BT) \tag{3b.10}$$

From this equation and a measurement of the carrier density n as a function of temperature it is possible to determine the "activation energy" $\Delta\epsilon_D$ of a donor if we know how N_c and g_D depend on temperature T. According to Eq.(3a.39) N_c is proportional to $T^{3/2}$. In Chap.3a the impurity spin degeneracy g_D has been introduced with a value of 2. This is true only at high temperatures. For the case of low temperatures (e.g. < 10 K for Sb in Ge) we have to consider, that what we normally label as the "impurity level" actually is the donor triplet state (parallel spins) and that there also is a singulet state (antiparallel spins) at an energy level which is lower than the triplet level by $\delta\epsilon = 0.57$ meV for Sb in Ge ($\Delta\epsilon_D = 9.6$ meV) and $\delta\epsilon = 4.0$ meV for As in Ge ($\Delta\epsilon_D = 12.7$ meV) [1]. This energy separation is of importance e.g. for the low-temperature ultrasonic attenuation which will be discussed at the end of Chap.6a. At present we notice that in Eq.(3b.10) we have to replace $\exp(\Delta\epsilon_D/k_BT)$ by $\frac{1}{4}\exp\{(\Delta\epsilon_D+\delta\epsilon)/k_BT\} + \frac{3}{4}\exp(\Delta\epsilon/k_BT)$ where 1/4 and 3/4 are the statistical weights of the singulet and the triplet states, respectively. Hence, Eq.(3b.10) becomes with $g_D = 2$

$$\frac{n(n+N_A)}{N_D-N_A-n} = \frac{N_c}{2} \cdot \frac{4}{\exp(\delta\epsilon/k_BT)+3} \exp(-\Delta\epsilon_D/k_BT) \tag{3b.11}$$

We can formally retain Eq.(3b.10) if we now define g_D as

$$g_D = \{\exp(\delta\epsilon/k_BT)+3\}/2 = g_D(T) \tag{3b.12}$$

For such low temperatures where $\delta\epsilon$ becomes important we can assume $n \ll N_A$ and $N_D - N_A$ and approximate Eq.(3b.10) by

$$\log(2ng_DT^{-3/2}) = \log\left(\frac{N_D-N_A}{N_A}\frac{2N_c}{T^{3/2}}\right) - \frac{\Delta\epsilon_D}{k_BT}\log e \tag{3b.13}$$

where log is to the base 10. According to Eq.(3a.31) $N_c/T^{3/2}$ is temperature independent. Hence a plot of the left-hand side of Eq.(3b.13) vs 1/T yields both $\Delta\epsilon_D$ and the compensation ratio $N_A/(N_D+N_A)$.

[1] H.Fritzsche, Phys.Rev.120 (1960) 1120.

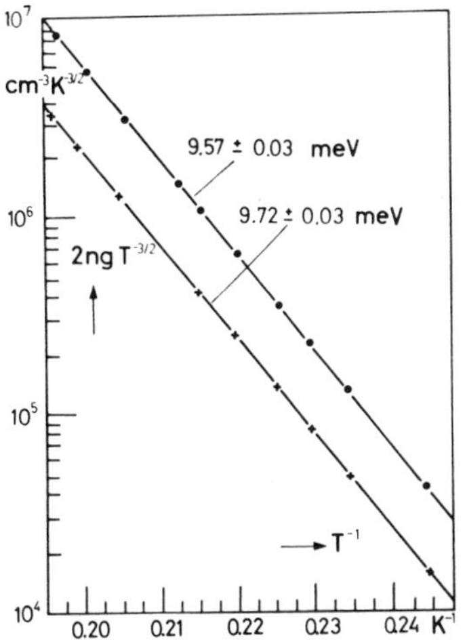

Fig.3.5 Experimental data of $2ng_DT^{-3/2}$ as a function of the reciprocal temperature for a determination of the energy separation between the conduction band and the donor triplet level derived from the unperturbed fourfold degenerate ground state of Sb in Ge (after Koenig et al., ref.2).

Fig.3.5 shows experimental data by Koenig et al. [2] of $2ng_DT^{-3/2}$ in a logarithmic scale vs T^{-1} for 2 samples of Sb-doped Ge in a temperature range between 4.0 and 5.1 K. The straight lines have slopes corresponding to $\Delta\epsilon_D = 9.57$ and 9.72 meV. The absolute values indicate $N_A/(N_D - N_A) = 0.036$ and 0.067, respectively. From this and the carrier density at high temperatures where it is $n = N_D - N_A$, the total impurity concentration $N_D + N_A$ could be determined. For some samples it turned out to be necessary to consider two types of donors with different energy levels, e.g. Sb and As.

For a p-type nondegenerate semiconductor, the equations (3b.3), (3b.4), and (3b.10) have to be replaced by

$$p = N_v \exp\{(\epsilon_v - \zeta)/k_BT\} \tag{3b.14}$$

$$\Delta\epsilon_A = \epsilon_A - \epsilon_v \tag{3b.15}$$

$$N_{A^-} = \frac{N_A}{\frac{g_A p}{N_v} \exp(\Delta\epsilon_A/k_BT) + 1} \tag{3b.16}$$

and

$$\frac{p(p + N_D)}{N_A - N_D - p} = \frac{N_v}{g_A} \exp(-\Delta\epsilon_A/k_BT) \tag{3b.17}$$

with quite similar meanings of the corresponding quantities. With careful experimental methods [3] it was possible to show in the temperature range 20-50 K that for In in Ge the high-temperature g-factor is 4 rather than 2 which takes account of the fact that at the valence band edge both light and heavy hole subbands are degenerate with one another.

In obtaining Eq.(3b.9) we have neglected the "minority carriers" which in this case are holes. These can be taken into account by calculating the product np from Eqs.(3b.3) and (3b.14) :

---

[2] S.H.Koenig, R.D.Brown III, and W.Schillinger, Phys.Rev.128 (1962) 1668, and literature cited there. See also D.Long, C.D.Motchenbacher, and J.Myers, J.Appl.Phys.30 (1959) 353; J.Blakemore: Semiconductor Statistics, Chap.3.2.4. Oxford: Pergamon. 1962.

[3] J.S.Blakemore, Phil.Mag.4 (1959) 560.

$$np = N_c N_v \exp(-\epsilon_G / k_B T) \tag{3b.18}$$

where the definition of the gap width

$$\epsilon_G = \epsilon_c - \epsilon_v \tag{3b.19}$$

has been used. The right-hand side of Eq.(3b.16) is denoted by n_i^2 where

$$n_i = \sqrt{N_c N_v} \exp(-\epsilon_G / 2k_B T) \tag{3b.20}$$

is the intrinsic concentration (see also Eq.(1c.2)). For the case of nearly but not exactly equal densities of electrons and holes one has to solve an equation of 4th power in either n or p which is obtained by combining Eqs.(3b.7), (3b.16), (3b.8), and (3b.18). The special case of nearly uncompensated n-type conductivity $n \gg n_i \gg p$ and N_A, yields an equation of only second order with the solution

$$n = N_{D^+} \approx \frac{N_D}{\frac{2n}{N_c} \exp(\Delta\epsilon_D / k_B T)} \approx \sqrt{\frac{1}{2} N_D N_C} \exp(-\Delta\epsilon_D / 2k_B T) \tag{3b.21}$$

In the exponential we find <u>half</u> the activation energy, $\frac{1}{2}\Delta\epsilon_D$. At somewhat lower temperatures, however, there is considerable freeze-out of carriers and hence $N_D > N_A \gg n \gg p$, and the concentration of ionized donors is nearly equal to that of acceptors since most of the electrons have just gone from donors to acceptors. From

$$\frac{1}{N_A} = \frac{1}{N_{D^+}} = \frac{1}{N_D}\left\{\frac{g_D n}{N_c} \exp(\Delta\epsilon_D / k_B T) + 1\right\} \tag{3b.22}$$

n is easily obtained

$$n = \frac{N_c}{g_D} \frac{N_D - N_A}{N_A} \exp(-\Delta\epsilon_D / k_B T) \tag{3b.23}$$

with the <u>full</u> activation energy $\Delta\epsilon_D$ in the exponential. Similar results are obtained for p-type conductivity. Fig.3.6 shows [4] experimental results obtained with p-type Si doped with 7.4×10^{14} acceptors/cm^3. At the transition from $n \gg N_D$ (at high temperature) to $n \ll N_D$ (at low temperature) there is a kink at 10^{11}/cm^3 which is the concentration of compensating donor impurities.

In an extrinsic semiconductor the Fermi level is close to the impurity level which determines the type of conductivity. This is clear from Eq.(3b.2) :

[4] N.B.Hannay: Semiconductors (N.B.Hannay,ed.) p.31. New York: Reinhold Publ.Co. 1959.

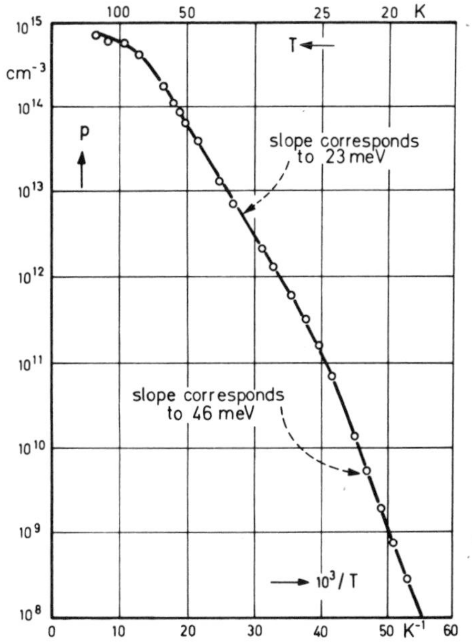

Fig.3.6 Hole concentration as a function of the reciprocal temperature for a silicon sample with a shallow acceptor of energy 46 meV and concentration N_A= 7.4 x $10^{14}/cm^3$, partly compensated by 1.0 x $10^{11}/cm^3$ donors (after F.J.Morin, published in ref.4).

In order to have an appreciable amount of impurities ionized (N_{D^+} of the order N_{Dx}), the difference $\zeta - \epsilon_D$ must be small compared to k_BT.

Quite interesting is the statistics of thermal ionization of double-donors and acceptors like e.g. Te : $Te^+ + e^- \leftrightharpoons Te^{++} + 2e^-$ which obviously is a double-donor [5]. In Chap.12b on photoconductivity, double-impurities will be discussed.

Fig.3.7 shows energy levels of impurities observed in Ge [6]. The numbers next to the chemical symbols are the energy intervals between the impurity levels and the nearest band edge in eVolt. Impurities known to yield either single-donors and acceptors have energy intervals of about 0.01 eVolt. These are denoted as "shallow-level impurities" or "hydrogen-type impurities". The latter notation is derived from the fact that the Rydberg energy $Ry = m_0 e^4/\{2(4\pi\kappa_0\hbar)^2\} = 13.6$ eV multiplied with the effective mass ratio, $m/m_0 = 0.12$, and divided by

[5] For the occupation probabilility of double donors see e.g. E.Spenke: Electronic Semiconductors, chap.VIII,1. New York: McGraw-Hill. 1965.

[6] For a review see e.g. E.M.Conwell, Proc.IRE 46 (1958) 1281; T.H.Geballe in ref.4, p.341 and 342.

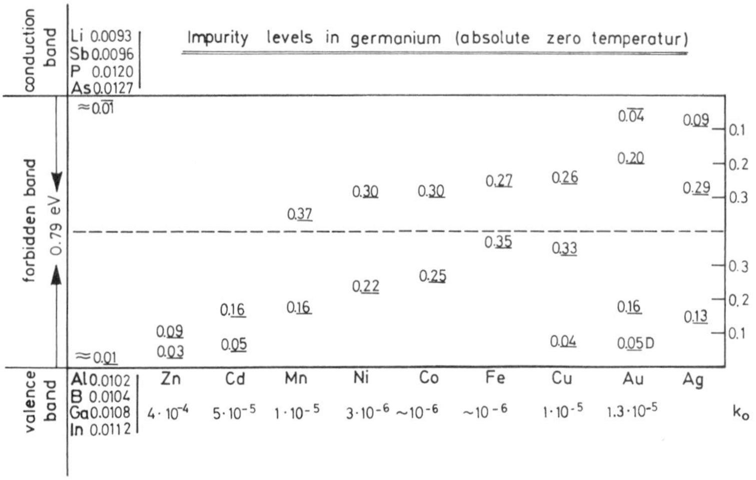

Fig.3.7 Impurity levels in germanium. All deep levels indicated are acceptors except for the lowest Au level. The numbers are energies in units of eV. Distribution coefficients k_o are given at the bottom of the figure (after R.Newman and W.W.Tyler: Solid State Physics (F.Seitz and D.Turnbull, eds.), vol.8, p.62. New York: Acad.Press. 1962.).

the square of the dielectric constant, $\kappa = 16$, yields an energy roughly equal to the energy interval of 0.01 eVolt given above. In this model the radius of the first electron orbit is the Bohr radius $a_B = 4\pi\kappa_o \hbar^2/m_o e^2 = 0.53$ Å multiplied by $\kappa/(m/m_o)$ which is about 70 Å and therefore much larger than the interatomic distance. This fact justifies the use of the macroscopic dielectric constant in the atomic model. In more rigorous quantum mechanical calculations the stress exerted by a possible misfit of a large impurity atom in the lattice has to be taken into account.

4. Charge and Energy Transport in a Non-degenerate Electron Gas

4a. Kinetic Theory of the Electron Gas

In the preceding chapters we have seen that a mobile charge carrier in a semiconductor has an effective mass m which is different from the free electron mass m_0. The effective mass takes care of the fact that the carrier is subject to the crystal field. In discussing the velocity distribution of the "gas" of carriers we found that the Fermi-Dirac distribution holds in general and that the Maxwell-Boltzmann distribution

$$f(v) \propto \exp(-mv^2/2k_BT) \tag{4a.1}$$

is an approximation of the former which is valid for "non-degenerate" semiconductors. Here the carrier density is small compared with the effective density of states N_c in the conduction band and N_v in the valence band, Eq.(3a.31). For these distributions no externally applied electric fields were assumed to be present. Instead, the calculations were based on the assumption of thermal equilibrium.

In this chapter we calculate the fluxes of charge and energy which are due to gradients of electric potential, of temperature, and of concentration of carriers. These fluxes are influenced by external magnetic fields and mechanical forces such as hydrostatic pressure and uniaxial stress. Very general relations between these fluxes and gradients were given by Onsager [1]. Rather than going into details of the thermodynamics of irreversible processes [2] we will take the simplified approach of the Boltzmann equation in the relaxation time approximation.

[1] L.Onsager, Phys.Rev. 37 (1931) 405; 38 (1931) 2265; see also e.g. A.C.Smith, J.F.Janak, and R.B.Adler: Electronic Conduction in Solids, Chap.3.5. New York: McGraw-Hill. 1967.

[2] S.R.de Groot: Thermodynamics of Irreversible Processes. Amsterdam: North-Holland Publ.Co. 1953.

4b. Electrical Conductivity and its Temperature Dependence

In the range of extrinsic conductivity the current density $\vec{j}$ which is due to an external electric field $\vec{E}$ is given by

$$\vec{j} = ne\vec{v}_d \tag{4b.1}$$

where $\vec{v}_d$ is the carrier drift velocity. The direction of $\vec{j}$ is from the positive to the negative pole while for electrons with $e < o$ the drift velocity $\vec{v}_d$ is opposite to $\vec{j}$. The absolute value of e has the value of the elementary charge which is also assumed to be the charge of the positive hole. Neglecting the thermal motion of the carriers the equation of motion is given by

$$d(m\vec{v}_d)/dt + m\vec{v}_d/\bar{\tau}_m = e\vec{E} \tag{4b.2}$$

where t is the time, $m\vec{v}_d$ the momentum of the carrier, and $\bar{\tau}_m$ is introduced as an average "momentum relaxation time". Relaxation means return to equilibrium. The second term describes the "friction" which the carriers experience on drifting through the crystal. This friction depends on the vibration of the individual atoms of the crystal lattice and therefore depends on crystal temperature. This is taken care of by a dependence of $\bar{\tau}_m$ on temperature.

In the steady state the first term vanishes and the drift velocity $\vec{v}_d$ is proportional to the electric field strength $\vec{E}$:

$$|\vec{v}_d| = \mu|\vec{E}| \tag{4b.3}$$

where the factor of proportionality, μ,

$$\mu = (|e|/m)\bar{\tau}_m \tag{4b.4}$$

is called the "mobility". If this is substituted in Eq.(4b.1) the current density $\vec{j}$ is given by

$$\vec{j} = \sigma\vec{E} \tag{4b.5}$$

where the conductivity σ is given by

$$\sigma = ne\mu = (ne^2/m)\bar{\tau}_m \tag{4b.6}$$

The inverse conductivity $1/\sigma = \rho$ is called "resistivity". The unit of resistivity is ohm-cm = Vcm/A. Eq.(4b.6) yields the unit of mobility as $cm^2/Vsec$ while Eq. (4b.4) yields the unit of mass as $V\,A\,sec^3/cm^2$ which is the same as 10^7 g.

In numerical calculations it is quite convenient to use the electromagnetic rather than the g-unit. The ratio*) e/m for $m=m_0$ has a value of $1.76 \times 10^{15}\ cm^2/Vsec^2$. A typical value of $\bar{\tau}_m$, which in certain simple cases is essentially the inverse vibration frequency of atoms in a crystal, has an order of magnitude value of 10^{-13} sec. This yields a mobility of 176 $cm^2/Vsec$. For electrons in n-type germanium m is only about one tenth of m_0 and $\bar{\tau}_m$ at room temperature some 10^{-13} sec. This yields a mobility which is in agreement in order of magnitude with the observed mobility of 3 900 $cm^2/Vsec$.

In looking again at Eqs.(4b.2)-(4b.5) it is obvious how the friction term leads to Ohm's law which states that σ is independent of the electric field strength. The notation "relaxation time" for $\bar{\tau}_m$ is clarified if we assume that at t=0 the electric field is switched off. The solution of Eq.(4b.2) is easily evaluated for this case. It is :

$$m\vec{v}_d = (m\vec{v}_d)_{t=0} \exp(-t/\bar{\tau}_m) \tag{4b.7}$$

The drift momentum decreases exponentially in a time $\bar{\tau}_m$ which is too short for us to be able to observe it directly. We shall see later on that in ac fields there is a phase shift between current and field which depends on the product of frequency and $\bar{\tau}_m$ (see Chap.4n).

In our model where the drifting electrons on their way occasionally collide with vibrating atoms of the crystal lattice, one would anticipate the existence of a mean free time between collisions. This "collision time" τ_c is of the order of τ_m. In Chap.6 this model will be refined and the various collision processes, appropriately called "scattering processes", will be discussed in detail. In general τ_m depends on the ratio of the carrier energy, ϵ, and the average energy of vibration of an atom which is of the order of k_BT. In special cases this relation takes the form

$$\tau_m = \tau_0 (\epsilon/k_BT)^r \tag{4b.8}$$

where the exponent r varies between -1/2 (for "acoustic deformation potential scattering") and +3/2 (for "ionized impurity scattering") and τ_0 is a factor of proportionality. For acoustic scattering the mean free path, l_{ac}, will be shown to be proportional to the inverse lattice temperature.

$$l_{ac} \propto T^{-1} \tag{4b.9}$$

Thus τ_0, which is given by

$$\tau_0 = l_{ac}/\sqrt{2k_BT/m} \tag{4b.10}$$

---

*) See table of constants at the end of the book.

where $\sqrt{2k_BT/m}$ is an average velocity of the carriers in equilibrium with the lattice, is proportional to $T^{-3/2}$.

The fact that τ_m in general is energy dependent requires a reformulation of its definition Eq.(4b.2). We have seen that there is a velocity distribution of carriers, f(v). The drift velocity $\vec{v}_d$ in an anisotropic material with an applied electric field in e.g. the z-direction will be given by its z-component

$$v_{dz} = \int_{-\infty}^{\infty} v_z f d^3v \Big/ \int_{-\infty}^{\infty} f d^3v \tag{4b.11}$$

where d^3v stands for $dv_x\, dv_y\, dv_z$. Similarly, $\bar{\tau}_m$ in Eq.(4b.2) is an <u>average</u> over the velocity distribution. Since τ_m is a function of velocity it would be appropriate to define it by the following equation which determines the relaxation of the distribution function $f(\vec{v})$.

$$\frac{df(v)}{dt} + \frac{f(v) - f_0(v)}{\tau_m(v)} = 0 \tag{4b.12}$$

where f_0 is the thermal-equilibrium distribution, i.e. without field. This is the "Boltzmann equation" in the relaxation time approximation. The "drift term" df/dt can be evaluated, if $dx/dt = v_x$; $dv_x/dt = (e/m)E_x$ etc. is taken into account:

$$\frac{df}{dt} = \frac{\partial f}{\partial t} + \frac{\partial f}{\partial x} v_x + \frac{\partial f}{\partial y} v_y + \frac{\partial f}{\partial z} v_z + \frac{\partial f}{\partial v_x}\frac{e}{m}E_x + \frac{\partial f}{\partial v_y}\frac{e}{m}E_y + \frac{\partial f}{\partial v_z}\frac{e}{m}E_z \tag{4b.13}$$

For dc fields with no temperature or concentration gradients in the crystal the first four terms on the right-hand side vanish. For $E_x = E_y = 0$ the Boltzmann equation becomes

$$\frac{\partial f}{\partial v_z}\frac{e}{m}E_z + \frac{f - f_0}{\tau_m} = 0 \tag{4b.14}$$

which can be written in the form

$$f = f_0 - \frac{e}{m}\tau_m \frac{\partial f}{\partial v_z} E_z \tag{4b.15}$$

We consider only small field intensities such that we may retain terms linear in E_z only. Therefore, in $\partial f/\partial v_z$ we may replace f by its equilibrium value f_0. If in a Legendre polynomial expansion of the derivative of the distribution function only the first two terms are retained, this is called the "diffusive approximation" of the distribution function.

$$f = f_0 - \frac{e}{m}\tau_m \frac{\partial f_0}{\partial v_z} E_z \tag{4b.16}$$

In an iterative process in which we replace f in $\partial f/\partial v_z$ by this value, a term in E_z^2 would be obtained. This E_z^2 term was supposed to be negligible at small field intensities. The case of high field intensities will be discussed in Chap.4m.

The calculation of the drift velocity from Eq.(4b.11) is straightforward. The first term in f, f_0, makes the integral in the numerator vanish since f_0 is an even function of v_z, $f_0(-v_z) = f_0(v_z)$.

$$\int_{-\infty}^{\infty} v_z f_0 d^3v = 0 \tag{4b.17}$$

Similarly, since

$$-\frac{\partial f_0}{\partial v_z} = v_z f_0 \frac{m}{k_B T} \tag{4b.18}$$

the second term in f makes the integral in the denominator vanish. Hence,

$$v_{dz} = \int_{-\infty}^{\infty} v_z [-(e/m)\tau_m(\partial f/\partial v_z)E_z] d^3v / \int_{-\infty}^{\infty} f_0 d^3v \tag{4b.19}$$

According to Eq.(4b.3) the mobility μ is defined by the ratio $|v_{dz}/E_z|$ and can be written in the form $(|e|/m)\langle\tau_m\rangle$ where the average momentum relaxation time $\langle\tau_m\rangle = \langle\tau_m(v)\rangle$ is given by

$$\langle\tau_m\rangle = \int_{-\infty}^{\infty} \tau_m(\vec{v}) v_z(-\partial f/\partial v_z) d^3v / \int_{-\infty}^{\infty} f_0 d^3v \tag{4b.20}$$

Later in this chapter we shall see that this kind of average is actually somewhat different from $\bar{\tau}_m$ defined by Eq.(4b.4).

We approximate $-\partial f/\partial v_z$ by $-\partial f_0/\partial v_z$ given for a Fermi-Dirac distribution function by

$$\partial f_0/\partial v_z = f_0(f_0 - 1) m v_z / k_B T \tag{4b.21}$$

and obtain

$$\langle\tau_m\rangle = (m/k_B T) \int_{-\infty}^{\infty} \tau_m v_z^2 f_0(1-f_0) d^3v / \int_{-\infty}^{\infty} f_0 d^3v \tag{4b.22}$$

We assume $\tau_m(\vec{v})$ to be an isotropic function $\tau_m(|\vec{v}|)$. Since $f_0(v)$ is an isotropic function we may then replace v_z^2 in the integral by $\frac{1}{3}v^2$. Furthermore, we assume the simple model of band structure and zero energy at the band edge, i.e. $\epsilon = mv^2/2$. More complex band structures will be considered in Chaps. 7 and 8. The volume element in v space is then given by $d^3v = 4\pi v^2 dv \propto \epsilon^{1/2} d\epsilon$. Hence Eq.(4b.22) becomes

$$\langle\tau_m\rangle = -\frac{2}{3} \int_0^{\infty} \tau_m \frac{\partial f_0}{\partial(\epsilon/k_B T)} (\epsilon/k_B T)^{3/2} d(\epsilon/k_B T) \Big/ \int_0^{\infty} f_0 (\epsilon/k_B T)^{1/2} d(\epsilon/k_B T) \tag{4b.23}$$

Let us now assume for $\tau_m(\epsilon)$ the power law given by Eq.(4b.8). For the general case of Fermi-Dirac statistics, $f_o(\epsilon)$ is given by Eq.(3a.25) and we obtain for $\langle\tau_m\rangle$ by partial integration :

$$\langle\tau_m\rangle = \tau_o \frac{2}{3}(r+3/2)\int_0^\infty \frac{x^{r+1/2}dx}{e^{x-\eta_n}+1} \Big/ \int_0^\infty \frac{x^{1/2}dx}{e^{x-\eta_n}+1} = \frac{4}{3\sqrt{\pi}}(r+3/2)!\,\tau_o F_{r+1/2}(\eta_n)/F_{1/2}(\eta_n) \quad (4b.24)$$

where the Fermi integrals are given by Eq.(3a.32) and η_n is the reduced Fermi energy. The factor $4/3\sqrt{\pi}$ is ≈ 0.752. For a non-degenerate electron gas where the Maxwell-Boltzmann distribution, Eq.(4a.1), is valid Eq.(4b.24) yields

$$\langle\tau_m\rangle = \frac{4}{3\sqrt{\pi}}(r+3/2)!\,\tau_o \quad (4b.25)$$

For nonpolar acoustic scattering the exponent $r = -1/2$, and $\langle\tau_m\rangle$ becomes

$$\langle\tau_m\rangle = \tau_o\, 4/3\sqrt{\pi} \approx 0.752\,\tau_o \quad (4b.26)$$

In a more general description it is convenient to introduce a wave vector $\vec{k}$ for the carrier by writing (see text after Eq.(2a.30))

$$\frac{\partial(\hbar\vec{k})}{\partial t} = e\vec{E} \quad (4b.27)$$

where $\hbar\vec{k}$ is the crystal momentum and $\hbar$ is Planck's constant divided by 2π. The carrier velocity is given by

$$\vec{v} = \hbar^{-1}\,\vec{\nabla}_k\epsilon \quad (4b.28)$$

In the Boltzmann equation the term $(\partial f/\partial v_z)eE_z/m$ is replaced by $e\hbar^{-1}(\partial f/\partial\epsilon)\cdot(\vec{\nabla}_k\epsilon\vec{E})$.The effective density of states N_c in the conduction band with zero energy at the band edge defined by Eq.(3a.30) is then given by

$$N_c = \frac{1}{4\pi^3}\int \exp(-\epsilon/k_BT)d^3k \quad (4b.29)$$

and the conductivity σ for $\vec{E}$ in z-direction by

$$\sigma = \frac{e^2}{4\pi^3\hbar^2}\int_{-\infty}^{\infty}\tau_m\left(\frac{\partial\epsilon}{\partial k_z}\right)^2\left(-\frac{\partial f_o}{\partial\epsilon}\right)d^3k \quad (4b.30)$$

Since this calculation does not assume an effective mass it is valid not only for the simple model of a parabolic band structure but also for a non-parabolic band structure.*)

*) For a first reading it may be profitable to continue with Chap.4c.

Now let us reconsider $\bar{\tau}_m$ defined by Eq.(4b.2). As has been suggested $\bar{\tau}_m$ is also an average of τ_m over the velocity distribution. We will now show that the equation of motion, Eq.(4b.2), is obtained from the Boltzmann equation. This will result in an expression for $\bar{\tau}_m$.

The Boltzmann equation (4b.12) with df/dt given by Eq.(4b.13) is multiplied by an arbitrary function [1] $Q(\vec{v})$ and integrated over velocity space

$$\int_{-\infty}^{\infty} Q(\vec{v}) \frac{\partial f}{\partial t} d^3v + \int_{-\infty}^{\infty} Q(\vec{v}) (\vec{v}\vec{\nabla}_r f) d^3v + \frac{e}{m}\int_{-\infty}^{\infty} Q(\vec{v}) (\vec{E}\vec{\nabla}_v f) d^3v + \int_{-\infty}^{\infty} Q(\vec{v}) \frac{(f-f_0)}{\tau_m} d^3v = 0 \tag{4b.31}$$

We can normalize f in such a way that

$$n = \int_{-\infty}^{\infty} f d^3v \tag{4b.32}$$

is the carrier concentration. The average of Q is defined as

$$\langle Q\rangle = \frac{1}{n}\int_{-\infty}^{\infty} Q(\vec{v}) f d^3v \tag{4b.33}$$

With these definitions the first term in Eq.(4b.31) can be written in the form

$$\int_{-\infty}^{\infty} Q(\vec{v}) \frac{\partial f}{\partial t} d^3v = \frac{\partial}{\partial t}\int_{-\infty}^{\infty} Q(\vec{v}) f d^3v = \frac{\partial (n\langle Q\rangle)}{\partial t} = n\frac{\partial \langle Q\rangle}{\partial t} + \langle Q\rangle \frac{\partial n}{\partial t} \tag{4b.34}$$

Similarly, the second term becomes

$$\int_{-\infty}^{\infty} Q(\vec{v}) (\vec{v}\vec{\nabla}_r f) d^3v = \vec{\nabla}_r \int_{-\infty}^{\infty} Q(\vec{v}) f\vec{v} d^3v = \vec{\nabla}_r (n\langle Q\vec{v}\rangle) \tag{4b.35}$$

The third term is integrated by parts :

$$\frac{e}{m}\int_{-\infty}^{\infty} Q(\vec{v}) (\vec{E}\vec{\nabla}_v f) d^3v = -\frac{e}{m}\vec{E}\int_{-\infty}^{\infty} f\vec{\nabla}_v Q d^3v = -\frac{ne}{m}\vec{E}\cdot\langle\vec{\nabla}_v Q\rangle \tag{4b.36}$$

where it is assumed that f vanishes at the boundaries exponentially.

We obtain from Eqs.(4b.31) - (4b.36) the "continuity equation" by giving Q the value 1 (taking the "zeroth moment" of the Boltzmann equation) :

$$\frac{\partial n}{\partial t} + \operatorname{div}(n\vec{v}_d) = 0 \tag{4b.37}$$

where $\langle\vec{v}\rangle = \vec{v}_d$ and the last integral in Eq.(4b.31) vanishes since the total density of carriers is not changed by collisions. From Eqs.(4b.34) and (4b.37) $\partial n/\partial t$ can be eliminated giving us

[1] Part of this treatment is closely related to L.Spitzer: Physics of Fully Ionized Gases, appendix. New York: J.Wiley and Sons. 1962.

$$\int_{-\infty}^{\infty} Q(\vec{v}) \frac{\partial f}{\partial t} d^3v = n \frac{\partial <Q>}{\partial t} - \vec{\nabla}(n\vec{v}_d) \tag{4b.38}$$

Giving Q the value mv_z; $<Q> = mv_{dz}$. Assuming that there is no dependence on position in the crystal (homogeneous crystal, no temperature gradient, no electromagnetic wave) $\vec{\nabla}(n\vec{v}_d) = 0$ and we obtain

$$n \frac{d(mv_{dz})}{dt} - neE_z + m \int_{-\infty}^{\infty} v_z \frac{f}{\tau_m} d^3v = 0 \tag{4b.39}$$

This is the equation of motion, Eq.(4b.2), where $1/\bar{\tau}_m$ is given by

$$1/\bar{\tau}_m = \int_{-\infty}^{\infty} \frac{1}{\tau_m} v_z f d^3v / nv_{dz} \tag{4b.40}$$

For evaluating this integral we approximate the distribution function $f(\vec{v})$ by a Fermi-Dirac distribution shifted by the drift velocity $\vec{v}_d$, and for $|\vec{v}_d| << |\vec{v}|$ we expand this function :

$$f(\vec{v}) \approx f_0(\vec{v}-\vec{v}_d) \approx f_0 - mv_z v_{dz} \partial f_0 / \partial \epsilon \tag{4b.41}$$

Since both τ_m and f_0 are even functions of $\vec{v}$ the integral vanishes for the first term, f_0, and the remaining term yields after an integration by parts :

$$1/\bar{\tau}_m = \frac{2}{3} \int_0^{\infty} \frac{d}{dx} \left(\frac{x^{3/2}}{\tau_m(x)}\right) f_0 dx \Big/ \int_0^{\infty} f_0 x^{1/2} dx \tag{4b.42}$$

For τ_m given by Eq.(4b.8) we finally obtain

$$1/\bar{\tau}_m = \frac{4}{3\sqrt{\pi}} \left(\frac{3}{2} - r\right)! \frac{1}{\tau_0} F_{1/2-r}(\eta_n)/F_{1/2}(\eta_n) \tag{4b.43}$$

Comparing Eq.(4b.42) with Eq.(4b.23) we notice that

$$1/\bar{\tau}_m = <\tau_m^{-1}> \tag{4b.44}$$

Hence, for the simple case of nonpolar acoustic scattering in a non-degenerate electron gas, the error made by replacing $\mu = (e/m) <\tau_m>$ by

$$\mu \approx (e/m)\bar{\tau}_m = e/(m<\tau_m^{-1}>) \tag{4b.45}$$

is $2(4/3\sqrt{\pi})^2 = 32/9\pi \approx 1.13$ and can occasionally be neglected. In Chap.6a quantum mechanical calculation will show that Eq.(4b.44) with $\bar{\tau}_m$ defined by a scattering probability is a poor approximation.

If in addition to the force $e\vec{E}$ there is a Lorentz force acting upon the carriers, a term

$$-n < \vec{\nabla}_v \{ \frac{e}{m} [\vec{v}\vec{B}] Q(\vec{v}) \} > = - \frac{ne}{m} < \{ [\vec{v}\vec{B}] \vec{\nabla}_v Q(\vec{v}) \} > \tag{4b.46}$$

has to be added to the right-hand side of Eq.(4b.36). In Eq.(4b.46) we used the fact that in e.g. $[\vec{v}\vec{B}]_x = v_y B_z - v_z B_y$ there is no v_x and therefore this is a constant factor for the differentiation $\partial/\partial v_x$. If we put $Q = mv_i$ one obtains

$$-ne <[\vec{v}\vec{B}]_i> = -ne\, [\vec{v}_d \vec{B}]_i \tag{4b.47}$$

In Chap.4m we will discuss "hot carriers" and find the definition of an "energy relaxation time" τ_ϵ quite useful. If we let Q be the energy ϵ of a carrier

$$Q(\vec{v}) = \sum_i \frac{m}{2} v_i^2 = \epsilon \tag{4b.48}$$

this provides an opportunity for introducing τ_ϵ at this time. For the Lorentz force, instead of Eq.(4b.47), we obtain

$$- \frac{ne}{2} <[\vec{v}\vec{B}]_i v_i> \tag{4b.49}$$

and the summation over all 3 directions of space lets this term vanish :

$$\sum_i [\vec{v}\vec{B}]_i v_i = ([\vec{v}\vec{B}]\vec{v}) = 0 \tag{4b.50}$$

Introducing ϵ_L and τ_ϵ by the definitions

$$\epsilon_L = \frac{3}{2} k_B T = \frac{1}{n} \int_{-\infty}^{\infty} Q(\vec{v})\, f_0\, d^3v \tag{4b.51}$$

and

$$\frac{<\epsilon> - \epsilon_L}{\tau_\epsilon} = \frac{1}{n} \int_{-\infty}^{\infty} \frac{\epsilon}{\tau_m} (f - f_0)\, d^3v \tag{4b.52}$$

and again assuming that there is no dependence on position in the crystal, we obtain from Eqs.(4b.31) - (4b.38) the "energy balance equation" :

$$\frac{d<\epsilon>}{dt} - e(\vec{E}\vec{v}_d) + \frac{<\epsilon> - \epsilon_L}{\tau_\epsilon} = 0 \tag{4b.53}$$

This equation does not explicitly contain the magnetic field strength; it is contained implicitly by the definition of τ_ϵ in Eq.(4b.52) where f depends on the magnetic field strength. Also, for hot carriers, f will depend on the electric field strength $\vec{E}$, and so will τ_ϵ. Consequently the decrease of $<\epsilon>$ with time to its equilibrium value, ϵ_L, after switching off the electric field will be non-exponential and τ_ϵ may be called a "relaxation time" in only a less strict sense of the word. For the rare case that τ_m does not depend on velocity, Eq.(4b52) shows τ_ϵ to be equal to τ_m.

For the calculation of τ_ϵ according to Eq.(4b.52), let us approximate the distribution function f by a zero-field distribution function f_o at a "carrier temperature" T_e assumed to be slightly higher than the lattice temperature T, and expand this function relative to $T_e - T$:

$$f \approx f_o(\epsilon/k_B T_e) \approx f_o(\epsilon/k_B T) - \frac{T_e - T}{T} \cdot \frac{\epsilon}{k_B T} \cdot \frac{df_o(\epsilon/k_B T)}{d(\epsilon/k_B T)} \tag{4b.54}$$

Hence, $f - f_o$ in Eq.(4b.52) is given by the second term on the right-hand side of Eq.(4b.54). For a non-degenerate electron gas, $\langle\epsilon\rangle - \epsilon_L$ is given by $\frac{3}{2}k_B(T_e - T)$. We then find for $1/\tau_\epsilon$ after an integration by parts :

$$1/\tau_\epsilon = \frac{2}{3}\int_0^\infty \frac{d}{dx}\left(\frac{x^{5/2}}{\tau_m(x)}\right) f_o dx \Big/ \int_0^\infty f_o x^{1/2} dx \tag{4b.55}$$

Assuming τ_m to be given by Eq.(4b.8), we find for a non-degenerate electron gas

$$1/\tau_\epsilon = (4/3\sqrt{\pi})\,(5/2\text{-}r)!\;\tau_o^{-1} \tag{4b.56}$$

which together with Eqs.(4b.43) and (4b.44) yields for this case

$$1/\tau_\epsilon = (5/2\text{-}r)\,\langle\tau_m^{-1}\rangle \tag{4b.57}$$

In Chap.6 we shall see that even for the simple case of nonpolar acoustic scattering this result is not correct. The correct result will be obtained by a quantum mechanical calculation. However, if we notice that the quantity r characterizes the scattering process, and if we interpret the product $\tau_\epsilon\langle\tau_m^{-1}\rangle$ as being approximately the number of collisions necessary for a transfer of the energy, gained from the field $\vec{E}$, to the crystal lattice, we obtain from Eq.(4b.57) the plausible result that this number depends on the type of scattering mechanism.

Let us return to the conductivity as given by Eq.(4b.6) with $\bar{\tau}_m$ replaced by $\langle\tau_m\rangle$. There are always both electrons and holes present in a semiconductor. Since these carriers do not influence each other (except for electron-hole scattering which is not important at low carrier densities) the currents going in opposite direction are subtracted from each other to yield the total current. Due to the opposite charges that electrons and holes have, the conductivities which are always positive quantities, add up to the total conductivity

$$\sigma = |e|(p\mu_p + n\mu_n) = |e|(p + nb) \tag{4b.58}$$

where μ_p is the hole mobility, μ_n the electron mobility, and b the ratio μ_n/μ_p. For intrinsic conduction $n = p = n_i$ and therefore

$$\sigma = |e| n_i \mu_p (1+b) \qquad (4b.59)$$

Since n_i according to Eq.(3b.20) rises exponentially with temperature we obtain the temperature dependence of σ as shown by Fig.4.1. The conductivity is plotted in a logarithmic scale vs the inverse temperature. A nearly straight line is obtained in the intrinsic region. When the temperature is lowered σ enters the extrinsic region where the carrier concentration is constant and σ rises, for the simplest case $\propto \tau_0 \propto T^{-3/2}$. At still lower temperatures there is carrier freeze-out as described by Eqs. (3b.21) and (3b.23).

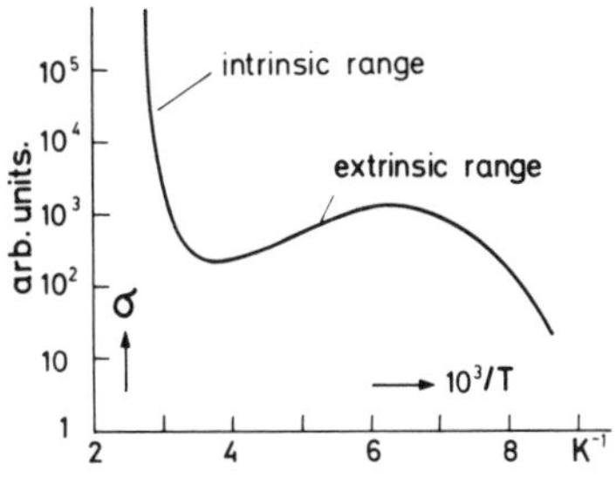

Fig.4.1 Schematic diagram of the conductivity as a function of the inverse temperature.

Methods of resistivity measurement are considered in Appendix C.

4c. Hall Effect in a Transverse Magnetic Field

Carriers which move perpendicular to the direction of a magnetic field or at an oblique angle, will be deflected from the direction of motion by the Lorentz force. This deflection causes a Hall voltage V_y in the experimental arrangement shown in Fig.4.2. This voltage is measured between side

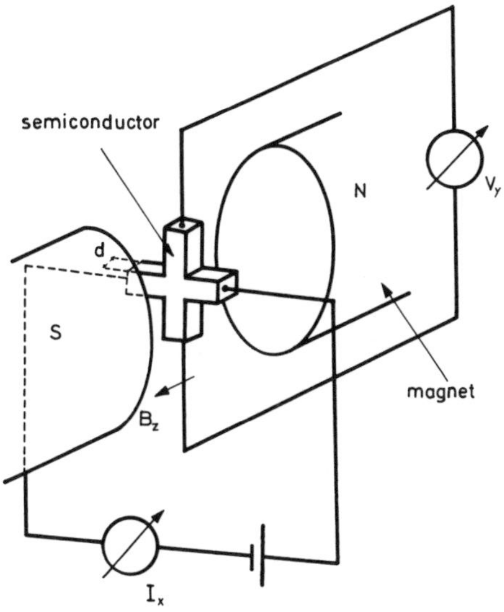

Fig.4.2 Hall arrangement. V_y is the Hall voltage.

arms of the filamentary sample which are opposite to each other. Its polarity is reversed by either reversing the current I_x or the magnetic induction B_z.

The Hall voltage is given by

$$V_y = R_H I_x B_z / d \tag{4c.1}$$

where d is the sample thickness in the direction of the magnetic field and R_H is called "Hall coefficient".

The Hall voltage is proportional to B_z if R_H in Eq.(4c.2) is assumed to be constant. This is true only for values of B_z which are small compared with the inverse mobility of the carriers as will be shown below. The Hall field E_y is given by

$$E_y = R_H j_x B_z \tag{4c.2}$$

where j_x is the current density. Usually the magnetic induction B_z of an electromagnet is given in units of kilogauss (kG). In semiconductor physics, however, the unit $Vsec/cm^2$ which is equal to 10^5 kG, is more convenient since quite often the dimensionless product of mobility, measured in units of $cm^2/Vsec$, and B_z will occur. The unit Vsec is sometimes called "Weber", and 1 $Weber/m^2$ equals $10^{-4} Vsec/cm^2$.

For a calculation*) of R_H we have to determine the energy distribution function $f(\epsilon)$ of the carriers under the influence of electric and magnetic fields. Without a magnetic field the distribution function is given by Eq. (4b.16) and can be modified (see Eq.(4b.28) and following text) to yield

$$f = f_0 - f_0' e \tau_m \hbar^{-1} (\vec{\nabla}_k \epsilon . \vec{E}) \tag{4c.3}$$

where f_0' stands for $\partial f_0 / \partial \epsilon$. When a magnetic field is present we retain essentially the form of f by introducing an unknown vector $\vec{G}$:

$$f = f_0 - f_0' \hbar^{-1} (\vec{\nabla}_k \epsilon . \vec{G}) \tag{4c.4}$$

In a later section of this chapter we will assume a temperature gradient to be present in the semiconductor in addition to electric and magnetic fields. Therefore it may be worthwhile to solve the Boltzmann equation for this more general case and adapt the solution to the special case of the Hall effect.

For this general case the Boltzmann equation in the relaxation time approximation, Eq.(4b.12) for $\partial f / \partial t = 0$, takes the form

$$(\vec{v}\vec{\nabla}_r f) + e\hbar^{-1} \{\vec{E} + [\vec{v}\vec{B}]\} \vec{\nabla}_k f + (f - f_0)/\tau_m = 0 \tag{4c.5}$$

---

*) Readers who want to skip this calculation continue after Eq.(4c.57).

Here we take for f its value as given by Eq.(4c.4). We first consider $\vec{\nabla}_k f$ in the second term of the Boltzmann equation

$$\vec{\nabla}_k f = \vec{\nabla}_k f_o - \vec{\nabla}_k \{f_o'(\vec{v}\vec{G})\} = f_o' \hbar\vec{v} - \vec{\nabla}_k \{f_o'(\vec{v}\vec{G})\} \tag{4c.6}$$

For a weak electric field $\vec{E}$, the deviation of f from the steady-state distribution f_o, which is essentially $\vec{G}$, is small. The product of $\vec{E}$ with this quantity which occurs in the second term of Eq.(4c.5) after $\vec{\nabla}_k f$ has been replaced by its value given by Eq.(4c.6), can therefore be neglected as a product of two small quantities (making it a second-order term). The product of the first term of Eq.(4c.6) with $[\vec{v}\vec{B}]$ vanishes because this triple product contains the same vector, $\vec{v}$, twice. This is also true for the second term on the right-hand side of the following equation :

$$-e\hbar^{-1}[\vec{v}\vec{B}]\vec{\nabla}_k\{f_o'(\vec{v}\vec{G})\} = -e\hbar^{-1}[\vec{v}\vec{B}]\, f_o'\vec{\nabla}_k(\vec{v}\vec{G}) - e\hbar^{-1}[\vec{v}\vec{B}]\, f_o''\hbar\vec{v}(\vec{v}\vec{G}) \tag{4c.7}$$

where f_o'' stands for $\partial^2 f_o/\partial\epsilon^2$. Hence, the second term in Eq.(4c.5) becomes

$$ef_o'(\vec{E}\vec{v}) - e\hbar^{-1} f_o'[\vec{v}\vec{B}]\vec{\nabla}_k(\vec{v}\vec{G}) \tag{4c.8}$$

To evaluate the last factor in Eq.(4c.8) we substitute for $\vec{v}$ its value $\hbar^{-1}\vec{\nabla}_k\epsilon$ and prove the following useful mathematical identity :

$$\vec{\nabla}_k(\vec{\nabla}_k\epsilon\vec{G}) = (\vec{G}\vec{\nabla}_k)\vec{\nabla}_k\epsilon + (\vec{\nabla}_k\epsilon\vec{\nabla}_k)\vec{G} + [\vec{\nabla}_k\epsilon[\vec{\nabla}_k\vec{G}]] \tag{4c.9}$$

The proof is easily done by expanding it in x-, y-, and z-components. E.g. the x-component is

$$\begin{aligned} &\frac{\partial}{\partial k_x}\left(\frac{\partial\epsilon}{\partial k_x}G_x + \frac{\partial\epsilon}{\partial k_y}G_y + \frac{\partial\epsilon}{\partial k_z}G_z\right) = \\ &= G_x\frac{\partial}{\partial k_x}\frac{\partial\epsilon}{\partial k_x} + G_y\frac{\partial}{\partial k_y}\frac{\partial\epsilon}{\partial k_x} + G_z\frac{\partial}{\partial k_z}\frac{\partial\epsilon}{\partial k_x} + \frac{\partial\epsilon}{\partial k_x}\frac{\partial G_x}{\partial k_x} + \\ &+ \underbrace{\frac{\partial\epsilon}{\partial k_y}\frac{\partial G_x}{\partial k_y}}_{\sim} + \underbrace{\frac{\partial\epsilon}{\partial k_z}\frac{\partial G_x}{\partial k_z}}_{\approx} + \frac{\partial\epsilon}{\partial k_y}\left(\frac{\partial G_y}{\partial k_x} - \underbrace{\frac{\partial G_x}{\partial k_y}}_{\sim}\right) - \frac{\partial\epsilon}{\partial k_z}\left(\underbrace{\frac{\partial G_x}{\partial k_z}}_{\approx} - \frac{\partial G_z}{\partial k_x}\right) \end{aligned} \tag{4c.10}$$

Evidently the terms marked ~~ cancel and so do the terms marked ≈≈ which proves the identity for the x-component. For the other components the proof is similar.

Now we assume that the vector $\vec{G}$ depends on $\vec{k}$ via ϵ and therefore is independent of the direction of $\vec{k}$. Thus the x-component of $[\vec{\nabla}_k\vec{G}]$ is given by

$$\frac{\partial G_z}{\partial k_y} - \frac{\partial G_y}{\partial k_z} = \frac{\partial G_z}{\partial\epsilon}\frac{\partial\epsilon}{\partial k_y} - \frac{\partial G_y}{\partial\epsilon}\frac{\partial\epsilon}{\partial k_z} = [\vec{\nabla}_k\epsilon\vec{G}']_x \tag{4c.11}$$

where $\vec{G}'$ stands for $\partial\vec{G}/\partial\epsilon$. The third term in Eq.(4c.9) is therefore

$$[\vec{\nabla}_k \epsilon\, [\vec{\nabla}_k \vec{G}]] = \vec{\nabla}_k \epsilon\, (\vec{\nabla}_k \epsilon \vec{G}') - (\vec{\nabla}_k \epsilon)^2 \vec{G}' \tag{4c.12}$$

The second term of this equation, $-(\vec{\nabla}_k \epsilon)^2 \vec{G}'$, can be written in the form

$$-\{\frac{\partial \epsilon}{\partial k_x}\frac{\partial \vec{G}}{\partial k_x} + \frac{\partial \epsilon}{\partial k_y}\frac{\partial \vec{G}}{\partial ky} + \frac{\partial \epsilon}{\partial k_z}\frac{\partial G}{\partial k_z}\} = -(\vec{\nabla}_k \epsilon\, \vec{\nabla}_k)\vec{G} \tag{4c.13}$$

which except for sign is identical with the second term in Eq.(4c.9). Hence Eq.(4c.9) becomes

$$\hbar \vec{\nabla}_k (\vec{v}\vec{G}) = (\vec{G}\vec{\nabla}_k)\, \vec{\nabla}_k \epsilon + \vec{\nabla}_k \epsilon\, (\vec{\nabla}_k \epsilon \vec{G}') \tag{4c.14}$$

If we multiply this with the vector product $[\vec{v}\vec{B}]$ as done in Eq.(4c.8), the second term vanishes since $\vec{v} \propto \vec{\nabla}_k \epsilon$ and the first term becomes

$$\hbar [\vec{v}\vec{B}]\, \vec{\nabla}_k (\vec{v}\vec{G}) = [\vec{v}\vec{B}]\, (\vec{G}\vec{\nabla}_k)\, \vec{\nabla}_k \epsilon = \vec{v}\, [\vec{B}\, (\vec{G}\vec{\nabla}_k)\, \vec{\nabla}_k \epsilon] \tag{4c.15}$$

Now the second term of the Boltzmann equation Eq.(4c.6) takes the form

$$e f_o' (\vec{v}\vec{E}) - e\hbar^{-2} f_o' (\vec{v}\, [\vec{B}\, (\vec{G}\vec{\nabla}_k)\, \vec{\nabla}_k \epsilon]) \tag{4c.16}$$

The first term of the Boltzmann equation, $\vec{v}\vec{\nabla}_r f$, is easily obtained from Eq. (4c.3):

$$\vec{v}\vec{\nabla}_r f = \vec{v}\vec{\nabla}_r f_o - \vec{v}\vec{\nabla}_r \{(f_o'(\vec{v}\vec{G}))\} = \vec{v}\vec{\nabla}_r f_o \tag{4c.17}$$

where the second term on the right-hand side is neglected since any position dependence of e.g. the temperature T will be assumed to be small.

In Chap.3 it is shown that the steady-state distribution of carriers, f_o, depends on the difference of carrier energy ϵ and the Fermi energy ζ relative to the energy $k_B T$. Therefore, we can express $\vec{\nabla}_r f_o$ by f_o':

$$\vec{\nabla}_r f_o = \frac{\partial f_o}{\partial \{(\epsilon - \zeta)/k_B T\}}\, \vec{\nabla}_r \frac{\epsilon - \zeta}{k_B T} = f_o' T \vec{\nabla}_r \frac{\epsilon - \zeta}{T} \tag{4c.18}$$

The behavior of "hot" carriers (see Chap.4m) can in a certain approximation be described by a function f_o which depends on an "electron temperature" T_e rather than the temperature T of the crystal lattice. In this case T in the last equation would have to be replaced by T_e.

In further discussing the Boltzmann equation it is convenient to introduce an "electrothermal field" $\vec{F}$ by

$$e\vec{F} = e\vec{E} + T\vec{\nabla}_r \frac{\epsilon - \zeta}{T} \tag{4c.19}$$

In this way we can combine the $\vec{E}$-dependent term in Eq.(4c.16) with the term proportional to $\vec{\nabla}_r f_o$ and obtain the Boltzmann equation in the form

$$ef'_0(\vec{v}\vec{F}) - e\hbar^{-2}f'_0(\vec{v}\,[\vec{B}(\vec{G}\vec{\nabla}_k)\,\vec{\nabla}_k\epsilon]) - f'_0(\vec{v}\vec{G})/\tau_m = 0 \tag{4c.20}$$

Since $f'_0\vec{v}$ is an arbitrary vector which is contained in each term, we find

$$e\vec{F} - e\hbar^{-2}[\vec{B}(\vec{G}\vec{\nabla}_k)\,\vec{\nabla}_k\epsilon] - \tau_m^{-1}\vec{G} = 0 \tag{4c.21}$$

It is possible to solve this equation for $\vec{G}$ if we introduce the $\epsilon(\vec{k})$-relation valid for the simple model of the band structure :

$$\epsilon = \epsilon_c + \frac{1}{2}\hbar^2\vec{k}\,m^{-1}\,\vec{k} \tag{4c.22}$$

where ϵ_c is the conduction band edge. We can simplify then

$$\hbar^{-2}(\vec{G}\vec{\nabla}_k)\,\vec{\nabla}_k\epsilon = m^{-1}\,\vec{G} \tag{4c.23}$$

Eqs.(4c.22) and (4c.23) hold also for the case that m is not a scalar but a tensor quantity with its main axes parallel to the coordinate axes :

$$m^{-1} = \begin{pmatrix} m_x^{-1} & 0 & 0 \\ 0 & m_y^{-1} & 0 \\ 0 & 0 & m_z^{-1} \end{pmatrix} \tag{4c.24}$$

where in general $m_x \neq m_y \neq m_z$. Eq.(4c.22) can be written in the form

$$\epsilon = \epsilon_c + \frac{1}{2}\hbar^2(k_x^2/m_x + k_y^2/m_y + k_z^2/m_z) \tag{4c.25}$$

Quite often also τ_m may be a tensor of the same form

$$\tau_m = \begin{pmatrix} \tau_x & 0 & 0 \\ 0 & \tau_y & 0 \\ 0 & 0 & \tau_z \end{pmatrix} \tag{4c.26}$$

With these tensors Eq.(4c.21) can be written in the form

$$e\vec{F} - e\,[\vec{B}m^{-1}\vec{G}] - \tau_m^{-1}\vec{G} = 0 \tag{4c.27}$$

where the product of the inverse mass tensor and $\vec{G}$ as given by Eq.(4c.23) has been introduced.

Solving Eq.(4c.21) for $\vec{G}$ requires the definition of another tensor

$$a = e^2 . \begin{pmatrix} \tau_y\tau_z/m_y m_z & 0 & 0 \\ 0 & \tau_x\tau_z/m_x m_z & 0 \\ 0 & 0 & \tau_x\tau_y/m_x m_y \end{pmatrix} \tag{4c.28}$$

In a spherical model a equals $(e\tau_m/m)^2$ which for an energy-independent τ_m would be the square of the mobility.

We shall prove that $\vec{G}$ turns out to be

$$\vec{G} = e\tau_m \frac{\vec{F} - e[\vec{B}.m^{-1}\tau_m \vec{F}] + a\vec{B}.(\vec{F}\vec{B})}{1 + (\vec{B}a\vec{B})} \tag{4c.29}$$

The numerator consists of 3 terms: The first one does not contain the magnetic induction; the second term is linear in $\vec{B}$ and takes care of the Hall effect and the third term as well as the denominator are quadratic in $\vec{B}$ and therefore yield the magnetoresistance.

For the proof of Eq.(4c.29) we begin with a mathematical identity

$$[\frac{a}{e^2}\vec{B}.\tau_m^{-1}\vec{G}] = m^{-1}\tau_m [\vec{B}.m^{-1}\vec{G}] \tag{4c.30}$$

This is easily shown to be an identity by writing in components, e.g. the x-component

$$\frac{\tau_x\tau_z}{m_x m_z} B_y \tau_z^{-1} G_z - \frac{\tau_x\tau_y}{m_x m_y} B_z \tau_y^{-1} G_y = \frac{\tau_x}{m_x}(B_y \frac{G_z}{m_z} - B_z \frac{G_y}{m_y}) \tag{4c.31}$$

Vectorial multiplication with $e^2\vec{B}$ yields

$$[\vec{B}[a\vec{B}.\tau_m^{-1}\vec{G}]] = e^2[\vec{B}.m^{-1}\tau_m [\vec{B}.m^{-1}\vec{G}]] \tag{4c.32}$$

which on the other hand is equal to

$$[\vec{B}[a\vec{B}.\tau_m^{-1}\vec{G}]] = a\vec{B}.(\vec{B}.\tau_m^{-1}\vec{G}) - \tau_m^{-1}\vec{G}(\vec{B}a\vec{B}) \tag{4c.33}$$

In the first term the factor $(\vec{B}.\tau_m^{-1}\vec{G})$ can be obtained by scalar multiplication of Eq.(4c.21) with $\vec{B}$ which yields $e.(\vec{F}\vec{B})$. The sum of $\tau_m^{-1}G(\vec{B}a\vec{B})$ as given by Eq.(4c.31) and $\tau_m^{-1}\vec{G}$ can now be written in the form

$$\tau_m^{-1}\vec{G}\{1 + (\vec{B}a\vec{B})\} = \tau_m^{-1}\vec{G} + a\vec{B}e(\vec{F}\vec{B}) - e^2[\vec{B}m^{-1}\tau_m [\vec{B}m^{-1}\vec{G}]] \tag{4c.34}$$

In the last term of this equation we replace $e[\vec{B}m^{-1}\vec{G}]$ as given by Eq.(4c.27) and obtain

$$+e[\vec{B}m^{-1}\vec{G}] - e^2[\vec{B}m^{-1}\tau_m \vec{F}] = e\vec{F} - \tau_m^{-1}\vec{G} - e^2[\vec{B}m^{-1}\tau_m \vec{F}] \tag{4c.35}$$

Hence, Eq.(4c.34) takes the form

$$\tau_m^{-1}\vec{G}\{1 + (\vec{B}a\vec{B})\} = e\vec{F} - e^2[\vec{B}.m^{-1}\tau_m \vec{F}] + a\vec{B}e(\vec{F}\vec{B}) \tag{4c.36}$$

Multiplication with $\tau_m/\{1 + (\vec{B}a\vec{B})\}$ yields Eq.(4c.29).

Now we are in a position to write the distribution function Eq.(4c.4) in its general form :

$$f = f_o - f_o' e\hbar^{-1}\vec{\nabla}_k e\tau_m \frac{\vec{F} - e[\vec{B}.m^{-1}\tau_m \vec{F}] + a\vec{B}(\vec{F}\vec{B})}{1 + (\vec{B}a\vec{B})} \tag{4c.37}$$

For the calculation of the conductivity without magnetic field, $\vec{B} = 0$ and

$\vec{F} = \vec{E}$, we obtain

$$f = f_o - f_o' e \tau_m (\vec{v}\vec{E}) \qquad (4c.38)$$

from which a conductivity tensor

$$\begin{pmatrix} \sigma & 0 & 0 \\ 0 & \sigma & 0 \\ 0 & 0 & \sigma \end{pmatrix} \qquad (4c.39)$$

is obtained where σ is equal to σ_o given by

$$\sigma_o = (ne^2/m) < \tau_m > \qquad (4c.40)$$

and the momentum relaxation time τ_m is averaged over a Maxwell-Boltzmann distribution function:

$$<\tau_m> = \frac{4}{3\sqrt{\pi}} \int_0^\infty \tau_m \cdot \left(\frac{\epsilon}{k_B T}\right)^{3/2} \exp\left(-\frac{\epsilon}{k_B T}\right) d\left(\frac{\epsilon}{k_B T}\right) \qquad (4c.41)$$

Except for the tensor notation, Eq.(4c.39), this is of course identical with the previous result Eq.(4b.23).

For the part of the distribution function Eq.(4c.37) linear in $\vec{B}$ we neglect the denominator and the first and third term in the numerator and obtain for the conductivity

$$\begin{pmatrix} 0 & \gamma B_z & -\gamma B_y \\ -\gamma B_z & 0 & \gamma B_x \\ \gamma B_y & -\gamma B_x & 0 \end{pmatrix} \qquad (4c.42)$$

where γ is equal to γ_o given by

$$\gamma_o = (ne^3/m^2) < \tau_m^2 > \qquad (4c.43)$$

For the evaluation of the Hall effect at an arbitrary magnetic induction, we will introduce quantities σ and γ defined as

$$\sigma = (ne^2/m^2) < \tau_m^2 / (1 + \omega_c^2 \tau_m^2) > \qquad (4c.44)$$

and

$$\gamma = (ne^3/m^2) < \tau_m^2 / (1 + \omega_c^2 \tau_m^2) > \qquad (4c.45)$$

where the "cyclotron frequency" ω_c has been introduced as given by

$$\omega_c = |e| B/m \qquad (4c.46)$$

The quantities σ and γ are obtained similarly as σ_0 and γ_0 except for retaining the denominator in the distribution function. The tensor Eq.(4c.42) applied to $\vec{E}$ yields e.g. an x-component of $\vec{j}$ given by

$$j_x = \gamma (B_z E_y - B_y E_z) = \gamma [\vec{E}\vec{B}]_x = (ne^2/m) <\tau_m [\vec{E}(e/m)\vec{B}\tau_m]_x /(1+\omega_c^2\tau_m^2)> \quad (4c.47)$$

which is also obtained from the vector-product term in Eq.(4c.37).

From the last term in the distribution function Eq.(4c.37) we obtain for the conductivity

$$\begin{pmatrix} -\beta B_x^2 & -\beta B_x B_y & -\beta B_x B_z \\ -\beta B_x B_y & -\beta B_y^2 & -\beta B_y B_z \\ -\beta B_x B_z & -\beta B_y B_z & -\beta B_z^2 \end{pmatrix} \quad (4c.48)$$

where β is given by

$$\beta = -(ne^4/m^3) < \tau_m^3/(1 + \omega_c^2 \tau_m^2) > \quad (4c.49)$$

For the case of a small magnetic field intensity $1/(1+\omega_c^2\tau_m^2)$ is replaced by $(1-\omega_c^2\tau_m^2)$ where $\omega_c^2 \propto B^2 = B_x^2 + B_y^2 + B_z^2$. As a consequence the tensor Eq.(4c.47) is replaced by

$$\begin{pmatrix} \beta_o (B_y^2 + B_z^2) & -\beta_0 B_x B_y & -\beta_0 B_x B_z \\ -\beta_0 B_x B_y & \beta_0 (B_x^2 + B_z^2) & -\beta_0 B_y B_z \\ -\beta_0 B_x B_z & -\beta_0 B_y B_z & \beta_0 (B_x + B_y^2) \end{pmatrix} \quad (4c.50)$$

where β_0 stands for

$$\beta_0 = -(ne^4/m^3) < \tau_m^3 > \quad (4c.51)$$

The tensor Eq.(4c.48) applied to $\vec{F}$ yields e.g. an x-component of $\vec{j}$ given by

$$j_x = -\beta B_x (B_x F_x + B_y F_y + B_z F_z) = -\beta B_x (\vec{F}\vec{B}) \quad (4c.52)$$

which is also obtained from the last term in Eq.(4c.37).

For a weak magnetic field the conductivity tensor is denoted by σ_w and is a combination of the tensors Eq.(4c.39), (4c.42), and (4c.50).

$$\sigma_w = \begin{pmatrix} \sigma_o + \beta_o (B_y^2 + B_z^2) & \gamma_0 B_z - \beta_0 B_x B_y & -\gamma_0 B_y - \beta_0 B_x B_z \\ -\gamma_0 B_z - \beta_0 B_x B_y & \sigma_0 + \beta_0 (B_x^2 + B_z^2) & \gamma_0 B_x - \beta_0 B_y B_z \\ \gamma_0 B_y - \beta_0 B_x B_z & -\gamma_0 B_x - \beta_0 B_y B_z & \sigma_0 + \beta_0 (B_x^2 + B_y^2) \end{pmatrix} \quad (4c.53)$$

Next we will discuss the Hall effect in a weak magnetic field. We choose a coordinate system such that the z-axis points in the direction of $\vec{B}$, i.e. $B_x = B_y = 0$. From $\vec{j} = \sigma_w \vec{E}$ we obtain an equation for E_z which is independent of B_z. Therefore we may assume $E_z = 0$. For the current components j_x and j_y we obtain

$$j_x = (\sigma_o + \beta_o B_z^2) E_x + \gamma_o B_z E_y \tag{4c.54}$$

and

$$j_y = -\gamma_o B_z E_x + (\sigma_o + \beta_o B_z^2) E_y \tag{4c.55}$$

For steady state the component j_y vanishes if the current through the voltmeter in Fig.4.2 measuring the Hall voltage V_y is small compared with the longitudinal current component I_x (in practice a high impedance voltmeter like a vacuum tube voltmeter or a solid state equivalent is used). The two sides of the sample perpendicular to the y-direction carry charges of opposite sign and of such a magnitude that the field between these charges which is the Hall field, counterbalances the Lorentz field. From $j_y = 0$ and Eq.(4c.55) we obtain for E_x:

$$E_x = \frac{\sigma_o + \beta_o B_z^2}{\gamma_o B_z} E_y \tag{4c.56}$$

We eliminate E_x from Eq.(4c.54) and obtain the Hall field E_y:

$$E_y = \frac{\gamma_o}{(\sigma_o + \beta_o B_z^2)^2 + \gamma_o^2 B_z^2} j_x B_z \tag{4c.57}$$

In comparing with Eq.(4c.2) we find for the Hall coefficient

$$R_H = \frac{\gamma_o}{(\sigma_o + \beta_o B_z^2)^2 + \gamma_o^2 B_z^2} \approx \gamma_o / \sigma_o^2 \tag{4c.58}$$

This approximation can be made since we have limited the calculation to the low-field case.

If we compare this equation with Eq.(4c.2) and take into account the definitions of γ_o and σ_o, Eqs.(4c.43) and (4c.40), respectively, we obtain for the Hall coefficient

$$R_H = r_H / ne \tag{4c.59}$$

where r_H stands for

$$r_H = \langle \tau_m^2 \rangle / \langle \tau_m \rangle^2 \tag{4c.60}$$

and is called the "Hall factor". In order to obtain an idea of the magnitude of the Hall factor we assume $\tau_m = \tau_0(\epsilon/kT)^r$ as discussed before, Eq.(4b.8), and obtain from Eq.(4b.25) for $<\tau_m>^2$:

$$<\tau_m>^2 = \{\frac{4}{3\sqrt{\pi}} \cdot \tau_0 \cdot (\frac{3}{2} + r)!\}^2 \qquad (4c.61)$$

Similarly by replacing in Eq.(4c.41) τ_m by τ_m^2, the average of τ_m^2 is obtained:

$$<\tau_m^2> = \frac{4}{3\sqrt{\pi}} \cdot \tau_0^2 \cdot (\frac{3}{2} + 2r)! \qquad (4c.62)$$

Hence, the Hall factor is given by

$$r_H = \frac{3\sqrt{\pi}}{4} \frac{(2r + 3/2)!}{\{(r + 3/2)!\}^2} \qquad (4c.63)$$

For acoustic deformation potential scattering where r equals -1/2, the Hall factor becomes

$$r_H = \frac{3\sqrt{\pi}}{4} \cdot \frac{\sqrt{\pi}/2}{1} = \frac{3\pi}{8} = 1.18 \qquad (4c.64)$$

while for ionized impurity scattering where r = + 3/2, we find

$$r_H = \frac{3\sqrt{\pi}}{4} \frac{2^{-5} \cdot 3^3 \cdot 5 \cdot 7\sqrt{\pi}}{2^2 \cdot 3^3} = \frac{315\pi}{512} = 1.93 \qquad (4c.65)$$

Obviously the order of magnitude of r_H is 1.

Due to the negative charge of the electron, $e < 0$, the Hall coefficient is negative for n-type conductivity while positive for p-type. Therefore the Hall effect is an important method for the determination of the type of conductivity. The case of intrinsic or nearly intrinsic semiconductors will be discussed later on.

From R_H the carrier concentration can be determined:

$$n \text{ or } p = r_H/R_H e \qquad (4c.66)$$

On the other hand, if R_H is known, the Hall effect is useful for an experimental determination for the magnetic induction $\vec{B}$. In contrast to the rotating solenoid it has no moving parts. There are many more device applications of the Hall effect. One is the multiplication of electrical signals which are used for generating j_x and B_z: The Hall voltage is determined by the product of both. A power meter may be constructed in this way.

The "Hall mobility" is defined by the product of the conductivity σ_0 and the Hall coefficient R_H:

$$\mu_H = R_H \sigma_o = (r_H/ne)\, ne\mu = r_H \mu \tag{4c.67}$$

The Hall mobility is different from the drift mobility by the Hall factor. Usually it is the Hall mobility which is measured rather than the drift mobility. From the observed temperature dependence of the Hall mobility one can get an idea of the most probable ϵ-dependence of τ_m from which the Hall factor r_H is calculated. In this way the drift mobility μ_H/r_H may be determined*). Since r_H is not too different from unity the error introduced by assuming an incorrect $\tau_m(\epsilon)$-dependence is not too large.

Another quantity of interest is the Hall angle θ_H which is given by

$$\tan\theta_H = E_y/E_x = \mu_H B_z = r_H \omega_c \langle\tau_m\rangle \tag{4c.68}$$

Only if the Hall angle is small, can one justify neglecting any dependence of r_H on $\vec{B}$ as was assumed in Eq.(4c.59). In any case of a determination of carrier concentration and mobility by the Hall effect the magnetic field intensity has to be small enough that a variation of $\vec{B}$ does not affect these quantities appreciably.

Because of the practical importance of the Hall effect we will give a numerical example. From Eqs.(4c.1) and (4c.59) the Hall voltage is given by

$$V_y = r_H I_x B_z/(ned) \tag{4c.69}$$

Assume a semiconductor with $r_H = 1.6$ and $n = 10^{14}/cm^3$. The sample width may be $d = 1$ mm $= 10^{-1}$ cm, the sample current 1 mA $= 10^{-3}$ A and the magnetic induction 1 kG $= 10^{-5}$ Vsec/cm^2. The Hall voltage is

$$V_y = 1.6 \times 10^{-3} \times 10^{-5}/(10^{14} \times 1.6 \times 10^{-19} \times 10^{-1})\,V = 10^{-2}\,V \tag{4c.70}$$

or 10 mV. Assume the semiconductor to be n-type germanium at room temperature with a Hall mobility of $\mu_H = 4 \times 10^3$ cm^2/Vsec (actually r_H is somewhat less than 1.6 in this case). The product $\mu_H B_z$ is 4×10^{-2} which is small compared to 1. Therefore the Hall voltage will vary linearly with B_z. We shall see, however, that the criterion $\mu_H B_z \ll 1$ may not be sufficient in the case of two types of carriers such as light and heavy holes which occur in most p-type semiconductors (Chap.8c; for the equivalent effect in the magnetoresistance see Eq.(4d.15) and Fig.4.5).

An experimental problem may arise from the difficulty to align perfectly the Hall arms of the sample. Let us assume a misalignment of 0.1 mm and calculate the voltage drop $I_x R_x$ along the sample filament of width 1 mm and thickness 1 mm

---

*) A direct method although rarely applicable is the Haynes-Shockley experiment (Chap.5b).

$$I_x R_x = \frac{10^{-3}\,\text{A} \times 10^{-2}\,\text{cm}}{10^{14}\,\text{cm}^{-3} \times 1.6 \times 10^{-19}\,\text{Asec} \times 4 \times 10^{3}\,\text{cm}^2/\text{Vsec} \times 10^{-2}\,\text{cm}^2} = $$

$$= 1.6 \times 10^{-2}\,\text{V} \tag{4c.71}$$

which is of the same order of magnitude as the Hall voltage. This misalignment voltage may be eliminated either by averaging $|V_y|$ over both magnetic field directions*) or by having two side arms at one side of the sample as shown in Fig.4.3. These arms are connected with the end contacts of a high-resistance potentiometer. Without magnetic field the potentiometer is set such that the voltage $V_y = 0$. With a magnetic field V_y is the true Hall voltage. An improved arrangement as first suggested by van der Pauw will be discussed in Appendix C.

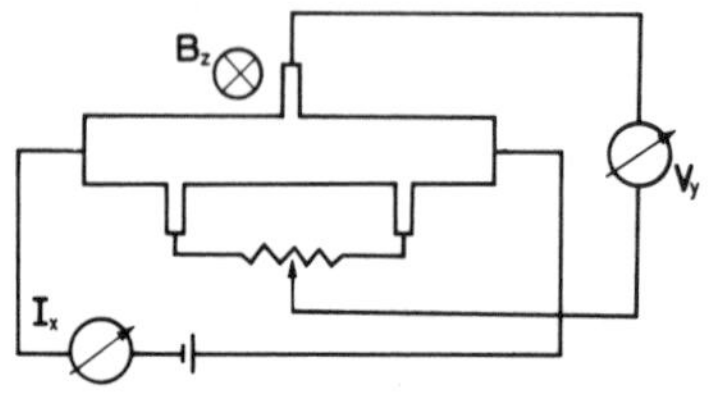

Fig.4.3 Hall sample shape which permits the electrical compensation of a longitudinal voltage drop when the magnet is switched off.

If both electrons and holes are present in comparable quantities the current densities of both carrier types add.

$$0 = j_y = \sigma_o E_y - |e| \cdot r_H (p\mu_p^2 - n\mu_n^2)\, E_x B_z \tag{4c.72}$$

and since in the weak-field approximation

$$E_y = R_H j_x B_z \approx R_H\, \sigma_o E_x B_z \tag{4c.73}$$

replacing E_y in Eq.(4c.72) and solving for R_H yields

$$R_H = \frac{|e|\, r_H}{\sigma_o^2}(p\mu_p^2 - n\mu_n^2) = \frac{r_H}{|e|}\,\frac{p\mu_p^2 - n\mu_n^2}{(p\mu_p + n\mu_n)^2} = \frac{r_H}{|e|}\,\frac{p - nb^2}{(p + nb)^2} \tag{4c.74}$$

where the ratio of mobilities,

$$b = \mu_n / \mu_p \tag{4c.75}$$

has been introduced. R_H changes sign at $p = nb^2$ rather than at the intrinsic concentration, $p = n$. E.g. for InSb at room temperature $b = 80$ and in intrinsic material the Hall effect is negative as in n-type. This causes the "Hall overshoot" of the p-type samples over the straight line characterizing the intrinsic conductivity shown [1, 2] in Fig.4.4. Similarly one finds an overshoot in the

[1] O.Madelung and H.Weiss, Z.Naturf.9a (1954) 527.

[2] N.I.Volokobinskaya, V.V.Galavanov, and D.N.Nasledov, Fiz.Tverd.Tela 1 (1959) 756; (Engl.: Sov.Phys.-Solid State 1 (1959) 687).

*) This is possible only for small electric field intensities such that R_H is independent of $\vec{E}$ (Chap.7f).

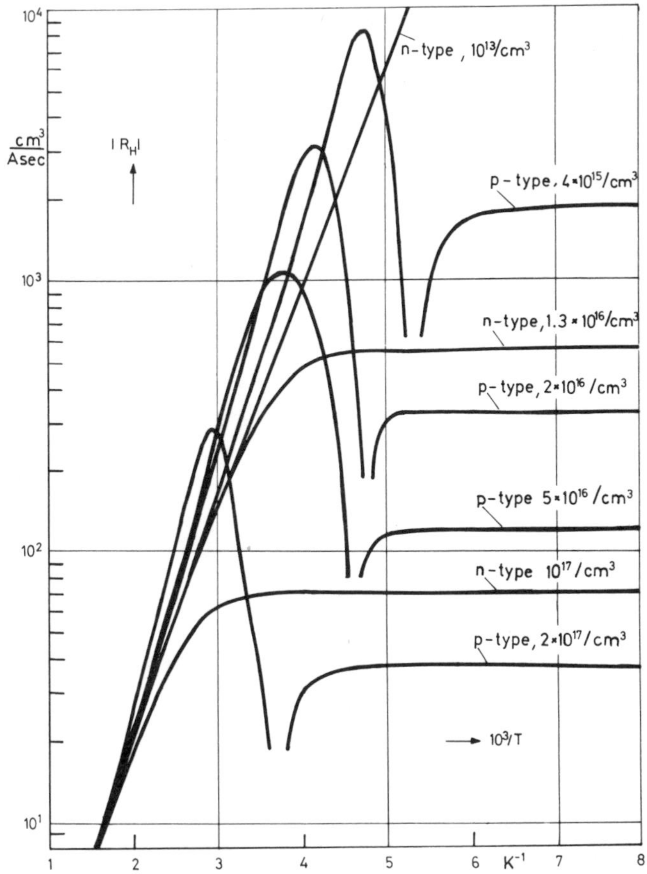

Fig.4.4 Hall coefficient of indium antimonide as a function of the reciprocal temperature (after Madelung and Weiss, ref.1, and Volokobinskaya et al., ref.2).

resistivity vs temperature curves of p-type samples : It is not at $p = n_i$ but at $p = n_i\sqrt{b}$ that the maximum of the resistivity occurs. For $b = 80$ this is nearly an order of magnitude larger than n_i and the maximum resistivity is by a factor of $(1+b)/2\sqrt{b} = 4.5$ times larger than the intrinsic conductivity.

In the final part of this section we consider the magnetic field dependence of the Hall coefficient. It is obvious from Eq.(4c.58) that for not too small magnetic field intensities there is a parabolic variation of R_H with B_z^2. The conductivity tensor valid for arbitrary magnetic fields is given in Appendix A.

For very strong magnetic fields we can approximate σ, γ, and β as given by Eqs.(4c.44), (4c.45), and (4c.49), respectively, in the following way :

$$\sigma \approx (ne^2/n\omega_c^2) <\tau_m^{-1}> \tag{4c.76}$$

$$\gamma \approx ne^3/m^2\omega_c^2 \tag{4c.77}$$

$$\beta \approx - (ne^4/m^3\omega_c^2) <\tau_m> \tag{4c.78}$$

The conductivity tensor σ_w is obtained from Eqs.(4c.39), (4c.42), and (4c.48):

$$\sigma_w = \begin{pmatrix} \sigma - \beta B_x^2 & -\beta B_x B_y + \gamma B_z & -\beta B_x B_z - \gamma B_y \\ -\beta B_x B_y - \gamma B_z & \sigma - \beta B_y^2 & -\beta B_y B_z + \gamma B_x \\ -\beta B_x B_z + \gamma B_y & -\beta B_y B_z - \gamma B_x & \sigma - \beta B_z^2 \end{pmatrix} \tag{4c.79}$$

For $B_x = B_y = E_z = 0$ the current components j_x and j_y are

$$j_x = \sigma E_x + \gamma B_z E_y \tag{4c.80}$$

and

$$j_y = - \gamma B_z E_x + \sigma E_y \tag{4c.81}$$

From $j_y = 0$ we obtain

$$E_x = (\sigma/\gamma B_z) E_y \tag{4c.82}$$

We eliminate E_x from Eq.(4c.80) and obtain the Hall field

$$E_y = \frac{\gamma}{\sigma^2 + (\gamma B_z)^2} j_x B_z = R_H j_x B_z \tag{4c.83}$$

We use Eqs.(4c76) - (4c.78) in order to obtain the Hall coefficient

$$R_H = \frac{ne^3/m^2\omega_c^2}{(ne^2/m\omega_c^2)^2 <\tau_m^{-1}> + (ne^3 B_z/m^2 m_c^2)^2} \approx 1/ne \tag{4c.84}$$

The approximation is valid for large magnetic field strengths, $(\mu_H B_z)^2 >> 1$. Assuming a value of 10 for $(\mu_H B_z)^2$ and a Hall mobility of e.g. 1 000 cm²/Vsec, a magnetic induction of 333 kG would be required which can be obtained in pulsed form only and by considerable effort. Few semiconductors have mobilities of more than 10^4 cm²/Vsec where the requirements on B_z are easier to meet. In this case, however, the carrier density is obtained without having to guess the unknown Hall factor. At very large magnetic field intensities, the carrier density n may be different from its low field value, though. In this case the energies ϵ_G, $\Delta\epsilon_A$ or $\Delta\epsilon_D$ are subject to changes induced by the magnetic field (Chap.9c) which can possibly be detected by optical methods.

4d. Magnetoresistance

The influence of a weak transverse magnetic induction B_z on the current density j_x parallel to an externally applied electric field of intensity E_x is obtained from Eqs.(4c.54) and (4c.55) with $j_y = 0$.

$$j_x = \{\sigma_o + \beta_o B_z^{\,2} + \frac{(\gamma_o B_z)^2}{\sigma_o + \beta_o B_z^2}\} E_x \qquad (4d.1)$$

It is common practice to introduce in this equation the resistivity ρ rather than the conductivity σ.

$$j_x = E_x / \rho_B \qquad (4d.2)$$

The subscript B has been used to denote the dependence on B_z. The relative change in resistivity, $\Delta\rho/\rho_B = (\rho_B - \rho_o)/\rho_B$, where $\rho_o = 1/\sigma_o$, is found from Eq.(4d.1)

$$\Delta\rho/\rho_B = -\,B_z^2\ \{(\beta_o/\sigma_o) + (\gamma_o/\sigma_o)^2\} = T_M\,(e <\tau_m> B_z/m)^2 \qquad (4d.3)$$

The magnetoresistance scattering coefficient T_M is defined by

$$T_M = (<\tau_m^3> \cdot <\tau_m> - <\tau_m^2>^2)/<\tau_m>^4 \qquad (4d.4)$$

For the case that $\tau_m(\epsilon)$ obeys a power law, Eq.(4b.8), T_M is given by

$$T_M = \frac{9\pi}{16} \cdot \frac{(3r+3/2)!(r+3/2)! - \{(2r+3/2)!\}^2}{\{(r+3/2)!\}^4} \qquad (4d.5)$$

The numerical value of T_M varies between 0.38 (for $r = -\,1/2$, acoustic deformation potential scattering) and 2.15 (for $r = +\,3/2$, ionized impurity scattering) and clearly depends stronger on the scattering mechanism than does the Hall factor r_H (see Eqs.(4c.64) and (4c.65)). If we introduce the drift mobility $\mu = e <\tau_m>/m$, the right-hand side of Eq.(4d.3) becomes $T_M(\mu B_z)^2$. The results obtained so far are valid only if*) $(\mu B_z)^2 \ll 1$. Therefore the proportionality of magnetoresistance to B_z^2 will be correct only as long as $\Delta\rho \ll \rho_B$.

In a strong transverse magnetic induction B_z the dependence of j_x on E_x is obtained from Eqs.(4c.80) and (4c.81).

---

*) μB_z is equal to the average of $\omega_c \tau_m$ over ϵ and in the distribution function $f(\epsilon)$ the latter product was assumed to be small, see text after Eq.(4c.49).

$$j_x = (\sigma + \gamma^2 B_z^2/\sigma)\, E_x \tag{4d.6}$$

Hence, the relative change in resistivity is given by

$$\Delta\rho/\rho_B = 1 - (\sigma + \gamma^2 B_z^2/\sigma)/\sigma_0 \tag{4d.7}$$

In the strong field approximation σ becomes much smaller than $\gamma^2 B_z^2/\sigma$ and can therefore be neglected. We insert for σ and γ their high-field values given by Eqs.(4c.76) and (4c.77):

$$\Delta\rho/\rho_B = 1 - \{<\tau_m><\tau_m^{-1}>\}^{-1} \tag{4d.8}$$

For the case that $\tau_m(\epsilon)$ obeys the power law mentioned earlier, we obtain for $\Delta\rho/\rho_B$

$$\Delta\rho/\rho = 1 - \frac{9\pi}{16\,(r+3/2)!(3/2-r)!} \tag{4d.9}$$

The numerical value is 0.116 for r = -1/2 and 0.706 for r = +3/2. For r = 0, where τ_m does not depend on energy and the averaging parenthesis may be left out, the magnetoresistance effect vanishes for any value of the magnetic field. This explains why in metals the magnetoresistance effect is so small: It is only the carriers close to the Fermi surface that contribute to the conduction process. These have nearly the same energy and the same value of τ_m, although for carriers in metals τ_m does depend on energy. However, the average of τ_m is essentially its value at the Fermi energy making it essentially a constant.

At low temperatures in strong magnetic fields quantum effects occur with the formation of "Landau levels". In degenerate semiconductors there is an oscillatory magnetoresistance which is known as "Shubnikov - de Haas effect" and which will be discussed in Chap.9b.

In intrinsic or nearly intrinsic semiconductors and in semimetals two types of carriers, electrons and holes, contribute to magnetoresistance. It is easily verified that σ, β, and γ have to be replaced by the sums of these contributions such as $\sigma_n + \sigma_p$ etc. For simplicity we omit the subscript o which denotes the weak field case, and obtain instead of Eq.(4d.3) for $\Delta\rho/\rho_B B_z^2$:

$$\frac{\Delta\rho}{\rho_B B_z^2} = -\frac{\beta_p + \beta_n}{\sigma_p + \sigma_n} - \left(\frac{\gamma_p + \gamma_n}{\sigma_p + \sigma_n}\right)^2 \tag{4d.10}$$

Since σ_0 (Eq.(4c.40)) and β_0 (Eq.(4c.51)) contain only even powers of e, σ_p and σ_n, and β_p and β_n have the same sign, while the opposite is true for γ_n and γ_p (Eq.(4c.45)). Assuming for simplicity the same scattering mechanism for electrons and holes, Eq.(4d.10) yields

$$\frac{\Delta\rho}{\rho_B B_z^2} = \frac{9\pi}{16}\left\{\frac{(3r+3/2)!}{\{(r+3/2)!\}^3}\frac{p\mu_p^3+n\mu_n^3}{p\mu_p+n\mu_n} - \left(\frac{(2r+3/2)!}{\{(r+3/2)!\}}\frac{p\mu_p^2+n\mu_n^2}{p\mu_p+n\mu_n}\right)^2\right\} \quad (4d.11)$$

We introduce as usual the ratio $b = \mu_n/\mu_p$. For an intrinsic semiconductor (n=p) and acoustic deformation potential scattering ($r = -1/2$) we obtain from Eq.(4d.11)

$$\frac{\Delta\rho}{\rho_B B_z^2} = \frac{9\pi}{16}\mu_p^2\left(\frac{1+b^3}{1+b} - \frac{\pi}{4}(1-b)^2\right) = \frac{9\pi}{16}\left(1-\frac{\pi}{4}\right)\mu_p^2\left(1+\frac{(\pi/2)-1}{1-\pi/4}b+b^2\right) \quad (4d.12)$$

while for large values of b it is reasonable to write this equation in the form

$$\frac{\Delta\rho}{\rho_B B_z^2} = \frac{9\pi}{16}\left(1-\frac{\pi}{4}\right)\mu_n^2\left(1+\frac{(\pi/2)-1}{1-\pi/4}b^{-1}+b^{-2}\right) \quad (4d.13)$$

A comparison of this equation with the last one reveals that it is the type of carrier with the higher mobility which determines galvanomagnetic effects such as magnetoresistance.

Considering again the case of semimetals where we may neglect the averaging procedure, both $9\pi/16$ and $\pi/4$ may be replaced by 1. We thus obtain from Eq.(4d.12)

$$\Delta\rho/\rho_B B_z^2 = \mu_n\mu_p \quad (4d.14)$$

It is in semimetals like e.g. bismuth and in intrinsic degenerate semiconductors where this type of behavior is found. The effect is several orders of magnitude larger than the magnetoresistance of metals with only one type of carrier. Before the invention of indium antimonide devices, bismuth spirals were used for magnetic field measurements.

Many p-type semiconductors contain heavy and light holes, denoted by subscripts h and l. Both contribute to the conduction processes. Here all carriers have the same sign of charge as contrasted to near-intrinsic semiconductors. We denote by b the mobility ratio μ_l/μ_h and by η the density ratio p_l/p_h. Assuming again a power law for $\tau_m(\epsilon)$ with an exponent $r = -1/2$, we obtain for the magnetoresistance

$$\frac{\Delta\rho}{\rho_B B_z^2} = \frac{9\pi}{16}\mu_p^2\left\{\frac{1+\eta b^3}{1+\eta b} - \frac{\pi}{4}\left(\frac{1+\eta b^2}{1+\eta b}\right)^2\right\} \quad (4d.15)$$

where $\mu_p \equiv \mu_h$ is the mobility of the majority of holes. In e.g. p-type germanium, $b = 8$ and $\eta = 4\ \%$ which yields for the factor between braces a value of 10.6 instead of the value $1 - \pi/4 = 0.215$ for $\eta = 0$. Therefore the one light hole out of 25 total holes raises $\Delta\rho$ by a factor of about 50 which is equivalent to a rise of sensitivity with respect to B_z by a factor of 7. The conductivity σ and the Hall coefficient R_H contain lower powers of b and therefore are much less sensitive to the light-hole contribution (σ is raised by a factor of 1.3 and R_H by a factor of 2). Experimental results of $\Delta\rho/\rho B^2$ obtained with p-type germanium at 205 K with $\vec{B}$ parallel to a <111> - direction are shown in Fig.4.5. For a quantitative analysis the effective mass anisotropy has to be considered [1]. The dashed line represents the heavy hole contribution. The observed Hall mobility (some 10^3 cm^2/Vsec) would suggest that in a one-type-of-carrier model, magnetoresistance is constant at fields up to a few kG. Obviously it is the light hole contribution which is large in magnetoresistance and comparatively small in the Hall mobility which makes the magnetoresistance vary strongly with $\vec{B}$ although $(\mu_H B_z)^2 \ll 1$ still holds. For more details see Chap.8c.

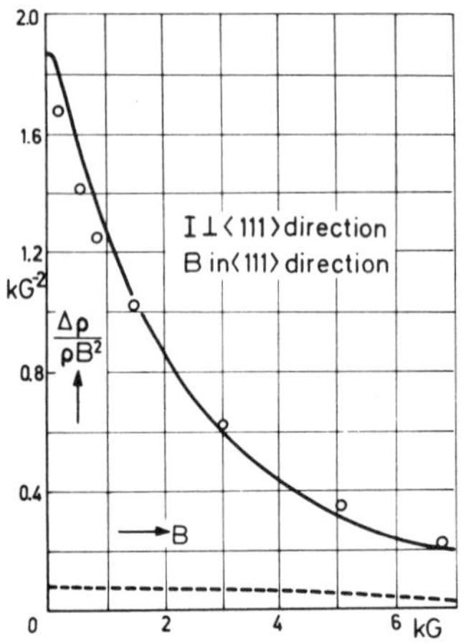

Fig.4.5 Experimental transverse magnetoresistance of p-type germanium at 205 K (B parallel to a <111> - direction). The dashed curve is calculated without inclusion of the high-mobility light holes (after Willardson et al., ref.1).

$\Delta\rho/\rho B^2$ is positive in the range of validity of the distribution function, Eq.(4c.37), since the Lorentz force deflects the carriers from the direction of the driving field and the drift velocity is only that component of the velocity in the direction of this field. It is mostly with "hot" carriers that a "negative magnetoresistance" may occur (see Chap.7f).

Longitudinal magnetoresistance where $\vec{B}$ and $\vec{F}$ are parallel in Eq.(4c.37), vanishes ($\Delta\rho=0$), if m^{-1}, τ_m, and a are scalar quantities. The case of non-vanishing longitudinal magnetoresistance will be discussed in Chaps. 7d and 8c.

4e Corbino Resistance

Measurements of the transverse magnetoresistance should be made with long filamentary samples, or else one obtains a "geometric" contribution to the magnetoresistance. This contribution is largest if the sample has the form of a circular disk with a central hole. Contacts are made at the hole and the circumference of the disk. The magnetic field is perpendicular to this "Corbino disk" [1, 2]. The arrangement is shown in Fig.4.6.

[1] R.K.Willardson, T.Harman, and A.C.Beer, Phys.Rev. 96 (1954) 1512.

Chap.4e:

[1] O.M.Corbino, Phys. Zeitschr. 12 (1911) 561.

[2] H.Welker and H.Weiss, Zeitschr. f. Physik 138 (1954) 322.

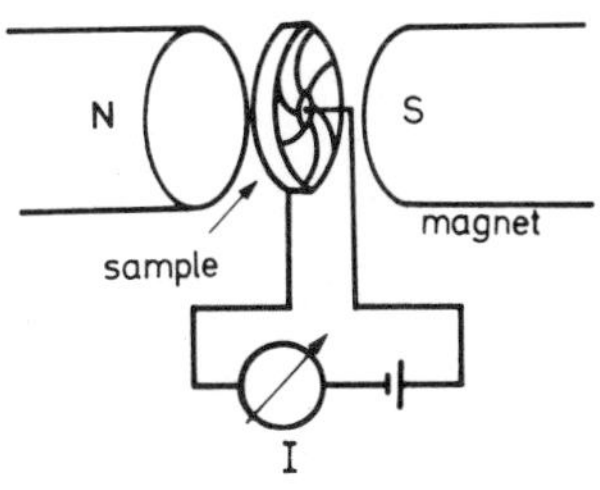

Fig.4.6 Corbino arrangement. The curves between the central hole and the circumference of the sample indicate the direction of current flow.

The relative change in resistance of the disk upon applying a magnetic field is

$$\frac{\Delta R}{R_o} = \frac{\Delta\rho}{\rho_o} + \frac{\mu_H^2 B_z^2}{1+\Delta\rho/\rho_o} \tag{4e.1}$$

where $\Delta\rho/\rho_o$ is the relative change in resistivity as measured with a long filamentary sample of the same material and μ_H is the Hall mobility. E.g. for n-type indium antimonide at room temperature in a field of 10 kG, $\Delta R/R_o = 17.7$ is obtained while $\Delta\rho/\rho_o$ is equal to only 0.48. There is no saturation of $\Delta R/R_o$ with increasing field intensity which makes this arrangement suitable for actual measurements of large magnetic field intensities. (From the experiment a rather linear relationship between ΔR and B_z is then obtained.)

The current lines in the sample are also shown in Fig.4.6. Because of radial symmetry the electric field $\vec{E}$ can have only radial components. Therefore along the x-axis the E_y-component vanishes. The current component j_y, however, does not vanish here as contrasted to conditions in magnetoresistance in filamentary samples. Hence, the resistivity cannot be defined by E_x/j_x but rather by

$$\rho_o + \Delta\rho = \rho_B = (\vec{E}\vec{j})/j^2 = E_x j_x/(j_x^2 + j_y^2) \tag{4e.2}$$

From Eqs.(4c.44) and (4c.45) with the notation

$$Z = 1/(1+\omega_c^2\tau_m^2) \tag{4e.3}$$

and putting $E_y = E_z = B_x = B_y = 0$ in the tensors (4c.39), the current densities take the form

$$j_x = (ne^2/m) <Z\tau_m> E_x \; ; \quad j_y = -(ne^3/m^2) <Z\tau_m^2> B_z E_x \tag{4e.4}$$

From Eq.(4e.2) we obtain for the resistivity ρ_B in a magnetic field

$$\rho_B = \frac{m <Z\tau_m>}{ne^2(<Z\tau_m>^2 + \omega_c^2 <Z\tau_m^2>^2)} \tag{4e.5}$$

The ratio of ρ_B and the zero-field resistivity

$$\rho_o = 1/\sigma_o = m/(ne^2<\tau_m>) \tag{4e.6}$$

is given by

$$\frac{\rho_B}{\rho_0} = 1 + \frac{\Delta\rho}{\rho_0} = \frac{<\tau_m><Z\tau_m>}{<Z\tau_m>^2 + \omega_c^2 <Z\tau_m^2>^2} \tag{4e.7}$$

On the other hand Eq.(4c.83) yields for R_H and thus for the product $\mu_H B_z$:

$$\mu_H B_z = \frac{R_H B_z}{\rho_0} = \frac{m<Z\tau_m^2>\omega_c}{ne^2(<Z\tau_m>^2 + \omega_c^2<Z\tau_m^2>^2)} \cdot \frac{ne^2<\tau_m>}{m} \tag{4e.8}$$

We obtain a simple expression for the following combination of terms :

$$1 + \frac{\Delta\rho}{\rho_0} + \frac{(\mu_H B_z)^2}{1+\Delta\rho/\rho_0} = \frac{<\tau_m><Z\tau_m>}{<Z\tau_m>^2 + \omega_c^2 <Z\tau_m^2>^2} +$$
$$+ \frac{<Z\tau_m^2>^2\,\omega_c^2<\tau_m>^2}{<\tau_m><Z\tau_m>(<Z\tau_m>^2 + \omega_c^2 <Z\tau_m^2>^2)} = \tag{4e.9}$$
$$= \frac{<\tau_m>}{<Z\tau_m>^2 + \omega_c^2 <Z\tau_m^2>^2}\left\{<Z\tau_m> + \frac{\omega_c^2<Z\tau_m^2>^2}{<Z\tau_m>}\right\} = \frac{<\tau_m>}{<Z\tau_m>}$$

This expression is obtained for the resistance ratio R_B/R_0 where R_B is given by $(E_x/j_x)_B$. From Eq.(4e.3) we obtain for this ratio

$$R_B/R_0 = <Z\tau_m>^{-1}/<\tau_m>^{-1} = <\tau_m>/<Z\tau_m> \tag{4e.10}$$

Denoting by ΔR the difference $R_B - R_0$ we obtain Eq.(4e.1) from Eqs.(4e.9) and (4e.10).

It may be of interest to solve Eq.(4e.1) for μ_H and evaluate μ_H from the observed data mentioned above for $B_z = 10$ kG :

$$\mu_H = \frac{1}{B_z}\sqrt{\left(1 + \frac{\Delta\rho}{\rho_0}\right)\left(\frac{\Delta R}{R_0} - \frac{\Delta\rho}{\rho_0}\right)} = 10^4\frac{\text{cm}^2}{\text{Vsec}}\sqrt{1.48 \times 17.2} = 50\,500\frac{\text{cm}^2}{\text{Vsec}} \tag{4e.11}$$

Experimental data at 1 kG are $\Delta\rho/\rho_0 = 0.014$ and $\Delta R/R_0 = 0.25$. At such a low field intensity the product $\mu_H B_z$ simply becomes $\sqrt{\Delta R/R_0}$ while for a filamentary sample it would be $r_H\sqrt{\Delta\rho/(\rho_B T_M)}$ which contains averages over $\tau_m(\epsilon)$. Therefore the low-field Corbino resistance may serve for a determination of the low-field Hall mobility as contrasted to magnetoresistance.

The second term on the right-hand side of Eq.(4e.1) is sometimes called the "geometric contribution" to magnetoresistance. Another, although smaller, contribution is obtained if the magnetoresistance is measured in the conventional way with a short thick sample rather than a long thin sample. The current lines and the equipotential lines for this case are shown in Fig.4.7.

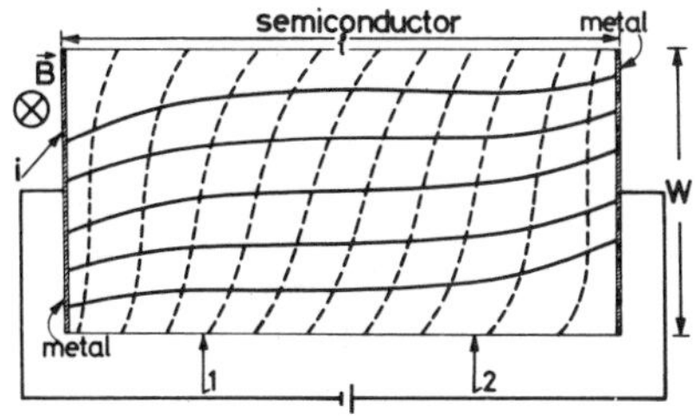

Fig.4.7
Current lines (full) and equipotential lines (dashed) in a bar-shaped sample in a magnetic field perpendicular to the plane of the drawing; the arrows marked 1 and 2 indicate potential probes (after H.Weiss: Semiconductors and Semimetals (R.K. Willardson and A.C.Beer, eds.), Vol.1, p.315. New York: Academic Press. 1966).

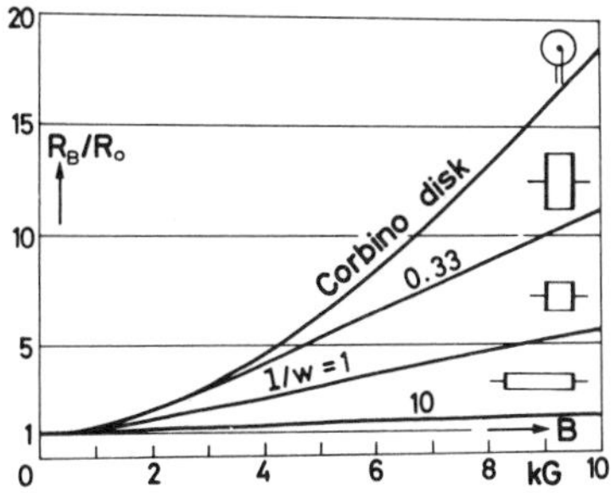

Fig.4.8
Relative resistance as a function of magnetic induction for four n-InSb samples of equal purity but different geometry (after ref.2).

Mainly near the end contacts of the sample the current lines do not go along the sample axis since there the contact surface short-circuits the Hall field. With l and w being sample length and width, respectively, Fig.4.8 shows [2] the magnetoresistance ratio as a function of B_z for various values of the ratio l/w. For a comparison, the Corbino resistance is plotted in the same figure. Only with high-mobility semiconductors is the geometric contribution to magnetoresistance important.

4f. Magnetoresistance in Inhomogeneous Samples

Quite often semiconductors are inhomogeneously doped. These inhomogeneities strongly influence magnetoresistance. How do they arise? The processes leading to inhomogeneous doping are still under investigation. When a single crystal is pulled from the melt in a Czochralski arrangement, depending on the velocity of pulling, zones of stronger and weaker doping are formed. These zones are called "striations" and are comparable in appearance to the annual rings of a tree [1]. In a direction perpendicular to the direction of pulling the relative magnetoresistance $\Delta R/R_0$ may be only one tenth of $\Delta R/R_0$ parallel to the direction of pulling [2].

The explanation of this phenomenon is based on the fact that the current lines are deflected from their original direction as they enter a zone of different carrier density due to a change in doping [3]. Fig.4.9 shows a transition from a slightly n-type part of a crystal (denoted "n") to a more heavily

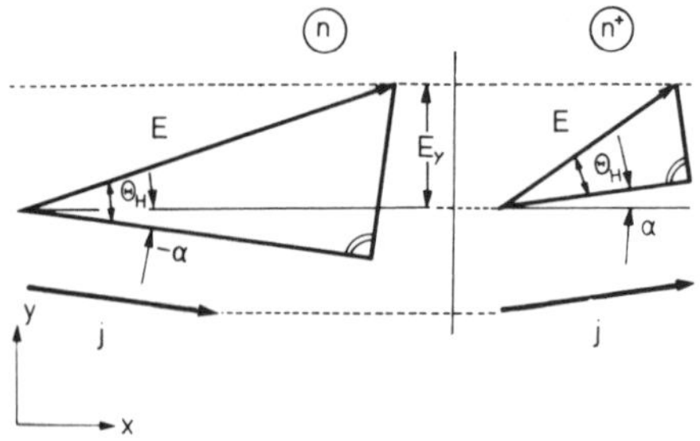

Fig.4.9 Rotation of the current lines in a transverse magnetic field by an angle α due to a discontinuity in the doping level (e.g. n - n^+ junction) (after H.Weiss: Semiconductors and Semimetals (R. K. Willardson and A. C.Beer, eds.), Vol.1, p.315. New York: Academic Press. 1966).

doped n-type part (denoted by "n^+"). For simplicity it is assumed that the Hall mobility is independent of n and $\vec{B}$. A coordinate system is chosen such that the x-direction is perpendicular to the n-n^+ barrier. The field component

[1] J. A. M. Dikhoff, Solid-State Electronics 1 (1960) 202; P. R. Camp, J. Appl. Phys. 25 (1954) 459.

[2] H. Rupprecht, R. Weber, and H. Weiss, Zeitschr.f.Naturf. 15a (1960) 783.

[3] C. Herring, J. Appl. Phys. 31 (1960) 1939; H. Weiss, J. Appl. Phys. Suppl. 32 (1961) 2064.

$E_x = j_x/(ne\mu)$ is proportional to $1/n$ since j_x is transmitted through the crystal by the externally applied voltage and hence is the same in both parts of the crystal. E_x is the same throughout the cross section of the crystal which can be written in the form $\partial E_x/\partial y = 0$. Since the magnetic field does not change with time we obtain from Maxwell's equations $[\vec{\nabla}_r \vec{E}]_z = 0$ which with $\partial E_x/\partial y = 0$ yields $\partial E_y/\partial x = 0$. This means that the Hall field E_y is also the same in both parts of the crystal. Therefore the field direction $\vec{E}$ in Fig.4.9 is different in the two parts. The Hall angle θ_H given by $\tan\theta_H = \mu_H B_z$ is the same, however, since μ_H is assumed to be the same. On the other hand in a homogeneous crystal $\tan\theta_H$ is E_y/E_x which is no longer true here. However, if we define θ_H as the angle between $\vec{E}$ and $\vec{j}$:

$$\tan\theta_H = [\vec{j}\vec{E}]/(\vec{j}\vec{E}) \tag{4f.1}$$

this definition would cover both the case of Fig.4.9 and that of a homogeneous crystal where $|\vec{j}| = j_x$ and hence $\tan\theta_H = E_y/E_x$. The current lines $\vec{j}$ in Fig.4.9 are deflected by the same angle α but in opposite directions on both sides of the barrier due to the magnetic field. The case of alternating n and n^+ regions is demonstrated by Fig.4.10. In the upper part of the figure the barriers are perpendicular to the average current density while in the lower part they are at an oblique angle to the average current density.

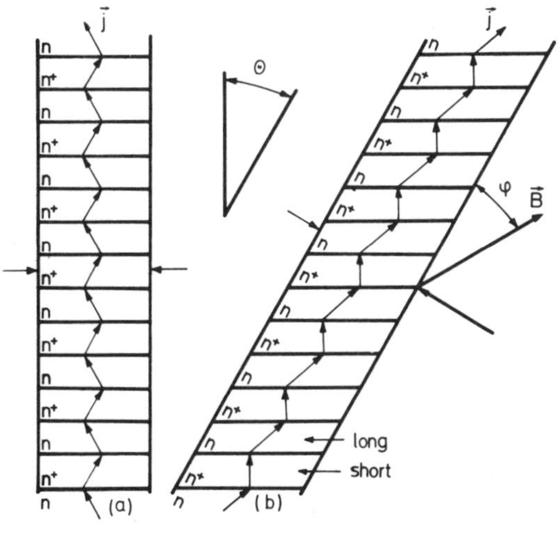

Fig.4.10 Current lines in a transverse magnetic field where the average current direction is perpendicular to the alternating n-n^+ junctions (a) or at an angle (b) (after H. Weiss: Semiconductors and Semimetals (R. K. Willardson and A. C. Beer, eds.), Vol.1, p.315. New York: Academic Press. 1966).

Herring neglected in his calculation [3] the energy dependence of the momentum relaxation time (τ_m = const.) and still obtained a magnetoresistance effect given by

$$\frac{\Delta\rho}{\rho_0} = \left(1 - \frac{1}{\langle n\rangle \cdot \langle n^{-1}\rangle}\right) \frac{\mu_H^2 B^2 \sin^2\varphi}{1+\mu_H^2 B^2 \cos^2\varphi} \tag{4f.2}$$

where $\langle n\rangle$ and $\langle n^{-1}\rangle$ are the values of n and 1/n, respectively, averaged over the whole crystal, and φ is the angle between the average current and the magnetic field. Let us for simplicity assume that $\langle n\rangle = \frac{1}{2}(n+n^+)$ and $\langle n^{-1}\rangle = \frac{1}{2}(1/n+1/n^+)$. For this case we obtain

$$1 - \frac{1}{\langle n\rangle \langle n^{-1}\rangle} = 1 - \frac{4nn^+}{(n^++n)^2} = \left(\frac{n^+-n}{n^++n}\right)^2 = \frac{1}{4}\left(\frac{\Delta n}{\langle n\rangle}\right)^2 \tag{4f.3}$$

if we introduce $\Delta n = n^+ - n$ and $\langle n\rangle = (n^++n)/2$. For a 10 % variation in n we obtain 0.25 % for this factor. The difference in doping will be larger because Δn is smoothed by the thermal diffusion of the carriers.

There may also be a variation of the direction of the average current density and therefore of φ throughout the crystal if the striations are not parallel to each other. A common method to make striations visible at the surface of the semiconductor crystal is by dyeing it with copper from a copper salt solution. In the galvanic process more copper will be deposited where the carrier density is higher than in other regions [1].

4g. Planar Hall Effect

In the Hall effect investigated so far the side arms of the sample pointed in a direction perpendicular to that of the magnetic field. In the arrangement shown in Fig.4.11 this is no longer the case. If the x-axis is still in the direction of current flow and the z-axis is in the direction of the Hall side arms, the magnetic induction $\vec{B}$ has components B_x and B_z in both of these directions. The "planar Hall field" E_z is given by

$$E_z = P_H \cdot j_x B_x B_z \tag{4g.1}$$

where P_H is the "planar Hall coefficient" measured in cm^5/wattsec^2. The observed voltage

$$V_z = P_H I_x B_x B_z / d \tag{4g.2}$$

where d is the thickness of the crystal perpendicular to $\vec{B}$; it is a maximum for $B_x = B_z = B/\sqrt{2}$ where the angle between $\vec{B}$ and $\vec{j}$ is 45°.

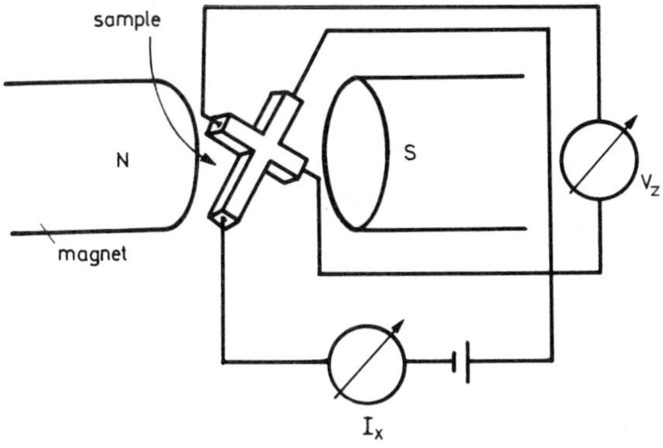

Fig.4.11 Arrangement for planar Hall effect.

$$V_z = P_H I_x B^2/2d \tag{4g.3}$$

The proportionality of V_z to B^2 immediately suggests a relation to magneto-resistance.

For a weak-field calculation of the coefficient P_H we derive the current density components from the conductivity tensor given by Eq.(4c.53) where we assume $B_y = 0$.

$$\left.\begin{aligned} j_x &= (\sigma_o\beta_o B_z^2)E_x + \gamma_o B_z E_y - \beta_o B_x B_z E_z \\ j_y &= -\gamma_o B_z E_x + (\sigma_o + \beta_o B_x^2 + \beta_o B_z^2) E_y + \gamma_o B_x E_z \\ j_z &= -\beta_o B_x B_z E_x - \gamma_o B_x E_y + (\sigma_o + \beta_o B_x^2) E_z \end{aligned}\right\} \tag{4g.4}$$

We solve $j_y = 0$ and $j_z = 0$ for E_x and E_y and insert these in the equation for j_x neglecting powers of B higher than the second. This equation is solved for E_z.

$$E_z = \{(\sigma_o\beta_o + \gamma_o^2)/\sigma_o^3\} j_x B_x B_z \tag{4g.5}$$

We compare this equation with Eq.(4g.1) and obtain for P_H:

$$P_H = \frac{1}{\sigma_o}\left\{\frac{\beta_o}{\sigma_o} + \left(\frac{\gamma_o}{\sigma_o}\right)^2\right\} \tag{4g.6}$$

Introducing the resistivity ρ_o for $1/\sigma_o$ and the transverse magnetoresistance Eq.(4d.3) for the term in braces, we find that the relationship between the planar Hall effect and the magnetoresistance suggested above is in fact :

$$P_H = -\rho_o \Delta\rho/\rho_B B^2 \tag{4g.7}$$

We estimate the order of magnitude of the planar Hall field relative to the Hall field Eqs.(4c.2) and (4c.59) by neglecting the factors r_H and T_M which makes $\Delta\rho/\rho_B B^2$ equal to the square of the mobility. We find that E_z is lower than E_y in the Hall arrangement by a factor $\mu_H B_x$ which should be $\ll 1$ in the weak field approximation. The planar Hall effect causes more interest in the many-valley model of the semiconductor where it will be investigated in more detail (Chap.7d).

4h. Thermal Conductivity, Lorenz Number, Comparison with Metals

Besides electric charge, carriers transport also energy. If we set up a temperature gradient $\vec{\nabla}_r T$ in a sample there will be a heat flow of density $\vec{w}$ which is determined by the thermal conductivity κ:

$$\vec{w} = -\kappa \vec{\nabla}_r T \tag{4h.1}$$

if there is no electric current. In metals where practically all of the thermal energy is transported by carriers, there is an electric field set up of the right amount to counteract the average carrier velocity in the direction of the temperature gradient. This is called the "thermoelectric field". Since some carriers have a higher than average velocity in this direction due to the Fermi-Dirac distribution function, there is still some energy transport by carriers which is proportional to the density of carriers and their mobility, hence proportional to the conductivity σ. The ratio

$$L = \kappa/(\sigma T) \tag{4h.2}$$

is denoted as "Lorenz number". The value calculated for a highly degenerate electron gas [1],

$$L = (\pi k_B/e)^2/3 = 2.45 \times 10^{-8}\,\text{watt-ohm/K}^2 \tag{4h.3}$$

is found experimentally to agree for most metals within 10 % at room temperature. The temperature independence of L is known as the "Wiedemann-Franz law". Since for metals $\sigma \propto 1/T$ to a good approximation, κ should be nearly independent of temperature.

The metal with the highest value of σ is silver; its value of κ at 273 K is 4.33 watt/cmK which is about 1 cal/cm sec K. Although heat is usually given in units of 1 cal = 4.19 wattsec, we prefer the electrical unit.

---

[1] For the evaluation of integrals of interest in the theory of a highly degenerate electron gas see e.g. A. C. Smith, J. F. Janak, and R. B. Adler: Electronic Conduction in Solids, appendix I. New York: McGraw-Hill. 1967.

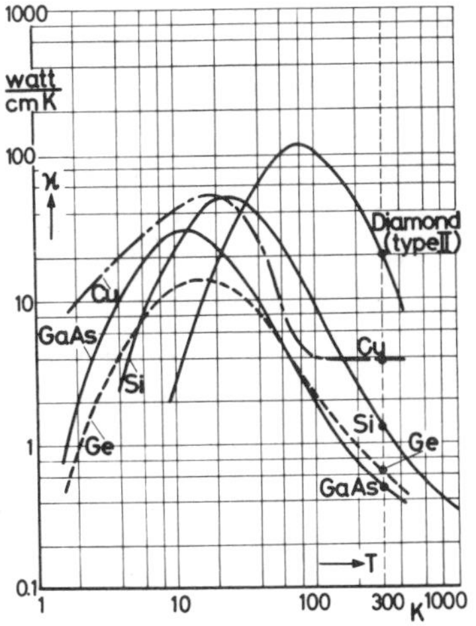

Fig.4.12 Measured thermal conductivity as a function of temperature for various semiconductors of high purity; for a comparison data for copper are also included in the diagram compiled by S. M. Sze: Physics of Semiconductor Devices. New York: J. B. Wiley and Sons. 1969 (after the data of Carruthers et al.; Holland; White; and Berman et al.).

In silicon the experimental value of κ is 1.4 watt/cm K (see Fig.4.12) which is about one third of that of silver, although the density of carriers is many orders of magnitude less. Even in insulators there is a thermal conductivity. This is entirely due to energy transport by thermal lattice waves. We will not consider the theory of this process [2]. It just may be worth mentioning that in an arrangement of coupled harmonic oscillators there is no interaction between lattice waves and hence no energy transfer. It is the anharmonicity of the lattice oscillators which determines the thermal conductivity in insulating crystals at high temperatures. E.g. in germanium the fact that it contains several isotopes plays an important role in the thermal conductivity [3].

In non-degenerate semiconductors the lattice contribution to thermal conductivity in almost all cases is orders of magnitude larger than the electronic contribution. Therefore we will not calculate the electronic contribution either but just comment upon the results of such calculations. Before we do this, we will discuss a typical experimental arrangement and some of the experimental results.

Methods of measuring κ were given by e.g. H. Weiss [4] and by J. Schröder [5] (for temperatures between 297 and 452 K). The arrangement by Weiss is shown in Fig.4.13. The filamentary sample of circular cross section has at the upper end a small electrically heated oven. The power consumption of this oven is measured. The lower end of the sample is cooled by a metal plate. In order to avoid convection, the arrangement is mounted in an evacuated cylinder (10^{-4} torr). Three more ovens are mounted outside the cylinder which, in order to avoid radiation losses, are kept at the same temperatures as the parts of the sample they face. Temperatures are measured by thermocouples (indi-

[2] A review is presented by J. Appel: Progress in Semiconductors. (A. F. Gibson and R. E. Burgess, eds.), Vol.5, p.142. London: Temple Press. 1960.

[3] T. H. Geballe and G. W. Hull, Phys.Rev. 110 (1958) 773.

[4] H. Weiss: Halbleiter und Phosphore. (M. Schön and H. Welker, eds.), p.497. Braunschweig: Vieweg. 1957.

[5] J. Schröder, Rev.Sci.Instr. 34 (1963) 615.

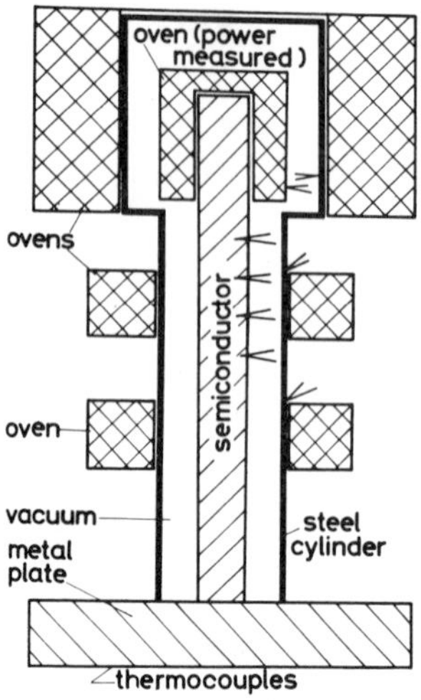

Fig.4.13 Cross section of an arrangement for measurements of thermal conductivity (after H. Weiss, ref.4).

cated in Fig.4.13 by <). The accuracy of the determination of κ is 10 % at normal and 20 % at high temperatures.

Results obtained [4] with n-type and p-type InSb are shown in Fig.4.14. κ is independent of doping. The lattice contribution to κ (dashed curve) is roughly proportional to 1/T. The electronic contribution is significant only above 500 K and agrees within a factor of 2 with values calculated for an intrinsic semiconductor (see below).

In Fig.4.15 the observed [6] thermal resistivity $1/\kappa$ of pure silicon is plotted versus temperature T. In a first approximation $1/\kappa$ is proportional to T equivalent to $\kappa \propto 1/T$ typical for lattice conduction processes.

Bismuth telluride, Bi_2Te_3, at room temperature has a thermal conductivity κ

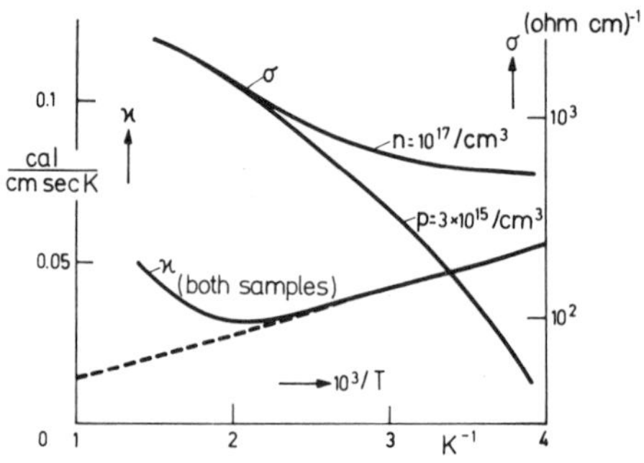

Fig.4.14 Dependence of electrical and thermal conductivities on the reciprocal of the temperature for n- and p-InSb (after H. Weiss, ref.4).

of 0.024 watt/cm K and, with suitable doping, an electric conductivity σ of 10^3(ohm-cm)$^{-1}$. The low value of κ and the high value of σ together with the possibility of obtaining it in n- and p-type form make it an interesting material for thermoelectric cooling and energy conversion applications. Its Lorenz number, $L = 8 \times 10^{-8}$ watt-ohm / K^2, is of the order of magnitude of that of metals, Eq.(4h.3). Since the material is degenerate for the value of σ quoted above, a comparison with L of metals is justified and reveals that the

[6] W. Fulkerson, J. P. Moore, R. K. Williams, R. S. Graves, and D. L. McElroy, Phys.Rev. 167 (1968) 765.

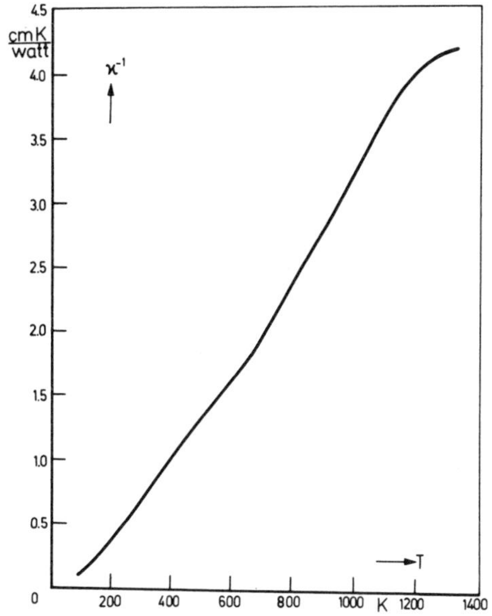

Fig.4.15 Thermal resistivity as a function of temperature for silicon (after W. Fulkerson et al., ref. 6).

electronic part of κ in Bi_2Te_3 should be about 30 %.

Finally we discuss results of a calculation of L for non-degenerate semiconductors. If we take for κ only its electronic part and assume a power law, ϵ^r, for $\tau_m(\epsilon)$, we obtain

$$L = (k/e)^2 \; (r+5/2) \tag{4h.4}$$

which is similar to Eq.(4h.3) except that $\pi^2/3$ is replaced by r+5/2. For acoustic deformation potential scattering (r = - 1/2) the latter factor becomes 2 while $\pi^2/3$ is about 3.29. Obviously degeneracy does not change the order of magnitude of L.

In intrinsic or nearly intrinsic non-degenerate semiconductors L becomes [7]

$$L = (k_B/e)^2 \,[r+5/2+ \{ 2(r+5/2) + \epsilon_G/k_BT\}^2 \, \sigma_n \sigma_p/\sigma^2] \tag{4h.5}$$

where the same scattering mechanism for electrons and holes is assumed and σ_n and σ_p are the electron and hole contributions to the electric conductivity $\sigma = \sigma_n + \sigma_p$. For intrinsic conduction, $\sigma_n \sigma_p/\sigma^2 = b/(1+b)^2$ where $b = \mu_n/\mu_p$, the

[7] For a calculation see e.g. O. Madelung: Grundlagen der Halbleiterphysik, p.117. Heidelberger Taschenbücher, Bd.71. Berlin - Heidelberg - New York: Springer. 1970.

mobility ratio. The term containing the band gap energy, ϵ_G, is due to generation of electron hole pairs at the hot sample end and recombination at the cold sample end. Each pair carries an energy of magnitude ϵ_G in addition to its energy of motion. The latter is the only energy converted to heat upon recombination in a zero-gap intrinsic semiconductor with equal mobilities of electrons and holes*) (b=1) and is taken care of by a term $(r+5/2)^2$ in addition to the "one-band" term r+5/2 in this case. For large-gap intrinsic semiconductors (which will occur at high temperatures only) the "ambipolar" part of L and therefore also κ will be quite large, although it may still be much smaller than the lattice contribution. This is the case e.g. in silicon at 800 K where κ calculated from Eqs.(4h.5) and (4h.2) is 10^{-3} watt/cm K while the observed value of κ is more than two orders of magnitude larger (4×10^{-1} watt/cm K).

One last remark about the "speed" of heat flow. The equation of continuity for $\vec{w}$ in one dimension is given by

$$\frac{\partial w_x}{\partial x} = -\frac{\partial}{\partial t}(c_v \rho T) \tag{4h.6}$$

where c_v is the specific heat per mass unit, ρ is the mass density, and the product $c_v\rho$ the specific heat per unit volume. This together with Eq.(4i.1) yields the "diffusion equation"

$$(\kappa/\rho c_v)\, \partial^2 T/\partial x^2 = \partial T/\partial t \tag{4h.7}$$

where we may introduce a "diffusion constant of temperature",

$$D_T = \kappa/\rho c_v \tag{4h.8}$$

which e.g. in silicon at room temperature has a value of

$$D_T = 1.4 \text{ watt/cm K} \,/\, (2.33 \text{ g cm}^{-3}\ 0.76 \text{ watt/gK}) = 0.79 \text{ cm}^2/\text{sec} \tag{4h.9}$$

(The product ρc_v is also the ratio of the specific heat per mole and volume per mole; for solids according to Dulong and Petit the former has a value of 25 wattsec/mole.K which may be useful for estimates of D_T).

A periodic solution of Eq.(4i.7) is given by

$$T \propto \exp(-x/l)\, \exp\{i\omega(t-x/v)\} \tag{4h.10}$$

where ω is the angular frequency, $v = \sqrt{2D_T\omega}$ is the velocity and $l = \sqrt{2D_T/\omega}$ is the range of the temperature wave. At a higher frequency the wave is faster

---

*) In this case there will be no thermoelectric field since electrons and holes travelling at the same speed neutralize each other.

but does not get as far. At a given minimum range l the maximum frequency is $\nu = D_T/\pi l^2$ which e.g. in a silicon sample of l = 1 mm length requires ν = 25 Hz and a transmission time of $1/\omega \approx 6 \times 10^{-3}$ sec. Contrary to these "slow" temperature waves acoustic lattice waves have a propagation velocity in solids of about 10^5 cm/sec which is nearly independent of frequency. It is mainly the "slowness" of temperature waves that thermal effects do not gain much interest in modern electronics while acoustoelectric effects have become increasingly interesting in recent years.

4i. Thermoelectric (Seebeck) Effect

The thermoelectric field mentioned in Chap.4h will be calculated now. In solving the Boltzmann equation for the case of a temperature gradient we have introduced an electrothermal field $\vec{F}$ given by Eq.(4c.19). Since $\vec{F}$ depends on the carrier energy ϵ we cannot take it outside the integral in calculating $\vec{j}$, but obtain instead

$$\vec{j} \propto \langle \tau_m e\vec{F} \rangle = e\vec{E}\langle \tau_m \rangle + T \cdot \langle \tau_m \vec{\nabla}_r(\epsilon-\zeta)/T \rangle \tag{4i.1}$$

Since the zero of energy is arbitrary we introduce an electron kinetic energy

$$\epsilon_n = \epsilon - \epsilon_c \tag{4i.2}$$

which is zero at the conduction band edge ϵ_c, and a hole kinetic energy

$$\epsilon_p = -(\epsilon - \epsilon_v) \tag{4i.3}$$

which is zero at the valence band edge. It is likewise convenient to introduce Fermi energies, ζ_n and ζ_p, by putting

$$\zeta_n = \zeta - \epsilon_c \tag{4i.4}$$

and

$$\zeta_p = -(\zeta - \epsilon_v) \tag{4i.5}$$

In non-degenerate semiconductors the Fermi level ζ is in the gap and both ζ_n and ζ_p are therefore negative ($\zeta_n + \zeta_p = -\epsilon_G$). If we subtract Eq.(4i.2) from Eq.(4i.4) we obtain

$$\epsilon_n - \zeta_n = \epsilon - \zeta \tag{4i.6}$$

Similarly Eqs.(4i.3) and (4i.5) yield

$$-(\epsilon_p - \zeta_p) = \epsilon - \zeta \tag{4i.7}$$

Except when considering simultaneous conduction by electrons and holes, we omit in this chapter for simplicity the subscripts n and p and, in the case of holes, also the negative sign in front of $\epsilon_p - \zeta_p$.

For a calculation of the thermoelectric field $\vec{E}$ we consider the case that there is no current flow through the conductor ("open circuit"). Eq.(4i.1) yields for $\vec{j} = 0$:

$$e\vec{E} = -\frac{T}{<\tau_m>} <\tau_m \vec{\nabla}_r \frac{\epsilon - \zeta}{T}> = \frac{1}{T}\left(\frac{<\tau_m \epsilon>}{<\tau_m>} - \zeta\right) \vec{\nabla}_r T + \vec{\nabla}_r \zeta \qquad (4i.8)$$

We introduce an "entropy transport parameter" S/e by [1]

$$S/e = \frac{1}{eT}\left(\frac{<\tau_m \epsilon>}{<\tau_m>} - \zeta\right) \qquad (4i.9)$$

which should not be confused with the entropy discussed in Chap.3. Now we have from Eq.(4i.8)

$$e\vec{E} = S\vec{\nabla}_r T + \vec{\nabla}_r \zeta \qquad (4i.10)$$

In principle a measurement of $\vec{E}$ involves the arrangement given in Fig.4.16.

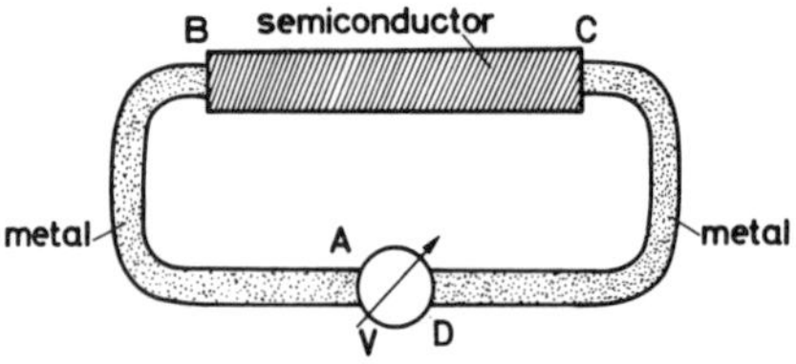

Fig.4.16 Arrangement for measurement of the Seebeck effect.

The filamentary sample has equal metal contacts at both ends, B and C, which are held at different temperatures, T_B and T_C. A voltmeter is connected at the end contacts. The voltage, V, indicated by the voltmeter is given by

$$V = \int_A^D (\vec{E}\, d\vec{r}) = \frac{1}{e}\int_A^D S(\vec{\nabla}_r T d\vec{r}) + \frac{1}{e}\int_A^D (\vec{\nabla}_r \zeta d\vec{r}) \qquad (4i.11)$$

A and D are the voltmeter contacts assumed to be at the same temperature and made from the same metal. Hence the second integral which is ζ/e between limits A and D, vanishes. Eq.(4i.11) is simplified to :

[1] M. W. Zemansky: Heat and Thermodynamics, 4th ed., p.192. New York: McGraw-Hill. 1957.

$$V = \frac{1}{e} \int_A^D S(\vec{\nabla}_r T d\vec{r}) \tag{4i.12}$$

From the path A → D we can split the part B → C which is in the semiconductor under investigation, while the rest, namely A → B plus C → D, is in the metal. Since the path through the voltmeter does not contribute to the integral and the parts C → D and A → B can thus be combined to give C → B which makes the integral the negative of that over B → C, we obtain

$$V = \frac{1}{e} \int_{T_B}^{T_C} (SdT)_{Semiconductor} - \frac{1}{e} \int_{T_B}^{T_C} (SdT)_{Metal} \tag{4i.13}$$

This is the thermoelectric force of the thermocouple consisting of the semiconductor and the metal. The effect is known as "Seebeck effect". We denote by $d\Theta/dT$ the "thermoelectric power" of a material :

$$\frac{d\Theta}{dT} = \frac{S}{e} = \frac{k_B}{e} \left(\frac{\langle \tau_m \epsilon / k_B T \rangle}{\langle \tau_m \rangle} - \frac{\zeta}{k_B T} \right) \tag{4i.14}$$

where $k_B/|e|$ is about*) 86 μV/K.

The "absolute thermoelectric force" Θ as a function of temperature T is given by

$$\Theta = \frac{1}{e} \int_0^T SdT \tag{4i.15}$$

From Eq.(4i.13) we obtain the observed voltage V in terms of $\Theta(T)$:

$$V = \{\Theta(T_C) - \Theta(T_B)\}_{Semiconductor} - \{\Theta(T_C) - \Theta(T_B)\}_{Metal} \tag{4i.16}$$

The Thomson effect, discussed in the following Chap.4j, offers a possibility for an experimental determination of $\Theta(T)$.

In the case that the momentum relaxation time $\tau_m(\epsilon)$ obeys a power law, ϵ^r, we obtain for the thermoelectric power of a non-degenerate n-type semiconductor

$$d\Theta/dT = -(k_B/|e|)(r+5/2 - \zeta_n/k_B T) \tag{4i.17}$$

while for the degenerate electron gas in a metal

$$d\Theta/dT = -(k_B/|e|)(r+3/2)(\pi^2/3)(k_B T/\zeta_n) \tag{4i.18}$$

*) For a more exact value see table of constants at the end of the book.

is obtained. At room temperature and below, $T \ll \zeta_n/k_B$, the thermoelectric power of a metal is much smaller than that of a non-degenerate semiconductor and in Eq.(4i.16) the term concerning the metal can therefore be neglected in most practical cases. In Fig.4.17 the thermoelectric power of copper between

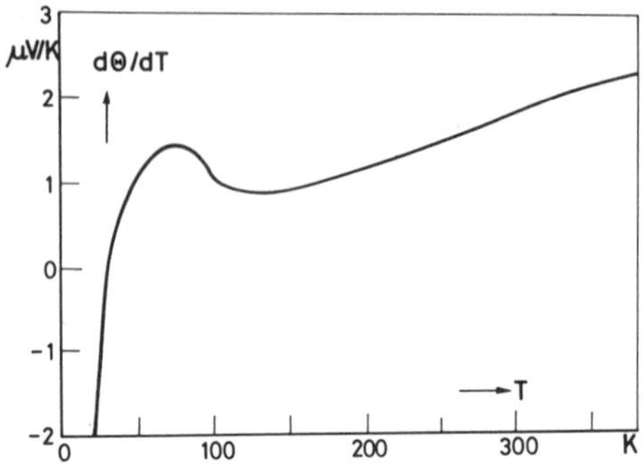

Fig.4.17 Thermoelectric power for high-purity copper (after A. V. Gold, D. K. C. McDonald, W. B. Pearson, and I. M. Templeton, Phil. Mag. 5 (1960) 765).

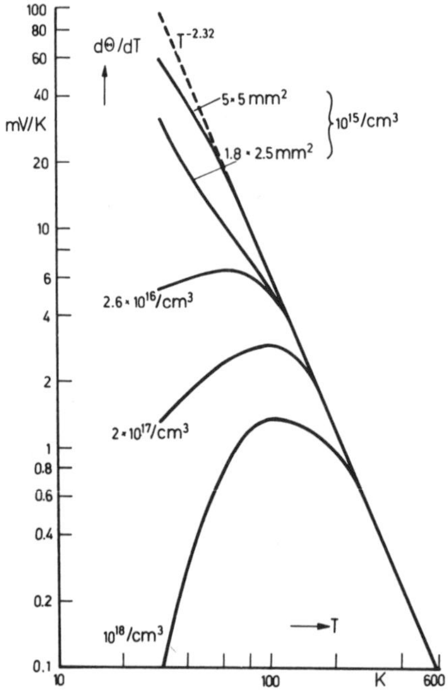

Fig.4.18 Thermoelectric power of p-type silicon for various concentrations of acceptors (boron) and compensating donors. Samples $10^{15}/cm^3$ are equal except for their cross sections (after T.H.Geballe and G.W. Hull, ref.4).

40 and 400 K is plotted which may serve for a typical example of a metal [2]. It is of the order of a few μV/K. The thermoelectric power of p-type silicon shown in Fig.4.18 is between 100 μV/K and 100 mV/K depending on doping and temperature [3].

Since $d\Theta/dT$ in Eq.(4i.14) depends on the first power of e, it is different in sign for electrons and holes. Fig.4.19 shows the product of T and $d\Theta/dT$ as

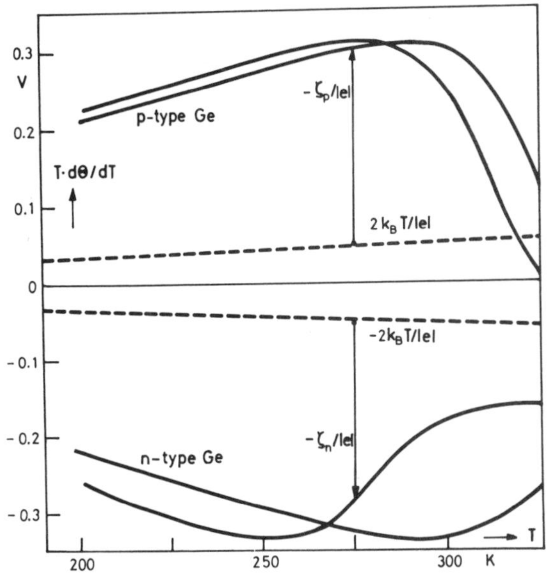

Fig.4.19 Thermoelectric power for n- and p-type germanium (experimental data and calculated curves). The vertical arrows are Fermi energies in units of the elementary charge.

a function of T for two n-type and two p-type germanium samples differing in doping [4]. The dashed lines represent $Td\Theta/dT = 2k_BT/e$ for both signs of e which is obtained from Eq.(4i.17) for $r = -1/2$ (acoustic deformation potential scattering) and $\zeta = 0$. Hence, the difference between an experimental curve and the dashed line with the same sign of e gives the negative Fermi energy in units of e. The vertical arrows in Fig.4.19 demonstrate these quantities for an

[2] A. V. Gold, D. K. C. McDonald, W. B. Pearson, and I. M. Templeton, Phil.Mag. 5 (1960) 765.
[3] T. H. Geballe and G. W. Hull, Phys.Rev. 98 (1955) 940.
[4] T. H. Geballe and G. W. Hull, Phys.Rev. 94 (1954) 1134.

arbitrarily chosen temperature of 275 K where the arrows have about equal lengths of nearly 0.3 V. Considering the fact that the n- and p-type samples are differently doped and the Fermi energies ζ_n and ζ_p are therefore not comparable in magnitude, the sum of ζ_n and ζ_p is about equal to the band gap energy, ϵ_G, which in germanium is about 0.7 eV. This indicates that to a first approximation the term $<\tau_m \epsilon/k_B T>/<\tau_m>$ in Eq.(4i.14) may be neglected and the Fermi energy determines the thermoelectric power.

A typical arrangement for a thermoelectric determination of the conductivity type of a semiconductor is shown in Fig.4.20. The sample is put on a cold metal plate while the hot point of a soldering iron is pressed against the top surface of the sample. A voltmeter is connected to the plate and the sol-

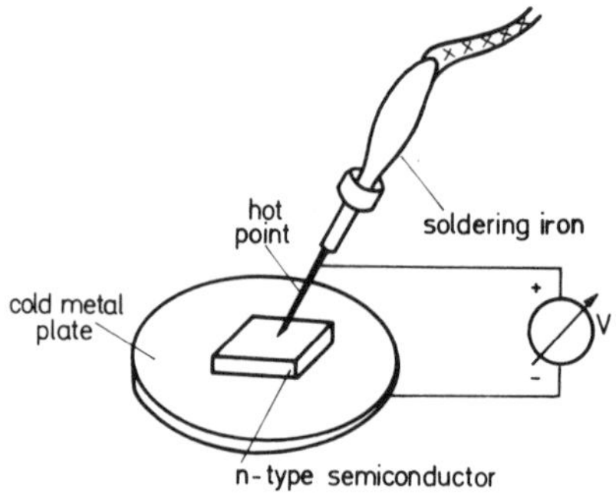

Fig.4.20 Hot-probe arrangement for a thermoelectric determination of the conductivity type of a semiconductor (for n-type the polarity of the voltage is shown; for p-type the polarity is reversed).

dering iron. For an n-type sample the polarity of the voltmeter is as indicated. For a p-type sample the polarity has to be reversed for normal deflection of the mV-meter. An arrangement where a hot plate and a cold point are used has been suggested [5] for high-resistivity material such as e.g. silicon carbide.

For intrinsic or near-intrinsic semiconductors both electrons and holes have to be considered. Since $\vec{j} = \vec{j}_p + \vec{j}_n = 0$, we obtain instead of Eq.(4i.8) for this case :

$$|e|\vec{E}(\sigma_p + \sigma_n) = -\frac{T}{<\tau_p>}<\tau_p \vec{\nabla}_r \frac{\epsilon_p - \zeta_n}{T}>\sigma_p - \frac{T}{<\tau_n>}<\tau_n \vec{\nabla}_r \frac{\epsilon_n - \zeta_n}{T}>\sigma_n \tag{4i.19}$$

which yields

$$\frac{d\Theta}{dT} = \frac{d\Theta_n}{dT}\cdot\frac{\sigma_n}{\sigma} + \frac{d\Theta_p}{dT}\cdot\frac{\sigma_p}{\sigma} \tag{4i.20}$$

[5] L. J. Kroko and A. G. Milnes, Solid State Electronics $\underline{8}$ (1965) 829.

where $\sigma = \sigma_n + \sigma_p$. For $\sigma_n = \sigma_p$ and $d\Theta_n/dT = -d\Theta_p/dT$ the thermoelectric power vanishes. Although this is an unrealistic case, it shows nevertheless that the thermoelectric power in the intrinsic temperature range of a semiconductor is smaller than that in the extrinsic range. In Fig.4.19 near the high-temperature ends of the curves, germanium becomes intrinsic and the absolute value of the thermoelectric power is decreased. For the curves marked "p-Ge" there even is an indication that a reversal in sign will occur at still higher temperatures which would be due to the fact that the electron mobility is higher than the hole mobility and hence $\sigma_n > \sigma_p$ in Eq.(4i.20).

In Fig.4.18 the curve representing the purest of the boron-doped silicon samples splits at its low temperature end; here the thermoelectric power depends on the geometric dimensions of the sample. Furthermore it is much larger than Eq.(4i.14) would suggest. An explanation was given by Herring [6] and Frederikse [7] by considering a non-equilibrium "phonon" distribution.

So far we have considered the electron "gas" as being in equilibrium with the oscillations of the atomic lattice. The interaction was described by the assumption of a momentum relaxation time τ_m. We shall consider scattering processes which determine the magnitude of τ_m in detail in Chap.6 and assume that the lattice waves are in equilibrium. In quantum theory lattice waves are quantized, the quanta being called "phonons". Besides the electron gas there is also the gas of "phonons", and the electron-Boltzmann equation and a "phonon-Boltzmann equation"*) are actually coupled integro-differential equations which have to be solved simultaneously in order to obtain an accurate description of conduction processes. Because of the mathematical difficulties involved we shall not attempt to do this but rather assume the equilibrium distribution of phonons as given by the well-known Planck equation. This is the procedure followed so far. However, in the present case of a temperature gradient in a semiconductor yielding the thermoelectric effect, phonons travel preferentially from the hot to the cold sample end. Due to the electron phonon interaction which will be discussed in detail in Chap.7h dealing with the acousto-electric effect, carriers are dragged by the phonons (in the acousto-electric effect carriers are transported by coherent sound waves, while the phonon drag effect is

[6] C. Herring: Halbleiter und Phosphore (M. Schön and H. Welker, eds.), p. 184. Braunschweig: Vieweg. 1958.

[7] H. P. R. Frederikse, Phys. Rev. 91 (1953) 491 ; 92 (1953) 248. For a review article on thermoelectric and thermomagnetic effects see L. Sosnowski: Semiconductors, Proc. Int. School of Physics XXII, p.436. London: Acad. Press. 1963.

*) There is no particle conservation for phonons, in contrast to electrons.

caused by incoherent waves of thermal conduction [6,7]). In this way more carriers are accumulated at the cold end than in the normal process described by Eq.(4i.14).

In a very simplified approach to the problem we consider a filamentary sample of a length equal to 2 phonon mean free paths, 2 l, with temperature T and energy density U(T) in the middle of the filament and temperature T-ldT/dx and energy density

$$U(T\text{-}l dT/dx) \approx U(T) - l(dT/dx)dU/dT \tag{4i.21}$$

at the hot end ($dT/dx < 0$) while for the cold end the minus signs should be replaced by plus signs. $dU/dT = c_v\rho$ is the specific heat per unit volume where ρ is the mass density. The power absorbed per unit volume is given by

$$\Delta U = U(T\text{-}l dT/dx) - U(T\text{+}l dT/dx) \approx -(dT/dx)\, 2lc_v\rho \tag{4i.22}$$

For simplicity the electron phonon interaction will be assumed to be strong enough to ensure a complete energy transfer to the carriers. With no current flow there will be an electric field E_x, and the force eE_x on the n2l carriers will counterbalance that due to the phonons :

$$n2leE_x = (dT/dx)\, 2lc_v\rho \tag{4i.23}$$

This yields a thermoelectric power of magnitude

$$\frac{d\Theta}{dT} = \frac{E_x}{dT/dx} = \frac{c_v\rho}{ne} \tag{4i.24}$$

The specific heat per mole of solids at high temperatures according to Dulong and Petit is 25 wattsec/mole.K and equals $c_v\rho$ times the volume per mole. The latter is 12 cm^3/mole in silicon which yields about 2 wattsec/Kcm^3 for $c_v\rho$. If the material contains e.g. $n = 10^{18}$ carriers/cm^3, the product ne = 0.16 watt-sec/V cm^3 and the ratio $c_v\rho/ne$ has an order of magnitude of 10 V/K. Since the specific heat at temperatures much lower than the Debye temperature is proportional to T^3, the same law should hold for $d\Theta/dT$.

In fact in Fig.4.18 there is an increase of $d\Theta/dT$ with T for the heavily doped samples. The order of magnitude, however, is only 10^{-3} V/K which can be explained by assuming that only a small fraction of the directional phonon power is transferred to the carriers. In fact, the effect is increased by decreasing the carrier concentration n as may be expected from Eq.(4i.24).

Herring [6] calculated the case of a weak electron phonon interaction. While the phonon wave velocity u_s is the sound velocity, the carriers will attain a drift velocity $v_d < u_s$ due to the phonon drag (for $v_d > u_s$ the wave would be amplified and the carriers would, in contrast, lose energy to the wave). The relative transferred energy, $-\Delta U/U$, will depend on the ratio v_d/u_s. The simplest assumption is to make $\Delta U/U$ equal to v_d/u_s. If we introduce the mobility μ by putting $v_d = \mu E_x$, take ΔU from Eq.(4i.22) and equate U to $\rho c_v T$ we obtain

$$(dT/dx)2l/T = \mu E_x/u_s \tag{4i.25}$$

This yields a thermoelectric power of magnitude

$$\frac{d\Theta}{dT} = \frac{E_x}{dT/dx} = 2\frac{u_s l}{\mu T} = 2\frac{u_s^2 \tau_{Ph}}{\mu T} \tag{4i.26}$$

where the ratio of the phonon mean free path, l, and the phonon velocity, u_s, is introduced as the phonon relaxation time, τ_{Ph}. While u_s is nearly independent of T and the product μT in silicon is roughly proportional to 1/T, a relaxation time τ_{Ph} strongly decreasing with T would explain the observation made on the purest sample (Fig.4.18). Values for τ_{Ph} of 4.1×10^{-9} sec at 20 K and 7.7×10^{-11} sec at 80 K obtained in this way seem to be reasonable [6]: At higher temperature the oscillation amplitude of the lattice atoms is increased and the anharmonicity of the oscillation more pronounced yielding a shorter mean free path between phonon-phonon collisions.

4j. Thomson and Peltier Effects

In contrast to the previous chapter we now permit an electric current density $\vec{j}$ to exist in addition to a temperature gradient $\vec{\nabla}_r T$ in the semiconductor sample. For this case Eq.(4i.10) must be replaced by

$$\vec{E} = \vec{j}/\sigma + \frac{1}{e}(S\vec{\nabla}_r T + \vec{\nabla}_r \zeta) \tag{4j.1}$$

which for $\vec{\nabla}_r T = \vec{\nabla}_r \zeta = 0$ yields Ohm's law and for $\vec{j} = 0$ yields Eq.(4i.10). We are interested in the heat generated in the sample which for thermal equilibrium is of course the well-known Joule heat. Let us first calculate the heat flow density $\vec{w}$. Carriers not only transport charge, e, but energy, ϵ, at the same time. In order to obtain $\vec{w}$ one might think of replacing e in $\vec{j}$ by ϵ. From Eqs.(4j.1) and (4i.9) we obtain $\vec{j}$ in the form

$$j = \frac{n}{m}\{\langle\tau_m e\rangle(e\vec{E} - \vec{\nabla}_r\zeta) - \langle\tau_m(\epsilon - \zeta)e\rangle T^{-1}\vec{\nabla}_r T\} \tag{4j.2}$$

The heat flow density $\vec{w}$ is given by

$$\vec{w} = \frac{n}{m}\{<\tau_m(\epsilon-\zeta)> (e\vec{E} - \vec{\nabla}_r\zeta) - <\tau_m(\epsilon-\zeta)^2> T^{-1}\vec{\nabla}_r T\} \qquad (4j.3)$$

A comparison of these two equations reveals that for obtaining $\vec{w}$ from $\vec{j}$, e has to be replaced by $\epsilon - \zeta$ except for e in the combination $e\vec{E}$ which is the driving force operating on the charged particle in an electric field $\vec{E}$. Why would one replace e by $\epsilon - \zeta$ rather than by just ϵ? It is known from thermodynamics that an increase in heat δQ is given by a change in internal energy dU minus a change in free energy (Helmholtz function) dF. The Fermi energy ζ was introduced by Eq.(3a.20) as the change in free energy, dF, with carrier concentration at constant temperature while ϵ is the change in U with n. Therefore, in the heat current density $\epsilon - \zeta$ is effective.

The signs in Eqs.(4j.2) and (4j.3) are typical for electrons with charge $e < 0$. In this case the Fermi energy ζ is given by Eq.(4i.4) and the difference $\epsilon - \zeta$ by Eq.(4i.6). For holes ($e > 0$) only the second term in $\vec{j}$ and the first term in $\vec{w}$ reverse sign since the energy $\epsilon - \zeta$ becomes negative according to Eq.(4i.7). Assuming for simplicity

$$\left.\begin{aligned} &<\tau_m\epsilon>_n/<\tau_m>_n = <\tau_m\epsilon>_p/<\tau_m>_p = k_BT(r+5/2) \\ &<\tau_m\epsilon^2>_n/<\tau_m>_n = <\tau_m\epsilon^2>_p/<\tau_m>_p = (k_BT)^2(r+7/2)(r+5/2) \end{aligned}\right\} \qquad (4j.4)$$

with the same scattering mechanism for both types of carriers and introducing

$$\sigma_n = ne^2<\tau_m>_n/m_n \; ; \quad \sigma_p = pe^2<\tau_m>_p/m_p \qquad (4j.5)$$

the current densities in the case of mixed conduction are given by

$$\vec{j} = \frac{\sigma_n + \sigma_p}{|e|}(|e|\,\vec{E} - \vec{\nabla}_r\zeta) + \left\{\frac{\sigma_n - \sigma_p}{|e|} k_BT\left(r+\frac{5}{2}\right) - \frac{\sigma_n\zeta_n - \sigma_p\zeta_p}{|e|}\right\}\frac{1}{T}\vec{\nabla}_r T \qquad (4j.6)$$

and

$$\vec{w} = -\left\{\frac{\sigma_n - \sigma_p}{e^2} k_BT\left(r+\frac{5}{2}\right) - \frac{\sigma_n\zeta_n - \sigma_p\zeta_p}{e^2}\right\}(|e|\,\vec{E} - \vec{\nabla}_r\zeta) - \left\{\frac{\sigma_n + \sigma_p}{e^2}(k_BT)^2 \left(r+\frac{7}{2}\right)\left(r+\frac{5}{2}\right) - 2\,\frac{\sigma_n\zeta_n + \sigma_p\zeta_p}{e^2} k_BT\left(r+\frac{5}{2}\right) + \frac{\sigma_n\zeta_n^2 + \sigma_p\zeta_p^2}{e^2}\right\}\frac{1}{T}\vec{\nabla}_r T \qquad (4j.7)$$

From these equations Eq.(4h.5) is easily obtained by putting $\vec{j} = 0$ and eliminating $\vec{E}$. Only the carrier contributions to $\vec{j}$ and $\vec{w}$ have been considered above. An additional term $+\kappa'\vec{\nabla}_r T$ on the right-hand side of Eq.(4j.6) would take care of the phonon drag contribution while additional terms $-\kappa_L\vec{\nabla}_r T$ and $-T\kappa'\vec{E}$ on the right-hand side of Eq.(4j.7) take care of heat conduction by phonons and the "electron drag" effect, respectively.

It may be interesting to investigate the diffusion of carriers in a temperature gradient. We shall see from Eq.(4j.2) that the terms depending on the Fermi level ζ account for this phenomenon. In $\vec{\nabla}_r\zeta$ we replace ζ by $\zeta_n + \epsilon_c$

according to Eq.(4i.4) while in $<\tau_m(\epsilon - \zeta)>$ we replace $\epsilon - \zeta$ by $\epsilon_n - \zeta_n$ according to Eq.(4i.6). From Eq.(3a.42) we obtain for $n >> N_{Dx}, N_{A-}$

$$n = N_c \cdot F_{1/2}(\eta_n) \tag{4j.8}$$

where $N_c \propto T^{3/2}$ and $\eta_n = \zeta_n / k_B T$. Since $\vec{\nabla}_r \zeta_n = (\partial \zeta_n / \partial T)\, \vec{\nabla}_r T$ we calculate $\partial n/\partial T$ from Eq.(4j.8) and solve for $\partial \zeta_n / \partial T$.

$$\frac{\partial n}{\partial t} = \frac{3}{2} \frac{N_c}{T} F_{1/2}(\eta_n) + N_c \left(\frac{1}{k_B T} \frac{\partial \zeta_n}{\partial T} - \frac{\zeta_n}{k_B T^2}\right) F_{-1/2}(\eta_n) \tag{4j.9}$$

$$\frac{\partial \zeta_n}{\partial T} = \frac{\zeta_n}{T} + \frac{k_B T}{N_c F_{-1/2}} \frac{\partial n}{\partial t} - \frac{3}{2} k_B \frac{F_{1/2}}{F_{-1/2}} \tag{4j.10}$$

where $\partial F_{1/2}(\eta_n)/\partial \eta_n = F_{-1/2}(\eta_n)$ has been applied. Since $(\partial n/\partial T)\vec{\nabla}_r T = \vec{\nabla}_r n$ we obtain for the current density from Eq.(4j.2) :

$$\vec{j} = n\,|e|\,\mu_n \vec{E} - \frac{<\tau_m e>}{m} k_B T \frac{F_{1/2}}{F_{-1/2}} \vec{\nabla}_r n + \frac{n}{m} \left(<\tau_m e> \frac{3}{2} k_B \frac{F_{1/2}}{F_{-1/2}} - \frac{<\tau_m \epsilon_n e>}{T}\right) \vec{\nabla}_r T \tag{4j.11}$$

The second term on the right-hand side is the diffusion current density usually written in the form $-e D_n \vec{\nabla}_r n$ where D_n is the diffusion coefficient for electrons. A comparison yields for D_n

$$D_n = \mu_n \frac{k_B T}{|e|} \cdot \frac{F_{1/2}(\eta_n)}{F_{-1/2}(\eta_n)} \approx \mu_n \frac{k_B T}{|e|} \tag{4j.12}$$

where we have introduced the mobility $\mu_n = ((|e|/m) <\tau_m>$. The approximation is valid for a non-degenerate electron gas.

This is known as the "Einstein relation". The ratio $F_{1/2}(\eta_n)/F_{-1/2}(\eta_n)$ equals about 3 for $\eta_n = 4$ and about 6.9 for $\eta_n = 10$.

For a calculation of heat transport in an n-type semiconductor we replace $e\vec{E} - \vec{\nabla}_r \zeta$ in Eq.(4j.3) from Eq.(4j.2) and remember that for $\vec{j} = 0$ the heat current density $\vec{w}$ is given by $-\kappa \vec{\nabla}_r T$. With S/e given by Eq.(4i.9) we obtain

$$\vec{w} = \vec{j} \cdot T \cdot (S/e) - \kappa \vec{\nabla}_r T = \Pi \cdot \vec{j} - \kappa \vec{\nabla}_r T \tag{4j.13}$$

where the "Peltier coefficient" Π has been introduced :

$$\Pi = T \cdot S/e = T \cdot d\Theta/dT \tag{4j.14}$$

The relation between the Peltier coefficient and the thermoelectric power is one of the "Onsager relations" and called the "second Kelvin relation".
The heat generated per unit volume and time, Q, is given by

$$Q = (\vec{j}\vec{E}) - (\vec{\nabla}_r \{\vec{w} - \zeta \vec{j}/e\}) \tag{4j.15}$$

where the second term is due to energy transport (see text after Eq.(4j.3)). Eliminating $\vec{E}$ and $\vec{w}$ from Eqs.(4j.1) and (4j.13), respectively, we find

$$Q = j^2/\sigma - \mu_{Th}(\vec{j}\vec{\nabla}_r T) + \vec{\nabla}_r(\kappa\vec{\nabla}_r T) \tag{4j.16}$$

where a "Thomson coefficient" μ_{Th} has been introduced by

$$\mu_{Th} = T\,d(S/e)/dT = Td^2\Theta/dT^2 \tag{4j.17}$$

The second term in Eq.(4j.16) is the Thomson heat while the first and third terms are Joule heat and heat transport by thermal conduction, respectively. The Thomson heat reverses sign on reversal of either $\vec{j}$ or $\vec{\nabla}_r T$.

A measurement of the Thomson heat at various temperatures allows for a determination of the absolute thermoelectric power which is important for metallic conductors. From Eq.(4j.17) the "first Kelvin relation"

$$\frac{d\Theta}{dT} = \int_0^T \frac{\mu_{Th}}{T}\,dT \tag{4j.18}$$

is obtained . If μ_{Th} as a function of T is known experimentally the integral may be calculated. This relation may also be obtained by thermodynamic arguments [1].

At a constant temperature throughout the semiconductor the Thomson heat at first sight seems to vanish since $\vec{\nabla}_r T = 0$. However, a reversible heat is still generated at places where the Fermi level changes with position such as at nn^+ junctions or a junction of two different materials ("heterojunction") if a current of intensity I flows through the junction. However, if we bring Q_{Th} into the form

$$Q_{Th} = -T\frac{d(S/e)}{dT}(\vec{j}\vec{\nabla}_r T) = -T\vec{j}\vec{\nabla}_r(S/e) \tag{4j.19}$$

and denote the junction cross section by A the heat generated per unit time is given by

$$-A\int_1^2 T\vec{j}\vec{\nabla}_r(S/e)dr = IT(S_1 - S_2)/e = \Pi_{(1\to2)}I \tag{4j.20}$$

and called the "Peltier heat" where the Peltier coefficient of the junction

$$\Pi_{(1\to2)} = \Pi_1 - \Pi_2 \tag{4j.21}$$

and Π of each conductor is defined by Eq.(4j.14). The Peltier heat is very large at a p-n junction with both sides degenerate. The difference in Fermi

---

[1] M. W. Zemansky: Heat and Thermodynamics, p.301. New York: McGraw-Hill. 1951. R. Becker: Theory of Heat, Chap.95. 2nd edition revised by G. Leibfried. Berlin-Heidelberg-New York: Springer. 1967.

levels on both sides is about ϵ_G, and if this is e.g. 1 eV, a current of 1 A will generate a Peltier heat of 1 watt.

In connection with the Peltier effect it has been suggested by Herring [2] that one consider a "vacuum semiconductor" which consists of two plane-parallel equal metal electrodes at temperatures T and T+dT, respectively, with T low enough that thermionic emission causes no appreciable space charge. The work function φ is assumed to be temperature-independent. We apply the Richardson equation and obtain for the equilibrium condition

$$AT^2 \exp(-|e|\,\varphi/k_BT) = A(T+dT)^2 \exp\{-|e|\,(\varphi+d\Theta)/k_B(T+dT)\} \tag{4j.22}$$

where the hotter electrode has gained a potential higher than the cooler one by an amount $d\Theta$ due to the emission of (negatively charged) electrons. Solving for $d\Theta/dT$ yields a thermoelectric power of magnitude

$$\left|\frac{d\Theta}{dT}\right| = 2\frac{k_B}{|e|} + \frac{\varphi}{T} \tag{4j.23}$$

This agrees with Eq.(4i.17) for $\zeta_n = -|e|\varphi$ and $r = -1/2$, the latter being typical for an energy-independent mean free path. If with both electrodes at the same temperature T a current is passed through the diode, the electrode which looses electrons suffers a heat loss. Both the incoming and the outgoing electrons each have a half-Maxwellian distribution and therefore each electron transports $2k_BT$ on the average. The heat loss per electron is $|e|\varphi+2k_BT$ on the average. This yields a Peltier coefficient

$$\Pi = \frac{|e|\,\varphi+2k_BT}{|e|} = T\cdot\frac{d\Theta}{dT} \tag{4j.24}$$

in agreement with Eq.(4j.14).

At low temperatures the Peltier effect may be subject to "electron drag" (see text after Eq.(4j.7)): Drifting carriers "drag" phonons which increase the heat flow.

Peltier cooling devices are occasionally used for laboratory purposes. For large-scale energy conversion thermoelectric devices with their low efficiency have not been able to compete with conventional devices even though they do not contain moving parts and therefore have a nearly unlimited lifetime. Much research has been devoted to the search for a suitable semiconductor material. A semiconductor is characterized by a "thermoelectric figure of merit"

$$Z = \frac{\sigma}{\kappa}\left(\frac{d\Theta}{dT}\right)^2 \tag{4j.25}$$

This combination of material constants can be explained by considering

---

[2] C. Herring: Halbleiter und Phosphore, p.184, appendix A. Braunschweig: Vieweg. 1958.

a voltage V_o applied to a bar-shaped sample of length l and cross section A. The current

$$I = V_o \sigma A/l \tag{4j.26}$$

through the sample causes a Peltier heat at the metal contacts of magnitude

$$\Pi I = T\frac{d\Theta}{dT} I = T\frac{d\Theta}{dT} V_o \sigma A/l \tag{4j.27}$$

neglecting the small contribution by the metal. The Peltier heat causes a temperature difference, ΔT, which can be used in a refrigerator. There is a heat loss by thermal conduction,

$$Q = \kappa \Delta T A/l \tag{4j.28}$$

which should be kept small by choosing a material with a low thermal conductivity κ. The ratio of Peltier heat and Q should be large :

$$\Pi I/Q = T(d\Theta/dT)\, V_o \sigma/(\kappa \Delta T) >> 1 \tag{4j.29}$$

A second effect of ΔT is the thermoelectric voltage which for no electric current, $I = 0$, would be simply

$$V = (d\Theta/dT)\, \Delta T \tag{4j.30}$$

We eliminate ΔT from Eqs.(4j.29) and (4j.30), and by introducing Z from Eq.(4j.25) we obtain

$$V_o Z >> V/T \tag{4j.31}$$

In this relation all the material constants form the "figure of merit" Z which should be as large as possible. In a more accurate calculation Joule heat and Thomson heat should be taken into account. Without going into more details we just give some data on a well-known thermoelectric semiconductor, bismuth telluride, Bi_2Te_3.

By appropriate doping Bi_2Te_3 can be obtained in n- and p-type form with the same absolute value of the thermoelectric power (but of course differing in sign) : $|d\Theta/dT| = 2 \times 10^{-4}$ V/K while the electrical conductivity, σ, equals 10^3(ohm-cm)$^{-1}$ and the thermal conductivity, κ, equals 1.5×10^{-2} watt/cm K. With these data one obtains from Eq.(4j.25) $Z = 3 \times 10^{-3}$/K which happens to be $= T^{-1}$ with T being room temperature where the data have been taken [3].

Hence, Eq.(4j.31) requires $V_o >> V$. Often thermocouples consisting of n- and p-type Bi_2Te_3 are connected to form a "battery".

[3] R. Bowers, R. W. Ure, Jr., J. E. Bauerle, and A. J. Cornish, J. Appl. Phys. 30 (1959) 930. A. F. Joffé: Semiconductor Thermoelements and Thermoelectric Cooling. London : Infosearch. 1958 ; for additional literature see e.g. H. P. R. Frederikse, V. A. Johnson, and W. W. Scanlon: Solid State Physics (K. Lark-Horowitz and V. A. Johnson, eds.), Vol. 6B, p.114. New York: Acad. Press. 1959.

4k. Thermomagnetic Effects

The thermomagnetic effect easiest to measure is the "Nernst effect". All others are more difficult to measure because the energy transport by carriers in semiconductors usually is many orders of magnitude smaller than the heat conduction by the crystal lattice and the voltages developed are correspondingly smaller.

A sample of the same shape as used for the Hall measurements may be used for Nernst measurements (Fig.4.2). A heat flow density w_x is transmitted (instead of an electric current I_x) by holding the sample ends at different temperatures and thus introducing a thermal gradient $\partial T/\partial x$. Just as in Hall measurements a transverse voltage V_y is developed between the side arms of the sample. The intensity of the "Nernst field" is given by

$$E_y = Q_N \cdot \frac{\partial T}{\partial x} B_z \tag{4k.1}$$

The "Nernst coefficient", Q_N, is measured in units of $cm^2/sec\ K$. It is taken as positive when the directions of $\vec{E}$, $\vec{B}$, and the temperature gradient are as indicated in Fig.4.21. The isothermal Nernst effect is subject to the condition $\partial T/\partial y = 0$. However, the adiabatic Nernst effect is what one normally observes. The difference between both effects is quite small, and therefore the isothermal effect is calculated from the observed adiabatic effect by adding to Q_N a small correcting term [1], $S_{RL}\, d\Theta/dT$, where S_{RL} will be given below by Eq.(4k.21).

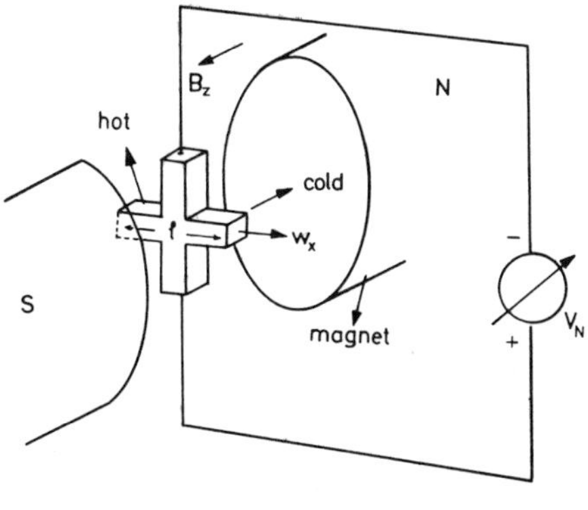

Fig.4.21 Nernst arrangement; $\vec{w}_x$ is the heat flow density. For the polarity of the Nernst voltage shown the Nernst coefficient is positive.

[1] E. H. Putley: The Hall Effect and Related Phenomena, p.84. London: Butterworth. 1960.

The occurrence of a transverse temperature gradient is due to the "Righi-Leduc effect":

$$\frac{\partial T}{\partial y} = S_{RL} \frac{\partial T}{\partial x} B_z \tag{4k.2}$$

where S_{RL} is measured in units of $cm^2/Vsec$. When the directions of $\vec{B}$ and of the temperature gradients are as indicated in Fig.4.22, $S_{RL} > 0$.

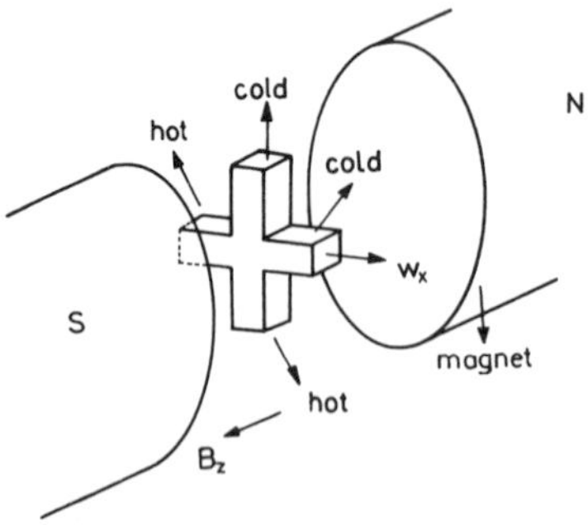

Fig.4.22 Righi-Leduc arrangement; $\vec{w}_x$ is the heat flow density. For the sign of the temperature gradient shown the Righi-Leduc coefficient is positive.

Hall measurements are occasionally subject to an error caused by the occurrence of the "Ettingshausen effect": The heat transported by carriers which are deflected from the $\vec{E}$-direction by a magnetic field generates a transverse temperature gradient

$$\frac{\partial T}{\partial y} = P_E \cdot j_x \cdot B_z \tag{4k.3}$$

where the "Ettingshausen coefficient" P_E is measured in units of $cm^3 K/watt$-sec. It is positive for the directions indicated in Fig.4.23. The Ettingshausen effect is called <u>isothermal</u> if $\partial T/\partial x = 0$. The temperature difference between the side arms of the sample yields a thermoelectric voltage between the metal-semiconductor contacts which adds to the Hall voltage. Since heat transport is a slow process in semiconductors (see Chap.4h) the Ettingshausen effect in Hall measurements can be eliminated by applying a low-frequency ac current instead of a dc current, and measuring the ac Hall voltage. The Ettingshausen effect may be appreciable (up to 10 % of V_{Hall}) in low-resistivity (e.g. 10^{-3} ohm-cm) semiconductors with low thermal conductivities (e.g. 5×10^{-2} watt/cm K). For certain laboratory applications Ettingshausen cooling shown by Fig. 4.24 may be of interest [2]. With a Bi/Sb alloy in a magnetic induction

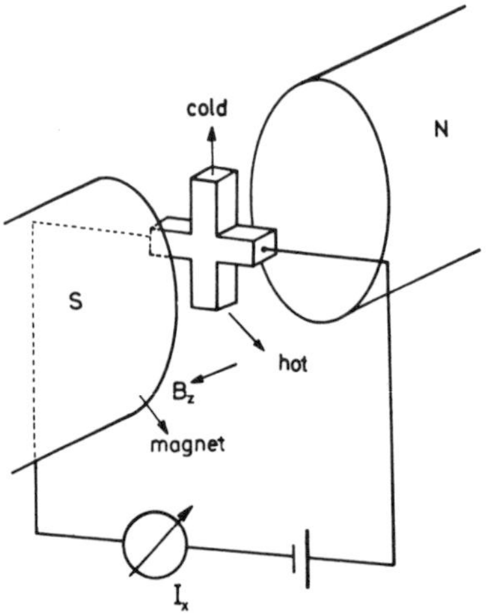

Fig.4.23 Ettingshausen arrangement; for the sign of the temperature gradient shown the Ettingshausen coefficient is positive.

[2] R. Wolfe, Semiconductor Products 6 (1963) 23; Sci. American 210 (1964) 70.

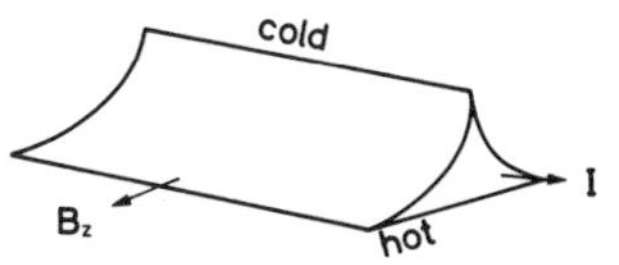

Fig.4.24 Sample shape for Ettingshausen cooling.

of 15 kG at a temperature of 156 K at the sample base, the temperature at the top of the sample was 102 K which is less by 54 K.

The temperature differences in the Righi-Leduc and the Ettingshausen effects are given by

$$\Delta T_y = S_{RL}\, \Delta T_x B_z b/l \tag{4k.4}$$

and

$$\Delta T_y = P_E I B_z/d \tag{4k.5}$$

respectively, while the Nernst voltage is given by

$$V_N = Q_N \Delta T_x B_z b/l \tag{4k.6}$$

where l is the length of the filamentary sample, d its thickness (in the $\vec{B}$-direction) and b its width (in the transverse direction).

A relation between Ettingshausen and Nernst effects,

$$P_E = Q_N T/\kappa \tag{4k.7}$$

called the "Bridgman relation" is obtained from thermodynamic arguments (the thermal conductivity, κ, includes the lattice contribution). It is convenient to measure the Nernst rather than the Ettingshausen effect and calculate P_E from Eq.(4k.7). Typical data e.g. for InSb with 8×10^{17} electrons/cm^3 at 600 K are [3] $Q_N = 0.3$ cm^2/sec K and $\kappa = 0.08$ watt/cm K. The Bridgman relation yields for this material $P_E = 2.25 \times 10^3$ cm^{3}K/wattsec. Assuming $j_x =$ $= 1$ mA/mm^2 $= 10^{-1}$ A/cm^2 and $B_z = 1$ kG $= 10^{-5}$ Vsec/cm^2 a temperature gradient $\partial T/\partial y = 2.25 \times 10^{-3}$ K/cm is calculated. For a sample width of 1 mm and assuming a thermoelectric power of 0.5 mV/K, a voltage of about 10^{-7} V is to be expected. This is many orders of magnitude smaller than the Hall voltage and in this case would be difficult to distinguish from the latter in an actual experiment.

Besides the thermomagnetic effects mentioned so far there is of course also an influence of the magnetic field on the thermoelectric effects. These effects have to be taken into account in measurements of e.g. the Nernst effect: Due to a misalignment which can hardly be avoided the side arms of the sample will have somewhat different temperatures if there is a temperature gradient along the sample filament. A thermoelectric voltage will occur in addition to the Nernst voltage. However, in contrast to the Nernst voltage it will keep its polarity if the magnetic field is reversed and can thus be eliminated.

Of the magneto-thermoelectric effects the easiest to calculate is the Corbino thermopower [4]. The experimental set-up is shown in Fig.4.25. A semiconducting Corbino disk is mounted on a heated metal rod. A thermoelectric

[3] H. Wagini, Z. Naturforsch. 19a (1964) 1541.

[4] A. C. Beer, J. A. Armstrong, and I. N. Greenberg, Phys. Rev. 107 (1957) 1506.

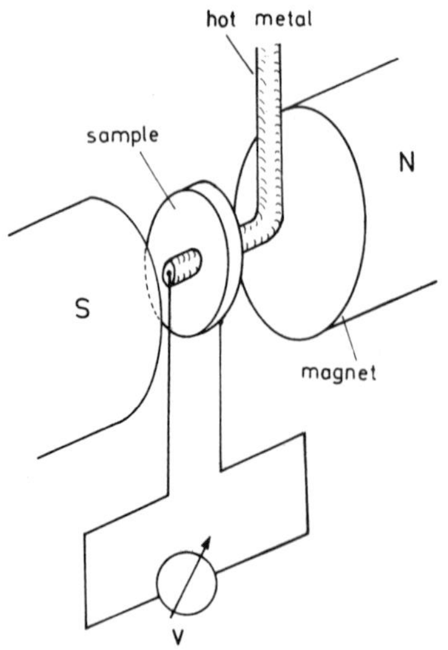

Fig.4.25 Corbino thermopower arrangement. The voltage is measured between the hot metallic rod and the circumference of the semiconducting disk. N and S are the magnet poles.

voltage is measured between the rod and the metal contact along the circumference of the disk. For small magnetic field intensities the change in thermoelectric power, $\Delta(d\Theta/dT)$, is proportional to B^2.

The saturation value of the transverse thermoelectric power with increasing magnetic field intensity may be used for a determination of a nonparabolicity of the energy bands [5].

For a calculation of the thermomagnetic effects in a non-degenerate semiconductor subject to a transverse magnetic field ($\vec{B} \perp \vec{j}$ and $\vec{w}$) we apply the equations given in Appendix A with $B_x = B_y = E_z = 0$. We obtain for the electrical current density in the case of a weak magnetic field for the isothermal Nernst effect ($\partial T/\partial y = \partial T/\partial z = \partial \zeta/\partial y = \partial \zeta/\partial z = 0$) :

$$j_x = \sigma(E_x + \frac{1}{|e|}\frac{d\zeta}{dx}) + \gamma B_z E_y + \sigma' \frac{1}{T}\frac{dT}{dx} \tag{4k.8}$$

$$j_y = -\gamma B_z(E_x + \frac{1}{|e|}\frac{d\zeta}{dx}) + \sigma E_y - \gamma' B_z \frac{1}{T}\frac{dT}{dx} \tag{4k.9}$$

We calculate E_y for $j_x = j_y = 0$ neglecting in Eq.(4k.8) the product of B_z and E_y because they are both small quantities:

$$\sigma(E_x + \frac{1}{|e|}\frac{d\zeta}{dx}) = -\sigma' \frac{1}{T}\frac{dT}{dx} \tag{4k.10}$$

$$\sigma^2 E_y = -\sigma'\gamma B_z \frac{1}{T}\frac{dT}{dx} + \sigma\gamma' B_z \frac{1}{T}\frac{dT}{dx} = (\sigma\gamma' - \sigma'\gamma) B_z \frac{1}{T}\frac{dT}{dx} \tag{4k.11}$$

The Nernst coefficient is defined by Eq.(4k.1). After eliminating E_y,

$$\sigma^2 T Q_N = \sigma\gamma' - \sigma'\gamma \tag{4k.12}$$

is obtained where σ, σ', γ, and γ' are given by Eqs.(A.17), (A.18), (A.20), and (A.21) of Appendix A, respectively, for B = 0:

[5] L. Sosnowski: Proc. Int. Conf. Phys. Semicond. Paris 1964, p.341. Paris: Dunod. 1964.

$$|e|\sigma^2 TQ_N = (\sigma_n + \sigma_p)\ \{(\sigma_n\mu_{Hn} + \sigma_p\mu_{Hp})k_BT(2r+5/2) - (\sigma_n\mu_{Hn}\zeta_n + \sigma_p\mu_{Hp}\zeta_p)\}$$
$$- (\sigma_n\mu_{Hn} - \sigma_p\mu_{Hp})\ \{(\sigma_n - \sigma_p)k_BT(r+5/2) - (\sigma_n\zeta_n - \sigma_p\zeta_p)\} \qquad (4k.13)$$

This can be simplified to yield

$$|e|\sigma^2 Q_N = k_B[(\sigma_n^2\mu_{Hn} + \sigma_p^2\mu_{Hp})r + \sigma_n\sigma_p(\mu_{Hn} + \mu_{Hp})\{3r+5 - (\zeta_n + \zeta_p)/k_BT\}] \qquad (4k.14)$$

We replace σ_n^2 by $\sigma_n(\sigma-\sigma_p)$, σ_p^2 by $\sigma_p(\sigma-\sigma_n)$, and $-(\zeta_n+\zeta_p)$ by ϵ_G :

$$Q_N = \frac{k_B}{|e|}\left\{\left(\frac{\sigma_n}{\sigma}\mu_{Hn} + \frac{\sigma_p}{\sigma}\mu_{Hp}\right)r + \frac{\sigma_n\sigma_p}{\sigma^2}(\mu_{Hn} + \mu_{Hp})\cdot(2r+5+\epsilon_G/k_BT)\right\} \qquad (4k.15)$$

For an extrinsic semiconductor where either σ_n or σ_p vanishes, the ambipolar term containing the factor $\sigma_n\sigma_p$ also vanishes. The sign of Q_N does not depend on the carrier type, but on the sign of r, in contrast to the Hall effect :

$$Q_N = \frac{k_B}{|e|}\mu_H r = \mu_H r \cdot 86\,\mu V/K \qquad (4k.16)$$

In the intrinsic region of temperature the ambipolar term dominates Q_N if electron and hole mobilities are about equal. Usually at temperatures somewhat below this temperature range lattice scattering dominates with $r < 0$ in the extrinsic range. Hence, a change of sign of Q_N can be expected at the transition from extrinsic to intrinsic behavior. At very low temperatures ionized impurity scattering dominates in semiconductors with not too large a dielectric constant. There is another sign reversal at the transition from ionized impurity scattering ($r = +3/2$) to lattice scattering ($r < 0$). This behavior is shown in Fig.4.26 where experimental results of Q_N obtained on highly doped p-type GaAs are plotted versus temperature. In the temperature range from 300 to 700 K lattice scattering is dominant while below this range it is ionized impurity scattering and above this range the material becomes intrinsic.

The Nernst voltage is usually of the order of μV since r in Eq.(4k.16) is of the order of unity and $\mu_H B \ll 1$. For $dT/dx = 10$ K/cm and a sample width of 1 mm the above order of magnitude for V_y is obtained.

The different behavior of the Hall and Nernst effects discussed above can be understood qualitatively by considering the drift motion of the carriers. In an electric field electrons and holes drift in opposite directions while in a temperature gradient they both drift in the same direction. Both types of carriers are deflected by a magnetic field to the same side of the sample in the first case (Hall effect) and to opposite sides in the second case (Nernst effect). In an extrinsic semiconductor due to the opposite charges of the carriers the direction of the Hall field depends on the charge of the carrier while the direction of the Nernst field does not. The sensitivity of the Nernst effect on the scattering mechanism is just as easy to understand. Carriers drifting from the

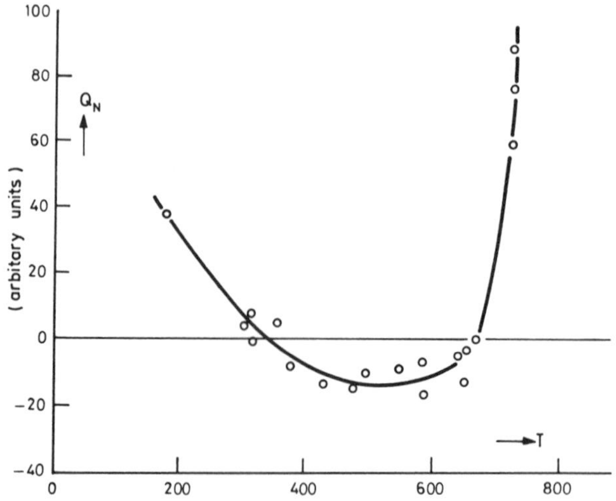

Fig.4.26 Nernst coefficient for p-type GaAs with 1.0 x 10^{17} holes/cm^3 as a function of temperature (after D. N. Nasledov, J. Appl. Phys. 32 Suppl. (1961) 2140).

hot to the cold end of the sample have to return after cooling off in order to maintain charge neutrality. Hence, there are two opposite flows of the same type of carrier, one of the "hot" carriers and one of the "cool" carriers. By a magnetic field hot carriers are deflected from their drift direction to one side of the sample while cool carriers are to the other side. Scattering decreases the influence of the magnetic field. For $\tau_m \propto \epsilon^r$ and $r > 0$ the hot carriers are less often scattered than cool carriers and the direction of the Nernst field is determined by the Lorentz force acting upon them. For $r < 0$ the same is true for the cool carriers instead of the hot carriers. Therefore the direction of the Nernst field depends on the sign of r.

In an intrinsic semiconductor with equal mobilities of both types of carriers there is no "return flow" of carriers since at the cold sample end electrons and holes recombine and release an energy of magnitude ϵ_G. It is this energy rather than the scattering processes which determines the Nernst effect in this case.

The Righi-Leduc effect is calculated from $j_x = j_y = w_y = 0$ where for a weak magnetic field

$$j_x = \sigma(E_x + \frac{1}{e}\frac{\partial \zeta}{\partial x}) + \frac{\sigma'}{T}\frac{\partial T}{\partial x} \tag{4k.17}$$

neglecting the product of small quantities B_z, E_y, $\partial\zeta/\partial y$, and $\partial T/\partial y$,

$$j_y = \sigma(E_y + \frac{1}{e}\frac{\partial\zeta}{\partial y}) + \frac{\sigma'}{T}\frac{\partial T}{\partial y} - \gamma B_z(E_x + \frac{1}{e}\frac{\partial\zeta}{\partial x}) - \gamma' B_z \frac{1}{T}\frac{\partial T}{\partial x} \qquad (4k.18)$$

$$-w_y = \sigma'(E_y + \frac{1}{e}\frac{\partial\zeta}{\partial y}) + \frac{\sigma''}{T}\frac{\partial T}{\partial y} - \gamma' B_z(E_x + \frac{1}{e}\frac{\partial\zeta}{\partial x}) - \gamma'' B_z \frac{1}{T}\frac{\partial T}{\partial x} + \kappa_1 \frac{\partial T}{\partial y} \qquad (4k.19)$$

and

$$\kappa_1 = \kappa + (\sigma'^2 - \sigma\sigma'')/\sigma T \qquad (4k.20)$$

is the lattice contribution to thermal conductivity κ. With the equations given in Appendix A the calculation of the Righi-Leduc coefficient S_{RL} defined by Eq.(4k.2) yields for an n-type semiconductor

$$S_{RL} = -s_{RL}\,\mu_H \left(\frac{k_B}{e}\right)^2 \frac{T\sigma}{\kappa} \qquad (4k.21)$$

where s_{RL} is the "Righi-Leduc factor" which for the case $\tau_m \propto \epsilon^r$ is given by

$$s_{RL} = r(r+2) + 5/2 \qquad (4k.22)$$

with values of 1.75, 2.5, and 7.75 for $r = -1/2$, 0, and $+3/2$, respectively; κ is the total thermal conductivity. If the lattice contribution to κ were negligible, the factor $(k_B/e)^2 T\sigma/\kappa$ would be the inverse Lorenz constant, given by $1/(r+5/2)$, and S_{RL} would be of the same order of magnitude as the Hall mobility μ_H. However, in most semiconductors the carrier contribution to κ is many orders of magnitude smaller than κ itself and S_{RL} is smaller than μ_H by this ratio. Assuming a ratio of e.g. 10^{-5}, a value of 2×10^{-2} for the product $\mu_H B_z$, and a temperature gradient $\partial T/\partial x = 10$ K/cm, we find $\partial T/\partial y = 2 \times 10^{-6}$ K/cm. Considering a sample width of 1mm and a thermoelectric power of 0.5 mV/K, a voltage of 10^{-10} V would have to be measured in order to observe the Righi-Leduc effect. Since this voltage is additive to the Nernst voltage which is many orders of magnitude larger, it could hardly be detected. Only in semiconductors with a large figure of merit Z, given by Eq.(4j.25) is the Righi-Leduc effect important.

The minus sign in Eq.(4k.21) characterizes an n-type semiconductor while a plus sign would be typical for a p-type semiconductor since μ_H contains the first power of the electronic charge while all other factors on the right-hand side of Eq.(4k.21) either contain e^2 or are independent of e.

Now we consider the Corbino thermopower mentioned above. In Chap.4e we noticed that due to the radial symmetry of the Corbino arrangement along the x-axis $E_y = 0$. The same is of course true for $\partial T/\partial y$ and $\partial\zeta/\partial y$. From $j_x = 0$ where j_x is given by

$$j_x = \sigma(E_x + \frac{1}{e}\frac{\partial\zeta}{\partial x}) + \sigma' \frac{1}{T}\frac{\partial T}{\partial x} \qquad (4k.23)$$

we obtain

$$E_x = -\frac{\sigma'}{\sigma T}\frac{\partial T}{\partial x} - \frac{1}{e}\frac{\partial \zeta}{\partial x} \tag{4k.24}$$

and for the thermoelectric power

$$\frac{d\Theta}{dT} = -\frac{\sigma'}{\sigma T} - \frac{\zeta}{e} \tag{4k.25}$$

where σ and σ' are given by Eqs.(A.3) and (A.4), respectively. Assuming an n-type semiconductor, the change of $d\Theta/dT$ as the result of the application of a weak magnetic field is given by

$$\Delta\frac{d\Theta}{dT} = \frac{\sigma_n k_B T(r+5/2) - \sigma_n \zeta_n - \sigma_n \mu_{Mn}^2 k_B T(3r+5/2)\, B^2 + \sigma_n \mu_{Mn}^2 \zeta_n B^2}{|e|\, T(\sigma_n - \sigma_n \mu_{Mn}^2 B^2)} - \frac{k_B T(r+5/2) - \zeta_n}{|e|\, T} \tag{4k.26}$$

where it is assumed that $\mu_{Mn}^2 B^2 << 1$. Therefore the denominator of the first fraction can be replaced by $|e| T\sigma_n$ if we multiply the first two terms in the numerator by $(1+\mu_{Mn}^2 B^2)$. We obtain the simple result

$$\Delta\frac{d\Theta}{dT} = -2r\frac{k_B}{|e|}\mu_{Mn}^2 B^2 = -(86\ \mu V/K)\frac{2r(3r+3/2)\,!}{\{(r+3/2)!\}^3}\frac{9\pi}{16}(\mu B)^2 \tag{4k.27}$$

where μ_{Mn}^2 is given by

$$\mu_{Mn}^2 = \mu^2(r_H^2 + T_M) = \mu_H^2 + T_M\mu^2 \tag{4k.28}$$

and T_M is given by Eq.(4d.4). For a p-type semiconductor the sign in Eq.(4k.27) is +. The r-dependent fraction in Eq.(4k.27) has values of + 1 for $r = -1/2$ and - 30 for $r = +3/2$: Its sign and magnitude depend strongly on the type of the scattering mechanism.

For the numerical example given in the discussion of the Nernst effect ($\mu = 10^4\ cm^2/Vsec$, B = 1 kG, $r = -1/2$) the value of $\Delta(d\Theta/dT)$ is 1.5 $\mu V/K$ which is of the same order of magnitude as the Nernst effect (of course, the geometry of the sample is different). As mentioned before the sign of $\Delta(d\Theta/dT)$ does not change when $\vec{B}$ is reversed.

An interesting result is obtained for the Corbino thermopower in the limit of a strong magnetic field. From Eqs.(4k.25), (A.27), and (A.28) we obtain for an n-type semiconductor

$$\Delta\frac{d\Theta}{dT} = -2r\frac{k_B}{|e|} = -2r\cdot 86\ \mu V/K \tag{4k.29}$$

For a p-type semiconductor the sign is +. The saturation value of the Corbino thermopower allows a direct determination of the exponent r for an assumed energy dependence of the momentum relaxation time given by $\tau_m \propto \epsilon^r$. If this assumption is not valid, we obtain instead of Eq.(4k.29) :

$$\Delta\frac{d\Theta}{dT} = -\frac{k_B}{|e|}\left(\frac{<\tau_m \epsilon/k_BT>}{<\tau_m>} - \frac{<\tau_m^{-1}\epsilon/k_BT>}{<\tau_m^{-1}>}\right) \qquad (4k.30)$$

where the integrals in the averages over the distribution function have to be evaluated numerically. Eqs.(4k.29) and (4k.30) are valid only if the condition $\mu_{Mn}^2 B^2 >> 1$ is fulfilled which for most semiconductors requires magnetic fields which can be obtained in pulsed form only, if at all. This poses a severe limit to the applicability of these equations.

4l. Piezoresistance

The change of the electrical resistivity upon the application of an external uniaxial stress or hydrostatic pressure is called "piezoresistance". Fig.4.27 shows the observed resistance of an n-type silicon sample as a function of hydrostatic pressure X transmitted by an electrically insulating liquid [1]. Up to X = 200 kbar (1 kbar = 10^3 bar = 10^9 dyn/cm^2 = 1 ton/0.981 cm^2) there is a slight linear decrease of the resistance with X in the semilog plot, which is followed by a drop of more than 6 orders of magnitude. This drop may be due to a phase transition of the silicon lattice and will not be discussed here. We focus our attention to the slight initial drop where $\Delta \log R \propto \Delta\rho/\rho \propto X$. Later on we will introduce a tensor π_{ik} of which two coefficients, π_{11} and π_{12}, will suffice to describe the relation between $\Delta\rho/\rho$ and X in the hydrostatic-pressure experiment:

$$\Delta\rho/\rho = -\Delta\sigma/\sigma = (\pi_{11} + 2\pi_{12})X \qquad (4l.1)$$

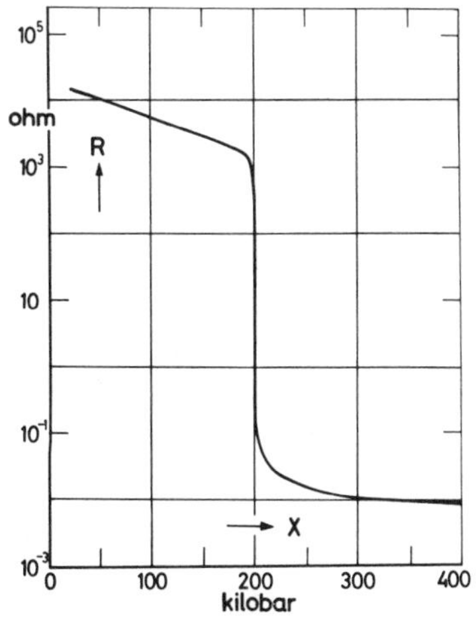

Fig.4.27 Resistance of an n-type silicon sample as a function of hydrostatic pressure (after ref.1).

These coefficients and another one, π_{44}, are needed for a description of uniaxial-stress experiments with possibly different directions of stress $\vec{X}$ and current $\vec{j}$ in a cubic semiconductor. The following table gives the combinations of the coefficients for some crystallographic directions of $\vec{X}$ and $\vec{j}$ where $\vec{X}$ is counted positive for a tension and negative for a compression.

---

[1] S. Minomura and H. G. Drickamer, J. Phys. Chem. Solids 23 (1962) 451.

Table 2 : Components of the piezoresistance tensor.

	$\vec{X}$	$\vec{j}$	$\Delta\rho/(\rho X)$
longit.	<100>	<100>	π_{11}
	<110>	<110>	$(\pi_{11}+\pi_{12}+\pi_{44})/2$
	<111>	<111>	$(\pi_{11}+2\pi_{12}+2\pi_{44})/3$
transv.	<100>	<010>	π_{12}
	<110>	$<1\bar{1}0>$	$(\pi_{11}+\pi_{12}-\pi_{44})/2$

All 3 components of the tensor can be determined from 3 measurements in different directions.

The complete piezoresistance tensor π_{ijkl} is given by

$$\Delta\rho_{ij}/\rho_o = -\Delta\sigma_{ij}/\sigma_o = \sum_{k,l=1}^{3} \pi_{ijkl} X_{kl} \tag{41.2}$$

where X_{kl} is the stress tensor and σ_{ij} is the conductivity tensor. It is well known from the theory of stress that X_{kl} is symmetrical and therefore has only 6 independent components which can be combined formally to give a vector in 6 dimensions [2]:

$$X_1 = X_{11};\; X_2 = X_{22};\; X_3 = X_{33};\; X_4 = X_{23};\; X_5 = X_{31};\; X_6 = X_{12}.$$

The same is true for the ρ_{ij} tensor (assuming no magnetic field). The piezoresistance tensor π_{ij} is then in 6 dimensions:

$$\Delta\rho_i/\rho_o = \sum_{j=1}^{6} \pi_{ij} X_j;\quad i = 1, 2, \cdots 6 \tag{41.3}$$

The "tensor of elastic constants", c_{ij}, in 6 dimensions is given by

$$X_i = \sum_{j=1}^{6} c_{ij} e_j \tag{41.4}$$

where the vector e_j contains the 6 components of the symmetrical "deformation tensor" which will be defined in Chap.7a. In Eq.(41.3) we substitute for X_j its value given by Eq.(41.4):

$$\Delta\rho_i/\rho_o = \sum_{j,k=1}^{6} \pi_{ij} c_{jk} e_k = \sum_{k=1}^{6} m_{ik} e_k \tag{41.5}$$

where we have introduced a tensor m_{ik} of "elastoresistance". For more details of these tensors see Paige [3], Smith [4], and Voigt [5].

[2] See e.g. C. Kittel: Introduction to Solid State Physics, 2nd ed., chap. 4. New York: J. Wiley and Sons, Inc. 1965.

[3] E. G. S. Paige: Progress in Semiconductors. (A.F.Gibson and R.E.Burgess, eds.),Vol. 8, p.159. London: Temple Press. 1964.

[4] C. S. Smith: Solid State Physics (F. Seitz and D. Turnbull, eds.), Vol.6, p.175. New York: Acad.Press. 1958.

[5] W. Voigt: Lehrbuch der Kristallphysik. Leipzig: Teubner. 1928.

An experimental set-up for measurements of the piezoresistance is shown in Fig.4.28. Assume a weight of 3 kg acting on an n-type silicon sample of

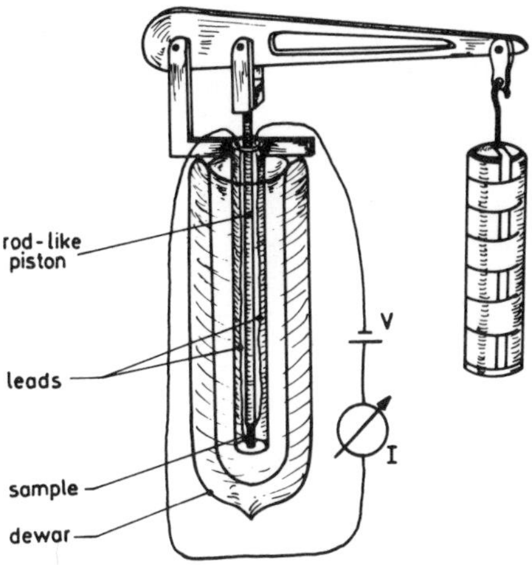

Fig.4.28 Arrangement for low-temperature measurements of the piezoresistance.

cross section 2 mm^2 via a lever with an arm ratio of 5:1. This results in a stress of 0.75 kbar and a relative resistance change, $\Delta\rho/\rho_o$, of 7.5 % at room temperature, (since $\pi_{11} = -102.2 \times 10^{-3}$/kbar) assuming $\Delta\rho/\rho_o \propto X$ up to this large value of $\Delta\rho/\rho_o$. In germanium the linear relationship is valid up to a stress of about 0.1 kbar.

Potential probes are used for the resistance determination. For a current density perpendicular to the applied stress, samples with side arms similar to those for Hall measurements are used. At least one transverse measurement is required for a determination of all 3 components of π_{ik}. The experimental data have to be corrected for changes in length and cross section of the sample under stress.

In making these corrections one has to take into account [6] the anisotropy of the resistivity.

Table 3 gives some data on germanium and silicon.

Table 3: Piezoresistance $\Delta\rho/(\rho X)$ in units of 10^{-3}/kbar at room temperature.

	ρ_o (ohm-cm)	π_{11}	π_{12}	π_{44}	$(\pi_{11}+2\pi_{12}+2\pi_{44})/3$	$(\pi_{11}+\pi_{12}+\pi_{44})/2$
n-Ge	16.6	-5.2	-5.5	-138.7	-96.6	-74.7
p-Ge	15.0	-10.6	+5.0	+98.6	+65.5	+41.5
n-Si	11.7	-102.2	+53.7	-13.6	-0.7	-31.1
p-Si	7.8	+6.6	-1.1	+138.1	+93.5	+71.8

[6] C. S. Smith, Phys. Rev. 94 (1954) 42.

The data in the next to last column of Table 3 represent the longitudinal piezoresistance in <111> direction. For n-type silicon it vanishes within experimental error. The longitudinal effect in <100> direction is given by π_{11} which for n-type germanium is comparatively small. The interpretation of these results requires the knowledge of the many-valley model of band structure which will be discussed in Chap.7.

The temperature dependence of the piezoresistance of n- and p-type silicon is plotted [7] in Fig.4.29. In n-type silicon the resistance change is proportional

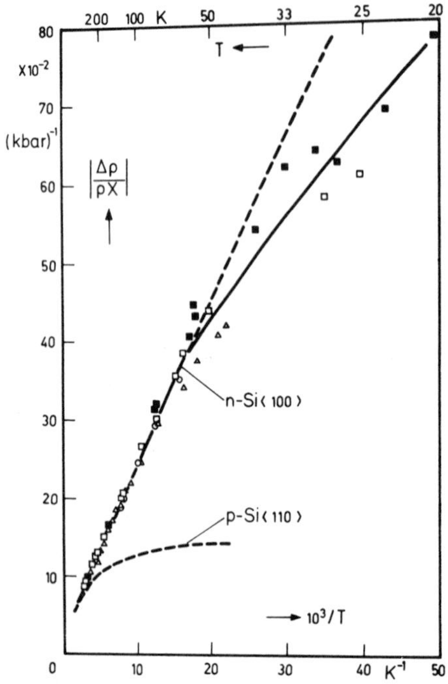

Fig.4.29 Longitudinal piezoresistance for n- and p-type silicon as a function of the reciprocal temperature (after ref.7).

to 1/T over a large range of the abscissa. This can be explained by a "repopulation of valleys" in the many-valley model (see Chap.7e) which yields a change in the "conductivity effective mass". In addition, "intervalley scattering" which determines τ_m in some semiconductors, is subject to change upon application of stress to the sample (see also Chap.7e).

According to Eq.(1c.3) the carrier concentration in intrinsic semiconductors is proportional to $\exp(-\epsilon_G/2k_BT)$ where the band-gap energy, ϵ_G, depends on the atomic distance and is changed by stress. This yields

$$\Delta\rho/\rho \propto (X/2k_BT)\, d\epsilon_G/dX \tag{41.6}$$

[7] F.J. Morin, T.H. Geballe, and C. Herring, Phys. Rev. 105 (1957) 525.

In this way a value for $d\epsilon_G/dX$ of 5 meV/kbar was determined from experimental data on intrinsic germanium at room temperature [8]. Eq.(2b.31) shows that the effective mass depends on the gap energy. Hence stress does also affect the mobility via the effective mass. This is observed with "direct" III-V compounds like e.g. InSb, InAs, GaAs, and InP where the conduction band minimum and the valence band maximum are at the same $\vec{k}$-value ($\vec{k} = 0$). Fig.4.30 shows the

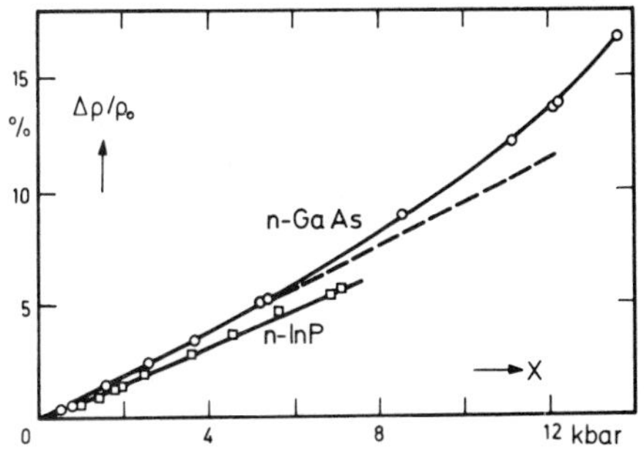

Fig.4.30 Piezoresistance for n-type GaAs and n-type InP as a function of hydrostatic pressure (after A. Sagar, quoted by R. W. Keyes in ref.8).

piezoresistance of n-GaAs and n-InP. E.g. for n-GaAs the initial slope of the curve is 0.96 %/kbar. Since the band gap energy is $\epsilon_G = 1.53$ eV and the compressibility $\kappa = \Delta V/(VX) = 1.38 \times 10^{-3}$/kbar, a value of 11 eV is found for the product $\epsilon_G \Delta\rho/(\rho X \kappa)$. For n-InSb the corresponding values are 6%/kbar, $\epsilon_G = 0.27$ eV, $\kappa = 2.3 \times 10^{-3}$/kbar, $\epsilon_G \Delta\rho/(\rho X \kappa) = 7$ eV. The value of this product is 9 eV for InAs and 8 eV for InP. The variation between 11 and 7 eV is small compared to that of the gap energy [8].

The mechanisms involved in the influences of stress on the resistance partly counterbalance each other, and the explanation of an observed stress dependence quite often is a difficult problem. We therefore will not consider here piezo-galvanomagnetic effects. The comparatively simple interpretation of a stress-induced shift of the optical absorption edge will be given in Chap.11b.

4m. Hot Electrons and Energy Relaxation Time

In the approximate solution of the Boltzmann equation, Eq.(4b.16), we have considered only small field intensities $\vec{E}$. For qualitative considerations this solution can be used for evaluating the term $\partial f/\partial v_z$ on the right-hand side of Eq. (4b.15). If this iteration is continued we obtain for f

$$f = f_0 - (\tau_m e E_z/m)\,\partial f_0/\partial v_z + (\tau_m e E_z/m)^2\,\partial^2 f_0/\partial v_z^2 + \cdots \qquad (4m.1)$$

[8] R. W. Keyes: Solid State Physics (F. Seitz and D. Turnbull, eds.), vol.11, p.172,Table V and p.176,Table VI. New York: Acad.Press. 1960.

For not too large values of E_z where the drift velocity is much smaller than the thermal velocity ($\approx\sqrt{k_B T/m}$) the series expansion converges. The drift velocity

$$v_{dz} = \int_{-\infty}^{\infty} v_z \left(-\frac{e}{m}\tau_m \frac{\partial f}{\partial v_z} E_z\right) d^3v \Big/ \int_{-\infty}^{\infty} f_0\, d^3v \tag{4m.2}$$

contains only odd powers of E_z if τ_m is independent of the direction of $\vec{v}$; integrals where an even function of v_z is averaged over $\partial f_0/\partial v_z$ or $\partial^3 f_0/\partial v_z^3$ etc. or an odd function averaged over $\partial^2 f_0/\partial v_z^2$ etc. vanish. The result for the drift velocity $v_{dz} = \mu E_z$ has the form

$$\mu E_z = \mu_0 (E_z + \beta E_z^3 + \cdots) \tag{4m.3}$$

and for the mobility

$$\mu = \mu_0 (1 + \beta E_z^2 + \cdots) \tag{4m.4}$$

where a coefficient β has been introduced; μ_0 is the zero-field mobility. The terms $\beta E_z^3 + \cdots$ represent a deviation from Ohm's law. If the series expansion may be terminated with the βE_z^3-term the carriers are considered to be "warm"; if one has to retain more terms the carriers are called "hot".

For a quantitative calculation of e.g. the coefficient β the scattering theory which will be treated in Chap.6 has to be applied. Often for hot carriers, and sometimes also for warm carriers, a Maxwell-Boltzmann distribution with an "electron temperature" T_e which is larger than the lattice temperature, is assumed in order to make the calculations simpler.

$$f \propto \exp(-\epsilon/k_B T_e) \tag{4m.5}$$

This distribution function has been justified by considering the energy gain per unit time of carriers from the field; it is the scalar product of the force eE and the drift velocity μE, namely $\mu e E^2$. In equilibrium this is equal to the energy loss by collisions

$$\mu e E^2 = -\langle \partial\epsilon/\partial t \rangle_{coll} \tag{4m.6}$$

If the energy gain of a particular carrier in the field direction is rapidly distributed in all other directions due to carrier-carrier interaction, the distribution given by Eq.(4m.5) is a good approximation. Its main advantage is, that integrals which contain this distribution function can be solved analytically. Since most of the experimentally observable quantities change only quantitatively if a more realistic distribution is used in the calculation, we will restrict analytical warm- and hot-carrier calculations to the Maxwell-Boltzmann distribution, Eq. (4m.5).

By Eq.(4b.52) we have introduced an energy relaxation time τ_ϵ. For the equilibrium case using Eq.(4m.5) and Eq.(4b.53) we obtain

$$\mu e E^2 = \frac{3}{2} k_B (T_e - T)/\tau_\epsilon \tag{4m.7}$$

A schematic representation of this equation is given in Fig.4.31 where the energy flow is demonstrated by arrows and the hatched areas indicate the energies contained in the carrier gas and in the crystal lattice.

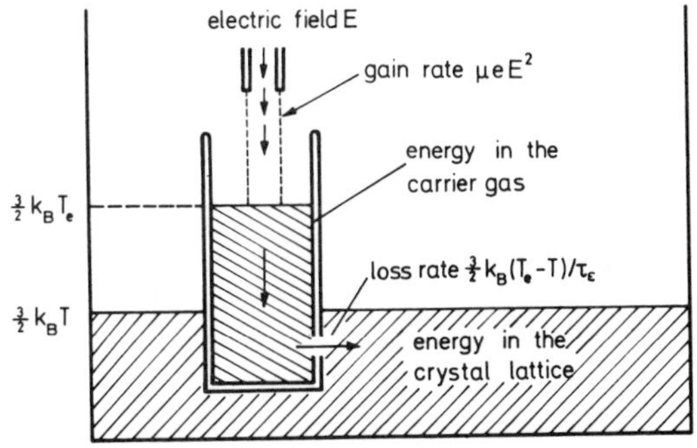

Fig.4.31 Schematic representation of the carrier energy balance indicating the energy gain from an applied electric field E and the energy loss to the crystal lattice. At equilibrium for $E \neq 0$ the mean carrier energy ($(3/2)k_BT_e$ for a non-degenerate carrier gas) is always larger than the mean thermal energy ($(3/2)k_BT$). Notice the small specific heat of the carriers and the large specific heat of the crystal lattice.

In general the momentum relaxation time τ_m depends on the carrier energy ϵ; the mobility $\mu = (e/m)\langle\tau_m\rangle$ is then a function of the electron temperature T_e; this function we will denote as $g(T_e)$. The ohmic mobility μ_0 is given by $g(T)$. A series expansion of the ratio μ/μ_0 yields

$$\mu/\mu_0 = 1 + (T_e - T)g'/g + \cdots \qquad (4m.8)$$

where g' stands for $\partial g/\partial T_e$ at $T_e = T$ and g stands for $g(T)$. For the case of warm carriers we terminate the expansion after the linear term and replace $T_e - T$ in Eq.(4m.7) by $(g/g')\,(\mu - \mu_0)/\mu_0$:

$$\mu e E^2 = \frac{3}{2}\,k_B\,(g/g')\beta E^2/\tau_\epsilon \qquad (4m.9)$$

where $(\mu - \mu_0)/\mu_0$ for warm carriers has been replaced by βE^2 according to Eq. (4m.3). In this approximation we may replace μ by μ_0 and obtain for the energy relaxation time

$$\tau_\epsilon = \frac{3}{2}\,(k_BT/e)\,\beta/\mu_0\,(d\ln g/d\ln T_e)_{T_e=T} = \frac{T}{7740\ \mathrm{K/V}}\cdot\frac{\beta}{\mu_0}\left(\frac{d\ln g}{d\ln T_e}\right)_{T_e=T} \qquad (4m.10)$$

For the case of $\tau_m \propto \epsilon^r$ the factor $d\ln g/d\ln T_e$ becomes $1/r$ which is of the order of magnitude unity. Assuming at $T = 77$ K (liquid nitrogen temperature) a mobility μ_0 of $10^4\ \mathrm{cm^2/Vsec}$ and a value of $10^{-4}\ \mathrm{cm^2/V^2}$ for $|\beta|$, we find τ_ϵ to be of the order of magnitude 10^{-10} sec. This is the time constant of a relaxation of the deviations from Ohm's law; it can be measured at a frequency of the order of magnitude $1/2\pi\tau_\epsilon \approx 1$ GHz which is in the microwave range of frequencies.

A method of measurement will be discussed in Chap.11o.

For a decrease of τ_m with ϵ, i.e. $r < 0$, the factor $d \ln g/d \ln T_e$ is negative. Since the time constant τ_ϵ has to be a positive quantity, the coefficient $\beta < 0$. Therefore in this case the mobility μ decreases with increasing electric field intensity $\vec{E}$. Similarly for an increase of τ_m with ϵ, i.e. $r > 0$, a positive sign of β can be deduced. Hence, for the limiting case of τ_m independent of ϵ one would expect β to vanish even though the expansion Eq.(4m.8) is not feasible. However, the present treatment is of a qualitative nature only and the scattering theory (Chap.6) has to be applied for a quantitative evaluation of β.

For a degenerate semiconductor we have to replace Eq.(4m.7) by

$$\mu e E^2 = \{<\epsilon(T_e)> - <\epsilon(T)>\}/\tau_\epsilon \tag{4m.11}$$

where $<\epsilon(T_e)>$ is given by

$$<\epsilon(T_e)> = \frac{3}{2} k_B T_e F_{3/2}(\zeta_n/k_B T_e)\, N_c(T_e)/n \tag{4m.12}$$

In this relation the function $F_{3/2}(x)$ is the Fermi-Dirac integral for $j = 3/2$, Eq.(3a.32), and $N_c(T)$ is given by Eq.(3a.31). For the case of two kinds of electrons of different effective mass but equal electron temperature T_e, which for $T_e = T$ has been given by Eq.(3a.38), we find

$$<\epsilon(T_e)> = \frac{3}{2n} k_B T_e [F_{3/2}(\zeta_n/k_B T_e) N_{c\Gamma}(T_e) + F_{3/2}(\{\zeta_n - \Delta_L\}/k_B T_e) N_{cL}(T_e)] \tag{4m.13}$$

and

$$\mu(T_e) = \{n_\Gamma(T_e)/n\}\, \mu_\Gamma(T_e) + \{n_L(T_e)/n\} \mu_L(T_e) \tag{4m.14}$$

where $N_{c\Gamma}(T)$, $N_{cL}(T)$, $n_\Gamma(T)$, and $n_L(T)$ are given by Eqs.(3a.39) and (3a.38), respectively. Eqs.(4m.13) and (4m.14) will be useful in Chap.11b.

For a quantitative calculation of the energy relaxation time τ_ϵ the scattering theory to be discussed in Chap.6 must be applied.

Experimental data on deviations from Ohm's law and on energy relaxation times are quite often obtained in extrinsic semiconductors in field regions where the carrier density does not change with field intensity. The current voltage characteristic then reflects the variation of the drift velocity with field intensity. First measurements on n-type Ge at 77, 193, and 298 K were made by Ryder and Shockley [1] and are shown in Fig.4.32 in a log-log plot where Ohm's law is represented by straight lines rising at an angle of 45°. Depending on lattice temperature deviations from these lines are significant at field strengths between 10^2 and 10^3 V/cm. Current saturation occurs above about 2 kV/cm at drift velocities of about 10^7 cm/sec nearly independent of lattice temperature.

A convenient way of showing the deviations from Ohm's law is a plot of the conductivity ratio σ/σ_0 where σ_0 is the zero-field conductivity. The full curves in Fig.4.33 are valid for a field applied in a <100> direction while the dashed

[1] E. J. Ryder and W. Shockley, Phys.Rev. 81 (1951) 139.

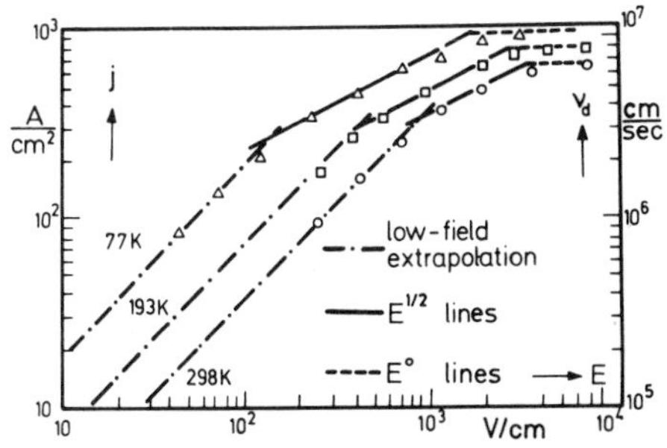

Fig.4.32 Current density in n-type germanium as a function of electric field intensity (after E. J. Ryder, ref. 4).

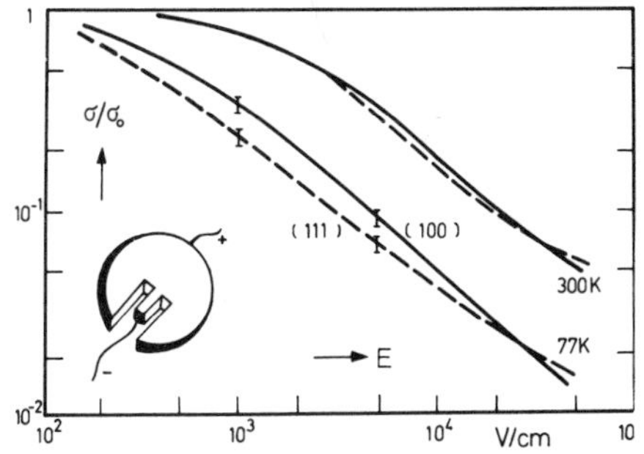

Fig.4.33 Conductivity of n-type germanium as a function of electric field intensity for two crystallographic directions of the applied field and for two temperatures. The inset shows the sample shape (after H. Heinrich and K. Seeger, ref.2).

curves are for a <111> direction. The inset in Fig.4.33 shows [2] the shape of the n-type Ge sample. The positive contact is large in order to prevent minority carrier injection (see Chap.5a). The data are corrected for the small voltage drop across the large-area part of the sample. The homogeneity of the field in the filamentary part of the sample has been questioned [3] even though the material was homogeneously doped.

At low temperatures (20 K in n-type Ge) positive deviations from the zero-field mobility may occur as indicated in Fig.4.34 [4]. These are considered to be due to ionized impurity scattering (see Chap.6f). For results obtained in n-type Si see Fig.7.14, in n-type GaAs see Fig.7.21 (schematic), for holes in Ge Fig.8.6, and for holes in Si Fig.8.9.

[2] H. Heinrich and K. Seeger, Verhandl.DPG (VI) 2 (1967) 26; K. Seeger, Acta Phys. Austriaca 27 (1968) 1.
[3] H. Heinrich, G. Bauer, and D. Kasperkovitz, phys. stat. sol. 28 (1968) K 51.
[4] E. J. Ryder, Phys. Rev. 90 (1953) 766.

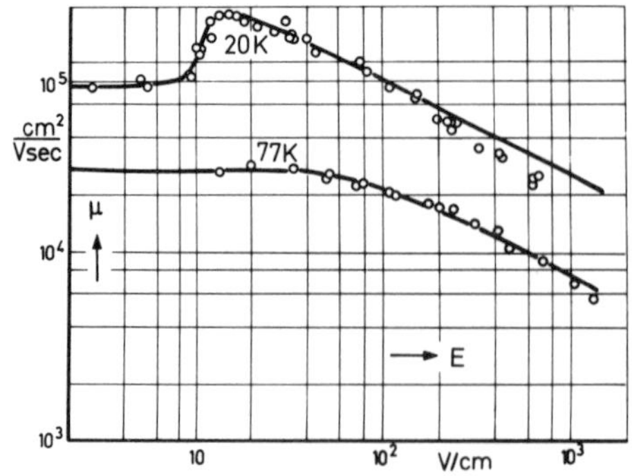

Fig.4.34 Mobility of n-type germanium as a function of electric field intensity at 20 and 77 K (after E. J. Ryder, ref. 4).

The range of warm electrons characterized by quadratic deviations from Ohm's law is shown in Fig.4.35 where μ/μ_0 is plotted vs E^2 at 77 K for N_D - N_A = 1.7 x 10^{16}/cm^3 (positive deviations) and 4 x 10^{14}/cm^3 (negative deviations) [5].

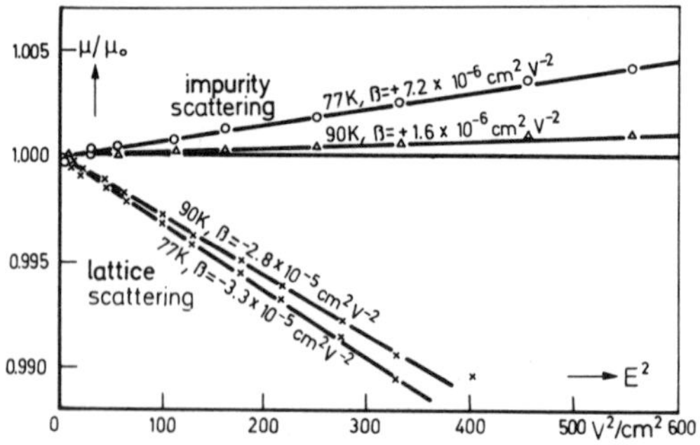

Fig.4.35 Mobility of two n-type germanium samples with different concentration of impurities versus the square of the applied field intensity at 77 and 90K. At higher field intensities than indicated here the experimental data deviate from the straight lines. Measurements were made by means of a pulsed bridge circuit (after J. B. Gunn, ref. 5).

The straight lines prove the square law for small deviations of μ/μ_0 from unity. The anisotropy of these deviations shown in Fig.4.36 will be explained in Chap. 7e (Fig.7.8a) [6]. The temperature dependence of β is quite strong, $\propto T^{-4.27}$ in pure n-Ge and $T^{-5.89}$ in n-Si [7]. For results obtained in n-InSb see Fig.7.11, for

---

[5] J. B. Gunn: Progress in Semiconductors (A. F. Gibson and R. E. Burgess, eds.), Vol. 2, p. 213. London: Temple Press. 1957.

[6] K. Seeger, Zeitschr. f. Physik 172 (1963) 68.

[7] P. Kästner, E. P. Röth, and K. Seeger, Zeitschr. f. Physik 187 (1965) 359.

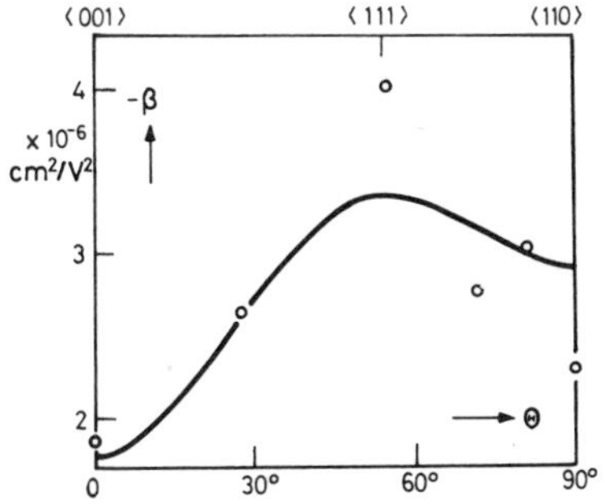

Fig.4.36 Observed values of $-\beta = -(\sigma - \sigma_0)/(\sigma_0 E^2)$ for n-type germanium at 194 K, as a function of the angle θ between the applied field $\vec{E}$ and the ⟨001⟩ direction in the $[1\bar{1}0]$ plane. The curve represents Eq.(7e.40) and was fitted to the data points by choosing suitable values for the parameters β_0 and γ of this equation (after K. Seeger, ref.6).

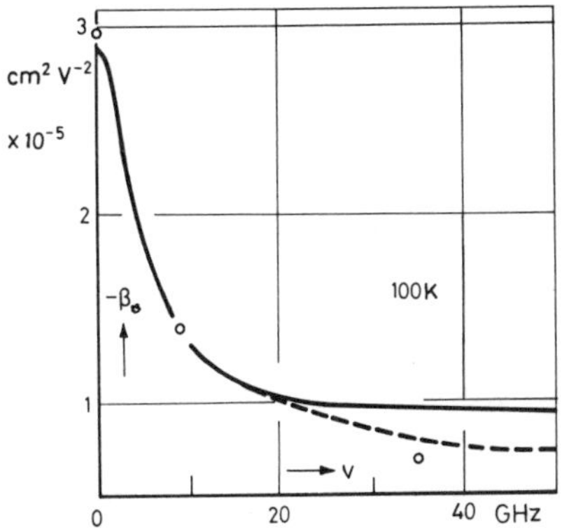

Fig.4.37 Frequency dependence of $-\beta_0$ (see caption of Fig.4.36) observed in the microwave range of frequencies in n-type germanium at 100 K. The point at $\nu = 0$ has been obtained by a dc method. The full line has been fitted with a two-parameter theory ($-\beta_0$ at $\nu = 0$ and τ_ϵ). For the dashed curve the known value of the momentum relaxation time has been taken into account (after K. Seeger and K. Hess, ref.8).

β as a function of the impurity concentration in n-Ge see Fig.6.8, and for the relation between T_e and E in n-GaSb see Fig.11.17.

Measurements of the frequency dependence of β in an arrangement similar to that for photoconductivity measurements (Fig.12.1) have been made in n-Ge at 100 K; the results shown in Fig.4.37 [6] can be explained by a combination of energy and momentum relaxation (dashed curve) [8] while energy relaxation alone [9] will be insufficient at the higher frequencies. The energy relaxation time τ_ϵ deduced from these and other data will be discussed in Chap.6h (Fig. 6.20). Measurements of τ_ϵ in n-GaAs and n-InAs are given in Fig.6.28. τ_ϵ of p-Te between 150 and 190 K is about 6 - 7 x 10^{-12} sec and nearly independent of temperature [10].

---

[8] K. Seeger and K. F. Hess, Zeitschr. f. Physik 237 (1970) 252.
[9] T. N. Morgan and C. E. Kelly, Phys. Rev. 137 (1965) A 1573.
[10] H. Kahlert, K. Hess, and K. Seeger, Solid State Comm. 7 (1969) 1149.

4n. High-Frequency Conductivity

The Boltzmann equation with df/dt given by Eq.(4b.13) contains a term $\partial f/\partial t$ which in the case of a sinusoidal time variation of the electric field intensity $\vec{E}$ depends on time t with the same frequency. Since $\partial f/\partial t$ is linear in $\vec{E}$ and $\partial f_0/\partial t = 0$ we have for $\vec{E} \propto \exp(i\omega t)$:

$$\partial f/\partial t = i\omega(f - f_0) \tag{4n.1}$$

Therefore

$$\partial f/\partial t + (\partial f/\partial v_x)eE_x/m + \cdots = -(f - f_0)/\tau_m \tag{4n.2}$$

may be written in the form

$$\{(\partial f/\partial v_x)eE_x/m + \cdots\}/(1 + i\omega\tau_m) = -(f - f_0)/\tau_m \tag{4n.3}$$

Hence the dc formulas developed so far may also be applied to the present case if we replace $\vec{E}$ by $\vec{E}/(1 + i\omega\tau_m) = \{\vec{E}/(1 + \omega^2\tau_m^2)\}(1 - i\omega\tau_m)$. In discussions of Maxwell's equations the imaginary part of this expression is found to be a contribution to the dielectric constant which is essential for the optical properties of semiconductors. The ac mobility μ_{ac} is then given by

$$\mu_{ac} = (e/m) \langle\tau_m/(1 + \omega^2\tau_m^2)\rangle \tag{4n.4}$$

For the special case of acoustic deformation potential scattering where $\tau_m = \tau_0(\epsilon/k_BT)^{-1/2}$ the averaging procedure can be performed analytically. For this purpose we introduce $\omega\tau_0 = q$; $q^2 + \epsilon/k_BT = y$ and obtain:

$$\mu_{ac}/\mu = \int_{q^2}^{\infty} \frac{(y-q^2)^2 \exp(-y+q^2)}{y}\, dy =$$

$$= \exp(q^2) \int_{q^2}^{\infty} y \exp(-y)dy - 2q^2 \int_{q^2}^{\infty} \exp(-y)dy + q^4 \int_{q^2}^{\infty} \exp(-y)dy/y =$$

$$= 1 - (\omega\tau_0)^2 - (\omega\tau_0)^4 \exp(\omega^2\tau_0^2)\mathrm{Ei}(-\omega^2\tau_0^2) \tag{4n.5}$$

For $\omega\tau_0 = 1$ the above is equal to 0.596 while without averaging $(1+\omega^2\tau_0^2)^{-1}$ – 0.500. For $\omega\tau_0 = 4$ the result Eq.(4n.5) is by about a factor of 2 larger than $(1 + \omega^2\tau_0^2)^{-1}$. $\sigma_{ac}/\sigma \propto \mu_{ac}/\mu$ as a function of $\omega\tau_0$ is shown in Fig.4.38; also shown is the contribution to the dielectric constant $-\Delta\kappa\kappa_0$ in units of $\sigma_{ac}\tau_0$ which has been calculated numerically. It seems worth mentioning that the integrals involved in the averaging procedures Eq.(4n.4) and a similar equation for $\Delta\kappa$ are the same as in Eqs.(A.3) to (A.11) of appendix A except that the cyclotron frequency, $\omega_c = eB/m$, is replaced by ω.

The momentum relaxation time is usually of the order of magnitude 10^{-12} to 10^{-13} sec. The product $\omega\tau_m$ equals unity at microwave frequencies. The maxi-

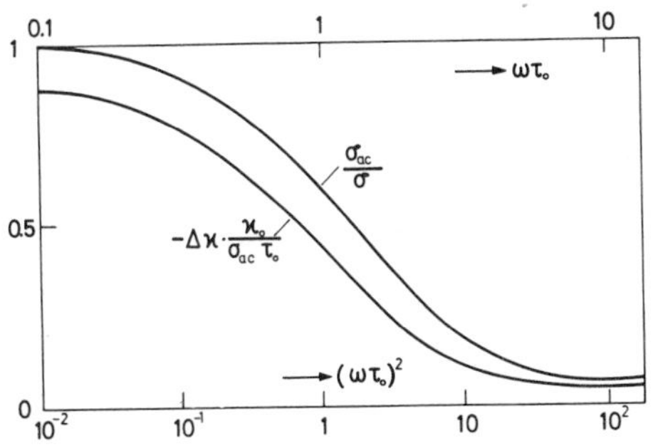

Fig.4.38 AC conductivity and contribution to the dielectric constant as a function of frequency.

mum sample dimension is usually about one wavelength or less, and it is therefore necessary to make measurements by waveguide techniques. Champlin and Glover [1] have shown that a number of experimental precautions be taken such as a close fit of the sample in the waveguide at all temperatures investigated.

At large amplitudes of the ac field carriers become warm or even hot and nonlinear terms develop which depend on the energy relaxation time τ_ϵ. For the simplified case of monoenergetic warm carriers we obtain from the momentum and energy balance equations [2] instead of Eq.(4n.4):

$$\mu_{ac} = \frac{e}{m} \cdot \frac{\tau_m}{1+\omega^2\tau_m^2} \left\{ 1 + \frac{\beta E_1^2}{4(1+\omega^2\tau_m^2)} \left(\frac{2}{1+\omega^2\tau_m^2} + \frac{1}{1+4\omega^2\tau_m^2} \right) \right\} \tag{4n.6}$$

where the coefficient β is given by Eq.(4m.4).

4o. Noise

The output of a device such as a resistor or a p-n structure when there is no signal input is called "noise" [1]. Even in the absence of current flow, the random thermal motion of carriers in a resistor produces noise across the terminals of the resistor which is known as "Johnson noise"; its power P is given by the Nyquist formula [2]

$$P = \langle \Delta V^2 \rangle \mathrm{Re}(1/Z) = 4k_B T \Delta f \tag{4o.1}$$

[1] K. S. Champlin and G. H. Glover, J. Appl. Phys. 37 (1966) 2355.
[2] K. Seeger and K. Hess, Zeitschr. f. Physik 237 (1970) 252.

Chap.4o.
[1] A. van der Ziel: Noise. New York: Prentice-Hall. 1954.
[2] H. Nyquist, Phys. Rev. 32 (1928) 110; see also e.g. R. Becker: Theory of Heat, 2nd edition revised by G. Leibfried, chap.85. Berlin - Heidelberg - New York: Springer. 1967.

where ΔV is the noise voltage, Z is the impedance, and Δf is the band width of the detector used for the noise measurement. For a band width of 100 kHz, a resistor of impedance 10^4 ohm at room temperature produces a rms noise voltage of 4 μV.

Besides fluctuations in the velocity of carriers, v_i, there are also fluctuations in the number of carriers, n, due to the detailed balance between generation and recombination of electrons and holes. If for simplicity we consider only one dimension, the current I is given by

$$I = e \sum_{i=1}^{n} v_i \approx \bar{n} e \bar{v} + e \sum_{i=1}^{\bar{n}} (v_i - \bar{v}) + e \bar{v} (n - \bar{n}) \tag{4o.2}$$

where a bar indicates the average value of the quantity. We denote $v_i - \bar{v}$ as Δv_i and $n - \bar{n}$ as Δn and obtain for the mean square of the fluctuation of the current, $\Delta I = I - \bar{n} e \bar{v}$,

$$\overline{(\Delta I)^2} = (e \sum_{i=1}^{\bar{n}} \Delta v_i)^2 + e^2 \bar{v}^2 \, \overline{\Delta n^2} \tag{4o.3}$$

where the first term on the right-hand side represents the Johnson noise and the second represents the "shot noise"; the latter depends on the average current $\bar{n} e \bar{v}$. There are many more noise sources such as the crystal surface [3] (surface recombination, surface states, see Chaps.5h and 14) and the metallic contacts.

For a p-n junction or a transistor a "noise figure" F is defined as the ratio of the total noise power and P given by Eq.(4o.1). Usually $10 \log_{10} F$ is quoted rather than F itself which is indicated by the addition of the symbol db ("decibel ") to the number. For a low-noise transistor a noise figure of 2 - 3 db is common.

If the noise power is independent of frequency (within a certain range of frequencies) the noise is called "white". The noise power of an etched Ge filament decreases with frequency f roughly as 1/f ("1/f noise") [4] in the range of 10 to 10^4 Hz. This can be explained by generation and recombination processes at localized energy levels in the band gap which are distributed over a considerable energy range [5].

Experimental investigations about the noise of hot carriers have been made in the range of low-temperature impact ionization (Chap.10a) by Lautz and Pil-

---

[3] A. L. McWhorter: Semiconductor Surface Physics (R. H. Kingston, ed.), p. 207. Philadelphia: Univ. of Pennsylvania Press. 1957; R. H. Kingston and A. L. McWhorter, Phys. Rev. 103 (1956) 534.

[4] T. G. Maple, L. Bess, and H. A. Gebbie, J. Appl. Phys. 26 (1955) 490; D. Sautter and K. Seiler, Zeitschr. f. Naturf. 12a (1957) 490.

[5] W. Shockley: Electrons and Holes in Semiconductors, p. 345. Princeton, N. J.: Van Nostrand. 1950; see also Chap.6n.

kuhn [6] and at liquid-nitrogen temperature by Erlbach and Gunn [7] and by Bryant [8]. While the break-down current increases by two orders of magnitude, $\overline{(\Delta I)^2}/\Delta f$ increases by a factor 10^8 up to a sharp maximum [6]. The idea underlying the Erlbach-Gunn experiment is to have more direct access to the electron temperature T_e by taking the noise power P as $4k_B T_e \Delta f$. A theoretical investigation by Price [9] shows, however, that this has to be done with some caution. Calculations for the simple case of acoustic phonon scattering yield a ratio of noise temperature to electron temperature which increases with field strength from unity up to about 1.14 in the high-field limit. Erlbach and Gunn measured noise temperatures at 420 MHz in n-type Ge at 77 K up to 3600 K at 1.4 kV/cm; it depends strongly on orientation. The calculations of Price, however, may not be applicable because of the strong influence of optical phonon scattering. Besides, Price has indicated a contribution to noise from intervalley scattering in a many-valley semiconductor [9].

---

[6] G. Lautz and M. Pilkuhn, Naturwiss. 47 (1960) 198 and 394; M. Pilkuhn, Thesis, T. H. Braunschweig 1960.

[7] E. Erlbach and J. B. Gunn: Proc. Int. Conf. Phys. Semic. Exeter 1962, p. 128. London: The Institute of Physics and the Phys.Soc.

[8] C. A. Bryant, Bull. Am. Phys. Soc. (Ser. II) 9 (1964) 62.

[9] P. J. Price: Fluctuation Phenomena in Solids (R. E. Burgess, ed.), Chap.8. New York: Acad. Press. 1964.

5. Carrier Diffusion Processes

The discussion of Eqs.(4j.8) - (4j.11) has shown that a temperature gradient in a conductor yields a concentration gradient $\vec{\nabla}_r n$ with the effect of a diffusion current $\vec{j} = -eD_n \vec{\nabla}_r n$ where D_n is proportional to the electron mobility due to the Einstein relation Eq.(4j.12). In this chapter we will investigate the diffusion of "injected" carriers in local variations in the type of doping, which is so typical for p-n junctions and transistors.

5a. Injection and Recombination

In Chap.1b we considered the transfer of an electron into a semiconductor from outside. Normally, however, electron-hole pairs rather than carriers of one type are "injected" into a semiconductor. The injection method, which is the easiest to understand, is pair generation by absorption of light quanta $\hbar\omega \geqslant \epsilon_G$ due to the "internal photoelectric effect" where ϵ_G is the gap energy. Let n_0 and p_0 be the dark-concentrations of electrons and holes, respectively. At a given light intensity these concentrations will be increased by $\Delta n = n\text{-}n_0$ and $\Delta p = p\text{-}p_0$, respectively. In an n-type semiconductor $n_0 \gg p_0$. At a not too low light intensity $n_0 \gg \Delta n = \Delta p \gg p_0$. The equality sign is valid for no recombination with impurities or trapped carriers. The holes, which are called "minority carriers" in this case, have been strongly increased in relative number while the electrons (called "majority carriers") have only been weakly increased:

$$\Delta p/p_0 \gg \Delta n/n_0 \tag{5a.1}$$

Obviously the contrary is true for a p-type semiconductor. Because of the strong relative increase of the minority carrier concentration the process is called "minority carrier injection".

Of course ionizing radiations other than light such as a beam of high-energy (several keV) electrons will act similarly. For practical purposes it is convenient to inject carriers by putting a metal pin on the semiconductor and applying a voltage of an appropriate magnitude and polarity between the pin and the semiconductor sample. The production of a p-n junction somewhere at the semiconductor surface by diffusion of impurities at an elevated temperature will also be as successful a tool as the previous method for injection: A metal contact placed on the p-n junction and a voltage of proper polarity applied to it will result in a very stable injection.

The electron hole pairs produced in this way gradually disappear by "recombination" such that a conduction electron becomes a bound valence electron. The binding energy, ϵ_G, is released in the form of either a photon or a quantum of lattice vibrations called a "phonon". In a special kind of semiconductor p-n junction called a "semiconductor-laser", even coherent recombination radiation is generated (Chap.13b). In most types of semiconductors, however, the phonon generation is predominant, especially in those semiconductors where there is an energy level for an impurity or a lattice defect near the middle of the gap. In this case the carrier may be trapped at this level and recombine in a second process. The surface of the crystal contains a large number of such states, up to $10^{15}/cm^2$ which is the order of magnitude of the concentration of surface atoms (Chap.14a).

If the carrier densities n and p are not too large, the recombination rate R' is proportional to the product np:

$$R' = Cnp = C(n_0 + \Delta n)(p_0 + \Delta p) \tag{5a.2}$$

where C is a factor of proportionality. In equilibrium with no injection ($\Delta n = = \Delta p = 0$) the recombination rate equals the rate of thermal generation, $Cn_0 p_0$, which does not concern us here and will be subtracted from R' to give R:

$$R = C\{(n_0 + \Delta n)(p_0 + \Delta p) - n_0 p_0\} \approx C(n_0 \Delta p + p_0 \Delta n) \tag{5a.3}$$

The product of two small quantities, $\Delta n \Delta p$, has been neglected in this approximation. This can be justified if Δn and Δp are much smaller than the majority carrier density. For the case $\Delta n \approx \Delta p$ in an n-type semiconductor $p_0 \Delta n \ll n_0 \Delta p$, and Eq.(5a.1) can be approximated by

$$R \approx Cn_0 \Delta p \tag{5a.4}$$

while in a p-type semiconductor $n_0 \Delta p \ll p_0 \Delta n$, and Eq.(5a.3) can be approximated by

$$R \approx Cp_0 \Delta n \tag{5a.5}$$

The result is that the recombination rate is proportional to the density of excess minority carriers. E.g. in an n-type semiconductor we denote the lifetime of an excess hole by τ_p which is defined by

$$R = -\frac{d\Delta p}{dt} = \frac{\Delta p}{\tau_p} \tag{5a.6}$$

where t is the time. The solution

$$\Delta p(t) = \Delta p(0) \cdot \exp(-t/\tau_p) \tag{5a.7}$$

is valid for the case of switching off the injection at $t = 0$. A comparison of

Eqs.(5a.4) and (5a.6) yields for the constant C :

$$C = \frac{1}{n_o \tau_p} \tag{5a.8}$$

Similarly for a p-type semiconductor we find

$$C = \frac{1}{p_o \tau_n} \tag{5a.9}$$

After this short introduction to injection [1] and recombination processes we turn our attention to the diffusion of excess carriers which are mostly generated at some part of the crystal surface.

5b. Diffusion and the Einstein Relation

If we increase the carrier concentration somewhere in the crystal by injection there will be a diffusion of the excess carriers throughout the crystal which is assumed to be homogeneous. The diffusion current $\vec{j}$ is proportional to the concentration gradient:

$$\vec{j} = -eD_n \vec{\nabla}_r n \tag{5b.1}$$

where D_n is the diffusion coefficient and the negative sign takes care of the fact that the carriers move away from regions of large concentrations. The dimension of D_n is cm^2/sec. In Eq.(5b.1) $e < 0$ for electrons, while for holes $e > 0$ and D_n and n are replaced by D_p and p, respectively.

In the presence of an electric field Eq.(5b.1) becomes (see Eq.(4j.11)):

$$\vec{j} = n|e|\mu_n \vec{E} - eD_n \vec{\nabla}_r n \tag{5b.2}$$

We introduce the electrostatic potential Φ by $\vec{E} = -\vec{\nabla}\Phi$. Assuming thermal equilibrium with no current flow in the crystal we obtain

$$0 = n|e|\mu_n \vec{\nabla}\Phi + eD_n \vec{\nabla} n \tag{5b.3}$$

which after division by neD_n,

$$0 = \frac{e\mu_n \vec{\nabla}\Phi}{|e|D_n} + \frac{\vec{\nabla} n}{n} \tag{5b.4}$$

can be integrated to yield

$$\text{const} = \frac{e\mu_n}{|e|D_n}\Phi + \ln n \tag{5b.5}$$

The zero-point of the potential Φ is arbitrary and can be chosen to give const = $\ln n_i$ where n_i is the intrinsic carrier density. Solving Eq.(5b.5) for n yields

[1] For "carrier extraction" see e.g. F. Stöckmann: Halbleiterprobleme, vol.VI, p. 279 (F. Sauter, ed.). Braunschweig: Vieweg. 1961.

$$n = n_i \exp\left(-\frac{e\mu_n}{|e|D_n}\Phi\right) \tag{5b.6}$$

If we introduce the Einstein relation for a non-degenerate electron gas, Eq.(4j.12),

$$D_n = \mu_n k_B T/|e| \tag{5b.7}$$

we obtain

$$n = n_i \exp(-e\Phi/k_B T) \tag{5b.8}$$

where $e < 0$ for electrons and $e > 0$ for holes. The non-homogeneous carrier distribution due to a potential barrier, $e\Phi$, (e.g. in a p-n junction) causes diffusion to take place.

This can also be derived using thermodynamic arguments. In fact if the electrostatic potential energy, $e\Phi$, is replaced by the gravitational potential energy, mgh, where m is the mass of air molecules, g the gravitational acceleration, and h the height above the surface of the earth, and n_i in Eq.(5b.8) is the density of air molecules on the earth, n is the density at height h. (Perrin's barometric formula.) The exponential in Eq.(5b.8) is frequently called "Boltzmann's factor". The treatment is valid only if the particles are independent from one another which is neither true in a liquid nor in a degenerate electron gas (Chap.4j).

At a temperature of 300 K the value of $k_B T/|e|$ is 25.9 mV; for a mobility of 1 000 $cm^2/Vsec$ D_n is 25.9 cm^2/sec. The Einstein relation allows the calculation of the diffusion coefficient from the value of the mobility. If the temperature dependence of the mobility is given by $\mu \propto T^{-1}$ the diffusion process is independent of the temperature.

We now assume that equal numbers of electrons and holes are injected into a semiconductor sample : $\Delta n = \Delta p$. The variation of the carrier densities with time and position is described by the continuity equation, Eq.(4b.37), where, however, the right-hand side is not zero but equal to the difference between the generation rate and the recombination rate. Let us denote the generation rate by G. The recombination rate is given by Eq.(5a.6). We thus obtain for the continuity equation

$$\frac{\partial \Delta n}{\partial t} + \frac{1}{e}(\vec{\nabla}_r \vec{j}_n) = G - \frac{\Delta n}{\tau_n} \tag{5b.9}$$

where $\vec{j}_n$ is given by

$$\vec{j}_n = \sigma_n \vec{E} - eD_n \vec{\nabla}_r n \tag{5b.10}$$

The electric field $\vec{E}$ may be an internal field due to different diffusion coefficients of electrons and holes. Eliminating $\vec{E}$ from Eq.(5b.10) and a corresponding equation for holes one obtains

$$(\vec{j}_n - |e|D_n \vec{\nabla} n)/\sigma_n = (\vec{j}_p + |e|D_p \vec{\nabla} p)/\sigma_p \tag{5b.11}$$

Since $\sigma_n \propto n\mu_n \propto nD_n$ and $\sigma_p \propto pD_p$ we find from this equation

$$\vec{j}_n/nD_n - \vec{j}_p/pD_p = |e|(\vec{\nabla}n/n + \vec{\nabla}p/p) \tag{5b.12}$$

Let us calculate the "ambipolar" diffusion by assuming that the total current density vanishes:

$$\vec{j} = \vec{j}_n + \vec{j}_p = 0 \tag{5b.13}$$

and that at the same time

$$\vec{\nabla}n = \vec{\nabla}p \tag{5b.14}$$

We introduce an "ambipolar diffusion coefficient" D by the definition

$$\vec{j}_n = |e| D \vec{\nabla}n \tag{5b.15}$$

for which a value of

$$D = \frac{(n+p)D_n D_p}{nD_n + pD_p} \tag{5b.16}$$

is obtained from Eqs.(5b.12), (5b.13), and (5b.14). In an n-type semiconductor $p \ll n$ and therefore $D \approx D_p$. Ambipolar diffusion is therefore determined by minority carriers. We introduce a "diffusion length" L_n by putting

$$\Delta n(x) = n(x) - n_0 = \Delta n(0) \exp(-x/L_n) \tag{5b.17}$$

and obtain

$$L_n = \sqrt{D_n \cdot \tau_n} \tag{5b.18}$$

from Eqs.(5b.9) and (5b.10) by assuming $\partial \Delta n/\partial t = (\vec{\nabla}_r \vec{E}) = G = 0$:

$$\Delta n/\tau_n = D_n (\vec{\nabla})^2 n = D_n \cdot \Delta n/L_n^2 \tag{5b.19}$$

Again taking a value of 25.9 cm^2/sec for D_n and a typical value of 100 μsec for the lifetime τ_n, a diffusion length of 0.5 mm is calculated. When the shortest distance to a surface is less than the diffusion length, calculated from the bulk lifetime, τ_n, surface recombination may be important (Eq.(5h.1)).

In our discussion on lifetime and diffusion length let us for completeness also introduce the "dielectric relaxation time" τ_d and the "Debye length" L_D. Consider a disturbance of the equilibrium distribution of majority carriers as e.g. in a plasma wave. The continuity equation for no generation and recombination and $\vec{j}_n = \sigma_n E$ is given by

$$\frac{\partial \Delta n}{\partial t} + \vec{\nabla}_r(\sigma \vec{E})/e = 0 \tag{5b.20}$$

The conductivity σ is assumed to be independent of position. Poisson's equation is

$$(\vec{\nabla}_r \vec{E}) = e\Delta n/\kappa\kappa_0 \tag{5b.21}$$

where κ is the relative dielectric constant and κ_0 the permittivity of free space. Eq.(5b.20) can be solved by integrating over time t:

$$\Delta n \propto \exp(-t/\tau_d) \tag{5b.22}$$

where the "dielectric relaxation time" τ_d is given by

$$\tau_d = \kappa\kappa_o/\sigma = 8.8 \times 10^{-14}\,\text{sec} \cdot \kappa\rho/\text{ohm-cm} \tag{5b.23}$$

and ρ is the resistivity of the semiconductor. N-type germanium ($\kappa = 16$) with an electron density of $10^{14}/\text{cm}^3$, which is equivalent to a resistivity of 15 ohm-cm, yields $\tau_d = 2.2 \times 10^{-11}$ sec at room temperature.

If in Eq.(5b.18) we introduce for τ_n the dielectric relaxation time τ_d and denote L_n by L_D, the "Debye length", we obtain

$$L_D = \sqrt{D_n\tau_d} = \sqrt{k_B T/e^2} \cdot \sqrt{\kappa\kappa_o/n} \tag{5b.24}$$

which at room temperature equals

$$L_D\,(300\,\text{K}) = \sqrt{(\kappa/n) \cdot 1.42 \times 10^4/\text{cm}} \tag{5b.25}$$

For the case of germanium mentioned above the Debye length L_D is 4.8×10^{-5} cm with on the average 1 730 germanium atoms and only 2 conduction electrons in one Debye length. In more heavily doped semiconductors inhomogeneities of doping will not seriously disturb the homogeneity of the majority carrier distribution in one Debye length.

Let us now return to minority carriers and consider simultaneous diffusion and drift in an externally applied electric field of intensity E_o. We introduce a new space coordinate

$$x' = x - \mu E_o t \tag{5b.26}$$

where μ is an "ambipolar drift mobility" defined by

$$\mu = |e|\,(n - p)\mu_n\mu_p/\sigma = (n - p)\mu_n\mu_p\,/\,(n\mu_n + p\mu_p) \tag{5b.27}$$

(note the minus-sign in the numerator, in contrast to the otherwise similar expression for the ambipolar diffusion coefficient, Eq.(5b.16).) E.g. for $p \ll n$ the ambipolar drift mobility is essentially equal to μ_p which is the minority mobility then.

The solution of Eq.(5b.9) for the present case is given by

$$\Delta n = \frac{\Delta n_o \exp(-t/\tau_n)}{2\sqrt{\pi D t}} \exp(-x'^2/4Dt) \tag{5b.28}$$

where Δn_o is the concentration of injected pairs. Diffusion profiles are given in Fig.5.1 at two different times, t_1 and t_2. During the course of the drift motion the distribution of the excess carriers over the space coordinate becomes broader due to diffusion. The half-width $\Delta_{1/2}$ of the pulse at a time t is given by

$$\Delta_{1/2} = 4\sqrt{Dt \ln 2} = 3.33\sqrt{Dt} \tag{5b.29}$$

E.g. for $D = 31\,\text{cm}^2/\text{sec}$ valid for electrons in p-type silicon, and $t = 60\,\mu$sec the half-width is 1.7 mm. For an electron drift of 1 cm a field intensity E_o of 14 V/cm is then required. Assume there is a collector at a distance d from the

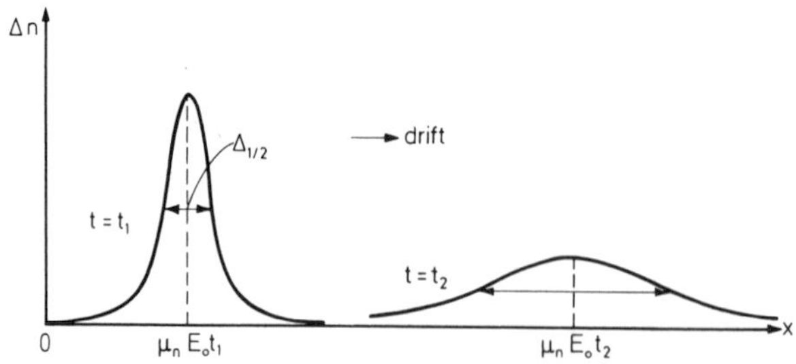

Fig.5.1 Diffusion profiles of carriers injected at x = 0 into a long filamentary sample at which an electric field E_o is applied, after elapsed time intervals t_1 and t_2 (after R. B. Adler, A. C. Smith, and R. L. Longini: Introduction to Semiconductor Physics. New York: J. Wiley and Sons. 1964).

point of injection which allows the determination of the drift time t and, from the total area under the pulse, also the lifetime τ. We can calculate the ambipolar mobility μ from

$$\mu = (\sqrt{1 + a^2} - a)\, d/E_o t \tag{5b.30}$$

where a stands for

$$a = (1 + \frac{2t}{\tau})\frac{n+p}{n-p}\frac{k_B T}{eE_o d} \tag{5b.31}$$

The a-dependent factor in Eq.(5b.30) is a consequence of the influence of carrier diffusion on the drift time.

An experiment of the kind described above has first been reported by J. R. Haynes and W. Shockley [1]. A schematic diagram of the arrangement is given in Fig.5.2. The ends of the filamentary sample are connected to a pulse generator.

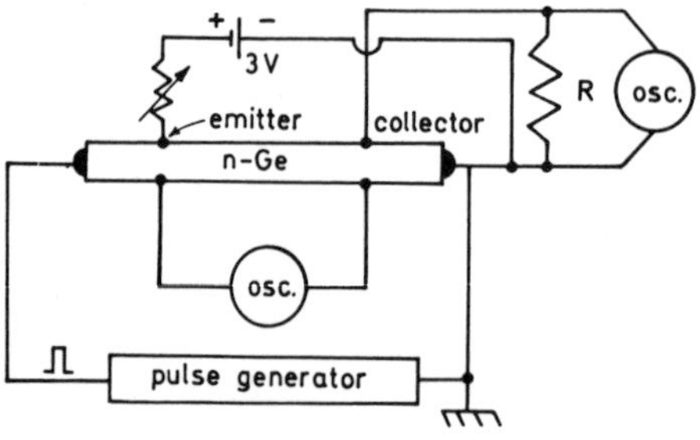

Fig.5.2 Haynes Shockley arrangement.

Holes are injected into the n-type germanium sample by an emitter contact mounted at one side of the sample. The applied pulsed field causes the carriers to drift towards the second side contact which is the "collector". In series with

[1] J. R. Haynes and W. Shockley, Phys. Rev. 81 (1951) 835.

the collector there is a resistor R of about 10^4 ohm, and the collector current is measured by observing with an oscilloscope the voltage drop across this resistor. The collector contact is rectifying and biased in the reverse direction (see Chap. 5c). The small reverse current strongly depends on the hole concentration in the vicinity of the contact area and is increased when the drifting carriers arrive at the contact.

The accuracy of the measurement is limited by the ratio of the half-width $\Delta_{1/2}$ and the drift distance $d = \mu E_0 t$. Since the field intensity E_0 is essentially the applied voltage V_0 relative to d, d becomes

$$d = \sqrt{\mu V_0 t} \tag{5b.32}$$

and the ratio $\Delta_{1/2}/d$ is given by

$$\Delta_{1/2}/d = 3.33 \sqrt{D/\mu V_0} \tag{5b.33}$$

which is proportional to $V_0^{-1/2}$. For highest accuracy the applied voltage should be as large as possible. Because of Joule heat generation the voltage has to be pulsed as is the case in Fig.5.2. At too large a field strength E_0 the mobility depends on E_0 ("hot electrons," Chap.4m).

By a comparison of the integrated pulses from different collectors mounted at various distances from the emitter, one can determine the lifetime τ using Eq. (5b.28): The ratio of the areas under the curves, giving the time variation of pulses obtained at average drift times t_1 and t_2, is equal to $\exp\{(t_2-t_1)/\tau\}$. The ratio of the pulse amplitudes, $\sqrt{t_2/t_1}\exp\{(t_2-t_1)/\tau\}$, is less suitable for a lifetime determination since t_2/t_1 is not as well-known experimentally as t_2-t_1.

The electron mobility observed with the Haynes-Shockley arrangement at various temperatures is shown in Fig.5.3. Below room temperature the mobility is proportional to $T^{-1.65}$. Above room temperature the semiconductor becomes intrinsic, and the ambipolar mobility, proportional to n-p, strongly decreases.

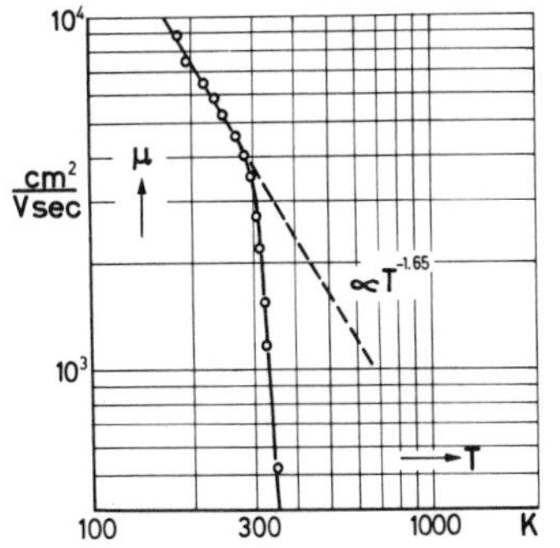

Fig.5.3 Observed drift mobility as a function of temperature (after J. P. McKelvey: Problems in Solid State Physics (H. J. Goldsmid, ed.), p. 82. New York: Acad.Press. 1968).

For electrons in silicon $\mu \propto T^{-2.5}$, for holes in n-type germanium and silicon $\mu \propto T^{-2.33}$ and $T^{-2.7}$, respectively. In Fig. 5.4 the room temperature drift mobilities of minority carriers in n- and p-type silicon as well as the drift mobilities of majority carriers (calculated from the Hall mobilities), are plotted vs the carrier densities. The carrier densities are about equal to the densities of ionized impurities. The decrease of mobility with increasing impurity concentration is due to ionized impurity scattering (Eq.6c.22). The differences in mobilities of one type of carrier in either n- or p-type material may be partly due to carrier-carrier interaction.

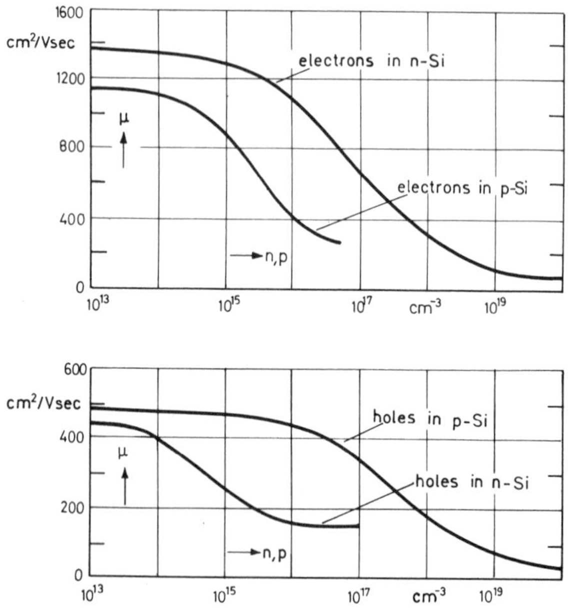

Fig.5.4 Mobility of majority- and minority carriers in n- and p-type silicon at room temperature, vs the majority-carrier concentration (after R. B. Adler, A. C. Smith, and R. L. Longini: Introduction to Semiconductor Physics. New York: J. Wiley and Sons. 1964).

5c. The p-n Junction

Consider a semiconducting single crystal which is partly doped with donors and partly with acceptors. In actual practice the transition between both parts will be gradual as indicated in Fig.5.5a by the dashed line ("graded junction").

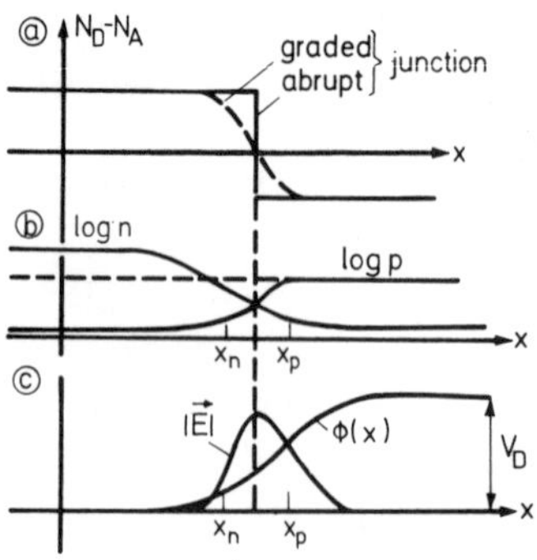

Fig.5.5 Abrupt p-n junction in thermal equilibrium.
a) Spatial distribution of donors and acceptors.
b) Spatial distribution of electrons and holes.
c) Potential distribution and electric field intensity; for a discussion of the energy band edges see Fig.5.9.

In order to simplify the calculation we will assume an "abrupt" junction where an n-type region containing a constant net donor concentration is next to a region with a constant net acceptor concentration as indicated by the full line. The idealized picture of the abrupt junction will give us the essential features of the physical behavior of the junction.

In a treatment of the problem by Schottky [1] Poisson's equation

$$\frac{d^2\Phi(x)}{dx^2} = \left\{ \begin{array}{l} |e|\, N_A/\kappa\kappa_o \text{ in the p-region} \\ -|e|\, N_D/\kappa\kappa_o \text{ in the n-region} \end{array} \right\} \text{ of the transition} \tag{5c.1}$$

is solved in one dimension neglecting the space charge due to the carriers which obviously is much less than that due to ionized impurities for the following reason: Since everywhere in the transition region for the case of no current flow the product $np = n_i^2$ is a constant, the sum $n + p = n + n_i^2/n$ has a minimum of value $2n_i$ at $n = n_i$ which has to be orders of magnitude less than both N_D and N_A; otherwise no junction would be possible by definition. With the minimum being so small there is a deficiency of carriers throughout the junction. Therefore the carriers do not contribute significantly to the space charge in the junction.

The potential Φ as a function of the space coordinate x is obtained by integration of Eq.(5c.1):

$$\Phi(x) = \left\{ \begin{array}{l} (|e|/2\kappa\kappa_o)N_A(x-x_p)^2 + \Phi_p \text{ in the p-region} \\ -(|e|/2\kappa\kappa_o)N_D(x-x_n)^2 + \Phi_n \text{ in the n-region} \end{array} \right. \tag{5c.2}$$

The constants of integration, Φ_p and Φ_n, are related to each other by the condition of continuity at $x = 0$:

$$(|e|/2\kappa\kappa_o)N_A x_p^2 + \Phi_p = -(|e|/2\kappa\kappa_o)N_D x_n^2 + \Phi_n \tag{5c.3}$$

The internal potential difference $\Phi_n - \Phi_p = V_D$ is called the "diffusion voltage" or sometimes the "built-in potential":

$$V_D = (|e|/2\kappa\kappa_o)(N_A x_p^2 + N_D x_n^2) \tag{5c.4}$$

A second boundary condition is that the field which is the first derivative of $\Phi(x)$ has also to be continuous at $x = 0$. This condition yields a relation between the second set of constants of integration, x_n and x_p:

$$-N_A x_p = N_D x_n \tag{5c.5}$$

From the parabolas given by Eq.(5c.2) it is obvious that the transition region extends into the p-region essentially for a distance x_p and into the n-region for a distance x_n. Since $x = 0$ between the two regions, one of the constants, x_n or x_p, is negative. The total width of the junction is thus given by

$$d = |x_n - x_p| \tag{5c.6}$$

From Eqs.(5c.4) and (5c.5) it is easy to calculate d. We add $N_A x_n$ on both sides

---

[1] W. Schottky, Zeitschr. f. Physik <u>118</u> (1942) 539.

of Eq.(5c.5) and multiply by $N_D(x_n - x_p)/(N_D + N_A)$:

$$\frac{N_D N_A}{N_D + N_A}(x_n - x_p)^2 = N_D x_n (x_n - x_p) = N_D x_n^2 + N_A x_p^2 \tag{5c.7}$$

where Eq.(5c.5) has again been used. Except for a factor $|e|/2\kappa\kappa_o$ the right-hand side of Eq.(5c.7) is V_D according to Eq.(5c.4). Hence, we obtain for $d = |x_n - x_p|$:

$$d = \sqrt{\frac{2\kappa\kappa_o}{|e|}(N_D^{-1} + N_A^{-1})V_D} \tag{5c.8}$$

(If in addition to the internal diffusion voltage we apply a bias V_B, d becomes

$$d = \sqrt{\frac{2\kappa\kappa_o}{|e|}(N_D^{-1} + N_A^{-1})(V_D + V_B)} \tag{5c.9}$$

and is proportional to $\sqrt{V_D + V_B}$).

How large is the diffusion voltage V_D? The potential energy $-|e|\Phi(x)$ for an electron as given by Eq.(5c.2) is large in the p-type region and small in the n-type region, the difference being $-|e|V_D$. For a non-degenerate electron gas we may assume Eq.(5b.8) with $\Phi = \Phi_n$ and $n = n_n$, characteristic of the n-type region, and a corresponding equation with $\Phi = \Phi_p$ and $n = n_p$, to be valid. The ratio of n_p and n_n for no external voltage is thus given by

$$n_p/n_n = \exp\{-|e|(\Phi_p - \Phi_n)/k_BT\} = \exp(-|e|V_D/k_BT) \tag{5c.10}$$

A similar equation holds for holes with $e = |e|$:

$$p_p/p_n = \exp\{-|e|(\Phi_p - \Phi_n)/k_BT\} = \exp(|e|V_D/k_BT) \tag{5c.11}$$

In equilibrium (no current flow) we have from Eq.(1c.6) the product

$$n_p p_p = n_n p_n = n_i^2 \tag{5c.12}$$

The intrinsic concentration n_i at a given temperature is a material constant; therefore the ratio n_p/n_n can be expressed as $n_i^2/n_n p_p$. In the case of complete ionization of impurities and no compensation the electron density n_n in the n-type region is equal to the net donor concentration N_D and likewise $p_p = N_A$. Solving Eq.(5c.10) this yields

$$V_D = \frac{k_BT}{|e|}\ln\frac{n_n}{n_p} = \frac{k_BT}{|e|}\ln\frac{N_D N_A}{n_i^2} = \frac{k_BT}{|e|}\left(\ln\frac{N_D}{n_i} + \ln\frac{N_A}{n_i}\right) \tag{5c.13}$$

It is convenient to introduce the decimal logarithm. For T equal to room temperature (300 K) we obtain

$$V_D = 59.6\text{ mV}\{\log(N_D/n_i) + \log(N_A/n_i)\} \tag{5c.14}$$

E.g. for a germanium sample where $n_i = 2.4 \times 10^{13}/\text{cm}^3$, $N_D = 2.4 \times 10^{16}/\text{cm}^3$, and $N_A = 2.4 \times 10^{14}/\text{cm}^3$ (the p-n junction as a rule is not symmetrical), a diffusion voltage V_D of 0.24 V is calculated from Eq.(5c.14).

A measurement of V_D as a function of temperature yields $n_i(T)$ from which according to Eq.(1c.3) the gap energy, ϵ_G , can be obtained. For this purpose we write Eq.(5c.13) in the form

$$|e| \cdot V_D = k_B T \ln(N_D N_A) - k_B T \ln n_i^2 = k_B T \ln(N_D N_A/C) - 3k_B T \ln T + \epsilon_G \quad (5c.15)$$

In a small temperature range the arguments of the logarithms may be taken as constants. V_D is then a linear function of T. The extrapolation of this line to $T = 0$ yields ϵ_G.

How can V_D be measured? The p-n junction may be considered to be a capacitor since as shown above there are fewer carriers of either sign in the transition region than outside this region. To a first approximation we consider the p-n junction to be a parallel plate condenser. If we denote the junction cross section by A, the capacity C is given by $C = \kappa\kappa_o A/d$. The variation of the free carrier contribution with the dielectric constant κ in the junction may be neglected. According to Eq.(5c.9) a plot of $1/C^2$ vs the bias V_B yields a straight line which can be extrapolated to cut the abscissa at $-V_D$.

For a numerical example we consider silicon at room temperature with $N_D \gg N_A$ and a resistivity ρ_p on the p-type side of the junction. If the unit of the capacity is the nanofarad (nF) we obtain

$$\left(\frac{A \cdot 33\ \mathrm{nF/cm^2}}{C}\right)^2 = \rho_p \cdot (V_D + V_B) \quad (5c.16)$$

Quite often this relationship between C and V_B cannot be verified by an experiment. The reason is that the junction is graded rather than abrupt. A more realistic calculation can be based on the assumption that the transition is "linear" with a constant gradient of the doping concentrations given by

$$d(N_D - N_A)/dx = \text{const.} \quad (5c.17)$$

In this case Poisson's equation becomes:

$$d^2\Phi(x)/dx^2 \propto -(|e|/\kappa\kappa_o)x \quad (5c.18)$$

The integration yields for the capacity of the junction

$$C \propto (V_D + V_B)^{-1/3} \quad (5c.19)$$

which yields a linear relationship between C^{-3} and V_B. In a typical diode the exponent of C is somewhere between −2 and −3. For a determination of the value of this exponent it is therefore useful first to plot the experimental data of C versus V_B on log-log paper. For a measurement of the capacity, a resonant circuit in an ac bridge is used with the ac voltage being much smaller than the dc voltage applied via a large series resistor [2] (see Fig.5.6). A p-n junction used as an electrically variable capacitor is called a "varactor diode"; this arrangement serves for the automatic fine adjustment of a resonant circuit.

Another device application is the high-energy particle counter shown in Fig.

[2] W. Shockley, Bell Syst. Tech. J. 28 (1949) 435.

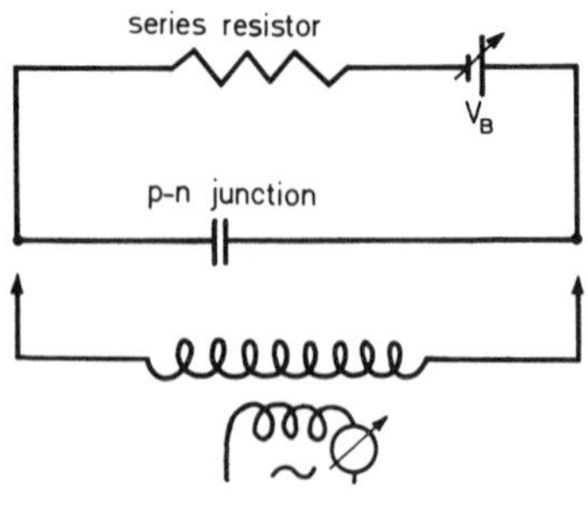

Fig.5.6 Arrangement for measurements of the p-n junction capacity as a function of the bias voltage V_B; the reverse biased p-n junction is shown as a capacitor. The meter in the primary coil indicates resonance when the frequency is varied.

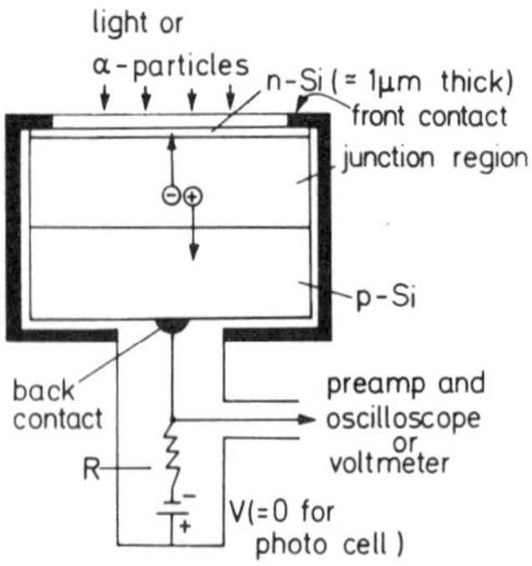

Fig.5.7 Reverse biased p-n junction used for the detection of high-energy particles or light. The junction region may be made intrinsic by lithium drift compensation (Chap.1c).

5.7. E.g. 8.78 MeV α-particles of a thorium C' source have a range of 60 μm in silicon. The junction width d in silicon with $N_D \gg N_A$ at room temperature according to Eq.(5c.16) is given by

$$d = 0.33\ \mu\text{m}\sqrt{\rho_p/\text{ohm-cm}}\ \sqrt{(V_D + V_B)/V} \tag{5c.20}$$

Assuming typical values of ρ_p = 1 000 ohm-cm and V_B = 40 V we obtain d = 65 μm which for this purpose is chosen larger than the range of the α-particles. Each α-particle generates electron hole pairs by impact ionization with an energy consumption of 3.5 eV per pair which is more than three times the gap energy ϵ_G = 1.1 eV of silicon. (For a discussion of impact ionization in semiconductors see Chap.10.) If all the electron hole pairs are separated by the field, the p-n junction acting as a capacitor obtains an additional charge of $(8.78\text{x}10^6/3.5)$ $\text{x}1.6\text{x}10^{-19}$ Asec = $4\text{x}10^{-13}$Asec. If we assume a junction area A=1 cm^2 and a bias voltage V_B = 40 V $\gg V_D$ we obtain from Eq.(5c.16) a capacity value of 165 pF. Hence the pulse height across the junction is $4\text{x}10^{-13}$ Asec/165 pF = 2.4 mV per α-particle. At a larger bias voltage the capacity is smaller and the pulse height larger. An upper limit for the bias voltage is given by the onset of electrical breakdown which eventually may result in permanent damage of the p-n junction. However, already at much lower bias there is a "dark current" through the junction which causes noise. The dark current and the noise level increase with bias. Hence, there is an optimum bias depending on the particular type of detector and the bandwidth of the amplifier used in connection with the detector. The average life of a particle detector is limited by radiation damage. Instead of material particles light quanta may also be detected, possibly without even using a bias ("photocell" Chap.5i).

So far the calculation did not take into account any current through the junction (aside from e.g. the few carriers generated by radiation). Now we will investigate the rectifying property of a p-n junction by considering the diffusion and recombination of carriers in the junction. We first consider the case that the generation rate G can be neglected and any recombination is small (diffusion length large compared to Debye length). Another case to be investigated later is that of a strong generation of carriers which dominates the reverse current of the junction diode.

For equilibrium ($\partial\Delta n/\partial t = 0$) and no generation ($G = 0$) the continuity equation (5b.9) becomes for holes ($e > 0$)

$$\frac{1}{e}(\vec{\nabla}_r \vec{j}_p) = -\frac{p-p_n}{\tau_p} \tag{5c.21}$$

To a first approximation the field current $\sigma\vec{E}$ may be neglected in Eq.(5b.10) if the applied voltage is much less than the diffusion voltage:

$$\vec{j}_p = -eD_p \vec{\nabla} p \tag{5c.22}$$

For this case the solution of Eq.(5c.21) is given by

$$p(x) - p_n = \{p(x_n) - p_n\} \exp\{-(x-x_n)/L_p\} \tag{5c.23}$$

where the diffusion length $L_p = \sqrt{D_p \tau_p}$ has been introduced and the boundary condition $p(\infty) = p_n$ has been taken into account.

We now apply to the junction a voltage V with a polarity opposite of the reverse bias V_B of Eq.(5c.9): $V = -V_B$ ("forward bias"). While for $V = 0$ the hole concentration at $x = x_n$, $p(x_n)$, is equal to p_n, it is now raised by a "Boltzmann factor"

$$p(x_n)/p_n = \exp(|e|\, V/k_B T) \tag{5c.24}$$

This is valid for not too large a voltage V such that $p(x_n) \ll p_p$. We substitute for $p(x_n)$ in Eq.(5c.23) its value given by Eq.(5c.24):

$$p(x) - p_n = p_n \{\exp(|e|V/k_B T) - 1\} \exp\{-(x-x_n)/L_p\} \tag{5c.25}$$

Eq.(5c.22) yields for $\vec{j}_p$ in x-direction (we omit the subscript x):

$$j_p = (|e|\, p_n D_p/L_p) \{\exp(|e|\, V/k_B T) - 1\} \exp\{-(x-x_n)/L_p\} \tag{5c.26}$$

The diffusion length of holes, L_p, is assumed to be much larger than the transition region $(x_p - x_n)$; therefore the last exponential may be omitted:

$$j_p = (|e|\, p_n D_p/L_p) \{\exp(|e|V/k_B T) - 1\} \tag{5c.27}$$

A similar expression is obtained for the electron current density j_n. The total current density $j = j_p + j_n$ is given by*) [3]

$$j = j_s \{\exp(|e|V/k_B T) - 1\} \tag{5c.28}$$

---

[3] C. T. Sah, R. N. Noyce, and W. Shockley, Proc.IRE 45 (1957) 1228; S. M. Sze: Physics of Semiconductor Devices, p.104. New York: J. Wiley and Sons. 1969.

*) The application of this type of law to actual j-V characteristics yields a factor between 1/2 and 1 in the argument of the exponential, see ref.3.

where j_s stands for

$$j_s = |e| (p_n D_p/L_p + n_p D_n/L_n) = |e| n_i^2 (N_A^{-1} \sqrt{D_n/\tau_n} + N_D^{-1} \sqrt{D_p/\tau_p}) \qquad (5c.29)$$

The dimension of $\sqrt{D/\tau}$ is that of a velocity. E.g. in germanium where $D_n = 100\ cm^2/sec$, $D_p = 49\ cm^2/sec$, $\tau_n = \tau_p = 100\ \mu sec$ we obtain $\sqrt{D_n/\tau_n} = 10^3\ cm/sec$ and $\sqrt{D_p/\tau_p} = 700\ cm/sec$; $n_i = 2.4x10^{13}/cm^3$; assuming as before $N_D = 2.4x10^{16}/cm^3$ and $N_A = 2.4x10^{14}/cm^3$, we obtain $j_s = 3.8x10^{-4}\ A/cm^2$.

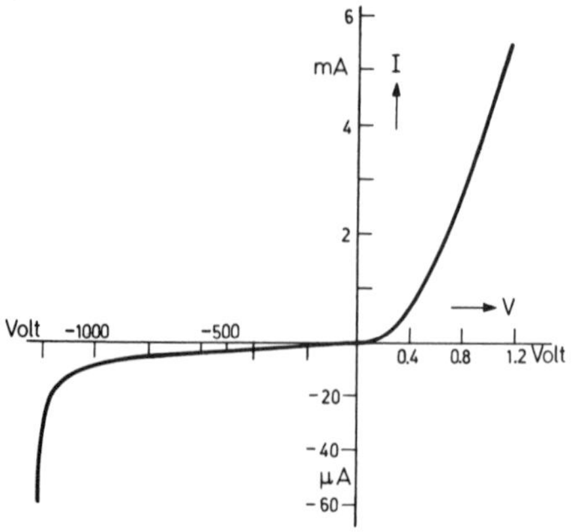

Fig.5.8 Current voltage characteristics of a germanium p-n junction (after W. Pietenpol, Phys. Rev. 82 (1951) 120).

Fig.5.8 shows the current voltage characteristics of a germanium p-n junction. In the "forward" direction ($V > 0$) the current increases exponentially in agreement with Eq.(5c.28). At higher current densities the series resistance which is due to the part of the sample outside the p-n junction, causes the characteristics to deviate from an exponential behavior towards values of V larger than at lower current densities. In the reverse direction Eq.(5c.28) suggests a saturation of the current density at a value of j_s. Fig.5.8 shows the reverse characteristics to be very flat at voltages below 10^3 V (note the difference in scale between forward and reverse direction) and a strong increase beyond this voltage which is typical of an electrical breakdown as we discussed earlier.

The calculated j_s-value, Eq.(5c.29), is in agreement with observations made on germanium diodes but not with those on silicon diodes. In silicon n_i^2 is about 6 orders of magnitude smaller than in germanium and j_s should be smaller by the same factor. This is not the case, however, and besides, the reverse current does not saturate. This is due to the thermal generation of electron hole pairs in the junction region at centers with energy levels near the middle of the band gap. Let us denote by f_t the occupation probability of such a center by an electron : $1 - f_t$ is then the probability of occupation by a hole. We assume the rate of electron generation, G, to be proportional to f_t:

$$G = a\,f_t \tag{5c.30}$$

where a is a factor of proportionality. In equilibrium G must be equal to the recombination rate of electrons, $n(1-f_t)/\tau_n$, which is proportional to the electron density, to the probability that a center is occupied by a hole, and inversely proportional to the lifetime of excess electrons:

$$n(1-f_t)/\tau_n = a\,f_t \tag{5c.31}$$

For an intrinsic semiconductor $n = n_i$, and the Fermi level is near the middle of the gap. To a good approximation we may assume the Fermi level to be at the energy level of the recombination center which yields $f_t = 1/2$ and from Eq. (5c.31) $a = n_i/\tau_n$. Hence the generation rate is given by

$$G = f_t n_i/\tau_n \tag{5c.32}$$

Under non-equilibrium conditions the increase of electron density per unit time is given by

$$\frac{\partial n}{\partial t} = \frac{n_i}{\tau_n} f_t - \frac{n(1-f_t)}{\tau_n} \tag{5c.33}$$

A similar equation for holes is easily obtained:

$$\frac{\partial p}{\partial t} = \frac{n_i}{\tau_p}(1-f_t) - \frac{pf_t}{\tau_p} \tag{5c.34}$$

The condition for pair generation

$$\frac{\partial n}{\partial t} = \frac{\partial p}{\partial t} \tag{5c.35}$$

yields for f_t:

$$f_t = \frac{n\tau_p + n_i\tau_n}{(n+n_i)\tau_p + (p+n_i)\tau_n} \tag{5c.36}$$

and for the rate of pair generation

$$\frac{\partial n}{\partial t} = \frac{\partial p}{\partial t} = \frac{n_i^2 - np}{(n+n_i)\tau_p + (p+n_i)\tau_n} \tag{5c.37}$$

In equilibrium, $\partial n/\partial t = 0$, we obtain $n_i^2 = np$ which in Eq.(1c.2) has been obtained by the law of mass action.

Since the electrons and holes usually are separated by the electrical field before they can recombine, n and p will be small in the junction region ($\ll n_i$) and thus can be neglected in Eq.(5c.37). Hence we obtain

$$\frac{\partial n}{\partial t} + \frac{\partial p}{\partial t} = \frac{n_i}{\tau} \tag{5c.38}$$

where

$$\tau = \frac{1}{2}(\tau_n + \tau_p) \tag{5c.39}$$

is an average lifetime. This yields for the current density j_s

$$j_s = |e| \left(\frac{\partial n}{\partial t} + \frac{\partial p}{\partial t}\right) d = |e| n_i d/\tau \tag{5c.40}$$

since the total reverse current will be proportional to the volume of the junction region and therefore the current density proportional to the width d. E.g. for silicon at room temperature $n_i = 1.3 \times 10^{10}/cm^3$. Assuming as above $d = 60\ \mu m$ and $\tau = 100\ \mu sec$ we obtain from Eq.(5c40) $j_s = 1.3 \times 10^{-7} A/cm^2$ which is 3 orders of magnitude larger than j_s from Eq.(5c.29) and in agreement with observations [3]. For germanium it would be somewhat smaller.

Since d in Eq.(5c.40) increases with the applied bias voltage, j_s also increases and there is no current saturation as observed in silicon.

5d. Quasi-Fermi Levels

The current voltage characteristics of a p-n junction were calculated without considering the energy distribution of the carriers*). The current was calculated from thermal diffusion of the carriers with diffusion constants which are related to the carrier mobilities by the Einstein relation. The voltage applied to the junction changes the potential difference between both sides of the junction. Depending on the polarity of the applied voltage this change in potential difference causes more or less carriers to diffuse to the opposite side than in thermal equilibrium.

The potential distribution is shown in Fig.5.9a. In going from the p- to the n-side of the junction both the conduction band edge and the valence band edge are lowered by $|e|V_D$ with no applied voltage, by $|e|(V_D - V)$ with a forward bias voltage V and by $|e|(V_D + V_B)$ with a reverse bias voltage V_B. In Fig.5.9b we consider the case of a reverse bias voltage V_B in more detail. In the regions $x > L_n$ and $x < -L_p$ the equilibrium carrier densities are hardly affected by the junction. There the densities of both carrier types can be described by the same Fermi level, denoted by ζ_p^* on the p-side and by ζ_n^* on the n-side (these should not be confused with ζ_p and ζ_n given by Eqs.(4i.5) and (4i.4), respectively) since any voltage applied to the junction will cause an additional potential drop in the region of the junction width, $-x_p \leqslant x \leqslant x_n$, while the field outside this region, though necessary for a current flow, is negligibly small to a first approximation. As we approach the junction from either side, the majority carrier densities are not seriously perturbed from their equilibrium values by the current flow, but the minority carriers are. Hence there is no longer a unique Fermi level for both carrier types. Therefore we denote ζ_p^* and ζ_n^* as "quasi-Fermi levels" (sometimes called "imrefs"). In Fig.5.9b ζ_p^*, ζ_n^*, and the intrinsic Fermi level ζ_i which is given by Eq.(3a.37), are shown. ζ_i is close to the middle of the gap, depending on the effective mass ratio, but independent of doping, and serves as a reference level

---

*) The reader who is not familiar with the contents of Chap.3 may find it profitable to continue with Chap.5e and leave the present chapter for a later reading.

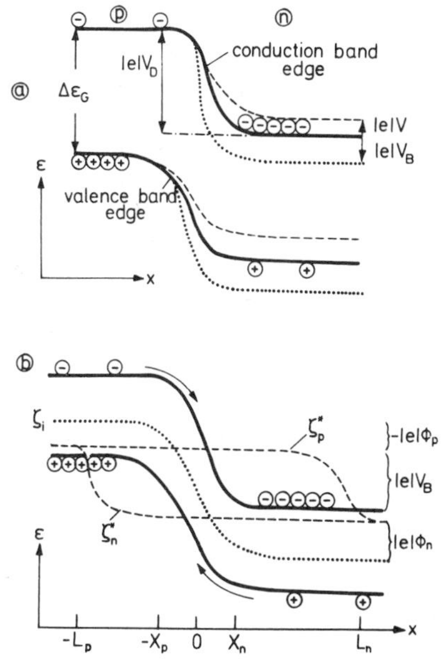

Fig.5.9a Energy band edges in a p-n junction at thermal equilibrium (full curves), with forward bias applied (dashed), and with reverse bias applied (dotted).

Fig.5.9b Quasi-Fermi levels in a reverse biased p-n junction.

throughout the junction. The constants Φ_n and Φ_p defined by Eq. (5c.2) are also indicated. Far away from the junction they are related to the quasi-Fermi levels by

$$|e|\,\Phi_n = \zeta_n^* - (\zeta_i)_{\text{n-side}} \tag{5d.1}$$

$$|e|\,\Phi_p = \zeta_p^* - (\zeta_i)_{\text{p-side}} \tag{5d.2}$$

Due to the potential barrier across the junction, electrons on the p-side slide down the potential hill, thus being converted from minority carriers to majority carriers. The same holds for the hole transfer from the n-side to the p-side (see arrows in Fig.5.9b). In the case of germanium it is exactly this charge transfer which causes the reverse current (in silicon thermal generation of electron hole pairs in the junction is predominant). At an average distance of a diffusion length away from the junction the excess majority carriers recombine with minority carriers; consequently at distances greater than the diffusion length the thermal equilibrium distribution is present.

The decrease in minority carrier densities in the regions contiguous to the junction is described by a decrease of ζ_n^* on the p-side and an increase of ζ_p^* on the n-side, by an amount $|e|\,V_B$ as shown in Fig.5.9b. For no applied bias V_B the Fermi level $\zeta = \zeta_n^* = \zeta_p^*$ goes straight through the whole crystal.

It is obvious from Fig.5.9b that the intrinsic Fermi level ζ_i drops across the junction by an amount

$$|e|\,\Phi_n + |e|\,V_B - |e|\,\Phi_p = |e|\,(V_B + V_D) \tag{5d.3}$$

where $V_D = \Phi_n - \Phi_p$ is the diffusion voltage. The drop of ζ_i is equal to the drop of the conduction band edge. Hence, the electron density in the p-type region, n_p, is lower than in the n-type region, n_n, by a factor of $\exp\{-|e|(V_D+V_B)/k_BT\}$. In comparison with Eq.(5c.10) which holds for $V_B = 0$, we find the sum of V_D and V_B to be the quantity which determines the carrier density ratio and therefore also the width of the junction given by Eq.(5c.9). A treatment of n - n$^+$ and p - p$^+$ junctions has been given by J. B. Gunn [1] and will not be discussed here.

[1] J. B. Gunn, J. Electron. Control 4 (1958) 17.

5e. The Transistor

A power amplifying p-n structure is called a "transistor". A typical example is the p-n-p transistor shown in Fig.5.10. The middle part of the single crystal is n-type; it is called the "base". The adjacent p-type parts are the emitter and the collector which have been introduced in the discussion of the Haynes-Shockley experiment, Chap.5b.

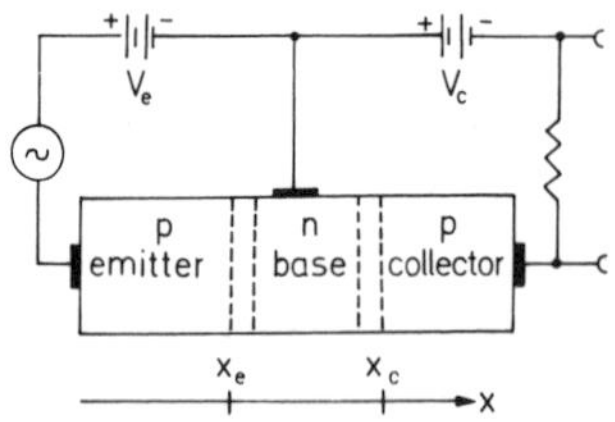

Fig.5.10 p-n-p transistor (schematic diagram)

For the emitter base junction we obtain from Eqs.(5c.21) and (5c.22):

$$D_p\, d^2p/dx^2 = (p-p_b)/\tau_b \qquad (5e.1)$$

where we introduced the symbols p_b and τ_b for p_n and τ_n, respectively, in the base. By

$$L_b = \sqrt{D_p \tau_b} \qquad (5e.2)$$

we denote the base diffusion length. The solution of Eq.(5e.1) for the region of the base, $x_e \leqslant x \leqslant x_c$, where $w = x_c - x_e$ is the base width, is given by

$$p(x) = p_b + \frac{p_e - p_b}{\sinh(w/L_b)} \sinh\frac{x_c - x}{L_b} + \frac{p_c - p_b}{\sinh(w/L_b)} \sinh\frac{x - x_e}{L_b} \qquad (5e.3)$$

where we denote by p_e the hole density $p(x_e)$ in the emitter and by p_c the hole density $p(x_c)$ in the collector. The boundary conditions $p(x_c) = p_c$; $p(x_e) = p_e$ are fulfilled. Due to the voltages V_e and V_c applied between the emitter and base, and the collector and base, respectively, the following relations exist between p_b, p_e, and p_c:

$$p_e = p_b \exp(|e|V_e/k_BT) \qquad (5e.4)$$

and

$$p_c = p_b \exp(|e|V_c/k_BT) \qquad (5e.5)$$

We substitute for p_e and p_c in Eq.(5e.3) their values given by Eqs.(5e.4) and (5e.5) and calculate j_p at $x = x_e$:

$$j_p(x_e) = -|e| D_p \frac{dp}{dx}\bigg|_{x_e} = \frac{|e|D_p p_b}{L_b} \cdot \frac{\cosh(w/L_b)\{\exp(|e|V_e/k_BT)-1\} - \{\exp(|e|V_c/k_BT)-1\}}{\sinh(w/L_b)} \qquad (5e.6)$$

This is the hole contribution to the emitter current density. For a calculation of the electron contribution we consider the ratio of electron densities in the p-type emitter, n_e, and in the n-type base, n_b, as a function of the emitter voltage V_e:

$$n_e/n_b = \exp(-|e|V_e/k_BT) \qquad (5e.7)$$

We find for the electron contribution to the emitter current density an equation similar to Eq.(5c.27) which represents a diode characteristic

$$j_n(x_e) = (|e|n_e D_n/L_e)\{\exp(|e|V_e/k_BT)-1\} \tag{5e.8}$$

where we introduced an electron diffusion length, $L_e = \sqrt{D_n \tau_e}$, in the emitter. The total emitter current density is given by

$$j_e(x_e) = j_p(x_e) + j_n(x_e) =$$

$$= \frac{|e|D_p p_b}{L_b \sinh(w/L_b)} \left\{\cosh\frac{w}{L_b}(\exp\frac{|e|V_e}{k_BT}-1)-(\exp\frac{|e|V_c}{k_BT}-1\right\}+\frac{|e|D_n n_e}{L_e}(\exp\frac{|e|V_e}{k_BT}-1) \tag{5e.9}$$

Similarly, the total collector current density, j_c, can be calculated

$$j_c = \frac{|e|D_p p_b}{L_b \sinh(w/L_b)}\left\{\cosh\frac{w}{L_b}(\exp\frac{|e|V_c}{k_BT}-1)+\exp\frac{|e|V_e}{k_BT}-1\right\}-\frac{|e|D_n n_e}{L_e}(\exp\frac{|e|V_c}{k_BT}-1) \tag{5e.10}$$

The "current amplification factor" a is given by

$$a = (\frac{\partial j_c}{\partial j_e})_{V_c = \text{const}} = (\frac{\partial j_c/\partial V_e}{\partial j_e/\partial V_e})_{V_c = \text{const}} \tag{5e.11}$$

In this definition it is assumed that the emitter input is a small signal and that the output is obtained at the collector. We obtain $\partial j_c/\partial V_e$ from Eq.(5e.10):

$$(\frac{\partial j_c}{\partial V_e})_{V_c = \text{const}} = \frac{|e|D_p p_b}{L_b \sinh(w/L_b)} \frac{|e|}{k_BT}\exp(|e|V_e/k_BT) \tag{5e.12}$$

and $\partial j_e/\partial V_e$ from Eq.(5e.9)

$$(\frac{\partial j_e}{\partial V_e})_{V_c = \text{const}} = \left\{\frac{|e|D_p p_b \cosh(w/L_b)}{L_b \sinh(w/L_b)} + \frac{|e|D_n n_e}{L_e}\right\}\frac{|e|}{k_BT}\exp\frac{|e|V_e}{k_BT} \tag{5e.13}$$

a is obtained from Eq.(5e.11)

$$a = \left\{\cosh(w/L_b) + \frac{D_n L_b n_e}{D_p L_e n_b}\sinh(w/L_b)\right\}^{-1} \tag{5e.14}$$

Since all quantities are positive, a is <1 which means there is no current amplification in the circuit shown by Fig.5.10. We will, however, now show that there is a power amplification.

In order to have a as close to 1 as possible the base width should be much less than the diffusion length and the base heavily doped: $w \ll L_b$; $p_b \gg n_e$. Assuming $L_b \approx L_e$ and $D_n \approx D_p$ we obtain for a:

$$a = 1/\cosh(w/L_b) \approx 1 - w^2/2L_b^2 \tag{5e.15}$$

It is possible to have $a = 0.98$ in germanium with a lifetime, τ_b, of 50 μsec, a diffusion length of 0.5 mm and a base width of 0.1 mm. In actual practice an even smaller base width is used.

In the differentiations in Eqs.(5e.12) and (5e.13) the base width w was assumed constant. The theory of the p-n junction has shown, however, that the junction width d depends on the applied voltage. Since the emitter voltage, V_e, is quite small the error made by neglecting this dependence is not very large. The collector voltage V_c, however, is relatively large and therefore

$$\frac{\partial w}{\partial V_c} = \frac{\partial d}{\partial V_c} = \frac{\partial}{\partial V_c}\left(\sqrt{V_D + V_c} \cdot d_o / \sqrt{V_c}\right) \approx \frac{d_o}{2\sqrt{V_D V_c}} \tag{5e.16}$$

where d_o is the base collector junction width for $V_c = 0$. This allows us to calculate the inverse collector impedance, $1/r_c$:

$$\frac{1}{r_c} = \left(\frac{\partial I_c}{\partial V_c}\right)_{V_e = \text{const}} \approx \frac{I_c}{w} \cdot \frac{\partial w}{\partial V_c} = \frac{I_c d_o}{2w\sqrt{V_D V_c}} \tag{5e.17}$$

Typical values are $d_o/w = 10^{-2}$; $V_D = 0.25$ V; $V_c = 4$ V and $I_c = 2$ mA. From Eq. (5e.17) a collector impedance of value 10^5 ohm is calculated.

The inverse emitter impedance,

$$\frac{1}{r_e} = \left(\frac{\partial I_e}{\partial V_e}\right)_{V_c = \text{const}} \tag{5e.18}$$

is obtained from Eqs.(5e.13) and (5e.9) neglecting to a first approximation the -1 terms and the $\exp(|e|V_c/k_BT)$ in Eq.(5e.9):

$$\frac{1}{r_e} \approx I_e \cdot \frac{|e|}{k_BT} = \frac{I_e}{25.9\,\text{mV}} \tag{5e.19}$$

where the equality sign is valid for room temperature. Assuming the emitter current to be the same as the collector current which is 2 mA, the order of magnitude obtained for the value of r_e is 10 ohm which is 4 orders of magnitude smaller than the collector impedance. Hence the power amplification factor in this case is given by

$$I_c^2 r_c / I_e^2 r_e = r_c/r_e \approx 10^4 \tag{5e.20}$$

This is also the factor of voltage amplification since $\alpha \approx 1$.

The operating power of the transistor is essentially the product $I_c V_c$ which in our case is of the order of a few mwatt. In vacuum tubes used before the invention of the transistor in 1948 by Bardeen, Brattain, and Shockley [1], the filament current power was a hundred times larger. A fast computer with about 10^5 transistors, which nowadays is quite common, would have been practically impossible with vacuum tubes because of the enormous heat production at the short distances required for fast signal transfer.

---

[1] See e.g. the Nobel prize lectures by John Bardeen: Semiconductor Research Leading to the Point Contact Transistor. Nobel Lectures "Physics" 1942-1962, p.318-341. Amsterdam: Elsevier Publ. Co. 1964. Walter H. Brattain: Surface Properties of Semiconductors, ibid. p.377-384. William Shockley: Transistor Technology Evokes New Physics, ibid. p.344-374.

The characteristics of a typical transistor are shown [2] in Fig. 5.11.

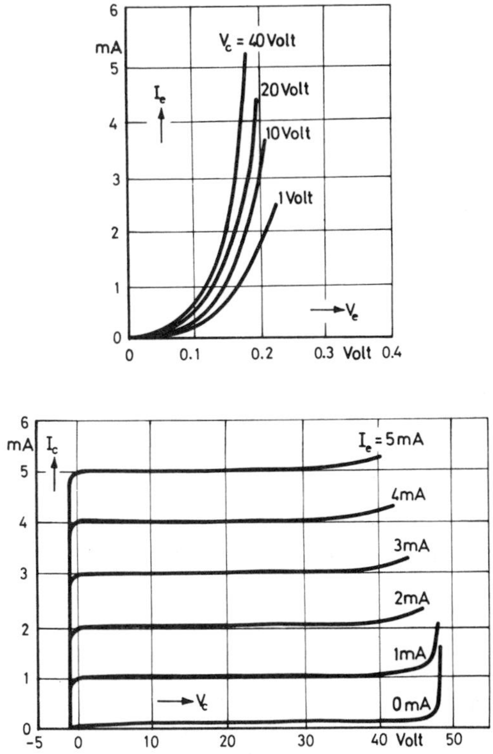

Fig. 5.11 Emitter and collector characteristics of a p-n-p transistor (after ref. 2).

The emitter characteristics, I_e vs V_e, are of course the forward characteristics while the collector characteristics, I_c vs V_c, are the reverse characteristics of a p-n junction diode. In the latter the emitter current I_e is a parameter. Up to quite large values of V_c, I_c is constant and equal to I_e, in agreement with $a \approx 1$. If the base is extremely thin and only lightly doped, it is possible that at a large collector voltage a galvanic connection of the collector and the emitter develops shorting the base. There is of course no amplification of signals in this case.

So far we investigated the electrical behavior of the p-n-p transistor by very formal mathematical methods. We will now consider qualitatively the behavior of individual electrons and holes in order to gain more "physical" understanding of transistor action.

---

[2] M. J. Morant: Introduction to Semiconductor Devices. Reading/Mass.: Addison-Wesley Publ. Co. Inc. 1964.

The potential distribution in a p-n-p transistor with no voltages applied is shown in Fig.5.12. In the heavily doped p-type emitter the Fermi level is locked at the acceptor levels, while in the heavily doped n-type base it is locked at the donor levels. The p-type collector is lightly doped. Since it may then be considered to be near-intrinsic the Fermi level is close to the middle of the gap (see Eq.(3a.37)). The potential well at the base is a barrier to the diffusion of holes from the emitter to the collector. The flow of holes across this barrier is controlled by the emitter-base voltage V_e as shown in Fig.5.11 (upper diagram), while the collector-base voltage, V_c, has little influence on I_c. The small difference of the current amplification factor α from unity is due to the loss of holes, on their way from the emitter through the base towards the collector, by recombination with electrons. In this rather qualitative consideration we have neglected the current contribution by electrons which may not be quite as apparent as the hole contribution. The reverse current of the base collector junction is due to electrons which are the minority carriers. Because the depletion region of this junction has a high impedance which manifests itself by flat collector characteristics, a small increase in collector current causes a large increase in collector voltage if the battery voltage is supplied via a series resistor having not too small a resistance.

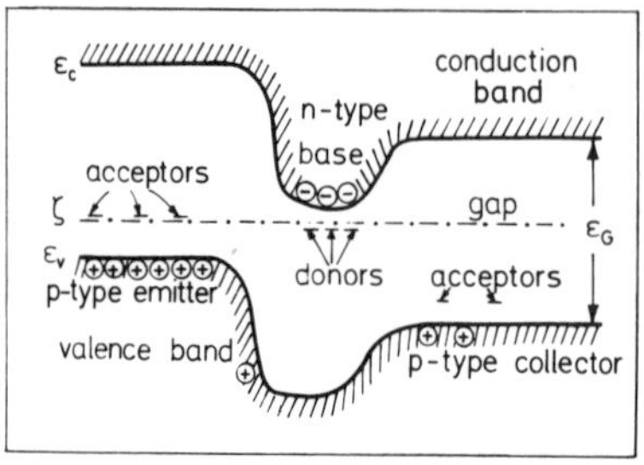

Fig.5.12 Energy band edges in a p-n-p transistor at thermal equilibrium.

The n-p-n transistor is similar to the p-n-p transistor except that the roles of electrons and holes are exchanged; it will not be discussed here.

Let us again consider the Haynes-Shockley experiment discussed in Chap.5b. The part of the germanium sample between the emitter and the collector contacts may be considered as a base. In contrast to the transistor discussed above the base is much longer than the diffusion length of minority carriers. This requires that the minority carriers drift rather than diffuse through the base; this is achieved by the application of a drift field. It is the action of emitter and collector which is the same as in a transistor. Similar to the Haynes-Shockley arrangement is the "filament transistor" where the metal-semiconductor contact close to the collector of the Haynes-Shockley type sample is used as a collector.

A feature of the Haynes-Shockley experiment is that the emitter and collector contacts are metal pins rather than p-n junctions. In the following chapter we shall explain how a metal-semiconductor contact under proper conditions behaves like a p-n junction. The type of transistor first invented made use of such contacts.

5f. The Metal-Semiconductor Contact

A metal-semiconductor contact can be used as a rectifier for an ac voltage like a p-n junction. In the early days of radio transmission a metal pin pressed against a natural crystal of lead sulfide was used as a rectifier. During World War II lead sulfide was replaced by silicon for rectification of microwave signals used in RADAR detection. Rectification is improved by "forming" which is a local heating of the extremely small contact area by a large current pulse of about one second duration. During the pulse metal atoms diffuse from the metal pin into the semiconductor thus forming a p-n junction. This method is not applicable in the case of large-area contacts such as used in "surface barrier counters" for high-energy particles in nuclear and high-energy physics [1]. For this purpose gold is evaporated on silicon to form a thin layer at one side of the crystal surface. By adsorption of oxygen molecules during a few days of exposure to room air an electrical dipole layer is formed which rejects majority carriers from the surface towards the bulk of the semiconductor thus forming a carrier depletion region at the surface called a "Schottky barrier".

We consider two cases of practical interest: in the first we entirely neglect the diffusion contribution to the current while in the second case the field contribution is neglected.

Let us consider the first case applicable to silicon and germanium. By n_b we denote the carrier density in the semiconductor bulk. If V is the voltage applied to the contact, and V_D is a diffusion voltage characterizing the band bending near the surface due to carrier rejection, the current density contribution due to holes flowing from the semiconductor into the metal, or electrons flowing in the opposite direction, is given by

$$j_1 = en_b \sqrt{k_B T/2\pi m} \exp\{-|e|(V_D - V)/k_B T\} \tag{5f.1}$$

The exponential represents the familiar "Boltzmann factor". We introduce the surface concentration of carriers, n_s, for no applied voltage ($V = 0$) given by

$$n_s = n_b \exp(-|e| V_D / k_B T) \tag{5f.2}$$

and obtain for j_1:

$$j_1 = en_s \sqrt{k_B T/2\pi m} \exp(|e|V/k_B T) \tag{5f.3}$$

This together with the contribution j_2 due to carriers going in the opposite direction yields the total current j:

$$j = j_1 + j_2 \tag{5f.4}$$

---

[1] See ref.1 of Chap.1c.

The magnitude of j_2 is easily obtained from the fact that $j = 0$ for no applied voltage, $V = 0$, and we obtain for the total current density:

$$j = e n_s \sqrt{k_B T / 2\pi m}\ \{\exp(|e|V/k_B T) - 1\} \tag{5f.5}$$

There is a saturation of the current density for a large reverse bias voltage.

For the second case (no field contribution) applicable to the selenium rectifier one obtains just as in the case of the p-n junction

$$j = j_s \{\exp(|e|V/k_B T) - 1\} \tag{5f.6}$$

where the current density for a large reverse voltage equals

$$j_s = |e| \mu_n n_s E_s \tag{5f.7}$$

and E_s is the electric field strength at the surface given by

$$E_s = \sqrt{(2 k_B T / e L_D^2)\ (V_D - V)} \tag{5f.8}$$

Since this factor depends on V there is no current saturation.

The properties of the semiconductor surface will be discussed in Chap.14.

5g. Various Types of Transistors

The first transistor was a point-contact transistor shown in Fig.5.13 in a cut-away view [1]. On top of a brass plug a germanium wafer 0.5 mm thick is soldered. Against the wafer two phosphor bronze springs (0.01 mm diameter) with sharp points are pressed at a distance of 0.02 mm which is much less than the diffusion length. The springs are welded to nickel mounting wires. After an electrical forming treatment one spring works as an emitter, the other as a collector while the plug is the base. Typical voltages and currents applied are : emitter 0.7 V, 0.6 mA ; collector 40 V, 2 mA. A large diffusion length and a good rectifying property are decisive for the qualification of a semiconductor as a transistor material. Prior to the invention of the germanium transistor, semiconductors used for rec-

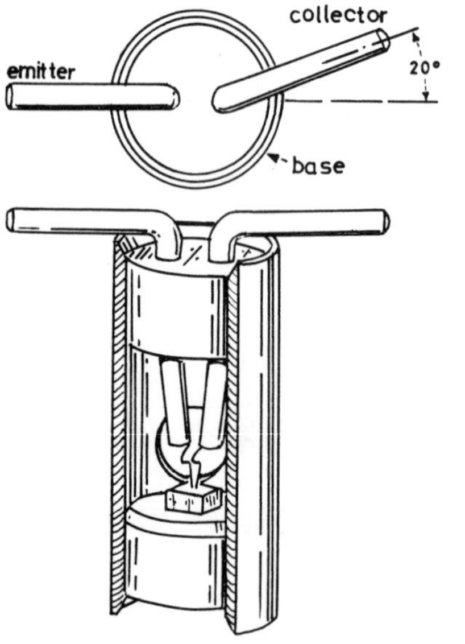

Fig.5.13 Cut-away view of a point-contact transistor (after W. Shockley, see ref.1.).

[1] J. Bardeen and W. H. Brattain, Phys.Rev.74 (1948) 230; W. Shockley: Electrons and Holes in Semiconductors. New York: Van Nostrand. 1950.

tification purposes were cuprous oxide and selenium. Lifetime and mobility in these semiconductors are too small, however, to yield transistor action. In a lead sulfide transistor a current amplification has not been observed [2]; the power amplification factor was 4 which is small compared to that in a germanium or silicon transistor. Even in these materials purification which nowadays is done by zone-refining [3] , is essential for obtaining the large diffusion length necessary for transistor action.

The alloy transistor [4] , usually a p-n-p type made of germanium, is shown in Fig.5.14. Two small pellets of indium metal are placed on either side of a semiconductor wafer of about 0.1 mm thickness. While the assembly is heated (to 820 K in the case of germanium) an alloying process takes place which results in the formation of the emitter and collector contacts. The base contact is a dot contact at one edge of the wafer. By making the collector somewhat larger than the emitter, good collection with a minimum of surface recombination around the emitter is guaranteed. The alloy process is not sufficiently controllable to yield the extremely small base width, required for a high-frequency transistor, which is of the order of 10^{-3} mm.

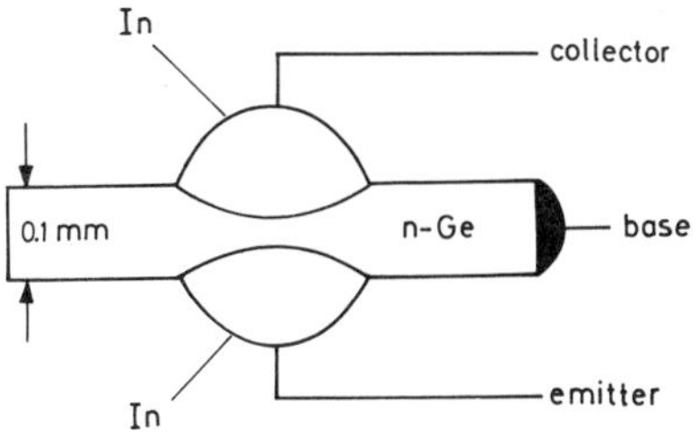

Fig.5.14 Schematic diagram of an alloy transistor.

In the years after 1960 the "planar transistor" [5] became the most frequently fabricated type of transistor. By thermal oxidation (for 1 hr at 1500 K or 10 hrs at 1200 K in steam) a thin (~ 1 μm) layer of silicon dioxide is grown on silicon which is free of holes and highly insulating ("quartz") [6]. With the help of the "photoresist process" small holes can be cut into this layer. For this process the surface of the silicon wafer should be flat ("planar"), otherwise imperfections result. The quartz layer is covered with a thin coat of "Kodak-Metal-Etch-Resist" which is exposed to light through a diaphragm. The exposed parts of the etch-resist solidify by polymerization due to the exposure while the unexposed parts can be removed by an organic solvent. The parts of the quartz layer not protected by the polymerized etch-resist are then etched away by

[2] H. A. Gebbie, P. C. Banbury, and C. A. Hogarth, Proc. Phys. Soc. (London) 63B (1950) 371.

[3] see e.g. H. Schildknecht: Zonenschmelzen. Weinheim/Germany: Chemie. 1964.

[4] M. Tanenbaum and D. G. Thomas, Bell Syst. Tech. J. 35 (1956) 1.

[5] J. A. Hoerni, IRE Electron Devices Meeting. Washington, D. C.: 1960.

[6] C. J. Frosch and L. Derrick, J. Electrochem. Soc. 104 (1957) 547; for further information on technology see e.g. A. S. Grove: Physics and Technology of Semiconductor Devices. New York: J. Wiley and Sons. 1967.

hydrofluoric acid. The remaining etch-resist is afterwards dissolved. The left over parts of the quartz layer are then used as a "mask" in a subsequent process of diffusion of impurities like e.g. boron into the n-type silicon substrate. On top of this structure another quartz layer is generated and perforated with holes smaller than the first ones but at the same locations. Through this second set of holes phosphorus is diffused into the silicon substrate to a depth that is less than the previous boron diffusion depth. Phosphorus overcompensates the existing boron doping and yields a heavily doped n-type layer. A cross section of the n-p-n structure so obtained is shown in Fig.5.15. At the position indicated by an arrow another hole is then etched through the second quartz layer. Fine metal wires are put through the holes and alloyed with the silicon thus serving as connections between the emitter (n-Si, phosphorous doped) and the emitter power supply, and between the base (p-Si, boron doped) and the base power supply, respectively. A third alloy contact at the back side of the n-type silicon wafer serves as the collector. Since the base is only a few μm thick the carrier diffusion time is quite small and the highest frequency which can be amplified is extremely high. Another advantage of this process is that the p-n junctions at the surface are protected against atmospheric influences by the grown oxide layers with the result of a smaller leakage current and better stability.

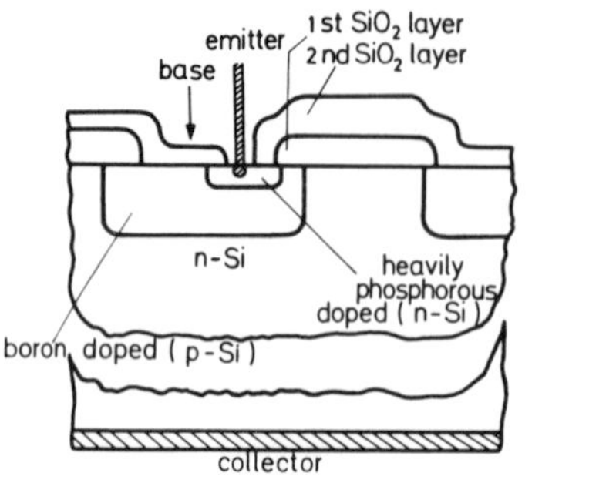

Fig.5.15 Silicon planar transistor (cross section)

With the help of the masking and diffusion techniques complete circuits with resistors, capacitors, transistors, and rectifiers, are fabricated in a single silicon wafer with very small requirements of space. These "integrated circuits" (I.C.) became important for use in fast computers. Up to 1000 transistors per mm^2 have been made on a single wafer. Precise photographic masks are required to define the microscopic areas to be exposed to etchants. In the "bipolar technology" discussed above from five to seven masking and etching steps are needed.

In a MOS technology only four such steps are required. A MOSFET is a "metal oxide semiconductor field effect transistor". A cross section is given in Fig.5.16. A voltage is applied between the electrodes labeled "source" and "drain". The circuit thus contains two p-n junctions one of which is biased in the reverse direction while the other is biased in the forward direction. The reverse biased junction has a high impedance. A positive potential is applied to the electrode called "gate" which is insulated from the semiconductor by a quartz layer. The positive potential at the gate induces a negative charge at the boundary between the semiconductor and the insulating quartz layer. Due to the induced negative charge the semiconductor although doped p-type converts into

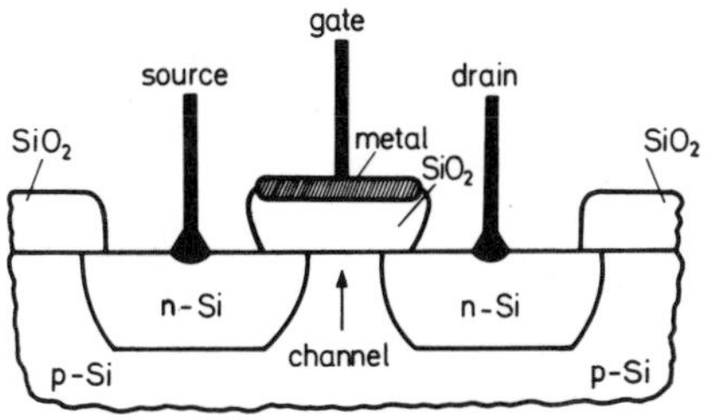

Fig.5.16 Metal oxide semiconductor field effect transistor (MOSFET), cross section.

n-type at the boundary thus forming an n-type channel between the two n-type source and drain regions. A strong current flows through the channel which can be controlled by the gate potential. The small leakage currents to the gate electrode are nearly independent of temperature. Since the breakdown field intensity in quartz is about 10^7 V/cm gate voltages up to 10^3 V can be used. The carrier mobility in the channel depends on the channel width which is less than the mean free path of carriers. The mobility then depends on the temperature T as $T^{-1.5}$ while the normal dependence of mobility on T in silicon is stronger [7] than the 1.5 power.

For a further treatment of semiconductor devices such as rectifiers and transistors see e.g. S. M. Sze: Physics of Semiconductor Devices. New York: J. Wiley and Sons. 1969.

5h. Dember-Effect and PEM-Effect

This and the next chapter are devoted to a further discussion of the diffusion of carrier pairs generated by the absorption of light. Details of the absorption process will not concern us here and will be discussed in Chap.11.

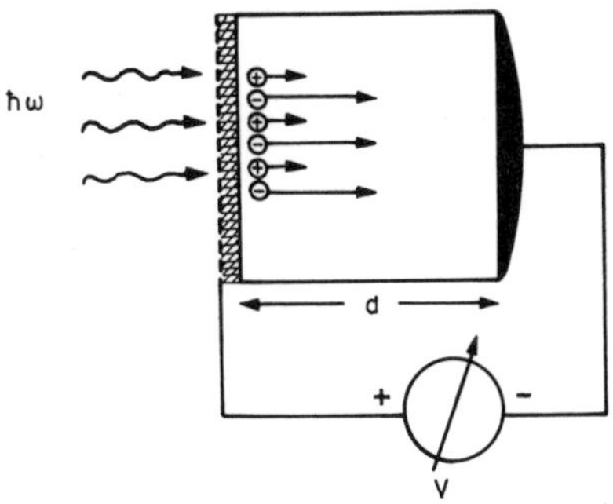

Fig.5.17 Dember arrangement.

Fig.5.17 shows a semiconductor sample with metal contacts on opposite sides which are connected by a voltmeter. One of the contacts is semitransparent to light of a frequency high enough to generate electron hole pairs close to the surface. Let us assume a penetration depth α^{-1} of the incident light in the semiconductor having a generation rate G for a given light in-

[7] O. Leistiko, A. S. Grove, and C. T. Sah, IEEE-Trans. ED-12 (1965) 248.

tensity (a = absorption coefficient). The product $a^{-1}G$ is denoted by G_o and measured in units $cm^{-2}\,sec^{-1}$. The electron hole pairs generated at one side of the sample diffuse towards the oppposite side. Since in general the diffusion coefficients of both carrier types are different, a voltage V is indicated by the voltmeter. The effect is called "photodiffusion" or "Dember effect" [1]. If one of the contacts is somewhat non-ohmic or even rectifying a photovoltaic effect is measured instead of the Dember effect; this will be discussed in the following chapter.

Recombination of carriers at the surface will be taken into account. A "surface recombination velocity" s is defined by considering a current density j_n perpendicular to the surface

$$j_n = |e|\, s\,\Delta n \tag{5h.1}$$

and a current density of holes, $j_p = -j_n$ where $\Delta n = \Delta p$ is the excess of carriers over the equilibrium values. s is the velocity of pair disappearance in the surface "drain".

The solution of the diffusion equation (5e.1) which we write in the form

$$D d^2 p/dx^2 = \Delta p/\tau \tag{5h.2}$$

is given by

$$\Delta p(x) = \frac{G_o L}{D}\,\frac{\cosh\{(x-d)/L\} - (L/L_{sd})\sinh\{(x-d)/L\}}{(1+L^2/L_{sd}L_{so})\sinh(d/L) + (L/L_{so}+L/L_{sd})\cosh(d/L)} \tag{5h.3}$$

where $L = \sqrt{D\tau}$; D is the ambipolar diffusion coefficient, Eq.(5b.16); τ is the lifetime of holes and electrons assumed to be equal; $L_{so} = D/s$ at the illuminated surface ($x = 0$); L_{sd} is the corresponding quantity at the opposite surface at $x = d$; and d is the sample dimension in the direction of light propagation, assumed to be $d \gg a^{-1}$. The current density

$$j_{nx} + j_{px} = \sigma E - |e|(D_p - D_n)\, dp/dx \tag{5h.4}$$

vanishes and the Dember voltage is given by

$$V = \frac{|e|(D_n - D_p)}{\sigma}\{\Delta p(0) - \Delta p(d)\} \tag{5h.5}$$

We obtain $\Delta p(0) - \Delta p(d)$ from Eq.(5h.3):

$$V = \frac{k_B T}{|e|}\,\frac{G_o L(\mu_n - \mu_p)\{\cosh(d/L) + (L/L_{sd})\sinh(d/L) - 1\}}{D(n\mu_n + p\mu_p)\{(L/L_{so}+L/L_{sd})\cosh(d/L) + (1+L^2/L_{sd}L_{so})\sinh(d/L)\}} \tag{5h.6}$$

For the simplifying case of $d \gg L$ and $L_{so} = L_{sd}$:

$$V = G_o\,\frac{k_B T}{|e|}\cdot\frac{\mu_n - \mu_p}{n\mu_n + p\mu_p}\cdot\frac{1}{\sqrt{D/\tau} + s} \tag{5h.7}$$

[1] H. Dember, Phys. Zeitschr. 32 (1931) 554; 856; 33 (1931) 207; J. Frenkel, Nature 132 (1933) 312; W. van Roosbroeck, J. Appl. Phys. 26 (1955) 380.

Obviously the method is applicable for a determination of the surface recombination velocity s if the mobilities and the lifetime of minority carriers are known. For a photon flux of $G_o = 10^{18}/cm^2$ sec, and an electron density of $2.6 \times 10^{13}/cm^3$ (and practically no holes), a hole mobility $\mu_p = \frac{1}{2}\mu_n$, a D/τ-ratio of $10^{16} cm^2/sec^2$ at room temperature, and neglecting surface recombination, a Dember voltage of 0.5 V would be obtained. A surface recombination velocity s of value 10^6 cm/sec which is typical of a sandblasted surface would reduce the effect to 500 μV. For an etched surface where s is of the order of magnitude of 10^2 cm/sec surface recombination would not appreciably influence V.

The application of a magnetic field perpendicular to the direction of light propagation yields the Hall equivalent of photodiffusion called the "photoelectromagnetic effect" (PEM-effect)*). The arrangement is shown in Fig.5.18. The polarity of the voltage V_y is reversed if B_z is reversed. The effect has been thoroughly treated theoretically by van Roosbroeck [2]. A simplified treatment neglecting surface recombination and assuming sample dimensions much larger than a diffusion length will be presented here. Only the case of a weak magnetic field in z-direction will be considered. The current density component in the y-("Hall-") direction is given by

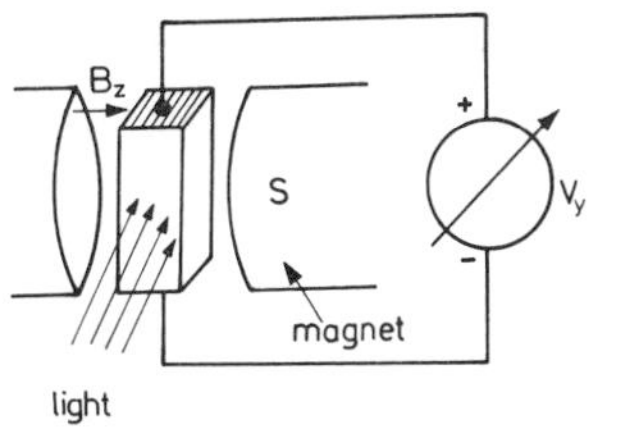

Fig.5.18 PEM arrangement (schematic).

$$j_y = \sigma E_y + (j_{nx}\mu_{Hn} - j_{px}\mu_{Hp})B_z \qquad (5h.8)$$

where the subscript H indicates the Hall mobilities. Since no current is assumed in the x-direction ($j_{px} = -j_{nx} = -|e|D\,dp/dx$) we obtain for the PEM short-circuit current density ($E_y = 0$) :

$$j_y = |e|\,D(\mu_{Hn} + \mu_{Hp})\,B_z \Delta p/L \qquad (5h.9)$$

where in the approximation mentioned above dp/dx has been replaced by $\Delta p/L$. Occasionally the PEM-effect is combined with a measurement of the photoconductivity to yield the minority carrier lifetime, τ. (Actually, the electrical injection method as applied in the Haynes-Shockley experiment yields more accurate results for τ, and the PEM-effect is more often used for a determination of the surface recombination velocity.) We apply an electrical field E_y and no magnetic field. The change in current density upon illumination of the sample, Δj_y, is given by

$$\Delta j_y = |e|\,(\mu_n + \mu_p)E_y \Delta p \qquad (5h.10)$$

[2] W. van Roosbroeck, Phys. Rev. 101 (1956) 1713.

*) A more appropriate name would be "photogalvanomagnetic effect", see e.g. O. Madelung: Handbuch der Physik, Vol.XX, p. 142 (S. Flügge, ed.). Berlin-Göttingen-Heidelberg: Springer. 1957.

and the ratio of photo and PEM current densities is given by

$$\frac{(\Delta j_y)_{\text{Photo}}}{(j_y)_{\text{PEM}}} = \frac{E_y}{B_z r_H} \frac{L}{D} = \frac{E_y}{B_z r_H} \sqrt{\frac{|e|\tau}{k_B T\mu}} \tag{5h.11}$$

where the Hall factor r_H has been assumed to be the same for electrons and holes and the same illumination is applied for the measurement of both effects.

The open-circuit PEM field E_y is obtained by replacing j_y in Eq.(5h.8) by $-\sigma E_y$ and averaging over the sample cross section :

$$E_y = -\frac{|e| D(\mu_{Hn} + \mu_{Hp}) B_z}{\sigma L} \cdot \frac{1}{d} \int_0^d \Delta p \, dx \tag{5h.12}$$

The integral in the approximation mentioned above is given by $\Delta p \cdot L$. The PEM voltage $V_y = E_y \cdot 1$ where l is the length of the crystal. Hence,

$$V_y = -\frac{|e| D(\mu_{Hn} + \mu_{Hp}) B_z}{\sigma} \cdot \frac{1}{d} \Delta p = -\frac{k_B T}{|e|} \mu_{Hp} B_z \left\{ \frac{b(b+1)(\eta+1)}{(\eta b+1)^2} \cdot \frac{\Delta p}{p} \right\} \tag{5h.13}$$

where $\eta = n/p$ and $b = \mu_n/\mu_p = \mu_{Hn}/\mu_{Hp}$ have been introduced. For η and $b \gg 1$ the factor in braces becomes $\Delta p/n = \Delta n/n$. Assuming for this factor a value of 10% and for $\mu_{Hp} B_z$ a value of 10%, the PEM voltage at room temperature is seen to be quite small, namely 260 μV. It is quite sensitive to inhomogeneities in doping since internal p-n, n^+-n, or p^+-p barriers yield a photovoltaic effect much larger than the PEM effect. The former will be described in the following chapter.

5i. Photovoltaic Effect

When light, x-rays, β-rays or other radiation, whose quantum energy exceeds a threshold of the order of the gap energy, ϵ_G, ionizes the region in or near a potential barrier, a photovoltaic potential is generated. The most important device application of the photovoltaic effect is the solar cell for conversion of solar radiation into electrical energy. At the surface of the atmosphere of the earth or at the moon the solar power density is 0.135 watt/cm². Assuming a conversion efficiency of 10% a 5 kwatt lunar vehicle motor requires 30 m² of effective solar cell surface for operation.

For equilibrium ($\partial \Delta n/\partial t = 0$) the continuity equation (5b.9) becomes for holes ($e > 0$)

$$\frac{1}{e} (\vec{\nabla}_r \vec{j}_p) = G - \frac{p - p_n}{\tau_p} \tag{5i.1}$$

which except for the term G is equal to Eq.(5c.21). With Eq.(5c.22) we obtain for holes at the n-side of the junction

$$-D_p(\vec{\nabla}_r)^2 p = G - \frac{p - p_n}{\tau_p} \tag{5i.2}$$

The boundary conditions are: $p = p_n$ and $G = 0$ at $x \gg x_n$; at $x = x_n$ as given by Eq.(5c.24). The p-side of the junction is assumed to be close to the illuminated surface as shown in Fig.5.7. An equation similar to Eq.(5i.2) holds for electrons at the p-side. For the boundary condition at the illuminated surface we assume for simplicity surface recombination to be very large so that we may assume G=0 at the p-side. Otherwise the conditions are the same as those given in Chap. 5c. For simplicity we assume that the incident radiation is converted into electron hole pairs only in the transition region, and the diffusion lengths, $L_p = \sqrt{D_p \tau_p}$, $L_n = \sqrt{D_n \tau_n}$, are much larger than the junction width $d = |x_p - x_n|$. The solution of Eq.(5i.2) is then given by

$$p(x) - p_n - G\tau_p = \{p(x_n) - p_n - G\tau_p\} \exp\{-(x - x_n)/L_p\} \tag{5i.3}$$

which yields for the hole component of the current density

$$j_p = \frac{|e| D_p p_n}{L_p} \{\exp(|e| V/k_B T) - 1\} - G \cdot |e| L_p \tag{5i.4}$$

where the relation $D_p \tau_p = L_p^2$ has been used. The electron component is similar except for the term $G|e|L_p$. Hence the total current density is given by

$$j = j_s \{\exp(|e| V/k_B T) - 1\} - j_L \tag{5i.5}$$

where

$$j_L = G \cdot |e| \cdot L_p = \frac{\Delta p}{\tau_p} |e| \sqrt{D_p \tau_p} = |e| \Delta p \sqrt{D_p/\tau_p} \tag{5i.6}$$

is the current component due to irradiation of the junction. A more complete expression for j_L which includes the finite penetration depth of light and a finite surface recombination vecolity is found in ref.[1].

The short-circuit current density $j = -j_L$. The open-circuit voltage V_{oc} is given by

$$V_{oc} = \frac{k_B T}{|e|} \cdot \ln(1 + j_L/j_s) \tag{5i.7}$$

The output power per front area is jV. It is a maximum when $\partial(jV)/\partial V = 0$; this yields an optimal current density

$$j = -\frac{j_L}{1 + \{1 - \exp(-|e| V/k_B T\} k_B T/(|e| V)} \tag{5i.8}$$

and a maximum output power per front area

$$|jV|_{max} = \frac{j_L V^2}{V + k_B T/|e|} \left(1 + \frac{j_s}{j_L}\right) \approx j_L V \tag{5i.9}$$

---

[1] O. Madelung: Handbuch der Physik, Vol. XX, p. 163. (S. Flügge, ed.) Berlin-Göttingen-Heidelberg : Springer. 1957.

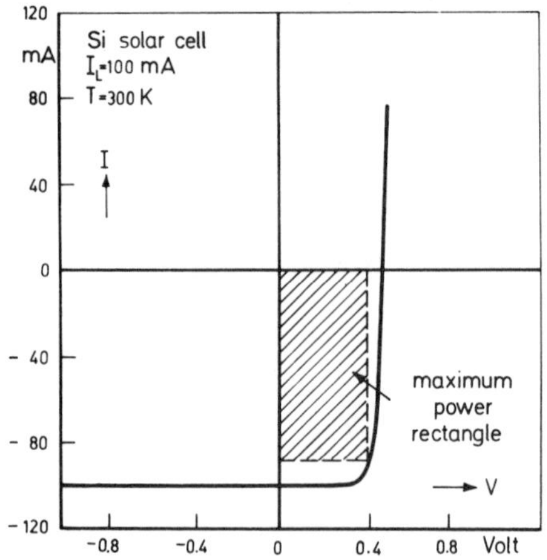

Fig.5.19 Current-voltage characteristics of a solar cell under illumination (after M. B. Prince, J. Appl. Phys. 26 (1955) 534).

In Fig.5.19 the current-voltage characteristics of a silicon p-n junction under illumination is shown. The shaded area represents the maximum output power of 35 mwatt. For the saturation current density j_s of silicon we apply Eq.(5c.40) assuming $\tau_n = \tau_p = \tau$ and obtain for the ratio j_L/j_s in Eq.(5i.7) a value of $L_p \Delta p/(dn_i)$. For $\Delta p \gg n_i$ and $L_p \gg d$ as assumed, we find $j_L/j_s \gg 1$. Even so, due to the logarithmic dependence of V_{oc} on j_L/j_s the voltage may not be much more than an order of magnitude larger than $k_B T/|e|$. This would be of about the same magnitude as the diffusion voltage V_D which is a few tenths of a volt at room temperature. Obviously the photovoltaic voltage is much larger than the PEM voltage discussed in the last chapter. (The influence of a weak magnetic field, necessary for the production of the PEM effect, on the photovoltaic voltage is negligible.)

Conversion efficiencies of over 16% have been reported for "heterojunction" solar cells consisting of p-type $Ga_{1-x}Al_xAs$ on a GaAs p-n junction where for x = 0.7 the greatest portion of the solar spectrum is utilized [2]. This type of heterojunction will be discussed in more detail in Chap.12b on semiconductor lasers. The improved efficiencies compared to conventional homojunction cells may be due to a reduction of both series resistance and surface recombination loss in the presence of the heavily doped $Ga_{1-x}Al_xAs$ layer (thickness 6 μm for x = 0.7, junction depth 0.8 μm). Values for V_{oc} of 1 V and for $-j_L$ of 21 mA/cm^2 for a solar input intensity of 98.3 mW/cm^2 at sea level have been found. The largest cell made with the reported high efficiency was about 0.15 cm^2 in size. Because of the large band gap of GaAs the cells are sensitive for photon energies not less than 1.4 eV and up to about 2.5 eV.

[2] J. M. Woodall and H. J. Hovel, Appl. Phys. Lett. 21 (1972) 379.

5j. Hot-Carrier Diffusion

In Chap.4n we have seen that the velocity distribution of carriers in a strong electric field is different from the distribution at thermal equilibrium. The average carrier energy is higher than the thermal energy. The question arises, what is the average energy of carriers in the strong electric field of a narrow p-n junction which is due to the built-in diffusion potential? Certainly, for the case of no current flow, the average energy of carriers must be the same everywhere in the crystal, including the junction region, for thermodynamic reasons. Then the mobility and the diffusion constant do not depend on the built-in field strength. In any case, both quantities are not uniquely determined by the local field strength but also depend on the local distribution of carriers. The problem first solved by Avak'yants [1] for a uniform barrier field has been treated quite generally by Stratton [2] and for the case of space-charge-limited current flow by Stratton and Jones [3].

The one-dimensional solution of the Boltzmann equation in the diffusive approximation is given by

$$f = f_0 - (eE_z/m)\tau_m \, \partial f_0/\partial v_z - \tau_m v_z \partial f_0/\partial z \tag{5j.1}$$

Then the current density j_z becomes

$$j_z = e\int v_z f d^3v = -(e^2 E_z/m)\int \tau_m v_z (\partial f_0/\partial v_z) d^3v - e\partial \int \tau_m v_z^2 f_0 d^3v/\partial z \tag{5j.2}$$

where it has been assumed that τ_m is independent of z and that z and v_z are the independent variables. The first intergal on the right-hand side treated in Chap.4b has a value of $n|e|\mu_n E_z$. The density n is given by the integral

$$n = \int f_0 d^3v = n(z) \tag{5j.3}$$

and depends on z since the distribution f_0 is a function of z in the junction region. If we define the diffusion coefficient D_n by

$$D_n = \int \tau_m v_z^2 f_0 d^3v / \int f_0 d^3v \tag{5j.4}$$

we obtain from Eq.(5j.2) for j_z:

$$j_z = n|e|\mu_n E_z - e\partial(nD_n)/\partial z \tag{5j.5}$$

and for the Einstein relation

$$|e|D_n/\mu_n = -m\int \tau_m v_z^2 f_0 d^3v / \int \tau_m v_z (\partial f_0/\partial v_z) d^3v \tag{5j.6}$$

[1] G. M. Avak'yants, Zh. Eksperim. i Teor. Fiz. 27 (1954) 333.

[2] R. Stratton, Phys. Rev. 126 (1962) 2002; J. Appl. Phys. 40 (1969) 4582; J. B. Gunn, J. Appl. Phys. 39 (1968) 4602; G. Persky and D. J. Bartelink, Phys. Rev. B1 (1970) 1614; V. L. Bonch-Bruevich, phys. stat. sol. 33 (1969) 911.

[3] R. Stratton and E. L. Jones, J. Appl. Phys. 38 (1967) 4596.

Only for the case of a Maxwell-Boltzmann distribution with an electron temperature T_e, where $\partial f_o/\partial v_z = -f_o m v_z/k_B T_e$, will the right-hand side of Eq.(5j.6) be equal to the familiar $k_B T_e$ (Eq.(4j.12)). The last term in Eq.(5j.5) can be split into two terms, $eD_n \partial n/\partial z + en(dD_n/dT_e)\partial T_e/\partial z$, where the second term represents the usual thermal diffusion.

From Eqs.(4b.31) and (4b.43) one can show that the energy balance equation, Eq.(4n.5), for the present case is given by [2]

$$j_z E_z = -n <\partial\epsilon/\partial t>_{coll} + \partial/\partial z \{-\kappa \partial T_e/\partial z - (j_z/e)\cdot <\tau_m \epsilon>/<\tau_m>\} \tag{5j.7}$$

where the term in braces is the energy flux in the positive z direction, $\kappa = \kappa(T_e)$ is the thermal conductivity of the carriers, and $<\tau_m \epsilon>/<\tau_m>$ is the average kinetic energy transported per carrier arising from the current flow.

Eqs.(5j.5), (5j.7), and Poisson's equation have been integrated numerically with the usual boundary conditions for a p-n junction: $n = n_b \exp(-eV_D/k_B T_e)$ at $z = d$; $n = n_b$ at $z = 0$, and in addition the requirement $T_e = T$ at $z = 0$ and d. For a small perturbation from equilibrium at $j_z = 0$ an analytical treatment is possible by expanding the z-dependent quantities as a power series in j_z. The applied voltage V becomes for this case

$$V = \int_o^d \{j_z/(n|e|\mu_n) - E_z(T_e - T)/T\}\, dz \tag{5j.8}$$

where the steady state values ($j_z = 0$) for n, μ_n, E_z, and T have to be used. From this equation the current-voltage characteristics can be determined for any specific barrier model.

Experimental results on D_n and D_p in <111> oriented silicon samples at room temperature have been obtained for field intensities between 6 and 50 kV/cm (Fig.5.20) [4]. Assuming a Maxwell-Boltzmann distribution, the Einstein

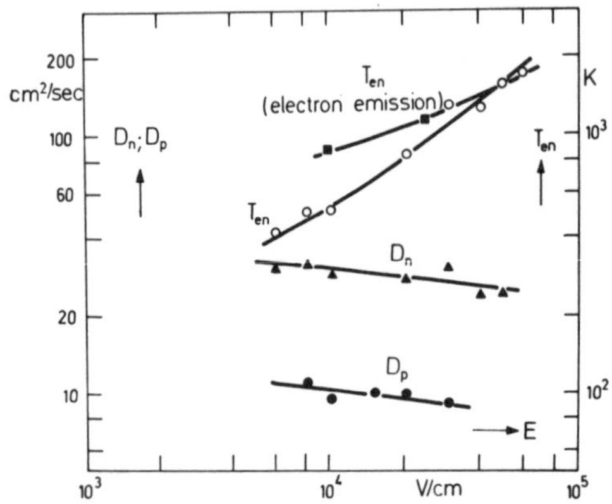

Fig.5.20 Carrier diffusion coefficients and hot carrier temperatures as a function of electric field intensity for <111> oriented silicon samples at room temperature (after ref.4).

[4] T. W. Sigmon and J. F. Gibbons, Appl. Phys. Lett. 15 (1969) 320.

relation in its simple form is applicable; this leads to the electron temperatures indicated in Fig.5.20. Plotted for comparison are electron temperatures obtained from experiments on thermal emission of electrons from a silicon p-n diode [5]. The temperatures obtained from the emission experiment are somewhat higher, but since the accuracy of the emission experiment in T_e is only about 30 % the agreement is quite good.

---

[5] E. A. Davies, J. Phys. Chem. Solids 25 (1964) 201.

6. Scattering Processes in a Spherical One-Valley Model

In Chap.4 we frequently assumed an energy dependence of the momentum relaxation time, $\tau_m \propto \epsilon^r$, where r is a constant, for the calculation of the galvanomagnetic, thermoelectric, thermomagnetic etc. effects. We will now treat the important scattering mechanisms and find the energy dependence of τ_m. For those cases where a power law is found the magnitude of the exponent r will be determined.

6a. Neutral Impurity Scattering

The equation of motion of a carrier in an electric field, Eq.(4b.2), contains a "friction term" which is essential for the establishment of a constant drift velocity at a given field intensity. From a microscopical point of view, friction is the interaction of carriers with imperfections of the crystal lattice such as impurities, lattice defects, and lattice vibrations. This interaction is called "scattering". The concept of a "scattering cross section" may be familiar from the theory of transmission of high-energy particles through matter [1] : The probability per unit time for a collision, $1/\tau_c$, is given by the density of scattering centers, N, the cross section of centers, σ_c, and the velocity v of the particle :

$$1/\tau_c = N\sigma_c v \tag{6a.1}$$

τ_c is called the "collision time"; it is the mean free time between collisions. For an explanation of this relation consider N parallel disks of area σ_c per unit volume. The particle moves perpendicular to the disks a distance vdt in the time interval dt and hits one of the disks with a probability $(N\sigma_c)vdt = dt/\tau_c$.

σ_c is obtained from a "differential cross section" $\sigma(\theta)$ by integrating over the solid angle $d\Omega = 2\pi \sin\theta\, d\theta$ where we assume the scattering center to be spherically symmetric:

$$\sigma_c = 2\pi \int_0^\pi \sigma(\theta) \sin\theta\, d\theta \tag{6a.2}$$

[1] see e.g. L. I. Schiff: Quantum Mechanics, 3rd ed., p. 110. New York: McGraw-Hill. 1968.

θ is the angle of deflection of the particle from its original direction of motion. Hence, after the collision the component of the particle velocity in the direction of its original motion is $v \cdot \cos\theta$. The relative change in this velocity component is therefore

$$\frac{v - v \cdot \cos\theta}{v} = 1 - \cos\theta \tag{6a.3}$$

and this is also the relative change in the corresponding momentum component. (The effective mass is assumed to remain constant during the scattering process.) Hence, the momentum-transfer cross section, σ_m, is given by a modification of Eq.(6a.2):

$$\sigma_m = 2\pi \int_0^\pi \sigma(\theta)(1 - \cos\theta) \sin\theta d\theta \tag{6a.4}$$

and the momentum relaxation time, τ_m, is then defined by

$$1/\tau_m = N\sigma_m v = Nv2\pi \int_0^\pi \sigma(\theta)(1 - \cos\theta) \sin\theta d\theta \tag{6a.5}$$

A very fundamental scattering process is the scattering of a conduction electron at a neutral impurity atom in the crystal lattice. A similar process is the scattering of low-energy electrons in a gas. This latter process has been treated quantum mechanically in great detail [2]; it had been observed before the development of quantum mechanics by Ramsauer [3]. The result of the theoretical treatment has been transferred to neutral-impurity scattering in crystals by Erginsoy [4].

The method applied here is that of "partial waves": The material wave of the electron is diffracted by the field of the impurity atom in such a way that it fits on smoothly to the undistorted wave function outside, which is thought to consist of the partial-wave functions of the plane wave without a scattering center present (expansion in Legendre functions of the scattering angle) and a scattered wave. Since this is a standard problem in text books on quantum mechanics [5] we shall not go into further detail here. The numerical calculation includes both electron exchange effects and the effect of the polarization of the atom by the incident electron. The result can be approximated by a total cross section

$$\sigma = 20\, a/k \tag{6a.6}$$

valid for electron energies of up to 25percent of the ionization energy of the impurity atom; a is the equivalent of the Bohr radius in an hydrogen atom

$$a = \kappa\kappa_0 \hbar^2/me^2 = \kappa a_B/(m/m_0) \tag{6a.7}$$

[2] H. S. W. Massey and E. H. S. Burhop: Electronic and Ionic Impact Phenomena, vol. 1, chap. 6.3. Oxford: Clarendon. 1969.
[3] C. Ramsauer, Ann.der Phys. 64 (1921) 513; 66 (1921) 545.
[4] C. Erginsoy, Phys. Rev. 79 (1950) 1013.
[5] E.g. L. I. Schiff: Quantum Mechanics, chap. 5. New York: McGraw-Hill. 1968.

where $a_B = 0.53$ Å is the Bohr radius, κ is the dielectric constant, and m is the "density-of-states effective mass" (see Chaps.7b and 8a) [6]; $k = 2\pi/\lambda = mv/\hbar$ is the absolute magnitude of the electron wave vector. It may be interesting to note that about the same cross section would have been obtained if the geometrical cross section πa^2 would have been multiplied by the ratio λ/a; there is, however, no simple explanation of this result. In fact, Eq.(6a.6) shows that the cross section varies inversely with the carrier velocity v while low-energy scattering by a perfectly rigid sphere or by a square-well potential has a cross section that is substantially independent of velocity [5]. For the carrier mobility μ we obtain

$$\mu = \frac{e}{20 a_B \hbar} \frac{m/m_o}{\kappa N^x} \tag{6a.8}$$

which is independent of temperature. In units of $cm^2/Vsec$ we find

$$\mu = \frac{1.44 \times 10^{22}\, cm^{-3}}{N^x} \frac{m/m_o}{\kappa} \tag{6a.9}$$

E.g. for electrons in Ge, where $m/m_o = 0.12$ and $\kappa = 16$, a mobility of $1.1 \times 10^3\, cm^2/Vsec$ is obtained assuming e.g. $10^{17}/cm^3$ neutral impurities. A large value of the dielectric constant favors neutral-impurity scattering, in contrast, as we shall see later on, to ionized-impurity scattering.

Neutral-impurity scattering is always accompanied by other scattering mechanisms such as ionized-impurity scattering and lattice scattering. At very low temperatures where impurities are neutral due to carrier freeze-out and where phonons have disappeared to a large extent making ionized-impurity scattering and lattice scattering unimportant, impurity band conduction dominates the conductivity (Chap.6n) and therefore no experimental data on mobility are available to compare with Eq.(6a.9).

However, both the linewidth of cyclotron resonance and the attenuation of ultrasonic waves have been shown to be determined at low temperatures by neutral impurity scattering. Although these effects will be discussed later on (Chaps.11k and 7h, respectively) we consider the results here as far as they can be explained by neutral-impurity scattering.

Fig.6.1 shows the inverse relaxation time obtained from the linewidth of cyclotron resonance, plotted vs the impurity concentration [7]. The data obtained for a shallow donor (Sb in Ge) are in good agreement with Erginsoy's formula (6a.8) where $\mu = (e/m)\tau_m$. However, for shallow acceptors (Ga and In in Ge) τ_m^{-1} is smaller by an order of magnitude and in agreement with a theory by Schwartz [8] which takes account of the fact that by a hydrogenic impurity a

---

[6] S. H. Koenig, R. D. Brown III, and W. Schillinger, Phys. Rev. 128 (1962) 1668.
[7] E. Otsuka, K. Murase, and J. Iseki, J. Phys. Soc. Japan 21 (1966) 1104.
[8] C. Schwartz, Phys. Rev. 124 (1961) 1468; M. Rotenberg, Ann. Phys. 19 (1962) 262.

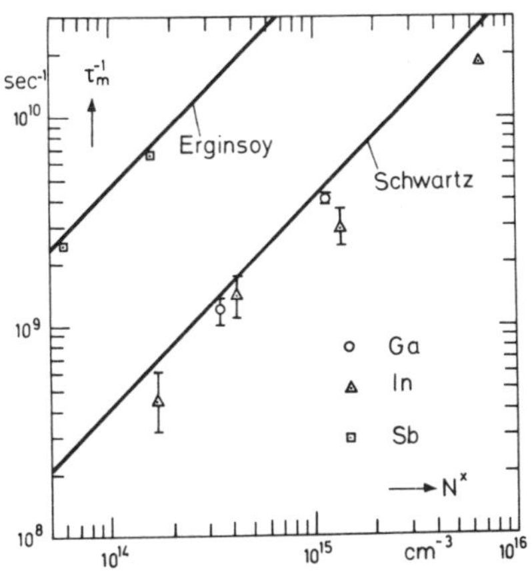

Fig.6.1 Concentration dependence of the inverse cyclotron relaxation time for neutral impurity scattering in germanium (after Otsuka et al., ref.7).

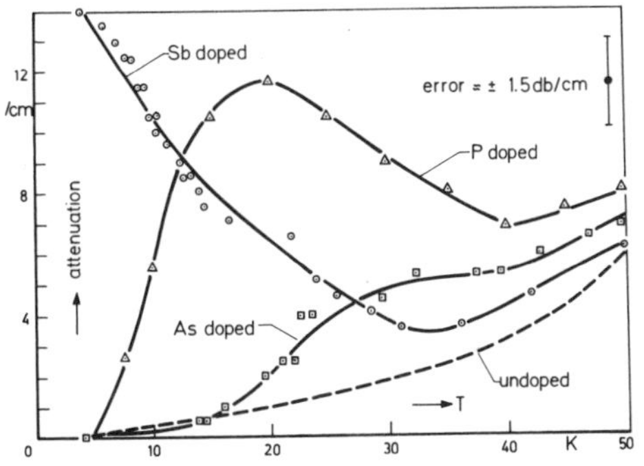

Fig.6.2 Attenuation of a transverse acoustic wave propagating in the <100> direction of doped and undoped germanium, as a function of temperature. The impurity density is about $3x10^{15}$/ cm^3 and the acoustic frequency is 9 GHz (after M. Pomerantz, ref.9).

positively charged particle (hole) is scattered differently than a negatively charged electron.

It may be surprising to find hydrogenic neutral impurities behave so similarly. In respect to effects other than carrier scattering this may be quite different.

E.g. Fig.6.2 shows [9] the attenuation of an ultrasonic wave of frequency 9 GHz in P doped, As doped, and Sb doped Ge, as a function of temperature. The attenuation at temperatures below 30 K is caused by the immobile electrons bound to the impurities. The dashed curve indicates the attenuation of undoped Ge. The different behavior of the impurities has been ascribed to the fact that the shear strain induced by the acoustic wave acts differently on singlet and triplet states of the electron bound to the impurity (see Chap.3b). The maximum attenuation has been found theoretically to occur at a temperature which is the ratio of the energy separation of the two states and 1.5 k_B. The known values of the energy separation divided by k_B are 3.7 K for Sb, 33 K for P, and 49 K for As. For As, the increasing lattice attenuation masks the decreasing impurity attenuation, and no maximum is observed therefore.

6b. Elastic Scattering Processes

While the calculation of the scattering cross section by a superposition of partial waves mentioned above results in series expansions which converge rapidly for $ka \ll 1$, the case of $ka \gg 1$ requires the method of Born's approximation. This method shall next be applied to the problem of electron scattering by ionized impurity atoms.

[9] M. Pomerantz, Proc. IEEE 53 (1965) 1438.

The scattering process is considered to be a small "perturbation" of the electron wave by the potential V(r) of the ionized impurity atom. Denoting by V the crystal volume, the ψ-function of the incoming electron having a wave vector $\vec{k}$,

$$\psi_n = V^{-1/2} \exp [i(\vec{k}_n \vec{r}) \tag{6b.1}$$

is the solution of the time-independent Schrödinger equation

$$H\psi_n = \hbar\omega_n \psi_n \,; \quad n = 0, 1, 2, 3, \cdots \tag{6b.2}$$

where H is the hamiltonian and $\hbar\omega_n$ are the eigenvalues of the unperturbed problem (i.e. no ionized impurity present) and the normalization of the ψ_n is given by

$$\int |\psi_n|^2 d^3r = \frac{1}{V} \int d^3r = 1 \tag{6b.3}$$

For the scattering process we have to solve the time-dependent Schrödinger equation of the perturbation problem

$$\{H + |e|\, V(r)\}\, \psi = i\hbar \partial \psi / \partial t \tag{6b.4}$$

where ψ is expanded in a series of the ψ_n

$$\psi = \sum_n a_n \psi_n \exp(-i\,\omega_n t) \tag{6b.5}$$

with unknown coefficients $a_n = a_n(t)$. Eq.(6b.4) yields

$$\sum_n a_n |e|\, V(r)\, \psi_n \exp(-i\omega_n t) = i\hbar \sum_n (da_n/dt)\, \psi_n \exp(-i\omega_n t) \tag{6b.6}$$

If this is multiplied by $\psi_m^* \exp(i\omega_m t)$, where m is an integer, and integrated over the crystal volume we obtain

$$\sum_n a_n H_{mn} \exp(i\omega_{mn} t) = i\hbar da_m/dt \tag{6b.7}$$

since the integral

$$\int \psi_m^* \psi_n d^3r = \frac{1}{V} \int \exp\{i(\vec{k}_n - \vec{k}_m)\vec{r}\} d^3r \tag{6b.8}$$

vanishes for $m \neq n$ and equals unity for $m = n$ due to the periodic boundary condition for the Schrödinger equation; H_{mn} and ω_{mn} are given by

$$H_{mn} = \int \psi_m^* |e|\, V(r)\, \psi_n d^3r \tag{6b.9}$$

and

$$\omega_{mn} = \omega_m - \omega_n \tag{6b.10}$$

The integration of Eq.(6b.7) yields for the coefficient $a_m = a_m(t)$

$$a_m(t) = -\frac{i}{\hbar} \sum_n H_{mn} \int_0^t a_n \exp(i\omega_{mn} t)\, dt \tag{6b.11}$$

where we have assumed V(r) and consequently H_{mn} to be time-independent neglecting the thermal motion of the ionized impurity.

The incoming electron is considered to be in an initial state k where $a_k = 1$ and all other a's vanish, and denoting the final state by k' we obtain from Eq.(6b.11)

$$|a_{k'}(t)|^2 = \hbar^2 |H_{k'k}|^2 t^2 \sin^2(\omega_{k'k} t/2)/(\omega_{k'k} t/2)^2 \tag{6b.12}$$

For times $t \to \infty$ (which are long enough that the scattering process has been completed) the function $\sin^2(\omega_{k'k} t/2)/(\omega_{k'k} t/2)^2$ has the properties of a δ-function: $2\pi\delta(\omega_{k'k} t) = (2\pi\hbar/t)\,\delta(\hbar\omega_{k'k})$. The transition probability S per unit time is then given by ("Golden Rule No. 2", [1])

$$S(\vec{k}, \vec{k}') = |a_{k'}(t)|^2/t = (2\pi/\hbar)|H_{k'k}|^2 \,\delta\,\{\epsilon(\vec{k}') - \epsilon(\vec{k})\} \tag{6b.13}$$

For transitions into or within a band we have to consider not only one single final state k' but a group of possible final states. Therefore we have to multiply S (1) by the density of states $g(\epsilon)d\epsilon$ given by Eq.(3a.28) (however, neglecting the factor of 2 since the electron spin is not changed by the scattering process), (2) by the probability $1 - f(\vec{k}')$ that the final state is not yet occupied, and (3) by the probability $f(\vec{k})$ that the initial state is occupied, and we have to integrate over phase space. This yields for the decrease of $f(\vec{k})$ with time t by the scattering process ("collision"):

$$\left(-\frac{\partial f(k)}{\partial t}\right)_{\text{coll, } k\to k'} = \frac{V}{(2\pi)^3} \int d^3k'\, S(\vec{k}, \vec{k}') f(\vec{k}) \{1 - f(\vec{k}')\} \tag{6b.14}$$

We still have to take into account scattering processes which go in the reverse direction:

$$\left(-\frac{\partial f(\vec{k})}{\partial t}\right)_{\text{coll, } k'\to k} = -\frac{V}{(2\pi)^3} \int d^3k'\, S(\vec{k}', \vec{k}) f(\vec{k}') \{1 - f(\vec{k})\} \tag{6b.15}$$

The total scattering rate consists of the sum of both contributions.

$$(-\partial f/\partial t)_{\text{coll}} = (-\partial f/\partial t)_{\text{coll, } k\to k'} + (-\partial f/\partial t)_{\text{coll, } k'\to k} \tag{6b.16}$$

For a non-degenerate semiconductor where the final states to a good approximation can be assumed unoccupied we may cancel the factors $1 - f(\vec{k}')$ and $1 - f(\vec{k})$ since they are approximately unity. For the equilibrium distribution $f_0(\vec{k})$ the time derivative $\partial f_0/\partial t = 0$. According to the principle of detailed balance [2]

$$S(\vec{k}, \vec{k}') f_0(\vec{k}) = S(\vec{k}', \vec{k}) f_0(\vec{k}') \tag{6b.17}$$

Hence, in Eq.(6b.15) we can eliminate $S(\vec{k}', \vec{k})$ and obtain for Eq.(6b.16)

$$(-\partial f/\partial t)_{\text{coll}} = V(2\pi)^{-3} \int d^3k'\, S(\vec{k}, \vec{k}') \{f(\vec{k}) - f(\vec{k}') f_0(\vec{k})/f_0(\vec{k}')\} \tag{6b.18}$$

For the simple model of band structure ($\epsilon = \hbar^2 k^2/2m$; $\vec{\nabla}_k \epsilon = (\hbar^2/m)\vec{k}$) we obtain from Eq.(4c.4)

$$f(\vec{k}) = f_0(\vec{k}) - (\hbar/m)(\partial f_0/\partial k) \cdot (\vec{k}\vec{G}) \tag{6b.19}$$

[1] E. Fermi: Nuclear Physics, p.142. Chicago: Univ. Chicago Press. 1950; "Golden Rule No. 1": l.c., p.148.

[2] See e.g. R. B. Adler, A. C. Smith, and R. L. Longini: Introduction to Semiconductor Physics, Sec. 1.5.3 and 3.3. New York: J. Wiley and Sons. 1964.

Introducing polar coordinates relative to the $\vec{k}$-direction we have

$$\left.\begin{aligned} d^3k' &= k'^2\,dk'\sin\theta\,d\theta\,d\varphi \\ (\vec{k}\vec{G}) &= kG\cos\vartheta \\ (\vec{k}'\vec{G}) &= k'G(\cos\vartheta\cos\theta + \sin\vartheta\sin\theta\cos\varphi) \end{aligned}\right\} \quad (6b.20)$$

Since ionized impurity scattering is highly elastic, $k = k'$ and since the equilibrium distribution $f_o(\vec{k})$ does not depend on the direction of $\vec{k}$, the ratio $f_o(\vec{k})/f_o(\vec{k}') = 1$ for elastic processes. We then find for the factor in braces in Eq.(6b.18):

$$f(\vec{k}) - f(\vec{k}') = -\frac{\hbar}{m}(\partial f_o/\partial k)kG[\cos\vartheta(1-\cos\theta) - \sin\vartheta\sin\theta\cos\varphi] \quad (6b.21)$$

$S(\vec{k}, \vec{k}')$ is independent of φ because of spherical symmetry. Therefore the integration in Eq.(6b.18) eliminates the last term in Eq.(6b.21). From Eqs.(6b.18), (6b.19), and (6b.21) we obtain

$$(-\partial f/\partial t)_{coll} = [f(\vec{k}) - f_o(\vec{k})]/\tau_m \quad (6b.22)$$

where

$$1/\tau_m = V(2\pi)^{-2} \int k'^2\,dk' \int_0^{\pi} S(\vec{k}, \vec{k}')(1-\cos\theta)\sin\theta\,d\theta \quad (6b.23)$$

is the inverse momentum relaxation time. The integration over k' is done in the Brillouin zone. A comparison with Eq.(6a.5) reveals that the differential cross section, $\sigma(\theta)$, is given by

$$\sigma(\theta) = (Vm)^2(2\pi\hbar)^{-3}k^{-1}\int S(\vec{k}, \vec{k}')\,k'd\epsilon' \quad (6b.24)$$

where one scattering center in the crystal volume ($N = 1/V$) has been assumed, the electron velocity v has been replaced by $\hbar k/m$, and $\epsilon' = \hbar^2 k^2/2m$. Due to the δ-function for the elastic scattering process, which according to Eq.(6b.13) is contained in S, the integral is easily solved:

$$\sigma(\theta) = \left\{\frac{Vm}{2\pi\hbar^2}\,|H_{k'k}|\right\}^2 \quad (6b.25)$$

6c. Ionized Impurity Scattering

Let us consider as a scattering center a singly ionized impurity atom of charge Ze fixed somewhere inside the crystal. In the classical picture the electron drift orbit is a hyperbola with the ion in one of its focal points depending on the sign of the electronic charge as shown in Fig.6.3. The distance p between the ion and the asymptote is called the "impact parameter". We introduce for convenience the distance

$$K = Ze^2/(4\pi\kappa\kappa_o mv^2) \quad (6c.1)$$

for which the potential energy equals twice the kinetic energy: κ is the relative dielectric constant and κ_o the permittivity of free space. The well-known Rutherford relation between impact parameter and scattering angle is given by

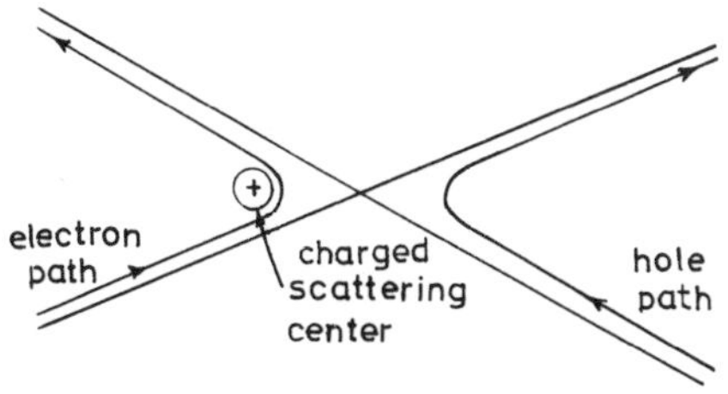

Fig.6.3 Coulomb scattering of an electron and a hole by a positive ion (after W. C. Dunlap, Jr.: An Introduction to Semiconductors. New York: J. Wiley and Sons. 1957.).

$$p = K \cot(\theta/2) \tag{6c.2}$$

Carriers deflected through an angle between θ and $\theta + d\theta$ into a solid angle $d\Omega$ have an impact.parameter of value between p and p + dp and therefore pass through a ring shaped area $2\pi p\,|dp|$ centered around the ion. The differential cross section is then obtained from

$$\sigma(\theta)\,d\Omega = 2\pi p\,|dp| = 2\pi p\,\frac{K d\theta/2}{\sin^2\theta/2} = 2\pi K^2\,\frac{\cot(\theta/2)\,d\theta/2}{\sin^2\theta/2} \tag{6c.3}$$

and since $d\Omega = 2\pi \sin\theta\, d\theta = 8\pi \sin^2(\theta/2)\cot(\theta/2)\,d\theta/2$,

$$\sigma(\theta) = \left(\frac{K/2}{\sin^2\theta/2}\right)^2 \tag{6c.4}$$

This differential cross section has thus been obtained from classical mechanics. It is based on the Coulomb potential of the impurity which has been assumed to extend to infinity. The calculation of τ_m from Eq.(6c.4) runs into difficulties: The integral has no finite value if it begins at zero scattering angle since $\sigma(0)=\infty$.

In practice the problem is solved by the fact that the Coulomb potential is not quite correct and a Yukawa potential

$$V(r) = -(Z|e|/4\pi\kappa\kappa_0 r)\exp(-r/L_D) \tag{6c.5}$$

is more adequate. The idea behind the cut-off distance L_D is that the electrostatic field of the individual ionized impurity is screened by the surrounding carrier gas. At high impurity concentrations the ionic space charge will also contribute to screening; we shall consider this contribution later on.

In the vicinity of an ionized impurity the density of carriers, n(r), will be different from the average carrier density, n:

$$n(r) = n\cdot\exp[-|e|V(r)/k_BT] \approx n[1-|e|V(r)/k_BT] \tag{6c.6}$$

The approximation is valid for a small screening effect. Solving Poisson's equation for spherical symmetry,

$$d^2[rV(r)]/dr^2 = -r|e|[n(r)-n]/\kappa\kappa_0 \tag{6c.7}$$

with n(r) given by Eq.(6c.6) and assuming that for $r \ll L_D$ the potential V(r) is given by the Coulomb potential, yields Eq.(6c.5) with L_D being the Debye length given by Eq.(5b.24).

The quantum mechanical calculation of the cross section based on the Yukawa potential has an analytical result, in contrast to a calculation based on classical mechanics. It is obvious from Eq.(6b.25) that the first step then will be the calculation of the hamiltonian matrix element, $H_{k'k}$, from Eq.(6b.9):

$$H_{k'k} = \frac{|e|}{V} \int V(r) \exp\{i(\vec{k}-\vec{k}')\vec{r}\} d^3 r \tag{6c.8}$$

For the evaluation of the integral we introduce

$$c = |\vec{k}-\vec{k}'| r \tag{6c.9}$$

$$\cos\varphi = (\vec{k}-\vec{k}')\vec{r}/c \tag{6c.10}$$

and

$$z = c \cdot \cos\varphi ; \quad -c \leqslant z \leqslant c \tag{6c.11}$$

Hence, the integral becomes

$$\int_0^\infty V(r)\, 2\pi r^2\, dr/c \int_{-c}^{c} \exp(iz)\, dz = 4\pi \int_0^\infty V(r) r^2 \,[\sin(c)/c]\, dr \tag{6c.12}$$

From Eq.(6c.5) we now obtain for the matrix element

$$\begin{aligned} H_{k'k} &= -\frac{Ze^2}{V\kappa\kappa_o |\vec{k}-\vec{k}'|} \int_0^\infty \exp(-r/L_D) \sin(|\vec{k}-\vec{k}'| r) dr = \\ &= -\frac{Ze^2}{V\kappa\kappa_o} \cdot \frac{1}{|\vec{k}-\vec{k}'|^2 + L_D^{-2}} \approx -\frac{Ze^2}{V\kappa\kappa_o 4k^2} \cdot \frac{1}{\sin^2(\theta/2) + (2kL_D)^{-2}} \end{aligned} \tag{6c.13}$$

where V is the crystal volume and the approximation is valid for the elastic scattering process considered here where $|\vec{k}| \approx |\vec{k}'|$ and

$$|\vec{k}-\vec{k}'| \approx 2k \sin(\theta/2) \tag{6c.14}$$

Eq.(6b.25) yields for the differential cross section

$$\sigma(\theta) = \left(\frac{K/2}{\sin^2(\theta/2) + \beta^{-2}}\right)^2 \tag{6c.15}$$

where

$$\beta = 2k L_D \tag{6c.16}$$

has been introduced and the distance K given by Eq.(6c.1). A comparison of the Eqs.(6c.15) and (6c.4) reveals that in contrast to the previous calculation $\sigma(\theta)$ remains finite even for a zero scattering angle.

The calculation of τ_m from Eq.(6a.5) is straightforward with the result

$$\tau_m = \epsilon^{3/2} N_I^{-1}\, 16\pi \sqrt{2m}\, (\kappa\kappa_o/Ze^2)^2\, [\ln(1+\beta^2) - \beta^2/(1+\beta^2)]^{-1} \tag{6c.17}$$

where N_I is the total concentration of ionized impurities in the crystal. Except for very low values of the carrier velocity $v \propto k \propto \beta$, the term in brackets is nearly constant and τ_m can be said to obey a power law

$$\tau_m \propto \epsilon^{3/2} \tag{6c.18}$$

with an exponent of +3/2.

For the averaging procedure, assuming a non-degenerate electron gas,

$$<\tau_m> = (4/3\sqrt{\pi}) \int_0^\infty \tau_m (\epsilon/k_BT)^{3/2} \exp(-\epsilon/k_BT) d\epsilon/k_BT \tag{6c.19}$$

we replace ϵ in β by that value for which the integrand $(\epsilon/k_BT)^{3/2}\exp(-\epsilon/k_BT)$ is a maximum; this is for $\epsilon = 3k_BT$. We then denote β by β_{BH} where B and H are the initials of the names of Brooks and Herring [1] to whom this calculation is due.

$$\beta_{BH} = 2\frac{m}{\hbar}\left(\frac{2}{m} \cdot 3k_BT\right)^{1/2} L_D \tag{6c.20}$$

or

$$\beta_{BH} = \left(\frac{\kappa}{16}\right)^{1/2} \cdot \frac{T}{100\,K} \cdot \left(\frac{m}{m_o}\right)^{1/2} \left(\frac{2.08 \times 10^{18}\,cm^{-3}}{n}\right)^{1/2} \tag{6c.21}$$

having a numerical value of e.g. 1 for n-type Ge ($\kappa = 16$; $m/m_o = 0.12$) with $n = 2.5 \times 10^{17}\,cm^{-3}$ at 100 K. The Debye length is then 55 Å while the average nearest distance between two ionized impurities is 159 Å assuming no compensation.

The mobility $\mu = (e/m)<\tau_m>$ is then given by

$$\mu = \frac{2^{7/2}(4\pi\kappa\kappa_o)^2 (k_BT)^{3/2}}{\pi^{3/2} Z^2 e^3 m^{1/2} N_I [\ln(1+\beta_{BH}^2) - \beta_{BH}^2/(1+\beta_{BH}^2)]} \tag{6c.22}$$

which in units of $cm^2/Vsec$ is

$$\mu = \frac{3.68 \times 10^{20}\,cm^{-3}}{N_I} \frac{1}{Z^2} \left(\frac{\kappa}{16}\right)^2 \left(\frac{T}{100\,K}\right)^{1.5} \frac{1}{(m/m_o)^{1/2} [\log(1+\beta_{BH}^2) - 0.434\,\beta_{BH}^2/(1+\beta_{BH}^2)]} \tag{6c.23}$$

and the log is to the base 10.

Historically the Brooks-Herring calculation was preceded by a calculation by Conwell and Weisskopf [2] based on Eq.(6c.4) with the requirement of a minimum scattering angle θ_{min}. This angle was obtained from Eq.(6c.2) and a maximum impact parameter p_{max} taken as half the average distance $N_I^{-1/3}$ between adjacent ionized impurity atoms. The calculation arrived at a formula similar to Eq.(6c.22) except that the term in brackets was replaced by $\log(1+\beta_{CW}^2)$ where

$$\beta_{CW} = \frac{1}{Z}\frac{\kappa}{16}\frac{T}{100\,K}\left(\frac{2.35 \times 10^{19}\,cm^{-3}}{N_I}\right)^{1/3} \tag{6c.24}$$

[1] H. Brooks: Advances in Electronics and Electron Physics (L. Marton, ed.), Vol.7, p. 85. New York: Acad. Press Inc. 1955; H. Brooks, Phys. Rev. 83 (1951) 879.

[2] E. Conwell and V. F. Weisskopf, Phys. Rev. 77 (1950) 388.

does not depend on the carrier concentration but on the ionized impurity concentration. Since the BH and CW results are different only in the logarithmic terms they yield about the same values of the mobility for concentrations up to about $10^{18}/cm^3$ beyond which most semiconductors become degenerate and the calculations given here are no longer valid. At constant temperature the mobility depends on N_I as shown in Fig.6.4 where $n = N_I$, $Z = 1$, $\kappa = 16$, and $m = m_0$ have been assumed.

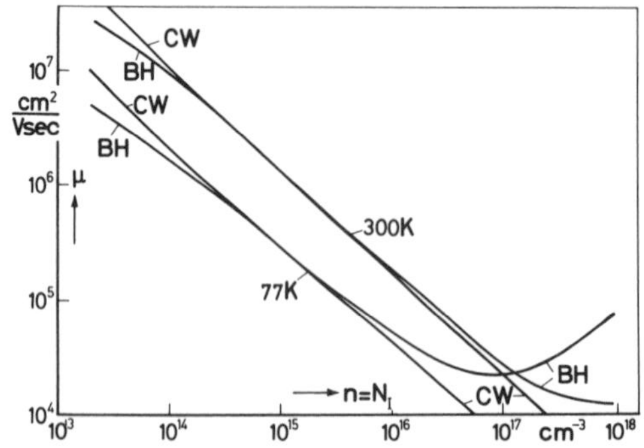

Fig.6.4 Dependence of the mobility on impurity concentration N_I at 77 and 300 K calculated according to Conwell and Weisskopf (CW) and Brooks and Herring (BH) for a hypothetical uncompensated semiconductor with effective mass equal to the free electron mass, dielectric constant 16, and impurity charge Ze = e.

At first sight it may be surprising to find that it is difficult to observe experimentally a $\mu \propto T^{3/2}$ behavior over a wide temperature range. At high temperatures lattice scattering is predominant while at low temperatures carriers freeze out at impurity levels thus neutralizing impurities and making N_I a function of T. A discussion of experimental results on $\mu(T)$ will be given after a treatment in which impurity and lattice scattering are considered simultaneously (Chap.6f).

The problem of shielding of ionized impurities by carriers in the range of carrier freeze-out will now be made plausible in a simplified treatment of the problem. The freeze-out has the effect of an increased density of carriers in the vicinity of an ionized impurity of opposite charge. For simplicity let us discuss only the combination of electrons and donors. The electron density which enters the Brooks-Herring formula will now be denoted by n'. It is larger than the average electron density in the semiconductor, n. The increase n'−n should be proportional to the occupancy N_{Dx}/N_D of the donors where $N_D = N_{D+} + N_{Dx}$ is the total donor concentration, N_{Dx} the concentration of neutral donors, and N_{D+} that of ionized donors. In a partially compensated n-type semiconductor there are also N_{A-} ionized acceptors. Because of charge neutrality

$$N_{A-} + n = N_{D+} + p \approx N_{D+} \tag{6c.25}$$

where p is negligibly small. Hence

$$n' - n \propto (N_D - N_{A^-} - n)/N_D \tag{6c.26}$$

For the case of a near-complete neutralization of the donors the increase $n'-n$ becomes small again since there are only few positive ions which disturb the random distribution of electrons in the crystal. Therefore $n'-n$ should also be proportional to the probability of ionization of donors, $N_{D^+}/N_D \approx (N_{A^-} + n)/N_D$. If we divide $n'-n$ by N_D, we may to a good approximation assume

$$(n'-n)/N_D = \{(N_D - N_{A^-} - n)/N_D\} \cdot (N_{A^-} + n)/N_D \tag{6c.27}$$

which yields Brooks' formula [1]:

$$n' = n + (N_D - N_{A^-} - n)(N_{A^-} + n)/N_D \tag{6c.28}$$

The total concentration of ionized impurities, N_I, is of course

$$N_I = N_{D^+} + N_{A^-} \tag{6c.29}$$

For a calculation of n' one has to take into account the fact that the activation energy of donors, $\Delta\epsilon_D$, depends somewhat on N_I. A more refined treatment of scattering reveals that the repulsive scattering process of a carrier at an ionized impurity of the same sign has a cross section different from that of the attractive scattering process where the carrier and ion are oppositely charged. For a discussion on the validity of Born's approximation see e.g. Schiff [3] and Blatt [4].

6d. Acoustic Deformation Potential Scattering of Thermal Carriers

When an acoustic wave propagates in a crystal lattice the atoms oscillate about their equilibrium positions. For small amplitudes A_s this oscillation is harmonic and for an angular frequency ω_s and wave vector $\vec{q}_s$ (subscript s for "sound") can be represented by

$$\delta\vec{r} = \vec{A}_s \exp[\pm i\{(\vec{q}_s\vec{r}) - \omega_s t\}] \tag{6d.1}$$

At present we shall not discuss the oscillation spectrum of a crystal but consider only long-wavelength acoustic waves where the sound velocity $u_s = \omega_s/q_s$ is a constant. The wavelength $2\pi/q_s$ is much longer than the interatomic distance and the crystal can be treated as a continuous medium. The difference in displacement between two adjacent atoms (average distance a) can be approximated by

[3] L. I. Schiff: Quantum Mechanics, 3rd ed., p. 325. New York: McGraw-Hill. 1968.

[4] F. J. Blatt, J. Phys. Chem. Solids 1 (1957) 262; F. J. Blatt: Solid State Physics (F. Seitz and D. Turnbull, eds.) Vol. 4, p. 199. New York: Acad. Press. 1957. See also D. Long, C. D. Motchenbacher, and J. Myers, J. Appl. Phys. 30 (1959) 353.

$$|\delta\vec{r}(a)-\delta\vec{r}(o)| = (\vec{\nabla}_r\delta\vec{r})a \qquad (6d.2)$$

where according to Eq.(6d.1) the periodic dilation $(\vec{\nabla}_r\delta\vec{r})$ is given by

$$(\vec{\nabla}_r\delta\vec{r}) = \pm i(\vec{q}_s\,\delta\vec{r}) \qquad (6d.3)$$

Since $\vec{q}_s$ is a vector in the direction of wave propagation and the product $(\vec{q}_s\delta\vec{r})$ vanishes for $\delta\vec{r}$ transverse to this direction, we shall consider here only longitudinal waves and use a subscript l instead of s.

$$\delta\vec{r} = \vec{A}_l \exp[\pm i\{(\vec{q}_l\vec{r})-\omega_l t\}] \qquad (6d.4)$$

The scattering of conduction electrons by acoustic waves requires the theorem of the deformation potential put forward by Bardeen and Shockley in 1950 [1]. For a principle treatment consider the variation of the valence band edge with the lattice constant calculated e.g. for diamond (Fig.2.9c). This variation can be taken as linear for a small change in the lattice spacing as it occurs in an acoustic wave. The change in energy of a hole at a transition from one atom to an adjacent atom, $\delta\epsilon_h$, is therefore proportional to $|\delta\vec{r}(a)-\delta\vec{r}(o)|$ or, according to Eq.(6d.2), proportional to $(\vec{\nabla}_r\delta\vec{r})$ in an acoustic wave:

$$\delta\epsilon_h = \epsilon_{ac,v}(\vec{\nabla}_r\delta\vec{r}) \qquad (6d.5)$$

where $\epsilon_{ac,v}$ is a factor of proportionality denoted as the "deformation potential constant" of the valence band. A similar relation applies for the electrons at the conduction band edge:

$$\delta\epsilon_e = \epsilon_{ac,c}(\vec{\nabla}_r\delta\vec{r}) \qquad (6d.6)$$

where $\epsilon_{ac,c}$ is the corresponding constant of the conduction band. Since the energies of electrons and holes are counted positive in opposite directions, Fig.2.9c shows that both constants have equal sign but in general different magnitudes.

Bardeen and Shockley proved that for a perturbation treatment of the interaction between the electron and the acoustic wave it is correct to take $\delta\epsilon = \epsilon_{ac}(\vec{\nabla}_r\delta\vec{r})$ for the perturbing potential (appendix B of ref.1). Since $\delta\epsilon \propto \exp\{i(\vec{q}_l\vec{r})\}$ the matrix element, Eq.(6c.8), is given by

$$|H_{k'k}| = \frac{\epsilon_{ac}q_l A_l}{V}\left|\int \exp\{i(\vec{k}-\vec{k}'\pm\vec{q}_l)\vec{r}\}\,d^3r\right| \qquad (6d.7)$$

For the quantum mechanical description it is more adequate to consider acoustic phonons of energy $\hbar\omega_l$ and momentum $\hbar q_l$ than acoustic waves. If the condition of momentum conservation

$$\vec{k}' = \vec{k} \pm \vec{q}_l \qquad (6d.8)$$

is fulfilled where the plus and minus signs refer to phonon absorption and emission, respectively, the integrand in Eq.(6d.7) is independent of r, and the crystal volume V cancels:

$$|H_{k'k}| = \epsilon_{ac}q_l A_l \qquad (6d.9)$$

[1] J. Bardeen and W. Shockley, Phys. Rev. 80 (1950) 72.

Eq.(6d.8) is a special case of a more general condition where the right-hand side contains also a lattice vector of the reciprocal lattice space given by Eq.(2d.2). Because of the periodicity of the crystal lattice and the Laue equation (see text before Eq.(2d.1)) this additional vector would not impair the result Eq.(6d.9). Such scattering processes are known as "Umklapp processes" [2]. Since we are dealing with scattering processes where both $\vec{k}$ and $\vec{k}'$ are relatively small we can neglect the vector of the reciprocal lattice.

For the vibration amplitude A_l in Eq.(6d.9) we now have to find its quantum mechanical equivalent. Since we investigate harmonic oscillations this is the matrix element of the space coordinate for a transition from the Nth vibrational state either to the N−1 state or to the N+1 state corresponding to the absorption or emission of a phonon, respectively:

$$A_l \to \left|\int \psi_{N\pm 1}\, x\, \psi_N\, d^3r\right| = \begin{cases} \{N\hbar/2M\omega_l\}^{1/2} & \text{for } N\to N-1 \\ \{(N+1)\hbar/2M\omega_l\}^{1/2} & \text{for } N\to N+1 \end{cases} \qquad (6d.10)$$

In the Nth state the oscillator energy is $(N+1/2)\hbar\omega_l$ consisting of N phonons. Since the crystal contains very many such oscillators we can replace N by the average number of phonons at a temperature T of the crystal which according to Planck is given by

$$N \to N_q = [\exp(\hbar\omega_l/k_BT) - 1]^{-1} \qquad (6d.11)$$

Taking for V now the volume of a unit cell of the crystal we can replace the oscillator mass M by the product ρV where ρ is the mass density. We finally obtain for the matrix element of the hamiltonian

$$|H_{k\pm q,k}| = \epsilon_{ac}\, q_l\, \{(N_q + 1/2 \mp 1/2)\hbar/2\rho V\omega_l\}^{1/2} \qquad (6d.12)$$

The acoustic phonon energy $\hbar\omega_l$ involved here is small compared with the thermal energy k_BT. Therefore N_q can be approximated by $k_BT/\hbar\omega_l \gg 1$ and since $N_q + 1 \approx N_q$ in this approximation we obtain the same matrix element for phonon absorption and emission:

$$|H_{k',k}| = \epsilon_{ac}\, q_l \{k_BT/2\rho V\omega_l^2\}^{1/2} = \epsilon_{ac}\{k_BT/2Vc_l\}^{1/2} \qquad (6d.13)$$

where the longitudinal elastic constant $c_l = \rho\omega_l^2/q_l^2 = qu_l^2$ has been introduced. (In Chap.7 we shall treat the tensor character of ϵ_{ac}; in brief we notice that for a <100> direction of wave propagation in a cubic lattice $c_l = c_{11}$, for a <110> direction $c_l = (c_{11} + c_{12} + c_{44})/2$, and for a <111> direction $c_l = (c_{11} + 2c_{12} + 4c_{44})/3$ while for other directions the waves are not to be strictly longitudinal having velocities between the extremes at <100> and <111>; c_{11}, c_{12}, and c_{44} are components of the elasticity tensor.)

The matrix element is independent of the electron energy and of the scattering angle. Since it is nearly the same for phonon emission and absorption, we can

[2] R. Peierls, Ann. Phys. (Leipzig) (5) 12 (1932) 154; see also A. H. Wilson: The Theory of Metals, p. 255 and 298. London: Cambridge Univ. Press. 1965.

take care of both processes simply by a factor of 2 in the scattering probability S (Eq.(6b.13)):

$$S \approx \frac{2\pi}{\hbar} |H_{k'k}|^2 [\delta\{\epsilon(\vec{k}')-\epsilon(\vec{k})+\hbar\omega_l\} + \delta\{\epsilon(\vec{k}')-\epsilon(\vec{k})-\hbar\omega_l\}]$$
$$\approx 2\frac{2\pi}{\hbar} |H_{k'k}|^2 \delta\{\epsilon(\vec{k}')-\epsilon(\vec{k})\} \qquad (6d.14)$$

The calculation of the momentum relaxation time according to Eq.(6b.23) yields since $k^2 dk = m^2 v\hbar^{-3} d\epsilon$:

$$1/\tau_m = v/l_{ac} \qquad (6d.15)$$

where the mean free path

$$l_{ac} = \pi\hbar^4 c_l/(m^2 \epsilon_{ac}^2 k_B T) \qquad (6d.16)$$

has been introduced which is independent of the carrier velocity. This proves that for the energy dependence of τ_m

$$\tau_m \propto \epsilon^{-1/2} \qquad (6d.17)$$

a power law with an exponent $-1/2$ is valid.

The mean free path decreases with increasing temperature since at higher temperatures more phonons are excited and therefore more "scattering centers" exist. The dependence on the effective mass arises from the density of states, $k^2 dk$. In Chap.7 for a more complex band structure a density-of-states effective mass will be defined, and it is of course this mass which enters here. For the simple model of band structure ($\epsilon \propto k^2$) considered at present there is only one kind of effective mass.

The calculation of the mobility is straightforward:

$$\mu = \frac{2\sqrt{2\pi}}{3} \frac{e\hbar^4 c_l}{m^{5/2}(k_B)^{3/2}\epsilon_{ac}^2} \qquad (6d.18)$$

which in units of $cm^2/Vsec$ is given by

$$\mu = 3.06\times10^4 \frac{c_l/10^{12}\,dyn\,cm^{-2}}{(m/m_o)^{5/2}(T/100\,K)^{3/2}(\epsilon_{ac}/eV)^2} \propto T^{-3/2} \qquad (6d.19)$$

E.g. for n-type Ge at T = 100 K a mobility of $3\times10^4\, cm^2/Vsec$ is calculated ($c_l = 1.56\times10^{12}\, dyn/cm^2$; $m/m_o = 0.2$; $\epsilon_{ac} = 9.5$ eV). However, at this and higher temperatures the contribution of optical deformation potential scattering cannot entirely be neglected and modifies the temperature dependence of the mobility: $\mu \propto T^{-1.67}$. For a determination of the exponent it is useful to plot μ vs T on log-log paper.

Since the mobility μ is proportional to $m^{-5/2}$, carriers with a small effective mass have a high mobility (e.g. light holes in Ge at low temperatures where acoustic phonon scattering dominates over optical phonon scattering).

6e. Acoustic Deformation Potential Scattering of Hot Carriers

We have so far considered acoustic deformation potential scattering to be essentially an elastic process. It may be interesting to see what the average energy loss per unit time of a carrier to the crystal lattice actually is [1]. According to definition

$$\langle -d\epsilon/dt\rangle_{coll} = \int \epsilon(\vec{k})\,(\partial f/\partial t)_{coll}\, d^3k / \int f d^3k \qquad (6e.1)$$

where $(-\partial f/\partial t)_{coll}$ is given by Eq.(6b.16). By partial integration Eq.(6e.1) can be manipulated into the form

$$\langle -d\epsilon/dt\rangle_{coll} = \int (-d\epsilon/dt)_{coll}\, f(\vec{k}) d^3k / \int f(\vec{k}) d^3k \qquad (6e.2)$$

where

$$(-d\epsilon/dt)_{coll} = V(2\pi)^{-3} \int [\epsilon(\vec{k}) - \epsilon(\vec{k}')]\, S(\vec{k}, \vec{k}')\, [1 - f(\vec{k}')] d^3k' \qquad (6e.3)$$

is obtained from Eq.(6b.16). For the scattering probability $S(\vec{k}, \vec{k}')$ we need the exact value

$$S(\vec{k}, \vec{k}') = \frac{2\pi}{\hbar} \cdot \frac{\epsilon_{ac}^2\, \hbar q}{2\rho V u_l} [N_q\, \delta\{\epsilon(\vec{k}') - \epsilon(\vec{k}) - \hbar u_l q\} + (N_q + 1) \delta\{\epsilon(\vec{k}') - \epsilon(\vec{k}) + \hbar u_l q\}] \qquad (6e.4)$$

rather than the approximate value given by Eq.(6d.14); we have replaced ω_l by $u_l q_l$ and for simplicity omitted the subscript l of q_l. The integration over d^3k' can be replaced by one over $d^3q = -2\pi q^2\, dq\, d(\cos\theta)$. The arguments of the δ-function are for the simple model of band structure

$$\epsilon(\vec{k}+\vec{q}) - \epsilon(\vec{k}) - \hbar u_l q = \frac{\hbar^2}{2m} (2kq \cos\theta + q^2 - 2mu_l q/\hbar) = \frac{\hbar^2 q}{2m} (q - q_\beta) \qquad (6e.5)$$

and

$$\epsilon(\vec{k}-\vec{q}) - \epsilon(\vec{k}) + \hbar u_l q = \frac{\hbar^2}{2m} (-2kq \cos\theta + q^2 + 2mu_l q/\hbar) = \frac{\hbar^2 q}{2m} (q - q_\alpha) \qquad (6e.6)$$

where the constants

$$q_\beta = -2k \cos\theta + 2mu_l/\hbar; \quad q_\alpha = 2k \cos\theta - 2mu_l/\hbar \qquad (6e.7)$$

have been introduced. Let us first integrate over q and afterwards over $\cos\theta$. Due to the δ-function*) the factor $\epsilon(\vec{k}) - \epsilon(\vec{k}')$ in Eq.(6e.3) becomes $+\hbar u_l q$ for the case of phonon absorption and $-\hbar u_l q$ for emission. Since

$$\delta[(\hbar^2 q/2m)(q - q_{\alpha,\beta})] = (2m/\hbar^2 q)\, \delta(q - q_{\alpha,\beta}) \qquad (6e.8)$$

we obtain from Eq.(6e.3)

[1] R. F. Greene, J. Electronics and Control 3 (1957) 387.

*) Some properties of the δ-function have been listed e.g. by L. I. Schiff: Quantum Mechanics, 3rd ed., p. 57. New York: McGraw-Hill. 1968. A useful equation is $\delta\{f(x)\} = \sum_i \delta(x - x_i)/|df/dx|_{x=x_i}$ where $f(x_i) = 0$.

$$(-d\epsilon/dt)_{coll} = -\frac{m\epsilon_{ac}^2}{2\pi\hbar\rho}\int_{-1}^{+1} d(\cos\theta)[q_\beta^3 N_{q_\beta}\{1-f(\vec{k}+\vec{q}_\beta)\} - q_a^3(N_{q_a}+1)\{1-f(\vec{k}+\vec{q}_a)\}] \quad (6e.9)$$

The limits of integration are provided by $q_\beta \geqslant 0$ and $q_a \geqslant 0$. For the first term we replace $d(\cos\theta)$ by $-(1/2k)dq_\beta$, for the second term by $+(1/2k)dq_a$. In the first integral the upper limit of integration would result from $\cos\theta=+1$ to be $q_\beta = -2k + 2mu_l/\hbar \approx -2k$ (the approximation is valid since the electron velocity $\hbar k/m$ even at low temperatures is large compared with the sound velocity u_l). Since $q_\beta \geqslant 0$ we find for the upper limit $q_\beta = 0$. The lower limit (from $\cos\theta = -1$) is denoted by $q_{\beta m}$:

$$q_{\beta m} = 2k + 2mu_l/\hbar \quad (6e.10)$$

In the second integral the upper limit (from $\cos\theta = +1$) is denoted by q_{am}:

$$q_{am} = 2k - 2mu_l/\hbar \quad (6e.11)$$

while the lower limit $-2k-2mu_l/\hbar$ would be negative and therefore is replaced by zero. Hence, Eq.(6e.9) yields

$$(-d\epsilon/dt)_{coll} = \frac{m\epsilon_{ac}^2}{2\pi\hbar\rho 2k}\int_{q_{\beta m}}^{0} dq q^3 N_q[1-f(\vec{k}+\vec{q})] + \int_0^{q_{am}} dq q^3 (N_q+1)[1-f(\vec{k}-\vec{q})]\} \quad (6e.12)$$

This can be manipulated into a more convenient form

$$(-d\epsilon/dt)_{coll} = -\frac{m\epsilon_{ac}^2}{4\pi\rho\hbar k}\left\{\int_{q_{am}}^{q_{\beta m}} dq q^3 N_q[1-f(\vec{k}+\vec{q})] + \right.$$
$$\left. + \int_0^{q_{am}} dq q^3 [N_q\{f(\vec{k}-\vec{q}) - f(\vec{k}+\vec{q})\} - \{1-f(\vec{k}-\vec{q})\}]\right\} \quad (6e.13)$$

For calculating the difference $f(\vec{k}-\vec{q})-f(\vec{k}+\vec{q})$ we expand

$$f(\epsilon \pm \hbar u_l q) \approx f(\epsilon) \pm \hbar u_l q \cdot df/d\epsilon \quad (6e.14)$$

As in Chap.4m, let us assume for $f(\epsilon)$ a Fermi-Dirac distribution function with an electron temperature T_e:

$$f(\epsilon) = \{1 + \exp[(\epsilon - \zeta)/k_B T_e]\}^{-1} \quad (6e.15)$$

where

$$df/d\epsilon = (f-1)/k_B T_e \quad (6e.16)$$

is valid. Except for the difference $f(\vec{k}-\vec{q})-f(\vec{k}+\vec{q})$, the functions $f(\vec{k} \pm \vec{q})$ may be approximated by $f(\epsilon)$; as before N_q is approximated by $k_B T/\hbar u_l q$:

$$(-d\epsilon/dt)_{coll} = -\frac{m\epsilon_{ac}^2}{4\pi\rho\hbar k}\left\{\frac{k_B T}{\hbar u_l}(1-f)\int_{q_{am}}^{q_{\beta m}} q^2 dq + \int_0^{q_{am}} q^3 dq \left[\frac{2T}{T_e} f(1-f) - (1-f)\right]\right\} \quad (6e.17)$$

In the approximation valid for $\hbar k/m \gg u_l$, the value of the first integral is $2^4 m u_l k^2/\hbar$ while in the second integral the upper limit can be replaced by $2k$:

$$(-d\epsilon/dt)_{coll} = (m\epsilon_{ac}^2 k^3/\pi\rho\hbar)\,(f-1)\,[(4mk_BT/\hbar^2k^2) + (2T/T_e)f - 1] \qquad (6e.18)$$

The averaging procedure according to Eq.(6e.2) yields by partial integration

$$<-d\epsilon/dt>_{coll} = \frac{2m\epsilon_{ac}^2}{\pi^{3/2}\hbar\rho}\left(\frac{2mk_BT_e}{\hbar^2}\right)^{3/2} 2\,\frac{T_e-T}{T_e}\,\frac{F_1(\eta)}{F_{1/2}(\eta)} \qquad (6e.19)$$

where the Fermi integrals are given by Eq.(3a.32) and $\eta = \zeta/k_BT_e$ is the reduced Fermi energy.

This expression is simplified

$$<-d\epsilon/dt>_{coll} = 4mu_l^2 <\tau_m^{-1}> (T_e-T)/T \qquad (6e.20)$$

by introducing the average reciprocal momentum relaxation time

$$<\tau_m^{-1}> = <l_{ac}^{-1}(2\epsilon/m)^{1/2}> = \frac{2m^2\epsilon_{ac}^2 k_BT}{\pi^{3/2}\hbar^4\rho u_l^2}\left(\frac{2k_BT_e}{m}\right)^{1/2} F_1(\eta)/F_{1/2}(\eta) \qquad (6e.21)$$

Eq.(6e.20) is valid for both degeneracy and nondegeneracy. For the latter case $<\tau_m^{-1}>$ is given by $(2/l_{ac})(2k_BT_e/\pi m)^{1/2} = 8/(3\pi<\tau_m>)$. Therefore

$$\mu = \frac{8}{3\pi}\cdot\frac{e}{m<\tau_m^{-1}>} \qquad (6e.22)$$

With the energy balance equation, Eq.(4m.5), and $\mu = \mu_o(T/T_e)^{1/2}$ according to Eq.(6d.17), we can solve Eq.(6e.20) for T_e/T:

$$T_e/T = \frac{1}{2}\left(1 + \sqrt{1 + \frac{3\pi}{8}\cdot\left(\frac{\mu_oE}{u_l}\right)^2}\right) \qquad (6e.23)$$

For warm carriers ($T_e - T \ll T$) the electron temperature increases with the square of E:

$$T_e/T = 1 + (3\pi\mu_o^2/32u_l^2)E^2 \qquad (6e.24)$$

while for hot carriers ($T_e \gg T$) the increase is linear:

$$T_e/T = (3\pi/32)^{1/2}\mu_o E/u_l \qquad (6e.25)$$

In Fig.6.5 the T_e vs E relationship is plotted. Eliminating T_e from the Eqs.(6e.23) and (4m.5) the field dependence of the mobility is obtained in the form

$$E^2 = \left(\frac{32}{8\pi}u_l^2/\mu_o^2\right)(\mu_o/\mu)^2\,[(\mu_o/\mu)^2 - 1] \qquad (6e.26)$$

In the warm carrier approximation the coefficient β is found to be

$$\beta = \frac{\mu-\mu_o}{\mu_oE^2} = -\frac{3\pi}{64}(\mu_o/u_l)^2 = -0.147(\mu_o/u_l)^2 \qquad (6e.27)$$

while for hot carriers the ratio μ/μ_o equals [2]

[2] W. Shockley, Bell Syst. Tech. J. 30 (1951) 990.

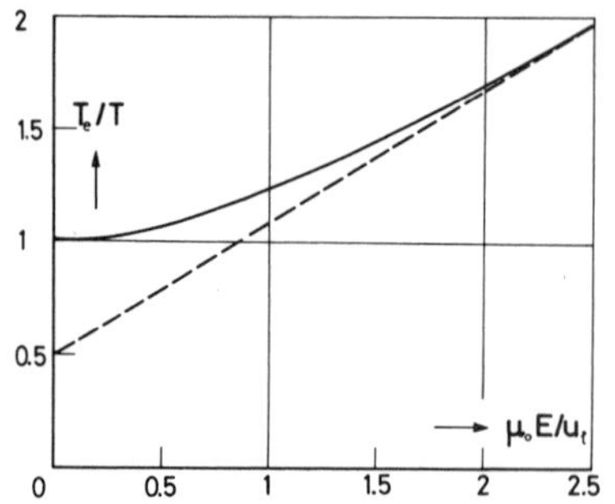

Fig.6.5 Electron temperature, in units of the lattice temperature, as a function of the electric field strength, in units of the ratio of sound velocity to zero-field mobility.

$$\mu/\mu_o = (32/3\pi)^{1/4} (u_l/\mu_o E)^{1/2} = 1.81 (u_l/\mu_o)^{1/2} E^{-1/2} \tag{6e.28}$$

and the drift velocity $v_d = \mu E$ increases proportional to $E^{1/2}$. Experimental results have been shown e.g. in Fig.4.34, where at T = 20 K for field intensities high enough that ionized impurity scattering can be neglected, acoustic phonon scattering should indeed be predominant.

The field intensity, where deviations from Ohm's law become significant, is u_l/μ_o, according to Eq.(6e.28). Typical values for u_l and μ_o are $5\text{x}10^5$ cm/sec and $5\text{x}10^3$ cm^2/Vsec, respectively, which yield a "critical" field intensity of 10^2 V/cm.

Let us finally calculate the energy relaxation time of warm carriers. This is defined by Eq.(4m.10) where the left-hand side for non-equilibrium conditions is $<-d\epsilon/dt>_{coll}$. For $T_e - T \ll T$ we find for the energy difference by a series expansion

$$<\epsilon(T_e)> - <\epsilon(T)> \approx (T_e - T)\frac{\partial}{\partial T_e}\int \epsilon f d^3k \Big/ \int f d^3k = \frac{3}{2} k_B (T_e - T)\,\partial(T_e F_{3/2}/F_{1/2})/\partial T_e \tag{6e.29}$$

For the differentiation we take into account that the carrier concentration $n \propto T_e^{3/2} F_{1/2}$ should be independent of T_e and therefore

$$0 = \frac{dn}{dT} \propto \frac{3}{2} T_e^{1/2} F_{1/2} + T_e^{3/2} \frac{dF_{1/2}}{d\eta} \cdot \frac{d\eta}{dT} \tag{6e.30}$$

which yields

$$T_e\, d\eta/dT_e = -(3/2) F_{1/2}/F_{-1/2} \tag{6e.31}$$

Therefore, the differentiation becomes simply

$$\frac{\partial}{\partial T_e}(T_e F_{3/2}/F_{1/2}) = F_{3/2}/F_{1/2} + (1 - F_{3/2} F_{-1/2}/F_{1/2}^2) T_e d\eta/dT_e = \frac{5}{2}\frac{F_{3/2}}{F_{1/2}} - \frac{3}{2}\frac{F_{1/2}}{F_{-1/2}} \tag{6e.32}$$

From Eqs.(4m.10), (6e.20), and (6e.29) we now obtain for the product $\tau_\epsilon \langle\tau_m^{-1}\rangle$:

$$\tau_\epsilon \langle\tau_m^{-1}\rangle = \frac{3\,k_B T}{8\,m u_l^2}\left(\frac{5}{2}\frac{F_{3/2}}{F_{1/2}} - \frac{3}{2}\frac{F_{1/2}}{F_{-1/2}}\right) \tag{6e.33}$$

which for the case of non-degeneracy is

$$\tau_\epsilon \langle\tau_m^{-1}\rangle = \frac{3\,k_B T}{8\,m u_l^2} \gg 1 \tag{6e.34}$$

This product is roughly the number of collisions necessary for the relaxation of energy. E.g. for n-Ge at 100 K where $u_l = 5.4\times10^5$ cm/sec, $m/m_o = 0.2$, the value of the product is about 10^2. Obviously a large number of collisions is necessary to relax the carrier energy while the momentum distribution is already relaxed after one collision, i.e. after a time of the order of magnitude of $\langle\tau_m\rangle$.

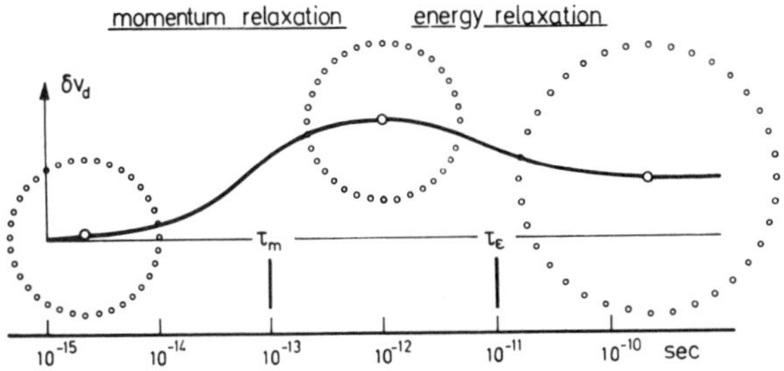

Fig.6.6 Schematic representation of the relaxation behavior of hot carriers (after Schmidt-Tiedemann, ref.3).

In Fig.6.6 a schematic representation of the relaxation behavior of warm and hot carriers is given [3]. Let us assume that an electric field E applied to a semiconductor sample is suddenly increased by a small amount δE such that the drift velocity of the carriers is increased by an amount δv_d. The change of δv_d with time is represented by the curve. The carrier distribution in $\vec{k}$-space at various times is symbolized by dotted circles with diameters which are a measure of the average carrier energy. The initial forward momentum of the carriers caused by the increase in field strength is randomized by collisions in a time of the order of magnitude $\langle\tau_m\rangle$ while the carrier energy does not yet change. The increase in carrier energy requires a larger time τ_ϵ. Since the momentum relaxation time $\langle\tau_m\rangle$ and therefore also the drift velocity decreases with increasing carrier energy, δv_d decreases somewhat during the period of energy relaxation. Now the relaxation is complete and the value of the drift velocity corresponds to the value of the applied field $E + \delta E$.

[3] K. J. Schmidt-Tiedemann: Festkörperprobleme (F. Sauter, ed.), Vol. 1, p. 122. Braunschweig: Vieweg. 1962.

Since deviations from Ohm's law and effects associated with it such as energy relaxation are best measurable at low temperatures where, however, ionized impurity scattering influences the mobility, we will discuss the combined lattice and impurity scattering mechanisms in the following chapter and consider experimental results in the light of this discussion.

6f. Combined Ionized Impurity and Acoustic Deformation Potential Scattering

If there are several scattering mechanisms active the corresponding scattering rates, which to a good approximation are the inverse momentum relaxation times, have to be added. For the case of combined ionized impurity and acoustic deformation potential scattering we have in the relaxation time approximation

$$\frac{1}{\tau_m} = \left(\frac{1}{\tau_m}\right)_{Ion} + \left(\frac{1}{\tau_m}\right)_{ac} \tag{6f.1}$$

Since $(1/\tau_m)_{ac} \propto \epsilon^{1/2}$ and $(1/\tau_m)_{Ion} \propto \epsilon^{-3/2}$, the ratio

$$(\tau_m)_{ac}/(\tau_m)_{Ion} = q^2 (\epsilon/k_B T)^{-2} \tag{6f.2}$$

where q^2 is a factor of proportionality given by

$$q^2 = 6\mu_{ac}/\mu_{Ion} \tag{6f.3}$$

and μ_{ac} and μ_{Ion} are the mobilities given by Eqs.(6d.18) and (6c.22), respectively. Hence τ_m is given by

$$\tau_m = (\tau_m)_{ac} (\epsilon/k_B T)^2 \{q^2 + (\epsilon/k_B T)^2\} \tag{6f.4}$$

For later convenience, let us calculate the mobility for a Maxwell Boltzmann distribution with an electron temperature T_e instead of the usual lattice temperature T. According to Eq.(4b.23) for the present case of non-degeneracy, the mobility is given by

$$\mu = \frac{4}{3\sqrt{\pi}} \frac{e}{m} \int_0^\infty \tau_m \exp(-\epsilon/k_B T_e)\, (\epsilon/k_B T_e)^{3/2}\, d(\epsilon/k_B T_e) \tag{6f.5}$$

With τ_m given by Eq.(6f.4) and $(\tau_m)_{ac} = \tau_0 (\epsilon/k_B T)^{-1/2}$, according to Eq.(6d.17), we find for the mobility

$$\mu = \mu_{ac} \int_0^\infty \frac{(\epsilon/k_B T)^{3/2}}{q^2 + (\epsilon/k_B T)^2} \exp(-\epsilon/k_B T_e)\, (\epsilon/k_B T_e)^{3/2}\, d(\epsilon/k_B T_e) \tag{6f.6}$$

where $\mu_{ac} = (4/3\sqrt{\pi})|e|\tau_0/m$ is the zero-field acoustic mobility given by Eq. (6d.18). We introduce for a parameter $\lambda = T/T_e$ and $q' = \lambda q$. Linearizing the denominator of the integrand yields

$$\mu = \mu_{ac}\lambda^{1/2}\,[1-\frac{q'^2}{2}\left\{\int_0^\infty \frac{\exp(-\epsilon/k_BT_e)\,d(\epsilon/k_BT_e)}{(\epsilon/k_BT_e)+iq'} + \int_0^\infty \frac{\exp(-\epsilon/k_BT_e)\,d(\epsilon/k_BT_e}{(\epsilon/k_BT_e)-iq'}\right\}] \quad (6f.7)$$

where i is the imaginary unit. The integrals are evaluated in the complex plane in terms of

$$-\mathrm{si}(q) = \int_q^\infty \frac{\sin t}{t}\,dt\,; \quad -\mathrm{Ci}(q) = \int_q^\infty \frac{\cos t}{t}\,dt \quad (6f.8)$$

and of the "auxiliary functions"

$$\left.\begin{aligned} f(q) &= \ \ \mathrm{Ci}(q)\,\sin(q) - \mathrm{si}(q)\,\cos(q) = 1/q + dg/dq \\ g(q) &= -\mathrm{Ci}(q)\,\cos(q) - \mathrm{si}(q)\,\sin(q) = -\,df/dq \end{aligned}\right\} \quad (6f.9)$$

with the result

$$\mu = \mu_{ac}\,\lambda^{1/2}\,[1-q'^2 g(q')] \quad (6f.10)$$

For thermal carriers ($\lambda = 1$) we find for the zero-field mobility [1]

$$\mu = \mu_{ac}\,[1-q^2 g(q)\,] \quad (6f.11)$$

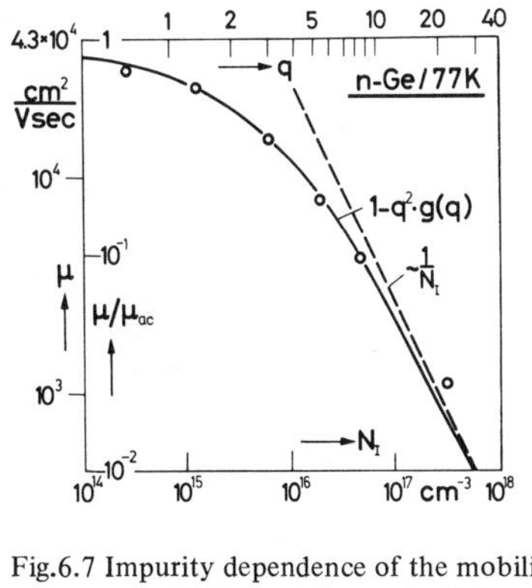

Fig.6.7 Impurity dependence of the mobility for acoustic deformation potential scattering, according to Eq.(6f.11).

The curve in Fig.6.7 represents the ratio μ/μ_{ac} as a function of q. The data points represent Hall mobility data observed in n-type Ge at 77 K for various impurity concentrations; they have been fitted to the curve by assuming $\mu_{ac} = 4.3\text{x}10^4$ cm^2/Vsec and a ratio $N_I/q^2 = 6.25\text{x}10^{14}$ cm^{-3}. For N_I the carrier concentration has been taken which means that impurity compensation has not been taken into account. In addition, optical phonon scattering known to be present in n-Ge at 77 K has been neglected. Therefore the values of the fitting parameters should not be taken too seriously.

In the energy balance ionized impurity scattering may be neglected because it is an elastic process. We shall consider here only the case of warm carriers since for hot carriers ionized impurity scattering (because of $(\tau_m)_{Ion} \propto \epsilon^{+3/2}$) becomes less important. For the warm carrier case the energy balance equation, Eq. (4m.5) with its right-hand side given in Eq.(6e.20), can be manipulated into the form

$$\mu e E^2 = -\,(\lambda-1)\,(32/3\pi)\,e u_l^2/\mu_{ac} \quad (6f.12)$$

If we take for μ its value given by Eq.(6f.11) and take for $-(3\pi/64)\,\mu_{ac}^2/u_l^2$ its

[1] P. P. Debye and E. M. Conwell, Phys. Rev. 93 (1954) 693.

value β_{ac} given by Eq.(6e.27) we obtain

$$[1-q^2 g(q)]E^2 = (\lambda-1)/(2\beta_{ac}) \quad (6f.13)$$

The factor $(\lambda-1)$ may be eliminated from this equation by means of Eq.(4m.7):

$$\beta E^2 = (\lambda-1)\left(\frac{1}{\mu}\cdot\frac{d\mu}{d\lambda}\right)_{\lambda=1} \quad (6f.14)$$

Solving for the ratio β/β_{ac} we find

$$\beta/\beta_{ac} = \{2[1-q^2 g(q)]/\mu(d\mu/d\lambda)_{\lambda=1} = (2/\mu_{ac})\,(d\mu/d\lambda)_{\lambda=1} \quad (6f.15)$$

The derivative of μ is easily calculated from Eq.(6f.9),

$$\frac{d\mu}{d\lambda} = \mu_{ac}\left[\frac{1-q'^2 g(q')}{2\sqrt{\lambda}} + \sqrt{\lambda}\cdot q\,\frac{d}{dq'}\,\{1-q'^2 g(q')\}\right] \quad (6f.16)$$

which for $\lambda = 1$ and $q' = q$ becomes

$$(2/\mu_{ac})\cdot\left(\frac{d\mu}{d\lambda}\right)_{\lambda=1} = 1-q^2 g-2q\,\frac{d}{dq}\,(q^2 g) = 1-5q^2 g-2q^3\,\frac{dg}{dq} \quad (6f.17)$$

From Eq.(6f.9) we finally obtain [2] for β/β_{ac}:

$$\beta/\beta_{ac} = 1 + 2q^2 - 5q^2 g - 2q^3 f \quad (6f.18)$$

For the case of a small density of ionized impurities, $q \ll 1$, we obtain from Eqs.(6f.11) and (6f.18), respectively:

$$\mu/\mu_{ac} \approx 1-q^2 \ln(1/1.781q) \quad (6f.19)$$

and

$$\beta/\beta_{ac} \approx 1-5q^2 \ln(1/1.781q) \quad (6f.20)$$

Both the zero-field mobility and β are reduced by impurity scattering, the latter more than the former.

For the case of a large density of ionized impurities, $q \gg 1$, we find

$$\mu/\mu_{ac} \approx 3!/q^2 - 5!/q^4 + 7!/q^6 - + \cdots \quad (6f.21)$$

and

$$\beta/\beta_{ac} = -3\mathrm{x}3!/q^2 + 7\mathrm{x}5!/q^4 - 11\mathrm{x}7!/q^6 + - \ldots \quad (6f.22)$$

Retaining only the first term in the expansion, μ becomes μ_{Ion} as shown by Eq. (6f.3) while β changes sign and is now inversely proportional to the ionized impurity density $N_I \propto q^2$. A comparison of the last two equations shows that β is more heavily affected by ionized impurity scattering than μ is. μ is reduced to $\frac{1}{2}\mu_{ac}$ by a q value of about 1.7 while β is reduced to $\frac{1}{2}\beta_{ac}$ already by $q = 0.5$.

Fig.6.8 (dashed curve) shows [3] the absolute value of β plotted vs $q^2 \propto N_I$. At large values of N_I the curve approaches the dashed straight line which in the

---

[2] M. S. Sodha, Phys. Rev. 107 (1957) 1266; K. Seeger, Zeitschr. f. Physik 156 (1959) 582.

[3] K. Seeger, Zeitschr. f. Physik 244 (1971) 439.

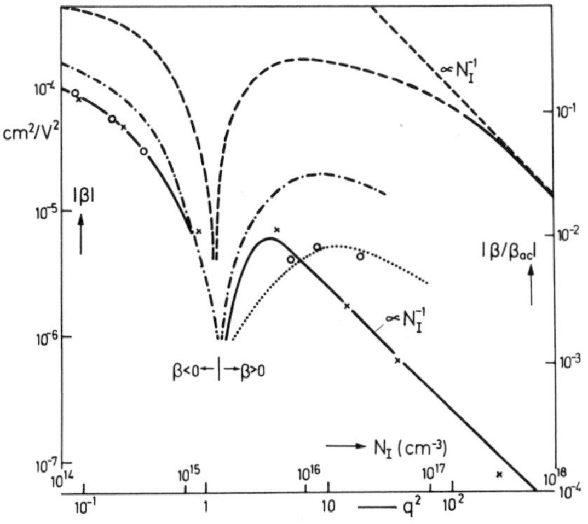

Fig.6.8 Absolute values of β vs ionized impurity concentration in n-Ge. Experimental data by Tschulena in <100> direction (crosses) and Gunn (circles). Calculations: according to Eq.(6f.18) (dashed curve and asymptote N_I^{-1}); by I. Adawi (Phys. Rev. 120 (1960) 118) using a variational method (dash-dotted curve); by I. Adawi using a Maxwell-Boltzmann distribution function (dotted curve). In all of Adawi's calculations acoustic and optical phonon scattering with a ratio of deformation potential constants, $D/\epsilon_{ac} = 0.4\ \omega_0/u_l$, are taken into account (after Seeger, ref.3).

log-log plot indicates a proportionality to N_I^{-1}. Although experimental data obtained on n-type Ge at 77 K by Gunn [4] (circles) and by Tschulena [5] (crosses) are much lower, the latter do show the expected proportionality to N_I^{-1}. Calculations, where optical phonon scattering (Chap.6h) in addition to the mechanisms discussed above, were taken into account agree qualitatively with observations (dash-dotted and dotted curves). In the limit of heavy doping a Maxwell-Boltzmann distribution function seems to be quite a good approximation because of the strong electron-electron interaction (see Chap.6k) while the doping is not yet so high that degeneracy would be important [6]. For this distribution Fig.6.9 shows [3] $\beta/\mu_0\mu_{I,o}$ as a function of temperature for various mechanisms for momentum and energy relaxation in the limit of heavy doping; μ_0 is the zero-field mobility in the pure sample and $\mu_{I,o}$ in the doped sample. E.g. the curve labeled ac/op represents calculations for acoustic mode momentum relaxation and optical mode energy relaxation in the pure sample. The latter is characterized by a

[4] J. B. Gunn, J. Phys. Chem. Solids 8 (1959) 239; see also Fig.4.35 and E. M. Conwell, Phys. Rev. 90 (1953) 769.
[5] G. Tschulena, Acta Phys. Austr. 33 (1971) 42.
[6] For the case of a degenerate semiconductor see ref.5, chap. 61, and G. Tschulena and R. Keil, phys. stat. sol. (b) 49 (1972) 191.

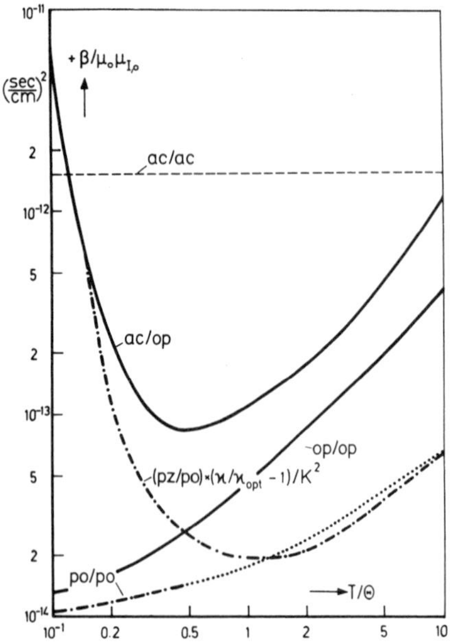

Fig.6.9 β relative to zero-field mobilities of a pure sample, μ_o, and of a sample with predominant ionized impurity scattering $\mu_{I,0}$, vs temperature, T, relative to Debye temperature, Θ, for various momentum and energy relaxation processes. E.g. "ac/op" means acoustic momentum relaxation and optical energy relaxation. The following values for the parameters have been chosen : $u_l = 5.4 \times 10^5$ cm/sec; $m/k_B\Theta = 10^{-14}$ sec^2/cm^2;$D/\epsilon_{ac} = 0.4\ \omega_o/u_l$ (after Seeger, ref. 3).

Debye temperature Θ. At $T/\Theta \approx 0.2$ valid for the present case of Ge at 77 K combined acoustic and optical mode momentum relaxation and optical mode energy relaxation would yield values of $\beta/\mu_o\mu_{I,0}$ well below the "ac/ac" horizontal line and thus yield values for β in agreement with observations. In any case, deviations from Ohm's law in homogeneously doped impure semiconductors are extremely small. It may be of interest to note that in metals such as copper and platinum no deviation from Ohm's law could be found up to field intensities of 2 kV/cm [7].

The case of hot carriers will briefly be discussed qualitatively on the basis of Eq.(6f.10). The influence of impurity scattering on the mobility is given by the term $q'^2 g(q')$ where $q' = q \cdot (T/T_e)$. As the carrier temperature rises to values $T_e \gg T$, q' becomes small even for large impurity densities, $q \gg 1$. Therefore we expect ionized impurity scattering to play no significant role in hot carrier experiments. Of course the field intensity required for carrier heating becomes

[7] H. Heinrich and W. Jantsch, Sol. State Comm. 7 (1969) 377.

larger as the mobility decreases due to impurity scattering, in agreement with the energy balance equation.

With increasing temperature the transition of the mobility from predominant ionized impurity scattering to lattice scattering is shown [8] in Fig.6.10.

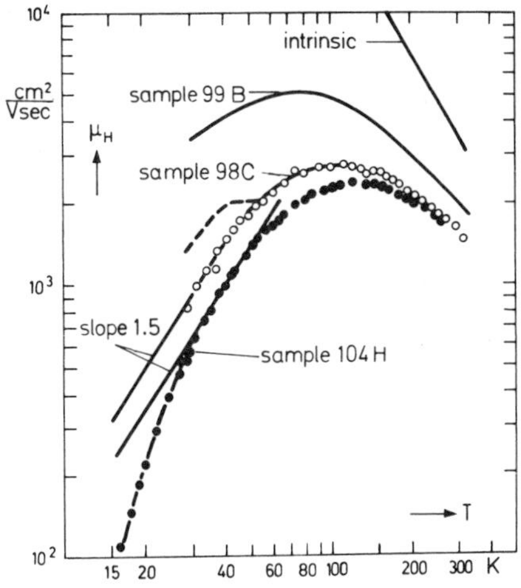

Fig.6.10 Hall mobility vs temperature for two n-type Zn-doped germanium samples (98C and 104H). Mobility data for intrinsic n-type Ge and for an uncompensated sample doped with 5×10^{16} Sb atoms/cm^3 are shown for comparison. For sample 98C, the dashed curve indicates the effect of exposure to light on mobility (after Tyler and Woodbury, ref.8).

The samples contain doubly charged zinc atoms and are very efficient in impurity scattering since the charge Z enters Eq.(6c.22) as Z^2. For samples 98C and 104H the charge of centers is not changing with temperature. The deviation of the observed mobility from the $T^{3/2}$ dependence in sample 104H at $T < 30$ K is attributed to impurity band conduction (Chap. 6n).

A method for obtaining the acceptor and donor concentrations in a semiconductor (actually p-type Si) by conductivity, Hall, and magnetoresistance measurements at 77 K in the range of validity of Ohm's law has been presented by Long et al.[9]. Brown and Bray [10] find that in n-type Ge at impurity concentrations of less than 10^{15}/cm^3 between 30 and 300 K the Brooks-Herring formula describes the ionized impurity scattering very well but overestimates the mobility for higher N_I or lower T. Eagles and Edwards [11] in an extension of calcula-

[8] W. W. Tyler and H. H. Woodbury, Phys. Rev. 102 (1956) 647.
[9] D. Long, C. D. Motchenbacher, and J. Myers, J. Appl. Phys. 30 (1959) 353.
[10] D. M. Brown and R. Bray, Phys. Rev. 127 (1962) 1593.
[11] P. M. Eagles and D. M. Edwards, Phys. Rev. 138 (1965) A 1706.

tions by Herring and Vogt [12] investigated the influence of ionized impurity scattering on the galvanomagnetic effects for the case of the many-valley model of band structure to be discussed in Chap.7.

6g. Piezoelectric Scattering

If a semiconductor crystal consists of dissimilar atoms such as e.g. SiC where the bonds are partly ionic (12 % in SiC) and the unit cell does not contain a center of symmetry (as in the zinc blende lattice or the trigonal lattice) carriers may be scattered by longitudinal acoustic waves due to "piezoelectric scattering" [1, 2]. Due to the oscillations of ions there arises a dipole moment per unit volume known as the "polarization" P. The dielectric displacement, D, given by

$$D = \kappa_o E + P \tag{6g.1}$$

vanishes, since there is no space charge because the spatial displacement of carriers is negligible in comparison to the spatial displacement of the ions. Therefore there is an ac electric field of intensity

$$E = -P/\kappa_o \tag{6g.2}$$

For a propagating wave of finite wavelength there will exist a strain $(\vec{\nabla}_r \delta\vec{r})$ where $\delta\vec{r}$ is the displacement of a lattice atom from its equilibrium position given by Eq.(6d.1). While for $(\vec{\nabla}_r \delta\vec{r}) = 0$ the dielectric constant κ is defined by the relation $D = \kappa\kappa_o E$, the dielectric displacement now contains a term which is proportional to $(\vec{\nabla}_r \delta\vec{r})$:

$$D = \kappa\kappa_o E + e_{pz}(\vec{\nabla}_r \delta\vec{r}) \tag{6g.3}$$

The factor of proportionality, e_{pz}, obtained from piezoelectric measurements is of the order of magnitude 10^{-5} Asec/cm^2; this value may be obtained by assuming a typical displacement of the atomic dimension of 10^{-8} cm for about 10^{22}/cm^3 elementary charges $e = 1.6 \times 10^{-19}$ Asec (one per atom). e_{pz} is called the "piezoelectric constant".

Eqs.(6g.2) and (6g.3) yield for $D = 0$:

$$E = -(e_{pz}/\kappa\kappa_o)\,(\vec{\nabla}_r \delta\vec{r}) \tag{6g.4}$$

In an acoustic wave having a propagation vector $\vec{q}$ the local variation of $(\vec{\nabla}_r \delta\vec{r})$ is given by Eq.(6d.3). Since E is proportional to $(\vec{\nabla}_r \delta\vec{r})$ the potential energy

---

[12] C. Herring and E. Vogt, Phys. Rev. 101 (1956) 944.

Chap.6g.

[1] A. R. Hutson, J. Appl. Phys. 32 Suppl. (1961) 2287.

[2] H. J. G. Meyer and D. Polder, Physica 19 (1953) 255.

$$\delta\epsilon = |e| \int E dr = |e| \frac{E}{q} \tag{6g.5}$$

depends also on $(\vec{\nabla}_r \delta\vec{r})$:

$$\delta\epsilon = (|e| e_{pz}/\kappa\kappa_o q)\,(\vec{\nabla}_r \delta\vec{r}) \tag{6g.6}$$

A comparison of this relation with Eq.(6d.6) for nonpolar acoustic scattering reveals that instead of the acoustic deformation potential constant, ϵ_{ac}, we now have $|e|\, e_{pz}/\kappa\kappa_o\, q$ which in contrast to ϵ_{ac} is not a constant but depends on

$$q = |\vec{k}' - \vec{k}| \approx 2|\vec{k}| \sin(\theta/2) = (2mv/\hbar) \sin(\theta/2) \tag{6g.7}$$

The absolute magnitude of the matrix element $H_{k'k}$ given by

$$|H_{k'k}| = \frac{|e|\, e_{pz}}{\kappa\kappa_o q} \left[\frac{k_B T}{2V c_l}\right]^{1/2} = \left[\frac{e^2 K^2 k_B T}{2V\kappa\kappa_o q^2}\right]^{1/2} \tag{6g.8}$$

is similar to Eq.(6d.12); we have introduced here the dimensionless "electromechanical coupling coefficient", K^2, defined by

$$K^2/(1-K^2) = e_{pz}^2/(\kappa\kappa_o c_l) \tag{6g.9}$$

which is approximately K^2 for $K^2 \ll 1$. E.g. in SiC with $e_{pz} = 10^{-5}$ Asec/cm^2, $\kappa = 10.2$, and $c_l = 1.8\text{x}10^{12}$ dyn cm^{-2} it is $K^2 = 6\text{x}10^{-4}$. For most polar semiconductors K^2 is of the order of magnitude 10^{-3}. The definition of K^2 given here is related to the power stored in a charged condenser which is proportional to the dielectric constant κ. Let us for simplicity denote the strain $(\vec{\nabla}_r \delta\vec{r})$ by S. The relation between tension, strain, and electric field strength in one dimension is given by

$$T = c_l S - e_{pz} E \tag{6g.10}$$

In a tensionfree crystal (T = 0) the strain induced by the electric field is

$$S = (e_{pz}/c_l) E \tag{6g.11}$$

If we eliminate S from Eq.(6g.3) we obtain for the ratio D/E

$$D/E = \kappa\kappa_o + e_{pz}^2/c_l \tag{6g.12}$$

We notice that the work necessary for charging the condenser in this case consists of an electric part $\propto \kappa\kappa_o$ and a mechanical part $\propto e_{pz}^2/c_l$ (If we can apply such a tension that there is no strain we have only the electric part). K^2 is now defined by the ratio of the mechanical work to the total work

$$K^2 = \frac{e_{pz}^2/c_l}{\kappa\kappa_o + e_{pz}^2/c_l} \tag{6g.13}$$

The calculation of the scattering probability from Eq.(6d.14) and of the momentum relaxation time τ_m from Eq.(6b.23) presents no problems:

$$\frac{1}{\tau_m} = \frac{V}{(2\pi)^3} \int 2\frac{2\pi}{\hbar} \cdot \frac{e^2 K^2 k_B T}{2V\kappa\kappa_o q^2}\, \delta\{\epsilon(k')-\epsilon(k)\} k^2 dk (1-\cos\theta) \sin\theta\, d\theta \cdot 2\pi \tag{6g.14}$$

where q^2 is approximately given by $4k^2 \sin^2(\theta/2)$ and $dk = \hbar^{-1}(m/2\epsilon)^{1/2} d\epsilon$. The integration yields for τ_m

$$\tau_m = \frac{2^{3/2} \pi \hbar^2 \kappa \kappa_0}{m^{1/2} e^2 K^2 k_B T} \epsilon^{1/2} \tag{6g.15}$$

We notice that for the energy dependence of τ_m

$$\tau_m \propto \epsilon^{+1/2} \tag{6g.16}$$

a power law with a positive exponent +1/2 is valid. Hence we expect for warm carriers positive values for the coefficient β.

For non-degenerate thermal carriers the mobility obtained from Eq.(6g.15) is given by

$$\mu = \frac{16\sqrt{2\pi}}{3} \cdot \frac{\hbar^2 \kappa \kappa_0}{m^{3/2} e \; K^2 (k_B T)^{1/2}} \propto T^{-1/2} \tag{6g.17}$$

and in units of $cm^2/Vsec$

$$\mu = 2.6 \frac{\kappa}{(m/m_0)^{3/2} K^2 (T/100\,K)^{1/2}} \tag{6g.18}$$

Assuming typical values $\kappa = 10$, $m/m_0 = 0.1$, $K^2 = 10^{-3}$, and $T = 100\,K$ a value of $8.25 \times 10^5\ cm^2/Vsec$ is found which is large compared with the mobility due to nonpolar acoustic scattering; the latter is a competing scattering process in semiconductors with partly ionic bonds.

Therefore it may be interesting to compare the mobility due to piezoelectric scattering (for this purpose denoted as μ_{pz}) with that due to acoustic deformation potential scattering (μ_{ac}) given by Eq.(6d.18). For the ratio μ_{pz}/μ_{ac} we find

$$\frac{\mu_{pz}}{\mu_{ac}} = 0.75 \frac{(\kappa/10)^2 (m/m_0)(T/100\,K)(\epsilon_{ac}/eV)^2}{(e_{pz}/10^{-5}\,Asec\,cm^{-2})^2} \tag{6g.19}$$

At a low enough temperature the mobility in a pure dislocationfree polar semiconductor may be determined by piezoelectric scattering rather than acoustic deformation potential scattering ($\mu_{pz} \ll \mu_{ac}$). A small effective mass m also favors piezoelectric scattering. In a typical case such as $\kappa = 10$; $m/m_0 = 0.1$; $T = 10\,K$; $\epsilon_{ac} = 5\,eV$; $e_{pz} = 10^{-5}\,Asec\,cm^{-2}$, we have $\mu_{pz}/\mu_{ac} \approx 0.2 \ll 1$ and piezoelectric scattering predominates. However, in polar semiconductors with purities available at present, ionized impurity scattering will probably dominate under these conditions. For this reason there are few practical applications for piezoelectric scattering. The energy transfer rate has been calculated by Kogan [3] and will not be discussed here since it has rarely been observed.

---

[3] Sh. M. Kogan, Fiz. Tverd. Tela 4(1962) 2474 (Engl. transl.: Sov. Phys.-Solid State 4(1963) 1813).

6h. The Phonon Spectrum of a Crystal

So far we have considered only long-wavelength acoustic phonons in electron scattering processes. They are characterized by a frequency-independent sound velocity. In crystals with more than one atom per unit cell one must also take into account "optical phonons". In this chapter we will derive the phonon spectrum of a crystal for which the simplified model of a one-dimensional chain will be assumed. In order to avoid surface effects we assume this chain to be ring-shaped and obtain a periodic boundary condition.

The chain may consist of N point-masses M which are connected by springs with a spring constant D. Let us consider mass j at x_j performing a harmonic oscillation; its equation of motion is given by

$$M d^2x_j/dt^2 = -D(x_j - x_{j-1}) - D(x_j - x_{j+1}) \tag{6h.1}$$

where the first term on the right-hand side is the force due to the spring connecting the mass under consideration with the mass at x_{j-1} and the second term is due to the spring on the other side of mass j. A harmonic solution of the differential equation is given by

$$x_j = x_o \exp(iqja - i\omega t) \tag{6h.2}$$

where a is the distance between adjacent masses at rest, ω is the angular frequency of the oscillation, and $q = 2\pi/\lambda$ where λ is the wavelength. Eq.(6h.1) yields for the solution given by Eq.(6h.2):

$$-M\omega^2 = D\{\exp(iqa) - 2 + \exp(-iqa)\} \tag{6h.3}$$

Since $\exp(iqa) + \exp(-iqa) = 2\cos(qa)$ we obtain

$$-M\omega^2/4D = \{\cos(qa) - 1\}/2 = -\sin^2(qa/2) \tag{6h.4}$$

Solving for $\omega > 0$ yields

$$\omega = \sqrt{D/M}\ |\sin(qa/2)| \tag{6h.5}$$

Since the N links of the chain are connected in a circle we have

$$x_{j+N} = x_j \tag{6h.6}$$

which results in

$$\exp(iqNa) = 1 \tag{6h.7}$$

or

$$q = n\,2\pi/Na\,; \quad n = 0, \pm 1, \pm 2, \cdots \tag{6h.8}$$

The difference between one q-value and the next is given by

$$\Delta q = 2\pi/Na \tag{6h.9}$$

which is smaller for a larger number of links.

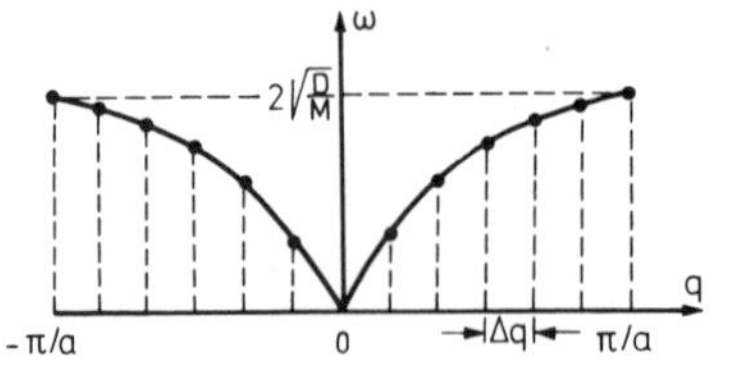

Fig.6.11 Quantized spectrum of oscillations of a circular chain.

Eq.(6h.5) is illustrated by Fig.6.11. In the language of quantum theory $\hbar\omega$ is the energy and $\hbar q$ the momentum of a phonon. The range $-N/2 \leqslant n \leqslant N/2$, which is equivalent to $-\pi/a \leqslant q \leqslant \pi/a$, is called the "first Brillouin zone" in phonon momentum space. For small values of q the sine-function in Eq.(6h.5) may be replaced by its argument. Dividing ω by q yields the sound velocity

$$u_s = \omega/q = a\sqrt{D/M} \tag{6h.10}$$

which is indeed independent of ω ("no dispersion"). However, at large values of q there is of course a dispersion.

Now we consider the case of two atoms per unit cell of length

$$a = r_1 + r_2 \tag{6h.11}$$

where $r_1 \neq r_2$; different spring constants $D_1 \neq D_2$ will also be assumed as illustrated by Fig.6.12. The equations of motion for this case are given by

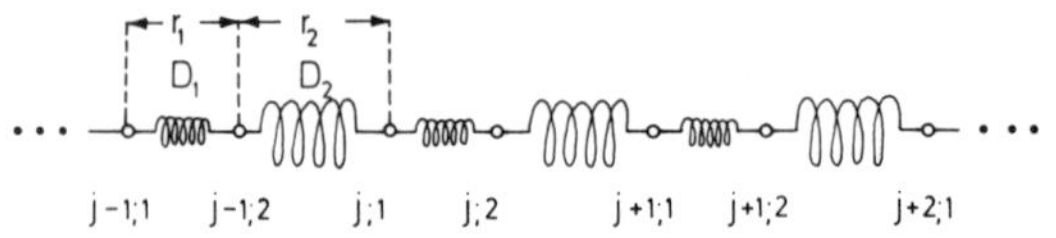

Fig.6.12 Chain with two kinds of springs in alternating succession.

$$\left.\begin{aligned} M d^2x_{j,1}/dt^2 &= -D_1(x_{j,1}-x_{j,2}) - D_2(x_{j,1}-x_{j-1,2}) \\ M d^2x_{j,2}/dt^2 &= -D_1(x_{j,2}-x_{j,1}) - D_2(x_{j,2}-x_{1+1,1}) \end{aligned}\right\} \tag{6h.12}$$

We try the harmonic solution

$$\left.\begin{aligned} x_{j,1} &= X_1 \exp(iqja-i\omega t) \\ x_{j,2} &= X_2 \exp\{iqj(a-r_2)-i\omega t\} \end{aligned}\right\} \tag{6h.13}$$

and for the sake of brevity introduce

$$Y = D_1 \exp(-iqjr_2) + D_2 \exp(-iqa-iqjr_2) \tag{6h.14}$$

Eqs.(6h.12) yield

$$\left.\begin{aligned} (D_1 + D_2 - M\omega^2)X_1 - YX_2 &= 0 \\ -Y^*X_1 + (D_1 + D_2 - M\omega^2)X_2 &= 0 \end{aligned}\right\} \tag{6h.15}$$

where Y^* is the complex conjugate of Y.

This is a linear homogeneous set of equations for X_1 and X_2. A solution exists if the determinant vanishes:

$$(D_1 + D_2 - M\omega^2)^2 - YY^* = 0 \qquad (6h.16)$$

This is solved for $M\omega^2$:

$$M\omega^2 = D_1 + D_2 \pm \sqrt{YY^*} = D_1 + D_2 \pm \sqrt{D_1^2 + D_2^2 + 2D_1D_2\cos(qa)}$$
$$= (D_1 + D_2) \cdot [1 \pm \sqrt{1 - \{4D_1D_2/(D_1 + D_2)^2\}\sin^2(qa)} \qquad (6h.17)$$

The two signs of the square root indicate two branches of the phonon spectrum, an acoustic and an "optical" branch. Temporarily we assume $D_1 = D_2$ which we denote by D:

$$\omega = \sqrt{2D/M}\ \sqrt{1 \pm \cos(qa/2)} \qquad (6h.18)$$

For the minus-sign we simplify

$$\omega = \omega_0\ |\sin(qa/4)| \qquad (6h.19)$$

where we have introduced

$$\omega_0 = 2\sqrt{D/M} \qquad (6h.20)$$

Eq.(6h.19) is similar to Eq.(6h.5) except that the unit cell is now twice as large as before. Eq.(6h.19) represents the acoustic mode.

For the plus-sign we simplify

$$\omega = \omega_0\ |\cos(qa/4)| \qquad (6h.21)$$

For q=0 the angular frequency is ω_0 which is called the "optical phonon frequency". At $q = \pi/a$ the optical mode is degenerate with the acoustic mode, the common frequency being $\omega_0/\sqrt{2}$. For reasons which will be given in Chap.7e this is called the "intervalley phonon frequency". Both branches of the phonon spectrum are shown in Fig.6.13.

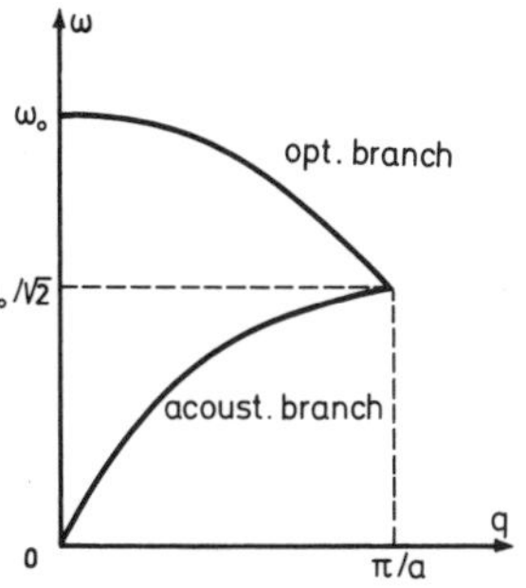

Fig.6.13 Phonon spectrum for a one-dimensional crystal with two equal point-masses per unit cell.

In obtaining Eq.(6h.18) we have assumed $D_1 = D_2$. In general this will not be the case, and in a compound semiconductor the atomic masses will also be different, with the effect that the two branches are not degenerate at the edge of the first Brillouin zone, $q = \pi/a$; there is a gap between the acoustic and the optical branches of the phonon spectrum as shown in Fig.6.14.

If there are more than 2 atoms in a unit cell there is more than one optical branch. In the one-dimensional model considered so far there are only longitudinal oscillations of the atoms. In a three-dimensional crystal there are in addition two transverse oscillations per longitudinal oscillation which are degenerate with each other in a cubic crystal.

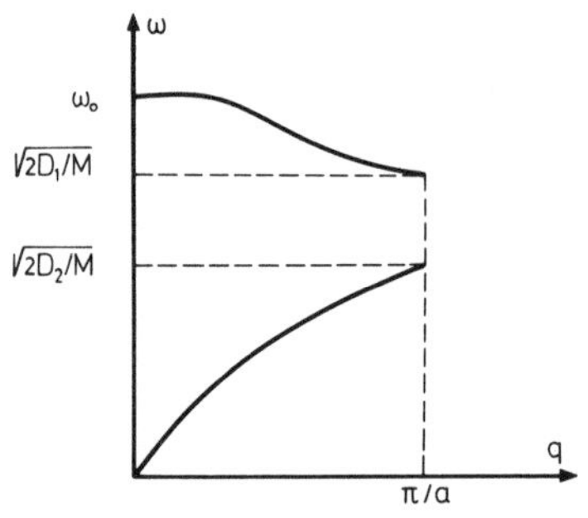

Fig.6.14 Phonon spectrum for a one-dimensional crystal with two equal point-masses per unit cell alternatingly connected by two different springs. (A similar spectrum is obtained when the masses are different.)

The cubic unit cell of the diamond and zinc blende lattices is shown in Fig.2.12. It contains a central atom and 4 atoms on the cube edges. The atoms on the cube edges are each shared by only 4 adjacent cubes because each atom makes only 4 bonds; therefore these atoms have to be weighted by a factor 1/4. The total number of atoms in the unit cell is then given by $1 + 4/4 = 2$.

Besides Raman spectroscopy, which will briefly be discussed in Chap.11g, neutron spectroscopy [1] provides a very satisfactory experimental method for obtaining the phonon spectrum of a crystal. The required high neutron flux is obtained from a nuclear reactor. The flux is filtered through a beryllium plate held at 4.2 K. "Cold neutrons" which are slow and whose wavelength is larger than twice the lattice constant are transmitted through the plate without diffraction; faster neutrons suffer multiple diffractions in the microcrystallites of the plate and are finally absorbed by the beryllium atoms with a cross section for absorption which is inversely proportional to the velocity and therefore large at 4.2 K. When the cold neutrons pass the warm semiconductor crystal they in turn become "warm" due to phonon absorption. By means of a time-of-flight spectrometer the velocity distribution of the neutrons is determined from which the energy and momentum of the absorbed phonons is obtained. Results for germanium [2] and gallium arsenide [3] are illustrated in Figs.6.15 and 6.16. The optical phonon temperature is also called "Debye temperature" *). For germanium it has a value of $\hbar\omega_0/k_B = 430$ K and for GaAs a value of 417 K. Due to the ionic binding character in compound semiconductors, atomic oscillations in the optical phonon mode may be excited by electromagnetic radiation (see Chap. 11g). This explains the name "optical phonon". For an optical phonon tempera-

[1] R. A. Cowley: Modern Solid State Physics (R. H. Enns and R. R. Haering, eds.) Vol.2, p.43. New York: Gordon and Breach. 1969; L. S. Kothari and K. S. Singwi: Solid State Physics (F. Seitz and D. Turnbull, eds.) Vol.8, p. 108. New York: Academic Press. 1969.

[2] I. Pelah, C. M. Eisenhauer, D. J. Hughes, and H. Palevsky, Phys. Rev. 108 (1957) 1091.

[3] B. N. Brockhouse, H. Palevsky, D. J. Hughes, W. Kley, and E. Tunkelo, Phys. Rev. Lett. 2 (1959) 258.

*) Actually this Debye temperature may be somewhat different from the one determined from specific heat measurements.

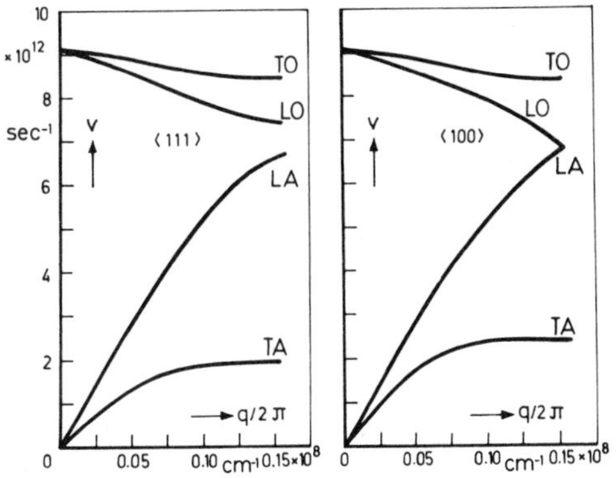

Fig.6.15 Lattice vibrational spectrum of germanium at 300 K in the ⟨111⟩ and ⟨100⟩ directions determined by the scattering of cold neutrons. The transverse branches are doubly degenerate (after B. N. Brockhouse and P. K. Iyengar, Phys. Rev. 111 (1958) 747).

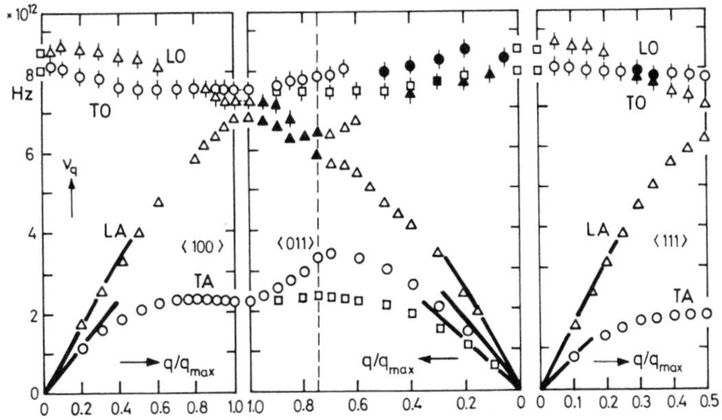

Fig.6.16 Lattice vibrational spectrum of gallium arsenide at 296 K. The solid points denote frequencies of undetermined polarization. The dashed line indicates the zone boundary in the ⟨011⟩ direction. The solid lines represent the slopes of the corresponding velocities of sound, calculated from the elastic constants (after G. Dolling and J. L. T. Waugh: Lattice Dynamics (R. F. Wallis, ed.) p. 19. London: Pergamon Press. 1965).

ture of e.g. 400 K a resonant wavelength of 32 μm in free space is calculated which is in the far infrared spectrum.

The calculation given above has been done for the model of a one-dimensional crystal. The curves in Figs.6.14 and 6.15 for a three-dimensional crystal can of course only in a crude approximation by represented by Eqs.(6h.19) and (6h.21).

Fig.6.15 shows that for a compound semiconductor with partly ionic bonds the longitudinal optical phonon frequency, ν_l, is somewhat larger than the transverse optical phonon frequency, ν_t, at $\vec{q} = 0$. In Chap.11g we shall see that the index of refraction, which is the square root of the dielectric constant for low frequencies, $\sqrt{\kappa}$, is also larger than for optical frequencies, where it is $\sqrt{\kappa_{opt}}$; κ is the static dielectric constant. We shall prove now the Lyddane-Sachs-Teller relation [4] for polar modes

$$\omega_l/\omega_t = \sqrt{\kappa} \,/\sqrt{\kappa_{opt}} \tag{6h.22}$$

which states that the ratio of the two phonon frequencies at q = 0 equals the ratio of the refractive indices.

The mechanical vibration of a polar crystal lattice, the dielectric polarization, $\vec{P}$, and the electric field $\vec{E}$ are interrelated by the following set of equations

$$\vec{P} = \gamma_{11}\vec{E} + \gamma_{12}\vec{w} \tag{6h.23}$$

$$-d^2\vec{w}/dt^2 = \gamma_{22}\vec{w} - \gamma_{12}\vec{E} \tag{6h.24}$$

where the γ_{ik} are coefficients, which will be determined later, and $\vec{w}$ is a "reduced displacement" given by

$$\vec{w} = \delta\vec{r}\,\sqrt{MN_u} \tag{6h.25}$$

N_u is the number of lattice cells per cm^3 and M is the reduced atomic mass given by

$$1/M = 1/M^+ + 1/M^- \tag{6h.26}$$

where M^+ and M^- are the masses of the positive and negative ions, respectively. $\delta\vec{r}$ is their relative displacement.

In Chap.7h we shall treat the interaction of carriers with coherent acoustic waves in polar lattices with quite similar equations, Eqs.(7h.17) and (7h.18). While for acoustic waves and no electric field, $\vec{E}= 0$, the polarization $\vec{P}$ is proportional to the strain, $S = (\vec{\nabla}_r \delta\vec{r})$, for optical modes $\vec{P}$ is proportional to $\delta\vec{r}$ itself as given by Eqs.(6h.23) and (6h.25). The essential feature of the optical vibration at $\vec{q} = 0$, i. e. at infinite wavelength, is that the positively charged atoms move as a body against the negatively charged atoms, in contrast to the long acoustic vibrations where atoms in a macroscopically small section move practically unison. In Chap.6k this feature of optical vibrations will also account for the $\delta\vec{r}$-proportionality of the interaction matrix element $H_{k'k}$.

The proportionality of $\vec{P}$ to $\vec{E}$ in Eq.(6h.23) for $\vec{w} = 0$ does not require much comment. Because of the inertia of the lattice atoms the case of $\vec{w} \approx 0$ is effective for an ac field at optical frequencies where the optical dielectric constant

[4] R. H. Lyddane, R. G. Sachs, and E. Teller, Phys. Rev. 59 (1941) 673; the proof follows closely M. Born and K. Huang: Dynamical Theory of Crystal Lattices, p.82. Oxford: Clarendon Press. 1954.

κ_{opt} enters this relation according to

$$\vec{P} = \kappa_o(\kappa_{opt} - 1)\vec{E} \tag{6h.27}$$

and therefore Eq.(6h.23) for $\vec{w} = 0$ yields

$$\gamma_{11} = \kappa_o(\kappa_{opt} - 1) \tag{6h.28}$$

Eq.(6h.23) for $\vec{E} = 0$ results in a polarization due to the relative reduced displacement $\vec{w}$ of the positively charged and negatively charged atoms. The factor of proportionality, γ_{12}, is also found in Eq.(6h.24), where a static displacement ($\vec{w}$ for $d^2\vec{w}/dt^2 = 0$) is shown to be caused by the application of a dc electric field $\vec{E}$:

$$\vec{w} = (\gamma_{12}/\gamma_{22})\vec{E} \tag{6h.29}$$

The fact that in both equations, (6h.23) and (6h.24), the same coefficient γ_{12} is found will become clear later on; it is due to the factor $\sqrt{MN_u}$ in the definition of $\vec{w}$ by Eq.(6h.25). If we eliminate $\vec{w}$ from the Eqs.(6h.29) and (6h.23) the polarization $\vec{P}$ as a function of the applied field $\vec{E}$ becomes

$$\vec{P} = (\gamma_{12}^2/\gamma_{22} + \gamma_{11})\vec{E} \tag{6h.30}$$

Since at present we deal with static conditions this relation is usually described by the static relative dielectric constant

$$\vec{P} = \kappa_o(\kappa - 1)\vec{E} \tag{6h.31}$$

where κ_o is the permittivity of free space. A comparison of both equations yields

$$\gamma_{12}^2/\gamma_{22} + \gamma_{11} = \kappa_o(\kappa - 1) \tag{6h.32}$$

If we take γ_{11} as given by Eq.(6h.28) and solve for γ_{12}^2 we find

$$\gamma_{12}^2 = \gamma_{22}\,\kappa_o(\kappa - \kappa_{opt}) \tag{6h.33}$$

Eq.(6h.24) for $\vec{E} = 0$ is the equation of motion for the atomic lattice vibration which is the well-known result of Hook's law and Newton's second law. The coefficient γ_{22} is the square of the angular frequency of the vibration.

For waves we consider only the field and the polarization in the direction of the wave propagation. Hence, for a transverse mode of oscillation (subscript t) $\vec{E} = 0$ and

$$\gamma_{22} = \omega_t^2 \tag{6h.34}$$

while for a longitudinal mode of oscillation (subscript l)

$$0 = \vec{D} = \kappa_o\vec{E} + \vec{P} \tag{6h.35}$$

where $\vec{P}$ is given by Eq.(6h.23):

$$0 = \kappa_o\vec{E} + \gamma_{11}\vec{E} + \gamma_{12}\vec{w} \tag{6h.36}$$

We solve this equation for $\vec{E}$ and by eliminating $\vec{E}$ from Eq.(6h.24) we obtain the equation of motion for the longitudinal mode

$$-d^2\vec{w}/dt^2 = \{\gamma_{22} + \gamma^2{}_{12}/(\kappa_o + \gamma_{11})\}\vec{w} \tag{6h.37}$$

The factor in braces is the square of the angular frequency of the vibration:

$$\gamma_{22} + \gamma_{12}^2/(\kappa_o + \gamma_{11}) = \omega_l^2 \tag{6h.38}$$

From Eqs.(6h.38) and (6h.34) we obtain for the ratio ω_l^2/ω_t^2 taking into account Eqs.(6h.33) and (6h.28):

$$\omega_l^2/\omega_t^2 = 1 + \gamma_{12}^2/\{\gamma_{22}(\kappa_o + \gamma_{11})\} = 1 + \kappa_o(\kappa - \kappa_{opt})/\{\kappa_o + \kappa_o(\kappa_{opt} - 1)\} = \kappa/\kappa_{opt} \tag{6h.39}$$

This yields Eq.(6h.22).

By using the method of electrostatics we have tacitly assumed a Coulomb interaction between charges in the lattice although in a lattice wave of course the Coulomb interaction is retarded.

Eq.(6h.24) can be obtained from a hamiltonian

$$H = \frac{1}{2}\left(\frac{d\vec{w}}{dt}\right)^2 + \frac{1}{2}\gamma_{22}\vec{w}^2 - \gamma_{12}(\vec{w}\vec{E}) - \frac{1}{2}\gamma_{11}\vec{E}^2 + (\vec{E}\vec{P}) + \frac{1}{2}(\kappa_o\vec{E}^2 + \vec{B}^2/\mu_o) \tag{6h.40}$$

where μ_o is the permeability of free space, by the Hamilton equations

$$d\dot{\vec{w}}/dt = -\vec{\nabla}_w H; \quad d\vec{w}/dt = \vec{\nabla}_{\dot{w}} H \tag{6h.41}$$

where the dot as usual means the time derivative. If for $\vec{P}$ in the hamiltonian its value given by Eq.(6h.23) is introduced, ω_t^2 for γ_{22}, and γ_{11} as given by Eq. (6h.28) we find for H:

$$H = \frac{1}{2}(d\vec{w}/dt)^2 + \frac{1}{2}\omega_t^2\vec{w}^2 + \frac{1}{2}(\kappa_o\kappa_{opt}\vec{E}^2 + \vec{B}^2/\mu_o) \tag{6h.42}$$

The first two terms are the mechanical lattice vibration energy per unit volume while the last term for a transverse mode represents radiative energy density. The time derivative of H,

$$\begin{aligned} dH/dt = {} & d\vec{w}/dt\,(d^2\vec{w}/dt^2 + \gamma_{22}\vec{w} - \gamma_{12}\vec{E}) + d\vec{E}/dt(\vec{P} - \gamma_{12}\vec{w} - \gamma_{11}\vec{E}) + \\ & + \vec{E}d(\kappa_o\vec{E} + \vec{P})/dt + (\vec{B}/\mu_o)d\vec{B}/dt \end{aligned} \tag{6h.43}$$

simplifies considerably by taking into account Eqs.(6h.23) and (6h.24):

$$dH/dt = \vec{E}\,d(\kappa_o\vec{E} + \vec{P})/dt + (\vec{B}/\mu_o)\,d\vec{B}/dt \tag{6h.44}$$

where $\kappa_o\vec{E} + \vec{P} = \vec{D}$ is the dielectric displacement vector and $\vec{B}/\mu_o$ the vector of the magnetic field strength. Eq.(6h.44) is the well-known rate of change of the energy density in a non-magnetic material. The Eqs.(6h.40 - 44) make it clear that the starting point of the calculation, Eqs.(6h.23) and (6h.24), with the same coefficient γ_{12} is essentially correct.*)

Let us finally derive a useful relation between $\vec{P}$ and $\delta\vec{r}$ for the longitudinal optical mode from Eqs.(6h.35), (6h.36), and (6h.25):

$$\vec{P} = -\kappa_o \vec{E} = \kappa_o \{\gamma_{12}/(\kappa_o + \gamma_{11})\} \delta\vec{r} \sqrt{MN_u} \tag{6h.45}$$

Introducing for γ_{12} and γ_{11} their values as given by Eqs.(6h.33) and (6h.28) and using the Lyddane-Sachs-Teller relation, Eq.(6h.39), we obtain

$$\vec{P} = \omega_l \{\kappa_o (\kappa_{opt}^{-1} - \kappa^{-1}) MN_u\}^{1/2} \delta\vec{r} \tag{6h.46}$$

The product MN_u is sometimes approximated by the mass density ρ. Since the polarization is a dipole moment, $e_C\delta\vec{r}$, per unit volume which here is the volume of the lattice cell, N_u^{-1},

$$P = e_C \delta\vec{r}/N_u^{-1} \tag{6h.47}$$

we find for the "Callen effective charge" [5] for longitudinal optical modes

$$e_C = \omega_l \{\kappa_o (\kappa_{opt}^{-1} - \kappa^{-1}) M/N_u\}^{1/2} \tag{6h.48}$$

e_C is usually given in units of the elementary charge. E.g. in α-SiC frequencies of $\omega_l = (1.82 \pm 0.05) \times 10^{14}\,\mathrm{sec}^{-1}$ and $\omega_t = (1.49 \pm 0.01) \times 10^{14}\,\mathrm{sec}^{-1}$ have been measured which yield

$$\kappa_{opt}^{-1} - \kappa^{-1} = \{(\omega_l/\omega_t)^2 - 1\}/\kappa = (0.49 \pm 0.15)/\kappa \tag{6h.49}$$

Although the frequencies have been measured quite accurately by Raman spectroscopy the difference $(\omega_l/\omega_t)^2-1$ is not well known. From the static relative dielectric constant $\kappa = 10.2$, the mass density $\rho = 3.2\,\mathrm{g/cm^3}$, and the density of lattice cells $N_u = 1.05\mathrm{x}10^{23}/\mathrm{cm}^3$ a value $e_C/e = 0.40$ has been calculated for SiC. In e.g. GaAs the value is 0.17 only. The corresponding values of the "Szigeti effective charge" e_S [6]

$$e_S = e_C\, 3\kappa_{opt}/(\kappa + 2) \tag{6h.50}$$

known from the theory of ionic bonds, are 0.67 in SiC and 0.38 in GaAs.

6i. Inelastic Scattering Processes

So far we have delt with practically elastic scattering processes such as impurity scattering and acoustic phonon scattering. For optical phonon scattering processes, however, the phonon energy is of the order of magnitude of the ther-

---

[5] H. B. Callen, Phys. Rev. 76 (1949) 1394.
[6] B. Szigeti, Trans. Faraday Soc. 45 (1949) 155.

*) For getting acquainted with dimensional considerations the reader should verify these two equations dimensionally by taking VA sec^3/cm^2 for the mass unit (equivalent to 10^7 g).

mal energy of carriers and these processes have to be treated as inelastic. As a consequence, scattering rates from a state $\vec{k}$ to a state $\vec{k}' = \vec{k} + \vec{q}$ will be very different from those to states $\vec{k}' = \vec{k} - \vec{q}$.

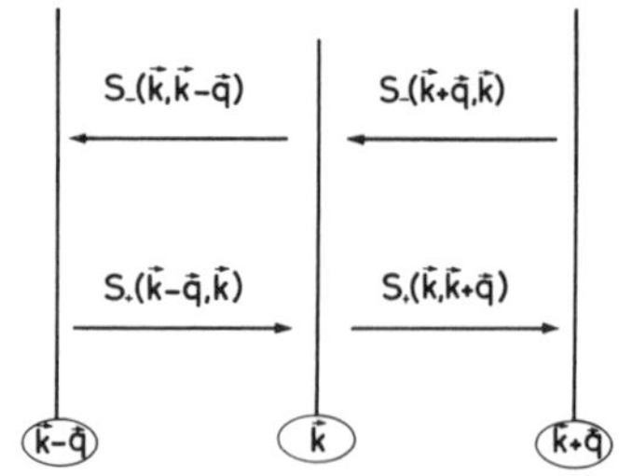

Fig.6.17 Schematic representation of electron transitions involving phonon emission and phonon absorption.

In Fig.6.17 all the four possible transitions from and to a state $\vec{k}$ are shown schematically. E.g. $S_-(\vec{k}, \vec{k}-\vec{q})$ represents the probability for the emission of a phonon of momentum $\hbar\vec{q}$ due to a change of the electron momentum from $\hbar k$ to $\hbar(\vec{k}-\vec{q})$. Hence, Eq.(6b.16) in full detail is now given by

$$(-\partial f(\vec{k})/\partial t)_{coll} = V(2\pi)^{-3} \int d^3q \{S_-(\vec{k}, \vec{k}-\vec{q}) f(\vec{k})[1-f(\vec{k}-\vec{q})] + S_+(\vec{k}, \vec{k}+\vec{q}) f(\vec{k})[1-f(\vec{k}+\vec{q})] - S_-(\vec{k}+\vec{q}, \vec{k}) f(\vec{k}+\vec{q})[1-f(\vec{k})] - S_+(\vec{k}-\vec{q}, \vec{k}) f(\vec{k}-\vec{q})[1-f(\vec{k})]\} \qquad (6i.1)$$

The subscripts + and − to S indicate phonon absorption and emission, respectively. The usual integration over $\vec{k}'$ has been replaced by an integration over $\vec{q}$ since $\vec{k}'$ is partly $\vec{k} + \vec{q}$ and partly $\vec{k} - \vec{q}$. For the case of a non-degenerate electron gas the factors in brackets can be omitted:

$$(-\partial f(\vec{k})/\partial t)_{coll} = V(2\pi)^{-3} \int d^3q \{[S_-(\vec{k}, \vec{k}-\vec{q}) + S_+(\vec{k}, \vec{k}+\vec{q})] f(\vec{k}) - S_-(\vec{k}+\vec{q}, \vec{k}) f(\vec{k}+\vec{q}) - S_+(\vec{k}-\vec{q}, \vec{k}) f(\vec{k}-\vec{q})\} \qquad (6i.2)$$

For inelastic processes we cannot approximate $f(\vec{k} \pm \vec{q})$ by $f(\vec{k})$ in order to be able to bring this equation into the form of Eq.(6b.22) and thus define a relaxation time τ_m. We shall prove, however, that in the special case where the matrix element $|H_{k'k}|$ is independent of $\vec{q}$ the $f(\vec{k} \pm \vec{q})$-factor in the last two terms in Eq. (6i.2) can be transformed into $f_0(\vec{k})$-factors by the integration over $\vec{q}$ and that even in this inelastic case τ_m can be defined; but otherwise a momentum relaxation time τ_m does not exist. Since indeed for optical deformation potential scattering the special case is realized we shall give the proof here.

In the diffusion approximation, Eq.(6b.19), $f(\vec{k})$ is given by

$$f(\vec{k}) = f_0(k) + f_1(k) \cos\vartheta \qquad (6i.3)$$

where for simplicity $-(\hbar kG/m)\partial f_0/\partial k$ has been denoted by $f_1(k)$. Since without a magnetic field $\vec{G}$ is proportional to the electric field $\vec{E}$, ϑ is the angle between $\vec{k}$ and $\vec{E}$; the direction of $\vec{E}$ is the polar axis. The angle between $\vec{k}'$ and $\vec{E}$ shall be denoted by ϑ'. Hence

$$f(\vec{k}) = f_0(k') + f_1(k') \cos\vartheta' \qquad (6i.4)$$

Let us first prove that

$$\int \cos\vartheta' \delta\{\epsilon(\vec{k}) - \epsilon(\vec{k}) - \hbar\omega_o\} d^3q = 0 \tag{6i.5}$$

where $\hbar\omega_o$ is the optical phonon energy which is taken as a constant. For the calculation of $\cos\vartheta'$ we introduce for $\vec{k}'$ polar coordinates with $\vec{k}$ being the polar axis:

$$\vec{k}' = \vec{k} \pm \vec{q} = (\pm q \sin\theta \cos\varphi, \pm q \sin\theta \sin\varphi, k \pm q \cos\theta) \tag{6i.6}$$

where θ is the angle between $\vec{k}$ and $\vec{k}'$ and φ is the azimuth. In this representation $\vec{E}$ is given by

$$\vec{E} = E(\sin\vartheta \cos\psi, \sin\vartheta \sin\psi, \cos\vartheta) \tag{6i.7}$$

and $\cos\vartheta'$ by

$$\cos\vartheta' = \frac{(\vec{k}'\vec{E})}{k'E} = \frac{\pm q \sin\theta \sin\vartheta \cos(\varphi-\psi) \pm q \cos\theta \cos\vartheta + k \cos\vartheta}{\sqrt{q^2 + k^2 \pm 2qk\cos\theta}} \tag{6i.8}$$

Since $d^3q = q^2 dq \sin\theta\, d\theta\, d\varphi$ we first perform the integration over φ which makes the $\cos(\varphi-\psi)$-term vanish and adds a factor of 2π to the other terms; the Dirac δ-function contained in the scattering probabilities S_- and S_+ in Eq. (6i.2)

$$\delta\{\epsilon(\vec{k} \pm \vec{q}) - \epsilon(\vec{k}) \mp \hbar\omega_o\} = \delta\{(\hbar^2/2m)(\pm 2kq\cos\theta + q^2) \mp \hbar\omega_o\} \tag{6i.9}$$

is independent of φ. The integrations over φ and θ are therefore easily performed (see p. 179, footnote):

$$\int_0^{\pi}\int_0^{2\pi} \cos\vartheta' d\varphi \delta\{\epsilon(\vec{k} \pm \vec{q}) - \epsilon(\vec{k}) \mp \hbar\omega_o\} \sin\theta\, d\theta = \frac{2\pi m}{\hbar^2 kq} \cos\vartheta \frac{A_\pm - q^2}{\sqrt{k^2 \pm 2m\omega_o/\hbar}} \tag{6i.10}$$

where for simplicity we have introduced the constant

$$A_\pm = 2k^2 \pm 2m\omega_o/\hbar \tag{6i.11}$$

For the final integration over q we consider only the q-dependent factors

$$\int_{q_1}^{q_2} \frac{A_\pm - q^2}{q} q^2 dq = \frac{1}{2}(q_2^2 - q_1^2)[A_\pm - \frac{1}{2}(q_1^2 + q_2^2)] \tag{6i.12}$$

Energy and momentum conservation, i.e. the vanishing argument of the δ-function Eq.(6i.9), and $q \geqslant 0$ yield for the lower limit

$$q_1 = \mp k \pm \sqrt{A_\pm - k^2} \tag{6i.13}$$

and for the upper limit

$$q_2 = +k + \sqrt{A_\pm - k^2} \tag{6i.14}$$

Fig.6.18 shows these limits in a diagram of q/k vs $\cos\theta$ for $\hbar\omega_o/\epsilon = 0.5$. With these values the integral given by Eq.(6i.12) vanishes which proves Eq.(6i.5) for both phonon absorption and emission. From Eq.(6i.4) in the form

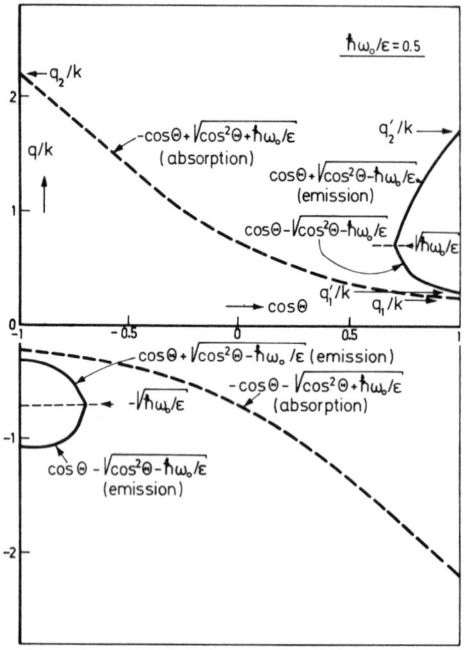

Fig.6.18 q/k obtained from energy and momentum conservation as a function of the cosine of the scattering angle, for phonon absorption and emission.

$$f(\vec{k} \pm \vec{q}) = f_0(\vec{k} \pm \vec{q}) + f_1(\vec{k} \pm \vec{q}) \cos\vartheta \qquad (6i.15)$$

and Eq.(6b.17) we can manipulate Eq.(6i.2) into a form where we have on the right-hand side the distribution functions in the form $f(\vec{k}) - f_0(\vec{k})$, and τ_m can be defined by Eq.(6b.22). The generalization of Eq.(6i.4) to an arbitrary number of spherical harmonics in the expansion

$$f(\vec{k}) = \sum_l f_l(k)\, P_l(\cos\vartheta) \qquad (6i.16)$$

has been given by Beer [1]

The result for $1/\tau_m$ is obtained from Eqs.(6b.22) and (6i.2):

$$\frac{1}{\tau_m} = \frac{V}{(2\pi)^3} \cdot \frac{2\pi}{\hbar} \{|H_{k+q,k}|^2 \int_{q_1}^{q_2} \frac{2\pi m}{\hbar^2 kq} q^2 dq + |H_{k-q,k}|^2 \int_{q_1'}^{q_2'} \frac{2\pi m}{\hbar^2 kq} q^2 dq\} \qquad (6i.17)$$

where for the first integral representing absorption of $\hbar\omega_0$ in a transition from $\vec{k}$ to $\vec{k} + \vec{q}$ the limits given by Eqs.(6i.13) and (6i.14) have been manipulated into the form

[1] A.C. Beer: Galvanomagnetic Effects in Semiconductors, in: Solid State Physics (F. Seitz and D. Turnbull, eds.), Suppl. 4, p. 286. New York: Acad. Press. 1963.

$$q_1 = -k + \sqrt{k^2 + 2m\omega_0/\hbar} = (a-1)k \tag{6i.18}$$

where

$$a = (1 + \hbar\omega_0 2m/\hbar^2 k^2)^{1/2} = (1 + \hbar\omega_0/\epsilon)^{1/2} \tag{6i.19}$$

corresponding to $\theta = 0$, and

$$q_2 = +k + \sqrt{k^2 + 2m\omega_0/\hbar} = (a+1)k \tag{6i.20}$$

corresponding to $\theta = \pi$, while for the second integral representing emission of $\hbar\omega_0$ in a transition from $\vec{k}$ to $\vec{k}-\vec{q}$ the limits are

$$q_1' = (1-b)k;\ q_2' = (1+b)k \tag{6i.21}$$

where b is given by

$$b = \mathrm{Re}(1 - \hbar\omega_0/\epsilon)^{1/2} \tag{6i.22}$$

and Re stands for "real part of" and assures that the emission term vanishes for $\epsilon < \hbar\omega_0$. The integrals are easily solved with the result

$$\frac{1}{\tau_m} = \frac{2^{1/2}Vm^{3/2}}{\pi\hbar^4}\left[|H_{k+q,k}|^2(\epsilon + \hbar\omega_0)^{1/2} + |H_{k-q,k}|^2\,\mathrm{Re}(\epsilon - \hbar\omega_0)^{1/2}\right] \tag{6i.23}$$

We shall see later that the second matrix element is different from the first one only by a factor $\exp(\hbar\omega_0/k_BT)$, hence :

$$\frac{1}{\tau_m} = \frac{2^{1/2}Vm^{3/2}}{\pi\hbar^4}|H_{k+q,k}|^2\left[(\epsilon + \hbar\omega_0)^{1/2} + \exp(\hbar\omega_0/k_BT)\,\mathrm{Re}(\epsilon - \hbar\omega_0)^{1/2}\right] \tag{6i.24}$$

The energy loss rate is easily obtained now since the energy gain or loss per collision is the optical phonon energy, $\pm\hbar\omega_0$, which is considered to be a constant:

$$\left(-\frac{d\epsilon}{dt}\right)_{coll} = -\frac{2^{1/2}Vm^{3/2}}{\pi\hbar^4}|H_{k+q,k}|^2\hbar\omega_0\left[(\epsilon + \hbar\omega_0)^{1/2} - \exp(\hbar\omega_0/k_BT)\,\mathrm{Re}(\epsilon - \hbar\omega_0)^{1/2}\right] \tag{6i.25}$$

In the absorption process the energy change is $+\hbar\omega_0$ while in the emission process it is $-\hbar\omega_0$.

6j. The Momentum Balance Equation and the Shifted Maxwellian

A method of calculating the field dependence of the mobility of warm and hot carriers with the simplifying assumption of a Maxwell-Boltzmann distribution with an electron temperature T_e consists of (1st) averaging τ_m over this distribution and calculating the mobility μ as a function of T_e, (2nd) calculating the relation between the field strength $\vec{E}$ and T_e from the energy balance equation

$$\mu e E^2 = \langle -d\epsilon/dt \rangle_{coll} \tag{6j.1}$$

and (3rd) eliminating T_e from the two relations.

In another method the first step is the calculation of an average of τ_m, $\bar{\tau}_m$, from

$$< -\frac{d(\hbar k_E)}{dt} >_{coll} = mv_d/\bar{\tau}_m \qquad (6j.2)$$

where the average on the left side is taken over a shifted distribution, Eq.(4b.41). The quantity $\hbar k_E$ is the component of the carrier momentum in the field direction which on the average is mv_d where v_d is the drift velocity. The mobility is then approximated by

$$\mu = (|e|/m)\bar{\tau}_m \qquad (6j.3)$$

Since $\mu E = v_d$ and $|e|E$ is the momentum gain per unit time from E, Eq.(6j.3) is obtained from Eq.(6j.2) and the momentum balance equation

$$<-d(\hbar k_E)/dt>_{coll} = |e| E \qquad (6j.4)$$

Otherwise the procedure is the same as above. The results are slightly different from those of the first method. Since the second method has been applied to the case of polar optical scattering [1] we shall consider it here briefly.

With k_E being given by

$$k_E = k \cos\vartheta \qquad (6j.5)$$

we obtain the component of the phonon wave vector in the field direction, q_E, from the Eqs.(6i.6) and (6i.7):

$$q_E = (\vec{q}\vec{E})/E = q \sin\theta \sin\vartheta \cos(\varphi-\psi) + q \cos\theta \cos\vartheta \qquad (6j.6)$$

The change of carrier momentum in the field direction is due to the gain or loss of a momentum $\hbar q_E$ at an absorption or emission process of a phonon, respectively, multiplied with the scattering probability (Fig.6.17) and integrated over phase space:

$$\{-d(\hbar k_E)/dt\}_{coll} = V(2\pi)^{-3} \int (\hbar q_E S_- - \hbar q_E S_+)\, d^3q \qquad (6j.7)$$

This is the quantum mechanical equivalent of the classical momentum balance equation, e.g. in the form of Eq.(4b.39).

Since the scattering probability S is independent of the azimuth φ, the integration over φ eliminates the term in Eq.(6j.5) which contains $\cos(\varphi-\psi)$. Hence, from Eq.(6j.7) the momentum loss rate

$$(-dk_E/dt)_{coll} = \cos\vartheta\, V(2\pi)^{-2} \int \cos\theta\, (S_- - S_+) q^3 dq \sin\theta\, d\theta \qquad (6j.8)$$

is obtained. Let us denote the right-hand side by $g(\epsilon)\cos\vartheta$ where

$$g(\epsilon) = V(2\pi)^{-2} \int \cos\theta\, (S_- - S_+) q^3 dq \sin\theta\, d\theta \qquad (6j.9)$$

Since the scattering probabilities S_+ and S_- contain δ-functions the integration over θ can be executed similarly to Eq.(6i.10)

$$\begin{aligned}\int_0^{\pi} \cos\theta\, \delta\{(\mp\hbar^2 kq/m)\cos\theta + (\hbar^2q^2/2m) \pm \hbar\omega_0\} \sin\theta\, d\theta &= \\ = \pm m/(2\hbar^2k^2) + m^2\hbar\omega_0/\hbar^4k^2q^2 &= \pm(1/4\epsilon)(1 \pm 2mk_B\Theta/\hbar^2q^2)\end{aligned} \qquad (6j.10)$$

[1] R. Stratton, Proc. Roy. Soc. (London) A 246 (1958) 406; J. Phys. Soc. Japan 17 (1962) 590.

where ϵ and $k_B\Theta$ have been introduced for $\hbar^2k^2/2m$ and $\hbar\omega_0$, respectively, and the upper signs pertain to phonon emission, the lower ones to absorption. For the function $g(\epsilon)$ we find

$$g(\epsilon) = V/(8\pi\hbar\epsilon)\{\int_{q_1}^{q_2}|H_{k+q,k}|^2(1-2mk_B\Theta/\hbar^2q^2)q^3dq + \int_{q'_1}^{q'_2}|H_{k-q,k}|^2(1+2mk_B\Theta/\hbar^2q^2)q^3dq\} \quad (6j.11)$$

where the limits are given by Eqs.(6i.18), (6i.20), and (6i.21).

Next we average Eq.(6j.7) over the shifted Maxwell-Boltzmann distribution function given by

$$f(\vec{k}) \propto \exp[-(\hbar\vec{k}-m\vec{v}_d)^2/2mk_BT_e] \quad (6j.12)$$

where the argument for not too large field intensities can be approximated by $-(\hbar^2k^2-2\hbar kmv_d\cos\vartheta)/2mk_BT_e$ and the exponential is approximated by

$$\exp(-\epsilon/k_BT_e)\exp(\hbar kv_d\cos\vartheta/k_BT_e) \approx (1+\hbar kv_d\cos\vartheta/k_BT_e)\cdot\exp(-\epsilon/k_BT_e) \quad (6j.13)$$

The distribution function

$$f(\vec{k}) = f_0(\epsilon) + f_1(\epsilon)\cos\vartheta \quad (6j.14)$$

therefore consists of a symmetric part $f_0(\epsilon) \propto \exp(-\epsilon/k_BT_e)$ and an asymmetric part $f_1(\epsilon)\cos\vartheta$ where $f_1(\epsilon)$ is given by

$$f_1(\epsilon) = f_0(\epsilon)v_d(2m\epsilon)^{1/2}/k_BT_e \quad (6j.15)$$

Since the averaging process

$$<-dk_E/dt>_{coll} = \int\cos\vartheta\, g(\epsilon)f(\vec{k})k^2dk\sin\vartheta\, d\vartheta\, d\psi / \int f(\vec{k})k^2dk\sin\vartheta\, d\vartheta\, d\psi \quad (6j.16)$$

contains the integrals

$$\int_0^\pi \cos\vartheta\, f(\vec{k})\sin\vartheta\, d\vartheta = \int_0^\pi \cos^2\vartheta\, f_1(\epsilon)\sin\vartheta\, d\vartheta = \frac{2}{3}f_1(\epsilon) \quad (6j.17)$$

and

$$\int_0^\pi f(\vec{k})\sin\vartheta\, d\vartheta = \int_0^\pi f_0(\epsilon)\sin\vartheta\, d\vartheta = 2f_0(\epsilon) \quad (6j.18)$$

this is simplified :

$$<-dk_E/dt>_{coll} = (2^{3/2}/3\pi^{1/2})v_d(m/k_BT_e)^{1/2}\int_0^\infty g(\epsilon)\cdot\exp(-\epsilon/k_BT_e)\cdot(\epsilon/k_BT_e)d(\epsilon/k_BT_e) \quad (6j.19)$$

For $1/\bar{\tau}_m$ we finally obtain from Eq.(6j.2)

$$1/\bar{\tau}_m = (2^{3/2}/3\pi^{1/2})\hbar/(mk_BT_e)^{1/2}\int_0^\infty g(\epsilon)\exp(-\epsilon/k_BT_e)(\epsilon/k_BT_e)\, d(\epsilon/k_BT_e) \quad (6j.20)$$

This quantity is somewhat different from $<\tau_m^{-1}>$ given by Eq.(6i.23). From Eq.(6j.3) we easily find the mobility as a function of the electron temperature T_e.

Next we calculate from Eq.(6j.1) the field strength E as a function of T_e

$$(-d\epsilon/dt)_{coll} = V(2\pi)^{-3}\int(\hbar\omega S_- - \hbar\omega S_+)d^3q \quad (6j.21)$$

If as for optical phonon scattering ω does not depend on q it can be taken outside the integral ($\omega = \omega_0$). Since the remainder on the right-hand side is just the difference of $1/\tau_m$ for emission und absorption, we find

$$(-d\epsilon/dt)_{coll} = \hbar\omega_0\{(1/\tau_m)_- - (1/\tau_m)_+\} \tag{6j.22}$$

which for optical deformation potential scattering yields Eq.(6i.25).

For the average only the symmetric part of the distribution function, $f_0(\epsilon)$, is relevant.

$$<-d\epsilon/dt>_{coll} = (2/\pi^{1/2})\hbar\omega_o \int_0^\infty \{(1/\tau_m)_- - (1/\tau_m)_+\} \exp(-\epsilon/k_B T_e)(\epsilon/k_B T_e)^{1/2} \, d(\epsilon/k_B T_e) \tag{6j.23}$$

From Eqs.(6j.1) and (6j.3) we find for the field intensity

$$E = (<-d\epsilon/dt>_{coll}\, m/\bar{\tau}_m e^2)^{1/2} \tag{6j.24}$$

and for the drift velocity

$$v_d = (<-d\epsilon/dt>_{coll}\, \bar{\tau}_m/m)^{1/2} \tag{6j.25}$$

both as functions of T_e where $\bar{\tau}_m$ and $<-d\epsilon/dt>_{coll}$ are given by Eqs.(6j.20) and (6j.23), respectively. If both functions of T_e are calculated one can finally obtain a plot of $v_d(E)$.

For polar optical scattering the main drawback of this method is due to the fact that the shifted Maxwell-Boltzmann distribution is a poor approximation to the real distribution: the real one is more spiked in the forward direction and cannot be represented with only one parameter T_e [2].

6k. Optical Deformation Potential Scattering

We will now investigate scattering processes of carriers by longitudinal optical phonons in nonpolar crystals. The optical phonon angular frequency which in Chap.6h has been denoted by ω_l, will now be denoted by ω_0. For the phonon energy, $\hbar\omega_0$, a temperature Θ is introduced by the relation

$$\hbar\omega_0 = k_B\Theta \tag{6k.1}$$

Θ is known as the "Debye temperature" since ω_0 is the highest phonon frequency and in his well-known theory of the specific heat Debye postulated a cut-off of the phonon spectrum at ω_0.

It was mentioned in Chap.6h (text after Eq.(6h.26)) that in the long-wavelength optical mode of vibration the positively charged atoms move as a body against the negatively charged atoms with the effect that a carrier in a transition

---

[2] E. M. Conwell: High Field Transport in Semiconductors, Solid State Physics, (F. Seitz, D. Turnbull, and H. Ehrenreich, eds.), Suppl. 9, chap. V. 7. New York: Acad. Press. 1967.

from a positively charged atom to a negatively charged atom changes its energy by an amount proportional to the atomic displacement:

$$\delta\epsilon = D \cdot \delta r \tag{6k.2}$$

The factor of proportionality, D, is the "optical deformation potential constant" of the band edge (in units of eV/cm) [1]. For simplicity we have omitted subscripts e or h for electrons or holes, respectively. Eq.(6k.2) is in contrast to the corresponding Eqs.(6d.5) and (6d.6) for the interaction between carriers and acoustic modes of vibration. Otherwise, however, the result of a calculation of the hamiltonian matrix element [2]

$$|H_{k\pm q,k}| = D\{(N_q + 1/2 \mp 1/2)\hbar/2\rho V\omega_o\}^{1/2} \tag{6k.3}$$

is quite similar to Eq.(6d.12). It is independent of q since the optical mode spectrum is stationary at $\vec{q} = 0$ and N_q depends on ω_o only:

$$N_q = \{\exp(\hbar\omega_o/k_B T) - 1\}^{-1} = \{\exp(\Theta/T) - 1\}^{-1} \tag{6k.4}$$

Since this is exactly the case treated in Chap.6i we can apply Eq.(6i.24) for a calculation of the momentum relaxation time τ_m:

$$\frac{1}{\tau_m} = \frac{m^{3/2} D^2 N_q}{2^{1/2}\pi\rho\hbar^2 k_B\Theta}\{(\epsilon + k_B\Theta)^{1/2} + \exp(\Theta/T)\,\mathrm{Re}(\epsilon - k_B\Theta)^{1/2}\} \tag{6k.5}$$

and Eq.(6i.25) for the energy loss rate:

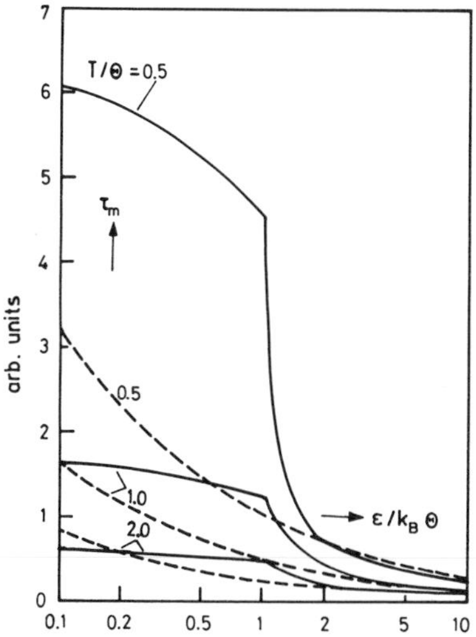

Fig.6.19 Momentum relaxation time as a function of energy for optical phonon scattering.

$$-\left.\frac{d\epsilon}{dt}\right|_{coll} = -\frac{m^{3/2} D^2 N_q}{2^{1/2}\pi\rho\hbar^2} \cdot \{(\epsilon + k_B\Theta)^{1/2} - \exp(\Theta/T)\,\mathrm{Re}(\epsilon - k_B\Theta)^{1/2}\} \tag{6k.6}$$

The dependence of τ_m on carrier energy ϵ is illustrated in Fig.6.19. At $\epsilon \gtrsim k_B\Theta$ the emission of optical phonons dominates $\tau_m(\epsilon)$ which is more pronounced at a lower lattice temperature. A kink similar to the one at $\epsilon = k_B\Theta$ should also occur at $\epsilon = 2k_B\Theta$, $3k_B\Theta$ etc., when two or three etc. optical phonons are emitted simultaneously. However, these processes were not incorporated in the theory given above, since the probability for a many-particle process is comparatively small. The

[1] H. J. G. Meyer, Phys. Rev. 112 (1958) 298.
[2] W. Shockley, Bell. Syst. Tech. J. 30 (1951) 990; W. A. Harrison, Phys. Rev. 104 (1956) 1281.

dashed curves show τ_m for acoustic phonon scattering where $\epsilon_{ac}/u_l = D/\omega_o$ and otherwise the same constants were assumed. The high-energy tails of the dashed curves merge with those of the full curves.

The momentum relaxation time decreases with increasing energy which for carrier heating has the consequence that deviations from Ohm's law are negative. The zero-field mobility $\mu_o = (e/m) \langle\tau_m\rangle$ where $\langle\tau_m\rangle$ is given by Eq.(6c.19) is readily evaluated for a non-degenerate carrier gas where a Maxwell-Boltzmann distribution is valid.

$$\mu_o = \frac{4\sqrt{2\pi}\, e\hbar^2 \rho (k_B\Theta)^{1/2}}{3m^{5/2} D^2} f(T/\Theta) \qquad (6k.7)$$

The function $f(T/\Theta)$ is given by

$$f(T/\Theta) = (2z)^{5/2} (e^{2z}-1) \int_0^\infty \frac{y^{3/2}\, e^{-2zy}\, dy}{\sqrt{y+1} + e^{2z}\, \mathrm{Re}\sqrt{y-1}} \qquad (6k.8)$$

where $z = \Theta/2T$ and $y = \epsilon/k_B\Theta$. The function is shown in Fig.6.20. The mobility

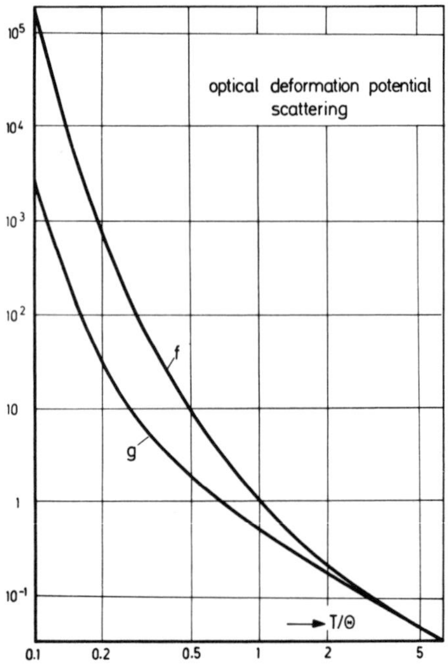

Fig.6.20 Functions $f(T/\Theta)$ and $g(T/\Theta)$ given by Eqs.(6k.8) and (6k.23), respectively.

which is proportional to the function decreases continuously with increasing temperature. Its numerical value in units of $cm^2/Vsec$ is given by

$$\mu = 2.04\text{x}10^3 \frac{(\rho/\text{g cm}^{-3})\,(\Theta/400\text{ K})^{1/2}}{(m/m_o)^{5/2}\,(D/10^8\text{ eV cm}^{-1})} f(\Theta/2T) \tag{6k.9}$$

The energy loss rate for hot carriers, Eq.(6k.6), will now be averaged over a Maxwell-Boltzmann distribution with an electron temperature T_e :

$$f(\epsilon) \propto \exp(-\epsilon/k_B T_e) \tag{6k.10}$$

For the average of the first term in braces of Eq.(6k.6) we obtain

$$\begin{aligned}\int_o^\infty \sqrt{\epsilon + k_B\Theta}\,\sqrt{\epsilon}\,\exp(-\epsilon/k_BT_e)d\epsilon &= (k_BT_e)^2 \int_0^\infty \sqrt{(\epsilon/k_BT_e)+(\Theta/T_e)} \\ &\cdot \sqrt{\epsilon/k_BT_e}\,\exp(-\epsilon/k_BT_e)\,d(\epsilon/k_BT_e) \\ &= (k_BT_e)^2(\Theta/2T_e)\exp(\Theta/2T_e)\,K_1(\Theta/2T_e)\end{aligned} \tag{6k.11}$$

where according to Appendix B, Eq.(B.4), K_1 is a modified Bessel function. For the average of the second term in braces of Eq.(6k.6) we obtain by introducing the variable $\xi = (\epsilon/k_BT_e) - \Theta/T_e$ and omitting for the moment the factor $\exp(\Theta/T)$:

$$\begin{aligned}\int_{k_B\Theta}^\infty \sqrt{\epsilon - k_B\Theta}\,\sqrt{\epsilon}\,\exp(-\epsilon/k_BT_e)d\epsilon &= (k_BT_e)^2\exp(-\Theta/T_e)\,\cdot \\ \int_o^\infty \sqrt{\xi}\sqrt{\xi+\Theta/T_e}\;e^{-\xi}\,d\xi &= (k_BT_e)^2(\Theta/2T_e)\exp(-\Theta/2T_e)K_1(\Theta/2T_e)\end{aligned} \tag{6k.12}$$

We have to divide both terms by the normalizing factor

$$\int_0^\infty \sqrt{\epsilon}\,\exp(-\epsilon/k_BT_e)d\epsilon = (k_BT_e)^{3/2}\sqrt{\pi}/2 \tag{6k.13}$$

For the sake of brevity we introduce as above $z = \Theta/2T$ and $\lambda = T/T_e$ and obtain for the average of Eq.(6h.25):

$$-\left\langle\frac{d\epsilon}{dt}\right\rangle_{coll} = \frac{m^{3/2}D^2(k_B\Theta)^{1/2}}{\pi^{3/2}\hbar^2\rho}\;\frac{(\lambda z)^{1/2}K_1(\lambda z)}{\sinh(z)}\sinh\{(1-\lambda)z\} \tag{6k.14}$$

An energy relaxation time τ_ϵ which is independent of T_e can only be defined rigorously for the case of warm electrons where the hyperbolic sine may be replaced to a good approximation by its argument:

$$\sinh\{(1-\lambda)z\} \approx (1-\lambda)z \approx z(T_e-T)/T \tag{6k.15}$$

and $\lambda = 1$ otherwise. From the definition of τ_ϵ given by Eqs.(4n.5) and (4n.6) we obtain for the warm electron case:

$$\tau_\epsilon = \frac{3\pi^{3/2}}{4}\cdot\frac{\hbar^2\rho(k_B\Theta)^{1/2}}{m^{3/2}D^2}\cdot\frac{\sinh(z)}{z^{5/2}K_1(z)} \tag{6k.16}$$

where $z = \Theta/2T$. For the case of the many-valley model to be discussed in Chap. 7.f, $m^{3/2}$ has to be replaced by $m_t m_l^{1/2}$ where m_t and m_l are the transverse and longitudinal effective masses. The numerical value of τ_ϵ in sec is given by

$$\tau_\epsilon = 1.6\text{x}10^{-12}\,\frac{(\rho/\text{g cm}^{-3})\,(\Theta/400\,\text{K})^{1/2}}{(m/m_0)^{3/2}\,(D/10^8\,\text{eVcm}^{-1})^2}\;\frac{\sinh(z)}{z^{5/2}\,K_1(z)} \qquad (6k.17)$$

where $z = \Theta/2T$.

The dependence of τ_ϵ on T/Θ is shown in Fig.6.21. It has a minimum near

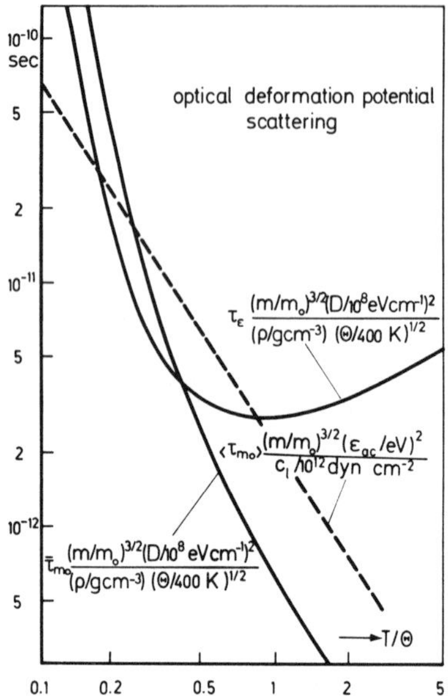

Fig.6.21 Energy relaxation time and momentum relaxation time averaged over a Maxwell-Boltzmann distribution, for optical deformation potential scattering; for comparison the momentum relaxation time for acoustic deformation potential scattering is included (dashed curve).

$T = \Theta$. At high temperatures both the absorption and the emission of optical phonons become very strong so that the result is the difference of two large quantities; hence there is no simple answer to the question why τ_ϵ rises proportionally to $\sqrt{T/\Theta}$ for $T \gg \Theta$.

Experimental results on $\tau_\epsilon(T)$ in n-type germanium obtained with various microwave methods [3–6] are shown in Fig.6.22. Probably the most accurate results are those of Hess [3] obtained by the method of harmonic mixing (see

[3] K. Hess and K. Seeger, Zeitschr. f. Physik 218 (1969) 431; 237 (1970) 252; K. Hess, thesis, Univ. Wien, Austria. 1970.
[4] T. N. Morgan and C. E. Kelly, Phys. Rev. 137 (1965) A 1573.
[5] K. Seeger, Zeitschr. f. Physik 172 (1963) 68.
[6] A. F. Gibson, J. W. Granville, and E. G. S. Paige, J. Phys. Chem. Solids 19 (1961) 198.

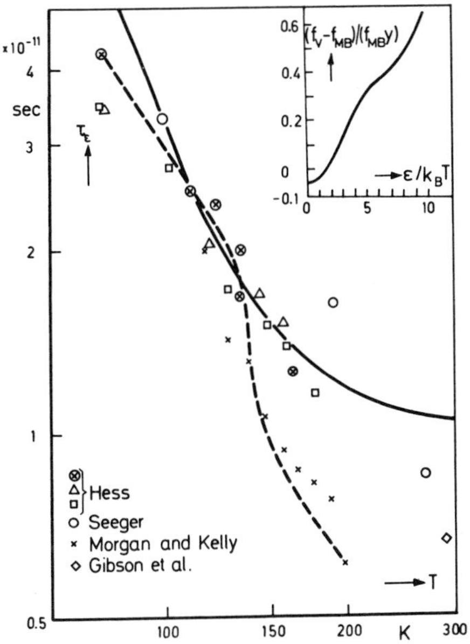

Fig.6.22 Observed and calculated energy relaxation time as a function of temperature for n-type germanium. The full curve has been calculated with a Maxwell-Boltzmann distribution, f_{MB}, with an optical deformation potential constant of $D = 4.8 \times 10^8$ eV/cm. The dashed curve was obtained for $D = 8 \times 10^8$ eV/cm by a variational method first applied by Adawi [8] for a calculation of $\beta = (\mu - \mu_0)/\mu_0 E^2$; the relative deviation of the variational-method-distribution f_V from f_{MB} is shown in the inset where y stands for $(3\pi/16)\mu_{ac}^2 E^2/u_l^2$ and μ_{ac} and u_l are the acoustic zero-field mobility and sound velocity, respectively. Acoustic phonon scattering was taken into account with a ratio of deformation potential constants, $D/\epsilon_{ac} = 0.4\, \omega_0/u_l$ (after Hess and Seeger, ref. 3, Morgan and Kelly, ref.4, Gibson et al., ref. 6).

Chap.4n). The full curve has been calculated from Eq.(6k.17) with $c_l = 1.56 \times 10^{12}$ dyn cm^{-2}, $m/m_0 = 0.2$, $\Theta = 430$ K characteristic for n-type Ge, and $D = 4.8 \times 10^8$ eV/cm. From infrared-absorption a value of 4×10^8 eV/cm has been obtained for D while from various kinds of electrical measurements values between 5 and 9×10^8 eV/cm have been found [7]. Since the Maxwell-Boltzmann distribution used here is only a crude approximation to the true distribution the agreement of D with the previous data is better than to be expected. Adawi [8] used a variational method for a calculation of β in n-type Ge, and by the same

[7] E. M. Conwell, Solid State Physics (F. Seitz, D. Turnbull, and H. Ehrenreich, eds.) Suppl. 9, p. 171. New York: Acad. Press. 1967.

[8] I. Adawi, Phys. Rev. 120 (1960) 118.

method both the distribution function shown in Fig.6.20 (inset) and τ_ϵ (dashed curve) have been calculated by Hess [9]. The agreement with the experimental data is not quite as good as when the Maxwell-Boltzmann distribution was used in the calculation. Earlier experimental data were obtained by Morgan and Kelly [4] (indicated by crosses), Seeger [5] (circles), and Gibson, Granville, and Paige [6] (diamond) by different microwave methods. The influence of acoustic phonon energy relaxation (estimated relaxation time $3\text{x}10^{-10}$ sec $\sqrt{100\,\text{K/T}}$) may be predominant below T = 50 K.

Measurements have also be made [3] on n-type Si and p-type Ge and Si, some also under uniaxial pressure. In contrast to n-Ge τ_ϵ is anisotropic in these materials which can probably be understood by considering the band structure.

The energy relaxation time discussed so far was that of warm carriers. For hot carriers the approximation Eq.(6k.14) is no longer valid and τ_ϵ will therefore depend on T_e and hence on the electric field intensity $\vec{E}$. In spite of the non-exponential time dependence of the energy a relaxation time defined by Eqs. (4n.5) and (4n.6) still remains useful. Fig.6.23 shows τ_ϵ as a function of T_e/T for $T = \Theta/6$, calculated with the assumption of a Maxwell Boltzmann distribution. There is a flat minimum near $T_e = 2T$ and an increase of τ_ϵ with $(T_e/T)^{1/2}$ for large values of the electron temperature T_e. Measurements of τ_ϵ in n-type Ge in the hot electron region have been made by Gibson, Granville, and Paige [6] at room temperature. While near $E = 1$ kV/cm a power law $\tau_\epsilon \propto T_e^{-1/4}$ could be fitted to the experimental curve, a power law $\tau_\epsilon \propto T_e^{+1/2}$ was more appropriate near $\vec{E} = 5$ kV/cm, and values of the exponent between $-1/4$ and $+1/2$ were found in the intermediate range of $\vec{E}$.

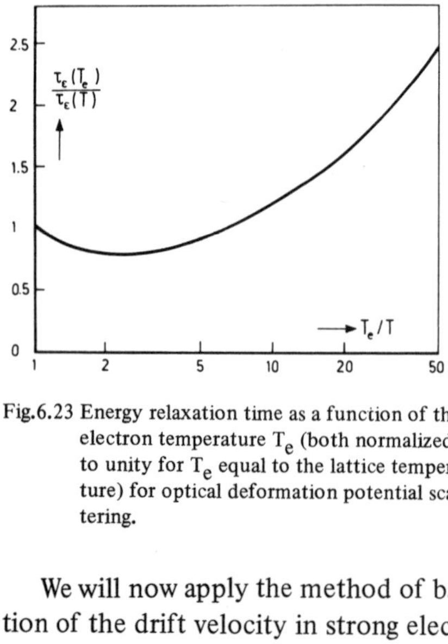

Fig.6.23 Energy relaxation time as a function of the electron temperature T_e (both normalized to unity for T_e equal to the lattice temperature) for optical deformation potential scattering.

We will now apply the method of balance equations (Chap.6j) for a calculation of the drift velocity in strong electric fields. Let us first evaluate the function $g(\epsilon)$ given by Eq.(6j.11). Since the hamiltonian matrix elements, $H_{k\pm q,k}$, do not depend on q (Eq.(6k.3) the integrations are straightforward.

$$g(\epsilon) = \frac{V}{8\pi\hbar\epsilon}\,\frac{\hbar^2 D^2 N_q}{2\rho V k_B\Theta}\left\{\frac{q_2^2-q_1^2}{2}\left(\frac{q_2^2+q_1^2}{2}-\frac{2mk_B\Theta}{\hbar^2}\right)\right.$$
$$\left.+\,e^{\Theta/T}\,\frac{q_2'^2-q_1'^2}{2}\left(\frac{q_2'^2+q_1'^2}{2}+\frac{2mk_B\Theta}{\hbar^2}\right)\right\} \qquad (6k.18)$$

[9] K. Hess (unpublished).

$$g(\epsilon) = \frac{m^2 D^2 N_q}{\pi\hbar^3 \rho k_B\Theta} \{\epsilon^{1/2}(\epsilon+k_B\Theta)^{1/2} + \epsilon^{\Theta/T}\, \epsilon^{1/2}\, \mathrm{Re}(\epsilon-k_B\Theta)^{1/2}\} \tag{6k.19}$$

For the calculation of $1/\bar{\tau}_m$ according to Eq.(6j.20) we introduce the parameter $t = \Theta/2T_e$ and the variable $\xi = \epsilon/k_BT_e$ for the first term in $g(\epsilon)$ and $\xi = (\epsilon-k_B\Theta)/k_BT_e$ for the second term

$$1/\bar{\tau}_m = \frac{2m^{3/2} D^2 N_q}{3\pi^{3/2}\hbar^2\rho(k_B\Theta)^{1/2}}\, t^{3/2}\{e^t[K_2(t) - K_1(t)] + e^{(\Theta/T)-t}[K_2(t)+K_1(t)]\} \tag{6k.20}$$

where K_1 and K_2 are modified Bessel functions given in Appendix B. With N_q given by Eq.(6d.11), we can write

$$1/\bar{\tau}_m = \frac{2m^{3/2} D^2 (\lambda z)^{3/2}}{3\pi^{3/2}\hbar^2\rho(k_B\Theta)^{1/2}\sinh(z)} [\cosh\{(1-\lambda)z\}\, K_2(\lambda z) + \sinh\{(1-\lambda)z\} K_1(\lambda z)] \tag{6k.21}$$

where as usual $z = \Theta/2T$ and $\lambda = T/T_e$. For the zero-field mobility, Eq.(6j.3) for $\lambda = 1$, we find in the present approximation

$$\mu_o = \frac{4\sqrt{2\pi}\, e\hbar^2\rho(k_B\Theta)^{1/2}}{3m^{5/2}D^2}\, g(T/\Theta) \tag{6k.22}$$

where the function $g(T/\Theta)$ (not to be confused with $g(\epsilon)$)

$$g(T/\Theta) = 9\pi 2^{-7/2} z^{-3/2} \sinh(z)/K_2(z) \tag{6k.23}$$

for comparison with $f(T/\Theta)$ in Eq.(6k.7) has been plotted in Fig.6.20. Especially at low temperatures the values of the mobility obtained by the method of balance equations are somewhat smaller than by the normal procedure.

The calculation of the field strength and the drift velocity, both as functions of λ, according to Eqs.(6j.24) and (6j.25), where the energy loss rate is given by Eq.(6k.14), yields

$$v_d/v_{ds} = \frac{2^{1/2}\coth^{1/2}(z)}{\lambda^{1/2} z^{1/2}\,[1+\coth\{(1-\lambda)z\}\, K_2(\lambda z)/K_1(\lambda z)]^{1/2}} \tag{6k.24}$$

and

$$\mu_o E/v_{ds} = \frac{2^{1/2}\coth^{1/2}(z)\lambda\, \sinh\{1-\lambda)z\}\, K_1(\lambda z)}{z^{1/2} K_2(z)} \cdot [1+\coth\{(1-\lambda)z\}\frac{K_2(\lambda z)}{K_1(\lambda z)}]^{1/2} \tag{6k.25}$$

where

$$v_{ds} = \left\{\frac{3k_B\Theta}{4m\coth(z)}\right\}^{1/2} \tag{6k.26}$$

is the saturation value of the drift velocity for large field intensities. In the range of saturation the drift energy of a carrier $mv_{ds}^2/2$ is about equal to $k_B\Theta = \hbar\omega_o$, the optical phonon energy, since at high electron temperatures phonon emission processes determine the drift velocity. v_{ds} decreases with increasing temperature, but only slowly (13 % between $T \ll \Theta$ and $T = \Theta/2$).

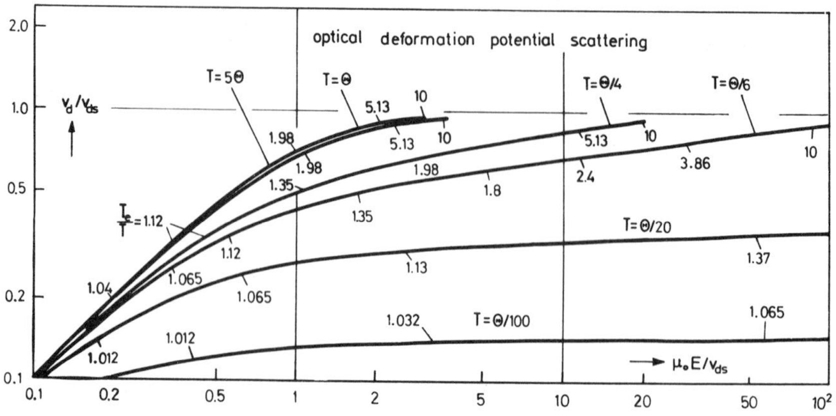

Fig.6.24 Normalized drift velocity vs normalized electric field intensity, with T/Θ as a parameter, for optical deformation potential scattering. Numbers given along the curves are values of T_e/T (T = lattice temperature, T_e = carrier temperature, Θ = Debye temperature).

Fig.6.24 shows a log-log plot of the calculated drift velocity vs electric field strength for various values of the lattice temperature, relative to the Debye temperature, T/Θ. Ohm's law is demonstrated by the initial rise at an angle of 45°. At large field strengths the drift velocity saturates; the saturation value which is temperature dependent through $z = \Theta/2T$ in Eq.(6k.26), has been used in Fig. 6.24 for normalization of both v_d and $\mu_o E$ where μ_o is the zero-field mobility. The numbers along the curves are the electron temperatures relative to the lattice temperature. At low temperatures saturation is approached only very gradually while the characteristics deviate from the 45° straight line already at very small field intensities. How small these field intensities are can be estimated from a calculation of the coefficient β characteristic for warm electrons.

From Eq.(4n.3) we find

$$\beta E^2 = (1-\lambda)\{d(\bar{\tau}_{mo}/\bar{\tau}_m)/d\lambda\}_{\lambda=1} \tag{6k.27}$$

where the factor $(1-\lambda)$ can be eliminated by means of the warm-carrier energy balance equation

$$\mu_o |e| E^2 = (1-\lambda)\, 3k_B T/2\tau_\epsilon \tag{6k.28}$$

Hence, β is given by

$$\beta = (2\mu_o |e| \tau_\epsilon/3k_B T)\{d(\bar{\tau}_{mo}/\bar{\tau}_m)/d\lambda\}_{\lambda=1} \tag{6k.29}$$

Differentiating the λ-dependent factors in Eq.(6k.21) and afterwards equating λ to unity yields

$$\left(\frac{d}{d\lambda}\frac{\bar{\tau}_{mo}}{\bar{\tau}_m}\right)_{\lambda=1} = \frac{3}{2} + z(K_2' - K_1)/K_2 = -(0.5 + 2zK_1/K_2) \tag{6k.30}$$

where K_2' is the derivative of K_2 which is given in Appendix B. With Eq.(6k.22) for μ_0 and (6k.16) for τ_ϵ we find the dimensionless quantity

$$-\beta k_B \Theta / m\mu_0^2 = \frac{(K_2 - K_1)(K_2 + 4zK_1)}{6zK_1K_2} \tag{6k.31}$$

where K_1 and K_2 are functions of $z = \Theta/2T$. In Fig.6.25 this quantity has been plotted vs T/Θ. It increases with temperature. However, since μ_0 strongly decreases with temperature as shown in Fig.6.20, $-\beta$ itself also strongly decreases (for about a factor of 10^5 in going from $T/\Theta = 0.1$ to 1). For an estimate of the order of magnitude for $-\beta$ at e.g. $T/\Theta = 0.2$ remember that $k_B\Theta/m$ is approximately the square of the saturation drift velocity, v_{ds}. Assuming typical values of $v_{ds} = 10^7$ cm/sec and $\mu_0 = 2\times10^4$ cm^2/Vsec, a value of about 10^{-7} cm^2/V^2 is found for $-\beta$. This is three orders of magnitude lower than the value observed e.g. in n-type Ge at 77 K ($T/\Theta \approx 0.2$) where both optical and acoustic deformation potential scattering determine the energy balance while the momentum balance is dominated by acoustic deformation potential scattering alone [10]; the different influence of the two scattering mechanisms is explained by the large number of acoustic phonons of low energy on one hand and the small number of optical phonons of large energy on the other hand.

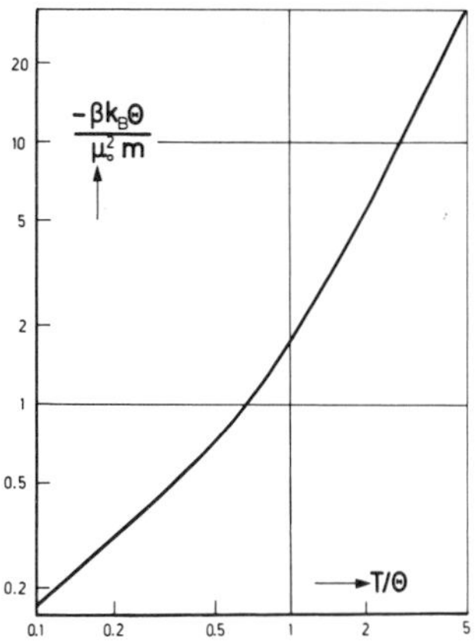

Fig.6.25 $-\beta k_B\Theta/(m\mu_o^2)$ as a function of the lattice temperature T normalized to the Debye temperature Θ, for optical deformation potential scattering.

Although e.g. in n-type germanium saturation has been observed and v_{ds} has the calculated order of magnitude, another theory in which the existence of "electric domains" (inhomogeneous field distribution) in the semiconductor is postulated is more likely to explain the observations. This will be discussed in Chap.7g.

The ratio $\tau_\epsilon/\bar{\tau}_{mo}$ is roughly the number of collisions required for the relaxation of energy. Its value is

$$\tau_\epsilon/\bar{\tau}_{mo} = (T/\Theta) K_2(\Theta/2T)/K_1(\Theta/2T) \tag{6k.32}$$

and contains as a material parameter only the Debye temperature Θ. At temperatures $T \gg 0.42\,\Theta$, the energy relaxation is much slower than the momentum relaxation which favors the randomization of the drift momentum as discussed in connection with Fig.6.6.

[10] T. N. Morgan, Bull. Am. Phys. Soc. (Ser.2) 2 (1959) 265; J. Phys. Chem. Solids 8 (1959) 245; see also E. M. Conwell: Solid State Physics (F. Seitz, D. Turnbull, and H. Ehrenreich, eds.), Suppl. 9, p.166. New York. Acad. Press. 1967.

6l. Polar Optical Scattering

In polar semiconductors the interaction of carriers with the optical mode of lattice vibrations is known as "polar optical scattering". As for piezoelectric scattering the potential energy of a carrier is given by Eq.(6g.5),

$$\delta\epsilon = |e| E/q \tag{6l.1}$$

where the electric field strength E is due to the polarization P of the longitudinal optical lattice vibration given by Eq.(6h.47)

$$E = -P/\kappa_o = -N_u e_C \delta r/\kappa_o \tag{6l.2}$$

This yields for $\delta\epsilon$:

$$\delta\epsilon = (-|e| N_u e_C/\kappa_o q)\delta r \tag{6l.3}$$

which except for the q-dependence of the factor of proportionality between $\delta\epsilon$ and δr is very similar to Eq.(6k.2). Hence, in analogy to Eq.(6k.3) we find for the hamiltonian matrix element

$$|H_{k\pm q,k}| = (|e| N_u e_C/\kappa_o q) \{(N_q + 1/2 \mp 1/2)\hbar/2\rho V\omega_o\}^{1/2} \tag{6l.4}$$

Two constants characteristic for an energy band of a semiconductor are convenient. One is defined by

$$E_o = \frac{|e| m}{4\pi\kappa_o^2 \rho\hbar\omega_o} (N_u e_C)^2 \tag{6l.5}$$

which according to Eq.(6h.48) can be written in the form

$$E_o = \frac{|e| m k_B \Theta}{4\pi\kappa_o \hbar^2} (\kappa_{opt}^{-1} - \kappa^{-1}) \tag{6l.6}$$

and in units of kV/cm has a numerical value of

$$E_o = 16.3(m/m_o)(\Theta/K)(\kappa_{opt}^{-1} - \kappa^{-1}) \tag{6l.7}$$

It is called "effective field strength". The second constant is the dimensionless "polar constant" α defined by

$$\alpha = \frac{\hbar|e|E_o}{2^{1/2} m^{1/2} (\hbar\omega_o)^{3/2}} = \frac{1}{137}\sqrt{\frac{mc^2}{2k_B\Theta}}\left(\frac{1}{\kappa_{opt}} - \frac{1}{\kappa}\right) = 397.4\sqrt{\frac{m/m_o}{\Theta/K}}\left(\frac{1}{\kappa_{opt}} - \frac{1}{\kappa}\right) \tag{6l.8}$$

where $1/137 = e^2/(4\pi\kappa_o\hbar c)$ is the fine structure constant and c is the velocity of light. E.g. in n-type GaAs $E_o = 5.95$ kV/cm and $\alpha = 0.067$.

With these constants the square of the matrix element, Eq.(6l.4), becomes

$$|H_{k\pm q,k}|^2 = \frac{2\pi\hbar^2|e|E_o}{Vmq^2}\left(N_q + \frac{1}{2} \mp \frac{1}{2}\right) = \frac{2^{3/2}\pi\hbar\alpha(\hbar\omega_o)^{3/2}}{Vm^{1/2}q^2}\left(N_q + \frac{1}{2} \mp \frac{1}{2}\right) \tag{6l.9}$$

We have shown in Chap.6i that in the present case, where the matrix element depends on q and the scattering process is inelastic ($\hbar\omega_0$ is of the order of magnitude of the average carrier energy or even larger), a momentum relaxation time τ_m does not exist strictly speaking. Even so, let us for an approximation in Eq. (6i.2) replace $f(\vec{k} \pm \vec{q})$ by $f_0(\vec{k})$. Now we obtain for $1/\tau_m$ given by Eq.(6i.17)

$$\frac{1}{\tau_m} = \frac{V}{(2\pi)^3} \cdot \frac{2\pi}{\hbar} \cdot \frac{2^{3/2}\pi\hbar a(\hbar\omega_0)^{3/2} N_q}{V m^{1/2}} \left\{ \int_{q_1}^{q_2} \frac{2\pi m}{\hbar^2 k q^3} q^2 dq + \exp(\Theta/T) \int_{q_1'}^{q_2'} \frac{2\pi m}{\hbar^2 k q^3} q^2 dq \right\} \tag{6l.10}$$

The solution of the integral is given by

$$\frac{1}{\tau_m} = a\omega_0 (\hbar\omega_0/\epsilon)^{1/2} N_q \{\ln\left|\frac{a+1}{a-1}\right| + \exp(\Theta/T) \ln\left|\frac{1+b}{1-b}\right|\} \tag{6l.11}$$

where a and b are defined by Eqs.(6i.19) and (6i.22), respectively.

At low temperatures $T \ll \Theta$ where $\epsilon \ll \hbar\omega_0$, $b \approx 0$, $N_q \approx \exp(-\Theta/T)$, and $1/\tau_m$ is approximated by

$$1/\tau_m \approx 2a\omega_0 \exp(-\Theta/T) \tag{6l.12}$$

The reciprocal momentum relaxation time is then essentially the product of the polar constant and the availability of an optical phonon for absorption. The mobility is simply $(e/m)\tau_m$:

$$\mu = [|e|/\{2m a\omega_0\}] \exp(\Theta/T) \tag{6l.13}$$

which in units of $cm^2/Vsec$ is given by

$$\mu = 2.6 \times 10^5 \frac{\exp(\Theta/T)}{a(m/m_0)(\Theta/K)} \tag{6l.14}$$

for $T \ll \Theta$. E.g. in n-type GaAs where $\Theta = 417$ K, $m/m_0 = 0.072$, $a = 0.067$, we calculate a mobility at 100 K of 2.2×10^5 $cm^2/Vsec$. This is an order of magnitude larger than the highest mobilities observed in this material. At this and lower temperatures the dominant scattering mechanism in compound semiconductors of even the highest purity available at present is impurity scattering. This unfortunately prevents a useful comparison of Eq.(6l.14) with experimental data.

The difficulty in calculating τ_m mentioned above can be overcome by e.g. the use of variational methods [1]. The mobility calculated in this way [2] for low temperatures, $T \ll \Theta$, agrees with the value given by Eq.(6l.14) while for high temperatures the latter has to be multiplied by $8\sqrt{T/9\pi\Theta}$. In the intermediate range numerical methods have been applied. By variational methods different momentum relaxation times $\tau_m(\epsilon)$ are obtained for conductivity, Hall effect, and thermoelectric effect. Fig.6.26 shows the exponent r of an assumed power law

---

[1] D. J. Howarth and E. H. Sondheimer, Proc. Roy. Soc. (London) A219 (1953) 53.
[2] H. Ehrenreich, J. Appl. Phys. 32 Suppl. (1961) 2155.

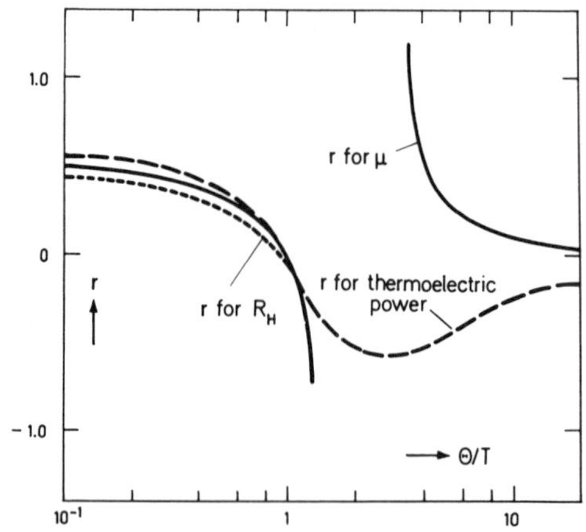

Fig.6.26 Exponent r of the energy dependence of the effective momentum relaxation time, $\tau_m \propto \epsilon^r$, for polar scattering. Different values of r apply for the mobility μ, the Hall coefficient R_H, and the thermoelectric power (Eq.(4i.17)) (after H. Ehrenreich, ref. 2).

$\tau_m \propto \epsilon^r$, as a function of Θ/T for these effects [2]. For the case of conductivity, r becomes infinite at $T \approx \Theta$ which demonstrates the difficulty in obtaining a function $\tau_m(\epsilon)$ valid for all values of ϵ.

For a calculation of the energy loss rate we multiply the first term on the right-hand side of Eq.(61.11) by $-\hbar\omega_o = -k_B\Theta$ and the second term by $+k_B\Theta$ since these represent phonon absorption and emission, respectively.

$$\left(-\frac{d\epsilon}{dt}\right)_{coll} = \frac{a(k_B\Theta)^{5/2}}{\hbar\sqrt{\epsilon}} N_q \left\{ -\ln\left|\frac{a+1}{a-1}\right| + e^{\Theta/T} \ln\left|\frac{1+b}{1-b}\right| \right\} \tag{61.15}$$

For a non-degenerate electron gas we average the energy loss rate over the Maxwell-Boltzmann distribution with an electron temperature T_e.

$$\begin{aligned} \langle -d\epsilon/dt \rangle_{coll} = 2\pi^{-1/2} a \hbar^{-1} (\Theta/T_e)^{3/2} N_q \{ -\int_0^\infty \ln\left|\frac{a+1}{a-1}\right| \exp(-\epsilon/k_B T_e) d\epsilon \\ + e^{\Theta/T} \int_{k_B T}^\infty \ln\left|\frac{1+b}{1-b}\right| \exp(-\epsilon/k_B T_e)\, d\epsilon \} \end{aligned} \tag{61.16}$$

The first integral on the right-hand side is easily solved (Eq.(B.3) of Appendix B) by introducing the variable $\xi = \epsilon/k_B T_e$ and the parameter $t = \Theta/(2T_e)$:

$$k_B T_e \int_0^\infty \ln\left|\frac{(1+2t/\xi)^{1/2}+1}{(1+2t/\xi)^{1/2}-1}\right| e^{-\xi} d\xi = k_B T_e\, e^t K_o(t) \tag{61.17}$$

For the second integral with the lower limit $k_B\Theta$ the variable $\xi = (\epsilon - k_B\Theta)/k_B T_e$ is introduced:

$$k_B T_e e^{\Theta/T} \int_0^\infty \ln \left| \frac{1+\{1-2t/(\xi+2t)\}^{1/2}}{1-\{1-2t/(\xi+2t)\}^{1/2}} \right| e^{-(\xi+2t)} d\xi = k_B T_e e^{(\Theta/T)-2t} e^t K_o(t) \qquad (61.18)$$

Since $t = \lambda z$ where $\lambda = T/T_e$, the average loss rate is finally obtained in the form [3] :

$$\langle -d\epsilon/dt \rangle_{coll} = 2^{3/2} \pi^{-1/2} a \hbar^{-1} (k_B\Theta)^2 (\lambda z)^{1/2} K_o(\lambda z) \sinh\{(1-\lambda)z\}/\sinh(z) \qquad (61.19)$$

For the energy relaxation time for warm carriers, where $\sinh\{(1-\lambda)z\} \approx (1-\lambda)z$ and $\lambda \approx 1$ otherwise, we find

$$\tau_\epsilon = \frac{3\pi^{1/2}}{2^{7/2} a\omega_o} \frac{\sinh(z)}{z^{5/2} K_o(z)} = \frac{0.47}{a\omega_o} \cdot \frac{\sinh(\Theta/2T)}{(\Theta/2T)^{5/2} K_o(\Theta/2T)} \qquad (61.20)$$

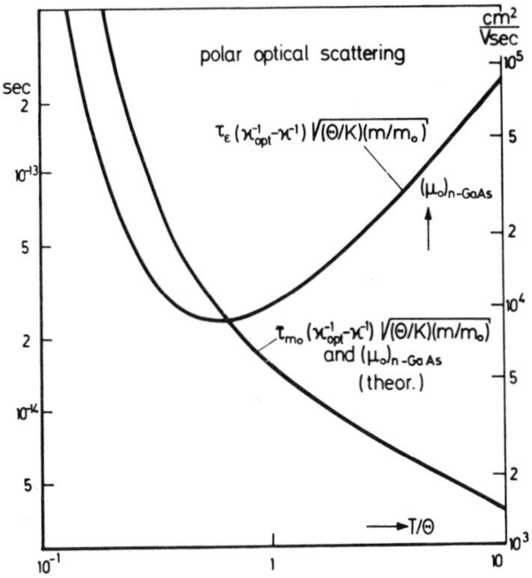

Fig.6.27 Energy relaxation time τ_ϵ for warm carriers and momentum relaxation time $\bar{\tau}_{mo}$ as a function of lattice temperature for polar optical scattering; for n-type gallium arsenide the curve for $\bar{\tau}_{mo}$ together with the right ordinate scale yields the mobility μ_o calculated according to Eq.(61.23) where $(\kappa_{opt}^{-1} - \kappa^{-1}) \sqrt{(\Theta/K)(m/m_o)} = 0.067$; $\Theta = 417$ K.

In Fig.6.27 the product

$$\tau_\epsilon (\kappa_{opt}^{-1} - \kappa^{-1}) \left(\frac{\Theta}{K}\right)^{1/2} \left(\frac{m}{m_o}\right)^{1/2} = 9.05\text{x}10^{-15} \sec \frac{\sinh(\Theta/2T)}{(\Theta/2T)^{5/2} K_o(\Theta/2T)} \qquad (61.21)$$

has been plotted as a function of T/Θ. There is a minimum at a temperature of about half the Debye temperature Θ (i.e. $\Theta/2T \approx 1$). At the minimum $\tau_\epsilon = 1.29/a\omega_o$ which indicates a suitable electromagnetic wavelength for the measurement of τ_ϵ of $\approx 1/a$ times the optical phonon wavelength. Since a for

[3] E. M. Conwell, Phys. Rev. 143 (1966) 657.

most semiconductors is of the order of magnitude 10^{-1} and the optical phonon wavelength is usually about 30 μm (Chap.11g) a suitable wavelength for measurements of τ_ϵ is about $\frac{1}{3}$ mm. Hence, τ_ϵ can be measured by sub-mm wave methods. Results obtained for n-GaAs and n-InAs are shown in Fig.6.28 [4-6]. The curves calculated according to Eq.(61.21) have been fitted to the experimental data by assuming suitable values for the polar constant α.

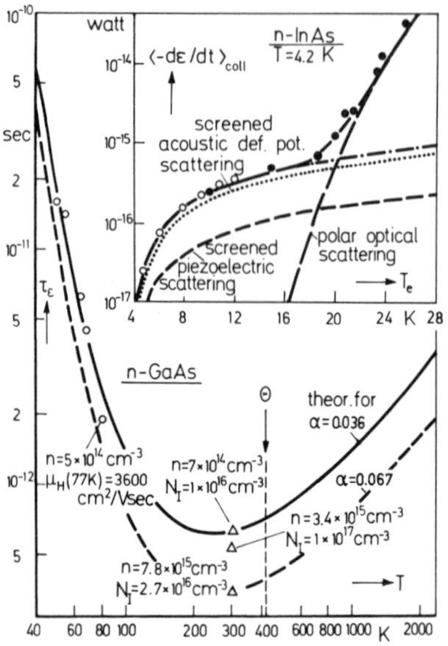

Fig.6.28 Warm-carrier energy relaxation time τ_ϵ observed in n-type GaAs by microwave (circles) and far-infrared techniques (triangles) as a function of lattice temperature (after Hess and Kahlert, ref. 4, and Kuchar et al., ref. 5). The inset shows τ_ϵ observed by the Shubnikov de Haas effect (open circles) and by mobility measurements and $\mu(E)eE^2$ (full circles) in n-type InAs as a function of electron temperature at a lattice temperature of 4.2 K (after Bauer and Kahlert, ref. 6).

Since the calculation of the drift velocity in strong electric fields by the method of balance equations [7] is very similar to the corresponding calculation for optical deformation potential scattering, Eqs.(6k.18)-(6k.31), except that now the matrix elements are proportional to q^{-2}, we will not go through details of the calculation but only give the results. The reciprocal average momentum relaxation time

$$1/\bar{\tau}_m = \frac{2^{5/2}\alpha\,\omega_o\,(\lambda z)^{3/2}}{3\pi^{1/2}\sinh(z)}\,[\cosh\{(1-\lambda)z\}K_1(\lambda z)+\sinh\{(1-\lambda)z\}K_0(\lambda z)] \tag{61.22}$$

where as usual $z = \Theta/2T$ and $\lambda = T/T_e$. The factor in brackets is different from the corresponding factor in Eq.(6k.21) only insomuch as the order of both of the modified Bessel functions is smaller by 1. For the zero-field mobility μ_0 we obtain in the present approximation

[4] K. Hess and H. Kahlert, J. Phys. Chem. Solids 32 (1971) 2262.
[5] F. Kuchar, A. Philipp, and K. Seeger, Solid State Comm. 11 (1972) 965.
[6] G. Bauer and H. Kahlert, Phys. Rev. B5 (1972) 566.
[7] R. Stratton, Proc. Roy. Soc. (London) A 246 (1958) 406.

$$\mu_o = \frac{3\pi^{1/2}}{2^{5/2}} \frac{|e|}{ma\omega_o} \frac{\sinh(z)}{z^{3/2} K_1(z)} = \frac{31.8\ cm^2/Vsec}{(\kappa_{opt}^{-1}-\kappa^{-1})\,(\Theta/K)^{1/2}\,(m/m_o)^{3/2}} \frac{\sinh(z)}{z^{3/2} K_1(z)} \qquad (61.23)$$

For high temperatures where $z \ll 1$ this is equal to the mobility obtained by variational methods [2] times a factor $32/9\pi \approx 1.13$ while for low temperatures ($z \gg 1$) it is lower than this mobility by a factor $3T/2\Theta$. The mobility calculated from Eq.(61.23) for electrons in n-type GaAs is shown in Fig.6.27. At room temperature a mobility of 7 800 $cm^2/Vsec$ is found which agrees with the experimental value.

In contrast to optical deformation potential scattering the drift velocity does not saturate at high field strengths. It is convenient, however, to introduce here also v_{ds} given by Eq.(6k.26). For the drift velocity v_d we find

$$v_d/v_{ds} = \frac{2^{1/2} \coth^{1/2}(z)}{\lambda^{1/2} z^{1/2} [1 + \coth\{(1-\lambda)z\} K_1(\lambda z)/K_o(\lambda z)]^{1/2}} \qquad (61.24)$$

where $\lambda = T/T_e$ and $z = \Theta/2T$, which is quite similar to the value for optical deformation potential scattering given by Eq.(6k.24). However, for small values of the argument, i.e. high electron temperatures, the ratio $K_1(\lambda z)/K_o(\lambda z) \approx 1/\{\lambda z \ln(2/\lambda z)\}$. Therefore,

$$(v_d)_{T_e \to \infty} \approx v_{ds} \{2 \ln(2/\lambda z)\}^{1/2} \to \infty \qquad (61.25)$$

Since this is a logarithmic dependence of the drift velocity on carrier temperature, there may be a region in the characteristics where the drift velocity "seems" to saturate.

The electric field strength E, in units of the material constant E_o, (Eq.61.6), is given by

$$E/E_o = 2^{3/2} (3\pi)^{-1/2} (\lambda z) \sinh\{(1-\lambda)z\} K_o(\lambda z) \cdot [1 + \coth\{(1-\lambda)z\} K_1(\lambda z)/K_o(\lambda z)]^{1/2}/\sinh(z) \qquad (61.26)$$

Fig.6.29 shows a diagram of v_d/v_{ds} vs $\mu_o E/v_{ds}$ which may be compared with the corresponding diagram for optical deformation potential scattering, Fig.6.24. At low temperatures ($T/\Theta \ll 1$) there is a plateau where the drift velocity is fairly constant over a range of field intensities, and finally a superlinear behavior which has been connected with "dielectric breakdown"[8]. At high temperatures ($T \gg \Theta$) no plateau is found and the breakdown characteristics develop right from the straight lines representing Ohm's law. Experimentally this breakdown has not yet been observed in semiconductors since at the high breakdown field

[8] H. Fröhlich and B. V. Paranjape, Proc. Phys. Soc. (London) B69 (1956) 21; see also E. M. Conwell: Solid State Physics (F. Seitz, D. Turnbull, and H. Ehrenreich, eds.) Suppl. 9, p. 200. New York: Academic Press. 1967. For "dielectric breakdown" see H. Fröhlich, Proc. Roy. Soc. (London) A160 (1937) 230; Adv. Physics 3 (1961) 325.

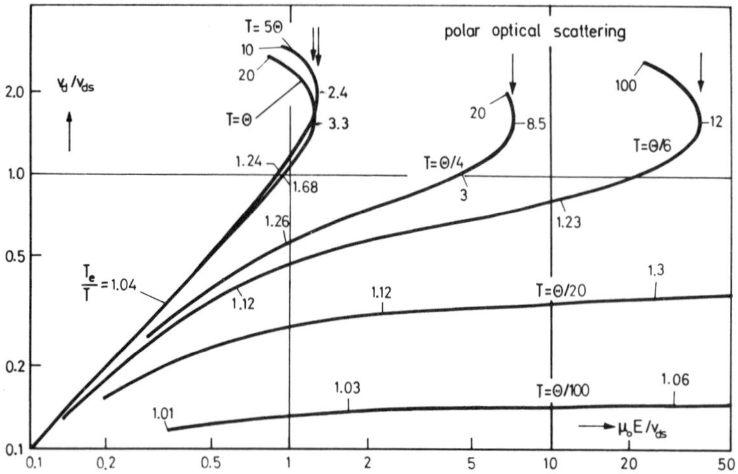

Fig.6.29 Drift velocity in units of v_{ds} as a function of the electric field intensity (in units of v_{ds}/μ_o) at various lattice temperatures (in units of the Debye temperature Θ) for polar optical scattering. The numbers next to the points are electron temperatures (in units of T). For various lattice temperatures the dielectric breakdown fields are indicated by arrows.

intensities other energy loss mechanisms such as intervalley scattering and impact ionization become important and prevent the high electron temperatures necessary for dielectric breakdown to occur.

There may be a chance to observe the onset of the superlinear characteristics in the warm-electron range of field intensities. The calculation of the coefficient β is similar to the corresponding calculation for optical deformation potential scattering, Eqs.(6k.27)–(6k.31), with the result [9]

$$-\beta = (m\mu_o^2/3k_B\Theta)\{4z-K_1(z)/K_o(z)\}$$
$$= (3\pi/16E_o^2)\sinh^2(z)\{4zK_o(z)-K_1(z)\}/\{z^3K_1^2(z)K_o(z)\} \quad (61.27)$$

In Fig.6.30 $|\beta|E_o^2$ is plotted vs T/Θ. At low temperatures, $T < 1.08\,\Theta$, β is negative, at higher temperatures positive. The positive values indicate the onset of breakdown. No experimental data of β have yet been reported for $T > \Theta$.

The number of collisions for energy relaxation is approximately given by the ratio

$$\tau_\epsilon/\bar{\tau}_{mo} = (T/\Theta)\,K_1(\Theta/2T)/K_o(\Theta/2T) \quad (61.28)$$

Except for the Debye temperature Θ it is independent of the particular material. At temperatures $T \gg 0.65\,\Theta$, the energy relaxation is much slower than the momentum relaxation. This indicates a rapid randomization of drift momentum which favors the formation of a drifted Maxwell-Boltzmann distribution.

The preceding calculations for all temperatures are, however, based on the

[9] R. Stratton, J. Phys. Soc. Japan 17 (1962) 590.

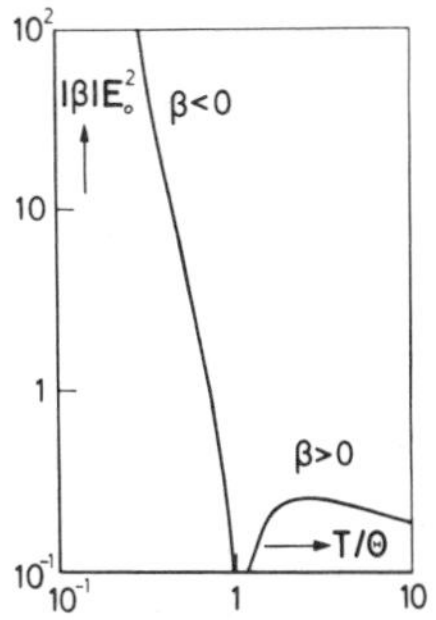

Fig.6.30 Dependence of $|\beta| E_0^2$ on lattice temperature T (in units of Θ) for polar optical scattering (after Stratton, ref.9).

assumption of a drifted Maxwell-Boltzmann distribution. For obtaining a better approximation of the true distribution function the variational method [2] and – more recently – the "Monte Carlo procedure" have been applied [10]. The latter is a computer simulation of the motion of a carrier in $\vec{k}$-space under the influence of both the applied electric field and the collisions. Sets of random numbers are applied for fixing both the scattering rate and the direction of carrier motion after each scattering event. Statistical convergence is obtained after some 10^4 events.

Another recently introduced method is the "iterative method" [11] modified by the concept of "self scattering" [12]. Let us assume that after n iterative steps we have arrived at a distribution function f_n. We then use this function to calculate the right-hand side of the Boltzmann equation which is determined by the scattering process. The distribution function on the left-hand side, which we denote as f_{n+1}, is then obtained by simple integration. "Self scattering" is an artificial non-physical process where the carrier in $\vec{k}$-space actually remains in position but which considerably simplifies the probability distribution for scattering. This method is well-suited e.g. for calculations of the ac conductivity in polar semiconductors [13].

In alkali halides where $\alpha \gtrsim 1$ (e.g. NaCl: $\alpha = 5.5$, AgBr : $\alpha = 1.6$) the concept of the "polaron" has been very fruitful. Because of the strong polar coupling, the lattice is polarized in the vicinity of a conduction electron and the polarization moves along with the electron. For this reason the electron effective mass, m_{pol}, is higher than without the polarization (where it would be m) and the ratio m_{pol}/m is a function of the polar coupling constant α. For the case of weak coupling, $\alpha \ll 1$, the polarization can be considered as a small perturbation and quantum mechanical perturbation theory yields [14]

$$m_{pol}/m = (1-\alpha/6)^{-1} \approx 1 + \alpha/6 \tag{61.29}$$

[10] T. Kurosawa, Proc. Int. Conf. Phys. Semicond. Kyoto 1966, J. Phys. Soc. Japan 21 Suppl. (1966) 424; W. Fawcett, A. D. Boardmann, and S. Swain, J. Phys. Chem. Solids 31 (1970) 1963.

[11] H. Budd, Proc. Int. Conf. Phys. Semicond. Kyoto 1966, J. Phys. Soc. Japan 21 Suppl. (1966) 420.

[12] H. D. Rees, J. Phys. Chem. Solids 30 (1969) 643; IBM J. Res. Develop. 13 (1969) 537.

[13] K. Seeger and H. Pötzl, Acta Phys. Austr., Suppl. X (1973) 341. Wien – New York: Springer. 1973; O. Zimmerl, thesis T. H. Wien 1972.

[14] For a brief review and further literature see e.g. C. Kittel: Quantum Theory of Solids, p. 137. New York: J. Wiley and Sons. 1963.

The average number of "virtual phonons" carried along by the electron is given by $a/2$. The conduction band edge is depressed by an amount $a\hbar\omega_0$. In a strong magnetic field, where Landau levels are formed in the energy bands (Chap.9b), the energy of the first of these levels is given by

$$\epsilon_{n=1} \approx \frac{3}{2}\hbar\omega_c(1-a/6) - a\hbar\omega_0 \tag{61.30}$$

where ω_c is the cyclotron frequency [15]. When an electron is strongly interacting with the polar lattice it is called a "polaron". (This should not be confused with the "polariton" mentioned in Chap.11g.) Hot carriers in polar semiconductors should accordingly be denoted as "hot polarons" [16]. At large carrier concentrations n the carrier-phonon interaction is screened, i.e. $a = a(n)$ [17].

6m. Carrier-Carrier Scattering

In a process where an electron is scattered by another electron ("e-e scattering") the total momentum of the electron gas is not changed. Hence electron-electron scattering as such has little influence on the mobility. However, since it is always combined with another scattering mechanism, which it may enhance, it can have quite an important influence. If e.g. an electron lost energy $\hbar\omega_0$ by optical phonon scattering and in $\vec{k}$-space is replaced due to e-e scattering by another electron with the same energy, the energy loss rate is enhanced.

The change in mobility by this process has been calculated for nonpolar semiconductors by Appel [1] and for polar semiconductors by Bate et al. [2]. Appel applied Kohler's variational method [3] which avoids the assumption of a relaxation time (the collision is inelastic!) and instead requires the rate of entropy production caused by all scattering processes involved to be a maximum for the steady state. For dominating ionized impurity scattering in a non-degenerate semiconductor, the mobility $\mu_I(0)$ is reduced to a value μ_I(e-e) by electron-electron scattering by a factor of about 0.6, while for extreme degeneracy there is no

[15] D. M. Larsen, Phys. Rev. 135 (1964) A 419; 144 (1966) 697; D. M. Larsen and E. J. Johnson, Proc. Int. Conf. Phys. Semic., Kyoto, 1966 (J. Phys. Soc. Japan 21, Suppl.) p. 443, Phys. Soc. Japan, Tokyo, 1966.

[16] M. Mikkor and F. C. Brown, Phys. Rev. 162 (1967) 848; J. W. Hodby, J. A. Borders, and F. C. Brown, J. Phys. C 3 (1970) 335.

[17] H. Ehrenreich, J. Phys. Chem. Solids 8 (1959) 130.

Chap.6m.

[1] J. Appel, Phys. Rev. 122 (1961) 1760.

[2] R. T. Bate, R. D. Baxter, F. J. Reid, and A. C. Beer, J. Phys. Chem. Solids 26 (1965) 1205.

[3] M. Kohler, Zeitschr. f. Physik 124 (1948) 772; 125 (1949) 679; E. H. Sondheimer, Proc. Roy. Soc. (London) A 203 (1950) 75.

reduction, which may be surprising at first glance. For the intermediate range Bate et al. find values for the reduction factor which are plotted vs the reduced Fermi energy, ζ_n/k_BT, in Fig.6.31. The upper scale gives values of the carrier density valid for n-type InSb at 80 K. There is a marked influence of electron-

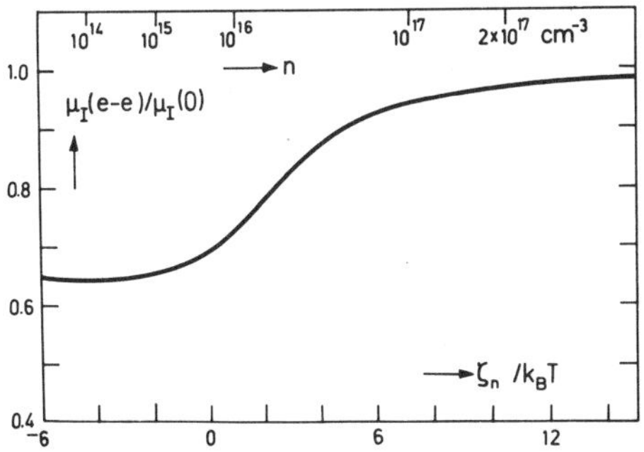

Fig.6.31 Electron-electron scattering correction for the mobility dominated by ionized impurity scattering, as a function of the reduced Fermi energy. The electron density given at the top scale is valid for the nonparabolic conduction band of InSb at 80 K (after Bate et al., ref.2).

electron scattering below $10^{17}/cm^3$ where $\zeta_n < 7 k_BT$. For small electron densities polar optical scattering will dominate over ionized impurity scattering and we have to consider the change of the polar optical scattering rate by electron-electron scattering. For even smaller electron densities ($n < 10^{14}/cm^3$) the influence of electron-electron scattering will vanish since it is proportional to n. (For impurity scattering in addition to electron-electron scattering, this is not quite so – see Fig.6.31 – because the impurity scattering rate is also proportional to n for $N_I = n$ and the ratio $\mu_I(e\text{-}e)/\mu_I(0)$ becomes independent of n.) When acoustic deformation potential scattering is predominant, electron-electron scattering has an even smaller effect. For n-type Ge containing 6.1×10^{14} shallow donors/cm^3 the maximum change is by a factor of 0.94; this maximum occurs at a temperature of 35 K [4].

In an intrinsic semiconductor electron-hole scattering may be of influence on the mobilities of both types of carriers. In hot carrier experiments this tends to keep electrons and holes at the same carrier temperature.

6n. Impurity Conduction and Hopping Processes

In heavily doped semiconductors one can no longer regard the impurity states as being localized. With increasing density of impurities the average distance between adjacent impurities becomes smaller and the energy levels of the impuri-

[4] T. P. McLean and E. G. S. Paige, J. Phys. Chem. Solids 16 (1960) 220.

ties broaden into bands. This is very similar to the process of forming energy bands in a crystal in the tight-binding approximation (Chap.2c) except that the impurities are distributed at random. (We will not discuss scattering by clusters [1].) These bands are at best half filled and therefore metallic conduction should be expected. However, the bands are narrow since the impurities are still relatively far apart, and therefore a very large effective mass and a low mobility are characteristic for "impurity conduction". In this respect impurity conduction is comparable to conduction in poor metals.

In n-type Ge impurity conduction prevails only at very low temperatures as shown [2] in Fig.6.32. In the curve labeled –1, normal conduction and carrier

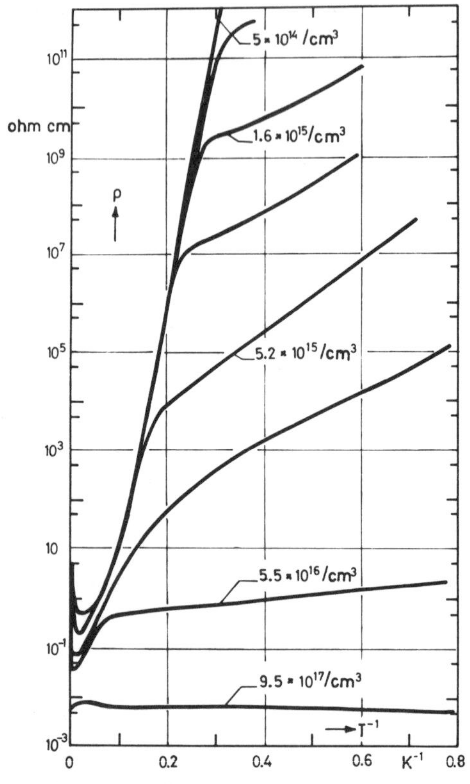

Fig.6.32 Resistivity of Sb-doped n-type Ge as a function of the reciprocal of the temperature (after H. Fritzsche, ref.2).

freeze-out persist up to the highest value of the resistivity. In the more heavily doped samples the curve flattens at the onset of impurity conduction. The Hall coefficient then drops markedly unless the doping is extremely high ($>10^{18}/cm^3$) and it remains independent of temperature down to the lowest temperature of about 2 K.

[1] L. R. Weisberg, J. Appl. Phys. 33 (1962) 1817.

[2] H. Fritzsche, J. Phys. Chem. Solids 6 (1958) 69.

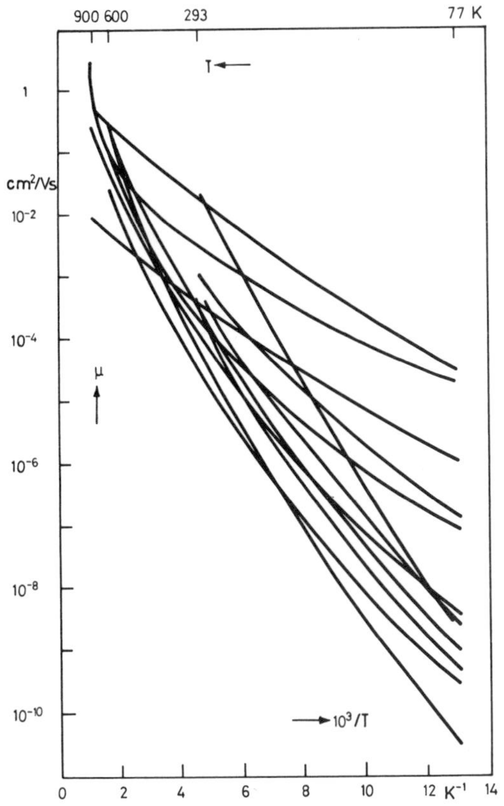

Fig.6.33
Temperature dependence of the mobility of various carbon-doped boron samples (after Klein and Geist, ref.3).

Fig.6.33 shows the mobility in heavily doped boron as a function of temperature [3]. The mobility increases with temperature exponentially with an average "activation energy" of 0.1 eV. Such a behavior, however, is more characteristic for "hopping processes" than for any type of metallic conduction: Carriers are thought to hop from one impurity atom to the next, and the Coulomb potential around the impurity atom is overcome by means of thermal energy.

The idea of hopping processes is supplemented in similar cases by the observation that the ac conductivity increases with frequency [4]. In heavily doped n-type Si at frequencies ν between 10^2 and 10^5 Hz and temperatures between 1 and 20 K an increase proportional to $\nu^{0.74} \cdots \nu^{0.79}$ has been found. Since hopping is a statistical process its probability will be given by $\int G(\tau) \exp(-t/\tau) d\tau$ where τ is the average time between two hops and $G(\tau)$ is a weight factor. The frequency

[3] W. Klein and D. Geist, Zeitschr. f. Physik 201 (1967) 411.
[4] M. Pollak and T. H. Geballe, Phys. Rev. 122 (1961) 1742.

dependence of the conductivity is then obtained by a Laplace transformation*):

$$\mathrm{Re}(\sigma) \propto \int_0^\infty G(\tau) \frac{\omega^2\tau^2}{1+\omega^2\tau^2} d\tau \qquad (6n.1)$$

If $G(\tau)$ is peaked at a value of $\tau \ll 1/\omega$, a dependence proportional to ω^2 would be expected; at $\tau \gg 1/\omega$ there would be no frequency dependence at all. Thus, by a suitable choice of the function $G(\tau)$ any exponent of ω between 0 and 2 over a certain range of frequencies can formally be explained. A more refined treatment applied to amorphous semiconductors has been given by Austin and Mott (ref. 9 of Chap.15b).

6o. Dislocation Scattering

Dislocations in a crystal lattice can be considered to belong to either one of two types shown in Fig.6.34: Edge-type or screw-type (spiral) [1]. For the for-

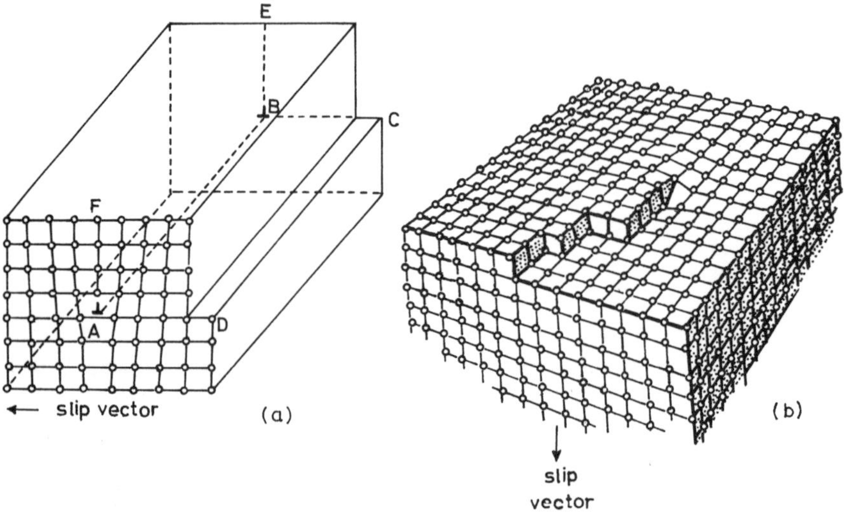

Fig.6.34 Schematic arrangement of atoms in a crystal containing (a) an edge-type dislocation, (b) a screw-type dislocation (after J. P. McKelvey: Solid-State and Semiconductor Physics. New York: Harper and Row. 1966.).

[1] J. Friedel: Dislocations. Oxford: Pergamon. 1964. S. Amelincks: The Direct Observation of Dislocations. Solid State Physics (F. Seitz and D. Turnbull, eds.) Suppl. 6. New York: Acad. Press. 1964.

*) An equivalent circuit for this problem is a condenser C in series with a resistor R: Application of a dc voltage results in a current first increasing in a step and then decreasing with time as $\exp(-t/\tau)$ where $\tau = RC$. The ac resistance is $R + 1/i\omega C$ and the real part of the conductance $R^{-1}\omega^2\tau^2/(1+\omega^2\tau^2)$.

mation of the edge-type dislocation an internal slip of the atomic arrays must occur perpendicular to the dislocation line A–B (slip vector B–C), while for the screw-type the slip vector is parallel to the dislocation line. The edge-type dislocation has been shown to introduce deep energy levels in germanium and other semiconductors. These levels have been considered as being due to dangling bonds which act as acceptors. However, experiments have shown that they can be occupied both by holes and electrons [3]. There is yet little agreement about the energy of the dislocation states even for the well-known semiconductor n-type Ge. Besides energy levels, also dislocation bands have been discussed [4].

It was suggested by Read [5] that dislocation lines are charged and surrounded by space charge cylinders. Carriers which move at an angle Θ relative to the cylinders are deflected and their mobility is reduced. From a simple mechanical model, involving specular reflection at the surface of impenetrable cylinders, a mean free path for the carriers, $l = 3/(8RN)$, was calculated where R is the cylinder radius and N is the number of dislocation lines per unit area. The scattering probability τ_m^{-1} was then assumed to be $(v/l) \sin \Theta$, where v is the carrier velocity, in addition to a term for lattice scattering. A further reduction in the mobility was assumed to be due to the reduction in crystal space available for the conduction electrons by the presence of the impenetrable cylinders.

A more satisfying treatment was given later by Bonch-Bruevich and Kogan [6]. They solved Poisson's equation in cylinder coordinates z, r, φ for a potential V(r)

$$\begin{aligned} d^2V/dr^2 + r^{-1}\, dV/dr &= (n - n_o)e/\kappa\kappa_o = n_o\{\exp(eV/k_BT) - 1\}e/\kappa\kappa_o \\ &\simeq n_o e^2 V/(\kappa\kappa_o k_B T) = V/L_D^2 \end{aligned} \tag{6o.1}$$

where n_o is the carrier density far away from a space charge cylinder and L_D is the Debye length given by Eq.(5b.24). A comparison of Eq.(6o.1) with Eq.(B.1) of appendix B shows that the solution is $V = AK_o(r/L_D)$ where A is a constant and K_o is a modified Bessel function of zero order. If Q is the charge of the dislocation line per unit length and $\vec{E} = -\vec{\nabla}_r V = AK_1/L_D$ is the electric field, we have

$$Q/\kappa\kappa_o = \int_0^\infty \{r^{-1} d(rE)/dr\} 2\pi r\, dr = 2\pi r E \Big|_0^\infty = 2\pi A \tag{60.2}$$

taking into account Eqs.(B.9),(B.13), and (B.16). Hence, the potential is given by

$$V(r) = (Q/2\pi\kappa\kappa_o) K_o(r/L_D) \tag{60.3}$$

By treating the present case of cylindrical symmetry similarly to the previous

---

[2] G. L. Pearson, W. T. Read, Jr., and F. J. Morin, Phys. Rev. 93 (1954) 93.
[3] W. Schröter, phys. stat. sol. 21 (1967) 211.
[4] W. Schröter and R. Labusch, phys. stat. sol. 36 (1969) 539.
[5] W. T. Read, Jr., Phil. Mag. 46 (1955) 111.
[6] V. L. Bonch-Bruevich and Sh. M. Kogan, Fiz. Tverd. Tela 1 (1959) 1221. [Engl. transl.: Sov. Phys. - Sol. State 1 (1959) 1118.]

case of spherical symmetry (Chap.6c), Pödör [7] was able to calculate the differential cross section

$$\sigma(\Theta) = \frac{2\pi m^2 e^2}{\hbar^4 k_t} \{ \int_0^\infty V(r) I_0(2k_t r \sin\Theta/2) r dr\}^2 = \frac{m^2 e^4 f^2}{8\pi\hbar^2 \kappa^2 \kappa_0^2 k_t^3 a^2} \cdot \frac{1}{(\sin^2\Theta/2 + \beta^{-2})^2} \tag{60.4}$$

and the reciprocal of the momentum relaxation time

$$\tau_m^{-1} = N v_t \int_0^{2\pi} (1 - \cos\Theta)\sigma(\Theta) d\Theta = \frac{N e^4 f^2 L_D}{8\kappa^2 \kappa_0^2 a^2 m^2} (v_t^2 + \hbar^2/4m^2\tau_d^2)^{-3/2} \tag{60.5}$$

where k_t and v_t are the components of $\vec{k}$ and $\vec{v}$ perpendicular to the dislocation lines, $I_0(t)$ is the zero-order Bessel function of the first kind, a is the distance between imperfection centers along the dislocation line and f is their occupation probability ($Q = ef/a$). Neglecting in the high temperature limit the second term in parenthesis, the mobility is given for a non-degenerate electron gas

$$\mu = \frac{30\sqrt{2\pi}\,\kappa^2 \kappa_0^2 a^2}{e^3 f^2 L_D m^{1/2}} \cdot \frac{(k_B T)^{3/2}}{N} \tag{60.6}$$

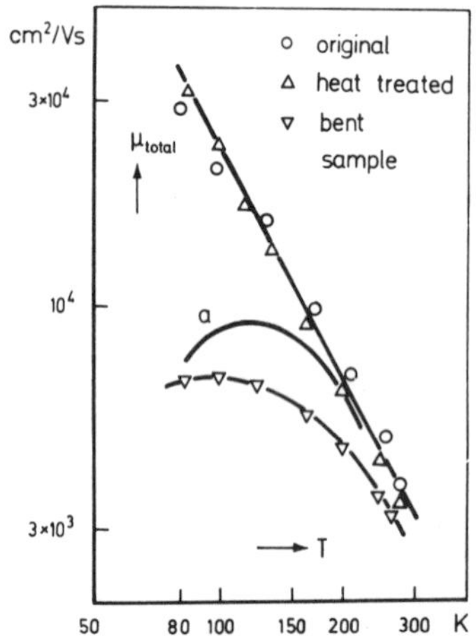

Fig.6.35 Mobility of electrons in n-type Germanium in the same sample before and after plastic bending. Curve (a) has been calculated for the bent sample (after Pödör, ref. 7).

Fig.6.35 (curve (a)) shows the temperature dependence of $(1/\mu + 1/\mu_l)^{-1}$ where μ_l is the lattice mobility, for $N = 10^7/cm^2$, $\kappa = 16$, $m = 0.3\, m_0$, and $a = 3.5$ Å. The straight line represents $\mu(T)$ for the undeformed sample ($N = 4x10^3/cm^2$), the lower curve for the bent sample. N has been calculated from the bending radius. The agreement between the lower curve and curve (a) is considered reasonable. Although the statement has not been made it may be assumed that measurements were made on the sample as a whole. Van Weeren et al. [8] have shown, however, that in bars cut from the bent sample at the neutral plane ("B") the mobility is different from that in bars cut above or below the neutral plane ("A"), that it is different in crystals bent about the <110> direction from that bent about the <211> direction and, of course, depends upon whether measurements are made with the current parallel or perpendicular to the bending axis. All of these cases are illustrated in Fig.6.36. As one might suppose the I-V characteristics for

[7] B. Pödör, Acta Physica Acad. Sci. Hung. 23 (1967) 393; phys.stat.sol. 16 (1966) K 167.

[8] J. H. P. van Weeren, R. Struikmans, and J. Blok, phys. stat. sol. 19 (1967) K 107.

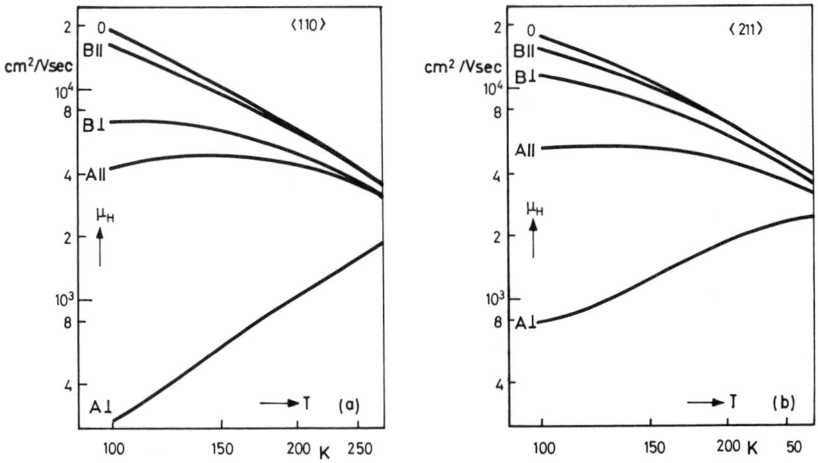

Fig.6.36 Hall mobility for n-type germanium in undeformed crystals and in crystals bent at 1103 K about the<110>(left) and the<211> direction (right diagram) (after Van Weeren et al., ref. 8).

hot carriers is different parallel and perpendicular to the dislocation lines [9].

[9] A. F. Gibson, J. Phys. Chem. Solids 8 (1959) 147.

7. Charge Transport and Scattering Processes in the Many-Valley Model

In Chap.2d, Figs.2.25 and 2.26, we have seen that the conduction bands of silicon and germanium near the band edges have constant energy surfaces which are either 8 half-ellipsoids or 6 ellipsoids of revolution; these correspond to 4 and 6 energy valleys, respectively. In these and many other semiconductors the "many-valley-model" of the energy bands has proved to be a fruitful concept for a description of the observed anisotropy of electrical and optical phenomena. Cyclotron resonance (Chap.11k) provides a direct experimental determination of the effective masses in each valley for any crystallographic direction.

In the present chapter we will calculate the most important galvanomagnetic effects in the many-valley model of energy bands, deal with intervalley transitions, and finally arrive at the high-field domain and the acousto-electric instabilities; the latter are considered also for piezoelectric coupling in a one-valley model since the calculation is similar to that for intervalley coupling in non-polar many-valley semiconductors.

7a. The Deformation Potential Tensor

By Eqs.(6d.5) and (6d.6) we have defined an "acoustic deformation potential constant", ϵ_{ac}, which in a crystal is actually a tensor. Let us first consider quite generally the 6 components of a deformation in a crystal [1]. In the undeformed crystal we assume a Cartesian coordinate system with unit vectors $\vec{a}$, $\vec{b}$, $\vec{c}$. By a small deformation these vectors are displaced. The displaced unit vectors are called $\vec{a}'$, $\vec{b}'$, $\vec{c}'$, where the origin of the coordinate system is taken at the same lattice point as before the deformation. The displacement is then simply a rotation described by the following system of equations:

$$\left.\begin{aligned} \vec{a}' &= (1+\epsilon_{xx})\vec{a} + \epsilon_{xy}\vec{b} + \epsilon_{xz}\vec{c} \\ \vec{b}' &= \epsilon_{yx}\vec{a} + (1+\epsilon_{yy})\vec{b} + \epsilon_{yz}\vec{c} \\ \vec{c}' &= \epsilon_{zx}\vec{a} + \epsilon_{zy}\vec{b} + (1+\epsilon_{zz})\vec{c} \end{aligned}\right\} \quad (7a.1)$$

[1] See e.g. C. Kittel: Introduction to Solid State Physics, chap. 4. New York: J. Wiley and Sons. 1965.

where the ϵ_{xx}, ϵ_{xy} etc. are assumed to be small quantities and their products being negligibly small. Therefore the cosines of angles between $\vec{a}'$ and $\vec{b}'$, $\vec{b}'$ and $\vec{c}'$, and $\vec{c}'$ and $\vec{a}'$ denoted as e_4, e_5, and e_6, respectively, are given by

$$e_4 = (\vec{a}'\,\vec{b}') = \epsilon_{yx} + \epsilon_{xy}\,;\quad e_5 = (\vec{b}'\,\vec{c}') = \epsilon_{zy} + \epsilon_{yz}\,;\quad e_6 = (\vec{c}'\,\vec{a}') = \epsilon_{zx} + \epsilon_{xz} \tag{7a.2}$$

With the notation

$$e_1 = \epsilon_{xx}\,;\quad e_2 = \epsilon_{yy}\,;\quad e_3 = \epsilon_{zz} \tag{7a.3}$$

we thus have given all 6 deformation components which have already been applied in Eq.(41.4). The volume V = 1 of a cube having edges $\vec{a}$, $\vec{b}$, and $\vec{c}$ is changed by the deformation to

$$1 + \delta V = (\vec{a}'\,[\vec{b}'\vec{c}']) \approx 1 + \epsilon_{xx} + \epsilon_{yy} + \epsilon_{zz} = 1 + e_1 + e_2 + e_3 \tag{7a.4}$$

Hence, the volume dilatation is given by

$$\delta V/V = e_1 + e_2 + e_3 \tag{7a.5}$$

The shift $\vec{r}' - \vec{r}$ of a mass point caused by the deformation can be described in the coordinate system $\vec{a}$, $\vec{b}$, $\vec{c}$ by coefficients u, v, w :

$$\vec{r}' - \vec{r} = u\vec{a} + v\vec{b} + w\vec{c} \tag{7a.6}$$

On the other hand $\vec{r}' = x\vec{a}' + y\vec{b}' + z\vec{c}'$ and $\vec{r} = x\vec{a} + y\vec{b} + z\vec{c}$ and therefore from Eq.(7a.1) we find

$$\left.\begin{aligned} u &= x\epsilon_{xx} + y\epsilon_{yx} + z\epsilon_{zx} \\ v &= x\epsilon_{xy} + y\epsilon_{yy} + z\epsilon_{zy} \\ w &= x\epsilon_{xz} + y\epsilon_{yz} + z\epsilon_{zz} \end{aligned}\right\} \tag{7a.7}$$

Each one of the 6 coefficients $e_1, e_2, \cdots e_6$ given by Eqs.(7a.2) and (7a.3) can now be expressed by derivatives $\partial u/\partial x$ etc.:

$$\begin{aligned} &e_1 = \partial u/\partial x;\quad e_2 = \partial v/\partial y;\quad e_3 = \partial w/\partial z; \\ &e_4 = \partial v/\partial x + \partial u/\partial y;\quad e_5 = \partial w/\partial y + \partial v/\partial z;\quad e_6 = \partial u/\partial z + \partial w/\partial x \end{aligned} \tag{7a.8}$$

The change of e.g. the conduction band edge, $\Delta\epsilon_c$, of a given valley due to a deformation may be expanded in a Taylor series:

$$\Delta\epsilon_c = \frac{\partial\epsilon}{\partial\epsilon_{xx}} e_1 + \frac{\partial\epsilon}{\partial\epsilon_{yy}} e_2 + \frac{\partial\epsilon}{\partial\epsilon_{zz}} e_3 + \frac{\partial\epsilon}{\partial\epsilon_{xy}} e_4 + \frac{\partial\epsilon}{\partial\epsilon_{yz}} e_5 + \frac{\partial\epsilon}{\partial\epsilon_{xz}} e_6 + \cdots \tag{7a.9}$$

We choose the x-axis of the coordinate system as the rotational axis of the ellipsoidal energy surface and introduce two constants [2]

$$\left.\begin{aligned} \Xi_d &= \frac{1}{2}\left(\frac{\partial\epsilon}{\partial\epsilon'_{yy}} + \frac{\partial\epsilon}{\partial\epsilon'_{zz}}\right) \\ \Xi_u &= \frac{1}{2}\left(2\frac{\partial\epsilon}{\partial\epsilon'_{xx}} - \frac{\partial\epsilon}{\partial\epsilon'_{yy}} - \frac{\partial\epsilon}{\partial\epsilon'_{zz}}\right) = \frac{\partial\epsilon}{\partial\epsilon'_{xx}} - \Xi_d \end{aligned}\right\} \tag{7a.10}$$

[2] C. Herring , Bell Syst. Tech. J. 34 (1955) 237.

where the deformation relative to the coordinate system of the ellipsoid is denoted by ϵ'_{ii}. We will show that Ξ_u is the deformation potential constant for a shear deformation along the symmetry axis of the valley and $\Xi_d + \frac{1}{3}\Xi_u$ is the deformation potential constant for a dilatation.

Let us first consider a conduction band valley in germanium located along a <111> axis. Symmetry considerations yield [3] $\partial\epsilon/\partial\epsilon_{xx} = \partial\epsilon/\partial\epsilon_{yy} = \partial\epsilon/\partial\epsilon_{zz}$ and $\partial\epsilon/\partial\epsilon_{xy} = \partial\epsilon/\partial\epsilon_{xz} = \partial\epsilon/\partial\epsilon_{yz}$. Hence, we obtain

$$\Delta\epsilon_c = (\partial\epsilon/\partial\epsilon_{xx})(e_1 + e_2 + e_3) + (\partial\epsilon/\partial\epsilon_{xy})(e_4 + e_5 + e_6) + \cdots \tag{7a.11}$$

A rotation of the x-axis into the symmetry axis of the ellipsoid may e.g. be described by the matrix

$$D_1 = \frac{1}{\sqrt{6}}\begin{pmatrix} \sqrt{2} & \sqrt{2} & \sqrt{2} \\ \sqrt{3} & -\sqrt{3} & 0 \\ 1 & 1 & -2 \end{pmatrix}; \quad D_1^{-1} = \frac{1}{\sqrt{6}}\begin{pmatrix} \sqrt{2} & \sqrt{3} & 1 \\ \sqrt{2} & -\sqrt{3} & 1 \\ \sqrt{2} & 0 & -2 \end{pmatrix} \tag{7a.12}$$

The ϵ-tensor with components $\epsilon_{xx}, \epsilon_{xy}, \cdots$ is thus related to the ϵ'-tensor with components $\epsilon'_{xx}, \epsilon'_{xy}, \cdots$ by the transformation

$$\epsilon = D_1^{-1}\,\epsilon' D_1 \tag{7a.13}$$

A calculation of the sum $e_4 + e_5 + e_6$ from this relation yields

$$e_4 + e_5 + e_6 = 2\epsilon'_{xx} - \epsilon'_{yy} - \epsilon'_{zz} = 2e'_1 - e'_2 - e'_3 \tag{7a.14}$$

The sum $e_1 + e_2 + e_3$ is obtained from the fact that the dilatation given by Eq. (7a.4) is invariant to the coordinate transformation:

$$e_1 + e_2 + e_3 = e'_1 + e'_2 + e'_3 \tag{7a.15}$$

The last two equations yield for $\Delta\epsilon_c$:

$$\Delta\epsilon_c = \left(\frac{\partial\epsilon}{\partial\epsilon_{xx}} + 2\frac{\partial\epsilon}{\partial\epsilon_{xy}}\right)e'_1 + \left(\frac{\partial\epsilon}{\partial\epsilon_{xx}} - \frac{\partial\epsilon}{\partial\epsilon_{xy}}\right)(e'_2 + e'_3) \tag{7a.16}$$

In the principal-axis system this is per definition:

$$\Delta\epsilon_c = (\partial\epsilon/\partial\epsilon'_{xx})e'_1 + (\partial\epsilon/\partial\epsilon'_{yy})e'_2 + (\partial\epsilon/\partial\epsilon'_{zz})e'_3 \tag{7a.17}$$

Due to rotational symmetry $\partial\epsilon/\partial\epsilon'_{zz} = \partial\epsilon/\partial\epsilon'_{yy}$. A comparison of Eqs.(7a.16) and (7a.17) yields

$$\frac{\partial\epsilon}{\partial\epsilon_{xx}} + 2\frac{\partial\epsilon}{\partial\epsilon_{xy}} = \frac{\partial\epsilon}{\partial\epsilon'_{xx}} \quad \text{and} \quad \frac{\partial\epsilon}{\partial\epsilon_{xx}} - \frac{\partial\epsilon}{\partial\epsilon_{xy}} = \frac{\partial\epsilon}{\partial\epsilon'_{yy}} \tag{7a.18}$$

From the definition, Eq.(7a.10), we find that

$$\partial\epsilon/\partial\epsilon'_{xx} = \Xi_u + \Xi_d \quad \text{and} \quad \partial\epsilon/\partial\epsilon'_{yy} = \Xi_d \tag{7a.19}$$

Solving Eqs.(7a.18) for $\partial\epsilon/\partial\epsilon_{xx}$ and $\partial\epsilon/\partial\epsilon_{yy}$ thus yields

[3] C. Herring and E. Vogt, Phys. Rev. 101 (1956) 944. Err.: ibid. 105 (1957) 1933.

$$\partial\epsilon/\partial\epsilon_{xx} = \Xi_d + \frac{1}{3}\Xi_u \quad \text{and} \quad \partial\epsilon/\partial\epsilon_{xy} = \frac{1}{3}\Xi_u \tag{7a.20}$$

We finally obtain for $\Delta\epsilon_c$

$$\Delta\epsilon_c = (\Xi_d + \frac{1}{3}\Xi_u)(e_1 + e_2 + e_3) + \frac{1}{3}\Xi_u(e_4 + e_5 + e_6) \tag{7a.21}$$

For the case of hydrostatic pressure, $e_4 = e_5 = e_6 = 0$ and $\Delta\epsilon_c \propto \delta V/V$ where the factor of proportionality is given by

$$\epsilon_{ac} = \Xi_d + \frac{1}{3}\Xi_u \tag{7a.22}$$

Experimental values of Ξ_u obtained from piezoresistance or from the acousto-electric effect in n-type Ge are between 15.8 and 19.3 eV. By the positive sign of Ξ_u we mean that by a uniaxial compression the energy of a <111>-valley is lowered relative to the three other valleys. Ξ_u is temperature dependent. Above 60 K it increases with temperature and reaches a value of 21 eV at 100 K [4]. At 77 K the ratio Ξ_d/Ξ_u is about -0.38 according to Herring and Vogt [3] and -0.45 according to Smith Jr. [5]. From these data values of -0.88 eV and -1.8 eV are calculated for $\Xi_d + \frac{1}{3}\Xi_u$. A hydrostatic pressure of 30 000 kg/cm² increases the <111> minima linearly by 0.03 eV while the <100> valleys are lowered such that both types of valleys have the same energy. The central valley at $\vec{k} = 0$ which without pressure is 0.15 eV above the <111> "satellite valleys" is raised by 0.21 eV by a pressure of 30 000 kg/cm² [6].

In silicon the lowest conduction band valleys are located on <100> and equivalent axes. For the <100> valley no transformation is required and due to symmetry $\partial\epsilon/\partial\epsilon_{zz} = \partial\epsilon/\partial\epsilon_{yy}$, $\partial\epsilon/\partial\epsilon_{xy} = 0$ etc.:

$$\Delta\epsilon_c = (\partial\epsilon/\partial\epsilon_{xx})e_1 + (\partial\epsilon/\partial\epsilon_{yy})(e_2 + e_3) \tag{7a.23}$$

Since $\epsilon'_{xx} = \epsilon_{xx}$ etc., Eq.(7a.10) yields

$$\Xi_d = \partial\epsilon/\partial\epsilon_{yy}\ ; \quad \Xi_d + \Xi_u = \partial\epsilon/\partial\epsilon_{xx} \tag{7a.24}$$

Therefore $\Delta\epsilon_c$ becomes

$$\Delta\epsilon_c = \Xi_d(e_1 + e_2 + e_3) + \Xi_u e_1 = (\Xi_d + \frac{1}{3}\Xi_u)(e_1 + e_2 + e_3) + \frac{1}{3}\Xi_u\{(e_1 - e_2) + (e_1 - e_3)\} \tag{7a.25}$$

$\Delta\epsilon_c$ for the other valleys is easily obtained by a transformation.

Experimental values for Ξ_u of silicon are between 8.5 and 9.6 eV. The deformation potential constant $\Xi_d \ll \Xi_u$ is not well known [7].

[4] S. H. Koenig: Proc. Int. School of Physics Vol.XXII, p. 515. New York: Acad. Press. 1961.

[5] J. E. Smith, Jr., Appl. Phys. Lett. 12 (1968) 233.

[6] E. G. S. Paige, Proc. in Semicond. 8 (1960) 158, Fig. 36.

[7] H. Heinrich and M. Kriechbaum, J. Phys. Chem. Solids 31 (1970) 927; K. Bulthuis, Philips Res. Repts. 23 (1968) 25.

7b. Electrical Conductivity

We consider first valley # ρ located in $\vec{k}$-space at $\vec{k}_\rho$ with its 3 main axes parallel to the coordinate axes. The effective masses in the directions of the axes are denoted by m_1, m_2, and m_3. It is convenient to introduce an "inverse mass tensor" with diagonal elements $\alpha_{ii} = m_\sigma/m_i$ where the "conductivity-effective mass" m_σ will be defined later (Eq.7b.18). The energy of a carrier is then given by

$$\epsilon = (\hbar^2/2m_\sigma)(\vec{k}-\vec{k}_\rho)\,\alpha\,(\vec{k}-\vec{k}_\rho) \tag{7b.1}$$

Since α is a diagonal tensor this relation is simplified to

$$\epsilon = (\hbar^2/2m_\sigma)w^2 \tag{7b.2}$$

by introducing a vector $\vec{w}$ with components

$$w_i = \sqrt{\alpha_{ii}}\,(k_i - k_{\rho i}). \tag{7b.3}$$

By this relation an ellipsoidal surface in $\vec{k}$-space is transformed in a spherical surface in $\vec{w}$-space [1].

In an electric field $\vec{E}$ and a magnetic induction $\vec{B}$ the distribution $f(\vec{k})$ of carriers is determined by the Boltzmann equation in the relaxation time approximation

$$(e/\hbar)(\vec{E} + \hbar^{-1}[\vec{\nabla}_k \epsilon \vec{B}])\,\vec{\nabla}_k f = -(f-f_0)/\tau_m \tag{7b.4}$$

Since for integrals in $\vec{k}$-space $dk_i = dw_i/\sqrt{\alpha_{ii}}$ and $fd^3k = fd^3w/\sqrt{\alpha_{11}\alpha_{22}\alpha_{33}} = gd^3w$, we define a function

$$g = f/\sqrt{\alpha_{11}\alpha_{22}\alpha_{33}} \tag{7b.5}$$

which for thermal equilibrium is denoted as

$$g_0 = f_o/\sqrt{\alpha_{11}\alpha_{22}\alpha_{33}} \tag{7b.6}$$

Since $\vec{\nabla}_k = \sqrt{\alpha}\,\vec{\nabla}_w$ we obtain from Eq.(7b.4) by dividing by $\sqrt{\alpha_{11}\alpha_{22}\alpha_{33}}$:

$$(e/\hbar)(\vec{E} + \hbar^{-1}[\sqrt{\alpha}\,\vec{\nabla}_w \epsilon \vec{B}])\sqrt{\alpha}\,\vec{\nabla}_w g = -(g-g_0)/\tau_m \tag{7b.7}$$

By introducing "effective field intensities"

$$E_i^* = \sqrt{\alpha_{11}}\;E_i \quad\text{and}\quad B_i^* = \sqrt{\alpha_{11}\alpha_{22}\alpha_{33}/\alpha_{ii}}\;B_i \tag{7b.8}$$

we bring Eq.(7b.7) into the form of Eq.(7b.4); the first term of Eq.(7b.7) becomes

$$(\vec{E}\sqrt{\alpha}\,\vec{\nabla}_w g) = \sum_i E_i\sqrt{\alpha_{ii}}\;\partial g/\partial w_i = \sum_i E_i^*\,\partial g/\partial w_i = (\vec{E}^*\vec{\nabla}_w g) \tag{7b.9}$$

and the second term

[1] C. Herring and E. Vogt, Phys. Rev. 101 (1956) 944. Err.: ibid. 105 (1957) 1933.

$$[\sqrt{a}\ \vec{\nabla}_w \epsilon \vec{B}]\ \sqrt{a}\ \vec{\nabla}_w g = \{\sqrt{a_{22}}\ (\partial\epsilon/\partial w_2)B_3 - \sqrt{a_{33}}\ (\partial\epsilon/\partial w_3)B_2\}\ \sqrt{a_{11}}\ \partial g/\partial w_1 + \text{cycl. permut.}$$
$$= \{(\partial\epsilon/\partial w_2)\ \sqrt{a_{11}\, a_{22}}\ B_3 - (\partial\epsilon/\partial w_3)\ \sqrt{a_{11}\, a_{33}}\ B_2\}\ \partial g/\partial w_1 + \text{cycl. permut.}$$
$$= ([\vec{\nabla}_w \epsilon \vec{B}^*]\ \vec{\nabla}_w g) \qquad (7b.10)$$

The Joule heat must be invariant to the transformation

$$(\vec{j}\vec{E}) = (\vec{j}^* \vec{E}^*) = (\vec{j}^* \sqrt{a}\ \vec{E}) = (\sqrt{a}\ \vec{j}^* \vec{E}) \qquad (7b.11)$$

This shows that $\vec{j}$ is transformed in the following way:

$$\vec{j} = \sqrt{a}\ \vec{j}^* \qquad (7b.12)$$

In $\vec{w}$-space the conductivity tensor $\sigma_w = \sigma_w(B^*)$ is defined by

$$\vec{j}^* = \sigma_w\ \vec{E}^* \qquad (7b.13)$$

Multiplication of this equation by $\sqrt{a}$ yields

$$\vec{j} = \sqrt{a}\ \sigma_w \vec{E}^* = \sqrt{a}\ \sigma_w\ \sqrt{a}\ \vec{E} = \sigma\vec{E} \qquad (7b.14)$$

This shows how σ is transformed:

$$\sigma = \sqrt{a}\ \sigma_w\ \sqrt{a} \qquad (7b.15)$$

We will consider here only the case of weak magnetic fields where the conductivity in a spherical valley, σ_w, is given by Eq.(4c.53) with $\vec{B}$ replaced by $\vec{B}^*$ and m by m_σ. The simple case of no magnetic field, $\vec{B}^* = 0$, will be considered first. There we have

$$\sigma = a\sigma_0 = a\ \frac{ne^2 < \tau_m >}{m_\sigma} \begin{pmatrix} 1 & 0 & 0 \\ 0 & 1 & 0 \\ 0 & 0 & 1 \end{pmatrix} \qquad (7b.16)$$

In a crystal lattice of cubic symmetry the conductivity is isotropic. Due to symmetry at least 3 valleys form the energy band. If there are 3 valleys their long axes are perpendicular to each other. Each one of the other two valleys first has to be transformed into the position of the first valley before the transformation equation Eq.(7b.3) can be applied. This is done by matrices

$$D_2 = D_2^{-1} = \begin{pmatrix} 0 & 1 & 0 \\ 1 & 0 & 0 \\ 0 & 0 & -1 \end{pmatrix} \quad \text{and} \quad D_3 = D_3^{-1} = \begin{pmatrix} 0 & 0 & 1 \\ 0 & -1 & 0 \\ 1 & 0 & 0 \end{pmatrix} \qquad (7b.17)$$

while, of course, D_1 is the unit matrix. By summation over all 3 valleys and taking into account that n, contained in σ_0 given by Eq.(4c.40), is the total carrier concentration assumed to be equally distributed among the 3 valleys (n/3 per valley), we find for an isotropic relaxation time $<\tau_m>$:

$$\sigma = \frac{1}{3} \Sigma\, a_{ii} \cdot \sigma_0 = \frac{ne^2 < \tau_m >}{m_\sigma}\ \frac{1}{3}\ \Sigma\, a_{ii} = \frac{ne^2 < \tau_m >}{m_\sigma} \qquad (7b.18)$$

For 6 valleys such as in silicon the same result is obtained as in the 3-valley model. Since $a_{ii} = m_o/m_i$ we finally obtain a "conductivity effective mass" m_σ given by

$$\frac{1}{m_\sigma} = \frac{1}{3}\left(\frac{1}{m_1} + \frac{1}{m_2} + \frac{1}{m_3}\right) \tag{7b.19}$$

The conductivity effective mass is also the mass in any particular valley for a direction which is symmetrical to all valleys, e.g. for the 3-valley model in <100> and equivalent directions:

$$\hbar^2 k^2/2m_\sigma = \hbar^2 k_1^2/2m_1 + \hbar^2 k_2^2/2m_2 + \hbar^2 k_3^2/2m_3 \tag{7b.20}$$

where $k_1^2 = k_2^2 = k_3^2 = \frac{1}{3} k^2$.

The "density-of-states effective mass" m_D is defined by

$$m_D = \sqrt[3]{m_1 m_2 m_3} \tag{7b.21}$$

since the density of states is proportional to $\int d^3k = \int d^3w/\sqrt{a_{11} a_{22} a_{33}}$.

We will now consider the case of germanium where there are 4 valleys located on the cube diagonals in $\vec{k}$-space as shown in Fig.2.26. The rotation matrix for the <111>-ellipsoid is given by Eq.(7a.11). Fig.7.1 shows the transformation of

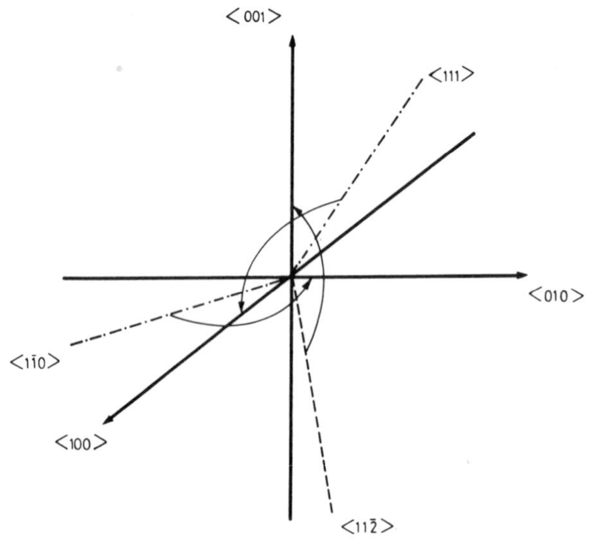

Fig.7.1 Rotation D_1 of axes of the Cartesian coordinate system.

axes into the <100>-, <010>-, and <001>-axes of the Cartesian coordinate system. It is important that the rotation does not include a mirror image, i.e. a transformation of a right-hand system into a left-hand system. The rotation matrices for the other three valleys have the same elements as D_1 except for sign and we simply give the signs :

$$D_2=\begin{pmatrix} + & + & - \\ + & - & 0 \\ - & - & - \end{pmatrix};\quad D_3=\begin{pmatrix} + & - & + \\ + & + & 0 \\ - & + & + \end{pmatrix}\quad \text{and}\quad D_4=\begin{pmatrix} - & + & + \\ + & + & 0 \\ - & + & - \end{pmatrix} \tag{7b.22}$$

Both the electric field $\vec{E}$ and the current density $\vec{j}$ are of course rotated in the same way, i.e. in Eq.(7b.14) we replace $\vec{j}$ by $D_\rho\vec{j}$ and $\vec{E}$ by $D_\rho\vec{E}$:

$$D_\rho\vec{j} = \sqrt{a}\,\sigma_w\,\sqrt{a}\,D_\rho\vec{E} \tag{7b.23}$$

Multiplication by D_ρ^{-1} yields $\vec{j} = \sigma\vec{E}$ with σ given by

$$\sigma = D_\rho^{-1}\sqrt{a}\,\sigma_w\,\sqrt{a}\,D_\rho \tag{7b.24}$$

For $\vec{B} = 0$, σ_w is diagonal and $\sqrt{a}\,\sigma_w\sqrt{a} = \sigma_0 a$. Let us introduce the following notations: The effective mass in the direction of the longitudinal axis of the rotational ellipsoid, m_l, and in the direction of the transverse axis, m_t, the anisotropy factor*) $K = m_l/m_t$, the ratio $(K-1)/(2K+1) = \lambda$, and the matrix

$$D_\rho^{-1} a D_\rho = \Lambda_\rho \tag{7b.25}$$

where e.g. Λ_1 is given by

$$\Lambda_1 = \frac{m_\sigma}{m_l}\cdot\frac{2K+1}{3}\cdot\begin{pmatrix} 1 & -\lambda & -\lambda \\ -\lambda & 1 & -\lambda \\ -\lambda & -\lambda & 1 \end{pmatrix} \tag{7b.26}$$

By a summation over all 4 valleys and taking into account that n, contained in σ_0 as given by Eq.(4c.40), is the total electron concentration which is equally distributed between the 4 valleys, we find

$$\sigma = \sigma_0\frac{1}{4}\sum_{\rho=1}^{4}\Lambda_\rho = \sigma_0\frac{m_\sigma}{4m_l}\cdot\frac{2K+1}{3}\,4 = \sigma_0 \tag{7b.27}$$

where for m_σ its value given by Eq.(7b.19) and $m_1 = m_2 = m_t$, $m_3 = m_l$ have been substituted :

$$m_\sigma = m_l\,3/(2K+1) \tag{7b.28}$$

In a model where 6 or 12 valleys are located on the <110> and equivalent axes [2], the conductivity σ_0 is also isotropic with an effective mass given by Eq. (7b.28), again assuming an isotropic relaxation time τ_m .

---

[2] M. Shibuya, Phys. Rev. 95 (1954) 1385; Physica 20 (1954) 971.
A. C. Beer: Solid State Physics, Suppl. 4, p. 228. (F. Seitz and D. Turnbull, eds.). New York: Acad. Press. 1963.

*) For simplicity we omit here the normal subscript m to K since the relaxation time has been assumed isotropic.

7c. Hall Effect in a Weak Magnetic Field

The terms in σ which are linear in B are given by Eq.(4c.42). As considered above we now have to replace $\vec{B}$ by $\vec{B}^*$ and m by m_σ. The Herring-Vogt transformation Eq.(7b.24) yields for <111> valleys such as in n-Ge :

$$\sigma = \frac{1}{4} \sum_{\rho=1}^{4} D_\rho^{-1} \sqrt{a} \begin{pmatrix} 0 & \gamma_o B_3^* & -\gamma_o B_2^* \\ -\gamma_o B_3^* & 0 & \gamma_o B_1^* \\ \gamma_o B_2^* & -\gamma_o B_1^* & 0 \end{pmatrix} \sqrt{a}\, D_\rho \tag{7c.1}$$

where γ_o is given by Eq.(4c.43) with m replaced by m_σ and $\vec{B}^*$ in valley #ρ has the components

$$B^*_{i(\rho)} = \sqrt{a_{11}a_{22}a_{33}/a_{ii}} \sum_{k=1}^{4} D_\rho^{ik} B_k \tag{7c.2}$$

E.g. for valley #1 which is in the <111> direction, $\vec{B}^*_{(1)}$ is given by

$$\vec{B}^*_{(1)} = \frac{m_\sigma}{m_l} \sqrt{\frac{K}{6}} \begin{pmatrix} \sqrt{2K}\,(B_1 + B_2 + B_3) \\ \sqrt{3}\,(B_1 - B_2) \\ B_1 + B_2 - 2B_3 \end{pmatrix} \tag{7c.3}$$

Let us introduce for a shorthand notation

$$\lambda' = (K + 2)/(K - 1) \tag{7c.4}$$

We obtain for the component σ_{12} of the conductivity tensor in the first valley

$$\begin{aligned} \sigma_{12}^{(1)} &= \frac{1}{4}\gamma_o \frac{m_\sigma}{m_l} \frac{\sqrt{K}}{6} (-2\sqrt{6}\, B_3^* + 2\sqrt{3K}\, B_1^*) \\ &= \frac{1}{4}\gamma_o \left(\frac{m_\sigma}{m_l}\right)^2 \frac{K}{3} (K-1)(B_1 + B_2 + \lambda' B_3) \end{aligned} \tag{7c.5}$$

$\sigma_{13}^{(1)}$ and $\sigma_{23}^{(1)}$ are similar except that the last factor is replaced by $(-B_1 - \lambda' B_2 - B_3)$ and by $(\lambda' B_1 + B_2 + B_3)$, respectively. For valley # 2 we find :

$$\vec{B}^*_{(2)} = \frac{m_\sigma}{m_l} \sqrt{\frac{K}{6}} \begin{pmatrix} \sqrt{2K}\,(B_1 + B_2 + B_3) \\ \sqrt{3}\,(B_1 - B_2) \\ -B_1 - B_2 - 2B_3 \end{pmatrix} \tag{7c.6}$$

and for the conductivity component σ_{12} :

$$\begin{aligned} \sigma_{12}^{(2)} &= \frac{1}{4}\gamma_o \frac{m_\sigma}{m_l} \frac{\sqrt{K}}{6} (-2\sqrt{6}\, B_3^* - 2\sqrt{3K}\, B_1^*) \\ &= \frac{1}{4}\gamma_o \left(\frac{m_\sigma}{m_l}\right)^2 \frac{K}{3} (K-1)(-B_1 - B_2 + \lambda' B_3) \end{aligned} \tag{7c.7}$$

The calculation for the other components and the other valleys is similar. By a summation over all 4 valleys we obtain

$$\sigma = \frac{1}{4}\,\gamma_0 \left(\frac{m_\sigma}{m_l}\right)^2 \frac{K}{3}\,(K-1)\,4\lambda' \begin{pmatrix} 0 & B_3 & -B_2 \\ -B_3 & 0 & B_1 \\ B_2 & -B_1 & 0 \end{pmatrix} \tag{7c.8}$$

The dependence of σ on $\vec{B}$ is the same as in the spherical one-valley model except that γ_0 is now replaced by a coefficient which we denote as γ_H:

$$\gamma_H = \gamma_0 \left(\frac{m_\sigma}{m_l}\right)^2 \frac{K}{3}\,(K-1)\lambda' = (ne^3/m_H^2) <\tau_m^2> \tag{7c.9}$$

where the "Hall effective mass",

$$m_H = m_l \sqrt{3/\{K(K+2)\}} \tag{7c.10}$$

has been introduced and the momentum relaxation time τ_m has been assumed to be isotropic.

For a calculation of the ratio of the Hall and drift mobilities, μ_H/μ, we may assume without loss of generality an electric field in the y-direction and a magnetic field in the z-direction. The Hall current density

$$\vec{j} = \gamma_H \begin{pmatrix} 0 & B_3 & 0 \\ -B_3 & 0 & 0 \\ 0 & 0 & 0 \end{pmatrix} \begin{pmatrix} 0 \\ E_2 \\ 0 \end{pmatrix} = \gamma_H \begin{pmatrix} E_2 B_3 \\ 0 \\ 0 \end{pmatrix} \tag{7c.11}$$

has then only a component in the "1" direction: $j_1 = \gamma_H E_2 B_3$. In addition we have of course the longitudinal current density $j_2 = \sigma E_2$ where in the weak-field approximation $\sigma = \sigma_0$ independent of B. The "Hall field" is given by

$$E_1 = j_1/\sigma = (\gamma_H/\sigma_0^2) j_2 B_3 \tag{7c.12}$$

and the "Hall coefficient" by

$$R_H = E_1/j_2 B_3 = \gamma_H/\sigma_0^2 = \frac{1}{ne} \frac{<\tau_m^2>}{<\tau_m>^2} \left(\frac{m_\sigma}{m_H}\right)^2 \tag{7c.13}$$

where the "Hall mass factor"

$$\left(\frac{m_\sigma}{m_H}\right)^2 = \frac{3K(K+2)}{(2K+1)^2} \tag{7c.14}$$

This quantity is unity for K = 1, 0.78 for K = 20, and approaches 0.75 for $K \to \infty$. Since $\mu_H = |R_H|\,\sigma$ and $\mu = \sigma/(n|e|)$, the Hall factor r_H becomes

$$r_H = \mu_H/\mu = (<\tau_m^2>/<\tau_m>^2)\,(m_\sigma/m_H)^2 \tag{7c.15}$$

For silicon the same results are obtained: The weak-field Hall coefficient is isotropic and proportional to a Hall mass factor given by Eq.(7c.14).

7d. The Weak-Field Magnetoresistance

The terms in σ which are quadratic in $\vec{B}$ are given by Eq.(4c.50). The Herring-Vogt transformation, Eq.(7b.24), yields for the case of n-Ge :

$$\sigma = \frac{1}{4}\sum_{\rho=1}^{4} D_\rho^{-1}\sqrt{a}\,\beta_0 \begin{pmatrix} B_2^{*2}+B_3^{*2} & -B_1^*B_2^* & -B_1^*B_3^* \\ -B_1^*B_2^* & B_1^{*2}+B_3^{*2} & -B_2^*B_3^* \\ -B_1^*B_3^* & -B_2^*B_3^* & B_1^{*2}+B_2^{*2} \end{pmatrix} \sqrt{a}\,D_\rho \qquad (7d.1)$$

where β_0 is given by Eq.(4c.51) with m replaced by m_σ.
Depending on the number ρ of the valley the B*-components have to be replaced by linear combinations of the B-components as given by Eqs.(7c.3), (7c.6), etc. The calculation although quite tedious is straightforward, and we merely state the result for the component which is quadratic in B [1].

$$\sigma = \beta_M \begin{pmatrix} B_2^2+B_3^2 & -B_1B_2 & -B_1B_3 \\ -B_1B_2 & B_1^2+B_3^2 & -B_2B_3 \\ -B_1B_3 & -B_2B_3 & B_1^2+B_2^2 \end{pmatrix} + \beta_M \frac{2(K-1)^2}{(2K+1)(K+2)} \begin{pmatrix} B_1^2 & 0 & 0 \\ 0 & B_2^2 & 0 \\ 0 & 0 & B_3^2 \end{pmatrix} \qquad (7d.2)$$

where

$$\beta_M = -(ne^4/m_l^3)\langle\tau_m^3\rangle K(K+2)(2K+1)/9 \qquad (7d.3)$$

The second term on the right-hand side of Eq.(7d.2) does neither occur in n-Si where the valleys are located on <100> and equivalent axes, nor in the spherical band model. For an investigation of this term let us consider the case where $\vec{j}$, $\vec{E}$, and $\vec{B}$ are parallel to the x-direction. The "longitudinal magnetoresistance" is then given by

$$\left.\frac{\Delta\rho}{\rho B^2}\right|_{longit} = \frac{j(0)-j(B)}{j(0)B^2} = -\frac{\beta_M}{\sigma_0}\,\frac{2(K-1)^2}{(2K+1)(K+2)} = \frac{e^2}{m_l^2}\,\frac{2K(K-1)^2}{3(2K+1)}\,\frac{\langle\tau_m^3\rangle}{\langle\tau_m\rangle} \qquad (7d.4)$$

where in σ both the terms quadratic and linear in $\vec{B}$ have been considered; the linear term, however, multiplied by $\vec{E} = (E,0,0)$ vanishes for $\vec{B} = (B,0,0)$. Introducing a "magnetoresistance mobility" μ_M given by Eq.(A.16) of appendix A with m_l substituted for m we find

$$(\Delta\rho/\rho B^2)_{longit} = \mu_M^2 \cdot \frac{2K(K-1)^2}{3(2K+1)} \qquad (7d.5)$$

For K = 20 which is typical for n-Ge the K-dependent factor has a value of 118 while it vanishes for K = 1; this shows that the effective-mass anisotropy strongly

[1] E. G. S. Paige: Progress in Semiconductors, Vol.8, p. 22. (A. F. Gibson and R. E. Burgess, eds.). London: Temple Press. 1960.

enters the magnetoresistance.

For the case of n-Si we obtain instead of Eq.(7d.2)

$$\sigma = \begin{pmatrix} \beta'_M (B_2^2 + B_3^2) & -\beta''_M B_1 B_2 & -\beta''_M B_1 B_3 \\ -\beta''_M B_1 B_2 & \beta'_M (B_1^2 + B_3^2) & -\beta''_M B_2 B_3 \\ -\beta''_M B_1 B_3 & -\beta''_M B_2 B_3 & \beta'_M (B_1^2 + B_2^2) \end{pmatrix} \tag{7d.6}$$

where the coefficients β'_M and β''_M are given by

$$\beta'_M = -(ne^4/m_l^3) < \tau_m^3 > K(K^2 + K + 1)/3 \tag{7d.7}$$

and

$$\beta''_M = -(ne^4/m_l^3) < \tau_m^3 > K^2 \tag{7d.8}$$

Since σ_{xx} is independent of B_1 there is no longitudinal magnetoresistance in the <100> direction, in contrast to the case of n-type Ge.

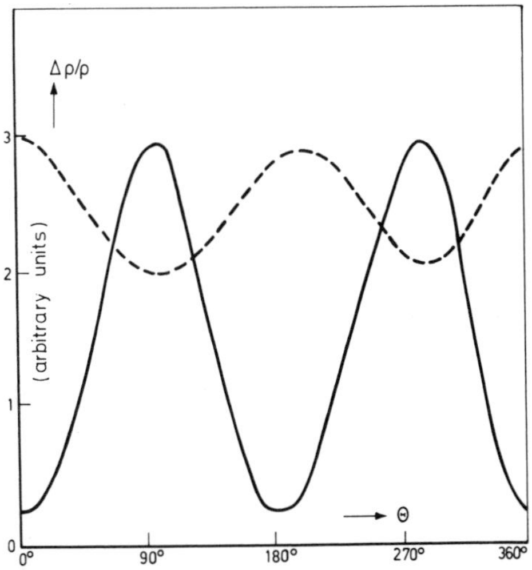

Fig.7.2 Magnetoresistance of n-type Ge (dashed curve) and n-type Si (full curve) as a function of angle between current (in the <100> direction) and magnetic field in a [010] plane.

Fig.7.2 shows the observed weak-field magnetoresistance for n-type Ge [2] and n-type Si [3] as a function of the angle θ between $\vec{B}$ and the <100> direction in the [010] plane where $\vec{j}$ is in the <100> direction. At $\theta = 0°$ the magnetoresistance is longitudinal; there it is a maximum for n-type Ge, and a minimum for n-type Si; in the latter case it should actually vanish as mentioned above.

[2] G. L. Pearson and H. Suhl, Phys. Rev. 83 (1951) 768.
[3] G. L. Pearson and C. Herring, Physica 20 (1954) 975.

The case of valleys located on <110> and equivalent axes will not be considered here [4].

In an experiment usually the current density $\vec{j}$ is kept constant and voltages are measured from which the Hall field and the longitudinal field intensities are determined. Instead of σ as a function of $\vec{E}$ and $\vec{B}$ we then need σ as a function of $\vec{j}$ and $\vec{B}$ for a comparison with experimental data. For the case of a weak magnetic field, the magnetoresistance in a cubic crystal is represented by the "inverted Seitz equation" [5] :

$$\Delta\rho/\rho B^2 = b + c(il + jm + kn)^2 + d(i^2 l^2 + j^2 m^2 + k^2 n^2) \tag{7d.9}$$

where (i,j,k) is a unit vector in the current direction and (l,m,n) is a unit vector in the direction of the magnetic field; b, c, and d are coefficients given by

$$b = -(\beta + \mu_H^2); \quad c = \mu_H^2 - \gamma \quad \text{and} \quad d = -\delta \tag{7d.10}$$

where β, γ, and δ are coefficients of the "normal Seitz equation"

$$\vec{j}/\sigma_0 = \vec{E} + \mu_H[\vec{E}\vec{B}] + \beta B^2 \vec{E} + \gamma(\vec{E}\vec{B})\vec{B} + \delta \begin{pmatrix} B_1^2 & 0 & 0 \\ 0 & B_2^2 & 0 \\ 0 & 0 & B_3^2 \end{pmatrix} \vec{E} \tag{7d.11}$$

μ_H is the Hall mobility ($= |R_H|\sigma_0$). The reference system for the unit vectors is given by the cubic-lattice vectors. For the longitudinal magnetoresistance we find from Eq.(7d.9) for the special case that both $\vec{j}$ and $\vec{B}$ are located in the $[1\bar{1}0]$-plane ($i = j = l = m = \sin\theta/\sqrt{2}$; $k = n = \cos\theta$) :

$$\Delta\rho/\rho B^2 = (b + c + d) - \frac{1}{2} d \sin^2\theta\,(3\cos^2\theta + 1) \tag{7d.12}$$

The same angular dependence will be obtained for $\Delta\rho/\rho E^2$ characteristic for warm carriers (Eq.(7e.41) and Fig.7.8a). For any direction of $\vec{j} \parallel \vec{B}$ in a [111]-plane we find

$$\Delta\rho/\rho B^2 = b + c + d/2 \tag{7d.13}$$

For the transverse magnetoresistance with $\vec{B}$ also in the [110] plane but with $\vec{j}$ in the <110> direction where $\vec{j} \perp \vec{B}$, we obtain

$$\Delta\rho/\rho B^2 = b + \frac{1}{2} d \sin^2\theta \tag{7d.14}$$

For the B-dependent Hall coefficient R_H an angular dependence similar to the one given by Eq.(7d.12) has been calculated where the θ-dependent term is proportional to B^2 [6].

---

[4] A. C. Beer: Solid State Physics (F. Seitz and D. Turnbull, eds.) Suppl. 4, Table VII, p.228. New York: Acad. Press. 1963.

[5] F. Seitz, Phys. Rev. 79 (1950) 372.

[6] H. Miyazawa, Proc. Int. Conf. Phys. Semic. Exeter (1962) p. 636. London: The Institute of Physics and The Physical Society. 1962.

With the shorthand notations

$$T'_M = \langle\tau_m^3\rangle\langle\tau_m\rangle/\langle\tau_m^2\rangle^2 \quad \text{and} \quad q = 1 + 3K/(K-1)^2 \tag{7d.15}$$

we obtain for n-type Si (<100> valleys) the relations

$$\left.\begin{aligned} b/\mu_H^2 &= T'_M (2K+1)(K^2+K+1)/\{K(K+2)^2\} - 1 \\ c/\mu_H^2 &= 1 - 3T'_M (2K+1)/(K+2)^2 \\ d/\mu_H^2 &= -T'_M (2K+1)(K-1)^2/\{K(K+2)^2\} \end{aligned}\right\} \tag{7d.16}$$

and

$$b + c = -d > 0 \quad \text{and} \quad q = -(\mu_H^2 + b)/d \tag{7d.17}$$

If the equation for q Eq.(7d.15) is solved for K we find two solutions indicated by subscripts + and – :

$$K_\pm = \frac{1}{2}(2q + 1 \pm \sqrt{12q-3})/(q-1) \tag{7d.18}$$

The solution $K_+ > 1$ yields prolate (cigar-shaped) ellipsoids for the constant-energy surfaces in $\vec{k}$-space while $K_- < 1$ yields oblate (disk-shaped) ellipsoids. If the scattering mechanism and therefore T'_M given by Eq.(7d.15) are known, one of the two K values may be excluded. E.g. Stirn and Becker [7] find for n-type AlSb, where like in n-type Si there are <100> valleys, that impurity scattering, for which Eq.(7d.15) yields $T'_M = 1.58 > 1$, is predominant at 77 K and therefore only the solution $K_+ > 1$ is physically significant.

If the direction of the current density $\vec{j}$ is the <100> direction and $\vec{B}$ is in any direction perpendicular to $\vec{j}$, Eq.(7d.9) results in $\Delta\rho/\rho^2$ being independent of the direction of $\vec{B}$. If this fact cannot be verified experimentally one may suspect that either a metallic contact at the semiconductor sample acts as a short circuit or that the crystal is inhomogeneous (remember that this is true only for the weak-$\vec{B}$ case).

Quite often the ratio $\Delta\rho/\rho B^2$ is denoted by M with a lower subscript indicating the crystallographic direction of the current density and an upper subscript indicating that of the magnetic field, such as e.g.

$$M_{110}^{\;001} = b \quad \text{and} \quad M_{110}^{\;1\bar{1}0} = b + \frac{1}{2}d$$

If the second M-value is smaller than the first one, $d < 0$, this is a strong indication for <100> valleys such as in n-type Si and n-type AlSb.

A criterion for <111> valleys such as in n-type Ge is given by

$$b = -c \quad \text{and} \quad d > 0 \tag{7d.19}$$

An experimental determination of b, c, and d for a single sample with side arms is possible by means of the "planar Hall effect" which has been discussed in Chap.4g. The coefficient P_H is given by $P_H = c/\sigma_0$ for a current in the <100> direction and a Hall field E_p in the <010> direction, while $P_H = (c+d)/\sigma_0$

[7] R. J. Stirn and W. M. Becker, Phys. Rev. 141 (1966) 621.

for $\vec{j} \parallel$ <110> and $\vec{E}_p \parallel$ <$\bar{1}$10>. If we thus determine c for $\vec{j} \parallel$ <100> and the coefficient b from the magnetoresistance for the same current direction, but $\vec{B} \parallel$ <010>, and the quantity b + c + d for $\vec{B} \parallel$ <100>, we can with only one sample determine b, c, and d. Another possibility is to measure c + d by the planar Hall effect for $\vec{j} \parallel$ <110>, b by the magnetoresistance for $\vec{B} \parallel$ <001>, and b + d/2 by the magnetoresistance for $\vec{B} \parallel$ <110>. Because of the finite thickness of the side arms of the sample the experimental data are subject to corrections [8].

So far we have assumed the momentum relaxation time τ_m to be isotropic. Herring and Vogt [9] and later Eagles and Edwards [10] consider the case of an anisotropic τ_m with components τ_l and τ_t in the longitudinal and the transverse direction, respectively, of a constant-energy ellipsoid of revolution in $\vec{k}$-space. The ratio $\tau_l/\tau_t = K_\tau$, and the ratio m_l/m_t, now denoted as K_m, yields for K the ratio

$$K = K_m / K_t \tag{7d.20}$$

since the conductivity is determined by the ratio m/τ_m. E.g. for b/μ_H^2 for n-type Si we obtain now [11] instead of Eq.(7d.16) :

$$b/\mu_H^2 = T'_M \frac{(\langle\tau_t^3 K_\tau^2\rangle/\langle\tau_t^3\rangle + K_m\langle\tau_t^3 K_\tau\rangle/\langle\tau_t^3\rangle + K_m^2)(\langle\tau_t K_\tau\rangle/\langle\tau_t\rangle + 2K_m)}{K_m(2\langle\tau_t^2 K_\tau\rangle/\langle\tau_t^2\rangle + K_m)^2} - 1 \tag{7d.21}$$

For lattice scattering τ_m seems to be fairly isotropic while for impurity scattering in n-type Ge $K_\tau = 11.2$ has been calculated [10]. In n-type Ge with $4 \times 10^{15}/cm^3$ impurities the ratio $K = K_m/K_\tau = 20$ at 4 K and K = 5 at 20 K; this temperature dependence has been attributed to a transition from lattice scattering at low temperatures where the impurities are neutral due to freeze-out of carriers, to a combination of ionized impurity scattering and lattice scattering at somewhat higher temperatures [12]. At 77 K the transition from lattice scattering to ionized impurity scattering may be obtained by a change of the impurity concentration N_I. K_τ as a function of N_I is shown in Fig.7.3 assuming saturation values of K_τ^i = 11 and 12 for K_τ in the limit of pure ionized impurity scattering, $N_I \to \infty$. The experimental data points have been determined from measurements of the microwave Faraday effect in n-type Ge at 77 K [13]; they are consistent with a value of 11 for K_τ^i which agrees within the accuracy of the experiment with the calculated value of 11.2 given above.

[8] C. Herring, T. H. Geballe, and J. E. Kunzler, Bell Syst. Tech. J. 38 (1959) 657.
[9] C. Herring and E. Vogt, Phys. Rev. 101 (1956) 944. Err.: ibid. 105 (1957) 1933.
[10] P. M. Eagles and D. M. Edwards, Phys. Rev. 138 (1965) A 1706.
[11] D. Long, Phys. Rev. 120 (1960) 2024.
[12] R. A. Laff and H. Y. Fan, Phys. Rev. 112 (1958) 317.
[13] H. O. Haller, thesis, Univ. Wien, Austria. 1972.
For further experimental data see L. J. Neuringer, Proc. Int. Conf. Semic. Physics, Paris 1964, p. 379. Paris: Dunod. 1964.

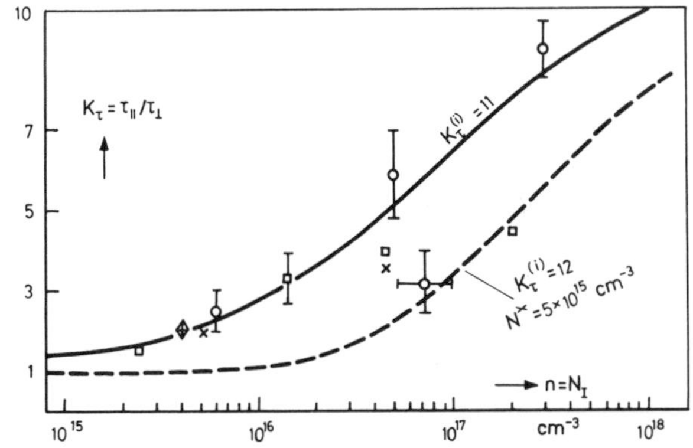

Fig.7.3 Dependence of the momentum relaxation time anisotropy K_τ on impurity concentration N_I in n-type Ge at 77 K. The full curve has been calculated for combined acoustic deformation potential scattering and ionized impurity scattering, the latter with an anisotropy of $K_\tau = 11$ for pure ionized impurity scattering. For the dashed curve, isotropic neutral impurity scattering has also been taken into account. Data points : by Haller, ref. 13: circles; from magnetoresistance measurements by H. Bruns, Zeitschr. f. Naturf. 19a (1964) 533: squares; by R. A. Laff and H. Y. Fan, Phys. Rev. 112 (1958) 317: diamond; by C. Goldberg and W. E. Howard, Phys. Rev. 110 (1953) 1035: crosses.

7e. Equivalent-Intervalley Scattering and Valley Repopulation Effects

In Chaps.7b - d we have assumed an equal distribution of carriers in all valleys which are at the same energy and therefore are denoted as "equivalent valleys". However, any field, pressure or temperature gradient etc. which introduces a preferential direction in an otherwise isotropic solid may lead to a "repopulation" of valleys since the equilibrium of the valley population is dynamic due to "intervalley scattering". Scattering within a valley is denoted as "intravalley scattering" [1]. Before considering repopulation effects let us investigate the intervalley scattering rate.

In a transition of an electron from one valley in $\vec{k}$-space to another, a large change in momentum, which is of the order of magnitude of the first-Brillouin-zone radius, is involved. This momentum may be taken up either by an impurity atom or by a phonon near the edge of the phonon Brillouin zone where the acoustic and optical branches are either degenerate or not too far from each other (Figs.6.13 and 6.14). We will not consider here the case of impurity scattering which is limited to low temperatures and very impure semiconductors. The phonon case is treated in a way similar to optical phonon scattering except that the optical phonon energy $\hbar\omega_0 = k\Theta$ is replaced by the somewhat lower

[1] C. Herring, Bell Syst. Tech. J. 34 (1955) 237.

"intervalley phonon" energy $\hbar\omega_i = k\Theta_i$ having a value between the optical and the acoustic phonon energy near the edge of the phonon Brillouin zone ($\Theta_i = 315$ K, $\Theta = 430$ K for Ge). Scattering selection rules have been considered by various authors [2]. In n-type Si two types of processes have been considered: "g scattering" occurs between a given valley and the valley on the opposite side of the same axis, e.g. between a <100> and a <$\bar{1}$00> valley, while in the "f process" a carrier is scattered to one of the remaining equivalent valleys, e.g. between a <100> and a <010> valley; both f and g scattering involve a reciprocal lattice vector ("Umklapp process", Chap.6d). g-type phonons have temperatures of about 190 and 720 K; phonons involved in f-scattering have a temperature of 680 K and take part in the repopulation of valleys which will be discussed below [3]. The inverse momentum relaxation time is given by [2]

$$1/\tau_i = w_2 \{\sqrt{(\epsilon/k_B\Theta_i)+1} + \exp(\Theta_i/T)\,\mathrm{Re}\sqrt{(\epsilon/k_B\Theta_i)-1}\}/\{\exp(\Theta_i/T)-1\} \quad (7e.1)$$

where w_2 is a rate constant. This equation corresponds to Eq.(6k.5) for optical deformation potential scattering. If we average $1/\tau_i$ over a Maxwell-Boltzmann distribution function $\propto \exp(-\lambda\epsilon/k_BT)$, where $\lambda = T/T_e$, we obtain

$$<1/\tau_i> = w_2\sqrt{2\lambda z/\pi}\,\cosh\{(1-\lambda)z\}K_1(\lambda z)/\sinh(z) \quad (7e.2)$$

where K_1 is a modified Bessel function (Appendix B) and $z = \Theta_i/2T$. For thermal carriers $\lambda = 1$. Taking into account acoustic deformation potential scattering in addition to intervalley scattering the total inverse momentum relaxation time is given by

$$1/\tau_m = 1/\tau_{ac} + 1/\tau_i \quad (7e.3)$$

where τ_{ac} stands for τ_m given by Eq.(6d.15). The mobility as a function of temperature T has been calculated by Herring [1]. The ratio μ/μ_o vs T/Θ_i where μ_o stands for

$$\mu_o = \mu_{ac}(T/\Theta_i)^{3/2} \quad (7e.4)$$

and $k_B\Theta_i = \hbar\omega_i$, is shown in Fig.7.4 for various values of the parameter w_2/w_1. The rate constant $w_1 = l_{ac}^{-1}(2k_B\Theta_i/m)^{1/2}\Theta_i/T$ where l_{ac} is the acoustic mean free path given by Eq.(6d.16). If an intervalley deformation potential constant D_i is introduced in analogy to the corresponding optical constant (Eq.(6k.5)) the ratio $w_2/w_1 = \frac{1}{2}(D_i u_l/\epsilon_{ac}\omega_i)^2$. It has been shown [3] that in n-Si f-type phonon scattering is weak and the observed temperature dependence of the mobility $\mu \propto T^{-2.5}$ is explained by a dependence similar to that shown in Fig.7.4 with

[2] H. W. Streitwolf, phys. stat. sol. 37 (1970) K 47, and references therein.

[3] D. Long, Phys. Rev. 120 (1960) 2024; M. Asche, B. L. Boichenko, V. M. Bondar and O. G. Sarbej, Proc. Int. Conf. Semic. Physics Moscow 1968, p. 793. Leningrad: Nauka 1968; W. A. Harrison, Phys. Rev. 104 (1956) 1281, has shown that optical intervalley scattering in n-Si is negligible.

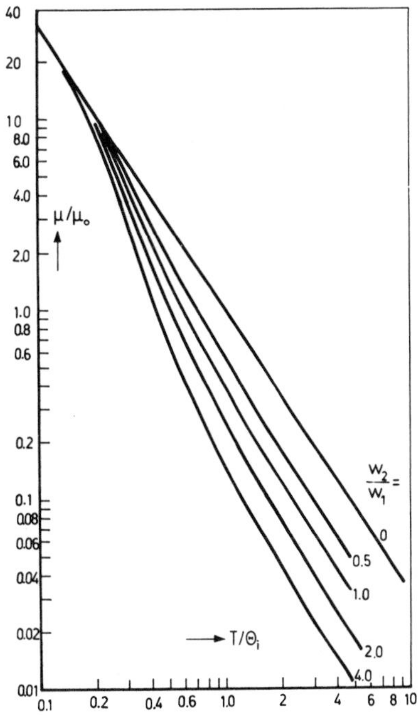

Fig.7.4 Influence of intervalley scattering on the temperature dependence of the mobility for various values of the coupling constants (after C. Herring, ref. 1).

$\Theta_i = 720\,K$ and $w_2/w_1 = 3$. For n-Ge w_2 has been determined experimentally from the acousto-electric effect (Chap. 7h).

We will now investigate the repopulation of equivalent valleys in a strong electric field. Fig.7.5 shows a two-valley model in $\vec{k}$-space with $\vec{E}$ at an oblique angle relative to the y-axis in co-ordinate space. In both valleys the electron effective mass is given by the direction of $\vec{E}$. Since the effective masses are proportional to the square of the length of the arrows in Fig.7.5, they are different for the two valleys except when $\vec{E}$ is at an angle of 45° relative to both valleys. Because of the different effective masses electrons are heated by the electrical field intensity $\vec{E}$ at a different rate. Therefore in the valley, whose longitudinal axis is at a larger angle relative to $\vec{E}$, the electron temperature T_e is higher than in the other valley. The transfer rate of carriers from a "hot valley" to a "cool valley" is larger than in the reverse direction according to Eq.(7e.2). Therefore the equilibrium population of a hot valley is smaller than that of a cool valley. The rate equation for an arbitrary number of valleys and a non-degenerate carrier gas is given by

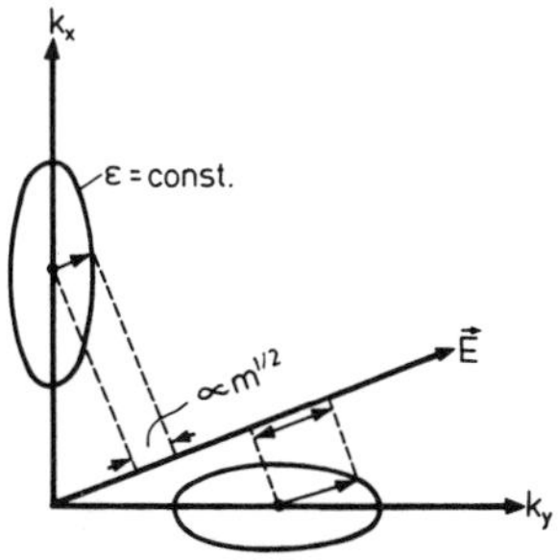

Fig.7.5 Two-valley model indicating different effective masses for a given field direction.

$$-dn_\rho/dt = n_\rho \sum_\sigma{}' \langle 1/\tau_{i,\rho\to\sigma}\rangle - \sum_\sigma{}' n_\sigma \langle 1/\tau_{i,\sigma\to\rho}\rangle \qquad (7e.5)$$

where the first term on the right-hand side represents the rate at which carriers are scattered out of valley $\#\rho$, the second term the scattering rate into this valley, and the prime at Σ indicates that the sum is taken over all valleys except $\#\rho$. Usually the simplifying assumption is made that the inverse time constant $\langle 1/\tau_{i,\rho\to\sigma}\rangle$ depends only on the temperature $T_{e\rho} = T/\lambda_\rho$ of the carrier emitting valley $\#\rho$. Of course, if several types of phonons with Debye temperature Θ_{i1}, Θ_{i2}, $\cdots \Theta_{is}$, $\cdots$ are involved in the intervalley processes, such as in n-type Si, a sum over all s appears on the right-hand side of Eq.(7e2) where z is replaced by $z_s = \Theta_{is}/2T$.

For a total number N of valleys, Eq.(7e.5) can now be written

$$-dn_\rho/dt = (n_\rho - n_\rho^{(0)})\,(N-1) < 1/\tau_{i\rho} > \quad (7e.6)$$

where the equilibrium carrier concentration $n_\rho^{(0)}$ is given by

$$n_\rho^{(0)} = \sum_\sigma{}' n_\sigma < 1/\tau_{i\sigma} > / \{(N-1) < 1/\tau_{i\rho} >\} \quad (7e.7)$$

Assuming for simplicity that the field direction is the longitudinal axis of a "cool" valley (subscript c) while all other N−1 valleys are equally "hot" (subscript h), we find for Eq.(7e.7)

$$n_c^{(0)} < 1/\tau_{ic} > = n_h^{(0)} < 1/\tau_{ih} > = n/\{< 1/\tau_{ic} >^{-1} + (N-1) < 1/\tau_{ih} >^{-1}\} \quad (7e.8)$$

where $n = n_c + (N-1)n_h$ is the total carrier concentration assumed to be field independent. These rate equations are useful in calculations of not only strong-dc-field effects but also of ac effects such as the acousto-electric effect (Chap.7h).

Intervalley scattering contributes also to the energy loss which is treated similarly to optical phonon scattering except that for the carrier energy change per collision $\hbar\omega_i$ is taken for $\hbar\omega_0$ and the deformation potential constant D in nonpolar semiconductors is replaced by a quantity D_{ij} [4]; for the effective mass m we take the product $m_t^{2/3} m_l^{1/3}$, valid for the valley into which the carrier is scattered. For intravalley scattering (both acoustic and optical) m is also replaced by $m_t^{2/3} m_l^{1/3}$ and, in addition, for acoustic scattering ϵ_{ac}^2 by $\frac{3}{4}\Xi_d^2\{1.31 + 1.61\,\Xi_u/\Xi_d + 1.01\,(\Xi_u/\Xi_d)^2\}$ in τ_m and by $\frac{2}{3}\Xi_d^2\{1 + 0.5\,(m_l/m_t)\,(1 + \Xi_u/\Xi_d)^2\}$ in $< -d\epsilon/dt >$ [5].

For a calculation of the current field characteristic under the influence of intervalley scattering let us call $n_0 = n/N$ the equilibrium carrier concentration in a given valley if there is no electric field. The equilibrium repopulation of the valley after application of an electric field is then given by

$$\Delta n_\rho = n_\rho^0 - n_0 \quad (7e.9)$$

where $\sum_\rho \Delta n_\rho = 0$ due to carrier conservation. For the case of no magnetic field we obtain for the current density

$$\vec{j} = \frac{1}{N}\sum_{\rho=1}^{N} (1 + \Delta n_\rho/n_0)\,\Lambda_\rho \sigma_w \vec{E} \quad (7e.10)$$

where Λ_ρ is given by Eq.(7b.25). E.g. let us consider the case of n-type Si with $\vec{E}$ in an arbitrary direction in the $[1\bar{1}0]$ plane. Valleys #1 and 2 are symmetric relative to $\vec{E}$ and therefore $\Delta n_1 = \Delta n_2$ while

$$\Delta n_3 = -2\Delta n_1 = -2\Delta n_2 \quad (7e.11)$$

[4] E. M. Conwell: High Field Effects in Semiconductors, Solid State Physics, Suppl. 9, p. 154 (F. Seitz, D. Turnbull, and H. Ehrenreich, eds.). New York: Acad. Press. 1967.

[5] Ref.4, p.115 and p.124 for acoustic and p.152 and p.155 for optical scattering.

For $\vec{E}$ in a <111> direction all valleys are symmetric relative to $\vec{E}$ and therefore $\Delta n_1 = \Delta n_2 = \Delta n_3 = 0$. Let us denote the conductivity in this direction by σ_0 S(E) where σ_0 is the ohmic conductivity which is isotropic in the cubic lattice of silicon*). We will calculate the current density for any direction of $\vec{E}$ in the $[1\bar{1}0]$ plane :

$$\vec{E} = E(\sin\theta/\sqrt{2}\,;\ \sin\theta/\sqrt{2}\,;\ \cos\theta) \tag{7e.12}$$

With the "effective field intensity" $E^*_{(\rho)}$ for valley #ρ defined by Eqs.(7b.8) and (7b.23) :

$$\vec{E}^*_{(\rho)} = \sqrt{a}\, D_\rho \vec{E} \tag{7e.13}$$

we obtain for the current density from Eq.(7e.10)

$$\vec{j} = \sigma_0 \frac{1}{N} \sum_{\rho=1}^{N} (1 + \Delta n_\rho / n_0)\, S(E^*_{(\rho)})\, \Lambda_\rho \vec{E} \tag{7e.14}$$

where Λ_ρ is given by Eq.(7b.25) and N = 3 for n-type Si. Let us calculate $E^*_{(\rho)}$ and Λ_ρ for the present case. For valley #1 the effective field intensity is given by

$$\vec{E}^*_{(1)} = E\sqrt{3/(2K+1)} \begin{pmatrix} \sin\theta/\sqrt{2} \\ \sqrt{K/2}\,\sin\theta \\ \sqrt{K}\,\cos\theta \end{pmatrix} \tag{7e.15}$$

and its magnitude by

$$E^*_{(1)} = E\sqrt{1 + \{(K-1)/(2K+1)\}\,(\tfrac{3}{2}\cos^2\theta - \tfrac{1}{2})} \tag{7e.16}$$

Because of symmetry $E^*_{(2)} = E^*_{(1)}$, while

$$\vec{E}^*_{(3)} = E\sqrt{3/(2K+1)} \begin{pmatrix} \cos\theta \\ -\sqrt{K/2}\,\sin\theta \\ \sqrt{K/2}\,\sin\theta \end{pmatrix} \tag{7e.17}$$

with a magnitude of

$$E^*_{(3)} = E\sqrt{1 + \{(K-1)/(2K+1)\}\,(1 - 3\cos^2\theta)} \tag{7e.18}$$

One can show that for a unit vector $\vec{h}_\rho$ from the origin to valley #ρ and a unit vector $\vec{e} = \vec{E}/E$ in the field direction

$$E^*_{(\rho)} = E\sqrt{1 + \{(K-1)/(2K+1)\} \cdot \{1 - 3(\vec{h}_\rho \vec{e})^2\}} \tag{7e.19}$$

is valid where in n-type Si $\vec{h}_1 = (1,0,0)$; $\vec{h}_2 = (0,1,0)$; $\vec{h}_3 = (0,0,1)$ and $\vec{e}$ is given for $\vec{E}$ in a $[1\bar{1}0]$ plane by Eq.(7e.12). The calculation of the Λ_ρ matrices is straightforward. Since $E^*_{(1)} = E^*_{(2)}$ one can add Λ_1 and Λ_2 :

---

*) This definition of σ_0 is different from Eq.(4c.39) insomuch as for m the conductivity effective mass m_σ is substituted.

$$(\Lambda_1+\Lambda_2)\vec{E} = E\,\frac{3}{2K+1}\begin{pmatrix}(K+1)\sin\theta/\sqrt{2}\\(K+1)\sin\theta/\sqrt{2}\\2K\cos\theta\end{pmatrix} \tag{7e.20}$$

Similarly we obtain for $\Lambda_3\vec{E}$:

$$\Lambda_3\vec{E} = E\frac{3}{2K+1}\begin{pmatrix}K\sin\theta/\sqrt{2}\\K\sin\theta/\sqrt{2}\\\cos\theta\end{pmatrix} \tag{7e.21}$$

Eq.(7e.14) shows that due to the factor $(1+\Delta n_\rho/n_o)S(E^*_{(\rho)})$ the direction of the current density is in general different from that of the electric field strength. This is true also if there were no repopulation of valleys. Repopulation enhances the hot-carrier effect. As shown by Fig.7.6 we introduce a longitudinal component j_l and a transverse component j_t of $\vec{j}$:

$$j_l = j\cos(\theta-\psi) = j\cos\psi\cos\theta + j\sin\psi\sin\theta = j_z\cos\theta + j_x\sqrt{2}\sin\theta \tag{7e.22}$$

and

$$j_t = j\sin(\theta-\psi) = j\cos\psi\sin\theta - j\sin\psi\cos\theta = j_z\sin\theta - j_x\sqrt{2}\cos\theta \tag{7e.23}$$

where

$$\vec{j} = j(\sin\psi/\sqrt{2};\ \sin\psi/\sqrt{2};\ \cos\psi) \tag{7e.24}$$

The calculation of j_l and j_t from Eqs.(7e.14) and (7e.20) – (7e.23) yields

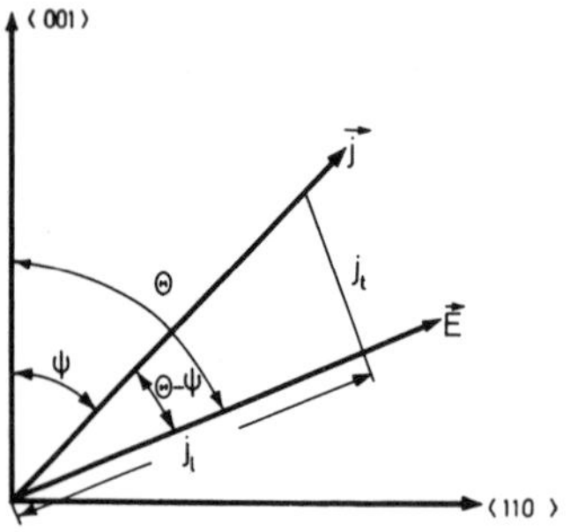

Fig.7.6 Longitudinal and transverse components of the current density in the [1$\bar{1}$0] plane.

$$j_l = \frac{\sigma_o E}{2K+1}\left\{2\left(K\cos^2\theta + \frac{K+1}{2}\sin^2\theta\right)\left(1-\frac{\Delta n_3}{2n_o}\right)S(E^*_{(1)}) + (\cos^2\theta + K\sin^2\theta)\left(1+\frac{\Delta n_3}{n_o}\right)S(E^*_{(3)})\right\} \tag{7e.25}$$

and

$$j_t = \frac{\sigma_o E}{2K+1}\cdot\frac{K-1}{2}\left\{(1-\Delta n_3/n_o)\,S(E^*_{(1)}) - (1+\Delta n_3/n_o)\,S(E^*_{(3)})\right\} \tag{7e.26}$$

One can write these equations in a form where they are also valid for n-type Ge with $\vec{E}$ in the [1$\bar{1}$0] plane :

$$j_l = \sigma_o E\frac{1}{N}\sum_{\rho=1}^{N}(E^*_{(\rho)}/E)^2(1+\Delta n_\rho/n_o)\,S(E^*_{(\rho)}) \tag{7e.27}$$

and

$$j_t = -\sigma_o E\frac{1}{N}\sum_{\rho=1}^{N}\frac{1}{2}(1+\Delta n_\rho/n_o)\,S(E^*_{(\rho)})\frac{d}{d\theta}(E^*_{(\rho)}/E)^2 \tag{7e.28}$$

The ratio j_t/j_l yields the "Sasaki angle" $\theta-\psi$ (Fig.7.6) due to the relation

$$\tan(\theta-\psi) = j_t/j_l \tag{7e.29}$$

A schematic diagram of the experimental arrangement used for a measurement of this ratio is shown in Fig.7.7; it is similar to the Hall arrangement (Fig.4.2) except that there is no magnetic field present. Usually voltages rather than current densities are measured and pulse techniques are used to minimize Joule heating effects. The ratio of field intensities $|E_t/E_l| = |j_t/j_l|$ is calculated from these data.

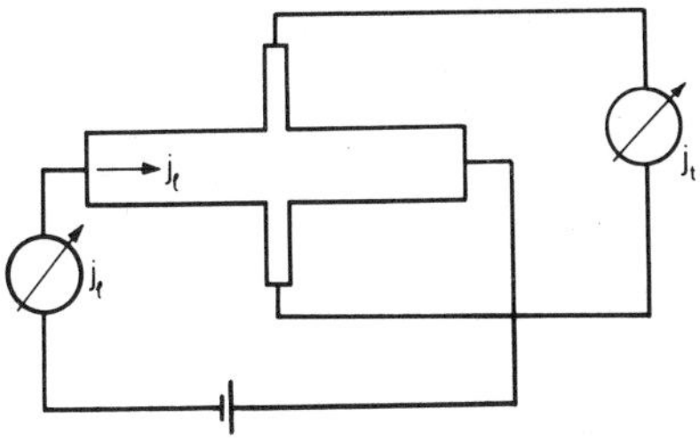

Fig.7.7 Experimental arrangement for Sasaki-Shibuya measurements.

Fig.7.8 shows j_l and j_t as a function of θ calculated from Eqs.(7e.27) and (7e.28) for the special case of "warm electrons". Curves for hot carriers are similar; experimental data for n-type Ge at 90 K and 750 V/cm are shown in Fig.7.9 [6]. The effect first predicated by Shibuya [7] and verified experimentally by Sasaki et al.[6] is known as the "Sasaki-Shibuya effect". The broken curve indicates the effect calculated by Shibuya for the case of no repopulation. It has been suggested by the authors that the discrepancy is due to repopulation. Theoretical considerations by Reik and Risken [8] suggest that in n-type Ge in the range of constant drift velocity the maximum repopulation is given by

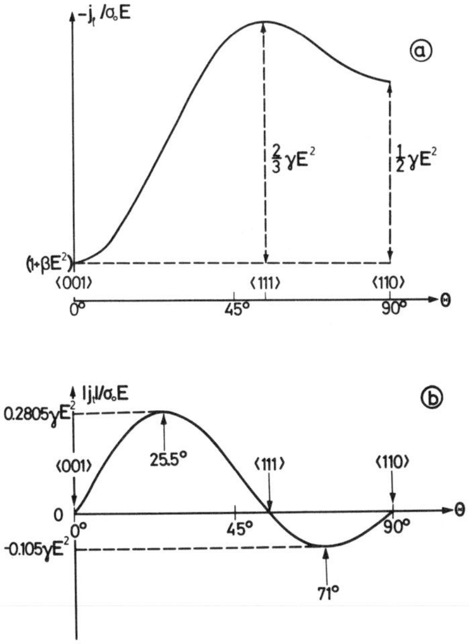

Fig.7.8 Longitudinal (a) and transverse (b) Sasaki-Shibuya current components as a function of the angle between field strength and the⟨001⟩direction. The numerical data are valid for warm electrons.

[6] W. Sasaki, M. Shibuya, and K. Mizuguchi, J. Phys. Soc. Japan 13 (1958) 456; W. Sasaki, M. Shibuya, K. Mizuguchi, and G. Hatoyama, J. Phys. Chem. Solids 8 (1959) 250.

[7] M. Shibuya, Phys. Rev. 99 (1955) 1189.

[8] H. G. Reik and H. Risken, Phys. Rev. 126 (1962) 1737.

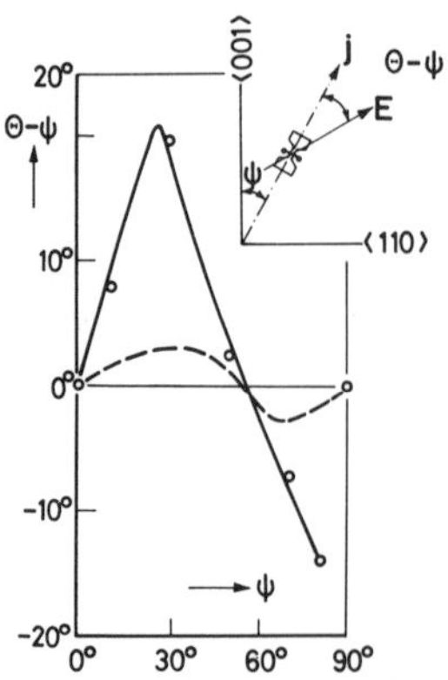

Fig.7.9 Theoretical (dashed curve) and experimental data (circles) of Sasaki angle vs the angle between the current density and the ⟨001⟩ direction valid for n-type Ge at 90 K (after Sasaki et al., ref.6).

$$(1+\Delta n_\rho/n_0)_{max} = (E/E^*_{(\rho)})/\frac{1}{4}\sum_{\rho=1}^{4}(E/E^*_{(\rho)}) \tag{7e.30}$$

This value has been verified [9] experimentally at high electric fields (Fig.7.10). The decrease of $\Delta n_\rho/n_0$ at even higher field strengths is probably due to carrier transfer to "upper valleys" located on ⟨100⟩ and equivalent axes ("nonequivalent intervalley scattering", Chap.7g). Any transfer between valleys by intervalley scattering reduces the temperature differences and therefore the repopulation.

Let us consider in more detail the case of warm carriers [10]. From Eq.(4n.3) we obtain for S(E):

$$S(E) = 1 + \beta E^2 \tag{7e.31}$$

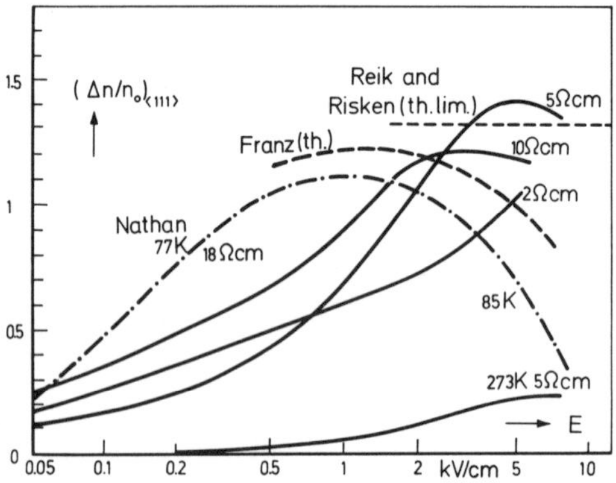

Fig.7.10 Relative increase in population of the "cool" valley at a ⟨111⟩ field direction for samples of different room temperature resistivities at 85 K (after Schweitzer et al., ref.9). Dash-dotted curve: Experimental data by M.I. Nathan, Phys. Rev. 130 (1963) 2201; horizontal dashed line: According to Eq.(7e.29). Dashed curve: Theory by W. Franz, phys. stat. sol. 3 (1963) 1260.

where $|\beta E^2| \ll 1$. Due to the low field intensity also $|\Delta n_\rho/n_0| \ll 1$. Therefore the product $(1+\Delta n_\rho/n_0)\,S(E^*_{(\rho)})$ can be approximated by $1+\Delta n_\rho/n_0+\beta E^{*2}_{(\rho)}$. Hence, Eq.(7e.27) yields for the longitudinal component

[9] D. Schweitzer and K. Seeger, Zeitschr. f. Physik 183 (1965) 207.
[10] K. J. Schmidt-Tiedemann, Philips Res. Repts. 18 (1963) 338.

$$j_l = \sigma_o E \frac{1}{N} \left\{ \sum_{\rho=1}^{N} (E^*_{(\rho)}/E)^2 + \sum_{\rho=1}^{N} (\Delta n_\rho / n_o)(E^*_{(\rho)}/E)^2 + \beta E^2 \sum_{\rho=1}^{N} (E^*_{(\rho)}/E)^4 \right\} \tag{7e.32}$$

Eq.(7e.19) yields for the first term on the right-hand side :

$$\frac{1}{N} \sum_{\rho=1}^{N} (E^*_{(\rho)}/E)^2 = 1 + (K-1)/(2K+1) \left\{ 1 - \frac{3}{N} \sum_{\rho=1}^{N} (\vec{h}_\rho \vec{e})^2 \right\} = 1 \tag{7e.33}$$

For the proof of the second equality remember that both $\vec{h}_\rho$ and $\vec{e}$ are unit vectors and the valley location has cubic symmetry. Next we consider the last term in Eq.(7e.32)

$$\frac{1}{N} \sum_{\rho=1}^{N} (E^*_{(\rho)}/E)^4 = 1 + 2(K-1)/(2K+1) \left\{ 1 - \frac{3}{N} \sum_{\rho=1}^{N} (\vec{h}_\rho \vec{e})^2 \right\} + (K-1)^2/(2K+1)^2 \frac{1}{N} \sum_{\rho=1}^{N} \left\{ 1 - 3(\vec{h}_\rho \vec{e})^2 \right\}^2 \tag{7e.34}$$

As in Eq.(7e.33) the second term on the right-hand side vanishes. The last sum is

$$\frac{1}{N} \sum_{\rho=1}^{N} \left\{ 1 - 3(\vec{h}_\rho \vec{e})^2 \right\}^2 = 1 - 2 + \frac{9}{N} \sum_{\rho=1}^{N} (\vec{h}_\rho \vec{e})^4 = -1 + 3(\cos^4\theta + \frac{1}{2}\sin^4\theta) = \frac{1}{2}(3\cos^2\theta - 1)^2 \tag{7e.35}$$

The second equality is true for $\vec{e}$ in a $[1\bar{1}0]$ plane with an angle θ relative to the $\langle 001 \rangle$ axis. We thus obtain from Eq.(7e.32)

$$j_l = \sigma_o E \left[1 + \frac{1}{N} \sum_{\rho=1}^{N} \frac{\Delta n_\rho}{n_o} \left(\frac{E^*_{(\rho)}}{E} \right)^2 + \beta E^2 \left\{ 1 + \left(\frac{K-1}{2K+1} \right)^2 \frac{1}{2} (3\cos^2\theta - 1)^2 \right\} \right] \tag{7e.36}$$

If in n-type Si the field is located in a $\langle 111 \rangle$ direction it is symmetric to all valleys and there is no repopulation, $\Delta n_\rho = 0$. Furthermore $3\cos^2\theta - 1 = 0$. If we put a subscript $\langle 111 \rangle$ on β for this direction we thus have

$$j_l = \sigma_o E (1 + \beta_{\langle 111 \rangle} E^2) \tag{7e.37}$$

For another direction of $\vec{E}$ in the $[1\bar{1}0]$ plane assuming no repopulation we thus obtain

$$j_l = \sigma_o E (1 + \beta E^2) \tag{7e.38}$$

with β given by

$$\beta = \beta_{\langle 111 \rangle} + \beta_{\langle 111 \rangle} \left(\frac{K-1}{2K+1} \right)^2 \frac{1}{2} (3\cos^2\theta - 1)^2$$

$$= \beta_{\langle 111 \rangle} + \frac{1}{3} \gamma \left\{ \frac{3}{2} (3\cos^2\theta + 1)(\cos^2\theta - 1) + 2 \right\} \tag{7e.39}$$

The dependence of β on θ is the result of the symmetry of the crystal lattice and thus cannot be changed by a repopulation. Therefore instead of the factor $\beta_{\langle 111 \rangle}(K-1)^2/(2K+1)^2$ we introduce a coefficient $\gamma/3$ which will depend on Δn_ρ (as will be shown in Eq.(7e.49) :

$$\beta = \beta_{\langle 111\rangle} + (\gamma/3)\,\{\frac{3}{2}\,(3\cos^2\theta + 1)\,(\cos^2\theta - 1) + 2\} \tag{7e.40}$$

If we denote β for $\theta = 0°$ by β_0 we find

$$\beta = \beta_0 - \gamma\,\frac{\sin^2\theta}{2}\,(3\cos^2\theta + 1) \tag{7e.41}$$

For any direction of $\vec{E}$ in a [111] plane one obtains $\beta = \beta_0 - \gamma/2$. The curve given in Fig.7.8a is obtained from Eq.(7e.41). Eq.(7e.41) is also valid for n-type Ge if the new constant is denoted as $-\gamma/2$ instead of $\gamma/3$; in n-type Ge there is no repopulation for $\vec{e} = (1,0,0)$. Experimental data for n-type Ge are shown in Fig. 4.36.

If we introduce γ in Eq.(7e.28) we obtain with a little algebra

$$j_t = \frac{1}{2}\,\gamma\sigma_0 E^3\,\frac{\sin 2\theta}{2}\,(3\cos^2\theta - 1) \tag{7e.42}$$

The ratio j_t/j_l for warm electrons may be approximated by $j_t/\sigma_0 E$:

$$j_t/\sigma_0 E = \tan(\theta - \psi) = \frac{1}{2}\,\gamma E^2\,\frac{\sin 2\theta}{2}\,(3\cos^2\theta - 1) \tag{7e.43}$$

The constant γ has been introduced in such a way that j_t is proportional to γ. The function is plotted in Fig.7.8b. It vanishes in the main symmetry directions of the crystal lattice which follows clearly from geometrical considerations: Assume a current direction at an angle θ relative to $\vec{E}$ with $\vec{E}$ in the direction of an N-fold symmetry axis; turn the crystal by an angle 360°/N around this axis; since this position cannot be distinguished from the former position $\vec{j}$ must have the same direction as before which is only possible for $\theta = 0°$.

Since the valleys are located along symmetry directions in $\vec{k}$-space it is not possible by the Sasaki-Shibuya effect to make measurements of $j_t \neq 0$ at maximum repopulation. Therefore the highest accuracy in Δn_ρ-determinations is obtained from measurements of the longitudinal current component [9].

For the determination of Δn_ρ for warm electrons in n-type Si we consider Eq.(7e.36) and the definition of γ:

$$\frac{1}{3}\sum_\rho \frac{\Delta n_\rho}{n_0}\,(\frac{E^*_{(\rho)}}{E})^2 + \beta E^2\,(\frac{K-1}{2K+1})^2\,\frac{1}{2}\,(3\cos^2\theta - 1)^2 = \frac{1}{3}\,\gamma E^2\,\frac{1}{2}\,(3\cos^2\theta - 1)^2 \tag{7e.44}$$

This can be simplified to yield

$$\frac{1}{3}\sum_\rho \frac{\Delta n_\rho}{n_0}\,(\frac{E^*_{(\rho)}}{E})^2 = \{-\beta\,(\frac{K-1}{2K+1})^2 + \frac{1}{3}\,\gamma\}\,E^2\,\frac{1}{2}\,(3\cos^2\theta - 1)^2 \tag{7e.45}$$

Since the factor $\frac{1}{2}\,(3\cos^2\theta - 1)^2$ on the right-hand side has been obtained from $\frac{1}{3}\sum_\rho (E^*_{(\rho)}/E)^4$, the simplest assumption for Δn_ρ is with 2 unknowns X and Y :

$$\Delta n_\rho/n_0 = X(E^*_{(\rho)}/E)^2 + Y \tag{7e.46}$$

This yields for Eq.(7e.45) :

$$X\{1+(\frac{K-1}{2K+1})^2\frac{1}{2}(3\cos^2\theta-1)^2\}+Y=\{-\beta(\frac{K-1}{2K+1})^2+\frac{1}{3}\gamma\}E^2\frac{1}{2}(3\cos^2\theta-1)^2 \tag{7e.47}$$

For $\cos\theta = 1/\sqrt{3}$ we find

$$X=-Y=\{-\beta_{\langle 111\rangle}+\frac{1}{3}\gamma(\frac{2K+1}{K-1})^2\}E^2 \tag{7e.48}$$

and for n-type Si therefore

$$\Delta n_\rho/n_0=\{-\beta_{\langle 111\rangle}+\frac{1}{3}\gamma(\frac{2K+1}{K-1})^2\}E^2\{(E^*_{(\rho)}/E)^2-1\} \tag{7e.49}$$

Similarly for n-type Ge :

$$\Delta n_\rho/n_0=\{-\beta_{\langle 100\rangle}-\frac{1}{2}\gamma(\frac{2K+1}{K-1})^2\}E^2\{(E^*_{(\rho)}/E)^2-1\} \tag{7e.50}$$

In both materials the largest repopulation is thus obtained for a field direction in the longitudinal direction of an energy ellipsoid, which is <111> in Ge and <100> in Si.

In conclusion we would like to point out that the existence of a transverse component of the current density is an important verification that the constant-energy surfaces are non-spherical.

7f. Warm- and Hot-Carrier Galvanomagnetic Phenomena

A magnetic field has always a cooling influence on warm and hot carriers: Due to the Lorentz force the carriers are deflected from the drift direction and thus gain less energy from the accelerating electric field than without a magnetic field. Before considering this phenomenon for the many-valley model let us briefly investigate the simpler case of the spherical one-valley model.

Fig.7.11 shows experimental results for the temperature dependence of the coefficient $|\beta|$ for warm electrons in n-InSb at $B=0$ and 27 kG [1]. In n-InSb we have a spherical valley at $\vec{k}=0$ and polar-optical scattering with the result of $\beta<0$ at temperatures between 65 and 90 K. By the application of a magnetic field of 27 kG the mobility of about $3\times10^5\ cm^2/Vsec$ is reduced to about 1/2 this value ($\mu B = 80 \gg 1$) and β is positive. Fig.7.12 is a schematic diagram of the mobility as a function of the electric field intensity with and without a strong magnetic field ($\mu B \gg 1$). For the simple case of predominant acoustic deformation potential scattering a negative magnetoresistance has been calculated for hot carriers taking into account the Hall field in addition to the applied electric field [2]:

---

[1] R. J. Sladek, Phys. Rev. 120 (1960) 1589.
[2] H. F. Budd, Phys. Rev. 140 (1965) A 2170.

$$\Delta\rho/\rho_o B^2 = -7.6\text{x}10^{-3}\mu_H^2(0) \qquad (7f.1)$$

where $\mu_H(0)$ is the zero-electric-field Hall mobility and the magnetic induction has been assumed to be $B \ll 1/\mu_H(0)$. For the limit of strong magnetic fields the calculation yields $\Delta\rho/\rho_o = -5.4\text{x}10^{-3}$.

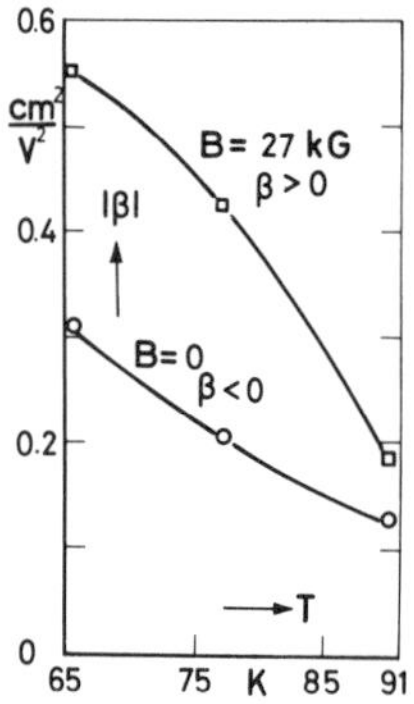

Fig.7.11 Observed temperature dependence of $\beta = \{\mu(E)-\mu_o\}/\mu_o E^2$ in n-type InSb in a strong magnetic field and in zero magnetic field (after Sladek, ref.1).

In many-valley semiconductors both a negative magnetoresistance and a polarity-dependent Hall effect and magnetoresistance have been observed ("polarity-dependence" means that by a reversal of $\vec{B}$ the Hall mobility μ_H and the magnetoresistance $\Delta\rho/\rho B^2$ are changed). Let us first consider the negative magnetoresistance which is caused by the electric-field-induced repopulation of valleys. (For $\vec{E}$-directions which are symmetric to the valleys and therefore do not cause carrier transfer, the magnetoresistance is positive.)

Fig.7.13 shows the observed weak-field transverse magnetoresistance in n-Si at 77 K as a function of $\vec{E} \parallel$ <100> [3]. The dash-dotted curve indicates the repopulation of the <100> valleys without a magnetic field as determined from the anisotropy of the j(E) characteristics. In the range of E values where $\Delta n/n_o \geqslant 0.5$ the magnetoresistance is found to be negative. For an explanation of this phenomenon we have to take into account that the magnetic field not only has a cooling effect but also generates a Hall field as already mentioned above; if this is added to the applied electric field $\vec{E} \parallel$ <100> the total field deviates from the longitudinal axis of the constant-energy ellipsoid in n-Si with the effect that the carrier effective mass becomes smaller, the carriers are more accelerated, and the conductivity is increased. Similar results have been observed in n-Ge for $\vec{j} \parallel$ <111> and $\vec{B} \parallel$ <1$\bar{1}$0> [4] in a field range where from j(E) measurements $\Delta n/n_o > 0.8$ has been found [5].

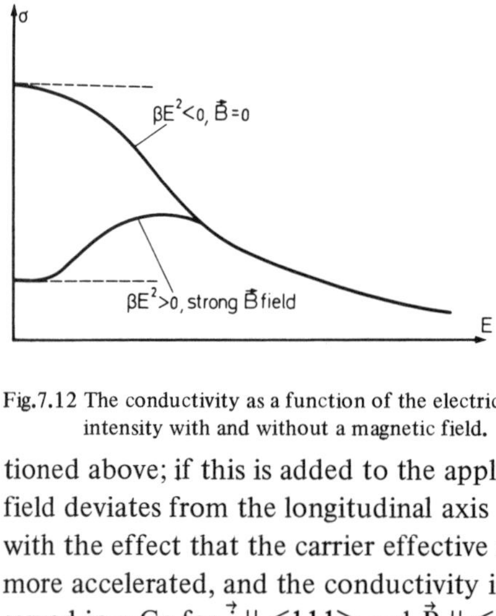

Fig.7.12 The conductivity as a function of the electric field intensity with and without a magnetic field.

[3] H. Heinrich and M. Kriechbaum, J. Phys. Chem. Solids 31 (1970) 927; M. Asche, V. M. Bondar, and O. G. Sarbej, phys. stat. sol. 31 (1969) K 143.
[4] E. A. Movchan and E. G. Miselyuk, Sov. Phys. Semicond. 3 (1969) 571.
[5] M. I. Nathan, Phys. Rev. 130 (1963) 2201.

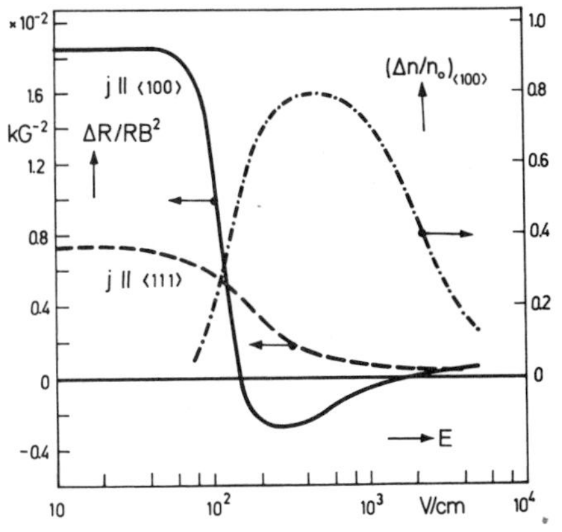

Fig.7.13 Observed weak-field transverse magnetoresistance of n-type Si at 77 K as a function of the electric field strength for two directions of the current density. The dash-dotted curve is an average of experimental values of the population of the ⟨100⟩ valleys determined by Kästner et al., Zeitschr. f. Physik 185 (1967) 359 (after Heinrich and Kriechbaum, ref.3).

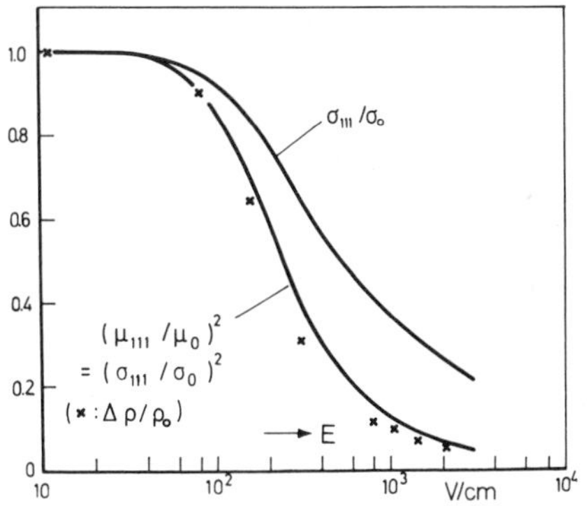

Fig.7.14 Observed conductivity $\sigma_{\langle 111\rangle}$ (normalized to the zero-field conductivity σ_0) of n-type silicon at 77 K in a ⟨111⟩ direction vs the electric field strength, and $(\sigma_{\langle 111\rangle}/\sigma_0)^2$. The crosses indicate observed values of the magnetoresistance (after Heinrich and Kriechbaum, ref.3).

If the current direction is symmetric to the valleys (as e.g. for $\vec{j} \parallel \langle 111\rangle$ in n-Si) the dependence of $\Delta\rho/\rho_0$ on $\vec{E}$ in weak magnetic fields is the same as that of μ^2 on E since $\Delta\rho/\rho_0 \propto \mu^2 B^2$. This is demonstrated by the experimental results shown in Fig.7.14 [3].

It is not surprising to find also the Hall coefficient in many-valley semiconductors to depend on the electric field strength. In Fig.7.15 R_H is plotted vs the applied E for 1.5, 4, and 6 kG in n-Ge ($\vec{B} \perp \vec{j} \parallel <1\bar{1}0>$) [4]. The maximum is simply explained by two-band conduction, Eq. (4c.74), with light and heavy electrons indicated by subscripts l and h, respectively.

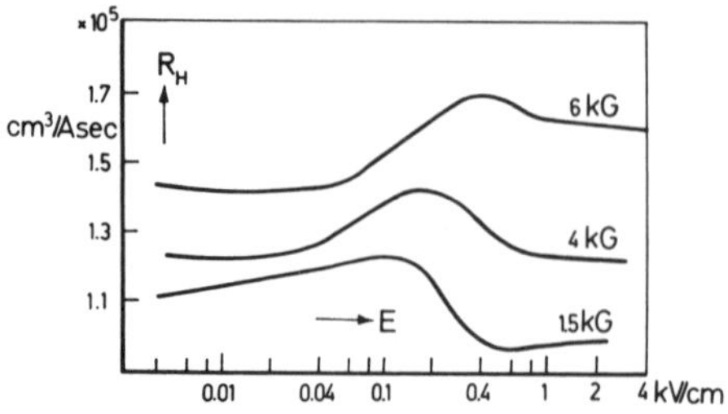

Fig.7.15 Hall coefficient of n-type germanium at 80 K as a function of the electric field strength for various magnetic fields (after Movchan and Miselyuk, ref. 4).

$$R_H = -\frac{r_H}{|e|} \cdot \frac{n_l \mu_l^2 + n_h \mu_h^2}{(n_l \mu_l + n_h \mu_h)^2} = R_H(0) \frac{(1+\eta)(1+\eta b^2)}{(1+\eta b)^2} \tag{7f.2}$$

where $R_H(0) = -r_H/\{|e|(n_l + n_h)\}$ and $b = \mu_h/\mu_l$ is the mobility ratio and $\eta = n_h/n_l$ the density ratio. With increasing population of the cool h valley at an increasing $\vec{E}$ field intensity there is a maximum of $\{R_H(E) - R_H(0)\}/R_H(0)$ of $(1-b)^2/4b$ at $\eta = 1/b$. For the 6 kG curve in Fig.7.15 the increase is 0.25 which yields $b = 0.4$ and $\eta = 2.5$ leading to a transfer rate $(n_h - n_o)/n_o$ of $(3-1/\eta)/(1+1/\eta) = 1.8$ where n_o is the zero-field equilibrium population (there are 4 valleys altogether). From the anisotropy of j(E) at B = 0 a maximum transfer rate of 1.1 has been found which is comparable in magnitude [5].

Let us now consider the polarity dependent Hall effect and magnetoresistance. For hot carriers in many-valley semiconductors even for magnetic fields $B \ll 1/\mu$ these effects may change in absolute magnitude at a sign reversal of $\vec{B}$ [6 - 9]. Since a similar effect is found for thermal carriers but in magnetic fields which are not small, and this effect is easily calculated we will first consider the Hall effect for this case.

For a magnetic field in $<0\bar{1}1>$ direction and a current in $<111>$ direction we expect the electric field to be (E_1, E_2, E_2) where

$$E_1 = E_L/\sqrt{3} + 2E_H/\sqrt{6} \text{ and } E_2 = E_L/\sqrt{3} - E_H/\sqrt{6} \tag{7f.3}$$

[6] M. Asche, Yu. G. Zav'yalov, and O. G. Sarbej, Pisma v JETP 13 (1971) 401 [Engl.: JETP Letters 13 (1971) 285] and private communication M. Asche.
[7] M. Asche, A. G. Maksimchuk, and O. G. Sarbej, phys. stat. sol. (b) 47 (1971) K 45.
[8] H. Heinrich and M. Kriechbaum, phys. stat. sol. (b) 50 (1972) K 45.
[9] M. Kriechbaum, H. Heinrich, and J. Wajda, J. Phys. Chem. Solids 33 (1972) 829.

i.e. to consist of the applied field $\vec{E}_L \parallel \vec{j}$ and the Hall field E_H<$2\bar{1}\bar{1}$>. The conductivity tensor for n-Si up to terms in B^2 applied to $\vec{E}$ as given above yields

$$\left.\begin{aligned}(\sigma_o + \beta'_M B^2)E_1 + \gamma_H \sqrt{2}\, BE_2 = j/\sqrt{3}\\ (-\gamma_H B/\sqrt{2})\, E_1 + \{\sigma_o + (\beta'_M + \beta''_M)\, B^2/2\}\, E_2 = j/\sqrt{3}\end{aligned}\right\} \quad (7f.4)$$

where $\vec{B} = (0;\ -B/\sqrt{2};\ B/\sqrt{2}\,)$ and Eqs.(7c.8) and (7d.6) have been taken into account. This set of equations can be solved for the ratio E_1/E_2:

$$\frac{E_1}{E_2} = \frac{\sigma_o - \gamma_H \sqrt{2}\, B + \frac{1}{2}(\beta'_M + \beta''_M)B^2}{\sigma_o + (\gamma_H/\sqrt{2}\,)B + \beta'_M B^2} \quad (7f.5)$$

Eq.(7f.3) yields for the Hall mobility $\mu_H = |E_H/(E_L B)|$:

$$\mu_H = (\sqrt{2}/B)\,(1 - E_1/E_2)/(2 + E_1/E_2) = \frac{\gamma_H + (\beta''_M - \beta'_M)\, B/3\sqrt{2}}{\sigma_o + (\beta''_M + 5\beta'_M)\; B^2/6} \quad (7f.6)$$

with $\sigma_o = (ne^2/m_l) < \tau_m > (2K+1)/3$ and γ_H, β'_M, and β''_M given by Eqs.(7c.9), (7d.7), and (7d.8), respectively, we finally obtain

$$\mu_H = \frac{\gamma_H}{\sigma_o} \cdot \frac{1 + \{(K-1)^2/(K+2)\}\, (B/3\sqrt{2}\,)e<\tau_m^3>/(m_l<\tau_m^2>)}{1 - \{5K(K^2 + 1.6K + 1)/(2K+1)\}\, (B^2/6)e^2<\tau_m^3>/(m_l^2<\tau_m>)} \quad (7f.7)$$

The result contains a term, which is linear in B and changes sign, when the polarity of $\vec{B}$ is changed. It is important for not too large field strengths B where the quadratic term can still be neglected, but B still larger than the range $B \ll 1/\mu_H$ considered in Chap.7c. Since the linear term is proportional to K−1 it vanishes in the spherical band model where K = 1. Similarly for the magnetoresistance a polarity dependence is found by taking into account powers of B larger than the second, in contrast to Chap.7d.

Fig.7.16 shows the transverse magnetoresistance of n-Si at 77 K for $\vec{j}$ at an angle of 24° relative to the <100> direction in the [110] plane and $\vec{B}$ also in this plane [6]. Curve 1 is parabolic indicating $\Delta\rho/\rho \propto B^2$ for E = 1 V/cm. Curve 3 for 260 V/cm is rather linear in a broad range of magnetic field intensities. Fig.7.17 shows the observed Hall mobility μ_H normalized to $(\mu_H)_{B\to 0}$ as a function of E for $\vec{B} \parallel$ <$0\bar{1}1$> (upper curve) and $\vec{B} \parallel$ <$01\bar{1}$> (lower curve) and $\vec{j} \parallel$ <111> [9]. Calculated results (not shown) are in qualitative agreement with these curves.

Fig.7.18 is a schematic diagram showing the <001>, <010>, and – in top view – <100> constant energy surface. For no magnetic field the full single arrows indicate the drift velocity, the double arrows the current density for each valley. The applied electric field is in the <011> direction. By the application of a magnetic field in <100> direction a Hall field E_H is generated which has been assumed in the drawing to be equal in magnitude to the applied electric field such that the total field is in the <010> direction. The dashed arrows show the drift velocities and current densities in the magnetic field. The drift velocity in the <010> valley is heavily reduced since the effective mass is increased. But because of the increase in population of this valley it yields the largest current contribution of all valleys.

At still higher field intensities carrier transfer to non-equivalent valleys occurs which will be considered in the following chapter.

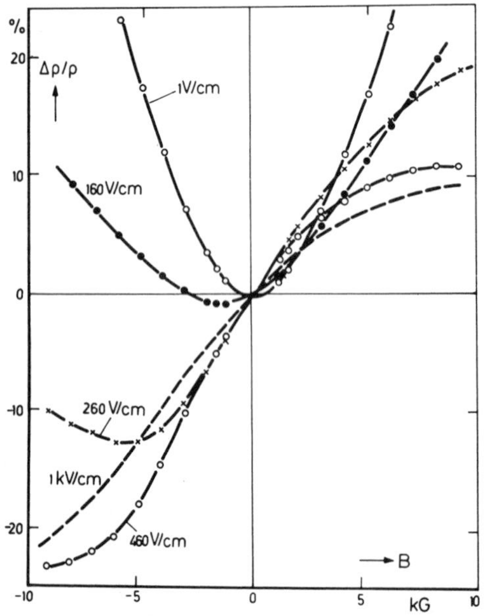

Fig.7.16 Dependence of $\Delta\rho/\rho$ on B for various electric field strengths; (orientations see text; after Asche et al., ref. 6).

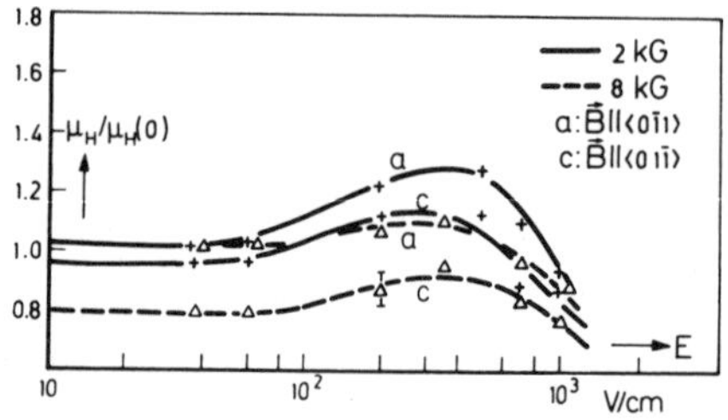

Fig.7.17 Normalized Hall mobility $\mu_H/\mu_H(0)$ for $\vec{j}$ ||<111> and $\vec{B}$ ||<0$\bar{1}$1> and <01$\bar{1}$>; B = 2 and 8 kG (after Kriechbaum et al., ref. 9).

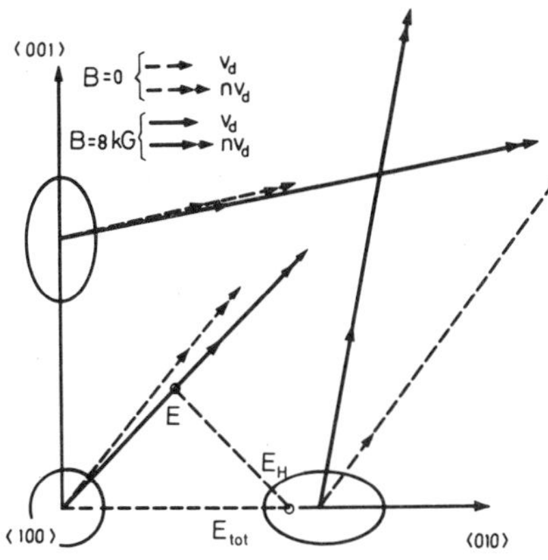

Fig.7.18
Schematic diagram showing the effect of a magnetic field on the drift velocity and current contribution for each valley (after Kriechbaum et al., ref.9).

7g. Non-equivalent Intervalley Scattering, Negative Differential Conductivity, and Gunn Oscillations

Besides the usual <111>-minima of the conduction band of germanium there are upper minima at $\vec{k} = 0$ and along the <100> and equivalent axes shown in Fig.2.26. Due to the strong curvature of the minimum at $\vec{k} = 0$, its density of states is very small and therefore the density of carriers which can be transferred to this valley is negligible. However, carriers are transferred to the "silicon-like" <100>-minima which are 0.18 eV above the <111> minima. Fig.7.19 shows experimental and theoretical results on the Hall coefficient R_H of n-type Ge at 200 K with j parallel to a <100> direction which is symmetric to all the normally occupied <111> valleys [1]. R_H increases strongly above 1 kV/cm. This is considered to be due to carrier transfer to the silicon-like valleys where the mobility is much smaller than in the normally occupied valleys. The dashed curves have been calculated for various mobility ratios b. If the carriers in the silicon-like valleys would be immobile, the Hall coefficient $R_H \propto 1/n_{\langle 111 \rangle}$ would increase as the number of carriers $n_{\langle 111 \rangle}$ in the normally occupied <111> valleys decreases. Further observations on n-type Ge will be discussed at the end of this chapter.

Let us now consider the simpler and more spectacular case of n-type GaAs where the central valley at $\vec{k} = 0$ is lowest in the conduction band and the silicon-like <100> "satellite" valleys are $\Delta\epsilon_{\Gamma X} = 0.36$ eV above the former as shown in Fig.7.20. The effective masses are $0.07\,m_0$ in the central valley and very probably $0.4\,m_0$ in each one of the 6 satellite valleys. ($0.4\,m_0$ is the value observed in n-type GaP where the <100> minima are lowest.) Since the effective density of states in one valley is proportional to $m^{3/2}$ according to Eq.(3a.31), we will con-

[1] H. Heinrich, K. Lischka, and M. Kriechbaum, Phys. Rev. B2 (1970) 2009.

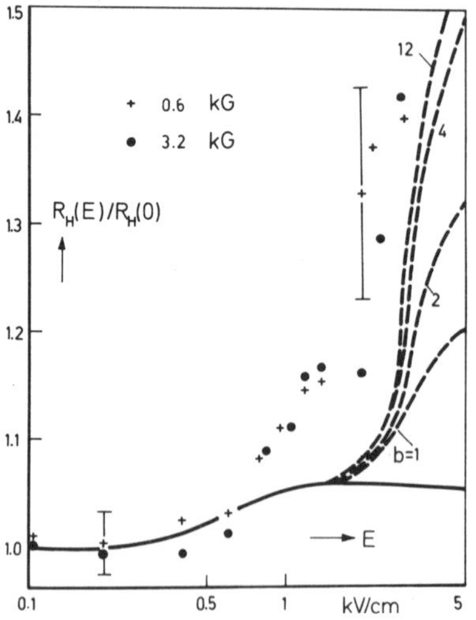

Fig.7.19 Hall coefficient, normalized to unity at zero electric field $\vec{E} = 0$, for n-type Ge at a lattice temperature of 200 K and $\vec{j} \parallel$ <100> direction, as a function of E. The full curve has been calculated for electrons only in <111> valleys. The dashed curves are valid for electron transfer to <100> valleys and non-equivalent intervalley scattering. The numbers at the dashed curves are the mobility ratios for both types of valleys assumed for the calculation.

sider from now on only one satellite valley instead of 6, with an effective mass of $(0.4^{3/2} \times 6)^{2/3}\, m_0 = 1.3\, m_0$.

Because of the small effective mass in the central valley electrons are heated very effectively by an electric field and gain an energy of magnitude $\Delta\epsilon_{\Gamma X} = 0.36$ eV at E ≈ 3 kV/cm. Before reaching this energy the mobility is not changed considerably due to polar-optical mode scattering (see Chap.6l). At E ≈ 3 kV/cm electrons are transferred to the satellite valleys where they are heavy and have a reduced momentum relaxation time because of scattering between these valleys and because of the higher density of states in these valleys. Therefore the mobility in the satellite valleys is estimated to be only 150 cm²/Vsec (80 cm²/Vsec observed in GaP where the X-valleys are lowest) while the mobility in the central Γ-valley is between 6 000 and 8 000 cm²/Vsec depending on purity; all data are valid for room temperature. As in the case of equivalent-intervalley scattering the momentum change at carrier transfer is supplied by an intervalley phonon which is either absorbed or emitted (Fig.7.20 and Chap.7e).

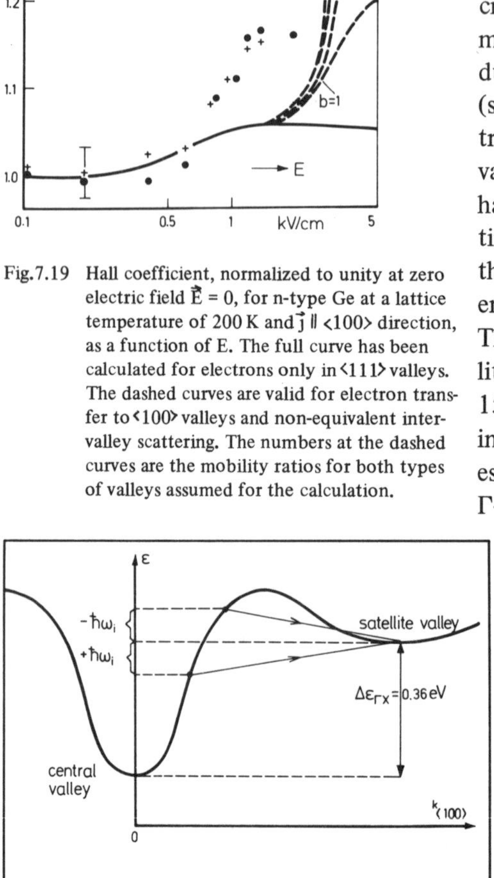

Fig.7.20 Central and satellite valleys in the conduction band of gallium arsenide (shown schematically). The full straight lines indicate electron transitions between the valleys with emission or absorption of an intervalley phonon.

These considerations lead to two possible drift velocity vs field characteristics plotted in Fig.7.21. At low field intensities the slope is the high mobility μ_1 in the central valley while at high field intensities the slope is the low mobility μ_2 in the satellite valleys assuming all carriers have been transferred. Curve #2 shows a range of negative differential conductivity

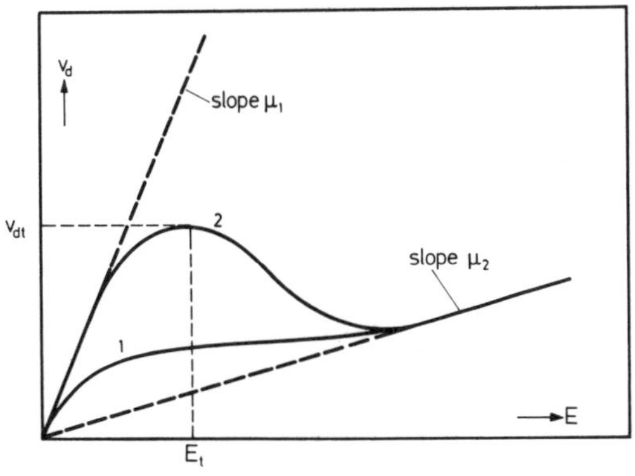

Fig.7.21 Drift velocity vs field characteristics for transitions from the central valley (mobility μ_1) to the satellite valley (mobility μ_2).

(ndc) while curve #1 everywhere has a positive slope. The type of curve observed depends on the ratio of the energy interval $\Delta\epsilon_{\Gamma X}$ to the thermal energy $k_B T$ which is 0.026 eV at room temperature. Since in GaAs this ratio is very high, a type-2-characteristic is realized in pure GaAs at room temperature. With increased density of impurities first the threshold field E_t increases and finally the ndc range disappears altogether.

Also in n-type GaSb the central valley (Γ) is lowest but the next-highest valleys are on <111> axes (L), $\Delta\epsilon_{\Gamma L} = 0.075$ eV above the central valley. The energy is only 3 times the thermal energy at room temperature. Consequently no ndc is observed. In InSb the energy $\Delta\epsilon_{\Gamma L}$ is 0.5 eV but the band gap $\Delta\epsilon_G$ is only 0.2 eV and except for very short times $\approx 10^{-9}$ sec there is impact ionization across the band gap rather than carrier transfer [2].

The idea of an ndc by electron transfer in GaAs and $GaAs_xP_{1-x}$ has first been suggested by Ridley and Watkins [3] and by Hilsum [4]. The generation of current oscillations at frequencies up to 10^3 GHz and the application for amplification purposes in this frequency range have been considered by these authors. Gunn [5] observed oscillations in n-type GaAs and InP at frequencies between 0.47 and 6.5 GHz by applying field intensities $\vec{E}$ of more than 3 kV/cm. Further experiments showed that the oscillations were indeed those which had been pre-

[2] J. E. Smith, M. I. Nathan, J. C. McGroddy, S. A. Porowski, and W. Paul, Appl. Phys. Lett. 15 (1969) 242.

[3] B. K. Ridley and T. B. Watkins, Proc. Phys. Soc. 78 (1961) 293; B. K. Ridley, Proc. Phys. Soc. 82 (1963) 954.

[4] C. Hilsum, Proc. IRE 50 (1962) 185.

[5] J. B. Gunn, Solid State Comm. 1 (1963) 88; J. B. Gunn: Plasma Effects in Solids (J. Bok, ed.), p. 199. Paris: Dunod. 1964.

dicted [6].

For a calculation of the ndc let us follow the simplified treatment given by Hilsum [4]. He assumes equal electron temperatures in the central valley (subscript 1) and the satellite valley (subscript 2). The ratio of electron densities, $\eta = n_1/n_2$, is given by the ratio of the densities of states, N_{c1}/N_{c2}, and the difference in valley energies, $\Delta\epsilon_{\Gamma X}$, by

$$\eta = (N_{c1}/N_{c2}) \exp(\Delta\epsilon_{\Gamma X}/k_B T_e) \tag{7g.1}$$

In later publications [7] calculations have been performed without making this simplifying assumption, with the result of a better reproduction of the experimental results. Eq.(7g.1) yields to a good approximation $n_1 = n\eta/(1+\eta)$ and $n_2 = n/(1+\eta)$ and a conductivity given by

$$\sigma = |e|\,(n_1\mu_1 + n_2\mu_2) = |e|\,n\mu_2(\eta b + 1)/(\eta + 1) \tag{7g.2}$$

where $b = \mu_1/\mu_2$ is the mobility ratio. For a voltage V applied to a filamentary sample of length L and area A, the current $I = \sigma AV/L$ vs V has a slope given by

$$\begin{aligned} dI/dV &= (A/L)\,(\sigma + V d\sigma/dV) = (A/L)\sigma\,\{1 + (E/\sigma)d\sigma/dE\} \\ &= \frac{A}{L}\, n|e|\mu_2\, \frac{\eta b + 1}{\eta + 1}\{1 - \frac{\eta(b-1)}{(\eta+1)(\eta b+1)}\, \frac{\Delta\epsilon_{\Gamma X} E}{k_B T_e^2}\, \frac{dT_e}{dE}\} \end{aligned} \tag{7g.3}$$

The sign of dI/dV depends strongly on dT_e/dE which has been calculated from Eq.(6l.26) and the energy balance Eq.(4m.6). The resulting current-field characteristic is shown in Fig.7.22. The threshold field strength of 3 kV/cm calculated from a shifted Maxwell-Boltzmann distribution is not changed appreciably by taking a distribution obtained as a solution of the Boltzmann equation.

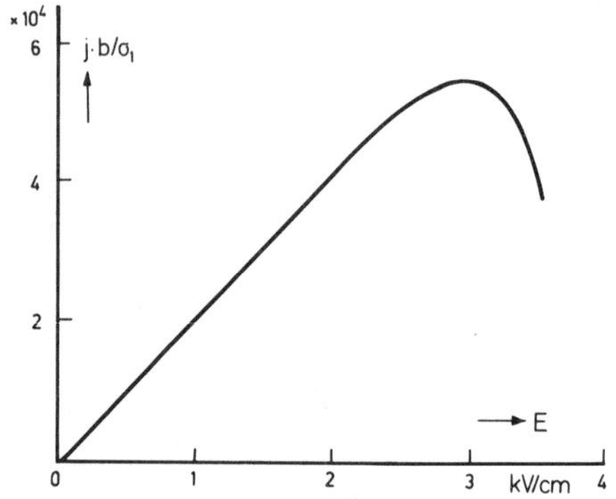

Fig.7.22 Current density vs. field characteristics calculated for n-type gallium arsenide at 373 K assuming polar optical scattering (after Hilsum, ref. 4).

[6] For a historical review see J. B. Gunn, Int. J. Sci. Technol. 46 (1965) 43.
[7] See e.g. E. M. Conwell: Solid State Physics (F. Seitz and D. Turnbull, eds.) Suppl. 9, p.254. New York: Acad. Press. 1967.

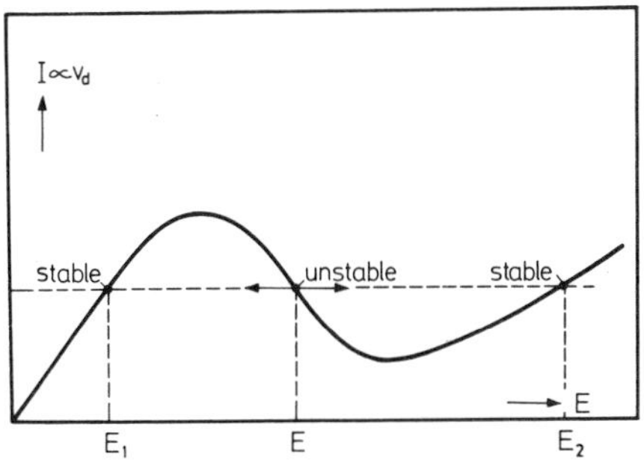

Fig.7.23 Instability in the region of negative differential conductivity.

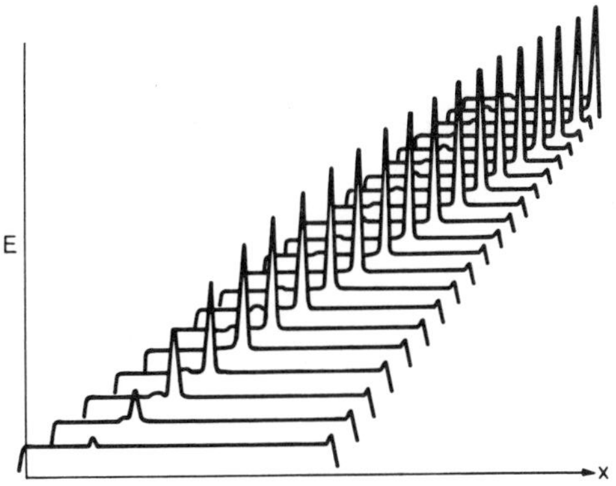

Fig.7.24 A high-field domain nucleates at an inhomogeneity of doping and moves to the anode of a "long" sample as it grows (after D. E. McCumber and A. G. Chynoweth, IEEE Trans. ED-13 (1966) 4.).

We will now consider the generation of oscillations in the range of ndc. From Fig.7.23 it can be shown that an operating point within the ndc range is not stable [3, 8]. At an average field intensity E two regions of length l_1 and l_2 with field intensities E_1 and E_2 are formed such that the total voltage $V = E_1 l_1 + E_2 l_2$. The high-field "domain" is not fixed locally but drifts with the electrons through the crystal. Inside the domain most of the electrons belong to the satellite valley and the drift velocity of these heavy electrons is the same as that of the light electrons outside the domain. The domain with the field strength E_2 indicated by Fig. 7.23 has a flat top. Such a flat-topped domain has not been ob observed in GaAs, however [8]. Fig.7.24 shows the actual field distribution in the sample taken after successive time intervals. The domain is nucleated at some inhomogeneity of doping close to the cathode and travels towards the anode while growing. After disappearing in the anode a new domain is generated. In this case the generation repetition rate is the oscillation frequency given by v_d/L where $v_d \approx 10^7$ cm/sec is the drift velocity and L is the sample length. For L = 0.1 mm a frequency of about 1 GHz is obtained which is in the microwave range. The generation of microwaves from a dc current in a piece of crystal, which is naturally sturdy and has a long lifetime compared to a vacuum tube, has important applications in RADAR and telecommunication systems; the generating crystal is called a "Gunn diode" (a competing device is the Read diode, see Chap.10b).

For a calculation of the growth rate of a high-field domain we begin with Poisson's equation which for simplicity is taken in one dimension

[8] J. E. Carroll: Hot Electron Microwave Generators, p. 105 ff. London: Arnold. 1970.

$$\partial E/\partial x = |e|\,(n - N_D)/\kappa\kappa_o \tag{7g.4}$$

Both the concentration of donors, N_D, and the dielectric constant, κ, are assumed constant while the carrier concentration n varies with the time, t, according to

$$n = n_o + \Delta n \cdot \exp(i\omega t) \tag{7g.5}$$

and similarly the current density

$$j = j_o + \Delta j \cdot \exp(i\omega t) \tag{7g.6}$$

and the field strength

$$E = E_o + \Delta E \cdot \exp(i\omega t) \tag{7g.7}$$

in a "small-signal theory" ($\Delta n \ll n_o$; $\Delta j \ll j_o$; $\Delta E \ll E_o$). For the current density we consider the drift term, the diffusion term, and the displacement term :

$$j = n|e|v_d - |e|D\partial n/\partial x + \kappa\kappa_o\,\partial E/\partial t \tag{7g.8}$$

The diffusion constant D is taken as field independent. The small-signal treatment yields

$$\Delta j = n_o\,|e|\,(\partial v_d/\partial E)\,\Delta E + |e|\,v_d\,\Delta n - |e|\,D\partial\Delta n/\partial x + i\omega\kappa\kappa_o\Delta E \tag{7g.9}$$

For Δn and its derivative we find from Eq.(7g.4) :

$$\partial\Delta E/\partial x = (|e|/\kappa\kappa_o)\Delta n \quad \text{and} \quad \partial^2\Delta E/\partial x^2 = (|e|/\kappa\kappa_o)\partial\Delta n/\partial x \tag{7g.10}$$

The left-hand sides become $-iq\Delta E$ and $-q^2\Delta E$, respectively, for a wave propagating as $\Delta E \propto \exp(-iqx)$ with a wave vector $\vec{q}$. This yields for Eq.(7g.9) :

$$\Delta j = \kappa\kappa_o\,(\tau_d^{-1} + i\omega - iqv_d + q^2 D)\Delta E \tag{7g.11}$$

where τ_d is a "differential dielectric relaxation time" given by

$$\tau_d^{-1} = (n_o\,|e|/\kappa\kappa_o)\,\partial v_d/\partial E \tag{7g.12}$$

Under constant-current conditions, $\Delta j = 0$, the electric waves propagating in the crystal have an angular frequency of

$$\omega = qv_d + i(\tau_d^{-1} + q^2 D) \tag{7g.13}$$

The imaginary part of ω represents a damping of the wave :

$$\exp(i\omega t) = \exp(iqv_d t)\,\exp\{-(\tau_d^{-1} + q^2 D)t\} \tag{7g.14}$$

The real part yields a phase velocity $\mathrm{Re}(\omega)/q = v_d$, the electron drift velocity. Since τ_d^{-1} is proportional to $\partial v_d/\partial E$, it is negative in the range of ndc. If the slope in this range is steep enough so that $-\tau_d^{-1} > q^2 D$, Eq.(7g.14) results in an exponential growth of the amplitude which finally saturates as shown in Fig.7.24. The pulse shape of the domain can be investigated by a Fourier analysis where in Eq.(7g.7) for each harmonic a corresponding term has to be taken. Diffusion counteracts the domain formation by its tendency to make the local distribution of electrons uniform.

If the diffusion may be neglected and τ_d and v_d can be approximated by their values averaged over the whole sample, we find ΔE by solving the differential equation obtained from Eqs.(7g.9) and (7g.10):

$$\Delta E = \Delta E(x) = \{\Delta jL/(\kappa\kappa_o v_d s)\}\ \{1-\exp(-sx/L)\} \tag{7g.15}$$

where the dimensionless complex quantity

$$s = (L/v_d)\,(\tau_d^{-1} + i\omega) \tag{7g.16}$$

has been introduced. The boundary condition $\Delta E(0) = 0$ is satisfied. The sample impedance $Z = Z(\omega)$ is given by

$$Z(\omega) = \frac{1}{A\Delta j}\int_0^L \Delta E dx = \frac{1}{A\kappa\kappa_o v_d s}\{L + (L/s)(e^{-s}-1)\} = \frac{L^2}{A\kappa\kappa_o v_d}\cdot\frac{e^{-e}+s-1}{s^2} \tag{7g.17}$$

The impedance vanishes at a maximum of $\mathrm{Im}(\omega)$ for $s = -2.09 \pm i\,7.46$, i.e. there is a current instability at a constant applied voltage. Inserting this value into Eq. (7g.16) yields a critical sample length

$$L_{crit} = 2.09 v_d(-\tau_d) = \frac{2.09\,\kappa\kappa_o v_d}{n_o|e|(-\partial v_d/\partial E)} \tag{7g.18}$$

For domain formation this is the minimum sample length. For n-type GaAs the product $n_o L_{crit}$ is determined by the maximum value of $-\partial v_d/\partial E$ which is 500 cm²/Vsec, by the average drift velocity, v_d, of $\approx 10^7$ cm/sec, and by the dielectric constant $\kappa = 13.5$. These data give

$$n_o L_{crit} = 3\mathrm{x}10^{11}/\mathrm{cm}^2 \tag{7g.19}$$

Assuming e.g. L = 0.3 mm this yields a minimum carrier concentration of $n_o = 10^{13}/\mathrm{cm}^3$. On the other hand, if the sample is too long the field at the cathode during domain formation is not sufficiently reduced to prevent the launching of a second domain before the first one has reached the anode.

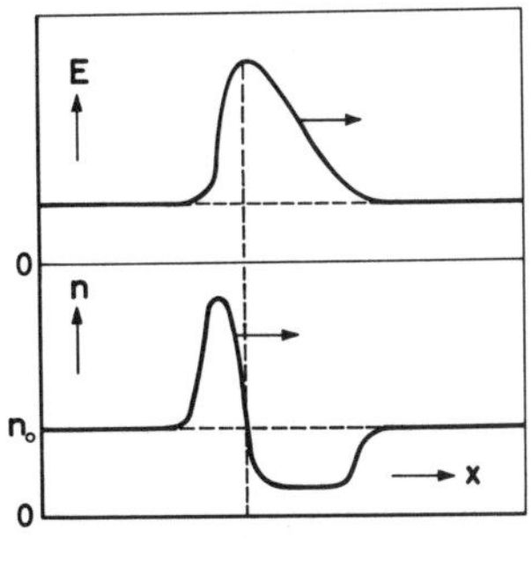

Fig.7.25 Schematic diagrams of the electric field and the electron density vs position for a high-field domain moving toward the right side (after P. N. Butcher, W. Fawcett, and C. Hilsum, Brit. J. Appl. Phys. 17 (1966) 841.).

Fig.7.25 shows the shape of a domain moving from left to right. At the leading edge there is a small deficiency of electrons while at the trailing edge there is an excess of electrons: The light Γ-electrons are caught by the leading edge where due to the strong electric field they are converted into heavy X-electrons. Because of their low mobility and in spite of the strong electric field they accumulate at the trailing edge of the pulse. If they should escape from the domain they are immediately reconverted into highly mobile Γ-electrons and catch up with the domain. In this way the electron accumulation at the trailing edge and the slope $\partial\Delta E/\partial x \propto \Delta n$

(see Eq.(7g.10)) can be understood. A numerical calculation which includes the diffusion of carriers has been published by Butcher and Fawcett [9]. The calculated domain shape has in fact been observed [10]; the comparison [11] yields an intervalley relaxation time, τ_i, of about 10^{-12} sec which is short compared with the drift time of 10^{-9} sec.

The explanation of the Gunn oscillations by non-equivalent intervalley electron transfer has been supported by two experiments, where the energy difference of the valleys was decreased either by hydrostatic pressure [12] on GaAs or by varying the composition x of a $GaAs_xP_{1-x}$ compound [13]. These effects are well known from measurements of the optical absorption edge and will be discussed in Chap.11b where e.g. Fig.11.13 shows the shift of the absorption edge in GaAs with hydrostatic pressure. The central valley moves up and the satellite valleys move down with pressure until at 60 000 atm they are at the same level of 0.49 eV above the central-valley position at zero pressure. The threshold field for Gunn oscillations decreases with pressure until a value of 1.4 kV/cm at 26 000 atm is reached. Beyond this pressure there are no more oscillations. With a decreasing energy separation $\Delta\epsilon_{\Gamma X}$ of the valleys, a lower electron temperature is sufficient for a carrier transfer to the satellite valleys. However, at too low values of $\Delta\epsilon_{\Gamma X}$ relative to k_BT, the ndc vanishes (curve #1 in Fig.7.20) and the oscillations stop.

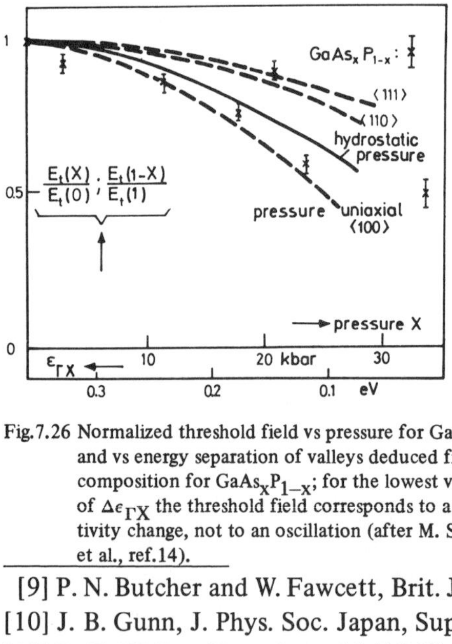

Fig.7.26 Normalized threshold field vs pressure for GaAs and vs energy separation of valleys deduced from composition for $GaAs_xP_{1-x}$; for the lowest value of $\Delta\epsilon_{\Gamma X}$ the threshold field corresponds to a resistivity change, not to an oscillation (after M. Shyam et al., ref.14).

Fig.7.26 shows the threshold field as a function of hydrostatic pressure [12] (full curve) and uniaxial pressure in 3 crystallographic directions [14] (dashed curves). The strongest decrease of the threshold field with uniaxial pressure is in the <100> direction, in agreement with the location of the satellite valleys on <100> and equivalent axes : The 2 valleys in the direction of the pressure move

[9] P. N. Butcher and W. Fawcett, Brit. J. Appl. Phys. 17 (1966) 1425.
[10] J. B. Gunn, J. Phys. Soc. Japan, Suppl. 21 (1966) 505.
[11] B. K. Ridley, Phys. Lett. 16 (1965) 105.
[12] A. R. Hutson, A. Jayaraman, A. G. Chynoweth, A. S. Corriel, and A. L. Feldman, Phys. Rev. Lett. 14 (1965) 639.
[13] J. W. Allen, M. Shyam, Y. S. Chen, and G. L. Pearson, Appl. Phys. Lett. 7 (1965) 78.
[14] M. Shyam, J. W. Allen, and G. L. Pearson, IEEE-Trans. ED-13 (1966) 63.

down and determine the threshold field, while the 4 others move slightly up [15]. Uniaxial pressure in a <111> direction is symmetric to all valleys, and since it acts in only one of the 3 directions of space its effect is only 1/3 of that of a hydrostatic pressure of equal magnitude.

In n-GaP the <100> valleys are lowest even without pressure as in the case of the conduction band of Si. If in $GaAs_xP_{1-x}$ a linear variation of $\Delta\epsilon_{\Gamma X}$ with composition x is assumed such that $\Delta\epsilon_{\Gamma X} = 0$ at x = 0.5 the pressure scale in Fig.7.26 can be supplemented by an energy scale $\Delta\epsilon_{\Gamma X}$ and the observed dependence of the threshold field on the composition [14] can be included in the diagram (crosses); there is good agreement with the observed pressure dependence.

For technical applications of the Gunn diode sinusoidal oscillations are preferred to pulse-shaped oscillations. In InP sinusoidal oscillations have been observed and explained by a 3-valley system with different deformation potential constants for the transitions [16]: An electron transfer $\Gamma \to X$ is followed by a transition $X \to L$ involving the emission of a phonon. These transitions have been considered fast compared with the final transition from the intermediate L-valleys to the original Γ-valley.

If the Gunn diode is mounted in a resonant circuit its frequency may be shifted by a factor of 2 without loss of efficiency. Diodes with nL-products of more than a few times $10^{12}/cm^2$ (see Eq.(7g.19)) show their largest output at a frequency which is somewhat longer than half the inverse transit time of the domain through the diode. If the nL-product is less than $10^{12}/cm^2$, the diode cannot oscillate but can be used for amplification purposes at a frequency of the inverse transit time [17]. In the "limited space charge accumulation oscillator" (LSA) the dc field $\vec{E}_0$ is supplemented by an ac field $\vec{E}_1 \cos(2\pi\nu t)$ parallel to $\vec{E}_0$ of so large an amplitude that for a small part of the period the differential mobility dv_d/dE is positive [18] as shown in Fig.7.27. At a sufficiently high frequency ν, nearly no space charge is generated: $n_0/\nu < 2\text{x}10^5/cm^3$ Hz. Due to the falling characteristic it is possible that the power P_{tot}, which an electron gains from the field $E_0 + E_1 \cos(2\pi\nu t)$ during a period $1/\nu$, is smaller than the power P_0 which it would gain from the dc field E_0 alone :

$$P_{tot} = e\nu \int_0^{1/\nu} Evdt < P_0 = e\nu E_0 \int_0^{1/\nu} vdt \qquad (7g.20)$$

[15] R. W. Keyes: Solid State Physics (F. Seitz and D. Turnbull, eds.), Vol. 11, p. 149. New York: Acad Press. 1960.

[16] C. Hilsum and H. D. Rees, Electr. Lett. 6 (1970) 277 and 310; H. C. Law and K. C. Kao, J. Appl. Phys. 41 (1970) 829; for energies in III-V compounds see C. Hilsum, Proc. Int. Conf. Semic. Phys. Paris (1964) p. 1127. Paris: Dunod. 1964.

[17] H. W. Thim, M. R. Barber, B. W. Hakki, S. Knight, and M. Uenohara, Appl. Phys. Lett. 7 (1965) 167.

[18] J. A. Copeland, J. Appl. Phys. 38 (1967) 3096.

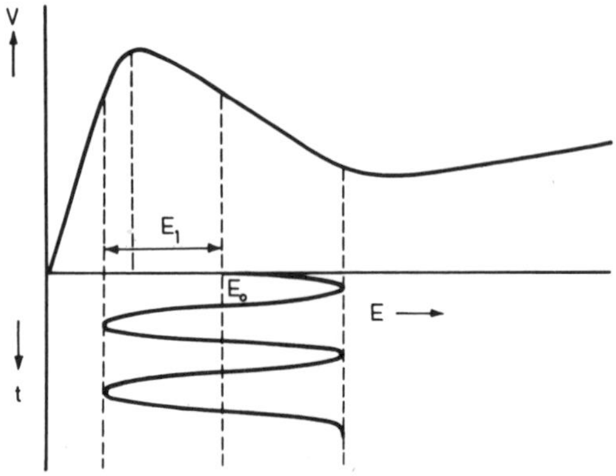

Fig.7.27 Field variation as a function of time in LSA oscillator (after G. A. Acket: Festkörperprobleme (O. Madelung, ed.), Vol. IX, p. 282. Oxford: Pergamon and Braunschweig: Vieweg. 1969.

i.e. the yield $(P_o - P_{tot})/P_o$ is positive. Maximum yields of 11 % at a zero-field mobility μ_o of 5 000 cm²/Vsec and of 23 % at μ_o = 9 000 cm²/Vsec have been calculated [19].

In another method of suppressing the free formation of Gunn domains the Gunn diode is coated with an insulating high-dielectric-constant material such as ferroelectric $BaTiO_3$ [20]. In a simplified picture according to Eq.(7g.12) the negative value of τ_d^{-1} can thus be reduced to a magnitude where amplification is possible but no free oscillation.

After the discovery of the Gunn effect in n-type GaAs and InP a few more semiconductors with a similar conduction band structure were shown to exhibit this type of oscillation: CdTe [21], InSb [2], $Ga_xIn_{1-x}Sb$ [22], and InAs [23]. It has been quite surprising to also find oscillations in n-type Ge [24], though much weaker, at temperatures below 130 K because there is nearly no difference in mobility between the lowest (L-type) valleys (3.9×10^4 cm²/Vsec at 77 K) and the upper (X-type) valleys (5×10^4 cm²/Vsec, observed in n-type Si at 77K). Nontheless a calculation by Paige [25] revealed a weak ndc above 2 - 5 kV/cm at

[19] I. B. Bott and C. Hilsum, IEEE-Trans. ED-14 (1967) 492.
[20] S. Kataoka, H. Tateno, M. Kawashima, and Y. Komamiya, 7th Int. Conf. Microwave Optical Generation and Amplification, Hamburg/Germany, 1968. Berlin: VDE Verlag. 1968.
[21] G. W. Ludwig, IEEE-Trans. ED-14 (1967) 547.
[22] J. C. McGroddy, M. R. Lorenz, and T. S. Plaskett, Solid State Comm. 7 (1969) 901.
[23] J. W. Allen, M. Shyam, and G. L. Pearson, Appl. Phys. Lett. 11 (1967) 253.
[24] J. C. McGroddy, M. I. Nathan, and J. E. Smith, Jr., IBM-J. Res. Develop. 13 (1969) 543.
[25] E. G. S. Paige, IBM-J. Res. Develop. 13 (1969) 562.

temperatures between 27 and 77 K where a value of 1×10^8 eV/cm is assumed for the acoustic deformation potential for the transition L → X with a phonon temperature $\Theta_i = 320$ K. This accounts for the weak oscillations observed for current directions of e.g. <100> or <110>, and the fact that there are no oscillations for a <111> direction; in the latter case there is a repopulation between L-valleys with the result of a lower mobility such that a later transfer to the X-valleys no longer results in an ndc range. For a current in a <100> direction the population of the L-type valleys is not changed but the transfer to the <100> valleys with their low mobility is preferred relative to the <010> and <001> valleys, while for a current in a <110> direction the populations are not too different from those for a <100> direction.

After the observation of a Gunn effect in n-type Ge the question arose why a constant-current range instead of an ndc is observed if the Gunn frequency is higher than the maximum frequency of the apparatus as has been the case in earlier measurements [24]. It has been supposed that the field distribution in the sample although static may be non-uniform with the formation of a flat-topped domain [8]. From Poisson's Eq.(7g.4) with $N_D = n_o$ and an average drift velocity $\langle v_d \rangle$, given by

$$n_o |e| \langle v_d \rangle = n(x) |e| v_d(E) , \tag{7g.21}$$

an explicit relation for $\partial E/\partial x$ is obtained :

$$\partial E/\partial x = (|e|\, n_o/\kappa\kappa_o) \, \{\langle v_d \rangle / v_d(E) - 1\} \tag{7g.22}$$

The $v_d(E)$-curve is shown in Fig.7.28a; $E(x)$ corresponding to two values of $\langle v_d \rangle$ is shown in Fig.28b. For a value of $\langle v_d \rangle$ below the threshold, E is uniform except for a region near the cathode where the material is heavily doped due to the

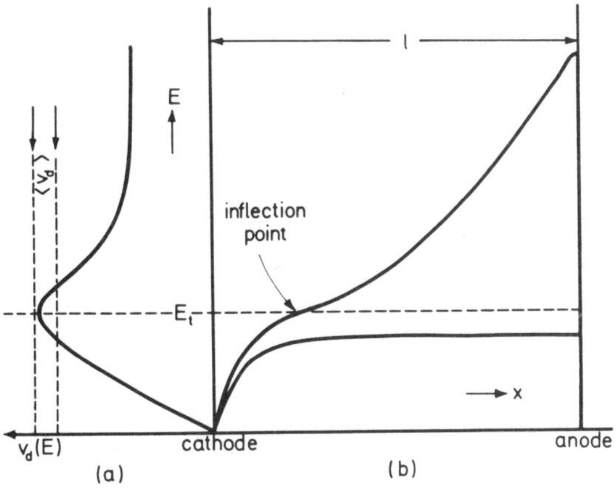

Fig.7.28 a Drift velocity vs field (similar to curve 2 in Fig. 7.21); b Resulting spatial dependence of electric field for two average drift velocities indicated in (a) (after McGroddy et al., ref. 24).

the formation of the metallic contact. For a value of $<v_d>$ above threshold, Max $\{v_d(E)\}$, Eq.(7g.22) yields $\partial E/\partial x > 0$ for all values of x, i.e. throughout the sample, with an inflection point at the threshold field strength E_t. If the product of carrier density and sample length satisfies the condition

$$n_o L > \kappa\kappa_o E_t / |e| \tag{7g.23}$$

the current density j rises only slightly with increasing total voltage across the sample as shown in Fig.7.29 once $j > n_o |e| \, \mathrm{Max}\{v_d(E)\}$ which is then denoted as the "saturation value" of j. Since small non-uniformities of doping in the sample produce gross non-uniformities of the field, it is not possible to deduce sign and magnitude of the differential conductivity near the saturation of j.

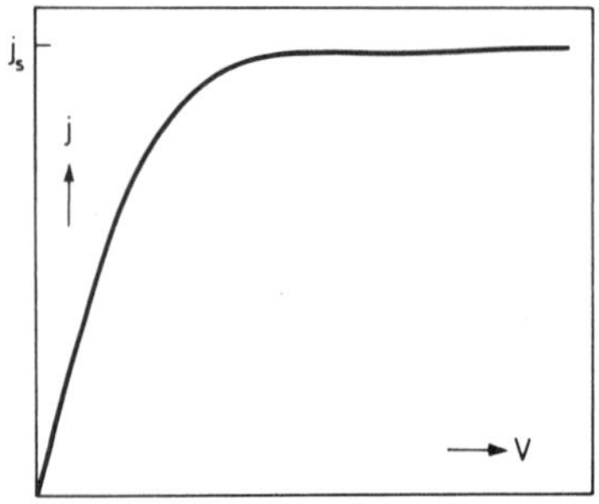

Fig.7.29 Current voltage characteristic resulting from situation shown in Fig.7.28, for the case of a "long" sample.

Let us finally consider Gunn oscillations in n-type Ge under uniaxial stress in a <111> direction for a current flowing in the $<11\bar{2}>$ direction which is perpendicular to the stress. The L-type valley whose longitudinal axis is in the stress direction is moved down while the 3 others are moved up. As shown in Fig.7.30, the effective masses in the current direction are: m_t in the low valley, 1.3 m_t in two of the upper valleys, and 6.4 m_t in the remaining upper valley where $m_t = 0.08\, m_o$ is the transverse mass. The threshold field as a function of stress is shown [26] in Fig.7.31 for electron concentrations between 10^{14} and $1.6 \times 10^{16}/cm^3$. The minimum in this dependence can be explained by assuming that at large stresses and correspondingly large energy differences high electron temperatures are necessary for electron transfer; at these electron temperatures the drift velocity in each valley is nearly kept constant due to the emission of energetic phonons [27].

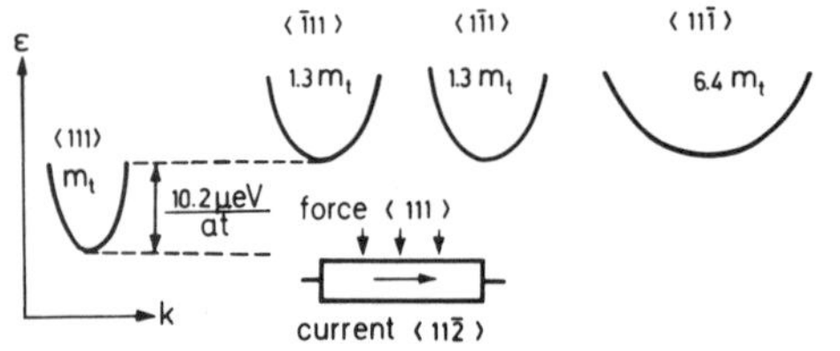

Fig.7.30 Valleys at the conduction band edge of germanium shifted by uniaxial stress in the <111> direction; effective mass values given in the current direction (after J. E. Smith, Jr., ref. 26).

In similar experiments with n-type Si the drift velocity was observed to saturate with E > 15 kV/cm at stresses larger than 5×10^3 atm but no oscillations have been observed. A Gunn effect bibliography has been compiled in refs. [28] and [8].

[26] J. E. Smith, Jr., Appl. Phys. Lett. 12 (1968) 233.
[27] J. E. Smith, Jr., J. C. McGroddy, and M. I. Nathan, Proc. Int. Conf. Semic. Physics Moscow 1968, p. 950. Leningrad: Nauka. 1968.
[28] T. K. Gaylord, P. L. Shah, and T. A. Rabson, IEEE-Trans. ED-15 (1968) 777, ED-16 (1969) 490.

Besides the Gunn effect there are many other effects which show the "voltage controlled" N-shaped characteristic shown in Fig. 7.21. Since these effects produce either amplitudes or oscillation frequencies that are much lower than in the Gunn effect, there are at present no device applications. In some homogeneously doped semiconductors a "current controlled" S-shaped characteristic is observed which is shown schematically in Fig.7.32; it leads to a formation of current filaments, j_1 and j_2. This behavior has been found e.g. during impact ionization in compensated n-type Ge (see Chap.10). For details on effects of this kind see e.g. Conwell [29] and Barnett [30].

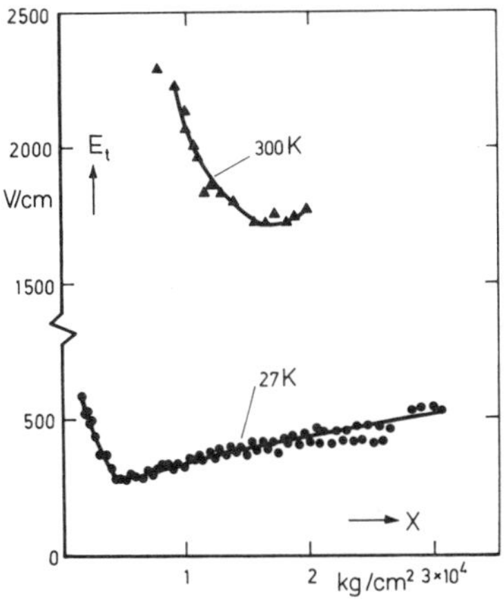

Fig.7.31 Threshold field vs compressive uniaxial stress in a ⟨111⟩ direction in n-type germanium at 27 and 300 K (after J. E. Smith, Jr., ref.26).

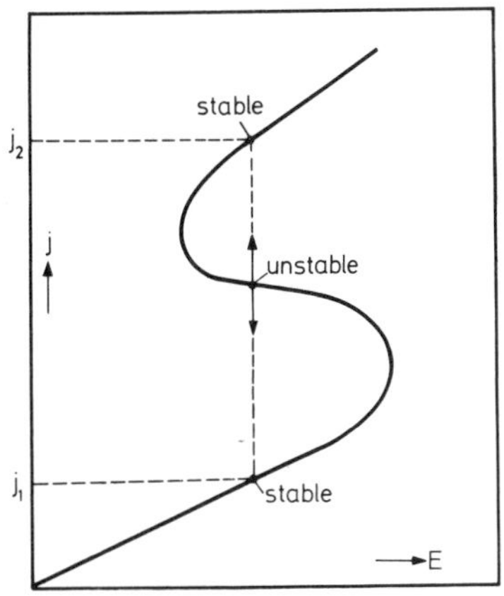

Fig.7.32 S-shaped current instability with a formation of current filaments.

7h. The Acousto-electric Effect

The appearance of a dc electric field along the direction of propagation of a travelling acoustic wave in a semiconductor is known as the "acousto-electric effect". It is due to the drag of carriers by the wave which is similar to the motion of driftwood toward a beach and to charge transport in a linear accelerator. Fig. 7.33 shows the sinusoidal variation of the potential energy $\Delta\epsilon_c$ in a crystal due to an acoustic wave at a certain time. Mobile carriers (represented by open circles) tend to bunch in the potential troughs as indicated by arrows along the curve. However, since the wave moves toward the right with a velocity u_s and a finite time τ_R would be required to reach equilibrium, the carriers can never

[29] Ref.7, p. 84 - 86.

[30] A. M. Barnett, IBM-J. Res. Develop. 13 (1969) 522.

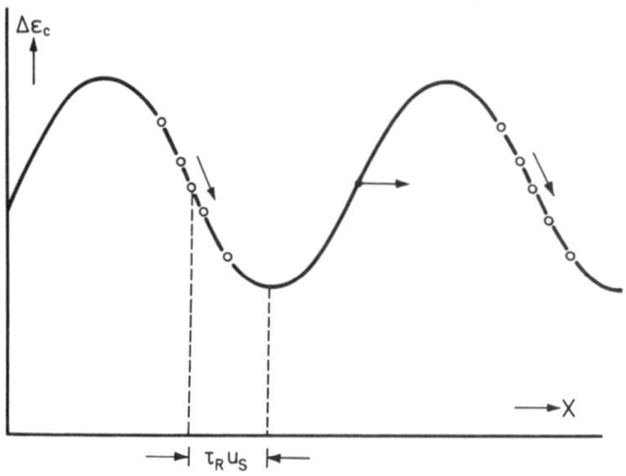

Fig.7.33 Spatial variation of the potential energy due to propagating acoustic wave and carrier bunching on the front slopes of the wave.

arrive at the troughs for the case of $\omega\tau_R \approx 1$ where ω is the angular frequency. Therefore, a higher carrier concentration is found on the front slopes of the wave than on the back slopes. A net average force on the carriers is therefore exerted in the propagation direction of the wave. The carriers finally accumulate on that sample face at which the acoustic wave emerges from the sample. This causes a dc field in the sample. If a dc current is allowed to flow, there is a net energy transfer from the acoustic wave to the electron gas and the wave is attenuated.

On the other hand, an acoustic wave is amplified by a dc current if the drift velocity of the carriers is slightly larger than the wave velocity. The carriers will then bunch on the back slopes. It has thus been possible to build acousto-electric amplifiers and also oscillators [1]. It may be interesting to note that the "phonon drag" mentioned in Chap.4j is different from the acousto-electric effect; the latter is produced by phase coherent acoustic waves rather than by incoherent phonons.

We will first consider the acousto-electric effect in non-polar many-valley semiconductors such as n-type Ge. In piezoelectric semiconductors such as CdS where the acousto-electric effect is much stronger, it does not rely upon a many-valley structure; but since the treatment to a large extent is very similar to the effect in n-Ge, we will consider it also here.

A shear wave obtained at frequencies of 20 or 60 MHz from a "Y cut" quartz oscillator is fed via a stopcock lubricant into an n-type Ge crystal in the <100> direction; the wave is polarized in the <010> direction [2]. Let us denote these

[1] J. D. Maines and E. G. S. Paige, J. Phys. C 2 (1969) 175.

[2] G. Weinreich, T. M. Sanders, Jr., and H.G. White, Phys. Rev. 114 (1959) 33; this and more recent papers have been reviewed by N. G. Einspruch: Solid State Physics (F. Seitz and D. Turnbull, eds.), Vol. 17, p. 243. New York: Acad. Press. 1965; M. Pomerantz, Proc. IEEE 53 (1965) 1438.

directions by x and y, respectively. For a material displacement

$$\delta y = \delta y_1 \sin(\omega t - qx) \tag{7h.1}$$

of amplitude δy_1 and angular frequency ω, we obtain a shear strain

$$\epsilon_{xy} = \frac{\partial \delta y}{\partial x} = -q\,\delta y_1 \cos(\omega t - qx) \tag{7h.2}$$

and a change in potential energy given by Eq.(7a.21):

$$\Delta\epsilon_c = \pm \frac{1}{3}\,\Xi_u\,\epsilon_{xy} \tag{7h.3}$$

where the sign depends on the valley in the conduction band of germanium shown in Fig.7.34. Hence, there are two classes of conduction band valleys

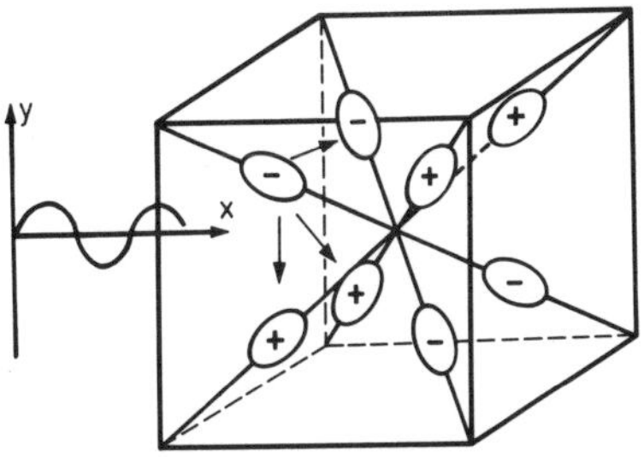

Fig.7.34 Separation of the constant-energy surfaces in $\vec{k}$-space for n-type germanium into two classes (+ and −) by a transverse acoustic wave propagating in real space with the indicated direction and polarization (for better visibility the half-ellipsoids have been replaced by full-ellipsoids). The cube does not represent the Brillouin zone.

which are indicated by subscripts + and −. Bunching of carriers from different classes of valleys takes place in different phases. It is convenient to introduce formally an "acoustic charge" e_{ac} given by

$$e_{ac} = \frac{1}{3}\,\Xi_u/\sqrt{c_t} \tag{7h.4}$$

where c_t is the elastic constant (denoted as c_{44} in elastic-tensor theory, Eq. (4m.4)) and an "acoustic potential" ϕ given by

$$\phi = -\sqrt{c_t}\; q\delta y_1 \cos(\omega t - qx) = \phi_1 \cos(\omega t - qx) \tag{7h.5}$$

In this notation Eq.(7h.3) becomes simply

$$\Delta\epsilon_c = \pm\, e_{ac}\,\phi \tag{7h.6}$$

Let us assume the material displacement and hence the wave amplitude to be small compared to the average displacement of the thermal vibration so that the carrier densities n_+ and n_- are given by

$$n_{\pm} = n_o \exp(\mp e_{ac}\phi/k_BT) \approx n_o(1 \mp e_{ac}\phi/k_BT) \tag{7h.7}$$

where n_o is the carrier density in each class of valleys for $\phi = 0$. For $\phi \neq 0$ we would have at equilibrium

$$n_+ - n_- = -2n_o e_{ac}\phi/k_BT \tag{7h.8}$$

Naturally, there is no equilibrium, and the non-equilibrium transfer rate $R_{+\to -}$ is given by

$$R_{+\to -} = -R_{-\to +} = \frac{2}{3}\left\langle\frac{1}{\tau_i}\right\rangle (n_+ - n_- + 2n_o e_{ac}\phi/k_BT) \tag{7h.9}$$

where τ_i is the intervalley scattering time (Chap.7e) and the factor 2/3 arises because in only two out of three intervalley transitions the electron changes the class (indicated by arrows in Fig.7.34). Particle conservation yields $n_+ + n_- = 2n_o$ and therefore from Eq.(7h.9) we find

$$R_{+\to -} = \frac{4}{3}\left\langle\frac{1}{\tau_i}\right\rangle \{n_+ - n_o(1 - e_{ac}\phi/k_BT)\} \tag{7h.10}$$

From the continuity equation

$$R_{+\to -} = \partial n_-/\partial t + \partial j_-/\partial x = -\partial n_+/\partial t - \partial j_+/\partial x \tag{7h.11}$$

the particle current densities given by

$$j_{\pm} = \mp e_{ac}\frac{\partial\phi}{\partial x} n_{\pm}\frac{D_n}{k_BT} - D_n\frac{\partial n_{\pm}}{\partial x} \tag{7h.12}$$

and the Einstein relation (between the carrier mobility and the diffusion coefficient D_n) we obtain after some algebra

$$n_+ - n_o = -n_o(1 + i\omega\tau_R)^{-1} e_{ac}\phi/k_BT \tag{7h.13}$$

where the frequency-dependent relaxation time

$$\tau_R = \left(\frac{4}{3}\left\langle\frac{1}{\tau_i}\right\rangle + q^2 D_n\right)^{-1} \approx \left(\frac{4}{3}\left\langle\frac{1}{\tau_i}\right\rangle + \frac{\rho\omega^2}{c_t} D_n\right)^{-1} \tag{7h.14}$$

has been introduced. Eq.(7h.14) shows that two processes contribute to the relaxation of the distribution, namely intervalley scattering and diffusion out of the bunches. The average force per particle, $\bar{F}$, exerted by the acoustic wave,

$$\bar{F} = \langle -e_{ac}(\partial\phi/\partial x)(n_+ - n_o)\rangle / \langle n_o\rangle \tag{7h.15}$$

is obtained from Eqs.(7h.5) and (7h.13):

$$\bar{F} = eE_o = \frac{\Xi_u^2 J}{9c_t u_s^2 k_BT}\cdot\frac{\omega^2\tau_R}{1+\omega^2\tau_R^2} \tag{7h.16}$$

where E_o is the acousto-electric field intensity, $u_s = \omega/q$ is the sound velocity and $J = 1/2\,(\phi_1^2 u_s)$ is the acoustic energy flux averaged over a period of the wave. For $J = 1$ watt/cm^2 at 60 MHz and 77 K, $\tau_i = 10^{-11}$ sec, $\Xi_u = 16$ eV, $c_t = 1.56\text{x}10^{12}$

dyn cm^{-2}, and $u_s = 5.4 \times 10^5$ cm/sec valid for n-type Ge, a field intensity of 1 μV/cm is calculated. Wave amplification by drifting carriers has been treated by e.g. Conwell [3].

From the experimental results [2] obtained at temperatures T between 20 and 160 K on 5 samples of different purity ($10^{14}-10^{16}/cm^3$), $<1/\tau_i>$ has been determined by means of Eqs.(7h.16) and (7h.14) and in Fig.7.35 plotted vs T. At the higher temperatures the data points for all the samples (and especially for the purer ones) fall on the dashed curve which has been calculated from Eq.(7e.2) with $\lambda=1$ (i.e. $T_e=T$), $\Theta_i = 315$ K, and $w_2 = 10^{11}$ sec^{-1}. Comparing data on the purest sample at low temperatures and 20 MHz with those at 60 MHz where $\omega^2 \tau_R^2 \approx 1$, the deformation potential constant $\Xi_u = 16$ eV has been determined to an accuracy of 10%. The deviation of the data from the dashed curve have been considered to be due to ionized and neutral impurity scattering.

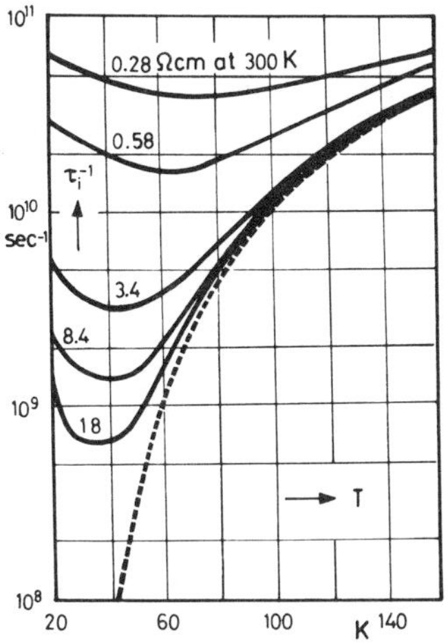

Fig.7.35 Temperature dependence of the intervalley scattering rate in n-type germanium (after Weinreich et al., ref.1).

Let us now investigate the amplification of longitudinal ultrasonic waves in piezoelectric semiconductors. This subject has been treated by White [4] and Hutson and White [5]. The coupling between the electrons and the lattice due to the piezoelectric effect has been discussed in Chap.6g. The dielectric displacement is given by Eq.(6g.3) where it is convenient to denote the strain $(\vec{\nabla}_r \delta\vec{r})$ by S :

$$D = \kappa\kappa_0 E + e_{pz} S \qquad (7h.17)$$

The stress T is given by Eq.(6g.10) :

$$T = c_l S - e_{pz} E \qquad (7h.18)$$

where the first term on the right-hand side represents Hook's law while the second term is due to the piezoelectric effect; c_l is the elastic constant at constant field. Let us denote the mean carrier density by n_0 and the instantaneous local density by $n_0 + fn_s$ where en_s is the space charge and only a fraction f of the space charge contributes to the conduction process; the rest is trapped at local states in the energy gap. Such traps are more important in piezoelectric semiconductors such as CdS or ZnO than they are in Ge or Si.

[3] E. M. Conwell: Solid State Physics (F. Seitz and D. Turnbull, eds.), Suppl. 9, p. 142. New York: Acad. Press. 1967.

[4] D. L. White, J. Appl. Phys. 33 (1962) 2547.

[5] A. R. Hutson and D. L. White, J. Appl. Phys. 33 (1962) 40.

The current density is given by

$$j = |e|\,(n_0 + fn_s)\mu E - eD_n \partial (n_0 + fn_s)/\partial x \tag{7h.19}$$

where $e < 0$ for electrons. The equation of continuity is

$$e\partial n_s/\partial t + \partial j/\partial x = 0 \tag{7h.20}$$

and Poisson's equation

$$\partial D/\partial x = en_s \tag{7h.21}$$

Differentiating the latter equation with respect to t yields

$$\partial^2 D/\partial x\partial t = -\partial/\partial x\{(\sigma + (e/|e|)\mu f\partial D/\partial x)E\} + D_n f\partial^3 D/\partial x^3 \tag{7h.22}$$

where we have taken Eq.(7h.20) into account. For an applied dc field E_0 and an an ac field of amplitude $E_1 \ll E_0$ which is due to an ultrasonic wave

$$\delta x = \delta x_1\, e^{i(\omega t - qx)} \tag{7h.23}$$

$$E = E_0 + E_1\, e^{i(\omega t - qx)} \tag{7h.24}$$

and similarly

$$D = D_0 + D_1\, e^{i(\omega t - qx)} = D_0 + (\kappa\kappa_0 E_1 - iqe_{pz}\delta x_1)\, e^{i(\omega t - qx)} \tag{7h.25}$$

where Eqs.(7h.17) and (7h.23) have been taken into account, we obtain

$$E_1 = i\frac{qe_{pz}\delta x_1}{\kappa\kappa_0}\,[1 - i/\{\omega\tau_d(\gamma - i\omega/\omega_D)\}]^{-1} \tag{7h.26}$$

where products of D_1 and E_1 which are second order terms have been neglected; the dielectric relaxation time τ_d given by Eq.(5b.23), a "diffusion frequency" $\omega_D = c_l/\rho D_n$, and a drift parameter $\gamma = 1 - (e/|e|)f\mu E_0/u_s$ have been introduced; μE_0 is the drift velocity; $u_s = \omega/q$ is the sound velocity. If we assume for the wave equation

$$\partial T/\partial x = \rho\partial^2\delta x/\partial t^2 = c_l\partial^2\delta x/\partial x^2 - e_{pz}\partial E/\partial x \tag{7h.27}$$

where Eq.(7h.18) has been taken into account, a solution of the form given by Eqs.(7h.23) and (7h.24), we find for the relation between δx_1 and E_1

$$-\rho\omega^2\delta x_1 = -c_l q^2\delta x_1 + iqe_{pz}E_1 \tag{7h.28}$$

This can be written in the usual form

$$-\rho\omega^2\delta x_1 = -c_l' q^2\delta x_1 \tag{7h.29}$$

by formally introducing a complex elastic constant c_l'. We obtain for c_l'/c_l by taking into account Eq.(7h.26)

$$c_l'/c_l = 1 - i(e_{pz}/q)E_1/\delta x_1 = 1 + K^2[1 - i/\{\omega\tau_d(\gamma - i\omega/\omega_D)\}]^{-1} \tag{7h.30}$$

where the electromechanical coupling coefficient $K^2 \ll 1$ given by Eq.(6g.9) has been introduced. The real part of c_l' is simply the product ρu_s^2 and since $K^2 \ll 1$

we can approximate $u_s = \mathrm{Re}\sqrt{c_1'/\rho}$ by

$$u_s = u_{so}\left\{1 + \frac{1}{2}K^2\left(1 - \frac{\frac{1}{\omega\tau_d}\left(\frac{1}{\omega\tau_d} + \frac{\omega}{\omega_D}\right)}{\gamma^2 + \left(\frac{1}{\omega\tau_d} + \frac{\omega}{\omega_D}\right)^2}\right)\right\} \tag{7h.31}$$

where $\sqrt{c_1/\rho} = u_{so}$ is the sound velocity for $K^2 = 0$. If we consider q in Eq.(7h.23) as complex with an imaginary part denoted by $-a = \mathrm{Im}(q)$, a positive value of a causes attenuation of the wave. The relation $\sqrt{\rho}\,\omega = \sqrt{c_1'}\,q$ obtained from Eq.(7h.29) thus yields for a in the same approximation as Eq.(7h.31)

$$a = -\omega\sqrt{\rho}\,\mathrm{Im}(1/\sqrt{c_1'}) = \frac{1}{2}\frac{K^2}{\tau_d u_{so}}\frac{\gamma}{\gamma^2 + \left(\frac{1}{\omega\tau_d} + \frac{\omega}{\omega_D}\right)^2} \tag{7h.32}$$

The attenuation constant is proportional to the drift parameter γ. Assuming $f = 1$ and a drift velocity μE_0 slightly larger than the sound velocity u_s and in the same direction, both γ and a are negative with the result of an amplification of the wave; for electrons ($e < 0$) the dc field $\vec{E}_0$ has to be negative, i.e. opposite to the direction of the wave.

The term $1/\omega\tau_d + \omega/\omega_D$ can be written as $1/\omega\tau_R$ where the frequency-dependent relaxation time τ_R has about the same form as τ_R given in Eq.(7h.14), namely

$$\tau_R = \left(\frac{1}{\tau_d} + \frac{\rho\omega^2}{\mathrm{Re}(c_1')}D_n\right)^{-1} \tag{7h.33}$$

The maximum relative reduction in the sound velocity occurs at $\gamma = 0$ and amounts to $\frac{1}{2}K^2/\{1 + (u_{so}/\omega L_D)^2\}$ which may not be appreciable since $K^2 \ll 1$; L_D is the Debye length given by Eq.(5b.24). However, the important fact is the negative attenuation coefficient. Its maximum value of $\frac{1}{8}K^2\sqrt{\omega_D/\tau_d}/u_{so} = K^2/8L_D$ is attained at $\gamma = -(2/u_{so})\sqrt{D_n/\tau_d}$. For CdS, values of $\gamma = -10.1$ and $-a = 5.6\times10^3$ db/cm have been calculated; even higher values of $-a$, namely 1.3×10^4 db/cm at $\gamma = -5.1$, are found in ZnO [6].

Figs.7.36 and 7.37 show experimental results of a vs frequency and vs γ, respectively, for CdS [7]. The curves have been calculated from Eq.(7h.32). The agreement between theoretical and experimental results is very good, although shear waves rather than longitudinal waves have actually been amplified in these experiments.

Fig.7.38 shows the current-voltage characteristics of CdS at various times between $\ll 1$ μsec and 80 μsec after application of the voltage [8]. The deviation

[6] N. I. Meyer and M. H. Jörgensen: Festkörperprobleme. Vol. X, p. 21. Oxford: Pergamon and Braunschweig: Vieweg. 1970.

[7] D. L. White, E. T. Handelman, and J. T. Hanlon, Proc. IEEE 53 (1965) 2157.

[8] J. H. McFee, J. Appl. Phys. 34 (1963) 1548.

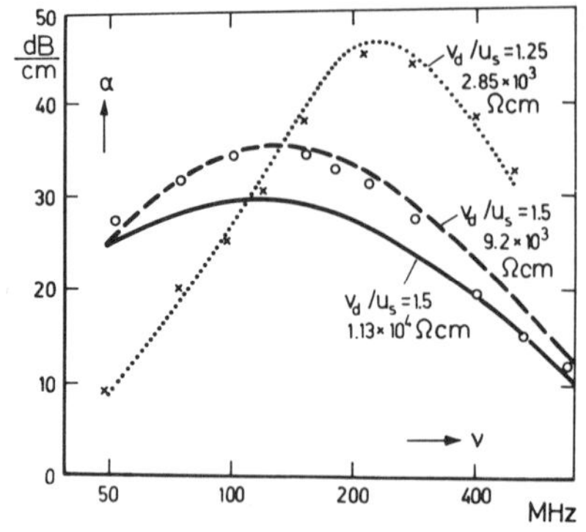

Fig.7.36 Electronic attenuation coefficient vs frequency for shear waves in photoconducting CdS; crosses and dotted curve: observed and calculated data, respectively, for $v_d/u_s = 1.25$ and $\rho = 2.85 \times 10^3$ ohm-cm; circles and full curve for $v_d/u_s = 1.5$ and $\rho = 1.13 \times 10^4$ ohm-cm; dashed curve calculated for $v_d/u_s = 1.5$ and $\rho = 9.2 \times 10^3$ ohm-cm (after D. L. White et al., ref. 7).

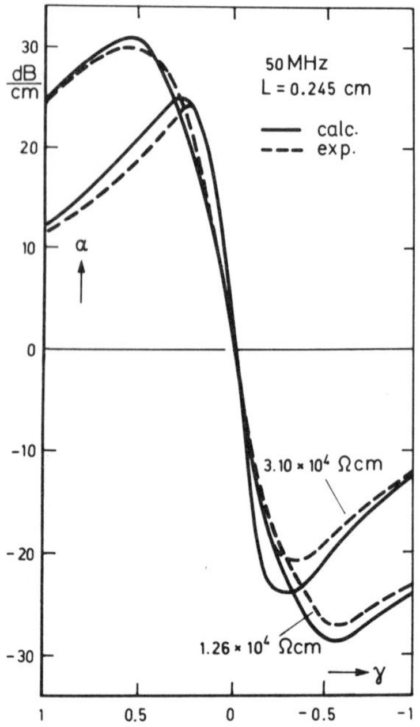

Fig.7.37 Electronic attenuation coefficient vs drift parameter for 50 MHz shear waves in photoconducting CdS at 300 K (after D. L. White, ref. 7).

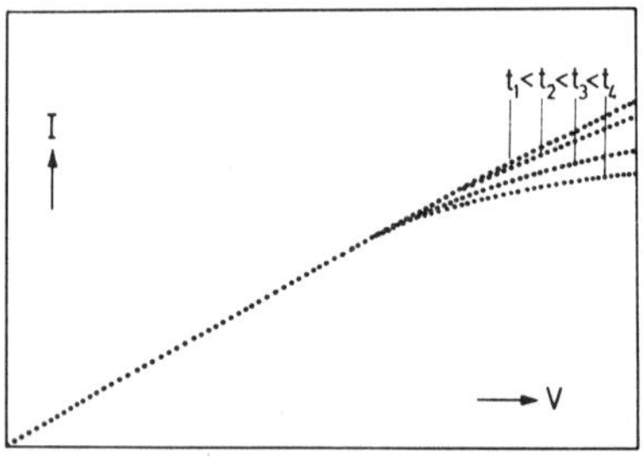

Fig.7.38 Current-voltage characteristic for cadmium sulphide at various time intervals after the application of the voltage (see Fig.7.39).

from Ohm's law at the larger time intervals is due to the generation of additional acoustic flux (acts by acoustic phonon scattering) from thermal lattice vibrations due to the acousto-electric effect. The output of a transducer (= ultrasonic microphone) attached to the sample clearly shows this generation [5] (Fig.7.39, lower trace). The trace in the middle of Fig.7.39 is the current through the sample. It shows oscillations which are caused by acousto-electric domains travelling back and forth in the crystal. These domains which are in some respects similar to Gunn domains have been shown to exist by observing the light scattered by them ("Brillouin scattering")*)

*) It is convenient to denote the scattering of visible or infrared light by acoustic phonons as "Brillouin scattering" while light scattering by optical phonons (see Chap.11g) or plasmas in solids is known as "Raman scattering": L. Brillouin, Ann. Phys. 17 (1922) 88; for a review see e.g. the introductory part of the paper by G. B. Benedek and K. Fritsch, Phys. Rev. 149 (1966) 647; R. W. Dickson, IEEE Trans. QE-3 (1967) 85; L. L. Hope, Phys. Rev. 166 (1968) 883.

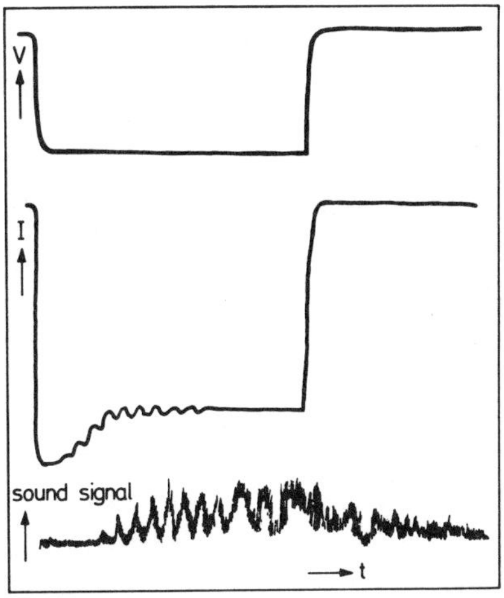

Fig.7.39 Buildup of ultrasonic flux in a CdS sample of resistivity $\rho = 3.3 \times 10^4$ ohm-cm; top trace: 1.5 kV voltage pulse (120 μsec long); middle trace: current pulse; bottom trace: signal from transducer tuned to 45 MHz (after McFee, ref. 8).

and by the modulation of microwaves transmitted through the sample which is shown in Fig.7.40 [9]. Present research is concerned with the kinetics of domain formation, nonlinear effects such as "parametric up-conversion" of frequencies at high acoustic power levels, and electrical instabilities due to the acousto-electric effect (for a review see e.g. ref. [6]).

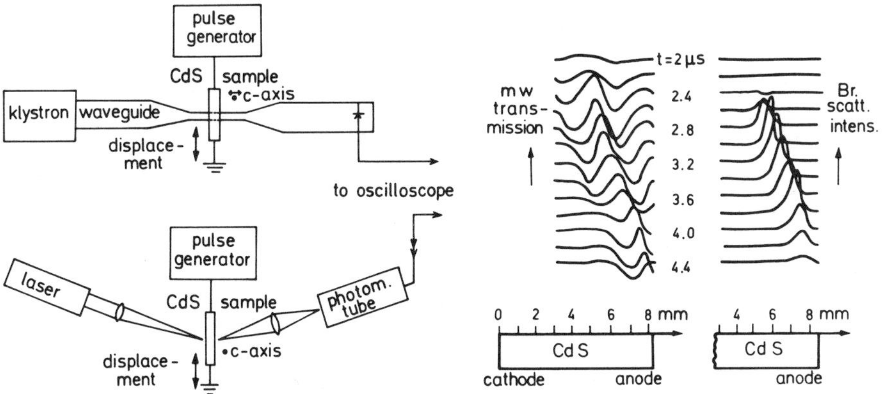

Fig.7.40 a Microwave transmission and Brillouin scattering arrangements for detection of acousto-electric domains.
b Microwave transmission and Brillouin scattering at various times after the application of the voltage. The negative peaks are due to space charge accumulation (after Wettling, ref. 9).

---

[9] W. Wettling: II-VI Semiconducting Compounds (D. G. Thomas, ed.), p. 928. New York: Benjamin. 1967; see also K. Hess and H. Kuzmany, Proc. Int. Conf. Phys. Semic. Warsaw 1972 (M. Miasek, ed.), p. 1233. Warsaw: PWN - Polish Scientific Publishers. 1972.

8. Carrier Transport in the Warped-Sphere Model

The valence bands of germanium, silicon, and the III-V compounds have an extremum at $\vec{k} = 0$ and are degenerate there. The constant-energy surfaces for this case are warped spheres which have already been discussed in Chap.2d (Figs. 2.29 and 2.30). In the zinc blende lattice typical for III-V compounds there is no center of inversion, in contrast to the diamond lattice.

8a. Energy Bands and Density of States

Although in Chap.2b we have considered the energy band model in general, we have not treated the warped-sphere model. However, it seems worthwhile to give here briefly an idea how it is developed from the Schrödinger equation without going into any details of the calculation*).

Since the valence band maximum is at $\vec{k} = 0$ the so-called $(\vec{k}.\vec{p})$ approximation is a convenient method of calculation. It is a perturbation approach to the problem. Substituting the wave functions $\psi(\vec{r}) = u(\vec{r}) \exp\{i(\vec{k}\vec{r})\}$, Eq.(2b.18), into the Schrödinger equation a solution of the form given by Eq.(2b.19) is found. Introducing the momentum operator $\vec{p} = -i\hbar\vec{\nabla}_r$ this can be written

$$\{p^2/2m + V(r)\}u(\vec{r}) + \{\hbar(\vec{k}\vec{p})/m\}u(\vec{r}) = (\epsilon - \hbar^2 k^2/2m)u(\vec{r}) \tag{8a.1}$$

For an energy band calculation at $\vec{k} = 0$ the term $\hbar(\vec{k}\vec{p})/m$ is treated as a small perturbation. For a short-hand notation let us introduce an eigenvalue

$$\epsilon' = \epsilon - \hbar^2 k^2/2m \tag{8a.2}$$

We will consider here only the diamond lattice where due to crystal symmetry at $\vec{k} = 0$ (= center of inversion) terms linear in $\vec{k}$ vanish and ϵ varies as $|\vec{k}|^2$ along any direction in $\vec{k}$-space. Hence, we have to solve the perturbation equation

$$H_{kp}\vec{u}(\vec{r}) = \{\hbar(\vec{k}\vec{p})/m\}\vec{u}(\vec{r}) = \epsilon'\vec{u}(\vec{r}) \tag{8a.3}$$

---

*) The following considerations have been adapted from L. Pincherle: Proc. Int. School of Physics XXII (R. A. Smith, ed.), p. 43. New York: Academic Press. 1963.

to second order where $\vec{u}(\vec{r})$ is given by a linear combination of atomic wave functions

$$\vec{u}(\vec{r}) = (\vec{a}yz + \vec{b}zx + \vec{c}xy)/r^2 \tag{8a.4}$$

where $\vec{a}$, $\vec{b}$, and $\vec{c}$ are unknown coefficients; spin-orbit coupling is neglected at present. For a determination of these coefficients $\vec{a} = \vec{a}(k_x, k_y, k_z)$ etc. from the set of linear equations (in components), Eq.(8a.3), is it sufficient to consider any physical problem which is proportional to the second order of a vectorial quantity, such as e.g. the magnetoresistance as a function of the magnetic field, as a model. An inspection of Eq.(4c.53) and neglecting the linear terms (i.e. $\gamma_0 = 0$) shows that the determinant of Eq.(8a.3) must have the form

$$\begin{vmatrix} Ak_x^2 + B(k_y^2 + k_z^2) - \epsilon & Ck_xk_y & Ck_xk_z \\ Ck_xk_y & Ak_y^2 + B(k_z^2 + k_x^2) - \epsilon & Ck_yk_z \\ Ck_xk_z & Ck_yk_z & Ak_z^2 + B(k_x^2 + k_y^2) - \epsilon \end{vmatrix} = 0 \tag{8a.5}$$

where A, B, and C are to be calculated by the usual methods of perturbation theory and ϵ' has been replaced by ϵ. We will not attempt to do such a calculation but just solve this equation for the $k_x k_y$ plane ($k_z = 0$) :

$$\{(Ak_x^2 + Bk_y^2 - \epsilon)(Ak_y^2 + Bk_x^2 - \epsilon) - C^2k_x^2k_y^2\}(Bk_x^2 + Bk_y^2 - \epsilon) = 0 \tag{8a.6}$$

which results in

$$\epsilon_\pm = \tfrac{1}{2}(A+B)(k_x^2 + k_y^2) \pm \sqrt{\tfrac{1}{4}(A-B)^2(k_x^2 - k_y^2)^2 + C^2k_x^2k_y^2} \tag{8a.7}$$

and

$$\epsilon = B(k_x^2 + k_y^2) \tag{8a.8}$$

The curves of constant energy consist of warped circles (Eq.(8a.7)) and normal circles (Eq.(8a.8) a set of which is shown in Fig.8.1.

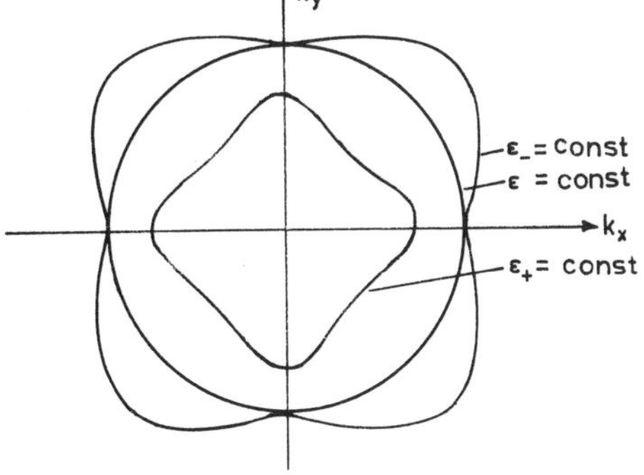

Fig.8.1 <100>cross section of the warped-sphere constant-energy surfaces.

Including now the spin-orbit interaction at $\vec{k} = 0$, each of the three states previously considered becomes double, and to the 6 x 6 matrix (in obvious short-hand notation

$$\begin{vmatrix} H_{kp} & 0 \\ 0 & H_{kp} \end{vmatrix}$$

we have to add a matrix which contains the Pauli spin matrices. In an approximation the total 6 x 6 matrix can be split into a

2 x 2 matrix and a 4 x 4 matrix. The solution of the 2 x 2 matrix is similar to Eq. (8a.8) except for a shift in energy which we denote by the symbol Δ:

$$\epsilon = A_o k^2 - \Delta \tag{8a.9}$$

where $k^2 = k_x^2 + k_y^2 + k_z^2$ and A_o is another coefficient. The 4 x 4 matrix yields an energy similar to Eq.(8a.7):

$$\epsilon_\pm = A_o k^2 \pm \sqrt{(B_o k^2)^2 + C_o^2 (k_x^2 k_y^2 + k_y^2 k_z^2 + k_z^2 k_x^2)} \tag{8a.10}$$

where B_o and C_o are two more coefficients [1]. For the split-off valence band given by Eq.(8a.9) the total angular momentum is $j = 1/2$ while for the heavy and light-hole bands given by the last equation it is $j = 3/2$.

Very often the anisotropy is small, i.e. $|C_o| \ll |B_o|$, and the square root can be expanded. It is then convenient to introduce the coefficients

$$B_o' = \sqrt{B_o^2 + \frac{1}{6} C_o^2} \tag{8a.11}$$

and

$$\Gamma_\pm = \mp C_o^2 / \{2B_o' (A_o \pm B_o')\} \tag{8a.12}$$

where the upper sign is valid for the light-hole band and the lower sign for the heavy-hole band. With these coefficients we obtain from Eq.(8a.10)

$$\epsilon_\pm = (A_o \pm B_o') k^2 \pm B_o' k^2 \left\{ \sqrt{1 \mp \frac{2(A_o \pm B_o')\Gamma}{B_o'} \left(\frac{k_x^2 k_y^2 + \cdots}{k^4} - \frac{1}{6} \right)} - 1 \right\} \tag{8a.13}$$

is expanded for small values of the second term of the radicand

$$\epsilon_\pm = (A_o \pm B_o') k^2 \{1 - \Gamma (\frac{k_x^2 k_y^2 + k_y^2 k_z^2 + k_z^2 k_x^2}{k^4} - \frac{1}{6})\} \tag{8a.14}$$

where higher powers of Γ have been neglected; since the second term is $\Gamma/6$ for $k_x = k_y = k_z = k/\sqrt{3}$, the series converges even for light holes in Si at 4 K where the factor $2(A_o + B_o')\Gamma/B_o' \approx 4$; consequently the approximation is acceptable. In the factor

$$\frac{k_x^2 k_y^2 + k_y^2 k_z^2 + k_z^2 k_x^2}{k^4} - \frac{1}{6} = \frac{1}{2} (\frac{2}{3} - \frac{k_x^4 + k_y^4 + k_z^4}{k^4}) \tag{8a.15}$$

which we will denote by $-q$, let us introduce polar coordinates

$$-q = \frac{1}{2} \{ \frac{2}{3} - \sin^4\theta\, (\cos^4\varphi + \sin^4\varphi) + \cos^4\theta \} \tag{8a.16}$$

With the notation

$$g_\pm = \{\hbar^2/2m_o)\, (A_o \pm B_o')/k_B T\}(1 + \Gamma q) \tag{8a.17}$$

we can write Eq.(8a.14) in the form

[1] G. Dresselhaus, A. F. Kip, and C. Kittel, Phys. Rev. 98 (1955) 368.

$$\epsilon_\pm / k_B T = -g_\pm k^2 \tag{8a.18}$$

Due to the negative sign the energy which is negative for holes yields a positive value for $g_\pm$. Eqs.(2b.32) for the heavy and light-hole bands have been obtained from this relation.

The hole concentration p is given by Eq.(3b.12). Let us now calculate the constant N_v by Eq.(4b.29) modified for the negative hole energy :

$$N_v = \frac{1}{4\pi^3} \int \exp(\epsilon/k_B T) d^3k = \frac{1}{4\pi^3} \int \exp(-gk^2) d^3k \tag{8a.19}$$

where d^3k, in polar coordinates, is given by

$$d^3k = k^2 dk \sin\theta d\theta d\varphi = \frac{1}{2} g^{-3/2} \sqrt{gk^2}\; d(gk^2) \sin\theta d\theta d\varphi \tag{8a.20}$$

For simplicity we omit the subscripts ±. The integration over gk^2 from 0 to ∞ yields a factor of $\sqrt{\pi}/2$. For the integration over θ and φ we expand $g^{-3/2}$ where g is given by Eq.(8a.17) :

$$g^{-3/2} = \left(\frac{\hbar^2}{2m_0} \frac{A_0 \pm B'_0}{k_B T}\right)^{3/2} \cdot \left(1 - \frac{3}{2}\Gamma q + \frac{15}{8}\Gamma^2 q^2 - \frac{35}{16}\Gamma^3 q^3 + - \cdots\right) \tag{8a.21}$$

The integration over φ from 0 to 2π and over θ from 0 to π yields for N_v:

$$N_v = 2\left(\frac{m_0 k_B T}{2\pi\hbar^2 (A_0 \pm B'_0)}\right)^{3/2} (1 + 0.05\Gamma + 0.0164\Gamma^2 + 0.000908\Gamma^3 + \cdots) \tag{8a.22}$$

which according to Eq.(3a.31) is given by $2(m_p k_B T/2\pi\hbar^2)^{3/2}$. Hence the "density-of-states effective mass", which we denote by m_d instead of m_p, is given by

$$m_d = \frac{m_0}{A_0 \pm B'_0} (1 + 0.03333\Gamma + 0.01057\Gamma^2 - 1.8\text{x}10^{-4}\Gamma^3 - 3\text{x}10^{-5}\Gamma^4 + \cdots) \tag{8a.23}$$

For a comparison we give the isotropic "conductivity-effective mass" m_σ which will be calculated in the following chapter,

$$m_\sigma = \frac{m_0}{A_0 \pm B'_0} (1 + 0.3333\Gamma - 0.02566\Gamma^2 - 9.5\text{x}10^{-4}\Gamma^3 \pm \cdots) \tag{8a.24}$$

In both equations the minus sign in the denominator is valid for heavy holes while the plus sign is valid for light holes. Assuming equal values of the momentum relaxation time for both types of holes the total conductivity is given by

$$\sigma = \{(p_h/m_h) + (p_l/m_l)\} e^2 \langle\tau_m\rangle \tag{8a.25}$$

where the subscripts h and l refer to heavy and light holes, respectively.

The experimental determination of the constants A_0, $|B_0|$, and $|C_0|$ from cyclotron resonance will be discussed in Chap.11k. The following Table 4 gives the values of these constants and of the conductivity-effective masses for Si and Ge at 4 K where m_s is the effective mass in the split-off valence band :

Table 4: Coefficients A_o, B_o, C_o, and conductivity effective masses (Si: ref. 1; Ge: ref. 2).

	A_o	$\lvert B_o \rvert$	$\lvert C_o \rvert$	m_h/m_o	m_l/m_o	m_s/m_o
Si	4.1	1.6	3.3	0.49	0.16	0.245
Ge	−13.27	8.62	12.4	0.29	0.0437	0.075

According to Eq.(8a.22) the ratio of constants N_v and therefore of the numbers of light and heavy holes is given by

$$p_l/p_h = (A_o - B_o')^{3/2} / \{A_o + B_o')^{3/2} (1 + 0.05\,\Gamma + \cdots)\} \tag{8a.26}$$

where the light hole band has been assumed isotropic. This ratio is 4 % for Ge and 16 % for Si.

Let us briefly discuss the influence of stress on the valence bands of germanium and silicon [3]. With the 6 deformation components $e_1 \cdots e_6$ given by Eqs. (7a.2) and (7a.3) and the deformation potential constants a, b, and d, the change in energy is given by

$$\Delta\epsilon_v = a(e_1 + e_2 + e_3) \pm \sqrt{\frac{1}{2} b^2\{(e_1 - e_2)^2 + (e_1 - e_3)^2 + (e_2 - e_3)^2\} + \frac{1}{4} d^2 (e_4^2 + e_5^2 + e_6^2)} \tag{8a.27}$$

where the split-off valence band has been neglected in the calculation. E.g. for Ge, a = 3.1 eV, b = 2.2 eV, and d = 4.5 eV while for Si $a = \Xi_d + \frac{1}{3}\Xi_u - 1.65$ eV, b = 2.2 eV, and d = 5.3 eV [4]. By a dilatation in the <111> direction the valence band in Ge is lowered by 4×10^{-6} eV/kg cm^{-2} given by the first term on the right-hand side of Eq.(8a.27). At low temperatures (≈ 4 K for $e_1 > 1/450$ in Ge) where the second term is much larger than the thermal energy $k_B T$ only the lower (in terms of hole energy) of the two subbands is populated. The constant-energy surface of this subband is an ellipsoid located at $\vec{k} = 0$. The semiconductor thus represents the ellipsoidal type and has been labelled "split p-germanium". The transport properties of this semiconductor have been reviewed by Koenig [5].

[2] B. W. Levinger and D. R. Frankl, J. Phys. Chem. Solids <u>20</u> (1961) 281.
[3] G. E. Pikus and G. L. Bir, Fiz.Tverd. Tela <u>1</u> (1959) 1642 [Engl.: Sov. Phys. Sol. State <u>1</u> (1959) 1502.].
[4] K. Bulthuis, Philips Res. Repts. <u>23</u> (1968) 25.
[5] S. H. Koenig: Proc. Int. School of Physics XXII (R. A. Smith, ed.), p. 515. New York: Acad. Press. 1963.

8b. The Electrical Conductivity

Since the number of holes in the split-off valence band is negligible, the conductivity is given by Eq.(8a.25) where the effective masses are given by Eq. (8a.24). For a proof of Eq.(8a.24) we consider the conductivity σ given by Eq. (4b.30) where τ_m is given by Eq.(4b.8) and the energy $|\epsilon|$ by Eq.(8a.18):

$$\tau_m = \tau_0 (gk^2)^r \tag{8b.1}$$

$(\partial\epsilon/\partial k_z)^2$ is calculated from Eq.(8a.13) and expanded for small values of Γ. The integration is similar to the one that led to Eq.(8a.22) and yields [1]:

$$\sigma = \frac{4}{3\sqrt{\pi}} \frac{e^2\tau_0}{m_0} (r + \frac{3}{2})! \; 2(\frac{m_0 k_B T}{2\pi\hbar^2})^{3/2} (A_0 \pm B_0')^{-1/2} \cdot (1 + 0.01667\,\Gamma + \cdots) \exp\{(\epsilon_v - \zeta)/k_B T\} \tag{8b.2}$$

For $\Gamma = 0$, $A_0 \pm B_0' = 1$, and an effective mass m_0, a conductivity $\sigma(m_0)$ would be obtained. Hence:

$$\sigma = \sigma(m_0)(A_0 \pm B_0')^{-1/2}(1 + 0.01667\,\Gamma + 0.041369\,\Gamma^2 + 9.0679\text{x}10^{-4}\,\Gamma^3 + 9.1959\text{x}10^{-4}\,\Gamma^4 + \cdots) \tag{8b.3}$$

The conductivity effective mass given by Eq.(8a.24) has been obtained from this relation. In this form the result is similar to that for a carrier of isotropic effective mass.

8c. Hall Effect and Magnetoresistance

To obtain a solution of the Boltzmann equation for carriers in electric and magnetic fields, discussed in Chap.4c, a quadratic $\epsilon(\vec{k})$-relation has been assumed. Since in the present case, Eq.(8a.10), we have a non-quadratic relation, the Boltzmann equation can be solved only for certain approximations. For weak magnetic fields Lax and Mavroides [1] calculated a series expansion in powers of $\vec{B}$. Let us briefly consider another method by McClure [2] valid for arbitrary magnetic field strengths. The carrier velocity given by Eq.(4b.28) is expanded along an orbit in $\vec{k}$-space ("hodograph") on a constant-energy surface intersected by a

[1] B. Lax and J. G. Mavroides, Phys. Rev. 100 (1955) 1650.

Chap.8c.

[1] B. Lax and J. G. Mavroides, Phys. Rev. 100 (1955) 1650.
[2] J. W. McClure, Phys. Rev. 101 (1956) 1642.

plane normal to $\vec{B}$, in powers of $\omega_c t$ where $\omega_c = \kappa e B(A_0 - B'_0)/m_0$ and t is the time period on the hodograph if no electric field is present. For a large deviation of the constant-energy surface from a sphere many more terms of the Fourier expansion have to be retained (up to $5\,\omega_c t$ for heavy holes).

Let us now consider the case of a weak magnetic field. The Boltzmann equation given by Eq.(4c.5) with $\vec{\nabla}_r f = 0$ and $\vec{v}$ given by Eq.(4b.28)

$$e\hbar^{-1}\{\vec{E} + \hbar^{-1}[\vec{\nabla}_k \epsilon \vec{B}]\,\vec{\nabla}_k f\} + (f - f_o)/\tau_m = 0 \qquad (8c.1)$$

is solved by assuming

$$f = f_o - a\partial f_o/\partial\epsilon \qquad (8c.2)$$

which is similar to Eq.(4c.3). The momentum relaxation time τ_m can be taken as a scalar quantity since the band extremum is at $\vec{k} = 0$ and τ_m has the symmetry of the cubic lattice. We thus obtain for Eq.(8c.1):

$$\frac{e}{\hbar}\{\vec{E} + \frac{1}{\hbar}[\vec{\nabla}_k \epsilon \vec{B}]\}\left(\frac{\partial f_o}{\partial\epsilon}\vec{\nabla}_k \epsilon - \frac{\partial f_o}{\partial\epsilon}\vec{\nabla}_k a - a\frac{\partial^2 f_o}{\partial\epsilon^2}\vec{\nabla}_k \epsilon\right) - \frac{a}{\tau_m}\frac{\partial f_o}{\partial\epsilon} = 0 \qquad (8c.3)$$

For the case of a weak electric field products of $\vec{E}$ with a and $\vec{\nabla}_k a$ can be neglected. The product $([\vec{\nabla}_k \epsilon \vec{B}]\,\vec{\nabla}_k \epsilon) = 0$ and therefore Eq.(8c.3) is simplified to:

$$a = \tau_m e\hbar^{-1}\{(\vec{E}\vec{\nabla}_k \epsilon) + \hbar^{-1}(\vec{B}[\vec{\nabla}_k \epsilon\, \vec{\nabla}_k a]) = \tau_m e\hbar^{-1}(\vec{E}\vec{\nabla}_k \epsilon) + \tau_m e(\vec{B}\vec{\Omega})a \qquad (8c.4)$$

where an operator

$$\vec{\Omega} = \hbar^2\,[\vec{\nabla}_k \epsilon\, \vec{\nabla}_k] \qquad (8c.5)$$

has been introduced. For small values of a we have to a first approximation

$$a_1 = \tau_m e\hbar^{-1}(\vec{E}\vec{\nabla}_k \epsilon) \qquad (8c.6)$$

and to a second approximation

$$a_2 = \tau_m e\hbar^{-1}(\vec{E}\vec{\nabla}_k \epsilon) + \tau_m e^2\hbar^{-2}(\vec{B}\vec{\Omega})\tau_m(\vec{E}\vec{\nabla}_k \epsilon) \qquad (8c.7)$$

etc. The current density $\vec{j}$ is obtained as usual (for the drift velocity see e.g. Eq. 4b.19)) to any approximation by substituting for a in the relation

$$\vec{j} = -\frac{e}{4\pi^3\hbar}\int \vec{\nabla}_k \epsilon\, a\frac{\partial f_o}{\partial\epsilon}\, d^3k \qquad (8c.8)$$

the quantities $a_1, a_2, \cdots$ given by Eq.(8c.6), (8c.7), etc. On the other hand $\vec{j}$ is given by its components

$$j_i = \sum_j \sigma_{ij} E_j + \sum_{jk} \sigma_{ijk} E_j B_k + \sum_{jkl} \sigma_{ijkl} E_j B_k B_l + \cdots \qquad (8c.9)$$

A comparison of coefficients yields

$$\sigma_{ij} = -\frac{e^2}{4\pi^3\hbar^2}\int \tau_m \frac{\partial\epsilon}{\partial k_i}\frac{\partial\epsilon}{\partial k_j}\frac{\partial f_o}{\partial\epsilon}\, d^3k \qquad (8c.10)$$

$$\sigma_{ijk} = \frac{e^3}{4\pi^3\hbar^4}\int \tau_m \frac{\partial\epsilon}{\partial k_i}\frac{\partial f_o}{\partial\epsilon}\sum_{r,s}\frac{\partial\epsilon}{\partial k_r}\frac{\partial}{\partial k_s}\left(\tau_m\frac{\partial\epsilon}{\partial k_j}\right)\delta_{krs}\,d^3k \tag{8c.11}$$

where δ_{krs} is the permutation tensor (= + 1 for a sequence 123 of krs, = −1 for a sequence 132, and = 0 for two equal subscripts).

$$\sigma_{ijkl} = -\frac{e^4}{4\pi^3\hbar^6}\int \tau_m \frac{\partial\epsilon}{\partial k_i}\frac{\partial f_o}{\partial\epsilon}\sum_{r,s,t,u}\frac{\partial\epsilon}{\partial k_r}\frac{\partial}{\partial k_s}\left[\tau_m\frac{\partial\epsilon}{\partial k_t}\frac{\partial}{\partial k_u}\left(\tau_m\frac{\partial\epsilon}{\partial k_j}\right)\right]\delta_{lrs}\,\delta_{ktu}\,d^3k \tag{8c.12}$$

If τ_m is a function of ϵ, independent of the direction of $\vec{k}$, we can put it before the operator $\vec{\Omega}$ in Eq.(8c.7) since

$$\vec{\Omega}\tau_m(\vec{E}\vec{\nabla}_k\epsilon) = (\vec{E}\vec{\nabla}_k\epsilon)\,\vec{\Omega}\tau_m + \tau_m\vec{\Omega}(\vec{E}\vec{\nabla}_k\epsilon) \tag{8c.13}$$

where

$$\vec{\Omega}\tau_m = \hbar^2[\vec{\nabla}_k\epsilon\,\vec{\nabla}_k\tau_m] = \hbar^2\frac{d\tau_m}{d\epsilon}[\vec{\nabla}_k\epsilon\,\vec{\nabla}_k\epsilon] = 0 \tag{8c.14}$$

Therefore in the integrands of Eqs.(8c.11) and (8c.12) τ_m can be taken outside the parenthesis and combined with the first factor τ_m to yield τ_m^2, τ_m^3, etc. The result of the integrations in the notation of Eq.(8b.3) is given by [3]

$$\sigma_{ijk} = \sigma_{ijk}(m_o)\,(A_o \pm B_o')^{1/2}\,(1-0.01667\,\Gamma + 0.017956\,\Gamma^2 - 0.0069857\,\Gamma^3 + - \cdots) \tag{8c.15}$$

$$\sigma_{xxyy} = \sigma_{xxyy}(m_o)\,(A_o \pm B_o')^{3/2}(1-0.2214\,\Gamma + 0.3838\,\Gamma^2 - 0.0167\,\Gamma^3 + 0.00755\,\Gamma^4 + - \cdots) \tag{8c.16}$$

$$\sigma_{xxxx} = -2\sigma_{xyyx} = \frac{16}{1155}\,\sigma_{xxyy}(m_o)\,(A_o \pm B_o')^{3/2}\,\Gamma^2 \cdot (1-0.4295\,\Gamma + 0.0188\,\Gamma^2 + 0.0103\,\Gamma^3 + - \cdots) \tag{8c.17}$$

$$\sigma_{xyxy} = -\sigma_{xxyy}(m_o)\,(A_o \pm B_o')^{3/2}\,(1-0.05\,\Gamma - 0.0469\,\Gamma^2 + 0.004\,\Gamma^3 - 0.00063\,\Gamma^4 + - \cdots) \tag{8c.18}$$

where $\sigma_{xxyy}(m_o) < 0$ and the existence of a longitudinal magnetoresistance is obvious from $\sigma_{xxxx} \neq 0$; this is a consequence of the warped energy surface. The magnetoresistance in the notation given in the text above Eq.(7d.19):

---

[3] J. G. Mavroides and B. Lax, Phys. Rev. 107 (1957) 1530; ibid. 108 (1957) 1648; see, however, footnote p. 875 of P. Lawaetz, Phys. Rev. 174 (1968) 867 and P. Lawaetz, thesis, Tech. Univ. Kopenhagen 1967, p. 138, where the statement is made that the value for σ_{xxyy} given by Mavroides and Lax is incorrect.

$$\left.\begin{aligned} M_{100}{}^{100} &= -\sigma_{xxxx}/\sigma_o \\ M_{110}{}^{110} &= -(\sigma_{xxxx}+\sigma_{xxyy}+\sigma_{xyyx}+\sigma_{xyxy})/2\sigma_o \\ M_{100}{}^{010} &= -(\sigma_{xxyy}/\sigma_o)-(\sigma_{xyz}/\sigma_o)^2 \\ M_{110}{}^{\bar{1}10} &= -(\sigma_{xxxx}+\sigma_{xxyy}-\sigma_{xyyx}-\sigma_{xyxy})/2\sigma_o \end{aligned}\right\} \quad (8c.19)$$

It may be sufficient to give these examples valid for certain principal crystallographic directions; for more details see refs. 3 and 4.

For a comparison of these calculations with experimental results obtained on p-type Ge [5] and p-type Si [6] the contributions by light and heavy holes have to be added in the calculation of the conductivity by Eq.(8a.27) and of the magnetoresistance by Eq.(4d.15). In the expression for the conductivity the light-hole contribution may be neglected due to the small number of these carriers while in the magnetoresistance the light holes have a very remarkable effect. The assumption of equal relaxation times for light and heavy holes can easily be justified: a given light hole is converted into a heavy hole by lattice scattering for two reasons, namely (1) energy is carried away by a phonon which makes this process more probable than the inverse process ($N_q + 1 > N_q$, see Eq.6d.12)), and (2) the heavy-hole density of states and therefore the transition probability are larger than for the inverse process. For the case of ionized impurity scattering, however, the relaxation times for light and heavy holes differ. In a range of temperatures, where light holes are scattered by ionized impurities and heavy holes by the lattice, a satisfactory explanation of experimental results is met with difficulty.

The observed temperature dependence of the Hall mobility in pure p-type Ge (e.g. $N_A = 1.44\times10^{13}/cm^3$; $N_D = 0.19\times10^{13}/cm^3$; [7]) between 120 and 300 K,

$$\mu_H = 2060\ cm^2/Vsec \times (T/300\,K)^{-2.3} \quad (8c.20)$$

has been explained by optical and acoustic phonon scattering with a value for the ratio of deformation potential constants, $Du_l/\epsilon_{ac}\omega_o$, of $\sqrt{3.8}$ by taking into account effective mass ratios of $m_h/m_o = 0.35$ and $m_l/m_o = 0.043$ and a concentration ratio of 23.2. The effective masses have been taken as temperature-independent although cyclotron resonance reveals an increase of the light-hole mass by 16 % between 10 and 100 K [8]: it does not enter appreciably into the

[4] A. C. Beer: Solid State Physics (F. Seitz and D. Turnbull, eds.), Suppl. 4, chap. 20b. New York: Acad. Press. 1963.
[5] G. L. Pearson and H. Suhl, Phys. Rev. 83 (1951) 768.
[6] G. L. Pearson and C. Herring, Physica 20 (1954) 975.
[7] D. M. Brown and R. Bray, Phys. Rev. 127 (1962) 1593.
[8] D. M. S. Bagguley and R. A. Stradling, Proc. Phys. Soc. (London) 78 (1961) 1078.

conductivity due to the small relative number of light holes. Between 50 and 80 K the mobility is reduced only by 10 - 20 % in comparison with the "acoustic" mobility and is proportional to $T^{-3/2}$. At lower temperatures the influence of ionized impurity scattering increases; very good agreement with the Brooks-Herring formula Eq.(6c.23) is obtained down to 20 K.

The influence of band structure on the galvanomagnetic properties, especially on magnetoresistance and, outside the range $\mu_H B \ll 1$, on the B-dependence of the Hall coefficient, R_H, is noteworthy. $R_H(B)$ has a minimum at a value of B which increases with increasing temperature. Experimental data of $R_H(B)$ are shown [9] in Fig.8.2. The curves are labelled with parameters q_h^2 and q_l^2 which are given by Eq.(6f.3), where the subscripts refer to heavy and light holes, respectively. There is qualitative agreement between the calculated and the observed data.

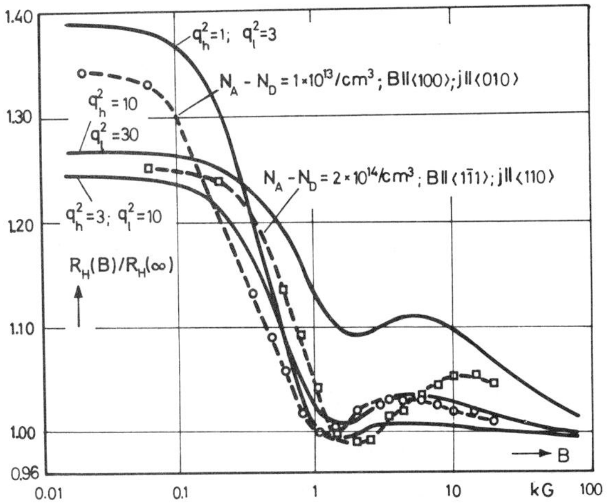

Fig.8.2 Hall coefficient of p-type germanium vs magnetic induction at 77 K observed for two values of the carrier density (dashed curves) and calculated (full curves) (after Beer and Willardson, ref. 9).

In the following table 5 the observed [5] magnetoresistance of p-type Ge at 77 K is compared with theoretical data [3] for various directions of $\vec{j}$ and $\vec{B}$ assuming an isotropic momentum relaxation time of 1.03×10^{-12} sec.

The longitudinal magnetoresistance in the <111> direction is smaller by an order of magnitude than in other directions. Therefore a small misalignment of the sample may cause a considerable error in the measurement, and the agreement between experimental and theoretical data is not as good as in directions where the magnetoresistance is large. In strong magnetic fields (about 100 kG) the longitudinal magnetoresistance saturates in a <111> direction but not in a <100> direction.

[9] A. C. Beer and R. K. Willardson, Phys. Rev. 110 (1958) 1286.

Table 5: Experimental and theoretical magnetoresistance of p-Ge at 77 K.

$(\Delta\rho/\rho B^2)_{\vec{j}}^{\vec{B}}$	experim.	$\times 10^{-3}/kG^2$	theor.
M_{100}^{010}	30.4		30.4
$M_{110}^{1\bar{1}0}$	27.0		27.4
M_{100}^{100}	0.14		0.093
M_{110}^{110}	2.0		3.25

Galvanomagnetic phenomena in p-type Si are different in some respects from those in p-type Ge. The energy Δ of the split-off valence band is only 0.04 eV in p-type Si; even so at room temperature where $k_BT = 0.026$ eV it is not appreciably populated because of its small density of states which is concurrent with its small effective mass. In band calculations even the small value of Δ given above leads to a non-parabolicity of the other two bands. Especially for the heavy hole band in the <110> direction in $\vec{k}$-space, the deviation from the linearity in Fig.8.3 is remarkable at energies of a few meV; cross sections of the heavy-hole band in a $[1\bar{1}0]$ plane for various energies relative to Δ (0.25, 0.95, and 1.5) have been shown in Fig.2.29b, p. 33. A calculation of the energy distribution of carriers in a non-parabolic spherical band has been given by Matz [10]. Asche and Borzeszkowski [11] calculated the temperature dependent mobility of heavy holes in this approximation as shown in Fig.8.4 for values of 0 and $\sqrt{2.5}$ for $Du_1/\epsilon_{ac}\omega_0$. In a <110> direction there is a swelling of the heavy-hole constant-energy surface with increasing energy which has been shown in Fig.2.29a, p. 33. For temperatures T > 100 K the consideration of optical deformation potential scattering in addition to acous-

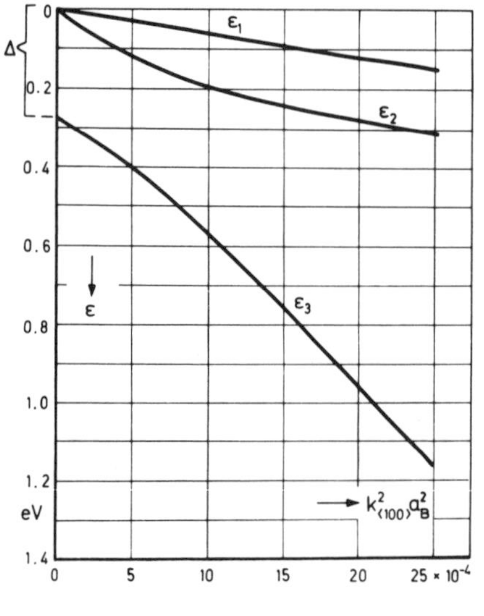

Fig.8.3 Valence band structure $\epsilon(k^2)$ for p-type germanium in a<100>direction; Δ = 0.29 eV; similar for p-type silicon where, however, Δ = 0.044 eV (after E. O. Kane, J. Phys. Chem. Solids 1 (1956) 82).

[10] D. Matz, J. Phys. Chem. Solids 28 (1967) 373.
[11] M. Asche and J. v. Borzeszkowski, phys. stat. sol. 37 (1970) 433.

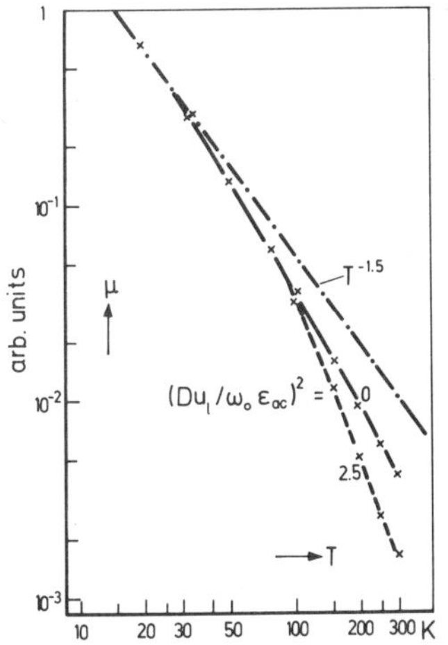

Fig.8.4 Calculated temperature dependence of the hole mobility in silicon taking band non-parabolicity into account, for two values of the coupling parameter (after Asche and Borzeszkowski, ref.11).

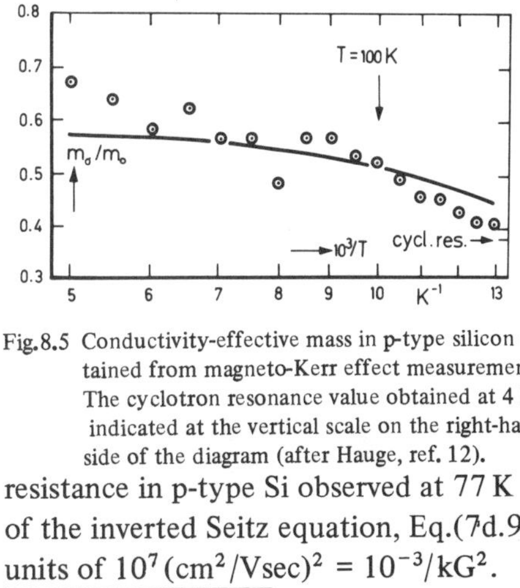

Fig.8.5 Conductivity-effective mass in p-type silicon obtained from magneto-Kerr effect measurements. The cyclotron resonance value obtained at 4 K is indicated at the vertical scale on the right-hand side of the diagram (after Hauge, ref. 12).

tic deformation potential scattering with a value of $\sqrt{2.5}$ for the ratio of the deformation potential constants leads to the observed $T^{-2.9}$ law (Fig.8.4). Between 50 and 100 K a $T^{-1.9}$ dependence is calculated for acoustic deformation potential scattering. The deviation of the exponent, −1.9, from the usual value of −1.5 is due to band non-parabolicity.

A consequence of the non-parabolicity of the heavy-hole band is an increase of the conductivity-effective mass with temperature because with increasing thermal energy k_BT the carriers are raised to less parabolic regions of the band. Fig.8.5 shows the conductivity-effective mass in p-type Si vs $10^3/T$ as determined from magneto-Kerr observations*) [12]. This dependence clearly demonstrates the band non-parabolicity [13]. Negative differential conductivity in non-parabolic bands has been discussed by Persky and Bartelink [14] as well as Fawcett and Ruch [15].

The B-dependent Hall coefficient R_H up to 90 kG has been found [16] to be anisotropic in p-type Si. This and the anisotropy of magnetoresistivity which increases with temperature has led to the conclusion that the band is non-parabolic as considered above. The weak-field magnetoresistance in p-type Si observed at 77 K by Long [17] yields for the coefficients of the inverted Seitz equation, Eq.(7d.9), b = 14.2, c = −10.6, and d = 0.8, all in units of $10^7\,(cm^2/Vsec)^2 = 10^{-3}/kG^2$.

[12] P. Hauge, thesis, The University of Minnesota, Minn., USA, 1967.

[13] For more recent calculations see e.g. M. Costato and L. Reggiani, Lett. Nuovo Cimento Ser. I, 3 (1970) 239.

[14] G. Persky and D. J. Bartelink, IBM-J. Res. Develop. 13 (1969) 607.

[15] W. Fawcett and J. G. Ruch, Appl. Phys. Lett. 15 (1969) 369.

[16] H. Miyazawa, K. Suzuki, and H. Maeda, Phys. Rev. 131 (1963) 2442.

[17] D. Long, Phys. Rev. 107 (1957) 672.

*) The magneto-Kerr effect is a change in ellipticity and polarization of a microwave upon reflection from a sample in a longitudinal magnetic field.

8d. Warm and Hot Holes

While only qualitative agreement between observed and calculated data has been found for galvanomagnetic phenomena in p-type Ge and Si, the situation is even more difficult in treating warm and hot holes. As already mentioned the time for a conversion of a light hole into a heavy hole is about one collision time, i.e. at a scattering process the conversion takes place with a large probability. Therefore the electron temperatures of both kinds of carriers may be taken equal.

Let us first consider very strong electric fields. The observed field dependence of the drift velocity of holes in p-type Ge is shown in Fig.8.6 for two values of

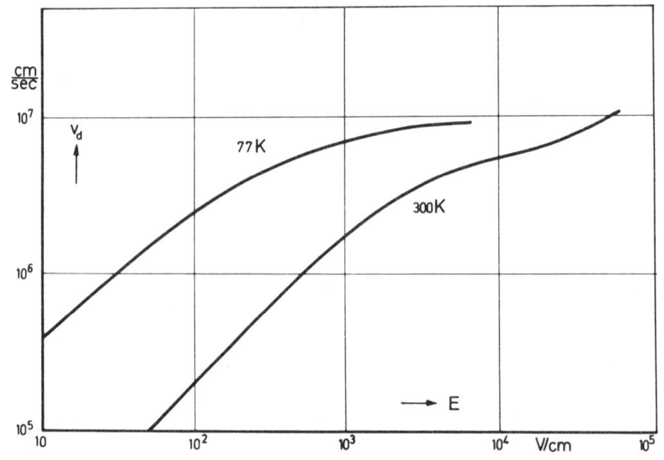

Fig.8.6 Drift velocity vs field characteristics in p-type germanium (after Brown and Bray; Prior; ref.1).

the lattice temperature [1]. At low temperatures there is a tendency toward saturation. The investigation of the Sasaki-Shibuya effect (Chap.7e, Fig.7.9) in p-type Ge as shown in Fig.8.7 yields the same sign as in n-type Ge: The deviation of the field direction from the current direction in both materials is towards the <111> direction [2]. A compensated sample behaves similarly as an uncompensated sample.

The energy distribution of holes in p-type Ge heated by a strong electric field has been investigated by the infrared absorption. This will be discussed in Chap. 11j. We will mention here only that the distribution is non-Maxwellian. An im-

[1] R. Bray and D. M. Brown: Proc. Int. Conf. Semic. Phys. Prague 1960, p. 82. Prague: Czech. Acad. Sciences. 1960; A. C. Prior, Proc. Phys. Soc. (London) 76 (1960) 465.

[2] W. E. K. Gibbs, J. Appl. Phys. 33 (1962) 3369.

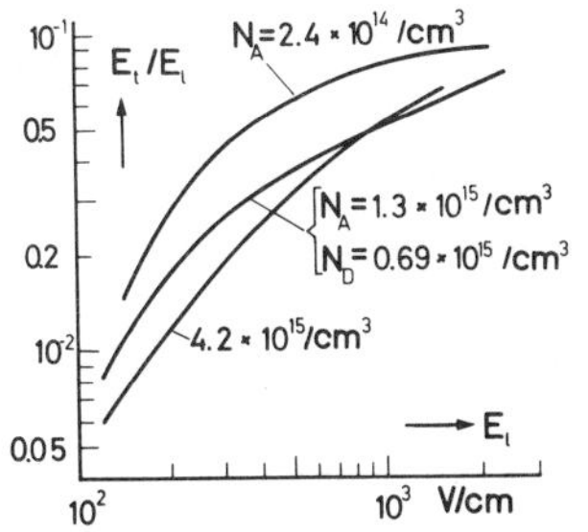

Fig.8.7 Sasaki-Shibuya effect in p-type germanium for various impurity concentrations at a lattice temperature of 77 K (after Gibbs, ref. 2).

poverishment of the hot-hole concentration at high energies relative to the Maxwellian distribution and the containment of the majority of the holes at energies below the optical phonon energy of 37 meV indicates strong optical phonon scattering: Carriers accelerated by the field to energies of more than this energy are scattered very efficiently to lower energies by optical phonon emission. This leads to a cyclical streaming motion in $\vec{k}$-space which is in contrast to the diffusion-type motion at a weak coupling between carriers and optical phonons [3]. A similar energy distribution has been calculated [4].

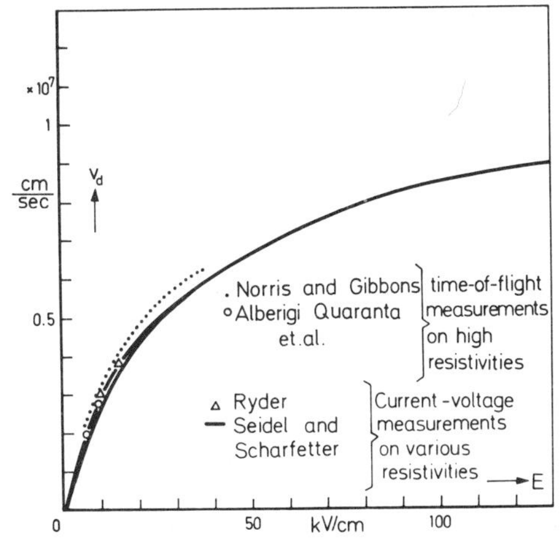

Fig.8.8 Drift velocity vs field characteristics for p-type silicon at room temperature (after Norris and Gibbons, Alberigi Quaranta et al., Ryder, and Seidel and Scharfetter, compiled by Seidel and Scharfetter, J. Phys. Chem. Solids 28 (1967) 2563).

The hole drift velocity observed in p-type Si in strong electric fields of strengths up to 100 kV/cm is plotted vs $\vec{E}$ in Fig.8.8. For $E < 10$ kV/cm at 77 K there is no anisotropy of v_d in pure samples where e.g. $\rho(300\,K) = 1500$ ohm-cm [5]. An explanation of the field dependence of the drift velocity at 77 K is given in terms of the band non-parabolicity: If the holes become hot, they become heavy at the same time and are less accelerated in the field. Hence the energy distribution is cut off at energies which are lower than the optical phonon energy, and in pure samples only acoustic phonon scattering is important. However, it is not clear

[3] W. E. Pinson and R. Bray, Phys. Rev. 136 (1964) A 1449;
A. C. Baynham and E. G. S. Paige, Phys. Lett. 6 (1963) 7.

[4] H. F. Budd, Phys. Rev. 158 (1967) 798.

[5] M. Asche and J. v. Borzeszkowski, phys. stat. sol. 37 (1970) 433.

if the light hole contribution may be neglected for this case. For temperatures $T > 150$ K the zero-field mobility is strongly influenced by optical phonon scattering as discussed above.

For the warm-hole range, where the current depends on $\vec{E}$ as given by Eqs. (7e.36) and (7e.38), the coefficients β_0 and γ have been measured both in p-type Ge and in p-type Si [6] but only qualitative agreement between theory and experiment has been obtained [7]. From the shape of the heavy-hole constant-energy surface it is obvious that the effective mass in a $\langle 100 \rangle$ direction is smaller than in a $\langle 111 \rangle$ direction, and the large current component due to the lighter $\langle 100 \rangle$-directed holes yields a rotation of the current vector away from the $\langle 111 \rangle$ direction ($\gamma > 0$).

---

[6] E. P. Röth, G. Tschulena, and K. Seeger, Zeitschr. f. Physik 212 (1968) 183.
[7] P. Lawaetz, thesis, Tech. Univ. Kopenhagen, 1967.

9. Quantum Effects in Transport Phenomena

In Chap. 2 we have learned how the quantization of the atomic energy levels results in the band structure of the crystalline solid. However, this is not the only domain of quantum mechanics in semiconductivity. Although most transport phenomena can be explained by the assumption of a classical electron gas, there are some which can be understood only by quantum mechanical arguments. In Chap. 9a we will treat phenomena which rely on the quantum mechanical "tunnel effect", while in Chaps. 9b-d the quantization of electron orbits in a strong magnetic field with the formation of "Landau levels" will be the basis for an understanding of the "oscillatory" behavior of transport phenomena.

9a. The Tunnel Diode

Although many scientists had observed an "anomalous" current-voltage characteristic in the forward direction of degenerate p-n junctions, it was not until 1958 that Esaki [1] gave an explanation in terms of a quantum tunneling concept. Since then the "Esaki diode" has become a useful device due to its ultrahigh-speed, low-power and low-noise operation [2].

The current-voltage characteristics is shown schematically in Fig. 9.1a. In the reverse direction the current increases as the voltage is increased. In the forward direction it shows a peak at a voltage V_P followed by an interesting region of "negative differential conductivity" between V_P and V_V. The various components of the current are plotted in Fig. 9.1b. Let us first consider the tunnel current component at temperatures near absolute zero.

Fig. 9.2a shows a simplified diagram of the equilibrium potential distribution in a p-n junction, which is degenerate on both sides. The Fermi level ζ is constant across the junction. On the n-side the conduction band is filled with electrons up to ζ, while on the p-side the valence band is filled with holes down to ζ (hatched

---

[1] L. Esaki, Phys. Rev. 109 (1958) 603.

[2] For a review see e.g. W. F. Chow: Principles of Tunnel Diode Circuits. New York: J. Wiley and Sons. 1964.

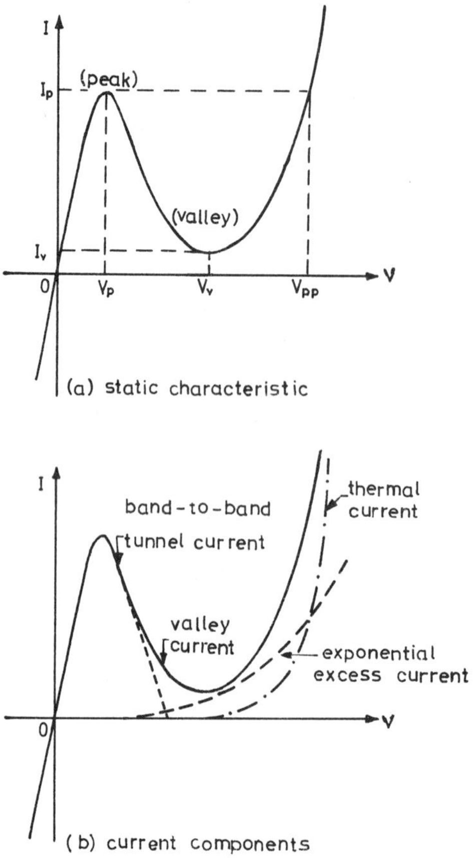

Fig.9.1a Static current-voltage characteristics of a typical tunnel diode (schematic diagram).
b The characteristics of Fig.9.1a decomposed into the current components (after S. M. Sze: Physics of Semiconductor Devices. New York: J. Wiley and Sons. 1969).

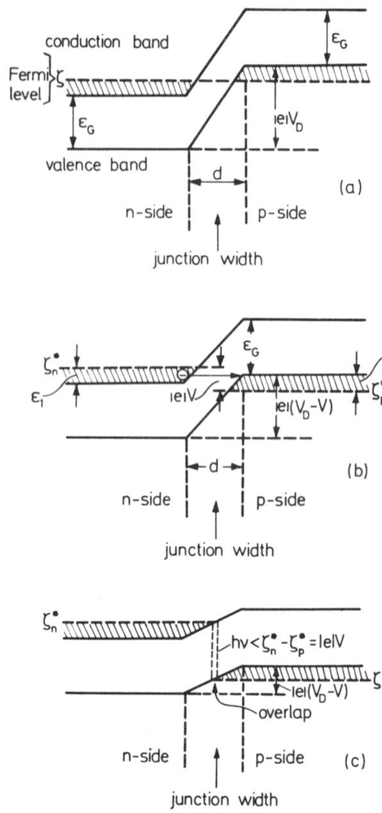

Fig.9.2a Degenerate p-n junction at thermal equilibrium.
b Degenerate p-n junction at a small forward bias; tunneling electron is indicated.
c Degenerate p-n junction at a large forward bias.

areas) In Fig.9.2b the potential in the junction upon application of a small forward bias V is given. Electrons from the conduction band of the n-side tunnel through the gap to the empty sites (or "holes") in the valence band of the p-side. Before we calculate the tunnel current let us consider the case of a large forward bias illustrated by Fig.9.2c. The overlap of the hatched areas in the vertical direction is of interest in the semiconductor laser to be discussed in Chap.13b; there is no overlap horizontally and consequently no tunneling of electrons. In this case we have only the usual thermal diffusion current indicated in Fig.9.1b.

The tunneling process is "direct" if in the energy-vs-momentum relationship of the carriers, $\epsilon(\vec{k})$, the conduction band edge is located at the same $\vec{k}$-value as the valence band edge, otherwise it is "indirect". In the indirect process the electron transfer from the conduction band edge to the valence band edge involves a change in electron momentum which may be supplied either by an impurity or a phonon ("impurity-assisted tunneling" and "phonon-assisted tunneling", respectively). We will consider here only direct tunneling.

The quantum mechanical calculation of the tunnel process through a potential barrier V_0 is similar to the calculation given in Eqs.(2b.1) - (2b.16). The incoming and reflected electron waves are described by

$$\psi = A_1 e^{ikx} + B_1 e^{-ikx} \quad \text{for } x < -w \tag{9a.1}$$

while the transmitted wave is given by

$$\psi = A_3 e^{ikx} \qquad \text{for } x > w \tag{9a.2}$$

and the wave inside the barrier is a solution of Eq.(2b.3):

$$\psi = A_2 e^{-\beta x} + B_2 e^{\beta x} \tag{9a.3}$$

The continuity condition for ψ and $d\psi/dx$ at $x = w$ and $x = -w$ yields for the transmission probability

$$T_t = |A_3/A_1|^2 = \{1 + \frac{1}{4}(\frac{\beta}{k} + \frac{k}{\beta})^2 \sinh^2 \beta w\}^{-1} \tag{9a.4}$$

which for $\beta w \gg 1$ can be approximated by

$$T_t = (\frac{4\beta k}{\beta^2 + k^2})^2 e^{-2\beta w} \approx e^{-2\beta w} \tag{9a.4}$$

The second approximation is valid for $k^2 \approx \beta^2 = 2m\hbar^{-2}(V_0 - \epsilon)$ where m is the reduced effective mass. So far we have assumed a rectangular potential barrier of width $d = 2w$ and height V_0. However, Fig.9.2b shows that the barrier is triangular and has a height of about ϵ_G, while the energy ϵ of the electron can be neglected. With β given by $\sqrt{2m\hbar^{-2}\epsilon_G}$ and w given by $d/2 \approx \epsilon_G/2|e|E$ where E is the electric field strength in the junction a calculation modified for the triangular barrier yields an additional factor of 2/3 in the argument of the exponential:

$$T_t = \exp(-\frac{4\sqrt{2m}\,\epsilon_G^{3/2}}{3|e|E\hbar}) \tag{9a.6}$$

For a parabolic barrier we obtain the same result except that the factor $4\sqrt{2}/3 = 1.88$ is replaced by $\pi/2^{3/2} = 1.11$. For simplicity we have neglected in this one-dimensional treatment the momentum perpendicular to the direction of tunneling. For phonon-assisted indirect tunneling ϵ_G in Eq.(9a.6) has to be replaced by $(\epsilon_G - \hbar\omega_0)$ where $\hbar\omega_0$ is the phonon energy [3]:

$$T_t = \exp(-\frac{4\sqrt{2m}\,(\epsilon_G - \hbar\omega_0)^{3/2}}{3|e|E\hbar}) \tag{9a.7}$$

The same expression for photon-assisted tunneling, where $\hbar\omega_0$ is replaced by the photon energy $\hbar\omega$, will be obtained in Chap.11e for light absorption in homogeneous semiconductors in an $\vec{E}$-field. For a calculation of the current, T_t is

[3] E. O. Kane, J. Appl. Phys. 32 (1961) 83; J. Phys. Chem. Solids 2 (1960) 181.

treated as a constant in the small voltage range involved.

At thermal equilibrium the tunneling current $I_{c\to v}$ from the conduction band to the valence band is given by

$$I_{c\to v} = AT_t \int_{\epsilon_c}^{\epsilon_v} f_c(\epsilon) g_c(\epsilon) \{1 - f_v(\epsilon)\} g_v(\epsilon) d\epsilon \tag{9a.8}$$

where $f_c(\epsilon)$ and $f_v(\epsilon)$ are the Fermi-Dirac distribution functions of electrons in the conduction band and valence band, respectively, $g_c(\epsilon)$ and $g_v(\epsilon)$ are the corresponding densities of states, and A is a constant. The reverse current is given by a corresponding expression :

$$I_{v\to c} = AT_t \int_{\epsilon_c}^{\epsilon_v} f_v(\epsilon) g_v(\epsilon) \{1 - f_c(\epsilon)\} g_c(\epsilon) d\epsilon \tag{9a.9}$$

and with no bias applied due to detailed balance $I_{c\to v} = I_{v\to c}$. When the junction is biased, the current can be approximated by [1]

$$I = I_{c\to v} - I_{v\to c} = AT_t \int_{\epsilon_c}^{\epsilon_v} (f_c - f_v)\, g_c g_v\, d\epsilon \tag{9a.10}$$

where in f_c and f_v the Fermi energies are replaced by the quasi-Fermi energies, ζ_n^* and ζ_p^*, and the applied voltage $V = (\zeta_n^* - \zeta_p^*)/e$. According to Eq.(3a.28) the densities of states vary as $\sqrt{\epsilon - \epsilon_c}$ and $\sqrt{\epsilon_v - \epsilon}$, respectively. For $2k_BT \geqslant \zeta_n^* - \epsilon_c$ and $\geqslant \epsilon_v - \zeta_p^*$ the Fermi-Dirac distribution can be linearized :

$$f_c(\epsilon) \approx \frac{1}{2} - (\epsilon - \zeta_n^*)/4k_BT; \quad f_v(\epsilon) \approx \frac{1}{2} + (\zeta_p^* - \epsilon)/4k_BT \tag{9a.11}$$

Hence we obtain from Eq.(9a.10)

$$I = A' \frac{|e|V}{4k_BT} (\epsilon_c - \epsilon_v)^2 = A' \frac{|e|V}{4k_BT} (\epsilon_1 + \epsilon_2 - |e|V)^2 \tag{9a.12}$$

for $\epsilon_1 + \epsilon_2 \geqslant |e|V$ where A' is another constant and according to Fig.9.2b $\epsilon_1 = \zeta_n^* - \epsilon_c$ and $\epsilon_2 = \epsilon_v - \zeta_p^*$ [4].

Fig.9.3 shows a comparison between experimental curves published by Esaki and points calculated according to Eq. (9a.12) where the constants A' and $\epsilon_1 + \epsilon_2$ have been fitted at the peaks of the curves [4]. Considering the approximation made in the calculation, the agreement is satisfactory.

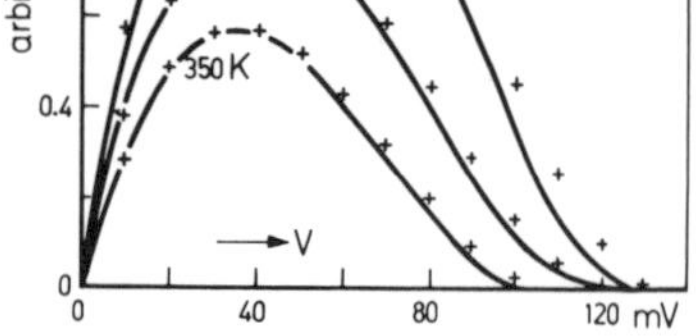

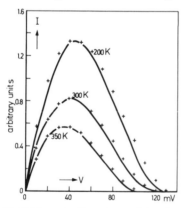

Fig.9.3 Current-voltage characteristics of a tunnel diode observed (crosses) and calculated (curves) for various temperatures (after Karlovsky, ref.4).

For voltages $V \geqslant (\epsilon_1 + \epsilon_2)/|e|$ the tunneling current should decrease to zero and only the normal diode current should flow; however, in practice there is a cur-

[4] J. Karlovsky, Phys. Rev. 127 (1962) 419.

rent in excess of the latter. This excess current is shown in Fig.9.1b. It is due to the fact that in a strong electric field the band edges are not as sharp as one might expect from $g(\epsilon) \propto \sqrt{\epsilon - \epsilon_c}$ but have "tails" [5] : $g(\epsilon) \propto \exp(-\text{const}.\epsilon^2)$ for $\epsilon < \epsilon_c$ (see Chap.11e). There is tunneling between the tails of the conduction and valence bands. Tunneling also occurs via impurity states in the energy gap.

If the p-n junction is not doped heavily enough so that both sides are just near degeneracy, there is no negative conductivity region but a region of high resistance in the forward direction. The current-voltage characteristics is then just opposite to that of a normal diode. Such a diode is called a "backward diode". It is useful for low-noise rectification of small microwave signals. For further details see e.g. Sze [6].

The scattering of tunneling electrons by impurities in the junction or in a "Schottky barrier" (metal degenerate-semiconductor junction with carrier depletion at the interface) where one or two phonons are emitted has been investigated by measuring the second derivative of the bias with respect to the current at very low temperatures. Fig.9.4 shows experimental results obtained by

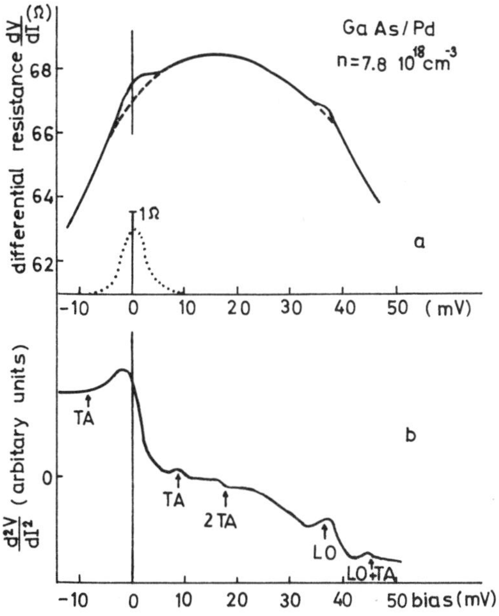

Fig.9.4 Derivatives of the current-voltage characteristics for a n-GaAs-palladium Schottky barrier at temperatures of 2 K (full) and 20 K (dashed); excess differential resistance shown by the dotted curve. Arrows indicate phonon energies (in units of the elementary charge) in GaAs (after Thomas and Queisser, ref.7)

[5] E. O. Kane, Phys. Rev. 131 (1963) 79.

[6] S. M. Sze: Physics of Semiconductor Devices. New York: J. Wiley and Sons. 1969.

Thomas and Queisser [7] on n-type GaAs/Pd Schottky barriers at 2 K. The type of phonons (transverse acoustic or longitudinal optical) which are emitted is indicated in Fig.9.4. Such measurements are denoted as "tunnel spectroscopy".

Tunneling through the gap has also been considered by Zener [8] in reverse biased non-degenerate p-n junctions: If the field $\vec{E}$ in the junction is very strong, the probability for tunneling given by Eq.(9a.6) is large enough that this process may lead to a breakdown in the reverse direction of the I-V characteristics. However, in almost all commercially available "Zener diodes" breakdown is due to impact ionization [9] rather than tunneling (Chap.10b).

9b. Magnetic Quantum Effects

In Chap.11k we will investigate the helical motion of a carrier in a magnetic field. The helix is centered around the direction of the field. The absorption of electromagnetic radiation in this case is called "cyclotron resonance". Here we consider the transport properties of a semiconductor in strong magnetic fields where the energy bands are converted into a series of energy levels known as "Landau levels" [1] due to the quantization of the orbits of the carriers.

The wave function $\psi(\vec{r})$ of an electron of energy ϵ in a magnetic induction $\vec{B} = [\vec{\nabla}_r \vec{A}]$ is obtained as the solution of the Schrödinger equation (Chap.2a)

$$\frac{1}{2m} (-i\hbar\vec{\nabla}_r + |e|\vec{A})^2 \psi = \epsilon\psi \tag{9b.1}$$

where the electron spin has been neglected. If we choose the z-direction of a Cartesian coordinate system as the direction of $\vec{B}$, a suitable gauge transformation yields for the vector potential $\vec{A}$:

$$\vec{A} = (0, Bx, 0) \tag{9b.2}$$

The wave function $\psi(\vec{r})$ is the product of $\exp(-ik_y y - ik_z z)$ and an unknown function $\varphi(x)$ for which we obtain the differential equation

$$d^2\varphi/dx'^2 + (2m/\hbar^2)(\epsilon' - \frac{1}{2} m\omega_c^2 x'^2)\varphi = 0 \tag{9b.3}$$

where $\epsilon' = \epsilon - \hbar^2 k_z^2/2m$, $x' = x - \hbar k_y/|e|B$, and $\omega_c = (|e|/m)B$.

This is the equation of motion for a harmonic oscillator having energy levels

---

[7] P. Thomas and H. J. Queisser, Phys. Rev. 175 (1968) 983; for reviews see e.g. P. Thomas: Fachberichte DPG, p. 114. Stuttgart: Teubner. 1968, and H. Zetsche, ibid. 1970, p. 172.

[8] C. Zener, Proc. Roy. Soc. (London) 145 (1934) 523; K. B. McAfee, E. J. Ryder, W. Shockley, and M. Sparks, Phys. Rev. 83 (1951) 650.

[9] S. L. Miller, Phys. Rev. 99 (1955) 1234.

Chapter 9b:

[1] L. D. Landau, Zeitschr. f. Physik 64 (1930) 629.

$\epsilon' = (n + \frac{1}{2})\hbar\omega_c$ where n is the quantum number. It yields for the electron energy:

$$\epsilon = (n + \frac{1}{2})\hbar\omega_c + \hbar^2 k_z^2/2m; \quad n = 0, 1, 2, \cdots \tag{9b.4}$$

The energy in a plane perpendicular to $\vec{B}$ is quantized, while the energy of motion in the direction of $\vec{B}$, $\epsilon_{kz} = \hbar^2 k_z^2/2m$, remains unaffected.

Let us now calculate the density of states. The number of allowed values of k_z, which is less than a given $|k_0| = 2\pi/\lambda_0$, is $L_z/(\lambda_0/2) = L_z k_0/\pi = L_z\sqrt{2m\epsilon_{k_0}}/\pi\hbar$ where L_z is the crystal dimension in the z-direction. By differentiation with respect to ϵ_{k_0} the number of allowed k_z values in the energy range $d\epsilon_{k_z}$ is found to be :

$$g(k_z)\,d\epsilon_{k_z} = \frac{L_z\sqrt{2m}}{2\pi\hbar\sqrt{\epsilon_{k_z}}}\,d\epsilon_{k_z} \tag{9b.5}$$

Assuming a crystal bounded by planes perpendicular to the x-, y-, and z-axes and having dimensions L_x, L_y, and L_z, k_y and k_z due to the periodicity of the Bloch functions have allowed values given by $k_y = n_y 2\pi/L_y$ and $k_z = n_z 2\pi/L_z$ where n_y and n_z are integers. There are $L_y/2\pi$ states per unit extent of k_y. We assume the electron wave function ψ to vanish at the boundaries $x = 0$ and $x = L_x$. x' has then values between 0 and L_x, and since ψ is centered around $x = \hbar k_y/|e|B$ (according to the definition of x'), k_y extends up to $L_x|e|B/\hbar$. The total number of values of k_y is thus given by

$$\mathrm{Max}(n_y) = (L_y/2\pi)\,\mathrm{Max}(k_y) = (L_y/2\pi)L_x|e|B/\hbar \tag{9b.6}$$

Multiplication of this expression with the right-hand side of Eq.(9b.5) and summation over all quantum numbers n yields for the density of states

$$g(\epsilon)d\epsilon = \sum_{n=0}^{n_{max}} \frac{V|e|B\sqrt{2m}}{(2\pi\hbar)^2\sqrt{\epsilon_{k_z}}}\,d\epsilon = \frac{V|e|B\sqrt{2m}}{(2\pi\hbar)^2}\sum_{n=0}^{n_{max}}\{\epsilon - (n + \frac{1}{2})\hbar\omega_c\}^{-1/2}\,d\epsilon \tag{9b.7}$$

where the crystal volume $V = L_x L_y L_z$ has been introduced and the sum is to be taken up to a maximum value, n_{max}, which makes $\epsilon \geqslant (n + \frac{1}{2})\hbar\omega_c$. The function $g(\epsilon)$ is shown in Fig.9.5a together with the density of states for the case of no magnetic field given by Eq.(3a.28) and now denoted by $g_0(\epsilon)$. Fig.9.5b shows the ratio $\int_0^\epsilon g(\epsilon)\,d\epsilon / \int_0^\epsilon g_0(\epsilon)\,d\epsilon$. For a degenerate semiconductor, which in zero field has a spherical constant-energy surface ("Fermi sphere"), the allowed $\vec{k}$-values in a magnetic field in the z-direction lie on the surfaces of a set of concentric cylinders whose axes are along the k_z-direction (Fig.9.6). As $\vec{B}$ is increased the radii of the cylinders become larger and the number of cylinders within a Fermi sphere of an energy, which is equal to the Fermi energy ζ, becomes smaller. For the case $\hbar\omega_c > \zeta$, only the innermost cylinder with quantum number $n = 0$ is available for accommodation of the carriers; this condition is called the "quantum limit".

In order to be able to measure quantum effects, $\hbar^2 k_z^2/2m$ in Eq.(9b.4) has to

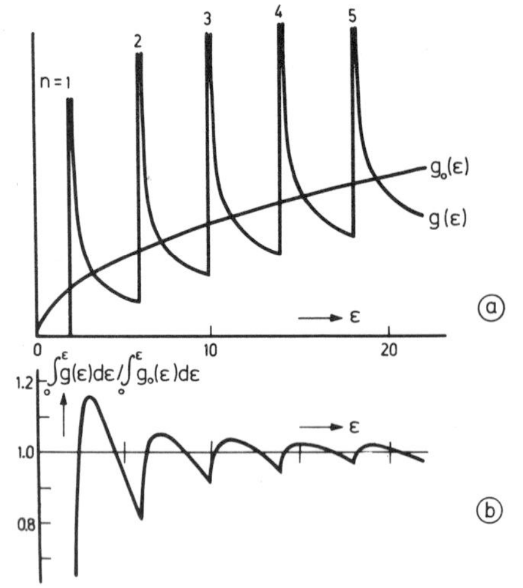

Fig.9.5a Density of states for no magnetic field, $g_0(\epsilon)$, and for $B \neq 0$, $g(\epsilon)$. The number of each Landau level is given in the figure.
b Ratio of the number of states below a given energy in a magnetic field relative to the same number without a field (after O. Madelung: Halbleiterprobleme (F. Sauter, ed.), Vol. V, p. 87. Braunschweig: Vieweg. 1960.

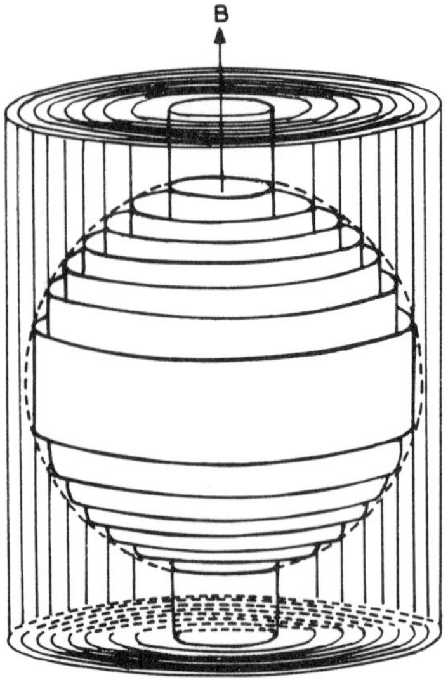

Fig.9.6 Magnetic quantization of a spherical constant energy surface in $\vec{k}$-space into discrete, concentric cylinders whose axes are parallel to the magnetic field (after R. G. Chambers, Can. J. Phys. 34 (1956) 1395.).

be much smaller than $\hbar\omega_c$, which is equivalent to the condition $k_B T \ll \hbar\omega_c$ or $T/K \ll B/7.46$ kG for $m = m_0$; even for small effective masses, temperatures down to a few K are required in order to observe quantum effects with available magnetic field intensities. At such low temperatures the carrier gas is degenerate and the average energy $= \zeta_n$ must be $\approx \hbar\omega_c$.

It is clear when the cylinders in Fig.9.6 move out of the Fermi sphere with increasing magnetic field strength that certain physical properties of the semiconductor show "oscillations" when plotted vs $\vec{B}$. Fig.9.7a shows [2] the longitudinal magnetoresistance $\Delta\rho/\rho$ of n-type InAs with a carrier density of $n = 2.5\times10^{16}/\mathrm{cm}^3$ as a function of B at various temperatures between 4.2 K and 12.5 K. These oscillations of the magnetoresistance are known as "Shubnikov -de Haas effect" [3]. A calculation for $\Delta\rho/\rho$ yields [4]:

$$\Delta\rho/\rho = \sum_{r=1}^{\infty} b_r \cos\left(\frac{2\pi\zeta_n}{\hbar\omega_c} r - \frac{\pi}{4}\right) \tag{9b.8}$$

---

[2] G. Bauer and H. Kahlert, Phys. Rev B5 (1972) 566 and Proc. Int. Conf. Semic. Physics, Cambridge, Mass., 1970 (S. P. Keller, J. C. Hensel, and F. Stern, eds.), p. 65. Oak Ridge/Tenn.: USAEC. 1970.

[3] L. Shubnikov and W. J. de Haas, Leiden Commun. 207a, 207c, 207d, 210a (1930).

[4] E. N. Adams and T. D. Holstein, J. Phys. Chem. Solids 10 (1959) 254.

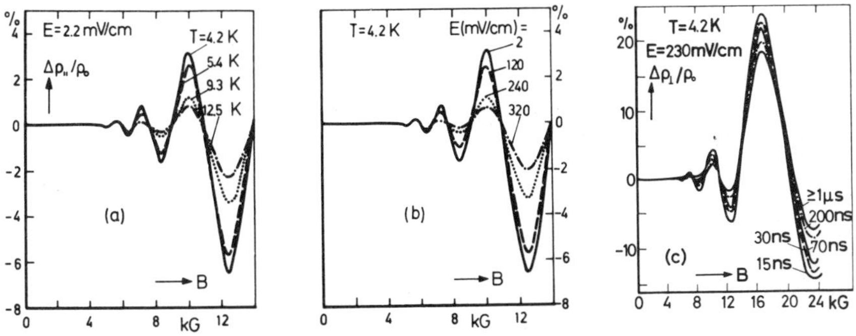

Fig.9.7a Shubnikov-de Haas effect in n-type InAs at various temperatures for a small electric field strength at long time intervals after the application of the electric field.
b Same as Fig.9.7a but for a lattice temperature T = 4.2 K and for various electric field strengths.
c Same as Fig.9.7a but for a lattice temperature T = 4.2 K, a strong electric field and various time intervals after the application of the electric field.

In a plot of $\Delta\rho/\rho$ vs $1/B \propto 1/\omega_c$ the period of the oscillation is independent of B; it depends on the carrier concentration via the Fermi energy ζ_n. The coefficients b_r rapidly decrease with increasing values of r and in practice only the r = 1 term is retained in the sum :

$$b_r = (-1)^r \sqrt{\frac{\hbar\omega_c}{2\zeta_n r}} \cos(r \frac{\pi}{2} gm/m_0) \frac{r\, 2\pi^2 k_B T/\hbar\omega_c}{\sinh(r\, 2\pi^2 k_B T/\hbar\omega_c)} \exp(-r\, 2\pi^2 k_B T_D/\hbar\omega_c) \tag{9b.9}$$

where g is the Landé factor of the carrier [5], and T_D is the "Dingle temperature" [6] which is a measure of the "natural line width" of the transitions between adjacent Landau levels. For a lifetime τ_c of a carrier in a state of energy ϵ the Heisenberg energy uncertainty

$$\Delta\epsilon = h/\tau_c \tag{9b.10}$$

divided by the level spacing $\hbar\omega_c$ yields an amplitude factor

$$\exp(-\Delta\epsilon/\hbar\omega_c) = \exp(-2\pi^2 k_B T_D/\hbar\omega_c) \tag{9b.11}$$

where

$$k_B T_D = \frac{1}{\pi}\hbar/\tau_c \tag{9b.12}$$

and since the hyperbolic sine in Eq.(9b.9) may be approximated by an exponential, the combination of this and the exponential given by Eq.(9b.11) yields $\exp\{-2\pi^2 k_B(T + T_D)/\hbar\omega_c\}$. The temperature T is primarily the carrier temperature T_e. For this reason it is possible to determine T_e for a degenerate electron gas, which is heated by a strong electric field $\vec{E}$,

[5] see e.g. L. I. Schiff: Quantum Mechanics, p. 441. New York: McGraw-Hill. 1968; for electrons in III-V compounds see M. Cardona, J. Phys. Chem. Solids 24 (1963) 1543.

[6] R. B. Dingle, Proc. Roy. Soc. (London) A 211 (1952) 517.

by means of the Shubnikov-de Haas effect [7]. Fig.9.7b shows oscillations for various values of $\vec{E}$ at a constant lattice temperature [2]. A comparison of Fig.9.7a and b reveals that the electron gas is heated to a temperature of e.g. 12.5 K by a field of 320 mV/cm while the lattice is at 4.2 K. Measurements at times elapsed after the application of the electric field between 10^{-8} and 10^{-6} sec shown in Fig. 9.7c yield a determination of the energy relaxation time which rises from 40 to 70 nsec in the electron temperature range from 4.2 to 12.5 K. For the evaluation of the data the Dingle temperature T_D (= 7.0 K for $\tau_c = 3.5 \times 10^{-13}$ sec in the present case) was assumed to be independent of $\vec{E}$ for the range of $\vec{E}$ values considered above: Since the mobility variation of the degenerate sample due to electron heating was less than 3 %, the influence of T_e on the collision time and therefore on the Dingle temperature was negligibly small.

Not only for degenerate semiconductors but also for semimetals such as gray tin (α-Sn) electron temperatures as a function of the applied electric field strength have been determined by this method: At E = 130 mV/cm the electron temperature equals 15 K at a lattice temperature of 4.2 K. No deviation from Ohm's law has been found at field strengths up to this value [8].

The lifting of spin degeneracy by a strong magnetic field adds another term to the right-hand side of Eq.(9b.4) :

$$\epsilon = (n + \frac{1}{2})\hbar\omega_c + \hbar^2 k_z^2/2m \pm \frac{1}{2} g\mu_B B \tag{9b.13}$$

where $\mu_B = e\hbar/2m_0 = 5.77\,\mu eV/kG$ is the Bohr magneton *) and g is the Lande factor. The ratio $g\mu_B B/\hbar\omega_c = (g/2)m/m_0$ is also a factor in the argument of the cosine in Eq.(9b.9); values between 0.33 (for n-InSb) and 0.013 (for n-GaAs) have been obtained from Shubnikov-de Haas measurements [5]. This yields for the Landé factor g values between −44 (for InSb) and + 0.32 (for GaAs). For reviews on further experimental work see ref.9.

The magnetooptical determination of g will be discussed in Chap.11d. Finally let us briefly consider the "de Haas-van Alphen effect" [10]: Oscillations occur in the magnetic susceptibility of metals, semimetals, and degenerate semiconductors. These oscillations are also periodic in 1/B and have an amplitude increasing

[7] R. A. Isaacson and F. Bridges, Solid State Comm. 4 (1966) 635.

[8] G. Bauer and H. Kahlert, Phys. Lett. 41A (1972) 351.

[9] L. M. Roth and P. N. Argyres: Semiconductors and Semimetals (R. K. Willardson and A. C. Beer, eds.), Vol.1., p. 159. New York: Acad. Press. 1966, and the article on thermomagnetic effects by S. M. Puri and T. H. Geballe, ibid., p. 203.

[10] W. J. de Haas and P. M. van Alphen, Leiden Commun. 208d, 212a (1930), and 220d (1933).

*) Strictly speaking the Bohr magneton is $\mu_0\mu_B$ where μ_0 is the permeability of free space.

with B. The magnetic moment is given by [11]

$$M = -(\partial F/\partial B)_{T = \text{const}} \tag{9b.14}$$

where F = U − TS is the free energy, U is the internal energy, and S is the entropy of the carriers (Chap.3a). The oscillatory part of the magnetic moment is given by

$$M_{osc} = \sum_{r=1}^{\infty} a_r \sin\left(\frac{2\pi\zeta}{\hbar\omega_c} r - \frac{\pi}{4}\right), \tag{9b.15}$$

where a_r is similar to b_r except that the factor $\sqrt{\hbar\omega_c/2\zeta r}$ contained in b_r has to be replaced by $3n(\hbar\omega_c)^{3/2}/\{r^{3/2}4\pi B\sqrt{2\zeta}\}$ and n is the carrier concentration. This effect has been investigated mostly in metals for a determination of the Fermi surface [12].

9c. Magnetic Freeze-out of Carriers

Eq.(9b.4) shows that the lowest Landau level (quantum number n = 0) with increasing magnetic field intensity rises by $\frac{1}{2}(e\hbar/m)B$. Besides the energy bands also the impurity levels are affected by the magnetic field. The hydrogen model of impurities has been calculated for the case of strong magnetic fields where $\frac{1}{2}\hbar\omega_c \gg$ the Rydberg energy [1]. When the ratio $\gamma = \hbar\omega_c/2Ry$ is $\gtrsim 1$, the originally spherical wave function of the hydrogen ground state becomes cigar shaped with the longitudinal axis in the direction of $\vec{B}$. The wave function is compressed, especially in the plane perpendicular to $\vec{B}$, the average distance of the electrons from the nucleus is reduced, and the ionization energy, defined as the difference in energy between the lowest bound state and the lowest state in the conduction band, is increased. For $\gamma \approx 2$ the ionization energy is doubled, but at greater values of γ the increase is sublinear. With increasing ionization energy the density of conduction electrons, n, decreases according to Eq.(3b.10) where, however, in the quantum limit for Maxwell-Boltzmann statistics N_c is given by $\sqrt{2\pi m k_B T}\, eB/h^2$ according to Eq.(9b.7) for $n_{max} = 0$ and Eq.(3a.36): (Consider that in e.g. InSb with its large g-factor for conduction electrons spin-splitting is about as large as Landau splitting and therefore the lowest level will contain electrons of only <u>one</u> direction of spin.)

---

[11] see e.g. R. A. Smith: The Physical Principles of Thermodynamics, p. 154. London: Chapman and Hall. 1952, and R. A. Smith: Wave Mechanics of Crystalline Solids, p. 384. London: Chapman and Hall. 1961.

[12] D. Shoenberg: The Physics of Metals, 1. Electrons (J. M. Ziman, ed.). Cambridge: Univ. Press. 1969; for a derivation of Eq.(9b.15) see e.g. J. M. Ziman: Principles of the Theory of Solids, chap. 9.7. Cambridge: Univ. Press. 1964.

Chap.9c:

[1] Y. Yafet, R. W. Keyes, and E. N. Adams, J. Phys. Chem. Solids 1 (1956) 137.

$$n = \sqrt{2\pi m k_B T}\,(|e|B/h^2)\exp\{(\zeta_n - \tfrac{1}{2}\hbar\omega_c)/k_B T\} = N_c \exp\{(\zeta_n - \tfrac{1}{2}\hbar\omega_c)/k_B T\} \tag{9c.1}$$

The decrease of n with increasing magnetic field strength is known as "magnetic freeze-out" of carriers.

There are few semiconductors where the condition $\gamma > 1$ is met with available magnetic field strengths. Even these semiconductors have not been obtained pure enough that the average spacing between neighboring impurity atoms is larger than about ten times the Bohr radius as has been assumed for the calculation (isolated hydrogen atom as a model) [1]. Hence, no quantitative agreement between theory and experiment is to be expected.

A semiconductor, which one might expect to show magnetic freeze-out at temperatures between 1 and 5 K, is n-type InSb. Fig.9.8 shows the results of field-dependent Hall constant measurements obtained [2] at 4.2 K with two samples of different zero-field carrier concentrations, $4\text{x}10^{14}/\text{cm}^3$ and $2\text{x}10^{16}/\text{cm}^3$. For the sample with the higher carrier concentration and consequently higher impurity concentration the overlap of the impurity wave functions persists to much higher magnetic fields than for the other sample. In Fig. 9.9 the Hall coefficient is plotted vs temperature for various values of B between 0.09 and 8.23 kG [3]. As should be expected from Eq.(3b.10) freeze-out increases with decreasing temperature. If the Hall coefficient R_H is assumed to be proportional to $\exp(\Delta\epsilon_D/k_B T)$ with a binding energy $\Delta\epsilon_D$, a dependence $\Delta\epsilon_D \propto B^{1/3}$ on the magnetic field intensity has been found [4]. In addition, there is a dependence of R_H on the carrier concentration n as shown in Fig.9.10 [5]: For a larger carrier concentration, a larger magnetic field strength is required to

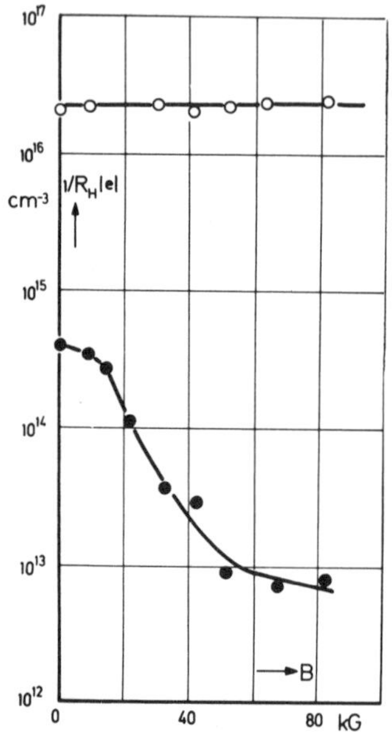

Fig.9.8 Magnetic freeze-out of carriers in n-type InSb at 4.2 K in a sample with about 10^{14} electrons/cm^3 (full circles) but not in a sample with about 10^{16} electrons/cm^3 (open circles) (after Keyes and Sladek, ref.2).

[2] R. W. Keyes and R. J. Sladek, J. Phys. Chem. Solids 1 (1956) 143.
[3] E. H. Putley, Proc. Phys. Soc. (London) 76 (1960) 802; J. Phys. Chem. Solids 22 (1961) 241.
[4] L. J. Neuringer, Proc. Int. Conf. Phys. Semicond. Moscow 1968, p. 715. Leningrad: Nauka. 1968.
[5] E. J. Fantner, thesis, Univ. Wien, Austria. 1973.

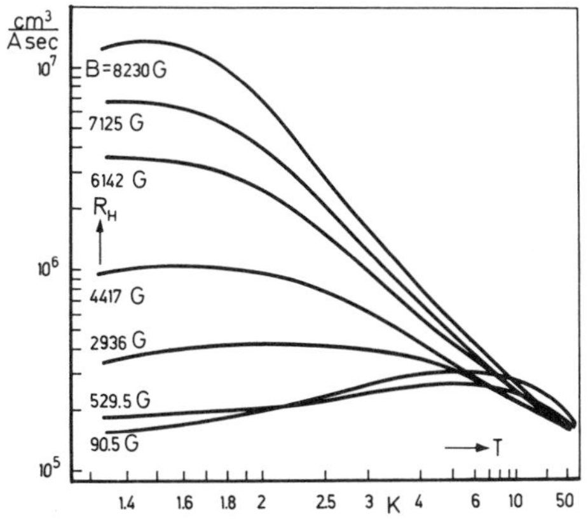

Fig.9.9 Dependence of the Hall coefficient in n-type InSb on temperature for various values of the magnetic induction (after E. H. Putley, ref.6).

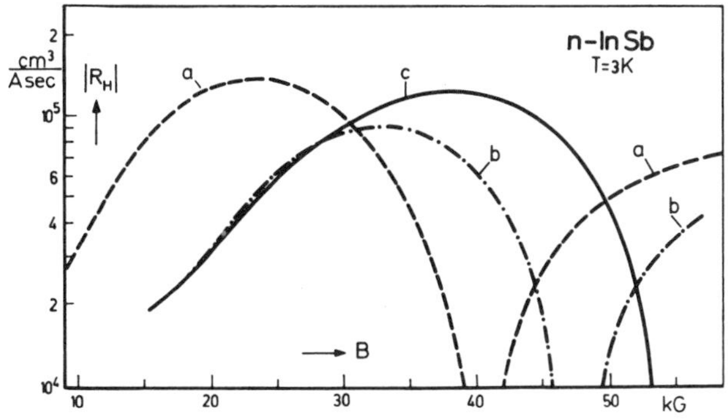

Fig.9.10 Magnetic field dependence of the Hall coefficient for n-InSb at T = 3 K for various values of the carrier concentration n : curve (a): n = 2.7x10^{14} cm^{-3}; Hall mobility at 4.2 K: μ_H = 5.5x10^4 cm^2/Vsec; (b): n = 6x10^{14} cm^{-3}, μ_H = 5.5x10^4 cm^2/Vsec; (c): n = 6x10^{14} cm^{-3}; μ_H = 3.5x10^4 cm^2/Vsec (after Fantner, ref. 5).

achieve the same freeze-out effect. At large magnetic field strengths there is a sign reversal of R_H. As yet there is no quantitative agreement and, concerning the sign reversal, not even qualitative agreement between experimental and theoretical data.

In n-InSb sub-mm waves (λ = 0.2 mm) are absorbed by the donors in a strong magnetic field ($\approx$ 7 kG) at 1.8 K (see Fig.9.11) and cause photoconductivity since the magnetically frozen-out electrons are lifted into the conduction band [6]. This effect has been applied for infrared detection. At even longer wave-

[6] E. H. Putley: Semiconductors and Semimetals (R. K. Willardson and A. C. Beer, eds.), Vol.1, p. 289. New York: Acad. Press. 1966.

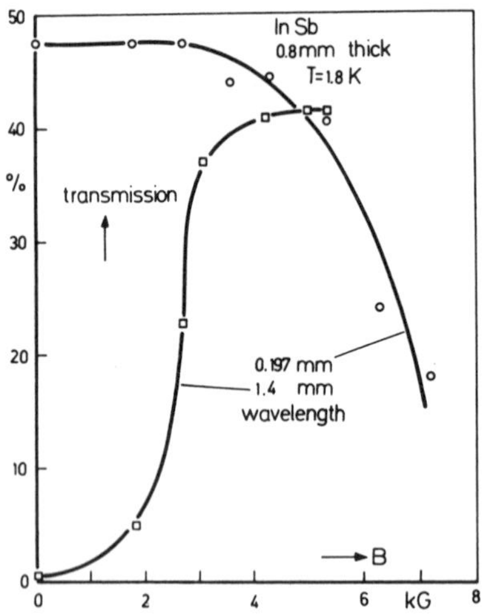

Fig.9.11 Microwave and far-infrared optical transmission of n-type InSb at 1.8 K (after E. H. Putley, ref. 6 and phys. stat. sol. 6 (1964) 571).

lengths (e.g. $\lambda = 1.4$ mm, Fig.9.10) where free-carrier absorption prevails (see Chap.11j) a reverse behavior has been observed and accounted for by the decreasing density of free carriers with increasing magnetic field.

9d. The Magnetophonon Effect

In contrast to the Shubnikov-de Haas oscillations, which require very low temperatures and a degenerate electron gas (see text after Eq.(9b.7)), there is another type of oscillations at higher temperatures which is due to longitudinal optical phonons of energy $\hbar\omega_0$ and which requires no degeneracy. These oscillations occur if the optical phonon energy is an integer multiple n of the Landau level spacing, $\hbar\omega_c$:

$$\hbar\omega_0 = n\hbar\omega_c\,; \quad n = 1, 2, 3, \cdots \tag{9d.1}$$

By absorption of a phonon an electron is transferred from a given Landau level $\# \rho$ to a level $\# \rho + n$. This effect is known as the "magnetophonon effect". It was predicted independently by Gurevich and Firsov [1] and by Klinger [2] and observed by Firsov et al.[3]. Their experimental data for longitudinal and transverse magnetoresistance obtained on n-type InSb at 90 K are displayed in Fig. 9.12. The inset shows the oscillatory part plotted vs 1/B. The period of the oscillation is constant and given by

$$\Delta(1/B) = e/(m\omega_0) \tag{9d.2}$$

[1] V. L. Gurevich and Yu. A. Firsov, Zh. Eksp. Teor. Fiz. 40 (1961) 199 [Engl.: Sov. Phys.-JETP 13 (1961) 137.].

[2] M. I. Klinger, Fiz. Tverd. Tela 3 (1961) 1342 [Engl.: Sov. Phys. Solid State 3 (1961) 974.].

[3] Yu. A. Firsov, V. L. Gurevich, R. V. Parfeniev, and S. S. Shalyt, Phys. Rev. Letters 12 (1964) 660.

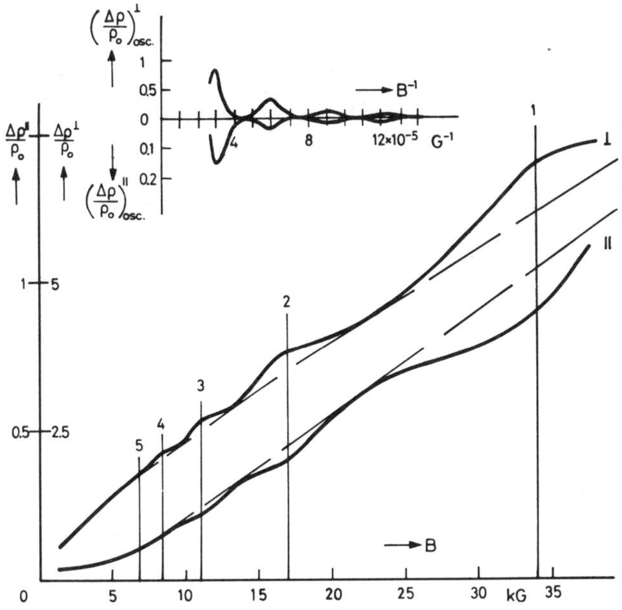

Fig.9.12 Transverse and longitudinal magnetoresistance for n-type InSb at 90 K. The inset shows the oscillatory part of the curves as a function of the inverse magnetic induction. The numbers at the vertical lines are values of the ratio of phonon frequency and cyclotron resonance frequency (after Firsov et al., ref.3).

in agreement with Eq.(9d.1). In n-type InSb at 34 kG, $m/m_0 = 0.016$ and $\omega_0 = 3.7\times10^{13}\,\mathrm{sec}^{-1}$, which yields $\Delta(1/B) = 3\times10^{-2}/\mathrm{kG}$. At resonance the transverse magnetoresistance shows maxima while the longitudinal magnetoresistance shows minima. The period is independent of the carrier density, in contrast to the Shubnikov-de Haas period (see Eq.(9b.7)). Since for large carrier densities [4] and consequently large impurity densities, ionized impurity scattering dominates over optical phonon scattering, no oscillations have been observed for densities of more than $5\times10^{15}/\mathrm{cm}^3$. At very low temperatures not enough optical phonons are available for the magnetophonon effect to occur, while at too high temperatures in spite of the strong magnetic field the condition $\omega_c\tau_m \gg 1$ necessary for the observation of Landau levels is not fulfilled. Because of this compromise the effect is quite small and requires refined electronic techniques such as double differentiation [5]. The longitudinal magnetophonon effect occurs

[4] A calculation for degenerate semiconductors has been presented by A. L. Efros, Fiz. Tverd. Tela 3 (1961) 2848 [Engl.: Sov. Phys. Solid State 3 (1962) 2079.].

[5] L. Eaves, R. A. Stradling, and R. A. Wood, Proc. Int. Conf. Semic. Physics, Cambridge, Mass., 1970 (S. P. Keller, J. C. Hensel, and F. Stern, eds.), p. 816. Oak Ridge/Tenn.: USAEC. 1970.

when two scattering processes such as inelastic scattering by an optical phonon and an elastic transition at an impurity site are operative [6].

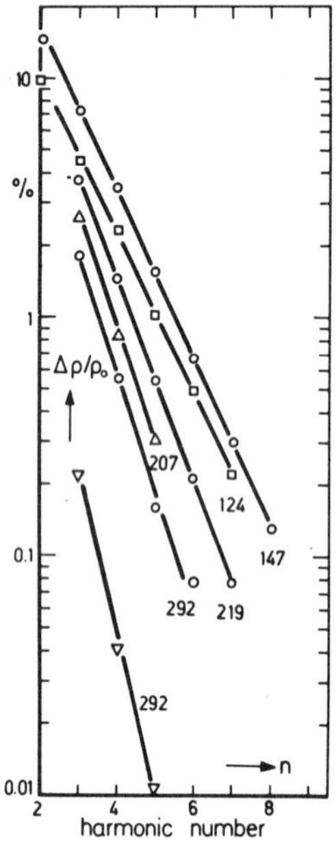

Fig.9.13 Observed amplitudes of the magnetophonon extrema in n-GaAs in the transverse configuration plotted vs harmonic number. The numbers at the curves are the temperatures in units of K (after Stradling and Wood, ref.7).

The amplitudes $\Delta\rho/\rho_0$ observed in the transverse configuration with n-type GaAs, are plotted vs the harmonic number in Fig.9.13 [7]. The straight lines in the semi-log plot suggest a proportionality of $\Delta\rho$ to $\exp(-\gamma\omega_l/\omega_c)$ with a constant γ which in GaAs is $\gamma = 0.77$ according to observations.

With an optical phonon frequency ω_0 known e.g. from Raman scattering, the magnetophonon effect serves for a determination of the effective mass m of the carriers. Fig.9.14 shows the electron and hole effective masses in Ge as a function of the angle between the magnetic field and the <100> direction in the $[1\bar{1}0]$ plane obtained from magnetophonon measurements. The effective hole masses and the transverse electron mass obtained at 120 K are significantly higher than those obtained from cyclotron resonance at 4 K (Fig.11.7b), while the longitudinal electron mass is in agreement with the cyclotron value [5]. The discrepancies are explained by the non-parabolicities of the bands: At the high temperatures of the magnetophonon measurements higher parts of the bands are occupied than at 4 K. In polar semiconductors due to the polaron effect (Chap. 61), the mass derived from magnetophonon oscillations is $1 + \alpha/3$ times the low-frequency mass and $1 + \alpha/2$ times the high-frequency or "bare" mass, where α is the polar coupling constant given by Eq.(61.8) and having a value of e.g. 0.06 in n-type GaAs [8]. The dependence of the magnetophonon mass m of n-type GaAs on temperature is shown [7] in Fig.9.15b; the variation of the optical phonon frequency with temperature, which is required for a determination of m, is plotted in Fig.9.15a. The values of m obtained after correcting for the band non-parabolicity are in agreement with those obtained by cyclotron resonance, Faraday rotation, and interband magneto-optic absorption.

While at temperatures below 60 K the magnetophonon effect in n-type GaAs is undetectable, electron heating by electric fields between 1 and 10 V/cm at

[6] V. L. Gurevich and Yu. A. Firsov, Zh. Eksp. Teor. Fiz. 47 (1964) 734 [Engl.: Sov. Phys.-JETP 20 (1964) 489.].

[7] R. A. Stradling and R. A. Wood, J. Phys. C 1 (1968) 1711.

[8] A. L. Mears, R. A. Stradling, and E. K. Inall, J. Phys. C 1 (1968) 821.

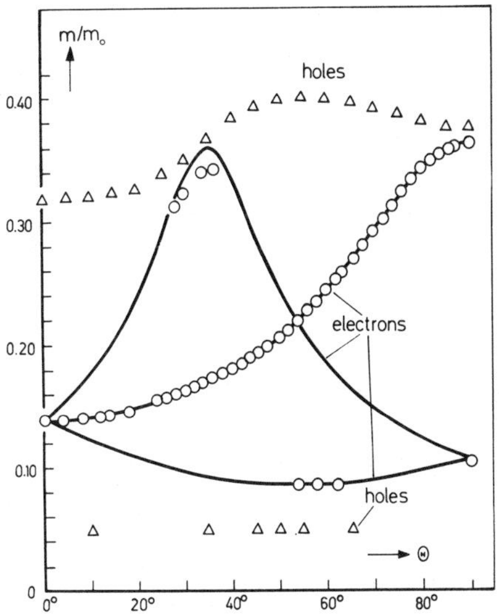

Fig.9.14 Variation of effective masses in Ge with angle between magnetic field and the <100> direction in the [110] plane, obtained from magnetophonon resonance (after Eaves et al., ref. 5).

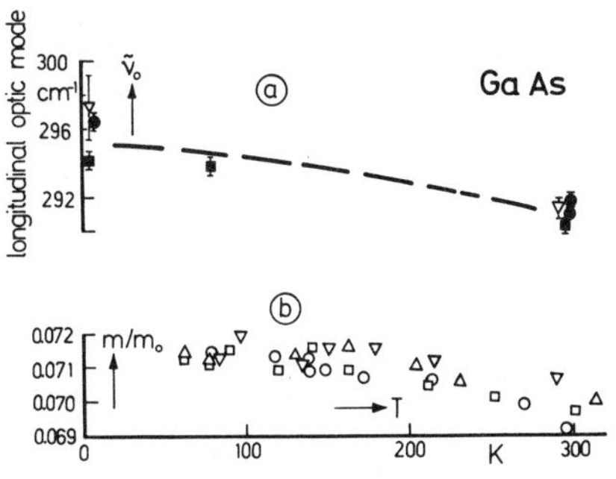

Fig.9.15 a Temperature dependence of the LO phonon frequency at $\vec{q} = 0$ for GaAs.
b Temperature dependence of the magnetophonon mass for n-type GaAs determined from the LO phonon frequency and the magnetophonon peaks (after Stradling and Wood, ref. 7).

lattice temperatures between 20 and 50 K results in a reappearance of the oscillation [9]. In Fig.9.16 the 1/B-values of the peaks are plotted [10] vs harmonic numbers for the series involving the emission of LO phonons with GaAs and CdTe. For a comparison, the high-temperature thermal-carrier peaks are also plotted. While the extrapolation of thermal peaks to $1/B = 0$ yields $n = 0$, the warm-carrier peaks are extrapolated to a negative value, $-1/2$, of n. Here the dominant energy loss mechanism is associated with the emission of an optical phonon and the capture of the warm electron by a donor of energy $\Delta\epsilon_D(B)$; at relatively low magnetic fields the donor energy may be approximated by $\Delta\epsilon_D(0) + \frac{1}{2}\hbar\omega_c$. Instead of Eq. (9d.1) we now have

$$\hbar\omega_o = \Delta\epsilon_D(0) + (n + 1/2)\hbar\omega_c \tag{9d.3}$$

A straight line through the $n \geqslant 7$ peaks of GaAs in Fig.9.15 fits this equation with $\Delta\epsilon_D(0) = 5.8$ meV which is consistent with a hydrogenic impurity, $\kappa = 12.5$, and $m/m_o = 0.0675$. The points corresponding to $n < 7$ deviate somewhat from this line since the dependence of $\Delta\epsilon_B$ on B becomes less strong as B is increased. In the warm-electron magnetophonon

[9] R. A. Reynolds, Solid State Electronics 11 (1968) 385.
[10] R. A. Stradling, L. Eaves, R. A. Hoult, A. L. Mears, and R. A. Wood, Proc. Int. Conf. Semic. Physics Cambridge/Mass. 1970 (S. P. Keller, J. C. Hensel, and F. Stern, eds.), p. 816. Oak Ridge/Tenn.: USAEC. 1970.

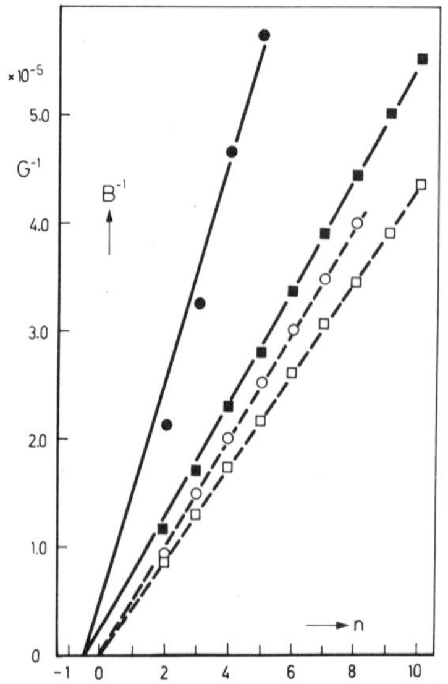

Fig.9.16 Inverse magnetic fields of the magnetophonon peaks vs harmonic numbers for the series involving the emission of LO phonons. Open squares: GaAs at 77 K, ohmic; full squares: GaAs at 20 K, high electric field; open circles: CdTe at 60 K, ohmic; full circles: CdTe at 14 K, high electric field (after Stradling et al., ref. 10).

effect the mobility and therefore the resistivity is determined by ionized impurity scattering and is very sensitive to a variation of the electron temperature T_e; the oscillation of T_e is caused by the oscillatory B-dependence of the probability of optical phonon emission from the high-energy tail of the distribution function. Consequently also the energy relaxation time τ_ϵ should depend on B in an oscillatory manner under magnetophonon conditions.

In CdTe for $E > 7$ V/cm additional peaks occur which agree with the field position of harmonics at higher temperatures under zero-field conditions and obey Eq. (9d.1).

In n-type InSb and GaAs at low temperatures (e.g. 11 K) and threshold fields of 0.07 V/cm and 0.8 V/cm, respectively, series of peaks are observed which have been attributed to a simultaneous emission of two oppositely directed transverse acoustic phonons, $\hbar\omega_t$, by electrons in the high-energy tail of the distribution.

$$2\hbar\omega_t = n\hbar\omega_c \; ; \quad n = 1, 2, 3, \cdots \tag{9d.4}$$

The phonons are thought to be at the X point in the phonon Brillouin zone and have energies of 5.15 meV in InSb and 9.7 meV in GaAs. In infrared lattice absorption (Chap.11 h) 2-phonon processes have also been observed. The matrix element for the 2-phonon emission has been estimated to be only 10^{-8} of that for the emission of a single LO phonon [10].

A magnetophonon Hall effect has also been measured, which is, however, much smaller than the magnetoresistance oscillations [11]. Oscillatory terms in ρ_{xy} arise from the second-order terms of magnitude $1/\omega_c^2 \tau_m^2$ which remain in the expression for the Hall coefficient in the high-field limit, Eq.(4c.84).

A magnetophonon effect in thermoelectric power in n-type InSb at 120 K is shown in Fig.9.17 [12]. The magnetic field is longitudinal. At least five maxima

[11] R. A. Wood, R. A. Stradling, and I. P. Molodyan, J. Phys. C 3 (1970) L 154.

[12] S. M. Puri and T. H. Geballe: Semiconductors and Semimetals (R. K. Willardson and A. C. Beer, eds.), Vol. 1, p. 203. New York: Acad. Press. 1966.

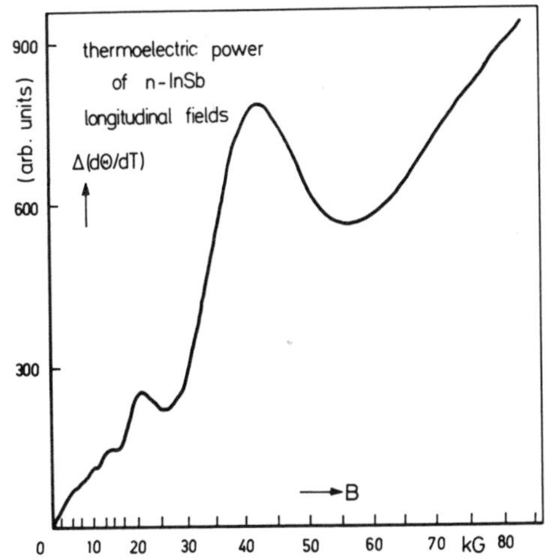

Fig.9.17 Variation of the thermoelectric power of n-type InSb with a longitudinal magnetic field at 120 K; peaks caused by magnetophonon resonance (after Puri and Geballe, ref. 12).

have been identified. The resonance at 42.5 kG can be correlated with a transition from the lowest to the next higher Landau level with the emission of a longitudinal optical phonon of energy 25 meV.

10. Impact Ionization and Avalanche Breakdown

Some aspects of impact ionization and avalanche breakdown in semiconductors are similar to the corresponding phenomena in gaseous discharges. Semiconductors may serve as model substances for gaseous plasmas since their ionic charges are practically immobile and therefore the interpretation of experimental data is facilitated. Impact ionization has been achieved both in the bulk of homogeneously doped semiconductors at low temperatures and in p-n junctions at room temperature. We will discuss these cases separately.

10a. Low-Temperature Impact Ionization in Homogeneous Semiconductors

Let us first consider impact ionization of shallow impurities in n-type Ge. Fig.10.1 shows typical I-V characteristics obtained at temperatures between 4.2 and 54.2 K [1]. At the lower temperatures most carriers are frozen-out at the impurities. Since the ionization energy is only about 10^{-2} eV, breakdown occurs already at fields of a few V/cm and persists until all impurities are ionized. At the highest temperature in Fig.10.1 all impurities are thermally ionized and hence there is no breakdown in the range of electric field intensities investigated. (At very high field strengths there is a tunnel effect across the band gap.) The onset of breakdown is more clearly shown in Fig.10.2 together with curves for the reciprocal of the Hall coefficient and the Hall mobility [2]. Already in the prebreakdown region there is a gradual increase of the carrier density with field intensity and a maximum in the mobility. At low field intensities ionized impurity scattering dominates the mobility. As the carrier energy is increased with increasing field intensity, the mobility increases also until lattice scattering becomes dominant and from there on the mobility decreases with field as a con-

---

[1] G. Lautz: Halbleiterprobleme (F. Sauter, ed.), Vol. VI., p. 21. Braunschweig: Vieweg. 1961.

[2] S. H. Koenig and G. R. Gunther-Mohr, J. Phys. Chem. Solids 2 (1957) 268; similar data have been obtained by N. Sclar and E. Burstein, J. Phys. Chem. Solids 2 (1957) 1.

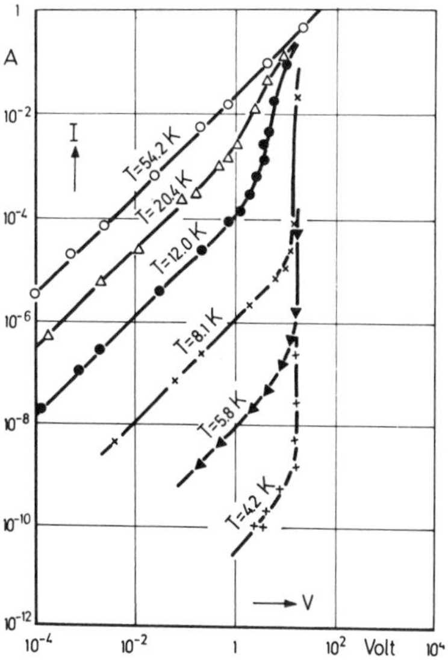

Fig.10.1 Current-voltage characteristics of n-type Ge at low temperature for various lattice temperatures (after G. Lautz, ref. 1).

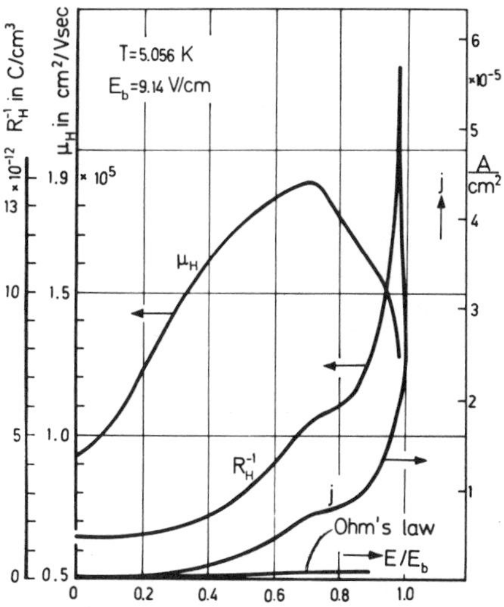

Fig.10.2 Impact ionization at low temperatures in n-type Ge (Sb doped; $N_D - N_A = 2.2 \times 10^{14}/cm^3$): Current density, reciprocal Hall coefficient, and Hall mobility vs electric field intensity. The extrapolated ohmic current-field characteristic is also shown (after S. H. Koenig et al., ref. 2).

sequence. Since lattice scattering is inelastic, in contrast to ionized impurity scattering, the increase of the carrier concentration with the field strength becomes weaker as lattice scattering becomes dominant which is indicated by a bump in the $R_H^{-1}(E)$ curve. The upper part of the characteristics has also been investigated in detail as shown in Fig.10.3 [3]. Curve (a) is for a carrier density of $n = 5.0 \times 10^{14}/cm^3$ at complete ionization and has been observed in various samples having a smallest cross dimension ranging from 0.5 to 5 mm. In contrast to earlier observations [4] no dependence on dimension has been found. Curve (b) represents data obtained from a sample with $n = 5.2 \times 10^{15}/cm^3$; both the breakdown field and the maximum conductivity are higher in this case. The decrease in conductivity is due to the above mentioned decrease in mobility at constant carrier density. The product of the low-field mobility and the

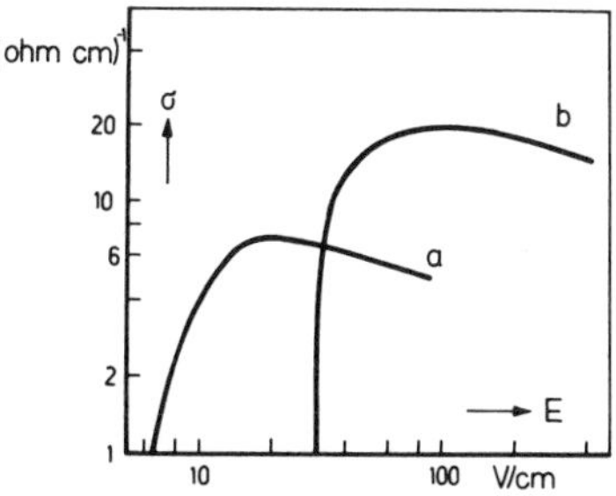

Fig.10.3 Electrical conductivity of n-type Ge with $N_D - N_A = 5 \times 10^{14}/cm^3$ (curve a) and $5.2 \times 10^{15}/cm^3$ (b) vs electric field strength at 4.2 K (after K. Baumann et al., ref. 3).

[3] K. Baumann, M. Kriechbaum, and H. Kahlert, J. Phys. Chem. Solids 31 (1970) 1163.

[4] A. Zylbersztejn and E. Conwell, Phys. Rev. Lett. 11 (1963) 417.

breakdown field intensity, which can be approximated by the drift velocity at the onset of breakdown, is about the same for all samples with shallow impurities irrespective of the impurity concentration [2].

For the impact ionization of shallow donors, the change of electron concentration n in an n-type semiconductor with time t at a given value of the electric field $\vec{E}$ and temperature T is given by [2]

$$dn/dt = A_T(N_D - N_A) + A_I n\{N_D - (N_A + n)\} - B_T n(N_A + n) - B_I n^2(N_A + n) \qquad (10a.1)$$

Here $A_T(T)$ and $A_I(E)$ are the coefficients for thermal and impact ionization processes, respectively; $B_T(T,E)$ is the coefficient for thermal recombination of a single electron with an ionized donor; $B_I(T,E)$ is the coefficient for the "Auger process" in which two electrons collide at an ionized donor, one being captured with the other taking off the excess energy. The Auger process is negligible for small values of n. $N_D - N_A$ is the concentration of uncompensated donors, $N_A + n$ is the concentration of ionized donors, and $N_D - (N_A + n)$ is the concentration of neutral donors. The concentrations of neutral acceptors and of holes have been assumed negligibly small.

The steady-state value of n which is denoted by n_o is easily obtained from Eq. (10a.1) for small values of n (neglect of Auger process and assuming $N_A \gg n$):

$$n_o = A_T(N_D - N_A)/\{B_T N_A - A_I(N_D - N_A)\} \qquad (10a.2)$$

Breakdown occurs at a field strength E_b for which the denominator vanishes:

$$B_T(T,E_b)N_A - A_I(E_b)(N_D - N_A) = 0 \qquad (10a.3)$$

Hence, at a given temperature the breakdown field strength is a unique function of the compensation ratio N_A/N_D. If we define a time constant

$$\tau = n_o/\{A_T(N_D - N_A)\} \qquad (10a.4)$$

Eq.(10a.1), subject to the simplifying assumptions that led to Eq.(10a.2), can be written in the form

$$-dn/dt = (n - n_o)/\tau \qquad (10a.5)$$

resulting in an exponential time dependence for a solution. Measurements of the recovery from breakdown yield a variation of τ with E at 4.2 K indicated by Fig. 10.4 (solid circles) [5]. The open circles show the variation of τ with temperature T for thermal electrons. The order of magnitude is 10^{-6} - 10^{-7} sec for a degree of compensation, $N_A/N_D = 5 \times 10^{12}/2 \times 10^{13} = 25\,\%$. Eq.(10a.4) shows that τ can be made short by adding compensating impurities. The capture cross section of about 10^{-12} cm^2 is an order of magnitude larger than the geometrical cross section and believed to be due to electron capture in a large, highly excited bound orbit,

[5] S. H. Koenig, Int. Conf. Solid State Physics, Brussels 1958 (M. Desirant, ed.), p. 422. London: Acad. Press. 1960.

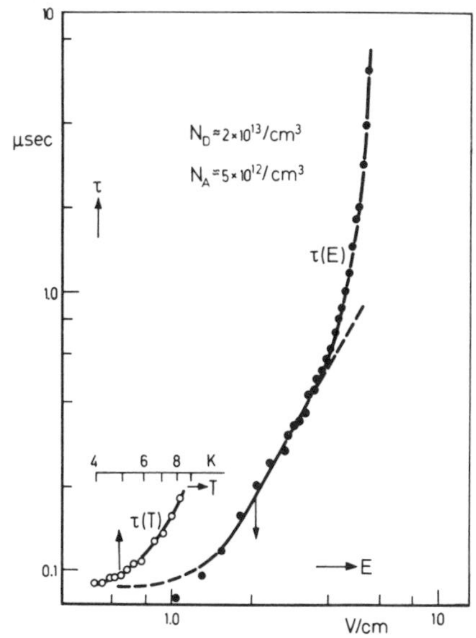

Fig.10.4 Variation of recombination time after breakdown with electric field strength at 4.2 K (full circles) and with temperature for thermal electrons (open circles) in n-type germanium (after S. H. Koenig, ref.5).

with a subsequent cascade of transitions to the ground state [5, 6].

In a many-valley semiconductor the breakdown field is anisotropic. Assuming that only the carriers in the hot valleys have sufficient energy for ionization at the onset of breakdown [7], we find for n-type Ge (valleys on <111> and equivalent axes) [8]:

$$E_b(\theta)/E_b(90^\circ) = \begin{cases} 1/\sqrt{1-(2-\cos^2\theta-\sqrt{2}\sin 2\theta)(1-K^{-1})/3} & \text{for } 0^\circ \leqslant \theta \leqslant 55^\circ \\ 1/\sqrt{1-\cos^2\theta\,(1-K^{-1})/3} & \text{for } 55^\circ \leqslant \theta \leqslant 90^\circ \end{cases} \tag{10a.6}$$

where as usual $\theta = 0^\circ$ indicates the <001> direction of $\vec{E}_b$, $\theta = 55^\circ$ the <111> direction, and $\theta = 90^\circ$ the <110> direction; $K = \mu_t/\mu_l$ is the ratio of transverse and longitudinal mobilities. Fig.10.5 compares experimental and theoretical data [9].

The impact ionization of deep-level impurities requires much larger field intensities than that of the shallow impurity levels discussed so far. E.g. Zn introduces a 0.033 and a 0.09 eV acceptor level in Ge (Zn^- and Zn^{--}, respectively). In Fig.10.6 shows the subsequent ionization of both levels at a temperature of 20.3 K [10]. Here as well as for shallow levels energy loss by lattice scattering

[6] M. Lax, J. Phys. Chem. Solids 8 (1959) 66.

[7] S. H. Koenig, R. D. Brown III, and W. Schillinger, Phys. Rev. 128 (1962) 1668.

[8] K. Seeger, Zeitschr. f. Physik 182 (1965) 510.

[9] N. Kawamura, Phys. Rev. 133 (1964) A 585; for n-Si see N. Kawamura, J. Phys. Chem. Solids 27 (1966) 919.

[10] A. Zylbersztejn, J. Phys. Chem. Solids 23 (1962) 297.

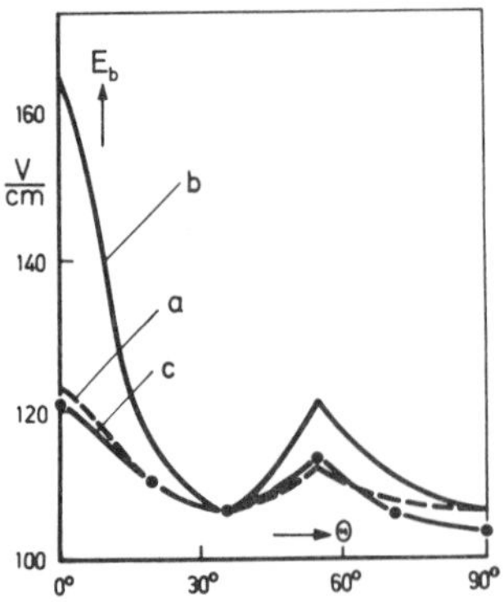

Fig.10.5 Dependence of low-temperature breakdown field on the angle between the field and the ⟨001⟩ direction in the $[1\bar{1}0]$ plane, calculated for energy loss by donor ionization (curve a), by acoustic deformation potential scattering (curve b), and experimental (curve c) in n-type Ge at 4.2 K (after N. Kawamura, ref.9).

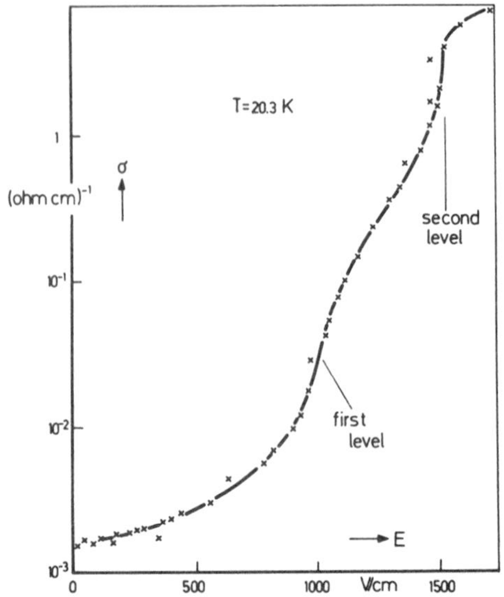

Fig.10.6 Conductivity of zinc-doped germanium vs electric field strength at 20.3 K. The two steps in the breakdown characteristic are consistent with the two acceptor levels of the zinc impurity (after A. Zylbersztejn, ref. 10).

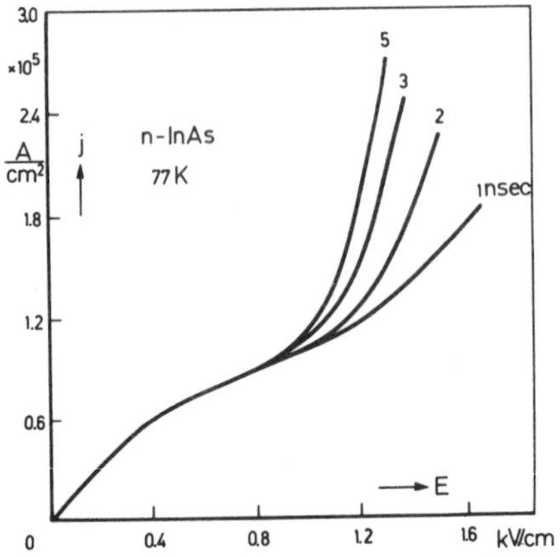

Fig.10.7 Current-field characteristics for n-type InAs at 77 K for various time intervals after the application of the voltage (after Bauer et al., ref.11).

completely dominates over the energy loss due to ionization.

In small-gap semiconductors such as InSb (ϵ_G = 0.2 eV), InAs (ϵ_G = 0.4 eV), and Te (ϵ_G = 0.35 eV), impact ionization of lattice atoms (i.e. not impurities) with the production of equal numbers of electrons and holes has been observed at temperatures between 4.2 and 300 K. In Fig. 10.7 experimental results obtained in n-type InAs at 77 K at various times after application of the field are shown [11]. From the development of the breakdown characteristics with time a "generation rate"

$$g(E) = (1/n_0)dn/dt \qquad (10a.7)$$

[11] G. Bauer and F. Kuchar, phys. stat. sol. (a) 13 (1972) 169.

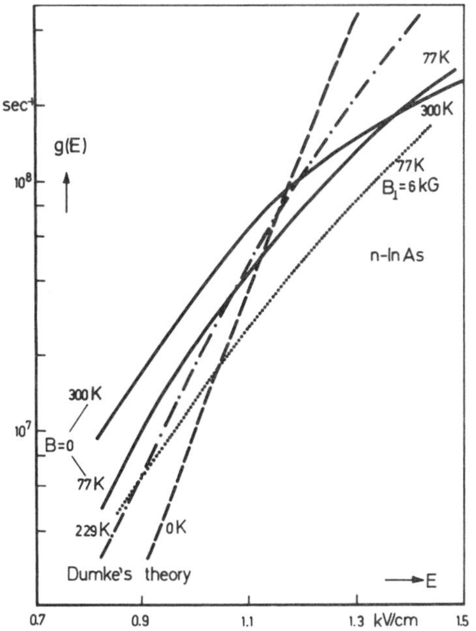

Fig.10.8 Generation rates: experimental (full curves and – in a transverse magnetic field of 6 kG – dotted) and calculated according to a theory by Dumke (ref. 12, dashed and dash-dotted curves) (after Bauer and Kuchar, ref. 11).

has been determined which is shown in Fig.10.8, together with the theoretical curves obtained by Dumke [12] for 0 and 229 K. The reciprocal of the generation rate is of the order of magnitude of 10^{-7} to 10^{-8} sec. For a transverse magnetic field of 6 kG at 77 K, where $\mu B \approx 2$, the generation rate is decreased by about 40% (Fig.10.8, dotted curve). This is interpreted by a cooling effect of the magnetic field on the hot carriers.

In many semiconductors such as p-type tellurium [13] at 77 K and compensated germanium [14] at 4.2 K regions of negative differential conductivity (n. d. c.) have been observed. Fig.10.9 shows V-I characteristics of indium-doped Ge compensated with antimony. Sample B is more heavily doped than sample A. The formation of current filaments in the breakdown region has been observed which is in agreement with the thermodynamic arguments given by Ridley [15] for S-shaped (current controlled) characteristics. The n. d. c. may be due to energy relaxation by deformation potential scattering in the magnetic field of the

---

[12] W. P. Dumke, Phys. Rev. 167 (1968) 783; see also R. C. Curby and D. K. Ferry, phys. stat. sol. (a) 15 (1973) 319.

[13] G. Nimtz, Proc. Int. Conf. Phys. Semic., Cambridge/Mass. 1970 (S. P. Keller, J. C. Hensel, and F. Stern, eds.), p. 396. Oak Ridge/Tenn.: USAEC. 1970.

[14] A. L. McWhorter and R. H. Rediker, Proc. Int. Conf. Phys. Semic., Prague 1960, p. 134. Prague: Czech. Acad. Sciences. 1960.

[15] B. K. Ridley, Proc. Phys. Soc. (London) 81 (1963) 996; A. M. Barnett, IBM-J. Res. Develop. 13 (1969) 522.

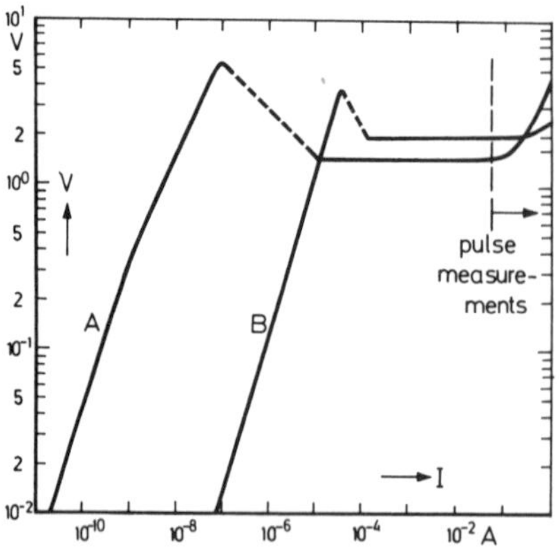

Fig.10.9 Voltage-current characteristics of indium-doped germanium strongly compensated with antimony, at 4.2 K. Sample B is more heavily doped than sample A. Sample thickness is 0.43 mm and contact diameter is 1 mm (after McWhorter and Rediker, ref. 14).

current [16] or due to momentum relaxation by impurity scattering. (In compensated material the impurity concentration is higher than normal.) For semiconductors with minority carriers having an effective mass much smaller than the majority carriers, one can imagine a process where the minority carriers generated in the avalanche are more easily accelerated by the field and, once they have sufficiently increased in number, dominate the ionization process.

In small-gap materials a "pinch-effect" has been observed after the formation of equal numbers of mobile positive and negative charges by the ionization process: The Lorentz force which arises due to the magnetic field of the current drives the carriers to the center of the sample thus forming a filamentary current [17]. The current density in the filament may be so high that the crystal lattice melts there and after switching off solidifies in polycrystalline form.

10b. Avalanche Breakdown in p-n Junctions

In the depletion layer of a reverse-biased p-n junction, electric field intensities up to 10^6 V/cm can be obtained with a negligible amount of Joule heating. Impact ionization across the band gap is then found even in large-gap semiconductors. In the junction characteristic shown in Fig.5.8 breakdown occurs at a voltage of 1200 V.

The relative increase in carrier density per unit length is called the "ionization rate":

$$a = (1/n)\, dn/dx \tag{10b.1}$$

The "multiplication factor" M is determined by the ionization rate:

$$1 - M^{-1} = \int_0^d a(E)\, dx \tag{10b.2}$$

where the assumption has been made that n_0 carriers have been injected at one end, x = 0, of the high-field region of width d and $n = Mn_0$ carriers have been

[16] R. F. Kazarinov and V. G. Skobov, Zh. Eksp. Teor. Fiz. 42 (1962) 1047. [Engl.: Sov. Phys. JETP 15 (1962) 726.]

[17] See e.g. B. Ancker-Johnson: Semiconductors and Semimetals (R. K. Willardson and A. C. Beer, eds.), Vol. 1, p. 379. New York: Acad. Press. 1966.

collected at x = d and the ionization rates for electrons and holes are equal. Eq.(10b.2) is obtained by considering the increase of the electron current component, j_n, between x and x + dx, which is due to ionization both by electrons and by holes (current component j_p):

$$dj_n/dz = \alpha j_n + \alpha j_p = \alpha j \qquad (10b.3)$$

The total current, j, is the same everywhere. At x = d the current consists only of electrons, i.e. there is no hole injection ($j_n = j$), while at x = 0, j_n is the injected electron current which equals $M^{-1}j$ (per definition of the multiplication factor M). With these boundary conditions the integration of Eq.(10b.3) yields Eq.(10b.2). Breakdown occurs when the integral on the right-hand side of Eq.(10b.2) equals unity, i.e. $M \to \infty$. For a measurement of M as a function of the reverse bias, carriers are usually injected by light [1, 2].

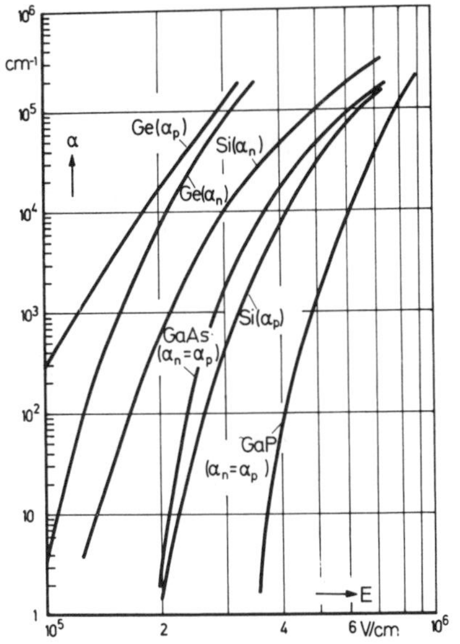

Fig.10.10 Observed ionization coefficient for avalanche multiplication vs electric field strength for electrons (n) and holes (p) in various semiconductors at room temperature (after Miller, Lee et al., Logan and Sze, and Logan and White, compiled by S. M. Sze, ref. 3).

Fig.10.10 shows experimental results of α vs E for Ge, Si, GaAs, and GaP [3]. The curves valid for Ge and Si can be represented by

$$\alpha \propto \exp(-\text{const}/E) \qquad (10b.4)$$

while those for GaAs and GaP are better represented by

$$\alpha \propto \exp(-\text{const}/E^2) \qquad (10b.5)$$

where the constants in the exponentials are fitting parameters. These forms of $\alpha(E)$ shall now be made plausible. A more complete calculation has been given by Baraff [4].

Let us consider [5] a collision where

[1] A. G. Chynoweth and K. G. McKay, Phys. Rev. 108 (1957) 29; for a review on charge multiplication phenomena see e.g. A. G. Chynoweth: Semiconductors and Semimetals (R. K. Willardson and A. C. Beer, eds.), Vol. 4, p. 263. New York: Acad. Press. 1968.

[2] C. A. Lee, R. A. Logan, R. L. Batdorf, J. J. Kleimack, and W. Wiegman, Phys. Rev. 134 (1964) A 761.

[3] S. M. Sze: Physics of Semiconductor Devices, p. 60. New York: J. Wiley and Sons. 1969.

[4] G. A. Baraff, Phys. Rev. 128 (1962) 2507; ibid. 133 (1964) A 26.

[5] This follows closely J. E. Carroll: Hot Electron Microwave Generators, p. 171 ff. London: Arnold. 1970.

the initial and final carrier velocities are v_i and v_f, respectively, and the velocities of the pair produced by the collision are v_n and v_p. For simplicity, let us consider a one-dimensional model where the incident particle is an electron. Momentum conservation yields

$$m_n v_i = m_n v_f + m_n v_n + m_p v_p \tag{10b.6}$$

and energy conservation gives

$$m_n v_i^2/2 = \epsilon_G + m_n v_f^2/2 + m_n v_n^2/2 + m_p v_p^2/2 \tag{10b.7}$$

A minimum of the incident kinetic energy,

$$\epsilon_i = m_n v_i^2/2 = (2m_n + m_p)/(m_n + m_p)\epsilon_G \tag{10b.8}$$

is obtained for $v_f = v_n = v_p$. Assuming the incident particle to be a hole yields the same value except that m_n and m_p are interchanged. For $m_n = m_p$ we find $\epsilon_i = 1.5\,\epsilon_G$. However, depending on the ratio m_n/m_p the factor in front of ϵ_G may have values between 1 and 2.

The number of carriers with energies $\epsilon > \epsilon_i$ is proportional to $\exp(-\epsilon_i/k_B T_e)$ if the carrier gas is non-degenerate with an electron temperature T_e. The latter has been determined from the energy balance equation

$$eEv_d = \frac{3}{2} \cdot \frac{k_B(T_e - T)}{\tau_\epsilon} \approx \frac{k_B}{\tau_\epsilon} T_e \tag{10b.9}$$

neglecting T and the factor 3/2 for a rough estimate. At the high field intensities of interest here, the drift velocity v_d is practically constant. Assuming a constant energy relaxation time τ_ϵ and considering the product $v_d \tau_\epsilon = l_o$ as a carrier-mean-free path for essentially optical phonon emission, the ionization rate a is determined by

$$a \propto \exp(-\epsilon_i/k_B T_e) = \exp(-\epsilon_i/eEl_o) \tag{10b.10}$$

Taking the quantity ϵ_i/el_o as a fitting parameter we arrive at Eq.(10b.4). On the other hand, assuming $v_d \tau_\epsilon$ to be proportional to E yields a dependence $a(E)$ as given by Eq.(10b.5).

Fig.10.11 shows the product al_o as a function of ϵ_i/eEl_o for various values of the ratio of the optical phonon energy $\hbar\omega_o$ and ϵ_i [3]. Experimental data obtained from Ge p-n junctions at 300 K have been fitted to the curve for which $\hbar\omega_o/\epsilon_i = 0.022$, assuming $\epsilon_i = 1.5\,\epsilon_G$, $\epsilon_G = 1.1$ eV for the direct energy gap at room temperature, and $\hbar\omega_o = 37$ meV. Values for l_o of 64 Å for electrons and 69 Å for holes are obtained. Since the drift velocity is about 6×10^6 cm/sec at high field strengths in Ge, values for τ_ϵ of about 10^{-13} sec are found which are, however, more than an order of magnitude lower than those obtained from microwave measurements [6]. Even so, the Baraff curve is a good fit to the experimental data.

[6] A. F. Gibson, J. W. Granville, and E. G. S. Paige, J. Phys. Chem. Solids 19 (1961) 198.

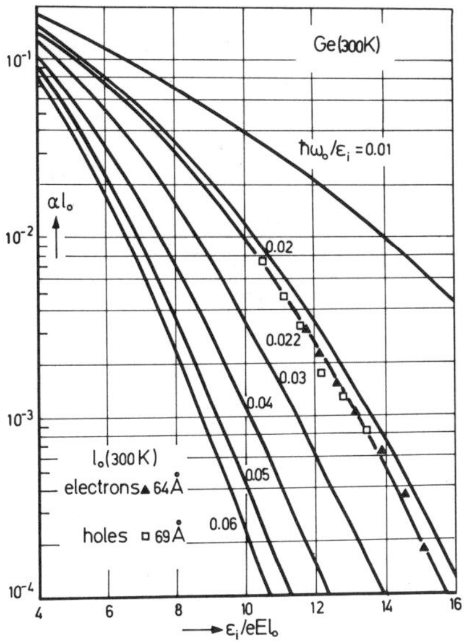

Fig.10.11 Baraff's plot calculated for various values of the ratio of the optical phonon energy and $\epsilon_i = 1.5\,\epsilon_G$, and data observed in germanium p-n junctions (after Baraff, ref. 4, and Logan and Sze, compiled by S. M. Sze, ref. 3).

In the range of impact ionization in diodes, light is emitted at an efficiency of 10^{-3} to 10^{-5} photons per electron crossing the junction. A typical spectrum obtained from GaAs at 77 K is shown in Fig.10.12 [7]. The maximum of the broad spectrum is near the energy gap. At the onset of breakdown light is emitted from various spots of the p-n junction rather than uniformly across the junction, and the current is noisy and concentrated on the same sites; these are denoted as "microplasmas". Dislocation-free junctions do not exhibit the phenomenon of microplasmas.

Impact ionization and a phase shift in ac current due to the carrier transit time through the p-n structure have been utilized for the generation of microwaves. The word "IMPATT" has been coined from "IMPact ionization Avalanche Transit Time" in order to characterize such devices. A special diode commonly known as the "Read diode" was the first device for this kind [8]. We can give here only a brief discussion of the physical principles of the Read diode and for further details refer to ref. 9.

Fig.10.13 shows the field distribution in a $p^+n\nu n^+$ structure where the ν region is essentially intrinsic (slightly n-type; for a slightly p-type region the letter π is used) [9] . In the high-field region of the p^+n junction impact ionization is assumed to occur. In the depletion region the field is assumed to be still high enough that the drift velocity v_d is nearly field-independent. Let us denote the admittance of the depletion region by Y such that the current density injected in the avalanche is given by $j_n = YV$ where V is the voltage across the whole diode. The charge travelling down the depletion region induces in the external circuit a capacitive current density $\kappa\kappa_0 \partial E/\partial t = (\kappa\kappa_0/L)\partial V/\partial t$ where L is the length of the depletion region. The external current at time t is induced from

[7] A. E. Michel, M. I. Nathan, and J. C. Marinace, J. Appl. Phys. 35 (1964) 3543.

[8] W. T. Read, Bell Syst. Tech. J. 37 (1958) 401; W. Shockley, Bell Syst. Tech. J. 33 (1954) 799.

[9] J. E. Carroll: Hot Electron Microwave Generators, p. 190. London: Arnold. 1970; H. Hartnagel: Semiconductor Plasma Instabilities, p. 100. London: Heinemann. 1969; ref. 3, p. 200.

carriers generated in the time interval between t and $t-\tau$ where $\tau = L/v_d$ is the carrier transit time. The delay between the injected charge and the induced current is therefore on the average $\tau/2$. For a sinusoidal variation of the current at a frequency ω, the induced current j is then given by

$$j \propto YV \exp(-i\omega\tau/2) \tag{10b.11}$$

If the phase factor of Y combined with the transit time phase $-\omega\tau/2$ yields a total phase between 90° and 270°, one can show that the current oscillates at the frequency ω even without any applied ac voltage. For a saturated drift velocity of $2\text{x}10^7$ cm/sec (as e.g. in Si) and a length $L = 10^{-3}$ cm, a frequency of about 5 GHz is obtained. Continuous-operation devices have been built with a power output of several watts. Frequencies as high as 300 GHz have been achieved.

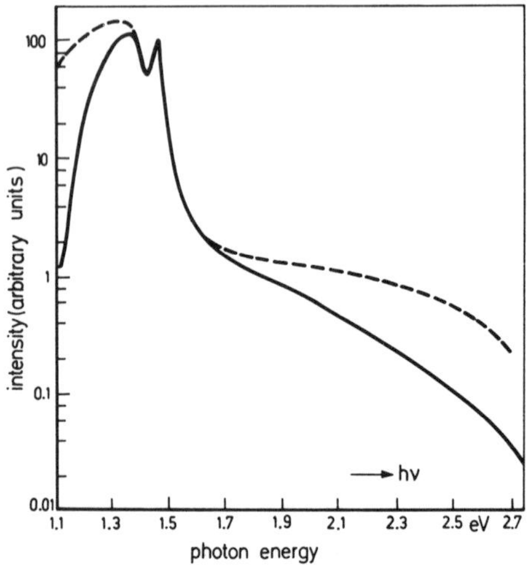

Fig.10.12 Emission spectrum from a reverse biased GaAs p-n junction at 77 K (original data: full curve; data corrected for spectral sensitivity of the detector: dashed curve) (after A. E. Michel et al., ref. 7).

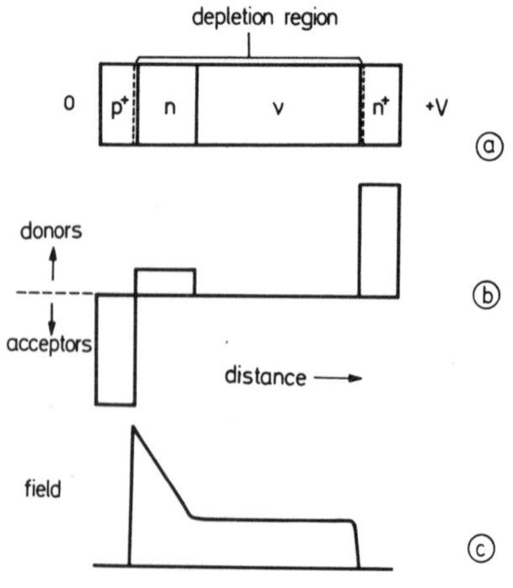

Fig.10.13 Read diode (a), impurity distribution (b), and field distribution (c) (after J. E. Carrol, ref. 9).

11. Optical Absorption and Reflection

The propagation of electromagnetic radiation through a semiconductor in general depends on temperature and external pressure and can be modified by electric and magnetic fields. Measurements of these effects provide information about band structure and energy levels in semiconductors. In Appendix D useful relations between reflection and transmission coefficients and the index of refraction and the extinction coefficient are given, together with the Kramers-Kronig relations. For numerical purposes it may be worth noting that a quantum energy $\hbar\omega$ of 1 eV is equivalent to a wavenumber $\tilde{\nu}$ of 8060 cm^{-1} and to a wavelength λ of 1.24 μm.

11a. Fundamental Absorption and Band Structure

Many transport properties of semiconductors require for an explanation the assumption of an energy gap ϵ_G between the valence band and the conduction band. Since in an ideal semiconductor there are no energy states within this gap the semiconductor should be transparent for light of angular frequencies less than a critical value, ω_e, given by

$$\hbar\omega_e = \epsilon_G \tag{11a.1}$$

if the absorption of light quanta is due only to the transfer of electrons from the valence band to the conduction band. This absorption is denoted as "fundamental absorption", and ω_e is the absorption edge.

This is true also for a non-ideal semiconductor if the density of carriers (called "free carriers" in optics) is not so large that there is metallic conduction and absorption by the carriers is of the same order of magnitude as the fundamental absorption.

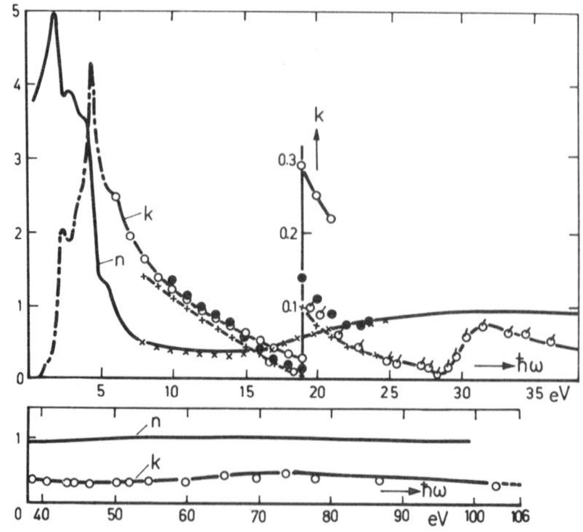

Fig. 11.1 Spectral dependence of the extinction coefficient k for germanium observed by Marton and Toots (crosses), Hunter (circles with partial slash), Sasaki (full circles), Philipp and Ehrenreich (open circles) and Philipp and Taft (dashed curve) (compiled by Lavilla and Mendlowitz, ref. 1). Refractive index n: experimental (crosses) by Marton and Toots (ref. 3) and calculated by a Kramers-Kronig analysis (full curve).

Fig. 11.1 shows the extinction coefficient, k, and the index of refraction, n, of germanium [1 - 4]. The extinction coefficient vanishes at quantum energies less than the energy gap of about 0.7 eV, rises to a peak at energies of a few eV and, as the energy increases to values which are in the soft-X-ray range, it decreases again. At the peak $k \approx n \approx 4$ and the imaginary part of the dielectric constant, $\kappa_i = 2nk$, is about 32.

For part of the spectrum the refractive index, n, has been calculated from $k(\omega)$ by means of the Kramers-Kronig relations (Appendix D). This calculation is confirmed by experimental data at energies between 8 and 20 eV, where n is less than 1. At low energies n has a value of about 4 corresponding to a real part of the dielectric constant of 16. In the X-ray range n approaches unity. Similar results are obtained for other semiconductors.

Fig. 11.2 shows the reflection coefficients r_∞ of germanium and silicon up to 12 eV [5], Fig. 11.3 the imaginary part, κ_i, of the dielectric constant of germa-

[1] R. E. Lavilla and H. Mendlowitz, J. Appl. Phys. 40 (1969) 3297.
[2] H. R. Philipp and E. A. Taft, Phys. Rev. 113 (1959) 1002.
[3] L. Marton and J. Toots, Phys. Rev. 160 (1967) 602.
[4] W. R. Hunter: Optical Properties and Electronic Structure of Metals and Alloys (F. Abeles, ed.), p. 136. Amsterdam: North-Holland Publ. Co. 1966.
[5] D. L. Greenaway and G. Harbeke: Optical Properties of Semiconductors. New York: Pergamon. 1968.

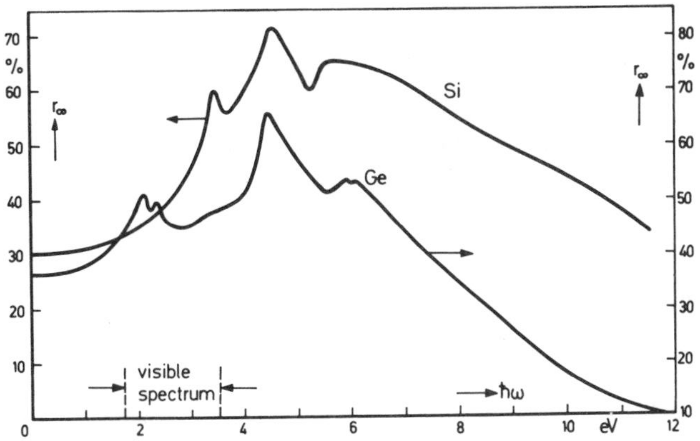

Fig. 11.2 Room temperature reflectivity of silicon and germanium (after Greenaway and Harbeke, ref. 5).

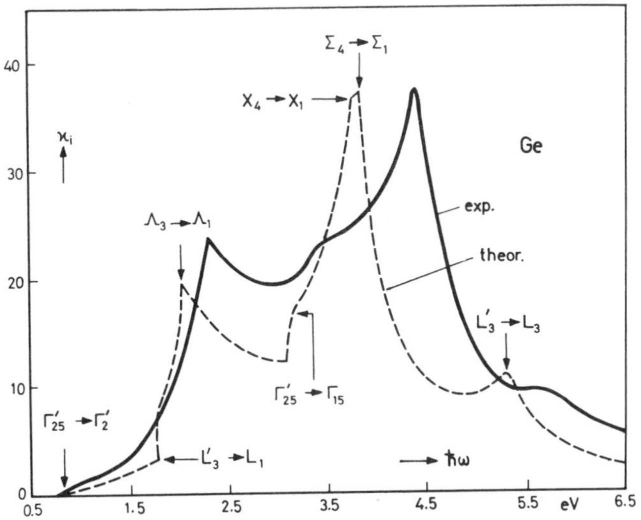

Fig. 11.3 Imaginary part of the dielectric constant for germanium. Solid curve: obtained from observed reflectance data. Dashed curve: calculated by Brust et al., ref. 6 .

nium [6]. The reflection coefficient r_∞ and the quantity κ_i have maxima at about the same energies. The maximum reflection is about 70 %. κ_i has also been calculated from the band structure shown by Fig.2.26 [7]; the result is indicated by the dash-dotted curve. There is qualitative agreement between the two curves and one can relate maxima in the $\kappa_i(\omega)$-relation to transitions in the energy band structure.

[6] Calculated curve by D. Brust, J. C. Phillips, and G. F. Bassani, Phys. Rev. Lett. 9 (1962) 94.

[7] M. Cardona and F. H. Pollak, Phys. Rev. 142 (1966) 530.

The most prominent maximum at about 4 eV is due to two transitions: One is from a valence band to the nearest conduction band at the edge of the first Brillouin zone in the $\langle 100 \rangle$ direction (denoted as $X_4 \to X_1$ in Fig.2.26); the other one, namely $\Sigma_4 \to \Sigma_1$, is a similar transition in the $\langle 110 \rangle$ direction and is not indicated in Fig.2.26 where only the $\langle 111 \rangle$ and $\langle 100 \rangle$ directions are shown. All transitions indicated, including peaks at 2 eV, 2.18 eV, and 6 eV, originate at the heavy hole band; at $\vec{k} = 0$, where the light and heavy hole bands are degenerate, transitions $\Gamma'_{25} \to \Gamma'_2$ (0.8 eV) and $\Gamma'_{25} \to \Gamma_{15}$ (3 eV) yield shoulders in the $\kappa_i(\omega)$-curve. The notation for points and axes in the first Brillouin zone by capitalized Greek and Latin letters has been introduced along with group theory (Chap.2d) and corresponding lattice symmetry considerations [8].

In the $\kappa_i(\omega)$-diagram the theoretical maxima occur at lower energies than the experimental ones. One might be tempted to improve the agreement by introducing more parameters in the band structure calculation. However, the fact that the present calculation includes only three adjustable parameters seems to be worth noting in view of the difficulties involved in such calculations.

The reflection spectra of germanium and silicon are very much alike, and so are the gross features of the band structure. Differences exist in the relative position of the conduction subbands: In Si Γ'_2 is practically identical with Γ_{15}, and therefore Γ'_2 is located below Λ_1 rather than above. The energy differences $L'_3 - L_1$ and $\Lambda_3 - \Lambda_1$ are larger than they are in Ge with the consequence that the lowest conduction band minimum is not at L_1 but at Δ_1. The reflectivity peak at 3.4 eV in Si is therefore related to a different transition than the 2.1 eV peak of the Ge curve; to what transition has not yet become clear: Possibly $\Gamma'_{25} \to \Gamma_{15}$ which is the direct transition from the top of the valence band to the conduction band at $\vec{k} = 0$, and also $\Gamma'_{25} \to \Gamma'_2$ and $L'_3 \to L_1$. An optical investigation of germanium-silicon alloys showed that the 3.4 eV peak becomes weaker with decreasing silicon content at about 75 atomic % Si [9]. Apparently for larger percentages of germanium a different transition determines the reflectivity at this energy.

We will choose InP as an example of a compound semiconductor and consider in Fig.11.4 the reflection coefficient r_∞ and in Fig.11.5 n and k [10]. Its band structure is similar to that of Ge. The correlation of the peaks to transitions in the band structure is again similar to that in Ge. At low temperatures a splitting of the $\Lambda_3 \to \Lambda_1$ peak of magnitude 0.14 eV has been observed which is considered due to the spin-orbit splitting of the valence band. For this peak a splitting of

[8] See e.g. J. Callaway: Energy Band Theory. New York: Acad. Press. 1964; for a short introduction see O. Madelung: Grundlagen der Halbleiterphysik. Berlin-Heidelberg-New York: Springer. 1970.

[9] J. Tauc and A. Abraham, J. Phys. Chem. Solids 20 (1961) 190.

[10] M. Cardona, J. Appl. Phys. 32 (1961) 2151 and 958; Zeitschr. f. Physik 161 (1961) 99.

0.28 eV has been found in Ge but has not been taken into account in the band structure shown in Fig.2.26. With increasing atomic weight the splitting energy increases and becomes 0.75 eV in AlSb and 0.8 eV in GaSb and InSb. For further details see ref. 11.

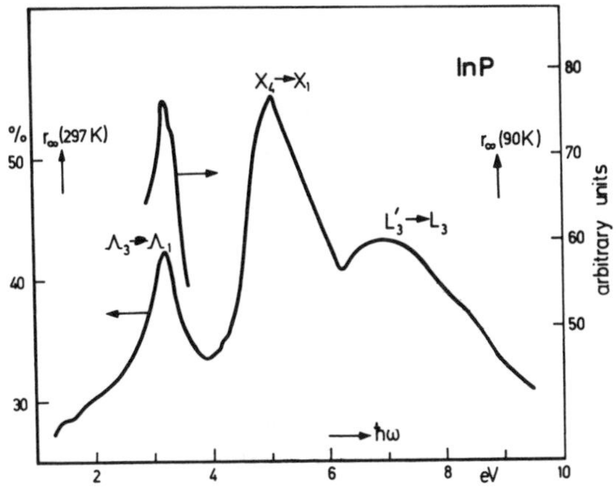

Fig.11.4 Room temperature reflectivity of indium phosphide (after Cardona, ref. 10).

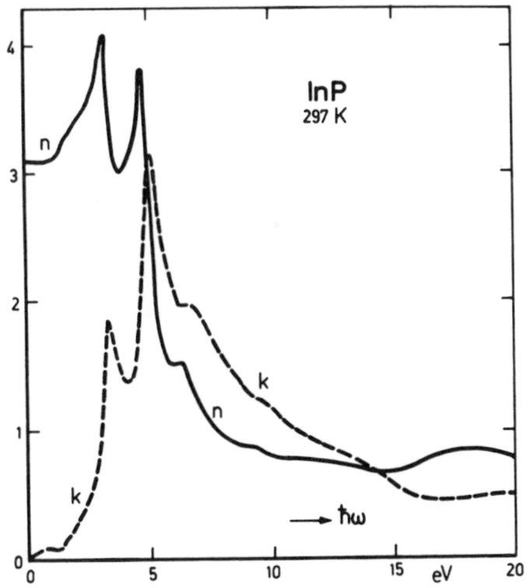

Fig.11.5 Refractive index and extinction coefficient of indium phosphide, obtained from the data in Fig.11.4 by a Kramers-Kronig analysis.

---

[11] M. Cardona: Semiconductors and Semimetals (R. K. Willardson and A. C. Beer, eds.), Vol. 3 , p. 125. New York: Acad. Press. 1967.

11b. Absorption Edge: Dependence on Temperature, Pressure, Alloy Composition, and Degeneracy

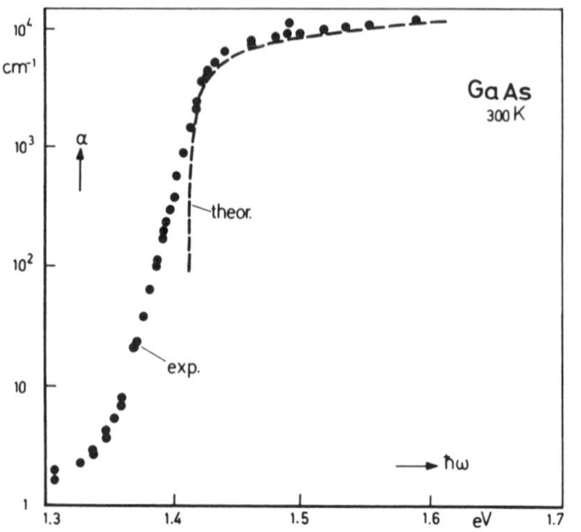

Fig. 11.6 Optical absorption edge of gallium arsenide at room temperature (after T. S. Moss and T. D. Hawkins, Infrared Physics 1 (1962) 111).

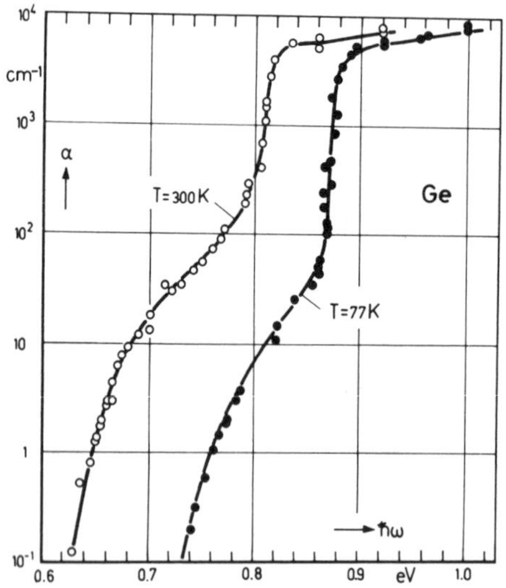

Fig. 11.7 Optical absorption edge of germanium at 77 and 300 K. The inflection at an absorption coefficient of 10^2 indicates the change from indirect to direct absorption processes (after R. Newman and W. W. Tyler: Solid State Physics (F. Seitz and D. Turnbull, eds.), Vol. 8, p. 49. New York: Acad. Press. 1959).

Let us now consider the edge of the fundamental absorption which according to Eq.(11a.1) is directly related to the energy gap. Figs. 11.6 and 11.7 show the absorption coefficient of GaAs and Ge, respectively, for various temperatures. In both cases the energy gap decreases with increasing temperature. Fig. 11.8 shows the temperature dependence of the gap for germanium. It is linear with a coefficient of -0.43 meV/K above 150 K. For semiconductors with energy gaps which are different by an order of magnitude this coefficient varies only within a factor of two: e.g. GaP ($\epsilon_G = 2.24$ eV at 300 K; -0.54 meV/K) and InSb ($\epsilon_G = 0.167$ eV at 300 K; -0.28 meV/K). Although in PbS, PbSe, and PbTe the gap increases with temperature, the normal behavior is a decrease.

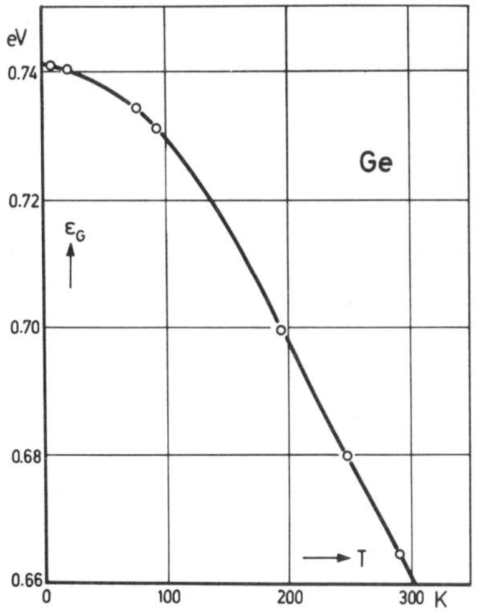

Fig. 11.8 Temperature dependence of the band gap in germanium (after G. G. Macfarlane, T. P. McLean, J. E. Quarrington, and V. Roberts, Phys. Rev. 108(1957) 1377).

A comparison of Figs. 11.6 and 11.7 reveals a small difference in the energy dependence of the absorption coefficients of GaAs and Ge. In Ge we observe a shoulder which is absent in GaAs. This shoulder is due to the fact that in Ge the lowest minimum of the conduction band is at the edge of the first Brillouin zone, while in GaAs it is at $\vec{k} = 0$. Since the valence band maximum in both cases is at $\vec{k} = 0$, a transition obeying Eq. (11a.1) in Ge involves a change in momentum while in GaAs it does not. This situation is shown schematically in Fig. 11.9. The transitions are called "indirect" and "direct", respectively. Both occur in Ge, but the indirect transition although it requires a smaller energy than the direct transition has a smaller probability and therefore it appears as a shoulder in Fig. 11.7. The conduction band of GaAs is indicated in Fig. 11.9 by a broken curve [1] (normalized to the Ge curve at $\vec{k} = 0$). In this case the direct transition requires the lowest possible energy, ϵ_G. The indirect transition indicated by a dash-dotted arrow may also occur but cannot be detected in the absorption spectrum because of its larger energy and smaller probability. Because of these transitions GaAs is called a "direct semiconductor" and Ge an "indirect semiconductor".

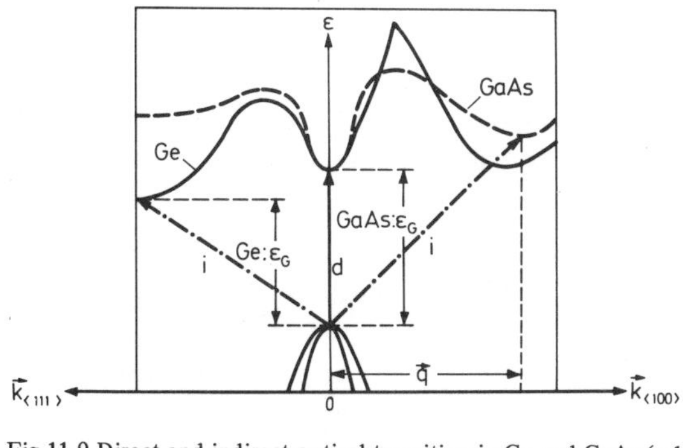

Fig. 11.9 Direct and indirect optical transition in Ge and GaAs (schematic); i indicates the indirect transition. The energy scales are normalized for equal direct gap in both semiconductors.

[1] For the complete band structure of GaAs see e.g. D. L. Greenaway and G. Harbeke: Optical Properties and Band Structure of Semiconductors, p. 67. New York: Pergamon. 1968.

Since the photon momentum is negligibly small*) momentum conservation in an indirect transition requires the co-operation of another particle. In the course of the transition this particle changes its momentum by an amount $\hbar\vec{q}$. The particle may be an intervalley phonon $\hbar\omega_i$ which is either generated or absorbed in the transition. Energy conservation requires in the case of absorption

$$\hbar\omega_e = \epsilon_G - \hbar\omega_i \qquad (11b.1)$$

instead of Eq.(11a.1). However, in general $\epsilon_G >> \hbar\omega_i$. Such a particle may also be an impurity atom or a dislocation. In these cases scattering processes can be considered elastic to a good approximation. The value of $\vec{q}$ has to be smaller than the reciprocal of the Debye length. In most cases of interest, however, elastic processes are less frequent than phonon interaction.

In Chap.11i a quantum mechanical treatment of optical transitions will be presented. The absorption of light is proportional to the transition probability which is essentially given by a matrix element times a delta-function integrated over all possible final states in the conduction band. To a first approximation the matrix element may be considered independent of $\vec{k}$ and $\vec{k}'$. Since $d^3k \propto \sqrt{\epsilon}\, d\epsilon$ and $d^3k' \propto \sqrt{-\epsilon_G - \epsilon'}\, d\epsilon'$ where spherical bands with zero energy at the conduction band edge have been assumed, the integration yields

$$\int \sqrt{\epsilon(-\epsilon_G - \epsilon')}\ \delta(\epsilon - \epsilon' - \hbar\omega \pm \hbar\omega_i)\, d\epsilon d\epsilon' =$$

$$\int_0^{\hbar\omega \mp \hbar\omega_i - \epsilon_G} \sqrt{\epsilon(-\epsilon + \hbar\omega \mp \hbar\omega_i - \epsilon_G)}\, d\epsilon = \frac{\pi^2}{8}(\hbar\omega \mp \hbar\omega_i - \epsilon_G)^2 \qquad (11b.2)$$

assuming $\hbar\omega \mp \hbar\omega_i > \epsilon_G$ (otherwise the integral = 0).

As in any phonon assisted transition process the matrix element includes the phonon distribution function, N_q. Hence, the absorption coefficient a is proportional to

$$a \propto N_q(\hbar\omega + \hbar\omega_i - \epsilon_G)^2 + (N_q + 1)(\hbar\omega - \hbar\omega_i - \epsilon_G)^2 \qquad (11b.3)$$

where $N_q = \{\exp(\hbar\omega_i/k_BT) - 1\}^{-1}$. While a direct transition is weakly temperature dependent due to $\epsilon_G = \epsilon_G(T)$ an indirect transition shows in addition an exponential dependence because of $N_q(T)$.

At low temperatures and for values of the absorption coefficient of the order of magnitude 10 cm^{-1}, structure is observed in the absorption spectrum which is correlated to longitudinal and transverse optical and acoustic intervalley phonons. For the case of GaP, where the lowest minimum of the conduction band is in <100> direction so that the band structure resembles that of silicon, absorption spectra are shown in Fig.11.10. The thresholds are sharpened by exciton effects to be discussed in Chap.11c. Except for this structure, the square root of the absorption coefficient is essentially a linear function of $\hbar\omega$ since phonon absorption can be neglected at low temperatures.

*) About 1/2000 of π/a for $\hbar\omega$ = 1eV and a lattice constant $a = 3\times10^{-8}$ cm.

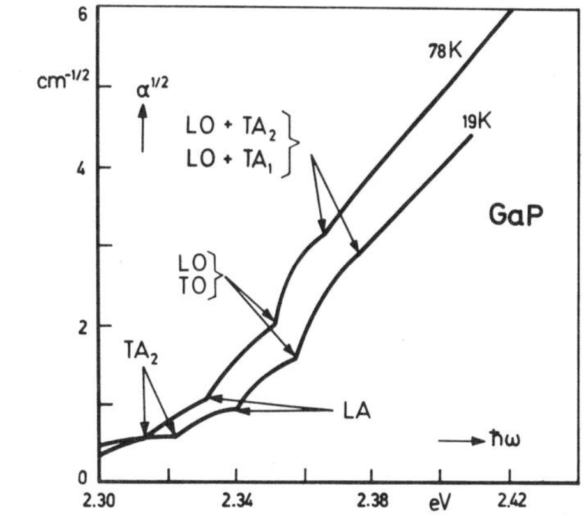

Fig.11.10
Optical absorption edge of gallium phosphide, showing threshold for the formation of free excitons with the emission of phonons (after M. Gershenzon, D. G. Thomas and R. E. Dietz, Proc. Int. Conf. Phys. Semicond., Exeter 1962, p. 752. London: The Institute of Physics and the Physical Society. 1962).

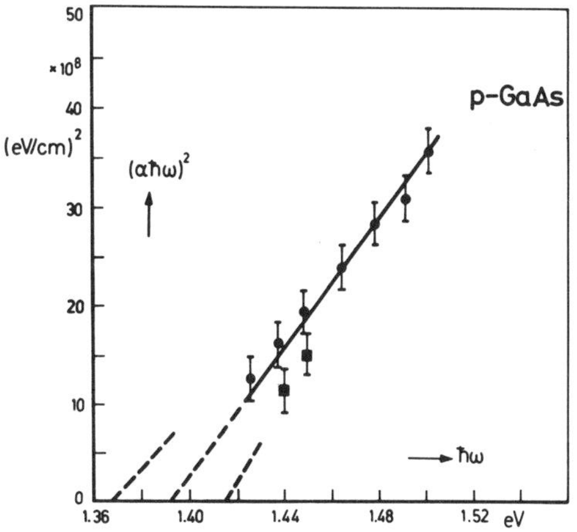

Fig.11.11 Optical absorption edge of gallium arsenide plotted for a determination of the threshold energy. Square points are valid for a second sample (after J. Kudman and T. Seidel, J. Appl. Phys. 33 (1962) 771).

The method usually applied for a determination of the absorption edge is illustrated in Fig.11.11. The square of the product of the absorption coefficient and the quantum energy is plotted vs the quantum energy for GaAs. A straight line can be drawn through the experimental data. Its intersection with the abscissa yields an energy gap of 1.29 eV with an accuracy of ± 0.02 eV. Even better accuracy is obtained by magneto-optical methods, however (see Chap.11d).

It will be shown in a quantum mechanical treatment (Chap.11i) that the absorption coefficient depends on frequency essentially according to a $\sqrt{\hbar\omega - \epsilon_G}$ law. The reason for plotting $(a\hbar\omega)^2$ as the ordinate rather than a^2 can easily be understood : Let R be the rate per unit volume for transitions of electrons from state $\vec{k}$ to state $\vec{k}'$, integrated over all final states, by absorption of a photon travelling with a velocity c/n in the medium of refractive index n. Assuming N photons in the volume V and the photon flux Nc/(nV), R is then given by the product of the photon flux and the absorption coefficient a :

$$R = N \frac{c}{nV} a \tag{11b.4}$$

The energy flux $\overline{S}$ is given by the product of the photon flux and photon energy $\hbar\omega$:

$$\overline{S} = N \frac{c}{nV} \hbar\omega \tag{11b.5}$$

On the other hand $\overline{S}$ is the time average of the Poynting vector $\vec{S}$ given by

$$\vec{S} = [\vec{E}\,\vec{H}] \tag{11b.6}$$

The vector potential $\vec{A}$ in a plane wave of wave vector $\vec{q}$ is given by

$$\vec{A} = A_0 \vec{a} \cos(\vec{q}\vec{r} - \omega t) = \frac{A_0}{2} \vec{a} \,\{\exp(i\omega t - i\vec{q}\vec{r}) + \exp(-i\omega t + i\vec{q}\vec{r})\} \tag{11b.7}$$

where A_0 is the amplitude and $\vec{a}$ a unit vector in the direction of $\vec{A}$. For non-magnetic media $\vec{E}$ and $\vec{H}$ are given in terms of $\vec{A}$ by

$$\vec{E} = -\partial\vec{A}/\partial t = \omega A_0 \vec{a} \sin(\vec{q}\vec{r} - \omega t) \tag{11b.8}$$

$$\vec{H} = [\vec{\nabla}_r \vec{A}]/\mu_0 = \{A_0 [\vec{a}\vec{q}]/\mu_0\} \sin(\vec{q}\vec{r} - \omega t) \tag{11b.9}$$

neglecting the scalar potential to be discussed below. The polarization vector $\vec{a}$ is directed in $\vec{E}$-direction. Eq.(11b.6) yields for the time-average of $|\vec{S}|$:

$$\overline{S} = |\vec{a}\,(\vec{a}\vec{q}) - \vec{q}|\, \omega A_0^2/2\mu_0 \tag{11b.10}$$

Assuming for simplicity a transverse electromagnetic wave having a propagation vector of magnitude $|\vec{q}| = \omega n/c$, we obtain for $\overline{S}$:

$$\overline{S} = \omega^2 n A_0^2/2\mu_0 c \tag{11b.11}$$

By eliminating N from Eqs.(11b.4) and (11b.5) we obtain for the product of the absorption coefficient and the quantum energy:

$$\alpha\hbar\omega = \frac{377 \text{ ohm}}{n} 2\hbar^2 R/A_0^2 \tag{11b.12}$$

where $\mu_0 c$ has been replaced by its numerical value of 377 ohm.

In Eq.(11b.8) we have neglected the scalar potential of the electromagnetic wave. It is well known from the theory of electromagnetic radiation that it can always be made to vanish by means of a gauge transformation*). The Hamiltonian of an electron of momentum $\hbar\vec{k}$ is then simply given by

$$H(\vec{k}, \vec{r}, t) = (\hbar\vec{k} - e\vec{A})^2/2m \tag{11b.13}$$

With this Hamiltonian it will be shown in Chap.11i that the quantum mechanical transition probability R is proportional to A_0^2 which justifies plotting the product $\alpha\hbar\omega$, as given by Eq.(11b.12), as a function of the frequency of absorbed light.

We have noticed the shift of the absorption edge with temperature. A shift may also be induced by hydrostatic pressure. Fig.11.12 shows this phenomenon for the case of GaAs. After an initial rise with a slope of 9.4 μeV/atm we find a maximum at 60 000 atm and a subsequent decrease with a slope of about − 8.7 μeV/atm. This behavior is explained by the assumption that at low pressures the

---

*) $\vec{A}' = \vec{A} - \vec{\nabla}_r f$; $V' = V + \partial f/\partial t$ where f(r, t) is an arbitrary function which can be chosen to yield V' = 0. The field intensities $\vec{E}$ and $\vec{H}$ are indifferent with respect to f; see e.g. J. A. Stratton: Electromagnetic Theory, p. 23. New York: McGraw-Hill. 1941.

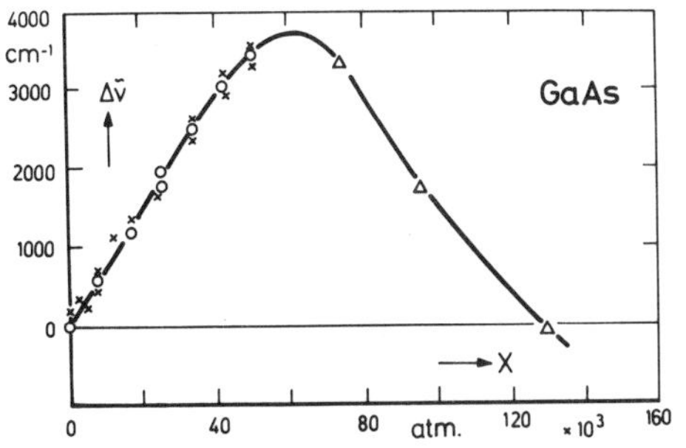

Fig. 11.12 Dependence of the absorption edge of gallium arsenide on hydrostatic pressure (after D. F. Edwards, T. E. Slykhouse, and H. G. Drickamer, J. Phys. Chem. Solids 11 (1959) 140).

conduction band edge is at $\vec{k} = 0$ (notation: Γ). It rises relative to the valence band maximum, also at $\vec{k} = 0$, with increasing pressure. However, there are satellite valleys the lowest of which are situated on $k_{\langle 100 \rangle}$ and equivalent axes (notation: X). These valleys fall with increasing pressure. At 60 000 atm the satellite valleys pass the central valley and then form the conduction band edge.

In germanium pressure coefficients of magnitude 5 μeV/atm for the <111> minimum, which at low pressure forms the conduction band edge, −2 μeV/atm for the <100> minimum, and 12 μeV/atm for the <000> minimum have been observed. In all semiconductors investigated so far the pressure coefficient of the <100> minimum has been found negative while those of the Γ- and L-valleys are always positive, the latter with values close to 5 μeV/atm in all semiconductors [2].

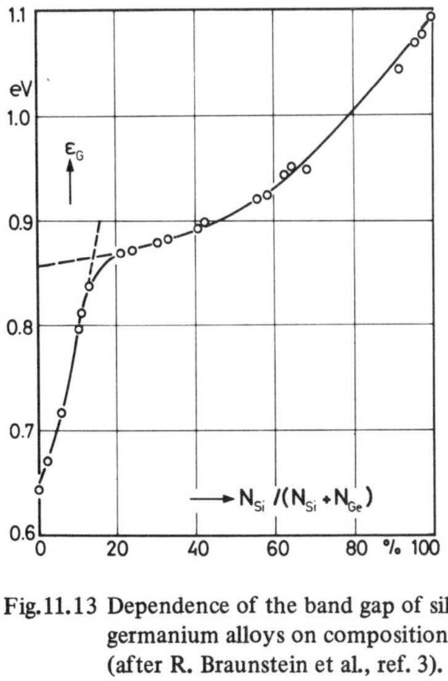

Fig. 11.13 Dependence of the band gap of silicon germanium alloys on composition (after R. Braunstein et al., ref. 3).

The absorption edge of alloys as a function of composition has some relation to the pressure dependence of its constituents. Fig. 11.13 shows the energy gap obtained from optical absorption in germanium silicon alloys [3]. In the germanium-rich alloys the conduction band edge is at L and rises with increasing silicon content. In the silicon-rich alloys the band edge rises at a slower pace and is formed by the <100> minimum. Apparently there is a cross-over of L- and X-minima at a silicon content of about 15 atomic %.

Finally we discuss the influence of carrier concentration n on the absorption edge. Such an influence has been observed in n-type semiconductors with low effective mass.

---

[2] For a compilation of data see e.g. E. J. Johnson: Semiconductors and Semimetals (R. K. Willardson and A. C. Beer, eds.), Vol. 3 , p. 200, Table II. New York: Acad. Press. 1967.

[3] R. Braunstein, A. R. Moore, and F. Herman, Phys. Rev. 109 (1958) 695.

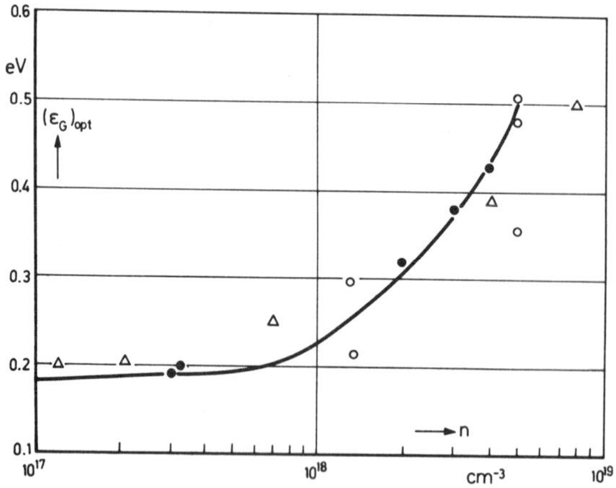

Fig. 11.14 "Burstein shift" of the apparent optical band gap of indium antimonide with increasing electron concentration n (after H. J. Hrostowski et al.: triangles; R. G. Breckenridge et al.: open circles; W. Kaiser and H. Y. Fan: solid points and calculated curve; ref. 4).

Fig. 11.14 shows observations in n-InSb [4]. For n larger than about $10^{18}/cm^3$ where the electron gas becomes degenerate, the optical absorption edge $(\epsilon_G)_{opt}$ rises with increasing values of n. The formula

$$(\epsilon_G)_{opt} - \epsilon_G = (1 + m_n/m_p)\,(\zeta_n - 4kT_e) \qquad (11b.14)$$

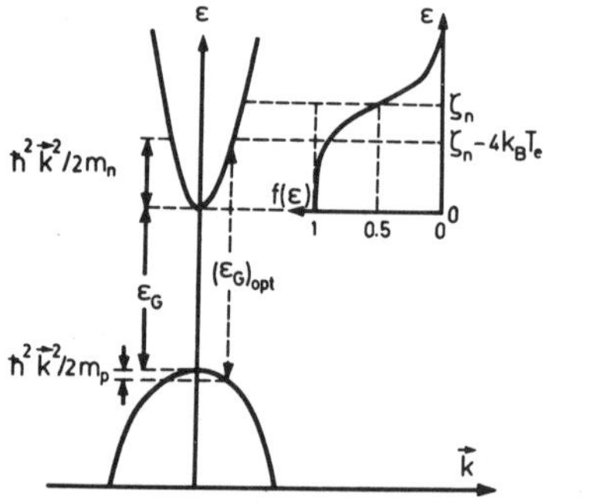

Fig. 11.15 Burstein shift in the energy band model of a direct semiconductor. In the diagram on the right-hand side the Fermi-Dirac distribution is illustrated.

correlates the Burstein shift with the Fermi energy ζ_n (for carriers in thermal equilibrium with the lattice the electron temperature T_e is equal to the lattice temperature) [5]. Let us assume for simplicity that all energy states in the conduction band up to an energy of

$$\hbar^2\vec{k}^2/2m_n = \zeta_n - 4kT_e \qquad (11b.15)$$

are occupied (since $1 - e^{-4} \approx 99\,\%$). Because of momentum conservation the minimum photon energy is not just $\epsilon_G + \hbar^2\vec{k}^2/2m_n$, but one has also to take into account the energy $\hbar^2\vec{k}^2/2m_p$ of the state in the valence band with the same $\vec{k}$-vector as illustrated by Fig. 11.15:

---

[4] W. Kaiser and H. Y. Fan, Phys. Rev. 98 (1955) 966; H. J. Hrostowski, G. H. Wheatley, and W. F. Flood, Phys. Rev. 95 (1954) 1683; R. G. Breckenridge, R. F. Blunt, W. R. Hosler, H. P. R. Frederikse, J. H. Becker, and W. Oshinsky, Phys. Rev. 96 (1954) 571.

[5] E. Burstein, Phys. Rev. 93 (1954) 632.

$$\hbar\omega = \epsilon_G + \frac{\hbar^2 \vec{k}^2}{2}\left(\frac{1}{m_n} + \frac{1}{m_p}\right) = \epsilon_G + \frac{\hbar^2 \vec{k}^2}{2m_n}\left(1 + \frac{m_n}{m_p}\right) \qquad (11b.16)$$

The electron energy $\hbar^2 \vec{k}^2/2m_n$ is given by Eq.(11b.15). Thus for the absorption edge $\hbar\omega_e = (\epsilon_G)_{opt}$ the value given by Eq.(11b.14) is found. For nonparabolic bands this relation has to be modified.

The Fermi energy depends on the electron temperature T_e and on the carrier concentration n. Shur [6] discovered that this dependence serves for a determination of the T_e-vs-$\vec{E}$-relation for hot carriers in degenerate semiconductors. Since no phonons are involved in a direct transition there is no temperature dependence other than through T_e. Therefore the result of one experiment at constant lattice temperature with increasing $\vec{E}$ can be compared with that of another experiment at increasing lattice temperature and $\vec{E} = 0$. In the second experiment $T_e = T$, and a T_e-vs-$\vec{E}$-relation is thus obtained empirically.

These experiments have been made e.g. with n-GaSb at 77 K [7]. Let us denote by $S(\hbar\omega)$ the product of the spectral sensitivity of the detector and the spectral distribution of the light intensity and by $T(\hbar\omega)$ the transmittivity of the degenerate sample. Polychromatic light causes a detector signal

$$U = \int_0^\infty T(\epsilon) S(\epsilon) d\epsilon \qquad (11b.17)$$

where $\epsilon = \hbar\omega$. If by the application of an electric field $\vec{E}$ the electron temperature is raised and the absorption edge shifted by $\Delta\epsilon$, the transmittivity is changed by $\Delta T = (dT/d\epsilon)\Delta\epsilon$ and the signal by

$$\Delta U \approx \Delta\epsilon \int_0^\infty \frac{dT}{d\epsilon} S(\epsilon) d\epsilon \propto \Delta\epsilon \qquad (11b.18)$$

provided that $|\Delta\epsilon| \ll 2S/(dS/d\epsilon)$. The maximum shift possible is $\zeta_n - 4kT_e$ where the degeneracy would be totally lifted. The change in the signal thus saturates at a value ΔU_s. In Fig.11.16b the quantity $(1 - \Delta U/\Delta U_s)$ has been plotted vs the

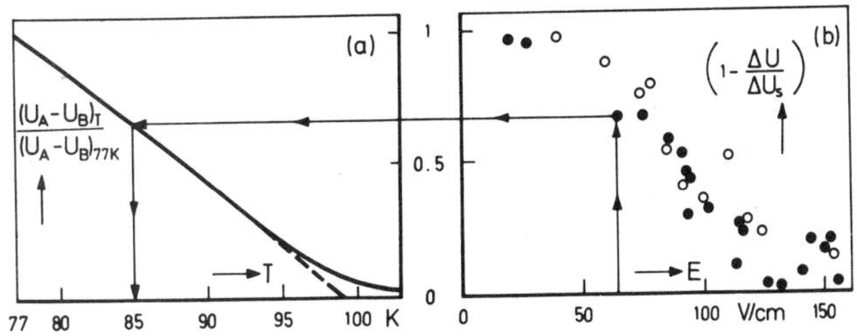

Fig.11.16a Temperature dependence of the difference in detector signal between sample A (n-type GaSb, n = $3.5\times10^{17}/cm^3$) and sample B (n = $0.68\times10^{17}/cm^3$) normalized to 77 K (full curve). The dashed curve represents calculated values of $\epsilon_B(T)/\epsilon_B(77\,K)$ where ϵ_B is the Burstein shift $(\epsilon_G)_{opt} - \epsilon_G$ given by Eq.(11b.14).
b Values for $1-\Delta U/\Delta U_s$ as a function of the electric field strength where the signal ΔU is given by Eq.(11b.18), for two samples (full and open circles). ΔU_s is the saturation value of ΔU for high field strengths (after Heinrich and Jantsch, ref. 7).

[6] M. S. Shur, Phys. Letters 29A (1969) 490.
[7] H. Heinrich and W. Jantsch, Phys. Rev. B4 (1971) 2504.

electric field strength. The saturation occurs at 150 V/cm for a sample with $n = 3.5\times10^{17}/cm^3$. For a more heavily doped sample the saturation field strength is higher. (Of course, then also a larger field intensity would be required to heat electrons to the same temperature because of stronger ionized impurity scattering.) In a second experiment the difference in signals $U_A - U_B$, from the degenerate sample A and a non-degenerate sample B ($n = 0.7\times10^{17}/cm^3$) has been measured as a function of temperature with no field applied. This difference, normalized to unity at the temperature of the first experiment (77 K), is plotted vs temperature in Fig.11.16a. Since according to Eq. (11b.18) a change in signal is proportional to a change in the Burstein shift which is also proportional to $U_A - U_B$, a relation between $\vec{E}$ and $T = T_e$ as indicated by the sequence of arrows can be established. The relation obtained is shown in Fig.11.17 by the data points. The curve has been calculated from the energy balance Eq. (4n.10) with an ohmic mobility $\mu = 1.77\times10^4\ cm^2/Vsec$ and an energy relaxation time $\tau_\epsilon = 1.6\times10^{-11}$ sec as determined from microwave harmonic mixing (Chap.11o). The agreement between the observed and the calculated data is satisfactory.

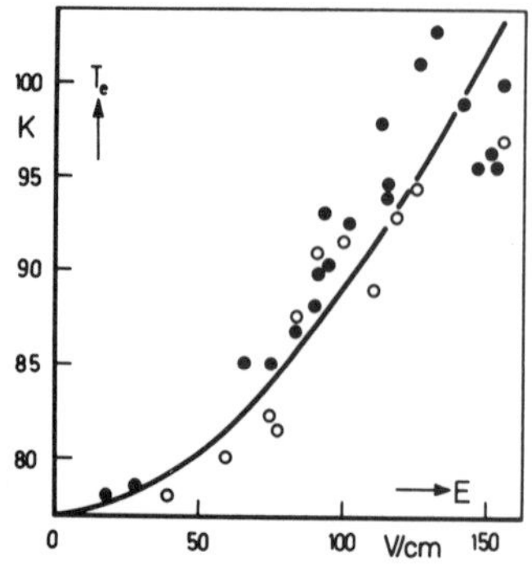

Fig.11.17 Electron temperature as a function of the electric field strength obtained in n-type GaSb from the Burstein shift ($n=3.5\times10^{17}/cm^3$).

11c. Exciton Absorption

So far we have considered the electron and the hole obtained as a pair by the absorption of a photon as being completely independent of each other. Actually, this is not always true. From atomic absorption spectra it is well known that besides the ionization continuum there are also discrete absorption lines due to excitations of the atoms. In a semiconductor such excitations can be described in a simplified manner by an electron and a hole bound to each other by Coulomb interaction. According to the hydrogen model the binding energy is given by

$$\Delta\epsilon_{exc} = -\frac{m_r e^4}{2\hbar^2(4\pi\kappa\kappa_0)^2}\cdot\frac{1}{n^2} = -\frac{\Delta\epsilon_{exc}^{(1)}}{n^2}\ ;\quad n = 1, 2, 3, \cdots \tag{11c.1}$$

where m_r is the reduced effective mass given by

$$m_r = 1/(m_n^{-1} + m_p^{-1}) \tag{11c.2}$$

and κ is the static dielectric constant. These excitations have been named "excitons". The ground-state exciton energy, $\Delta\epsilon_{exc}^{(1)}$, is $(m_r/m_0)\kappa^{-2}$ of the Rydberg energy, 13.6 eV; for values of e.g. $m_r/m_0 = 0.05$ and $\kappa = 13$ we find 4 meV. Hence, a few meV below the absorption edge a series of discrete absorption lines

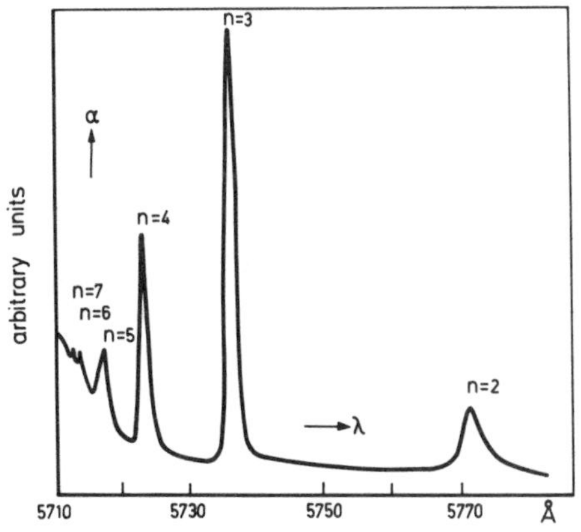

Fig.11.18 Exciton absorption spectrum of Cu_2O at 4 K (after Nikitine et al., ref. 1).

should be observed. Fig.11.18 shows such a spectrum found in Cu_2O [1]. Since the (even parity) exciton is "dipole forbidden" (Chap. 11i) the line spectrum begins with a quantum number n = 2. In allowed transitions observed e.g. in GaAs (Fig.11.19) usually only the n = 1 peak is observed while the rest of the discrete absorption lines is merged with the absorption edge [2]. The absorption spectrum above the edge is the ionization continuum which is different from the spectrum calculated without taking excitons into account, by a

factor of $$\frac{2\pi\sqrt{\Delta\epsilon_{exc}^{(1)}/(\hbar\omega-\epsilon_G)}}{1-\exp(2\pi\sqrt{\Delta\epsilon_{exc}^{(1)}/(\hbar\omega-\epsilon_G})}$$

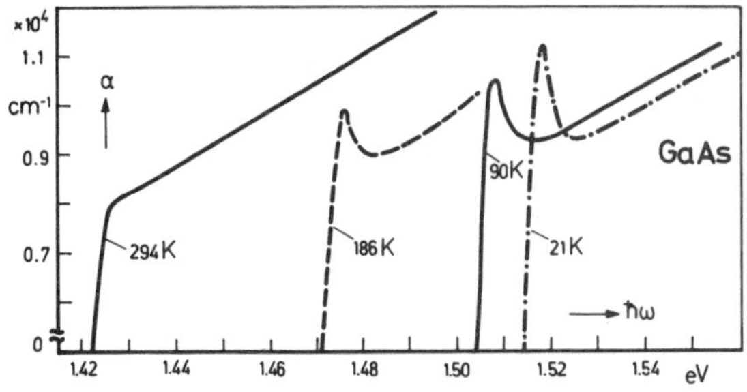

Fig.11.19 Observed exciton absorption spectra in GaAs at various temperatures between 21 and 294 K (after M. D. Sturge, ref. 2).

Fig.11.20 shows theoretical absorption spectra both including and neglecting the exciton effects. For $\Delta\epsilon_{exc}$ in GaAs, Fig.11.19 thus yields a value of 3.4 meV at 21 K. At room temperature the exciton peak is completely wiped out since the binding energy is readily supplied by phonons. Also in semiconductors with large carrier concentrations ($n > 2x10^{16}/cm^3$ in GaAs), in semimetals, and in metals no excitons exist because free carriers tend to shield the electron hole interaction.

[1] S. Nikitine, J. Bielmann, J. L. Deiss, M. Grossmann, J. B. Grun, J. Ringeissen, G. Schwab, M. Siesskind, and L. Wursteisen, Proc. Int. Conf. Physics of Semiconductors Exeter 1962. London: The Institute of Physics and the Physical Society. 1962.

[2] M. D. Sturge, Phys. Rev. 127 (1962) 768.

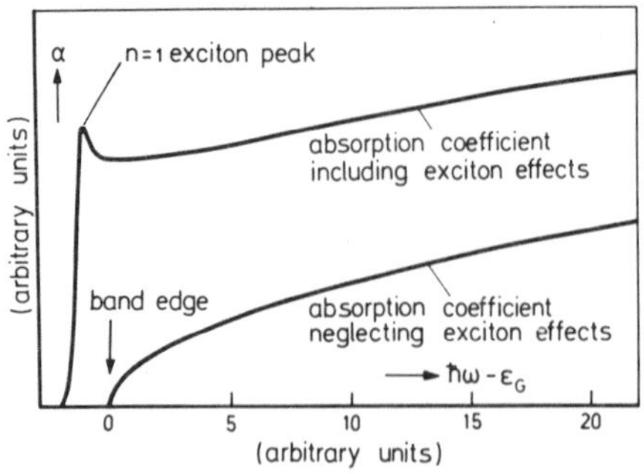

Fig.11.20 Calculated optical absorption coefficient for direct transitions in the simple band model neglecting and including the n = 1 exciton peak below the band edge (after M. D. Sturge, ref. 2).

Also neutral impurities cause a broadening of the exciton lines and, at large concentrations, cause their disappearance.

Since GaAs is a "direct semiconductor" (Chap.11b) we call the exciton a "direct exciton". In GaP which is an indirect semiconductor the absorption edge, shown in Fig. 11.10, is determined by "indirect excitons". The phonon involved in the indirect transition can either be absorbed or emitted. The absorption due to excitons should occur in steps; there should be twice as many steps as there are phonon branches. Actually shoulders are observed as shown by Fig.11.10. Similar spectra have been observed in Ge and Si.

At very low temperatures (e.g. 1.6 K in GaP) discrete exciton lines have been observed in indirect semiconductors such as GaP. These lines are interpreted by the assumption of excitons bound to neutral donors [3] ("bound excitons"). For sulphur in GaP the exciton binding energy of 14 meV is about 10% of the donor ionization energy.

From the hydrogen model discussed above we calculate the Bohr radius of the exciton:

$$a_{exc} = \frac{\kappa}{m_r/m_0} a_B \tag{11c.3}$$

where $a_B = \hbar^2/m_0 e^2 \approx 0.53$ Å is the Bohr radius of the hydrogen atom. E.g. for $m_r/m_0 = 0.05$ and $\kappa = 13$ we find $a_{exc} = 130$ Å which is about the same as the radius for a valence electron orbit in a shallow donor or acceptor. Carrier shielding of excitons occurs if the carrier concentration is larger than the concentration $\{2/(2\pi)^3\}4\pi|\vec{k}|^3/3$ corresponding to a Fermi wave vector $|\vec{k}| = 1/a_{exc}$.

Excitons with the rather large spatial extent given by Eq.(11c.3) are adequately described by the "Wannier model" [4]. Frenkel [5] considered excitons in a solid consisting of weakly interacting atoms such as rare-gas atoms. In this case the excitation is essentially that of a single atom or molecule; the interaction of nearest neighbors can be treated as a small perturbation. The radius of the "Frenkel exciton" is therefore at most a few Å which is of the order of the lattice constant. Such excitons have been discussed in some alkali halides and organic phosphors.

[3] M. Gershenzon, D. G. Thomas, and R. E. Dietz: Proc. Int. Conf. Phys. Semic., Exeter 1962, p. 752. London: The Institute of Physics and the Physical Society. 1962; J. J. Hopfield: Proc. Int. Conf. Phys. Semic., Paris 1964, p. 726. Paris: Dunod. 1964.

[4] G. H. Wannier, Phys. Rev. 52 (1937) 191.

[5] J. Frenkel, Phys. Rev. 37 (1931) 17; 1276.

If a magnetic field is applied perpendicular to the motion of an exciton, the Lorentz force tends to separate the negative electron from the positive hole. An electric field applied along the dipole axis of the exciton would have the same effect on the exciton, and in both cases the optical absorption shows a "Stark effect". It has been observed and measured in CdS by Thomas and Hopfield [6]. The experiment demonstrates that the exciton is created with a nonzero velocity. More detailed information on excitons may be obtained e.g. from the book by Dexter and Knox [7] and the article by Knox [8]

11d. Interband Transitions in a Magnetic Field

In Chap.9 it has been shown that by a strong magnetic field Landau levels are formed in the conduction and the valence bands. For a parabolic band (effective mass m) their energies are given by

$$\epsilon_n = (n + \frac{1}{2})\,\hbar\omega_c + \hbar^2 k_z^2/2m; \quad n = 0, 1, 2, \cdots \tag{11d.1}$$

where the magnetic field has been assumed in z-direction and ω_c is the cyclotron angular frequency

$$\omega_c = (e/m)\,B_z \tag{11d.2}$$

For n = 0 we find that the band edge moves with B_z such that the energy gap widens:

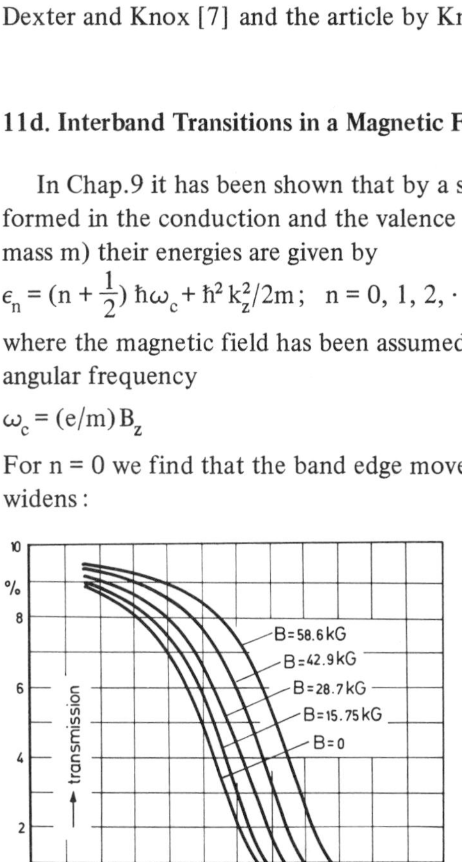

Fig. 11.21 Shift of the optical absorption edge of InSb with magnetic field at room temperature; due to the non-parabolicity of the conduction band the shift is nonlinear in B_z (after Burstein et al., ref. 1).

$$\epsilon_G(B) = \epsilon_G(0) + \frac{1}{2}\hbar(e/m_n + e/m_p)B_z = \epsilon_G(0) + \frac{e\hbar}{2m_r}B_z \tag{11d.3}$$

where m_r is the reduced mass given by Eq.(11c.2). A corresponding shift of the fundamental absorption edge has been observed. For parabolic bands this shift should be proportional to B_z. A nonlinear behavior may be due to band nonparabolicity.

Fig.11.21 shows the transmission of InSb at the fundamental absorption edge for various values of

[6] D. G. Thomas and J. J. Hopfield, Phys. Rev. 124 (1961) 657.
[7] D. L. Dexter and R. S. Knox: Excitons. New York: J. Wiley and Sons. 1965.
[8] R. S. Knox: Solid State Physics (F. Seitz and D. Turnbull, eds.), Suppl. 5. New York: Acad. Press. 1963.

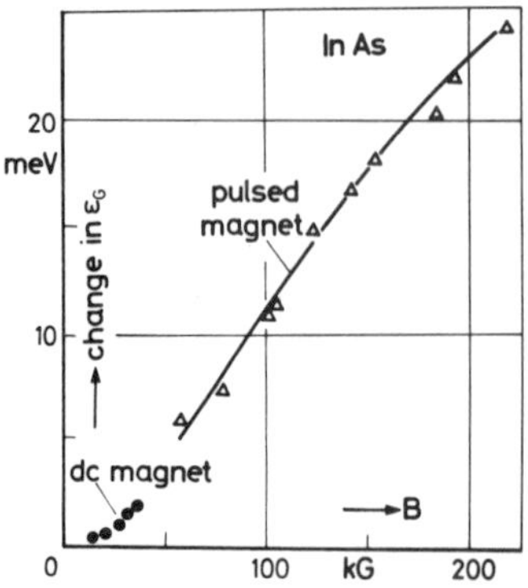

Fig. 11.22 Variation of band gap for indium arsenide with magnetic field (after Zwerdling et al., ref. 2).

the magnetic induction [1]. With increasing B_z the edge is shifted towards larger energies. In this case the shift is not proportional to B_z. This is explained by the fact that since $m_n \ll m_p$, the shift is mostly due to electrons in the conduction band which is non-parabolic. In Fig. 11.22 the magnetic shift of the conduction band edge in InAs is shown [2]. The nonlinear behavior at small magnetic fields cannot be explained in the simple Landau model. The nonlinear behavior at large magnetic fields is probably due to a nonparabolicity of the conduction band.

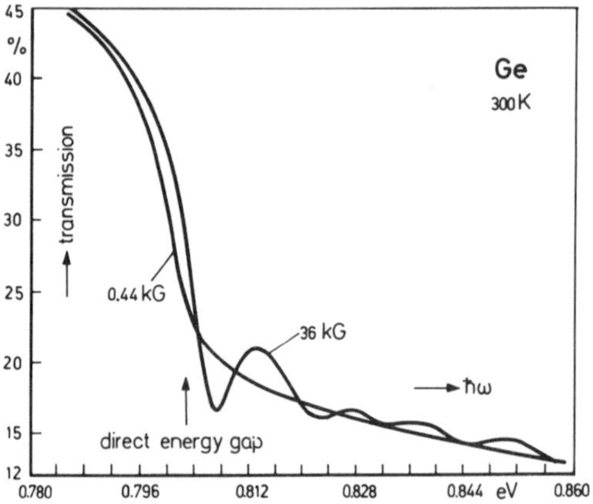

Fig. 11.23 Oscillatory transmission of germanium at room temperature in magnetic fields of 0.44 and 36 kG (after Zwerdling, ref. 3).

Besides the shift of the absorption edge there is an oscillatory behavior at somewhat higher photon energies. This is shown in Fig. 11.23 for the case of the direct transition at $\vec{k} = 0$ in Ge [3]. The minima of the transmission curve are plotted as a function of B_z in Fig. 11.24. For each minimum there is a linear relation between $\hbar\omega$ and B_z. From the slope of line # 1 the electron effective mass in the Γ-valley has been determined as (0.036 ± 0.002) eV at room temperature. The accuracy is better than that obtained by any other method.

At low temperatures where in Eq.(11d.1) the term $\hbar^2 k_0^2/2m$ is small compared with the first term, transitions between Landau levels of the valence and conduction bands have a fine structure. At large magnetic fields the Landau levels are split because of the two possible electron spins of quantum number $M = \pm 1/2$. The Zeeman splitting energy is given by

$$\Delta\epsilon_{\pm} = g M \mu_B B_z \tag{11d.4}$$

[1] E. Burstein, G. S. Picus, H. A. Gebbie, and F. Blatt, Phys. Rev. 103 (1956) 826.
[2] S. Zwerdling, R. J. Keyes, S. Foner, H. H. Kolm, and B. Lax, Phys. Rev. 104 (1956) 1805.
[3] S. Zwerdling, B. Lax, and L. M. Roth, Phys. Rev. 108 (1957) 1402.

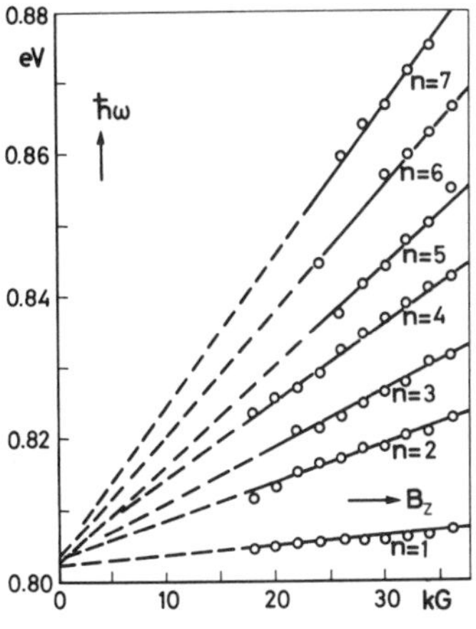

Fig. 11.24 Transmission minima of Fig. 11.23, as a function of the magnetic field (after S. Zwerdling and B. Lax, Phys. Rev. 106 (1957) 51).

where g is the Landé g-factor of the electrons and $\mu_B = e\hbar/2m_o = 5.8\ \mu eV/kG$ is the Bohr magneton. Fig. 11.25 shows the fine structure of the direct transition in Ge observed at 4.2 K, together with the theoretical line spectrum [4]. In the valence band both heavy and light holes have been considered. A rough correlation is found between the theoretical and the experimental intensities for all lines except the two lowest; these persist down to zero magnetic field and therefore have been attributed to the direct exciton transition. (An unintentional strain split the valence bands and caused two exciton transitions instead of one.) A value for the g-factor of $g = -2.5$ is found for Ge. For n-InSb a very large negative value of $g = -54$ is obtained in a similar way.

The influence of a magnetic field on the reflectivity at the absorption edge is illustrated by Fig. 11.26 [5]. From the simple model of a damped harmonic oscillator one can calculate the change in the index of refraction and the extinction coefficient and thus the change in reflectivity. This model serves for a qualitative understanding of the observed reflection spectra.

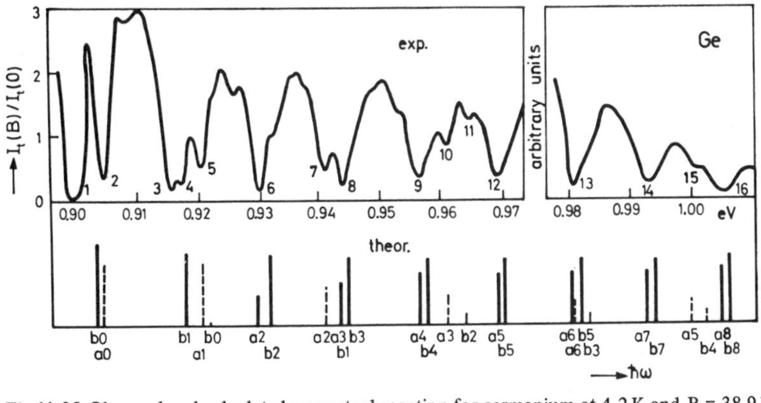

Fig. 11.25 Observed and calculated magnetoabsorption for germanium at 4.2 K and B = 38.9 kG parallel to the E-vector of the light wave (after Roth et al., ref. 4).

Besides direct also indirect transitions have been found in magnetic fields. E.g. Fig. 11.27 shows Landau transitions and an indirect exciton transition ob-

[4] L. M. Roth, B. Lax, and S. Zwerdling, Phys. Rev. 114 (1959) 90.
[5] G. B. Wright and B. Lax, J. Appl. Phys. Suppl. 32 (1961) 2113.

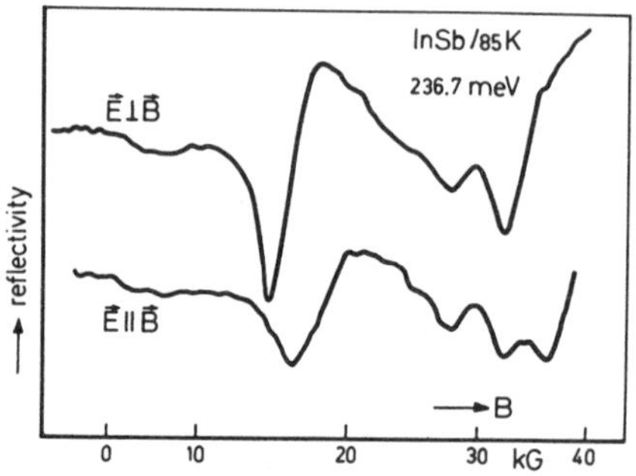

Fig.11.26 Interband magnetoreflections in InSb at 85 K for a photon energy of 236.7 meV (after Wright and Lax, ref. 5).

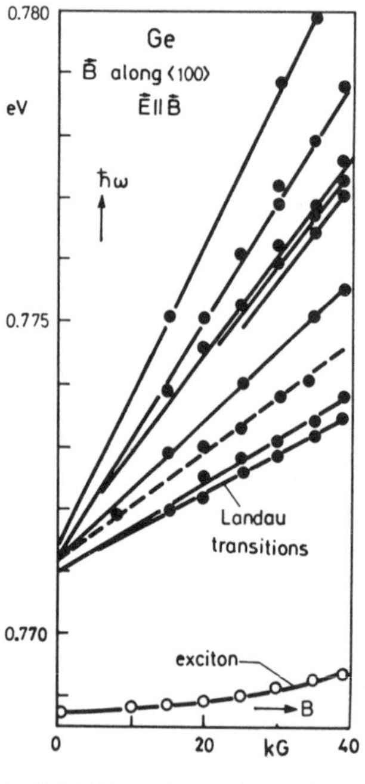

Fig.11.27 Transmission minima for indirect Landau transitions in Ge at 1.5 K as a function of the magnetic field (after Zwerdling et al., ref. 6).

served at 1.5 K in Ge [6]. After subtracting the energy of the emitted phonon from the results extrapolated to $B_z = 0$ an energy gap of $\epsilon_G = 0.744$ eV is obtained.

The exciton transition is linear with B_z only at high field intensities. At small field intensities a parabolic behavior is found. The indirect exciton in Ge shown in Fig.11.27 can be resolved as two lines, one with a binding energy of 2.1 meV, the other with 3.2 meV. The dependence on B_z is different for $B_z \parallel$ <100> and <111> directions; this can be explained by the different phonon spectra in various crystallographic directions and the fact that the g-factors for electrons and holes are actually tensors. From microwave spin resonance experiments the tensorial character of the g-factor has been shown to exist.

In a stronger magnetic field of 73.8 kG a further splitting of the indirect exciton lines has been found. This is due to the spin splitting of the Landau levels. Electronic g-factors of 1.8 ± 0.3 and 1.5 ± 0.3 have been determined from these experiments; they agree with calculated values. For such measurements the sample is only a few μm thick and must not be mounted on a substrate of different thermal expansion coefficient; otherwise on cooling the degeneracy of energy levels would be lifted by strain and an unintentional line splitting would be obtained in a magnetic field. Such precautions have been taken in the experiment discussed above.

The field dependence of the direct exciton in Ge at 77 K is shown in Fig.11.28 [7]. From the linear part of curve # 1 the reduced mass of the exciton has been determined as $0.33\ m_0$.

[6] S. Zwerdling, B. Lax, L. M. Roth, and K. J. Button, Phys. Rev. 114 (1957) 80.

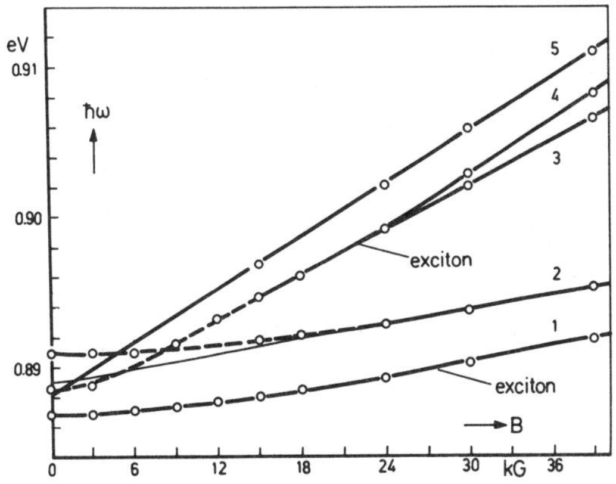

Fig.11.28 Transmission minima for direct exciton and Landau transitions in Ge at 77 K as a function of the magnetic field (after Lax and Zwerdling, ref. 7).

The convergence of curves # 3 - 5 yields a direct gap of 888.6 ± 0.4 meV at 77 K. A comparison with the room temperature results obtained from Fig.11.26 shows an increase with temperature of the effective mass in the Γ-minimum and a decrease of the band gap. The binding energy of the direct exciton is estimated as $13.6\,\text{eV} \times 0.033/16^2 = 1.7\,\text{meV}$ for a relative dielectric constant of 16 in Ge*). A more refined theoretical analysis yields a value of 1.5 meV. It also shows that for $p_{\pm}$-states there is no linear dependence on B_z if the effective masses of the electron and the hole are equal. This may be the case in Cu_2O where only a quadratic dependence has been observed.

11e. The Franz-Keldysh Effect (Electroabsorption and Electroreflectance)

We now investigate the effect of a strong electric field on light absorption in semiconductors at the fundamental absorption edge. Theoretical treatments of this effect have been given by W. Franz [1] and L. V. Keldysh [2]. It can best be described as photon-assisted tunneling through the energy barrier of the band gap (for tunneling effects see also Chap.9) and exists in insulators as well as semiconductors. In order to distinguish from hot-electron effects, "semi-insulating semiconductors" are best suited for an experimental investigation.

---

[7] B. Lax and S. Zwerdling, Progress in Semiconductors (A. F. Gibson and R. E. Burgess, eds.), Vol. 5. London: Temple Press. 1960.

Chap.11e:

[1] W. Franz, Z. Naturforsch. 13a (1958) 484.

[2] L. V. Keldysh, Zh. Eksp. Teor. Fiz. 34 (1958) 1138. [Engl.: Sov. Phys. JETP 7 (1958) 788.]

*) 13.6 eV = Rydberg energy, $0.033\,m_0$ = effective mass.

A simplified treatment considering an electron in vacuum will be presented first. Its potential energy in an electric field of intensity E in x-direction is given by $-|e|Ex$ where zero energy has been assumed at $x = 0$. For a quantum mechanical tunneling effect the Schrödinger equation

$$-(\hbar^2/2m_0)d^2\psi/dx^2 - |e|Ex\psi = \epsilon\psi \tag{11e.1}$$

for the ψ-function of the electron has to be solved. Here ϵ is the kinetic energy in x-direction which except for $\hbar^2(k_y^2 + k_z^2)/2m_0$ is the total kinetic energy. This equation is simplified to

$$d^2\psi/d\xi^2 = \xi\psi \tag{11e.2}$$

by introducing a dimensionless coordinate

$$\xi = -(x + \epsilon/|e|E)/l \tag{11e.3}$$

where l is an "effective length" given by

$$l = (\hbar^2/2m_0|e|E)^{1/3} \tag{11e.4}$$

A solution of Eq.(11e.2) for large positive values of ξ,

$$\psi \propto \xi^{-1/4}\exp(\pm \frac{2}{3}\xi^{3/2}) \tag{11e.5}$$

is easily verified. Let us denote this expression by y. The logarithmic derivative of y is given by

$$(\ln y)' = y'/y = \pm\xi^{1/2} - 1/(4\xi) \tag{11e.6}$$

since

$$\ln y = \pm\frac{2}{3}\xi^{3/2} - \frac{1}{4}\ln\xi \tag{11e.7}$$

From Eq.(11e.6) we obtain

$$y''/y = (y'/y)^2 \pm \frac{1}{2}\xi^{-1/2} + 1/(4\xi^2) = \xi \mp 1/(2\xi^{1/2}) + 1/(16\xi^2) \pm 1/(2\xi^{1/2}) + 1/(4\xi^2) \approx \xi \tag{11e.8}$$

The approximation is valid for large values of ξ where terms in ξ^{-2} are negligible. Since for a finite electron density ψ has to vanish at infinite values of ξ, only a negative value of the exponent in Eq.(11e.5) is physically significant for $\xi > 0$. Therefore the solution is essentially an exponential decrease with distance from the origin.

A solution for large negative values of ξ is given by

$$\psi \propto |\xi|^{-1/4}\sin(\frac{2}{3}|\xi|^{3/2} + \pi/4) \tag{11e.9}$$

where

$$\sin(\frac{2}{3}|\xi|^{3/2} + \pi/4) = \{\sin(\frac{2}{3}|\xi|^{3/2}) + \cos(\frac{2}{3}|\xi|^{3/2})\}/\sqrt{2} \tag{11e.10}$$

This solution is made plausible by replacing in Eq.(11e.5) ξ by $-\xi$, which makes the exponent purely imaginary. Since $\sin\varphi$ and $\cos\varphi$ are linear combinations of $e^{i\varphi}$ and $e^{-i\varphi}$ Eq.(11e.9) will be a solution of the homogeneous differential Eq.(11e.2) for this case. The complete solution is given by the Airy function [3].

[3] K. Tharmalingam, Phys. Rev. 130 (1963) 2204.

(In tabulated form the Airy function is found in M. Abramowitz and I. A. Stegun: Handbook of Mathematical Function, p. 475 - 477. New York: Dover Publ. 1965.)

We now consider a conduction electron in a semiconductor. The solution Eq. (11e.9) will be correct within the conduction band and Eq.(11e.5) in the forbidden gap if we replace in Eq.(11e.4) the free electron mass by the effective mass. The probability of finding an electron at x is given by $|\psi(x)|^2\,dx$. The absorption coefficient a is proportional to the probability integrated over the energy gap:

$$a \propto \int |\psi(\xi)|^2 d\xi \propto \int y^2 d\xi = y^2\xi - (dy/d\xi)^2 \qquad (11e.11)$$

where $d^2y/d\xi^2 = \xi y$ has been taken into account in the evaluation of the integral. For large positive values of ξ we will neglect any power of ξ and consider only the exponential dependence. In this approximation we obtain for $x = 0$ and $\epsilon = \hbar\omega - \epsilon_G$:

$$a \propto \exp(-\frac{4}{3}\xi^{3/2}) = \exp\{-\frac{4}{3}(\frac{\epsilon_G - \hbar\omega}{|e|\,E\,l})^{3/2}\} \qquad (11e.12)$$

With l as given by Eq.(11e.4) this becomes:

$$a \propto \exp(-\frac{4\sqrt{2m}\;(\epsilon_G - \hbar\omega)^{3/2}}{3|e|E\hbar}) \qquad (11e.13)$$

A similar expression for phonon-assisted tunneling has been obtained in Eq.(9a.7). At a given photon energy, $\hbar\omega$, the absorption increases with electric field intensity. This can be interpreted as a shift of the absorption edge to lower photon energies. For a shift of e.g. 10 meV a field intensity E is required which is obtained to a good approximation from

$$3|e|E\hbar/4\sqrt{2m} = (10^{-2}\,eV)^{3/2} \qquad (11e.14)$$

or, for $m \approx m_o$

$$E = \frac{4\text{x}10^{-3}}{3\hbar}\sqrt{2m_o|e|} \approx 5\text{x}10^4\,V/cm \qquad (11e.15)$$

A convenient method of measurement is to apply an ac voltage of e.g. 4 kV across a disc-shaped sample of thickness e.g. 1/3 mm perpendicular to a beam of monochromatic light passing through the sample and connect an amplifier to the light detector which is tuned to a frequency of twice the frequency of the ac voltage. The signal as a function of $\hbar\omega$ is proportional to the derivative of the absorption coefficient with respect to $\hbar\omega$. It is obvious that for an arbitrary function f

$$a = f(\frac{\epsilon_G - \hbar\omega}{E^{2/3}}) \qquad (11e.16)$$

yields

$$\frac{da}{dE} = \frac{2}{3}\,\frac{\epsilon_G - \hbar\omega}{E}\,\frac{da}{d(\hbar\omega)} \qquad (11e.17)$$

For not too strong absorption and for short samples, Eq.(D.16) yields an average transmission, $\langle T \rangle$, which is essentially proportional to $\exp(-ad)$ where d is the sample length. The change of $\langle T \rangle$ with E is proportional to the change of a

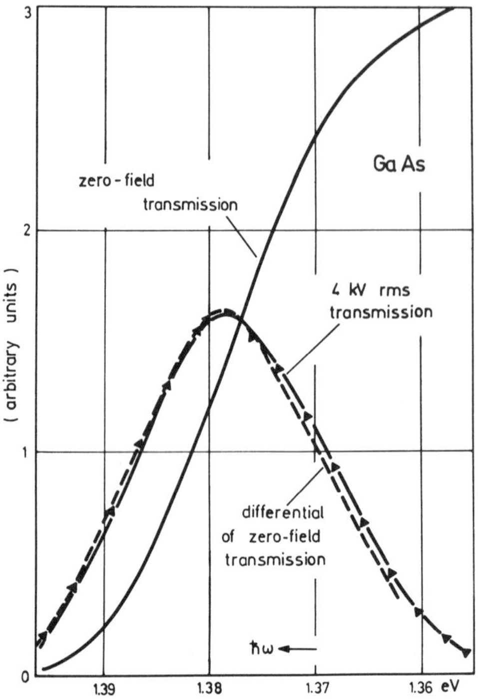

Fig.11.29 Optical transmission at the absorption edge (S-shaped full curve) and ac component of transmission when a strong ac electric field is applied to a semi-insulating GaAs sample (data points). The differential of the zero-field transmission is shown by the dashed curve ("Franz-Keldysh effect", after T. S. Moss, ref. 4).

with E. The data have to be corrected for Joule heating of the crystal resulting in a change of ϵ_G with lattice temperature.

Fig.11.29 shows results of $d\alpha/dE$ obtained in semi-insulating GaAs, together with the zero-field transmission [4]. The Franz-Keldysh effect has also been observed in other large-gap semiconductors such as CdS which in the form of pure or compensated crystals, often with deep-level impurities, show little Joule heating at the high voltage applied. Sometimes a large reverse bias applied to a semiconductor diode serves to generate the large electric field with not too much heat generation.

For photon energies larger than the band gap, Eq.(11e.9) predicts an oscillatory behavior of the absorption with E. Observations which could be interpreted in this way can also be explained by the assumption that an electric field destroys exciton absorption [5].

So far we have investigated the effect of a strong electric field on absorption of light with frequencies near the fundamental edge. It was shown in Chap.11a that within the range of fundamental absorption a measurement of reflection is more adequate. The effect of a strong electric field on reflectivity in this range has been observed by the differential method discussed above. The field has been applied perpendicular to the reflecting surface. The relative change in reflectivity, $\Delta r_\infty/r_\infty$, for a given magnitude of the electric field intensity observed in germanium at room temperature is shown in Fig.11.30 in the range of photon

[4] T. S. Moss, J. Appl. Phys. Suppl. 32 (1961) 2136.

[5] Y. Hamakawa, F. Germano, and P. Handler: Proc. Int. Conf. Semicond. Physics Kyoto 1966, p. 111, "Comment by the authors". Tokyo: The Physical Soc. Japan. 1966.

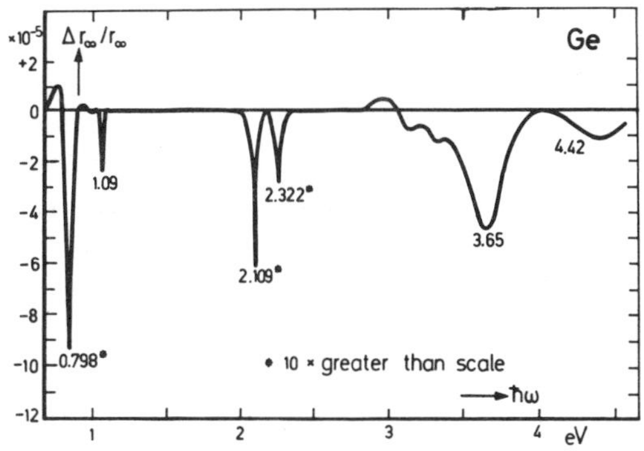

Fig.11.30 Relative change in reflectivity of germanium upon application of an electric field normal to the reflecting surface (after Seraphin and Hess, ref. 6).

energies, $\hbar\omega$, between 0.5 and 4.5 eV [6]. A comparison of this curve with Fig. 11.3 reveals the increased sensitivity of the electroreflectance method in determinations of "critical points" (van Hove singularities, see Chap.11i). Strongly temperature dependent dips in the electroreflectance spectrum may indicate excitons at band minima which do not form a band edge. However, exciton Stark effect and exciton quenching [7] by the strong electric field will also occur and have to be taken into account in the evaluation of data in terms of band structure.

11f. Impurity Absorption

The optical absorption spectrum of a semiconductor is modified in two ways by the presence of impurities :

1. There are transitions between the ground state and the excited states of a neutral impurity. The maximum probability for such a transition occurs at an energy $\hbar\omega$ which is of the order of magnitude of the ionization energy, $\Delta\epsilon_A$ for an acceptor and $\Delta\epsilon_D$ for a donor. Usually this energy is much less than the band gap energy. Such a transition may also occur even though the impurity cannot be ionized thermally ("electrically inactive impurity").

2. The transition energy may be near the fundamental edge as illustrated by Fig.11.31. A shallow acceptor level and a shallow donor level are each represented by a horizontal line. The half width of the line is given by $1/a_{imp}$ where the Bohr radius of the impurity atom, a_{imp}, is given by

$$a_{imp} = \frac{\kappa}{m/m_o} a_B \qquad (11f.1)$$

[6] B. O. Seraphin and R. B. Hess, Phys. Rev. Lett. 14 (1965) 138.

[7] D. L. Greenaway and G. Harbeke: Optical Properties and Band Structure of Semiconductors. New York: Pergamon. 1968.

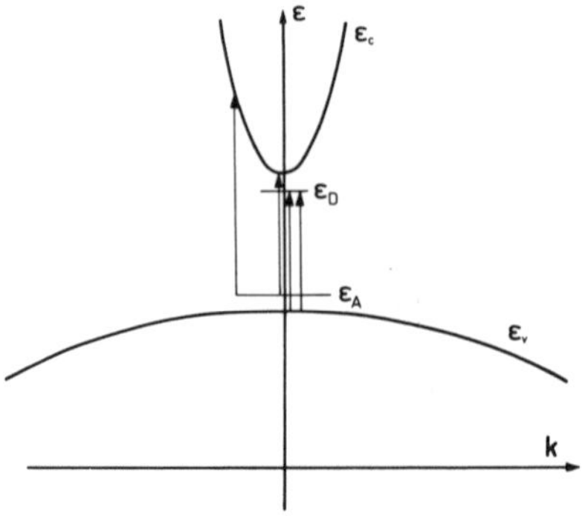

Fig.11.31 Schematic energy band diagram illustrating spectral spread of impurity absorption due to spread of k of impurity wave functions (after E. J. Johnson: Semiconductors and Semimetals (R. K. Willardson and A. C. Beer, eds.), Vol. 3, p. 153. New York: Acad. Press. 1967).

$a_B \approx 0.5$ Å is the Bohr radius of the hydrogen atom and m is the effective mass of an electron in the conduction band, if the impurity is a shallow donor, and of a hole in the valence band, if the impurity is a shallow acceptor. Except for the effective mass, the same considerations can be applied here which have been made in connection with the exciton (Eq.11b.3). For deep level impurities and for acceptors in group IV and III-V semiconductors the impurity wave function contains contributions from several bands. In an indirect semiconductor, transitions involve phonons as in band-to-band transitions.

E.g. let us consider a transition from a shallow acceptor to the conduction band. The transition probability is proportional to the density N_{A^-} of ionized acceptors which from Eqs.(3b.14) and (3b.16) is obtained as

$$N_{A^-} = \frac{N_A}{g_A \exp\{(\epsilon_A - \zeta)/k_B T\} + 1} \qquad (11f.2)$$

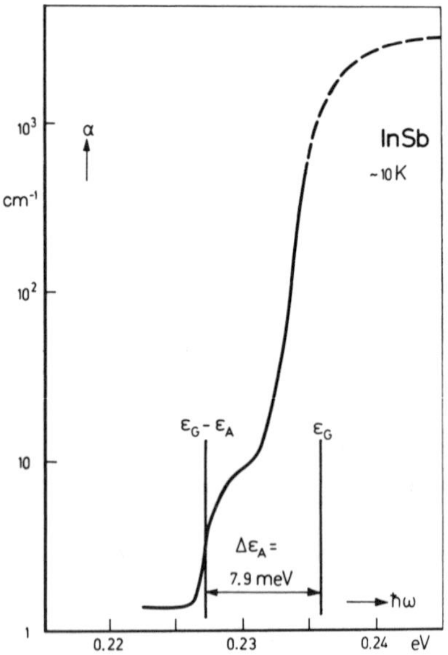

Fig.11.32 Impurity absorption in InSb at about 10 K (after Johnson and Fan, ref. 1).

For allowed transitions involving acceptors where $m_p \gg m_n$ the absorption coefficient α is given by the proportionality

$$\alpha \hbar\omega \propto \frac{N_A \sqrt{\hbar\omega - (\epsilon_G - \Delta\epsilon_A)}}{g_A \exp\{(\epsilon_A - \zeta)/k_B T\} + 1} \qquad (11f.3)$$

Except for the shift by $\Delta\epsilon_A$ the spectral variation is the same as for the fundamental absorption. However, the absorption by impurities is temperature dependent and proportional to the impurity concentration.

Experimental results on optical absorption by zinc or cadmium in InSb are shown in Fig.11.32 [1]. Assuming an energy gap, ϵ_G, of 235.7 meV at 10 K, an ionization energy, $\Delta\epsilon_A$, of 7.9 meV is found from Eq.(11f.3). (A more accurate value is obtained from magne-

[1] E. J. Johnson and H. Y. Fan, Phys. Rev. 139 (1965)A 1991.

toabsorption.) The influence of the impurity concentration on the optical absorption of p-GaSb at 10 K is illustrated by Fig. 11.33 [1]. The impurity concentrations in units of $10^{17}\,cm^{-3}$ measured at 300 K are 1.4 and 2.5 in sample # 1 and

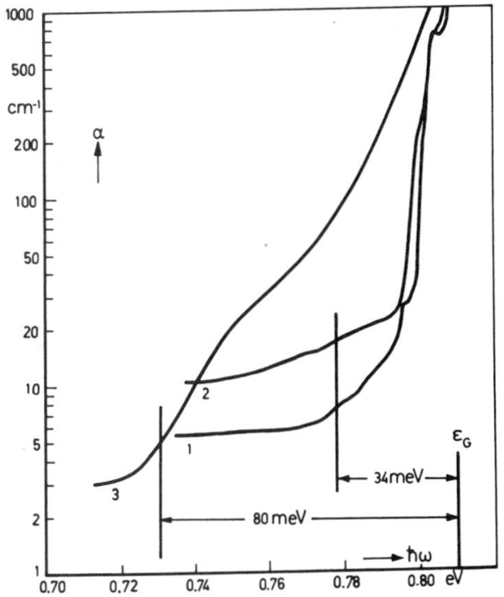

Fig. 11.33 Tail of fundamental absorption of GaSb influenced by different impurity concentrations: (1) and (2) undoped samples, (3) tellurium doped (after Johnson and Fan, ref. 1).

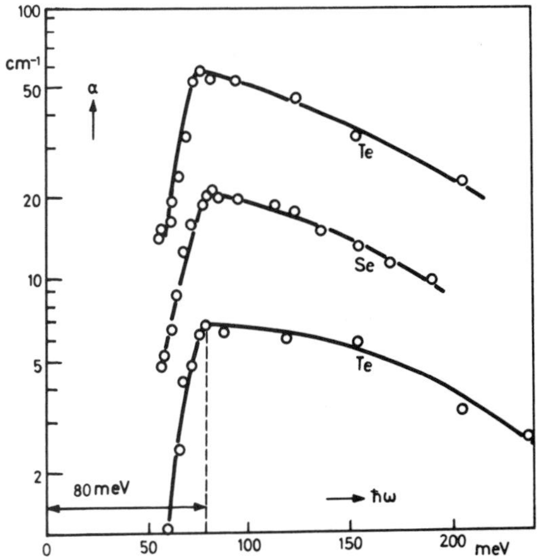

Fig. 11.34 Impurity absorption of GaSb doped with tellurium and selenium of various concentrations (after Johnson and Fan, ref. 1).

2, respectively, while in sample # 3 the acceptor impurities are partly compensated by tellurium which is a donor in GaSb. The effect of compensation is to raise the Fermi level to about 80 meV above the valence band edge such that a deep acceptor level, which has not been identified so far, is filled with electrons. Transition of these electrons into the conduction band by absorption of light quanta is indicated by the long tail of the fundamental absorption edge (Fig. 11.33, curve # 3). The deep acceptor level is also visible at low photon energies $\hbar\omega \approx 80$ meV (Fig.11.34) where holes are transferred from the acceptor level to the valence band [1]. This corresponds to case # 1 discussed above. Since the deep level in GaSb can also be obtained by bombarding the crystal with high-energy electrons it has been suspected that it is due to lattice defects rather than to impurities [2].

Low energy photon absorption by shallow donors and acceptors in silicon at 4.2 K is shown in Figs.11.35 and 11.36, respectively. Sharp bands are obtained in pure samples where line broadening by the interaction of neighboring impurity atoms is negligible. There are several sharp absorption bands at energies which are less than the ionization energy. The absorption band at a wave number of 316.4 cm^{-1} is due to an antimony contamination. The spectra of bismuth and arsenic impurities are very similar; the main difference is a shift in wave number of 140 cm^{-1} which is equivalent to a shift in energy of 17.3 meV. The similarity of the spectra is due to the fact that the valence electron of a shallow donor orbits with a radius which is several times larger than the lattice constant; the electron wave function is more determined by the host lattice than by the substitutional impurity itself. The same is true for holes bound to shallow acceptors. Morgan [3] has shown that the local strain fields generated by size differences between the substitutional impurity atom and the host atom it replaces, essentially explain the small differences in binding energies ϵ_B between various impurities. The agreement between the observed and calculated data is improved by taking into account the Stark effect due to the ionic electric fields. Observations of the donor → acceptor luminescent transition in GaP have shown that the misfit volume associated with the impurity is distributed over a range of at least four lattice parameters [4]. X-ray determinations of the lattice constant in

[2] R. Kaiser and H. Y. Fan, Phys. Rev. 138 (1956) A 156; E. B. Owens and A. J. Strauss: Proc. Conf. Ultrapurif. Semicond. Materials, p. 340. New York: Macmillan. 1961; D. Effer and P. J. Etter, J. Phys. Chem. Solids 25 (1964) 451.

[3] T. N. Morgan, Proc. Int. Conf. Phys. Semic., Cambridge/Mass. 1970 (S. P. Keller, J. C. Hensel, and F. Stern, eds.), p. 266. Oak Ridge/Tenn.: USAEC. 1970.

[4] T. N. Morgan: Proc. Int. Conf. Phys. Semic., Warsaw 1972 (M. Miasek, ed.), p. 989. Warsaw: PWN - Polish Sci.Publ. 1972; T. N. Morgan and H. Maier, Phys. Rev. Lett. 27 (1971) 1200.

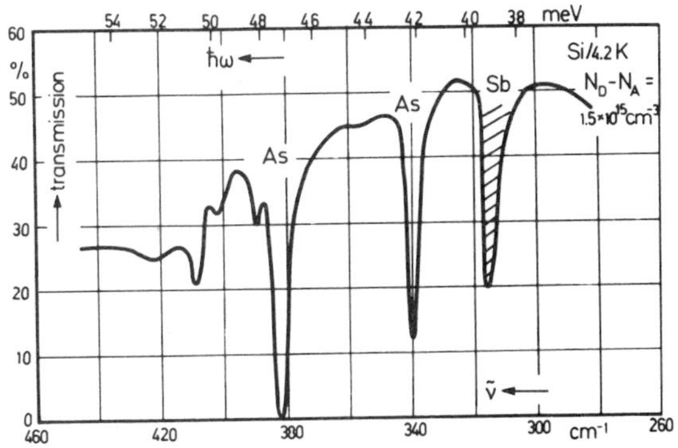

Fig.11.35 Optical transmission spectra of silicon doped with $1.5x10^{15}/cm^3$ arsenic atoms, at 4.2 K; the hatched band arises from antimony contamination (after H. J. Hrostowski: Semiconductors (N. B. Hannay, ed.). New York: Reinhold Publ. Co. 1959).

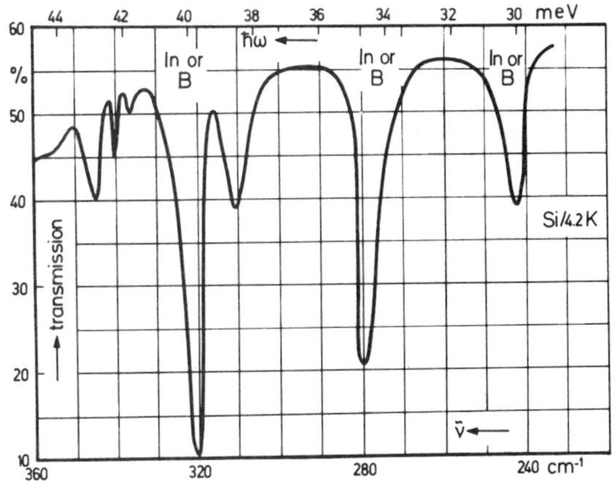

Fig.11.36 Optical transmission spectra of silicon doped with boron, at 4.2 K (after H. J. Hrostowski: Semiconductors (N. B. Hannay, ed.). New York: Reinhold Publ. Co. 1959).

Si have shown that there is a volume increase with doping [5]. At high doping levels in compensated semiconductors there is a formation of ion pairs $N_D^+ N_A^-$.

Optical absorption peaks at 64 and 141 meV in Si are due to an oxygen impurity which is electrically inactive. Even for $10^{18}/cm^3$ oxygen atoms the absorption coefficient at 4.2 K is only a few cm^{-1} as illustrated by Fig.11.37. If the more common 0^{16} isotope is replaced by the 0^{18} the absorption bands are shifted

[5] F. H. Horn, Phys. Rev. 97 (1955) 1521; V. T. Bublik, S. S. Gorelik, and A. N. Dubrovina, Fiz. Tverd. Tela 10 (1968) 2846 [Engl.: Sov. Phys.-Sol. State 10 (1969) 2247]; J. A. Baker, T. N. Tucker, N. E. Moyer, and R. C. Buschert, J. Appl. Phys. 39 (1968) 4365.

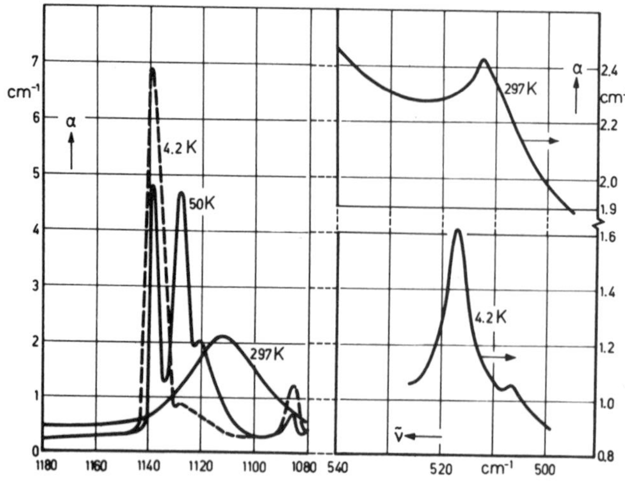

Fig.11.37 Oxygen absorption bands in silicon at 4.2, 50, and 297 K for a crystal containing oxygen enriched with 12% O^{18} and 1% O^{17}; absorption bands due to O^{16} are at 1106 and 515 cm^{-1} (after Hrostowski and Kaiser, ref. 6).

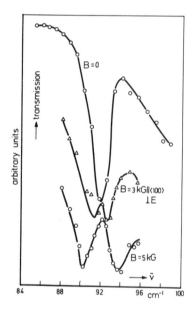

Fig.11.38 Zeeman splitting of a $2p_{\pm}$ phosphorous impurity level in germanium at 4.2 K (after Boyle, ref. 7).

which proves that the bands are associated with oxygen [6].

Valuable information about the assignment of spin states to transitions like those shown in Fig.11.35 and 11.36 is gained from magneto-optical observations. If a longitudinal magnetic field is applied along a <100> direction in a germanium crystal at 4.2 K, the Landau levels in the four valleys of the conduction band are left degenerate and Zeeman splitting of the bound states about a phosphorous donor may be investigated. The 1S → 2p, m = 0 line at 76.0 cm^{-1} remains unchanged while the 1S → 2p,m = ± 1 line at 91.5 cm^{-1} at low fields splits into a symmetrical doublet as shown in Fig.11.38 [7]. The magnitude of the splitting of this line yields a value for the transverse effective mass in Ge of (0.077 ± 0.005) m_0 which agrees well with the value of 0.082 m_0 obtained from cyclotron resonance. In the theory of linear Zeeman splitting [8]

[6] H. J. Hrostowski and R. H. Kaiser, Phys. Rev. 107 (1957) 966.
[7] W. S. Boyle, J. Phys. Chem. Solids 8 (1959) 321.
[8] J. M. Luttinger and W. Kohn, Phys. Rev. 97 (1955) 869.

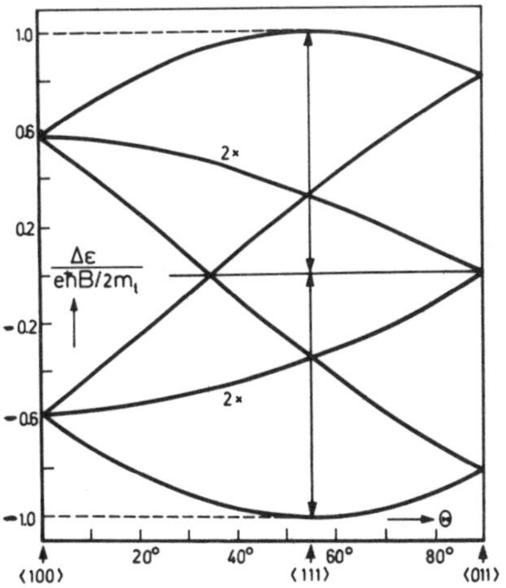

Fig. 11.39 Calculated dependence of the linear Zeeman splitting of donor $2p_\pm$ states in germanium on the angle between the magnetic field and the ‹100› direction in a [1$\bar{1}$0] plane (after R. R. Haering, ref. 9).

of donor $2p_\pm$ states in Ge we have to consider four ellipsoids of the conduction band and obtain for $\Delta\epsilon_\pm$ a dependence on the angle θ between $\vec{B}$ and the <100> direction in the [1$\bar{1}$0] plane, which is shown in Fig. 11.39 [9]. The symbol 2x indicates a two-fold degeneracy which is due to the fact that two of the four ellipsoids are symmetrical with respect to the [1$\bar{1}$0] plane. For a <111> direction the splitting for the ellipsoid with the longitudinal axis in this direction becomes

$$\Delta\epsilon_{p_\pm} = \pm \frac{e\hbar}{2m_t} |\vec{B}| = \pm \frac{\mu_B |\vec{B}|}{m_t/m_o} \tag{11f.4}$$

where m_t is the transverse effective mass and $\mu_B = 5.77\ \mu eV/kG$ is the Bohr magneton. The calculation neglects spin orbit interaction which is justified for strong magnetic fields.

For a spherical band the linear Zeeman splitting of donor $p_\pm$ states is given by

$$\Delta\epsilon_{p_\pm} = \pm\ \hbar\omega_c \tag{11f.5}$$

while for p_o and s states there is no splitting but only a quadratic shift.

For s states the shift is given by [10]

$$\Delta\epsilon_s = (\hbar\omega_c)^2/(8\Delta\epsilon_I) \tag{11f.6}$$

where $\Delta\epsilon_I$ is the ionization energy and a small magnetic field strength has been assumed ($\hbar\omega_c \ll \Delta\epsilon_I$).

In a strong magnetic field Landau levels are formed in the conduction band, and the photoionization continuum shows an oscillatory behavior. For the case of Ge doped with $3\times10^{15}\ cm^{-3}$ arsenic donors this is shown in Fig. 11.40. The correct Landau quantum numbers n assigned to the transmission minima may be obtained by observing the transmission as a function of $\vec{B}$ at a fixed wave number of e.g. 137 cm^{-1}. The wave number location of the successive minima in the transmission at two different magnetic fields is indicated in Fig. 11.41. The two straight lines extrapolated to negative values of n intersect an $n = -1/2$. Since the Landau energy is a linear function of $(n + 1/2)\hbar\omega_c$ where $\hbar\omega_c = (e/m)B$ one would expect that for $n = -1/2$ any dependence on $\vec{B}$ would disappear as it is the case. The intersection occurs at a wave number of 108 cm^{-1} which corresponds

[9] R. R. Haering, Can. J. Phys. 36 (1958) 1161.
[10] J. H. van Vleck: The Theory of Electric and Magnetic Susceptibilities, p. 178. London: Oxford University Press. 1932.

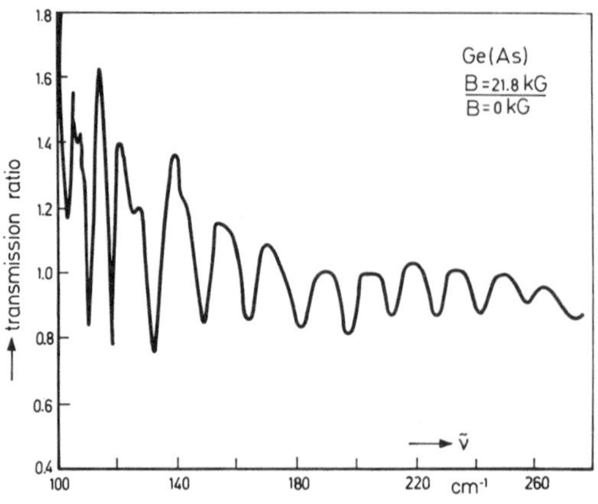

Fig.11.40
Transmission ratio at 4 K with and without a field of B = 21.8 kG in germanium, as a function of wave number. The absorption is due to arsenic impurities (after Boyle, ref. 7).

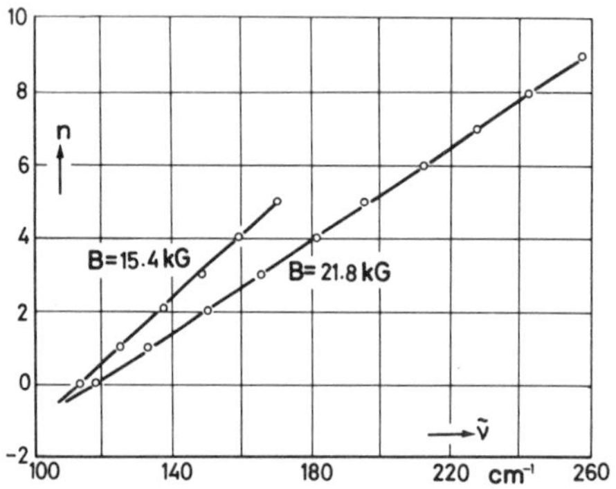

Fig.11.41
The wave number location of the successive transmission minima of Fig.11.40 (B = 21.8 kG) and of a similar plot for B = 15.4 kG (not shown) (after Boyle, ref. 7).

to an energy of 13.4 meV. It is incorrect to compare this energy with the donor binding energy of 12.8 meV, found from electrical measurements at various temperatures [11], as pointed out by Howard and Hasegawa [12]: Actually the transitions discussed so far are not transitions to free Landau states but to bound states whose energies are lowered relative to the free magnetic states by the Coulomb energy of the impurity ("bound Landau states"). These localized states are found at energies both between the Landau levels and between the zero-Landau level and the zero-field band edge. Later experiments by Boyle and Howard [13] confirmed this hypothesis. There is no simple way of obtaining the impurity ionization energy from magnetooptical data of this kind [14].

[11] T. H. Geballe and F. J. Morin, Phys. Rev. 95 (1954) 1805.

[12] R. E. Howard and H. Hasegawa, Bull. Am. Phys. Soc. 5 (1960) 178.

[13] W. S. Boyle and R. E. Howard, J. Phys. Chem. Solids 19 (1961) 181.

[14] B. Lax and S. Zwerdling: Progress in Semiconductors (A. F. Gibson and R. E. Burgess, eds.), Vol. 5 , p. 269. London: Temple Press. 1960.

11g. Lattice Reflection in Polar Semiconductors

In polar crystals transverse optical lattice vibrations have an electric dipole moment which can be excited by an ac electric field with frequencies in the far-infrared part of the electromagnetic spectrum. Within a small range of frequencies the reflectivity is close to 100%. By multiple reflections this band can therefore be filtered out of a white spectrum. The remaining light beams are called "reststrahlen".

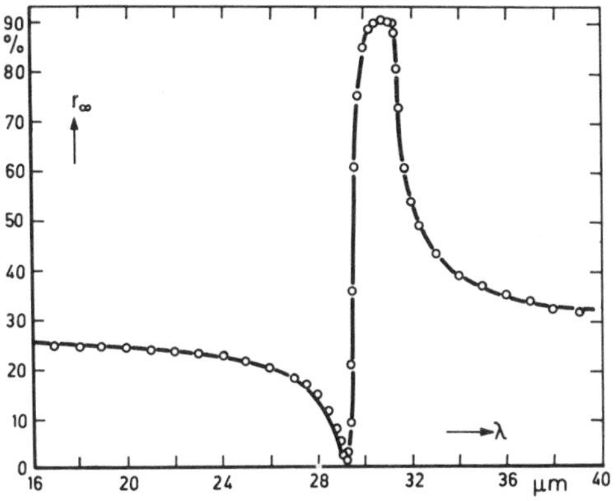

Fig. 11.42 Lattice reflection spectrum of AlSb; the points represent experimental data; the curve represents a calculated fit for a single classical oscillator (after Turner and Reese, ref. 1).

Fig. 11.42 shows the reflectivity of AlSb observed at room temperature in the spectral range from 18 to 40 μm [1]. The reststrahlen band is found between 29 and 31 μm.

We will treat the problem theoretically simply as a classical harmonic oscillator of eigenfrequency ω_e and damping constant γ. In an electric field of amplitude E_1 and angular frequency ω the equation of motion

$$m d^2x/dt^2 + m\gamma dx/dt + m\omega_e^2 x = eE_1 \exp(i\omega t) \tag{11g.1}$$

has a solution given by

$$x = \frac{e}{m} \cdot \frac{E_1 \exp(i\omega t)}{\omega_e^2 - \omega^2 + i\omega\gamma} \tag{11g.2}$$

Assuming n oscillators per unit volume, the dipole moment per unit volume is given by

$$\frac{ne^2}{m} \cdot \frac{E_1 \exp(i\omega t)}{\omega_e^2 - \omega^2 + i\omega\gamma} = \frac{\omega_p^2}{\omega_e^2 - \omega^2 + i\omega\gamma} \kappa_0 E_1 \exp(i\omega t) \tag{11g.3}$$

where we have introduced the "plasma angular frequency"

$$\omega_p = \sqrt{\frac{ne^2}{m\kappa_0}} \tag{11g.4}$$

For the complex relative dielectric constant $\kappa_r + i\kappa_i$ we obtain from Eqs.(6h.27) and (6h.31)

[1] W. J. Turner and W. E. Reese, Phys. Rev. 127 (1962) 126.

$$\kappa_r - i\kappa_i = \kappa_{opt} + \frac{\omega_p^2}{\omega_e^2 - \omega^2 + i\omega\gamma} \tag{11g.5}$$

where κ_{opt} is the dielectric constant at frequencies $\omega \gg \omega_e$ and ω_p. This equation can be solved for the real and imaginary parts:

$$n^2 - k^2 = \kappa_r = \kappa_{opt} + \omega_p^2 \frac{\omega_e^2 - \omega^2}{(\omega_e^2 - \omega^2)^2 + \omega^2\gamma^2} \tag{11g.6}$$

and

$$2nk = \kappa_i = \omega_p^2 \cdot \frac{\omega\gamma}{(\omega_e^2 - \omega^2)^2 + \omega^2\gamma^2} \tag{11g.7}$$

where the relation between κ_r, κ_i, the index of refraction, n (not to be confused with the density of harmonic oscillators), and the extinction coefficient, k, as given by Eq.(D.18) of Appendix D, has been introduced. Solving for n and k we have

$$n = \sqrt{\frac{1}{2}\left\{\kappa_r + \sqrt{\kappa_r^2 + \kappa_i^2}\right\}} \tag{11g.8}$$

and

$$k = \sqrt{\frac{1}{2}\left\{-\kappa_r + \sqrt{\kappa_r^2 + \kappa_i^2}\right\}} \tag{11g.9}$$

The reflectivity of an "infinitely thick" plate at normal incidence is given by:

$$r_\infty = \frac{(n-1)^2 + k^2}{(n+1)^2 + k^2} \tag{11g.10}$$

Fig.11.43 shows κ_r, κ_i, and r_∞ as functions of ω/ω_e with parameters arbitrarily chosen as $\kappa_{opt} = 12$; $(\omega_p/\omega_e)^2 = 3$; $\gamma/\omega_e = 0.05$. Near the resonance frequency ω_e there is a sharp maximum of κ_i and a broader peak of r_∞ while the frequency behavior of $\kappa_r - \kappa_{opt}$ resembles that of the differential of κ_i with respect to ω/ω_e.

The static dielectric constant, κ, is obtained from Eq.(11g.6) for $\omega = 0$:

$$\kappa = \kappa_{opt} + (\omega_p/\omega_e)^2 \tag{11g.11}$$

and in Fig.11.43 a value of 15 has been chosen. The reflectivity r_∞ has a minimum at a "reduced frequency" given by

$$\frac{\omega}{\omega_e} = \sqrt{\frac{\kappa - 1}{\kappa_{opt} - 1}} \tag{11g.12}$$

which for values of κ and $\kappa_{opt} \gg 1$ can be approximated by $\sqrt{\kappa/\kappa_{opt}}$. The Lyddane-Sachs-Teller relation, Eq.(6g.22), shows that this quantity is given by the ratio ω_0/ω_t where ω_0 and ω_t are the longitudinal and transverse frequencies for long-wavelength optical phonons, respectively. From Eq.(11g.12) we find

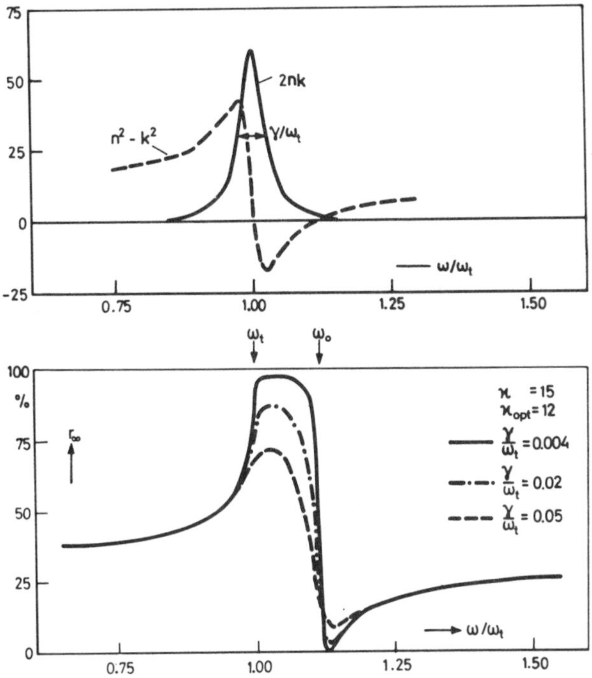

Fig.11.43 Calculated real (n^2-k^2) and imaginary (2nk) parts of the complex dielectric constant for a single classical oscillator, as a function of the frequency relative to the oscillator frequency (top), and calculated reflectivity for various degrees of damping (bottom) (after M. Hass: Semiconductor and Semimetals (R. K. Willardson and A. C. Beer, eds.), Vol. 3, p. 7. New York: Acad. Press. 1967).

that the eigenfrequency ω_e is the transverse frequency ω_t since the transverse oscillations are excited by the transverse electromagnetic wave. Hence, the reflectivity minimum occurs at the longitudinal optical phonon frequency ω_o.

The evaluation of the reflectivity spectrum of AlSb given in Fig.11.42 yields phonon energies of $\hbar\omega_o = 42.0$ meV and $\hbar\omega_t = 39.6$ meV. The ratio $\gamma/\omega_t = 5.9\ (\pm 0.5) \times 10^{-3}$ where $\omega_t/2\pi c = \tilde{\nu}_t = 318.8\ (\pm 0.5)\ \mathrm{cm}^{-1}$. From Raman scattering experiments [2] values of $\tilde{\nu}_t = 318.9\ (\pm 0.5)\ \mathrm{cm}^{-1}$ and $\tilde{\nu}_o = 339.9\ (\pm 0.5)\ \mathrm{cm}^{-1}$ have been obtained which are in close agreement with the reflectivity data. The comparatively large inaccuracy in γ is attributed to the difficulty in measuring the absolute magnitude of the maximum reflectivity. The damping constant γ is largely due to lattice anharmonicity.

The Raman spectrum of AlSb is shown in Fig.11.44 [2]. The lines are obtained from the incident laser light either by absorption or emission of phonons of wave numbers $\tilde{\nu}_t$ (strong lines) or $\tilde{\nu}_o$ (weak lines). (The laser wavelength of 1.06 μm obtained from a YAG:Nd^{3+} laser*) is subject to the condition that the

[2] A. Mooradian and G. B. Wright, Solid State Comm. 4 (1966) 431.
*) YAG = Yttrium Aluminium Garnet

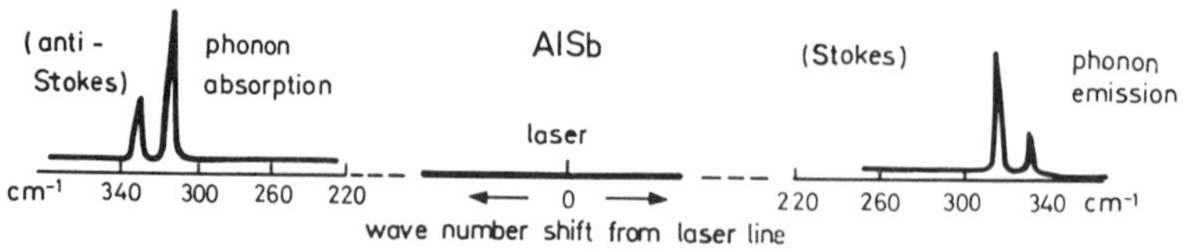

Fig. 11.44 Raman spectrum of AlSb at room temperature (after Mooradian and Wright, ref. 2).

crystal is transparent in this part of the spectrum [3]; for a distinction between Raman scattering and Brillouin scattering see Chap. 7h, footnote.) *)

11h. Multiphonon Lattice Absorption

In the elemental diamond-type semiconductors the lattice atoms are not charged and consequently there is no first-order electric moment ("infrared-active fundamental vibration"). However, this and all other semiconductors show a number of weak absorption bands which have been identified as being due to the interaction of the incident photon with several phonons simultaneously ("multiphonon absorption"). Because of their weakness the lines cannot be found experimentally in the reflectivity spectrum [1].

Fig. 11.45 shows the absorption spectrum of silicon at various temperatures between 20 and 365 K in the wavelength range from 7 μm (corresponding to about 150 mm^{-1}) to 20 μm (corresponding to about 30 mm^{-1}) [2]. Between 20 and 77 K there is not much change in the data. The photon energies of the peaks have been found to be combinations of 4 photon energies, namely the transverse (TO) and longitudinal optical (LO) phonon energies of 59.8 and 51.3 meV,

[3] For a review of Raman scattering in semiconductors see e.g. A. Mooradian: Festkörper-Probleme (O. Madelung, ed.), Vol. IX, p. 74. London: Pergamon and Braunschweig: Vieweg. 1969, where also surface scattering and Raman scattering by thermal and hot solid state plasmas are considered.

[4] K. Huang, Proc. Roy. Soc. A 208 (1951) 352; C. H. Henry and J. J. Hopfield, Phys. Rev. Lett. 15 (1965) 964; J. J. Hopfield, Phys. Rev. 112 (1958) 1555; U. Fano, Phys. Rev. 103 (1956) 1202.

*) In a lattice of strong polarity an electromagnetic wave is coupled to a sound wave and vice versa. A distinction between phonon and photon can no longer be made and one prefers then to denote the quantum of vibration by the word "polariton" [4].

Chap. 11h:

[1] For a review see W. G. Spitzer: Semiconductors and Semimetals (R. K. Willardson and A. C. Beer, eds.), Vol. 3, p. 17. New York: Acad. Press. 1967.

[2] F. A. Johnson, Proc. Phys. Soc. (London) B 73 (1959) 265.

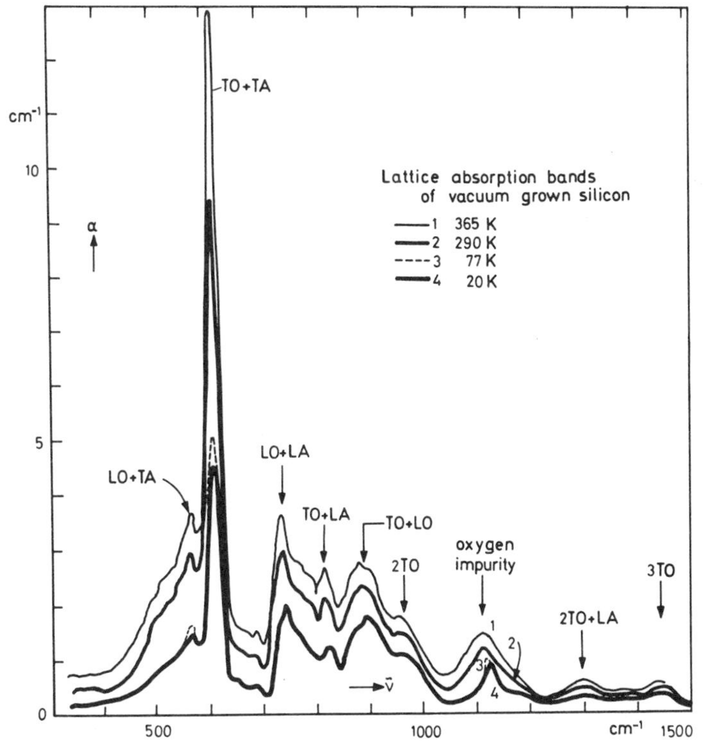

Fig.11.45 Lattice absorption due to multiphonon processes in vacuum grown silicon (after Johnson, ref. 2).

respectively, and the transverse (TA) and longitudinal acoustic (LA) phonon energies of 15.8 and 41.4 meV, respectively, all at the edge of the phonon Brillouin zone*). (For the case of Ge the phonon spectrum has been shown in Fig.6.15.) E.g. to the highest peak a combination of TO + TA could be assigned, while e.g. the small peak at 145 mm^{-1} is labeled 3 TO etc. A total of 10 peaks could be fitted quite well and their dependence on temperature agrees with calculations based on these phonon energies. The peak at 110 mm^{-1} is, however, due to an oxygen impurity which exists even though the crystal was vacuum grown (see Fig.11.37).

The phonon energies so obtained compare favorably well with those found from slow-neutron spectroscopy where in a <100> direction TO = 58.7 meV, LO = LA = 49.2 meV, and TA = 17.9 meV at $\vec{q} = \vec{q}_{max}$ [3] (Chap.6h).

Similar measurements have been made on other group-IV elements, on III-V compounds, on some II-VI compounds, and on SiC. An interesting case is that

[3] B. N. Brockhouse, Phys. Rev. Lett. 2 (1959) 256.

*) An average of the LO and TO frequencies has been denoted as "intervalley phonon frequency" in Chap.7e.

of silicon germanium alloys. The silicon- and germanium-like summation bands are not sensitive to the composition which suggests that there are pure Ge and Si conglomerates rather than a true disordered alloy [4]. In addition two new bands have been observed: one in the Ge-rich alloy and the other in the Si-rich alloy which are impurity bands since they are not present in the pure elements and grow in intensity as the composition is changed toward 50 % of either constituent. The shift of phonon frequencies with composition of the alloy is shown in Fig. 11.46.

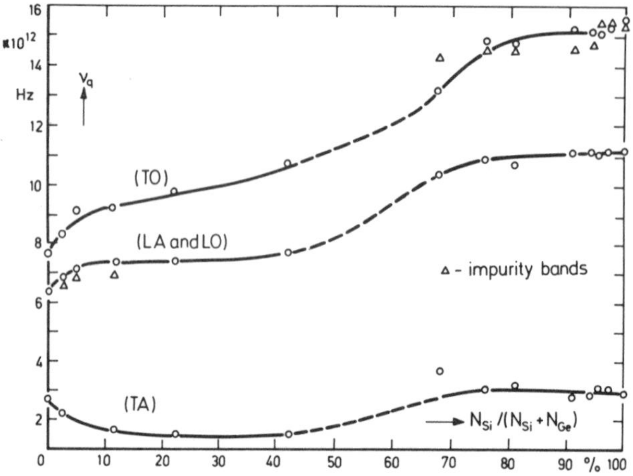

Fig. 11.46 Phonon frequencies in Ge-Si alloys as a function of composition (after Braunstein, ref. 4).

Let us finally consider an empirical relationship between the ratio $(LO/TO)^2$ and the square of the Szigeti effective charge, e_S, defined by Eq.(6h.50). Fig. 11.47 shows that there is a linear relationship which most materials obey to a good approximation [5]. For $e_S = 0$, as in the elemental semiconductors Ge and Si,

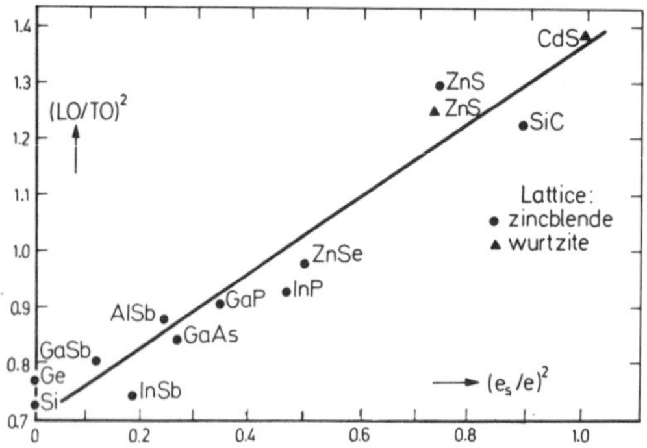

Fig. 11.47 The square of the ratio of longitudinal and transverse optical phonon frequencies as a function of the square of the Szigeti effective charge (in units of the elementary charge) (after Marshall and Mitra, ref. 5).

[4] R. Braunstein, Phys. Rev. 130 (1963) 879.
[5] R. Marshall and S. S. Mitra, Phys. Rev. 134 (1964) A 1019.

LO < TO at the zone edge. However, with increasing values of e_S the LO branch moves up relative to the TO branch such that for $(e_S/e)^2 > 0.5$, LO > TO at the zone edge. For the case of GaAs the phonon spectrum is shown in Fig.6.16.

Another linear relationship is one between the ratio $(LO/LA)^2$ and the ratio of the atomic masses of the constituents of a compound [5]. This is shown in Fig.11.48. The theory of the diatomic linear chain yields for the ratio of the optical and acoustic frequencies at the zone edge $\sqrt{D_1/D_2}$ for $M_1 = M_2$ (see Fig.6.14) and $\sqrt{M_1/M_2}$ for $D_1 = D_2$ where D_1 and D_2 are the spring constants and M_1 and M_2 are the atomic masses.

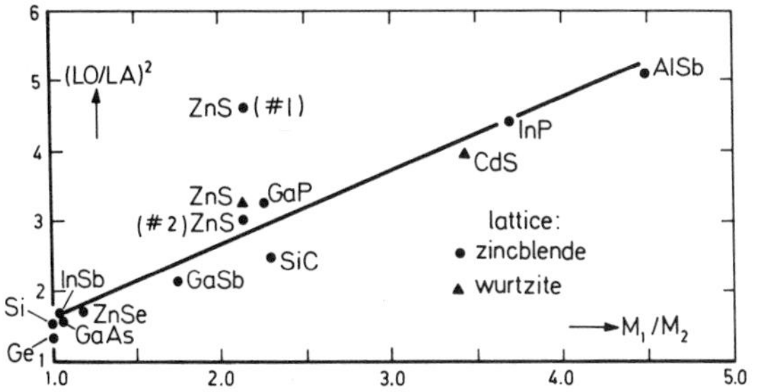

Fig.11.48 The square of the ratio of longitudinal optical and acoustic phonon frequencies as a function of the ionic mass ratio in compound semiconductors (after Marshall and Mitra, ref. 5).

11i. Quantum Mechanical Treatment of the Fundamental Optical Absorption Edge

In Chap.11b we have considered the observed optical absorption near the fundamental edge. Now we will treat this problem by means of a quantum mechanical perturbation theory where the electromagnetic wave is considered as a small perturbation of the electron wave.

The classical Hamiltonian of an electron in an electromagnetic field is given by Eq.(11b.13). The quantum mechanical commutation relation between e.g. the x-component of the momentum vector $\hbar\vec{k} = -i\hbar\vec{\nabla}_r$ and the same component of the vector potential $\vec{A}$ is easily found to be [2]

$$\hbar k_x A_x - A_x \hbar k_x = -i\hbar \partial A_x/\partial x \quad (11i.1)$$

[1] D. L. Greenaway and G. Harbeke: Optical Properties and Band Structure of Semiconductors. London: Pergamon. 1968; G. F. Bassani: Proc. Int. School of Physics (J. Tauc, ed.), Vol. XXXIV , p. 33. New York: Acad. Press. 1966; for optical processes in general see also W. Heitler: Quantum Theory of Radiation. Oxford: Clarendon. 1954.

[2] E.g. L. I. Schiff: Quantum Mechanics. New York. McGraw-Hill. 1968.

*) The reader who is not familiar with wave mechanics may find it profitable to reserve this chapter for a later reading. More comprehensive treatments have been given in ref.1.

and therefore in the Hamiltonian, $(\hbar k_x - eA_x)^2$ has to be replaced by

$$(\hbar k_x - eA_x)(\hbar k_x - eA_x) = \hbar^2 k_x^2 - 2eA_x \hbar k_x + ie\hbar \partial A_x/\partial x + e^2 A_x^2 \quad (11i.2)$$

For low light intensities the term $\propto A^2$ can be neglected in comparison to the linear terms. The term $\propto(\vec{\nabla}_r \vec{A})$ can be made to vanish by a suitable gauge transformation [3]. Hence, the time-dependent Schrödinger equation is given by

$$(H_0 + H_1)\psi = i\hbar \partial\psi/\partial t \quad (11i.3)$$

where the interaction between the electron and the radiation field is represented by the operator H_1:

$$H_1 = i(e\hbar/m_0)(\vec{A}\vec{\nabla}_r) \quad (11i.4)$$

In a perturbation treatment the matrix element for a transition of an electron from state $\vec{k}$ to state $\vec{k}'$ (e.g. valence band to conduction band) is given by

$$H_{k'k} = \int \psi_{k'}^* H_1 \psi_k \, d^3r \quad (11i.5)$$

The electron wave functions are the usual Bloch functions $u_x \exp\{i(\vec{k}\vec{r})\}$ (Chap. 2a). Since the momentum of the absorbed or emitted photon at optical frequencies is much smaller than the average electron momentum and no phonon is assumed to be involved, the transition in the energy-vs-momentum diagram is nearly vertical ($\vec{k}' \approx \vec{k}$).

With $\vec{A}$ given by Eq.(11b.7) we obtain for the matrix element

$$H_{k'k} = i(e\hbar A_0/2m_0) \int u_{k'}^* \{(\vec{a}\vec{\nabla}_r u_k) + i(\vec{a}\vec{k})u_k\} d^3r \quad (11i.6)$$

For an allowed transition the second term in the integrand is much smaller than the first term and can be neglected. It would vanish for an exactly vertical transition in the ϵ-vs-$\vec{k}$ diagram. With the definition of a momentum matrix element

$$\vec{p}_{k'k} = -i\hbar \int u_{k'}^* \vec{\nabla}_r u_k \, d^3r \quad (11i.7)$$

we can write for $H_{k'k}$:

$$H_{k'k} = -(eA_0/2m_0)(\vec{a}\vec{p}_{k'k}) \quad (11i.8)$$

The transition rate R defined by Eq.(11b.4) is given by the right-hand side of Eq.(6b.14) with the transition probability S(k,k') given by Eq.(6b.13)

$$R = \frac{2V}{(2\pi)^3} \int d^3k \, (2\pi/\hbar)(eA_0/2m_0)^2 (\vec{a}\vec{p}_{k'k})^2 \, \delta(\epsilon' - \epsilon - \hbar\omega) \quad (11i.9)$$

where V is the volume of the unit cell, a factor of 2 takes care of the possible change in spin in the photon absorption process, and the initial state has been assumed occupied and the final state vacant, i.e. f(k) = 1; f(k') = 0. Furthermore, induced emission of photons can be neglected because the band gap is usually large compared with $k_B T$.

For an evaluation of the integral which contains a delta function consider the

[3] See footnote Chap.11b.

footnote on p. 180. The argument of the delta function, $\epsilon' - \epsilon - \hbar\omega$, is a function of $\vec{k}$. Since the photon momentum is negligibly small, we have $k' = \vec{k}$ for the transition (assumed to be "direct") and we write for the delta function

$$\delta(\epsilon' - \epsilon - \hbar\omega) = \delta(k' - k)/|\vec{\nabla}_k(\epsilon' - \epsilon)|_{\epsilon' - \epsilon = \hbar\omega} \tag{11i.10}$$

Since the initial state is located in the valence band and the final state in the conduction band it is convenient to introduce subscripts v and c, respectively. Because there are points in the energy band diagram where $\vec{\nabla}_k \epsilon_c(\vec{k}) = \vec{\nabla}_k \epsilon_v(k)$ and therefore the denominator vanishes, there are large contributions to R at these "critical points". Compared with the delta function, the factor $(\vec{a}\vec{p}_{vc})^2$ is only a slowly varying function of $\vec{k}$ and can therefore be taken outside the integral. The product

$$f_{vc} = (2V/m_o \hbar\omega)(\vec{a}\vec{p}_{vc})^2 \tag{11i.11}$$

is called the dimensionless "oscillator strength" of the transition and the integral

$$J_{vc}(\omega) = \frac{2}{(2\pi)^3} \int \delta\{\epsilon_c(\vec{k}) - \epsilon_v(\vec{k}) - \hbar\omega\}\, d^3k \tag{11i.12}$$

the "joint density of states" (dimension $(eV)^{-1}\,cm^{-3}$). Now we obtain for the transition rate R

$$R = \frac{\pi e^2 A_o^2 \omega}{4m_o} f_{vc} J_{vc}(\omega) \tag{11i.13}$$

and from Eq.(11b.12) for the absorption coefficient

$$a = \frac{\mu_o c}{n} \frac{\pi e^2 \hbar}{2m_o} f_{vc} J_{vc}(\omega) \tag{11i.14}$$

where n is the refractive index. The factor $\mu_o c\pi e^2 \hbar/2m_o = 1.098 \times 10^{-16}\,eVcm^2$.

Now let us consider the joint density of states for the simple case of spherical energy surfaces at $\vec{k} = 0$ where

$$\hbar\omega = \epsilon_G + \hbar^2 k^2/2m_n + \hbar^2 k^2/2m_p = \epsilon_G + \hbar^2 k^2/2m_r \tag{11i.15}$$

where $m_r = (m_n^{-1} + m_p^{-1})^{-1}$ is a reduced effective mass. Let us for simplicity denote $\hbar\omega$ by ϵ. From Eq.(11i.12) we now obtain for J_{vc}

$$J_{vc} = \frac{2}{(2\pi)^3} \frac{d}{d\epsilon}\left(\frac{4\pi}{3} k^3\right) = \frac{1}{2\pi^2}\left(\frac{2m_r}{\hbar^2}\right)^{3/2} (\epsilon - \epsilon_G)^{1/2} \tag{11i.16}$$

The dependence of a on $(\hbar\omega - \epsilon_G)^{1/2}$ is just the observed frequency dependence for a direct transition. From this value of J_{vc} we finally obtain for a in units of cm^{-1}

$$a = \frac{2.7 \times 10^5}{n} (2m_r/m_o)^{3/2} f_{vc} \{(\hbar\omega - \epsilon_G)/eV\}^{1/2} \tag{11i.17}$$

Assuming e.g. $\hbar\omega - \epsilon_G = 10$ meV; $m_r = m_o/2$; $f_{vc} = 1$; and $n = 4$; we find $a = 0.67 \times 10^4\,cm^{-1}$. This is the observed order of magnitude to which a rises at the fundamental absorption edge.

It has not yet been become clear why the oscillator strength is of the order of magnitude 1, i.e. what $\vec{p}_{vc}$ defined by Eq.(11i.7) really means. However, since the Bloch wave functions in the conduction band and the valence band in general are not known explicitly except for their symmetry, we will only note that it depends on the polarization of the light and that selection rules for the transitions can be determined from these symmetry properties. As for atomic spectra, an f-sum rule has been shown to exist [4] which makes plausible that $f_{vc} \approx 1$ for allowed dipole transitions.

It is very interesting to investigate the function $J_{vc}(\omega)$ for various types of critical points. Taking Eq.(11i.10) into account we write for Eq.(11i.12)

$$J_{vc}(\omega) = \frac{2}{(2\pi)^3} \int \frac{dS}{|\vec{\nabla}_k \{\epsilon_c(\vec{k}) - \epsilon_v(\vec{k})\}|_{\epsilon_c - \epsilon_v = \hbar\omega}} \tag{11i.18}$$

where dS is an element of an energy surface $\epsilon_c(\vec{k}) - \epsilon_v(\vec{k}) = \hbar\omega$ in $\vec{k}$-space. Let us for simplicity assume a 2-dimensional $\vec{k}$-space and an energy surface [5]

$$\epsilon_c(\vec{k}) - \epsilon_v(\vec{k}) = \epsilon_G - A\{\cos(k_x a) + \cos(k_y a) - 2\} \tag{11i.19}$$

where $a > 0$ is a constant and ϵ_G is the energy gap at $\vec{k} = 0$. The van Hove singularities are the solutions of

$$\left.\begin{aligned} \partial(\epsilon_c - \epsilon_v)/\partial k_x &= Aa \sin(k_x a) = 0 \\ \partial(\epsilon_c - \epsilon_v)/\partial k_y &= Aa \sin(k_y a) = 0 \end{aligned}\right\} \tag{11i.20}$$

in the first Brillouin zone :

$$\vec{k} = (0,0);\ (0, \pm \pi/a);\ (\pm \pi/a, 0);\ (\pm \pi/a, \pm \pi/a) \tag{11i.21}$$

We expand the energy bands around these points according to $\cos x \approx 1 - x^2/2$:

$$\begin{aligned} &\text{minimum at } (0,0): \\ &\text{maximum at } (\pi/a, \pi/a): \epsilon - \epsilon_G = \\ &\text{saddle point at } (0, \pi/a): \end{aligned} \begin{cases} \frac{1}{2} Aa^2(\Delta k_x^2 + \Delta k_y^2) \\ -\frac{1}{2} Aa^2(\Delta k_x^2 + \Delta k_y^2) + 4A \\ \frac{1}{2} Aa^2(\Delta k_x^2 - \Delta k_y^2) + 2A \end{cases} \tag{11i.22}$$

The joint density of states at the extremal points is simply obtained *)

$$J_{vc}(\omega) = \frac{2}{(2\pi)^2} \left| \frac{d}{d\epsilon} \{\pi(\Delta k)^2\} \right| = \begin{cases} 1/\pi A a^2 \text{ for } \epsilon > \epsilon_G \text{ at } (0,0) \\ 1/\pi A a^2 \text{ for } \epsilon < \epsilon_G + 4A \text{ at } (\pi/a, \pi/a) \end{cases} \tag{11i.23}$$

[4] A. H. Wilson: The Theory ot Metals, p. 46. London: Cambridge Univ. Press. 1953.

[5] M. Cardona: Problems in Solid State Physics (H. J. Goldsmid, ed.), p. 425. London: Pion Ltd.; New York: Acad. Press. 1968.

*) This is again a result of the equation of the footnote on p. 180: $\int \delta\{\epsilon(k_x, k_y)\} dk_x dk_y = |(d/d\epsilon) \int dk_x dk_y|$.

and zero otherwise while for the saddle point $(0, \pi/a)$ it is

$$J_{vc}(\omega) = \frac{2}{(2\pi)^2}\int_L \frac{ds}{|\vec{\nabla}_k \epsilon|} = \frac{2}{(2\pi)^2}\, 4 \int_{\Delta k_{xm}}^{\Delta k_{xM}} \frac{d(\Delta k_x)}{|\partial\epsilon/\partial\Delta k_y|} = \frac{2}{\pi^2 A a^2}\int_{\Delta k_{xm}}^{\Delta k_{xM}} \frac{d(\Delta k_x)}{\Delta k_y} \qquad (11i.24)$$

where ds is an element of and L the complete constant-energy line on the two-dimensional $\epsilon(k_x, k_y)$ surface and Δk_{xm} and Δk_{xM} are the minimum and the maximum values, respectively, of Δk_x on this line. Solving Eq.(11i.22) ("saddle point") for Δk_y and evaluating the integral in Eq.(11i.24), we obtain

$$J_{vc}(\omega) = \frac{2}{\pi^2 A a^2}\{\ln(\Delta k_{xM} + \sqrt{\Delta k_{xM}^2 - 2(\epsilon-\epsilon_G-2A)/Aa^2}) - \ln\sqrt{2(\epsilon-\epsilon_G-2A)/Aa^2}\} \qquad (11i.25)$$

where $\Delta k_{xm} = \sqrt{2(\epsilon-\epsilon_G-2A)/Aa^2}$ has been taken into account. (This is found from Eq.(11i.22) for $\Delta k_y = 0$ and valid for a positive value of the radicand) and $\Delta k_{xM} \approx \pi/a$. Eq.(11i.25) shows that $J_{vc}(\omega)$ has a logarithmic singularity at the saddle point, where $\epsilon = \epsilon_G + 2A$. The singularity is shown in Fig.11.49. Such singularities have first been analysed by van Hove [6] and the points in the Brillouin zone where they occur have been identified by Phillips [7].

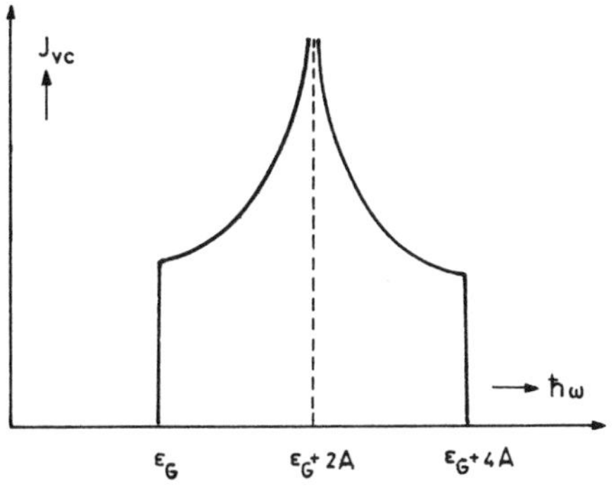

Fig.11.49 Joint density of states for a two-dimensional surface of constant energy.

In three dimensions the expansion around a critical point

$$\epsilon(\vec{k}) = \epsilon_c(\vec{k}) - \epsilon_v(\vec{k}) = \epsilon_0 + \frac{a^2}{2}\sum_{i=1}^{3} A_i(\Delta k_i)^2 \qquad (11i.26)$$

contains three coefficients A_i which may be positive or negative. Saddle points are characterized by either one negative and two positive coefficients (denoted "M_1") or two negative and one positive ("M_2"). For three positive coefficients we find a minimum ("M_0") where J_{vc} varies with $\epsilon = \hbar\omega$ as given by Eq. (11i.16) while for all coefficients being negative there is a maximum ("M_3") and $J_{vc} \propto (\epsilon_G - \hbar\omega)^{1/2}$. The functions $J_{vc}(\omega)$ are shown schematically in Fig.11.50. For a comparison, let us consider again the $\kappa_i(\omega)$ spectrum for Ge shown in Fig. 11.3. It is fairly obvious that e.g. the edge denoted by $\Lambda_3 \to \Lambda_1$ is of the M_1 type. In Fig.2.26 the transition has been indicated in the energy band structure. Although it is not possible to see the saddle point in this diagram one notices that $\vec{\nabla}_k \epsilon_c = \vec{\nabla}_k \epsilon_v$ for the $k_{\langle 111\rangle}$ direction indicated there. Another transition of interest to us is the direct transition from the valence band maximum to the conduction band minimum at $\vec{k} = 0$ which is of course of the M_0 type (Fig.11.3).

[6] L. van Hove, Phys. Rev. 89 (1953) 1189.
[7] J. C. Phillips, Phys. Rev. 104 (1956) 1263.

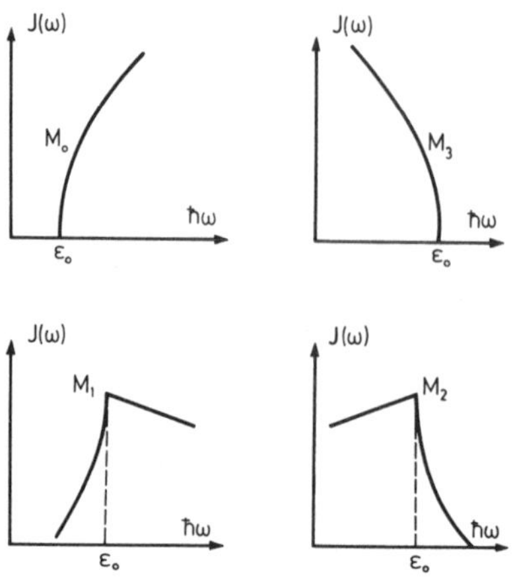

Fig.11.50 Joint density of states near critical points; the subscript of M indicates the number of negative coefficients in the expansion of the energy difference between the bands as a function of $\vec{k}$ (after Bassani, ref. 1).

11j. Free-Carrier Absorption and Reflection

Light absorption by free carriers (i.e. conduction electrons or holes) shall first be treated by the classical theory of the harmonic oscillator in the limit of vanishing binding energy $\hbar\omega_e$. For this case Eqs.(11g.6) and (11g.7) yield

$$n^2 - k^2 = \kappa_r = \kappa_{opt} - \omega_p^2/(\omega^2 + \gamma^2) \tag{11j.1}$$

and

$$2nk = \kappa_i = \omega_p^2\gamma/\{\omega(\omega^2 + \gamma^2)\} \tag{11j.2}$$

It will be shown below by a quantum mechanical treatment that the damping constant γ is essentially the inverse momentum relaxation time, τ_m^{-1}. For the case of small damping we substitute for n in Eq.(11j.2) its value $\sqrt{\kappa_{opt}}$ obtained from Eq.(11j.1) for $\gamma = 0$ and for frequencies ω which are well above the plasma frequency ω_p. The extinction coefficient k is then given by

$$k = \omega_p^2\gamma/(2\omega^3\sqrt{\kappa_{opt}}) \tag{11j.3}$$

and the absorption coefficient a by

$$a = k4\pi/\lambda = \frac{377\ \text{ohm}}{\sqrt{\kappa_{opt}}}\ \sigma_o/(\omega\tau_m)^2 \propto \lambda^2 \tag{11j.4}$$

where σ_o is the dc conductivity and γ has been replaced by τ_m^{-1}. E.g. for τ_m = $2\text{x}10^{-13}$ sec and a wavelength of $\lambda = 1\mu$m we find $\omega\tau_m \approx 10^2$. A refractive index of e.g. $\sqrt{\kappa_{opt}} = 4$ and a conductivity of e.g. $\sigma_o = 10^2\,(\text{ohm cm})^{-1}$ yield $a \approx 1\ \text{cm}^{-1}$ which is small compared with the fundamental absorption where $a \approx 10^4\ \text{cm}^{-1}$. However, in metals where $\sigma_o \approx 10^6\,(\text{ohm cm})^{-1}$ the free-carrier absorption coefficient has the same order of absolute magnitude as the fundamental absorption in semiconductors or is even larger (e.g. Ag at $\lambda = 0.6\,\mu$m and 300 K : k = 4.09, n = 0.05).

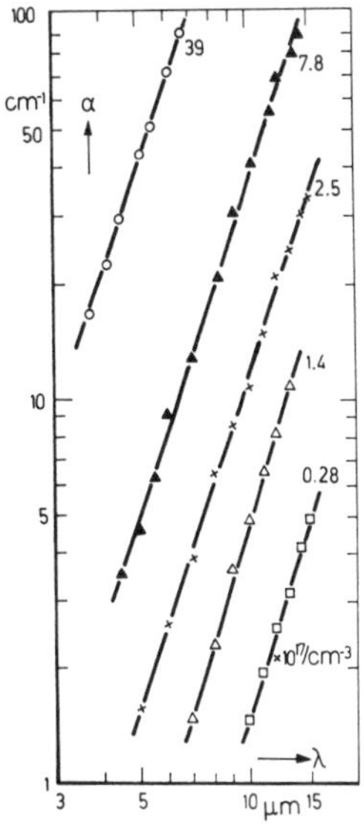

Fig. 11.51 Free-carrier absorption spectrum in n-type indium arsenide for various carrier densities at room temperature (after J. R. Dixon: Proc. Int. Conf. Phys. Semicond. Prague, 1960, p. 366. Prague: Czech. Acad. Sci. 1960).

Eq. (11j.4) suggests a dependence of α on the second power of the wavelength. Fig. 11.51 shows a log-log plot of α vs λ obtained in n-InAs at room temperature for various electron densities. Although a power law is found, the slope of the curves is rather compatible with a λ^3-dependence for α. This discrepancy is resolved by a quantum mechanical treatment of the problem which is different for the various kinds of electron phonon or electron impurity interactions. Fig. 11.52 shows that in fact an absorption or emission process for a photon of energy $\hbar\omega = \Delta\epsilon$ must involve a change in momentum $\hbar q$; this is much larger than the photon momentum itself. The momentum difference may be supplied by a phonon with a momentum $\hbar\vec{q} = \hbar(\vec{k} - \vec{k}')$ having an energy $\hbar\omega_q$. This phonon may either be absorbed or emitted.

Since 3 particles (electron, photon, and phonon) are involved in the absorption process we treat the problem with second-order perturbation theory*) where an "intermediate" or "virtual state" k' is assumed to exist intermediate between the initial state k and the final state k". The transition from k to k" may take place first by absorption of a photon $\hbar\omega$ (transition to state n) and then the subsequent absorption or emission of a phonon $\hbar\omega_q$, or by the reverse process. In addition, we will consider here also "induced emission" of light which is not contained in classical theory, again with either emission or absorption of a phonon. There are altogether 4 processes indicated in Fig. 11.52.

The transition probability S in second-order perturbation theory is given by [3]

[1] A. Yariv: Quantum Electronics, chap. 21.7. New York: J. Wiley and Sons. 1968.

[2] H. Y. Fan, W. Spitzer, and R. J. Collins, Phys. Rev. 101 (1956) 566; H. Y. Fan: Repts. Progr. Phys., Vol. 14., p. 119. London: The Physical Society. 1956.

[3] L. I. Schiff: Quantum Mechanics, p. 247. New York: McGraw-Hill. 1968.

*) The reader who is not familiar with quantum mechanics may prefer to reserve this treatment for a later reading. The calculation has been adapted from ref. 1. The method has been reported by Fan et al. [2].

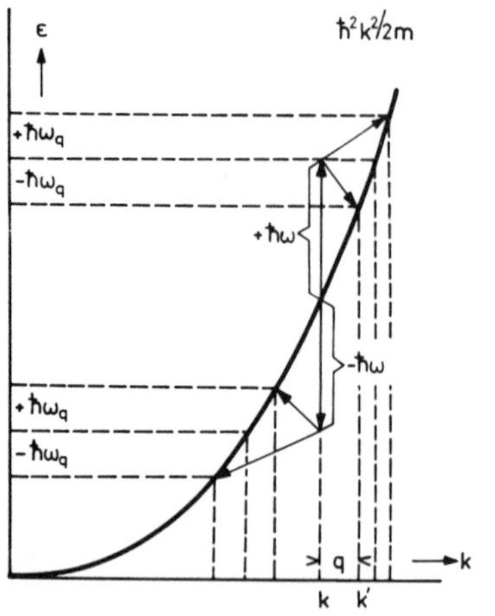

Fig. 11.52 Change in carrier momentum as a consequence of a change in energy for intravalley transitions by photon absorption or emission.

$$S_{\pm} = \frac{2\pi}{\hbar}\{|H_{k'k}|^2_{\pm}\,|H_{k''k'}|^2/\hbar^2\omega^2\}\,\delta(\epsilon' - \epsilon - \hbar\omega \pm \hbar\omega_q) \tag{11j.5}$$

where $\hbar\omega$ is the photon energy, $H_{k'k}$ is the matrix element of the hamiltonian for phonon emission or absorption (subscript + or −, respectively), and $H_{k''k'}$ is the corresponding matrix element for photon absorption. For the process of induced emission of photons the transition probability is denoted by $S'_{\pm}$; it differs from $S_{\pm}$ only by the sign of $\hbar\omega$ in the argument of the delta function.

The transition rate R is given by the right-hand side of Eq.(6b.14) except that we now have to integrate also over all initial states k where the spin degeneracy introduces a factor of 2 :

$$R_{\pm} = \frac{2V}{(2\pi)^6}\int d^3k\, d^3k'\, S_{\pm} f(k)\{1 - f(k')\} \tag{11j.6}$$

We will limit our treatment to a non-degenerate gas of carriers with an electron temperature T_e, where we can neglect f(k'), and for f(k) we write

$$f(k) = (n/N_c)\exp(-\epsilon/k_B T_e) = (n/2)\,(2\pi\hbar^2/mk_B T_e)^{3/2}\exp(-\epsilon/k_B T_e) \tag{11j.7}$$

where n is the carrier concentration. It is convenient to replace d^3k' in Eq.(11j.6) by d^3q where $\vec{q}$ is the phonon wave vector.

The matrix element $H_{k''k'}$ is given by the right-hand side of Eq.(11i.8) except that we now introduce for the momentum matrix element divided by the electron mass

$$\vec{p}_{k'k}/m_0 \approx \hbar(\vec{k}' - \vec{k})/m = \hbar\vec{q}/m \tag{11j.8}$$

neglecting as in Chap. 11i the photon momentum and taking into account the crystal potential by introducing the carrier effective mass, m. Hence, we find for $R_{\pm}$ from Eq.(11j.6) :

$$R_{\pm} = \frac{Ve^2 A_0^2 n}{2(2\pi)^5 \hbar\omega^2 m^2 N_c}\int |H_{k'k}|^2_{\pm}\,(\vec{a}\vec{q})^2\,\delta(\epsilon' - \epsilon - \hbar\omega \pm \hbar\omega_q)\exp(-\epsilon/k_B T_e)\,d^3k\,d^3q \tag{11j.9}$$

In general the matrix element $H_{k'k}$ depends on q, but not on k. Therefore, the integral over k can be performed rightaway. The integral of the delta function over θ and φ yields a factor :

$$\int\delta\{(\hbar^2/2m)(q^2 + 2kq\cos\theta) - \hbar\omega \pm \hbar\omega_q\}\sin\theta\, d\theta\, d\varphi = -2\pi m/\hbar^2 kq \tag{11j.10}$$

and the integral over $k^2\,dk$:

$$-2\pi m/\hbar^2 q \int_{k_{min}}^{\infty} \exp(-\hbar^2 k^2/2mk_B T_e) k dk = (2\pi m^2 k_B T_e/\hbar^4 q) \exp(-\hbar^2 k_{min}^2/2mk_B T_e) \quad (11j.11)$$

where the lower limit k_{min} is found from energy conservation (argument of the delta function = 0) for $\cos\theta = 1$:

$$k_{min}^2 = \left(\frac{m(\hbar\omega \mp \hbar\omega_q)}{\hbar^2 q} - \frac{q}{2}\right)^2 \quad (11j.12)$$

while $k_{max}^2 = \infty$ is obtained from $\cos\theta = 0$*). It is convenient to introduce the parameter

$$z_\pm = (\hbar\omega \pm \hbar\omega_q)/2k_B T_e \quad (11j.13)$$

(subscript + indicates phonon absorption, subscript − emission) and the variable

$$\xi = \hbar^2 q^2/4mk_B T_e \quad (11j.14)$$

We now obtain for Eq.(11j.11)

$$\int \delta(\epsilon' - \epsilon - \hbar\omega \pm \hbar\omega_q) \exp(-\epsilon/k_B T_e) d^3k = \pi\hbar^{-3} m^{3/2} (k_B T_e)^{1/2} \exp(z_\mp) \xi^{-1/2} \exp\{-\tfrac{1}{2}(\xi + z_\mp^2/\xi)\} \quad (11j.15)$$

For the integral over $d^3q = q^2 dq \sin\theta' d\theta' d\varphi'$ we introduce the angle θ' between the unit vector $\vec{a}$ of the light polarization and the phonon wave vector $\vec{q}$:

$$\int_0^\pi (\vec{a}\vec{q})^2 \sin\theta' d\theta' d\varphi' = \int_0^\pi q^2 \cos^2\theta' \sin\theta' d\theta' d\varphi' = \xi\, 16\pi m k_B T_e/3\hbar^2 \quad (11j.16)$$

and finally obtain for the transition rate $R_\pm$

$$R_\pm = \frac{Ve^2 A_0^2 n}{2(2\pi)^5 \hbar\omega^2 m^2 N_c} \frac{\pi m^{3/2}(k_B T_e)^{1/2}}{\hbar^3} \frac{16\pi m k_B T_e}{3\hbar^2} \frac{(4mk_B T_e)^{3/2}}{2\hbar^3} \cdot \int_0^\infty |H_{k'k}|_\pm^2 e^{z_\mp} \exp\{-\tfrac{1}{2}(\xi + z_\mp^2/\xi)\} \xi d\xi \quad (11j.17)$$

and from Eq.(11b.12) for the absorption coefficient

$$a_\pm = R_\pm 2\hbar\mu_0 c/(\sqrt{\kappa_{opt}} A_0^2 \omega) = \frac{\mu_0 c}{\sqrt{\kappa_{opt}}} \cdot \frac{2^{3/2}}{3\pi^{3/2}} \cdot \frac{ne^2 m^{1/2}(k_B T_e)^{3/2} V}{\hbar^5 \omega^3} \int_0^\infty |H_{k'k}|_\pm^2 e^{z_\mp} \exp\{-\tfrac{1}{2}(\xi + z_\mp^2/\xi)\} \xi d\xi \quad (11j.18)$$

The limits of integration have been extended to 0 and ∞ in view of the exponential factor in the integrand which vanishes at both limits. Note $\kappa_{opt} = \kappa_{opt}(\omega)$.

Let us first consider the simple case where the matrix element is independent of q and therefore of ξ so that we can take it outside the integral. According to Appendix B, Eq.(B.5), the integral is given by

*) In a crystal k_{max} is actually of the order of magnitude π/a where a is the lattice constant; since the integrand depends on k^2 exponentially and is very small for $k = \pi/a$, we replace π/a by ∞.

$$\int_0^\infty |H_{k'k}|^2_\pm e^{z_\mp} \exp\{-\frac{1}{2}(\xi + z_\mp^2/\xi)\}\xi d\xi = |H_{k'k}|^2_\pm e^{z_\mp} 2z_\mp^2 K_2(|z_\mp|) \qquad (11j.19)$$

where K_2 is a modified Bessel function.

Another case of interest is that of $H_{k'k} \propto 1/q$. For this case we obtain according to Eq.(B.4)

$$\int_0^\infty q^{-2} e^{z_\mp} \exp\{-\frac{1}{2}(\xi + z_\mp^2/\xi)\}\xi d\xi = \frac{\hbar^2}{4mk_BT_e} e^{z_\mp} 2|z_\mp| K_1(|z_\mp|) \qquad (11j.20)$$

The former case is realized for acoustic and optical deformation potential scattering while the latter case pertains to piezoelectric and polar optical scattering.

For ionized impurity scattering no phonon is involved, hence $z_\pm$ becomes $z = \hbar\omega/2k_BT_e$ and since $H_{k'k} \propto 1/q^2$ for screening neglected we find

$$\int_0^\infty q^{-4} e^{z} \exp\{-\frac{1}{2}(\xi + z^2/\xi)\}\xi d\xi = (\frac{\hbar^2}{4mk_BT_e})^2 e^z 2K_0(|z|) \qquad (11j.21)$$

Second-order perturbation theory is applicable since the "third particle" is the ionized impurity atom.

Let us first consider the case of acoustic deformation potential scattering. The matrix element given by Eq.(6d.13) is the same for phonon emission and absorption. Since the phonon energy is very small, $\hbar\omega_q \ll \hbar\omega$, Eq.(11j.13) yields $z_\pm \approx \hbar\omega/2k_BT_e$. Hence, for photon absorption (subscript a) we find

$$a_a = \frac{\mu_0 c}{\sqrt{\kappa_{opt}}} \cdot \frac{2^{5/2}}{3\pi^{3/2}} \cdot \frac{ne^2\epsilon_{ac}^2 m^{1/2} k_BT(k_BT_e)^{3/2}}{\hbar^5 c_l \omega^3} (\frac{\hbar\omega}{2k_BT_e})^2 \exp(\frac{\hbar\omega}{2k_BT_e}) K_2(\frac{\hbar\omega}{2k_BT_e}) \qquad (11j.22)$$

where a factor of 2 has been introduced to account for both phonon absorption and emission processes. We still have to treat the induced photon emission (subscript e). This process is obtained from Eqs.(11j.5), (11j.9), etc. simply by changing the sign of $\hbar\omega$. Since the argument of K_2 is always positive we obtain from Eq.(11j.19)

$$a_e \propto |H_{k'k}|^2_\pm e^{-z_\pm} 2z_\pm^2 K_2(|z_\pm|) \propto (\frac{\hbar\omega}{2k_BT_e})^2 \exp(-\frac{\hbar\omega}{2k_BT_e}) K_2(\frac{\hbar\omega}{2k_BT_e}) \qquad (11j.23)$$

where the right-hand side is of course valid for acoustic scattering. Since the emission term a_e has to be subtracted from a_a we find for the overall absorption $a = a_a - a_e \propto e^z - e^{-z} = 2\sinh(z)$:

$$a = \frac{\mu_0 c\, 2^{3/2} ne^2 \epsilon_{ac}^2 (mk_BT)^{1/2}}{\sqrt{\kappa_{opt}}\, 3\pi^{3/2}\hbar^3 c_l \omega} (\frac{T}{T_e})^{1/2} \sinh(\frac{\hbar\omega}{2k_BT_e}) K_2(\frac{\hbar\omega}{2k_BT_e}) \qquad (11j.24)$$

Let us first consider the case of thermal equilibrium, T_e=T, and high temperatures, $2k_BT \gg \hbar\omega$. Since for this case $\sinh(z) \approx z$ and $K_2(z) \approx 2/z^2$, we obtain

$$a \approx \frac{\mu_0 c\, 2^{7/2} ne^2 \epsilon_{ac}^2 m^{1/2}(k_BT)^{3/2}}{\sqrt{\kappa_{opt}}\, 3\pi^{3/2}\hbar^4 c_l \omega^2} \qquad (11j.25)$$

We simplify by introducing the dc mobility given by Eq.(6d.18) and the dc conductivity $\sigma_o = ne\mu$; neglecting a factor of $32/9\pi \approx 1.13$, we finally obtain Eq. (11j.4). This proves that we were justified in replacing the damping constant γ in the classical model by τ_m^{-1}. Because $\hbar\omega \ll 2k_BT$, it is also correct to neglect the induced emission in the classical treatment. The dependence of a on the carrier temperature T_e in this approximation is just as we expect from Eq.(6d.17)

$$a \propto \gamma = \langle \tau_m^{-1} \rangle \propto T_e^{-1/2} \qquad (11j.26)$$

i.e. there is no increase of the optical absorption with increasing electron temperature. Since for hot carriers according to Eq.(6e.25) T_e increases linearly with the applied dc field E, we find $a \propto E^{1/2}$ for the classical Drude model.

In the quantum limit, $\hbar\omega \gg 2k_BT_e$, the absorption becomes independent of T_e:

$$a \approx \frac{\mu_o c}{\sqrt{\kappa_{opt}}} \cdot \frac{2^{1/2}}{3\pi} \cdot \frac{ne^2\epsilon_{ac}^2 m^{1/2} k_B T}{\hbar^{7/2} c_l \omega^{3/2}} = a_{cl} \frac{(\pi\hbar\omega)^{1/2}}{8(k_BT)^{1/2}} \qquad (11j.27)$$

where a_{cl} is the classical zero-field absorption given by Eq.(11j.25). The absorption is stronger in the quantum limit than it is in the classical limit, but the wavelength dependence is less strong ($\propto \lambda^{3/2}$, compared with λ^2 in the classical limit). Fig.11.53 shows this dependence over a broad wavelength range for various temperatures T in a log-log plot. The $\lambda^{3/2}$ and λ^2 dependences are indicated by straight lines (dash-dotted and dashed, respectively), fitted to the 77 K curve at low and high values of λ, respectively.

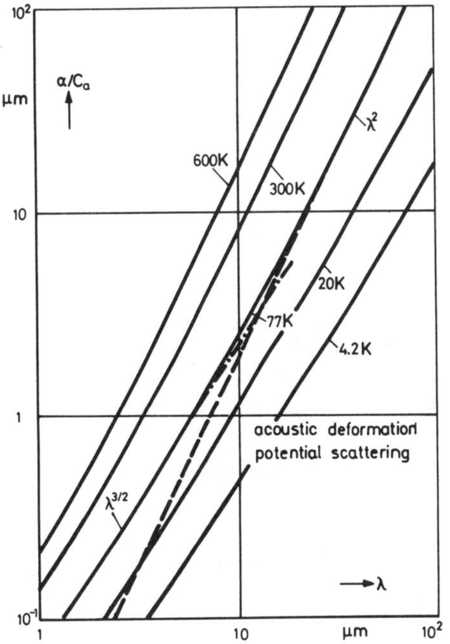

Fig. 11.53 Free-carrier optical absorption coefficient as a function of the wavelength of light for acoustic deformation potential scattering. C_a is a factor of proportionality.

The ratio of absorption in an applied field E and zero-field absorption is given by

$$\frac{a(E)}{a(o)} = \left(\frac{T}{T_e}\right)^{1/2} \cdot \frac{\sinh(\hbar\omega/2k_BT_e)\, K_2(\hbar\omega/2k_BT_e)}{\sinh(\hbar\omega/2k_BT)\, K_2(\hbar\omega/2k_BT)} \qquad (11j.28)$$

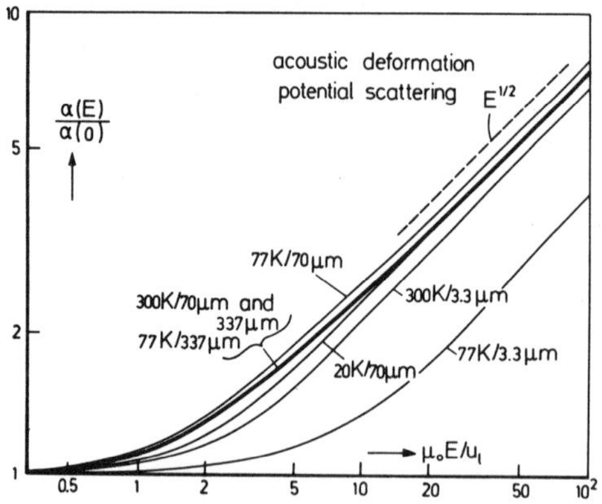

Fig. 11.54 Electric-field dependence of the free-carrier absorption for acoustic deformation potential scattering. The dashed curve shows an $E^{1/2}$ dependence.

In Fig. 11.54 the ratio $a(E)/a(0)$ is plotted vs $\mu_0 E/u_l$ where the $T_e(E)$ dependence as given by Eq.(6e.23) has been taken into account. An $E^{1/2}$ dependence is indicated by the dashed line. At high field strengths the carrier temperature is high enough that classical optics is applicable in any case.

Numerically the absorption cross section is given by

$$\frac{a}{n} = \frac{(4.693\ \text{Å})^2 (\lambda/\mu m) \sinh\{(7195.3/(\lambda/\mu m)(T_e/K)\} K_2\{7195.3/(\lambda/\mu m)(T_e/K)\}}{\sqrt{\kappa_{opt}}\,(m/m_0)^2 (\mu/cm^2 V^{-1} sec^{-1})(T_e/100\,K)} \tag{11j.29}$$

where $\mu = \mu(T_e) \propto T_e^{-1/2}$ is the acoustic mobility.

For piezoelectric scattering, the matrix element is proportional to $1/q$ according to Eq.(6g.8), and Eq.(11j.20) can be applied :

$$a = \frac{\mu_0 c\, 2^{1/2} n e^4 K^2 (k_B T)^{1/2}}{\sqrt{\kappa_{opt}}\; 3\pi^{3/2} \hbar^2 \kappa\kappa_0 m^{1/2} \omega^2} (T/T_e)^{1/2} \sinh(\hbar\omega/k_B T_e)\, K_1(\hbar\omega/2k_B T_e) \tag{11j.30}$$

In the classical limit the factor $\sinh(\hbar\omega/2k_BT_e)\cdot K_1(\hbar\omega/2k_BT_e) \approx 1$ and $a \propto \lambda^2$ while in the quantum limit it is $\approx (\pi k_B T_e/\hbar\omega)^{1/2}$ and $a \propto \lambda^{2.5}$ independent of electron temperature T_e. Except for the stronger wavelength dependence, this case is quite similar to acoustic deformation potential scattering. In contrast to the latter, however, a decreases with T_e for $T_e \geqslant \hbar\omega/3k_B = 4800\,K/(\lambda/\mu m)$ while below this range it increases as normal.

For numerical purposes we introduce the mobility μ given by Eq.(6g.17) dependent on $T_e \propto \langle\epsilon\rangle$ as given by Eq.(6g.16) and obtain for the absorption cross section a/n :

$$\frac{a}{n} = \frac{(0.771\ \text{Å})^2 (\lambda/\mu m)^2}{\sqrt{\kappa_{opt}}\,(\mu/cm^2 V^{-1} sec^{-1})(m/m_0)^2} \sinh\left(\frac{7195.3}{(\lambda/\mu m)(T/K)}\right) K_1\left(\frac{7195.3}{(\lambda/\mu m)(T/K)}\right) \tag{11j.31}$$

Because of the high mobilities found for piezoelectric scattering in Chap. 6 g, the optical absorption due to this process can usually be neglected.

Let us consider now optical deformation potential scattering. The matrix element is given by Eq.(6k.3). It is independent of q and therefore Eq.(11j.19) may be applied.*) ω_q is now the optical phonon frequency, ω_0. As before we will introduce $z = \hbar\omega_0/2k_BT$. Since $N_q + 1 = N_q e^{2z}$ we find for the 4 processes except for a common factor:

$$\left.\begin{aligned} &\alpha_{a+} \propto e^{z_+} z_+^2 K_2(z_+)\,; \quad \alpha_{a-} \propto e^{2z+z_-} z_-^2 K_2(|z_-|)\,; \\ &\alpha_{e+} \propto e^{-z_-} z_-^2 K_2(|z_-|)\,; \alpha_{e-} \propto e^{2z-z_+} z_+^2 K_2(z_+)\,; \end{aligned}\right\} \quad (11j.32)$$

and

$$\alpha = \alpha_{a+} + \alpha_{a-} - \alpha_{e+} - \alpha_{e-} \propto 2e^z \{\sinh(z_+ - z) z_+^2 K_2(z_+) + \sinh(z_- + z) z_-^2 K_2(|z_-|)\} \quad (11j.33)$$

For $2e^z N_q$ we write $1/\sinh(z)$ and finally obtain from Eq.(11j.18):

$$\alpha = \frac{\mu_o c}{\sqrt{\kappa_{opt}}} \cdot \frac{2^{3/2} n e^2 D^2 m^{1/2}}{3\pi^{3/2}\hbar^4 \rho\, \omega_o\, \omega^3} (k_B T_e)^{3/2} \frac{\sinh(z_+ - z)\, z_+^2 K_2(z_+) + \sinh(z_- + z)\, z_-^2 K_2(|z_-|)}{\sinh(z)} \quad (11j.34)$$

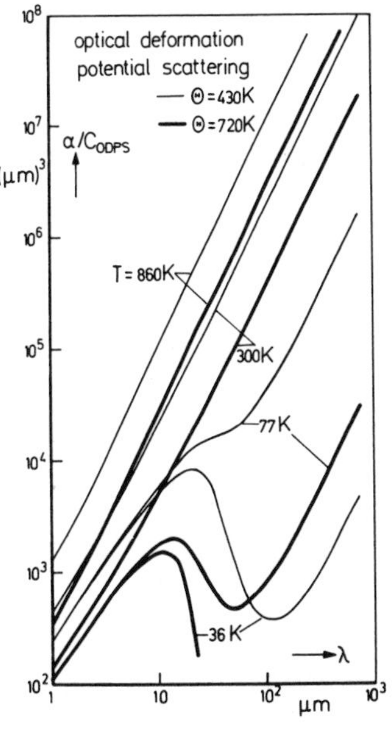

Fig. 11.55 Same as Fig. 11.53, but for optical deformation potential scattering.

For the case of thermal carriers, where $T_e = T$, we find $z_+ - z = z_- + z = \hbar\omega/2k_BT$. Introducing the mobility μ given by Eq.(6k.22), α/n becomes

$$\frac{\alpha}{n} = \frac{(4.27\times10^{-2}\,\text{Å})^2}{\sqrt{\kappa_{opt}}} \cdot \frac{(\theta/100\,\text{K})\,(\lambda/\mu\text{m})^3 \sinh(z_o)\{z_+^2 K_2(z_+) + z_-^2 K_2(|z_-|)\}}{(m/m_o)^2\,(\mu/\text{cm}^2\,\text{V}^{-1}\,\text{sec}^{-1})\,(\theta/2T)^3 K_2(\theta/2T)} \quad (11j.35)$$

where $z_o = 7195.3/(\lambda/\mu\text{m})\,(T/\text{K})$ and at present $z_\pm = z_o \pm \theta/2T$. In Fig. 11.55 the quantity

$$\lambda^3 \sinh(z_o) \frac{z_+^2 K_2(z_+) + z_-^2 K_2(|z_-|)}{(\theta/2T)^{3/2} \sinh(\theta/2T)} = \alpha/C_{ODPS} \quad (11j.36)$$

which defines a factor of proportionality, C_{OPDS}, is plotted vs λ for various values of lattice temperature T and Debye temperature θ. At long wavelengths, $\alpha \propto \lambda^2$. This is the same

*) Although for hot carriers the Maxwell-Boltzmann distribution may not be a good approximation of the true distribution it yields results in analytical form which should at least qualitatively be correct.

as for acoustic scattering. At low temperatures, however, there is a maximum of a at roughly $\{7200/(\theta/2)\}$ μm. This corresponds to $z_- = 0$. An inspection of the 4 components of a, given by Eq.(11j.32), reveals that the maximum is due to a_{a-}; at $\omega = \omega_0$ there is a resonance absorption where a photon is absorbed and subsequently the electron returns to its initial state by emitting an optical phonon of the same frequency. A comparison of Eq.(11j.34) with Eq.(11j.24) for $T_e = T$ and $\hbar\omega = \hbar\omega_0 \gg 2k_BT$ results in a ratio of the absorption coefficients for optical and acoustic deformation potential scattering

$$a_{opt}/a_{ac} = (4/\sqrt{\pi})\,(Du_l/\epsilon_{ac}\omega_0)^2\,\sqrt{T/\theta} \qquad (11j.37)$$

This is of the same order of magnitude as the ratio $(Du_l/\epsilon_{ac}\omega_0)^2$ which e.g. for n-Ge is about 0.2. In the quantum limit, $a_{opt} \propto \lambda^{1.5}$ for $\hbar\omega \gg \hbar\omega_0$.

For the case of hot carriers we obtain from Eq.(11j.34) for the ratio of absorption coefficients

$$\frac{a(E)}{a(0)} = \left(\frac{T_e}{T}\right)^{1/2} \frac{(z_0+z)^2 K_2(z_+)\sinh(z_+ - z) + (z_0 - z)^2 K_2(|z_-|)\sinh(z_- + z)}{\{(z_0+z)^2 K_2(z_0+z) + (z_0-z)^2 K_2(|z_0 - z|)\}\sinh(z_0)} \qquad (11j.38)$$

where $z_\pm = (\hbar\omega \pm \hbar\omega_0)/2k_BT_e$, $z_0 = \hbar\omega/2k_BT$, and $z = \hbar\omega_0/2k_BT$.

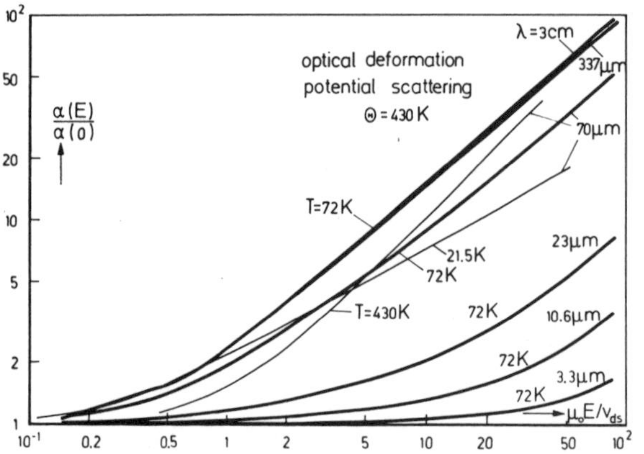

Fig.11.56
Same as Fig.11.54, but for optical deformation potential scattering.

Fig.11.56 shows this ratio as a function of E where the T_e-vs-E relationship is given by Eq.(6k.25). For a wavelength of 70 μm and a Debye temperature of 430 K the influence of lattice temperature is indicated: The curves cross for $\mu_0 E/v_{ds} \approx 4$. For the other curves the wavelength has been chosen for a parameter. At all field intensities the ratio $a(E)/a(0)$ increases with wavelength. At long wavelengths, however, there is a saturation: In Fig.11.56 there is practically no difference between 337 μm (which is the HCN laser wavelength) and 3 cm (microwave X-band). For a comparison with acoustic scattering shown in Fig.11.54 we have to take into account that the scales of E are different: For acoustic scattering E is related to the sound velocity while for optical scattering E is related to the saturation drift velocity which e.g. in n-Ge is about 20 times the acoustic sound velocity. The effects of both scattering mechanisms will be compared with each other in Fig.11.60.

Let us now consider polar optical scattering. The hamiltonian matrix element is given by Eq.(6 l.19) and proportional to 1/q. Hence, Eq.(11j.20) may be applied.*) In order to distinguish between the absorption coefficient a and the polar constant, the latter shall now be denoted as a_{pol}. The result is:

$$a = \frac{\mu_0 c}{\sqrt{\kappa_{opt}}} \cdot \frac{2a_{pol} n e^2 (\hbar\omega_0)^{3/2}}{3\pi^{1/2}\hbar^2\omega^3 m} (k_B T_e)^{1/2} \frac{\sinh(z_+ - z)\, z_+ K_1(z_+) + \sinh(z_- + z)|z_-| K_1(|z_-|)}{\sinh(z)} \tag{11j.39}$$

In Fig.11.57 the quantity

$$\lambda^3 \sinh(z_0)\{z_+ K_1(z_+) + |z_-| K_1(|z_-|)\}/\{(\theta/2T)^{3/2}\sinh(\theta/2T)\} = a/C_{POS} \tag{11j.40}$$

which defines a factor of proportionality, C_{POS}, is plotted vs λ for various values of lattice temperature T and Debye temperature Θ. At long wavelengths $a \propto \lambda^2$ while at short wavelengths $a \propto \lambda^{2.5}$. The plot is very similar to that for optical deformation potential scattering. For thermal carriers we introduce the mobility given by Eq.(6 l.23) and find for the absorption cross section

$$\frac{a}{n} = \frac{(3.02\times10^{-2}\,\text{Å})^2}{\sqrt{\kappa_{opt}}} \cdot \frac{(\theta/100\,\mathrm{K})(\lambda/\mu m)^3 \sinh(z_0)\{z_+ K_1(z_+) + |z_-| K_1(|z_-|)}{(m/m_0)^2 (\mu/\mathrm{cm^2 V^{-1} sec^{-1}})(\theta/2T)^2 K_1(\theta/2T)} \tag{11j.41}$$

For the case of hot carriers we obtain from Eq.(11j.39) for the ratio of absorption coefficients

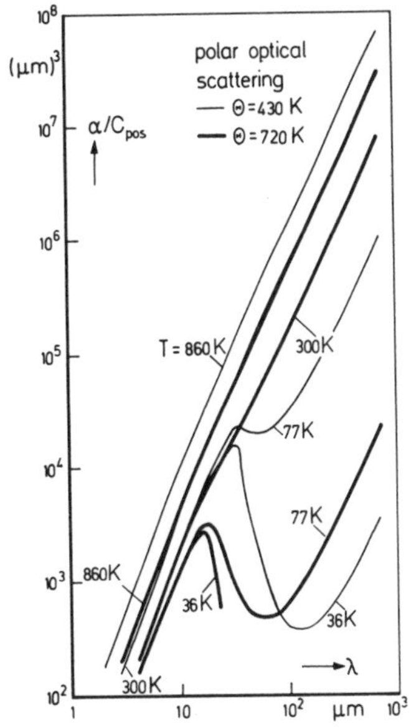

Fig.11.57 Same as Fig.11.53, but for polar optical scattering.

$$\frac{a(E)}{a(0)} = \left(\frac{T_e}{T}\right)^{1/2} \frac{(z_0+z) K_1(z_+) \sinh(z_+ - z) + |z_0 - z| K_1(|z_-|) \sinh(z_- + z)}{\{(z_0+z) K_1(z_0+z) + |z_0 - z| K_1(|z_0 - z|)\} \sinh(z_0)} \tag{11j.42}$$

where $z_\pm = (\hbar\omega \pm \hbar\omega_0)/2k_B T_e$, $z_0 = \hbar\omega/2k_B T$, and $z = \hbar\omega_0/2k_B T$. Fig.11.58 shows this ratio as a function of E where the T_e-vs-E relationship is given by Eq. (6 l.26). As in Fig.6.29 there is a breakdown field and an ambiguity in a(E). Otherwise, the curves are similar to those for optical deformation potential

*) See footnote on p. 381.

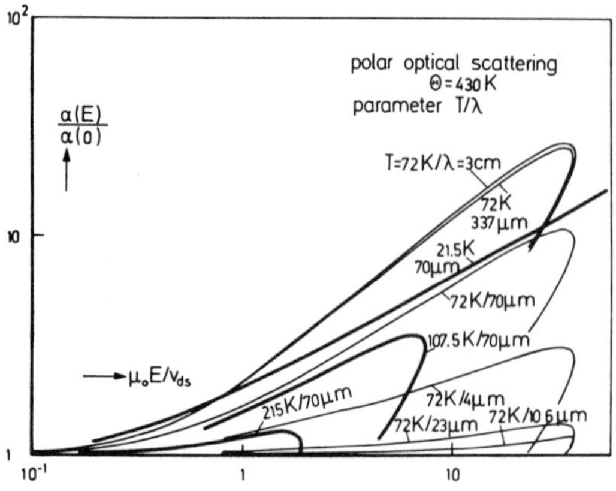

Fig.11.58
Same as Fig.11.54, but for polar optical scattering.

scattering shown in Fig.11.56. However, as for piezoelectric scattering, there is a range of electron temperatures T_e where a decreases with increasing T_e.

The present results for acoustic and optical deformation potential scattering have been extended to the many-valley model by Meyer [4]. Gurevich et al. [5], König [6], and König and Kranzer [7] treated the case of polar optical scattering.

Ionized impurity scattering has been treated by Wolfe [8]. For $H_{k'k}$ given by Eq.(6c.13) for $L_D^{-1} \approx 0$ and $|\vec{k}-\vec{k}'| = q$ we find from Eqs.(11j.18) and (11j.21) for the case of absorption

$$a_a = \frac{\mu_o c}{\sqrt{\kappa_{opt}}} \cdot \frac{n V^{-1} Z^2 e^6}{2^{3/2} 3\pi^{3/2} \hbar\kappa^2 \kappa_o^2 m^{3/2} \omega^3} (k_B T_e)^{-1/2} e^z K_o(|z|) \qquad (11j.43)$$

where $z = \hbar\omega/2k_BT_e$. The case of induced emission is obtained by changing the sign of $\hbar\omega$. As in the transition from Eq.(6c.13) to (6c.17) we replace V^{-1} by the impurity concentration, N_I. Hence, the total absorption coefficient is given by

$$a = \frac{\mu_o c}{\sqrt{\kappa_{opt}}} \frac{n N_I Z^2 e^6}{2^{1/2} 3\pi^{3/2} \hbar\kappa^2 \kappa_o^2 m^{3/2} \omega^3} (k_B T_e)^{-1/2} \sinh\left(\frac{\hbar\omega}{2k_B T_e}\right) K_o\left(\frac{\hbar\omega}{2k_B T_e}\right) \qquad (11j.44)$$

and the absorption cross section

$$\frac{a}{n} = \frac{(3.88\times10^{-2}\,\text{Å})^2}{\sqrt{\kappa_{opt}}} \frac{Z^2 N_I(\lambda/\mu m)^3}{10^{17}\,cm^{-3}} \frac{\sinh\left(\frac{7195}{(\lambda/\mu m)(T_e/K)}\right) K_o\left(\frac{7195}{(\lambda/\mu m)(T_e/K)}\right)}{\kappa^2 (m/m_o)^{3/2} (T_e/100\,K)^{1/2}} \qquad (11j.45)$$

[4] H. J. G. Meyer, Phys. Rev. 112 (1958) 298; J. Phys. Chem. Solids 8 (1959) 264; see also R. Rosenberg and M. Lax, Phys. Rev. 112 (1958) 843.
[5] V. L. Gurevich, I. G. Lang, and Yu. A. Firsov, Fiz. Tverd. Tela 4 (1962) 1252 [Engl.: Soviet Physics Solid State 4, (1962) 918].
[6] W. M. König, Acta Phys. Austr. 33 (1971) 275.
[7] D. Kranzer and W. M. König, phys. stat. sol. (b) 48 (1971) K 133.
[8] R. Wolfe, Proc. Phys. Soc. (London) A 67 (1954) 74.

where the impurity concentration N_I has been related to a typical value of $10^{17}/cm^3$. Since for an uncompensated semiconductor in the extrinsic range $N_I = n$, we observe an increase of a with n^2 rather than n as in the cases treated previously where phonons were involved.

As mentioned by Wolfe [8] the Born approximation in the perturbation treatment of ionized impurity scattering may not be applicable for certain cases. Different approximations have been derived from the idea that the process is essentially the inverse process of bremsstrahlung [4, 9] for which an exact expression has been given by Sommerfeld [10] where, however, the dielectric constant, κ, and the effective mass ratio, m/m_o, have to be introduced *). Denoting the ionization energy of the impurity by $\Delta\epsilon_I$, the Born approximation is valid for both $k_B T_e$ and $|k_B T_e \pm \hbar\omega| \gg \Delta\epsilon_I$. In the limit of classical optics where $\hbar\omega \ll \Delta\epsilon_I$ but still $k_B T_e \gg \Delta\epsilon_I$ the Sommerfeld equation yields Eq.(11j.44) where, however, the factor $\sinh(z) K_o(z)$ is replaced by the constant $\pi/2\sqrt{3} = 0.907$. For the intermediate case where $\hbar\omega \approx \Delta\epsilon_I$ and $k_B T_e \ll \Delta\epsilon_I$ we have to replace this factor by 1 while for $k_B T_e \approx \Delta\epsilon_I$ and $\hbar\omega \ll \Delta\epsilon_I$, Eq.(11j.44) is again valid. Fig.11.59 shows a plot of both $\lambda^3 \sinh(z) K_o(z)$ (full curves) and $\lambda^3(1-e^{-2z})$ (dash dotted) vs λ for $T_e = 40$, 77, and 300 K. A λ^3-dependence is indicated by the dashed straight line. Which one of the curves is applicable depends on the ionization energy of the impurity. In the quantum limit the Born approximation yields $a \propto \lambda^{3.5}$ independent of temperature.

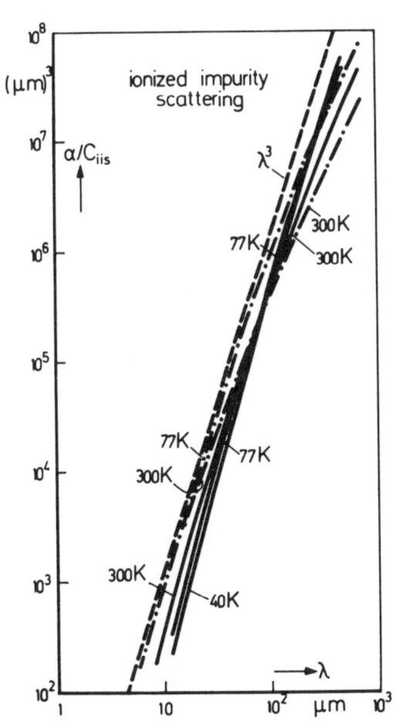

Fig. 11.59 Same as Fig. 11.53 but for ionized impurity scattering.

For a combination of several scattering mechanisms, the absorption coefficients a have to be added, because the scattering probabilities S are additive and a is proportional to S. This makes an absorption experiment easier to interpret than transport phenomena although in both cases the complicated $T_e = T_e(E)$ relation is involved. Regions of the band for $\epsilon < \epsilon_G$ may be explored which possibly cannot be reached by carriers in an experiment on transport phenomena [4]. On the other hand, absorption by impurities, excitons, or the crystal lattice may

[[9] S. Visvanathan, Phys. Rev. 120 (1960) 379.
[10] A. Sommerfeld, Ann. Physik 11 (1931) 257.

*) This has been overlooked in the definition of the interaction parameter in ref. 9; the reason for introducing κ is the same as in the hydrogene model for the shallow impurity.

cover part of the spectrum of free-carrier absorption.

The combination of acoustic and optical deformation potential scattering has been investigated in n-type Ge at 200 K for a wavelength of 10.6 μm (CO_2 laser radiation). For simplicity a spherical band model has been assumed. The $T_e(E)$ relation has been calculated from the energy balance equation in the form

$$E = \frac{\{[\mu_{ac}^{-1}(T_e) + \mu_{opt}^{-1}(T_e)] \cdot [-\langle d\epsilon/dt\rangle_{ac} - \langle d\epsilon/dt\rangle_{opt}]\}^{1/2}}{\mu_o(T)\,\{\mu_{ac}^{-1}(T) + \mu_{opt}^{-1}(T)\}} \tag{11j.46}$$

where $\mu_o(T)$ is the experimental zero-field mobility (7700 cm²/Vsec at 200 K) and $\mu^{-1} \approx \mu_{ac}^{-1} + \mu_{opt}^{-1}$ has been assumed for simplicity. For a parameter, the ratio $(Du_l/\epsilon_{ac}\omega_o)^2 = b$ has been introduced. Fig. 11.60 shows $\alpha(E)/\alpha(0)$ as a function

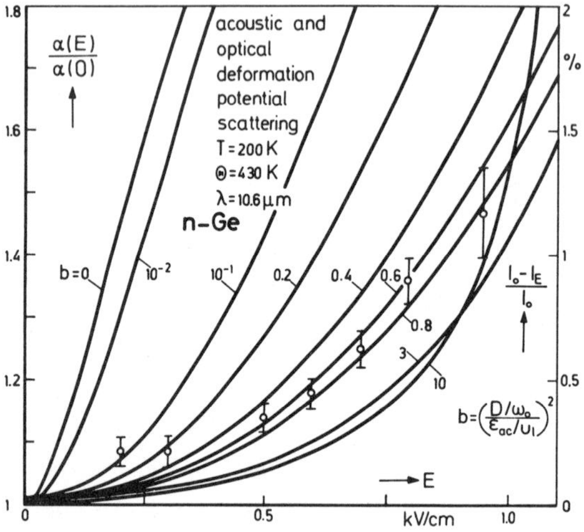

Fig. 11.60
Acoustic and optical deformation potential scattering, contributing both to optical absorption by hot carriers in n-Ge at 200 K and λ = 10.6 μm. Data points for $N_I \approx n = 5\times10^{15}/cm^3$.

of E. Pure acoustic scattering is obtained for b = 0 while optical scattering is found for b ≫ 1. The data points observed by Lischka [11] agree with b ≈ 0.6. Hence, the optical phonon contribution to scattering is weak in n-type Ge, as has already been found from galvanomagnetic effects (Chap. 6k). Since the carrier concentration was only $5\times10^{15}/cm^3$ ionized impurity scattering was negligibly small. For the sample thickness d = 0.24 cm ≪ $1/\alpha$, the ratio $\alpha(E)/\alpha(0)$ was simply obtained from the transmitted light intensities I_E and I_o with and without the applied dc electric field E, respectively, by the relation

$$\alpha(E)/\alpha(0) = 1 + \{(1-r_\infty^2)/(1+r_\infty^2)\}(I_o - I_E)/I_o\,\alpha(0)d \tag{11j.47}$$

[11] K. Seeger and K. Lischka, Verhandl. DPG (VI) 7 (1962) 586; prior to these measurements, a CO_2-laser modulation by carrier heating in n-type Ge at room temperature has been reported by K. H. Müller, G. Nimtz, and M. Selders, Appl. Phys. Lett. 20 (1972) 322.

In the present calculations for hot carriers an isotropic distribution function with an electron temperature T_e has been assumed. However, as has been calculated by Fomin [12], the average of the factor $(\vec{a}\vec{p}_{k'k})^2$ in Eq.(11i.9) with an anisotropic distribution of the carriers such as the expansion Eq.(6i.16) even for a spherical band model yields a dependence of the absorption on the angle between the light polarization, $\vec{a}$, and the field direction, $\vec{E}$, by a factor $1 + \frac{1}{5}(f_2/f_0) \cdot \{3\cos^2(\vec{a}, \vec{E}) - 1\}$, where f_2 and f_0 are coefficients of the expansion.

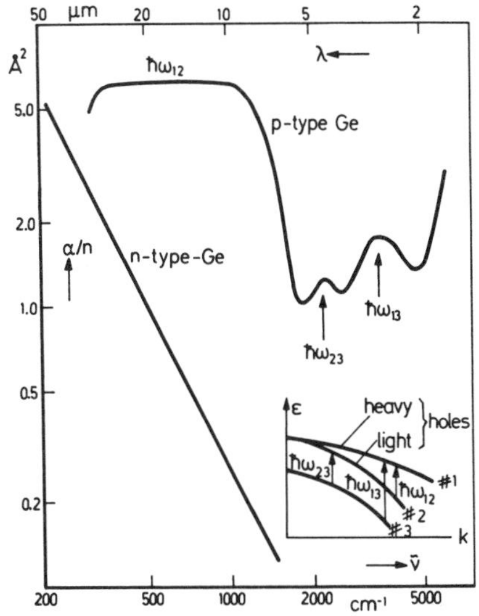

Fig. 11.61 Ratio of free-carrier absorption coefficient and the carrier density for p- and n-type germanium at room temperature. The lattice absorption independent of carrier concentration has been subtracted. The indicated transitions between valence subbands are those shown in the inset (after Kaiser et al., ref. 16).

For a determination of the distribution function, i.e. of the coefficients f_l ($l = 0, 1, 2, \cdots$), for hot carriers, direct interband transitions are more useful than the intraband transitions discussed so far which by necessity are indirect, i.e. involve phonons. The valence bands of most semiconductors consist of the light and heavy-hole and the split-off subbands and transitions between these have been shown to be nearly ideal for the present case [13-15]. Fig. 11.61 shows the absorption spectrum of p-type Ge and, for comparison, also of n-type Ge [16]. The inset indicates transitions between the subbands which cause the observed absorption maxima in p-type Ge. (The arrows indicate electron transitions.) E.g. the hole transition $1 \rightarrow 2$ by the absorption of the 10.6 μm radiation of a CO_2 laser (10 kW, pulsed) converts heavy to light holes with

[12] N. V. Fomin, Fiz. Tver. Tela 2 (1960) 605 [Engl.: Sov. Phys. Solid State 2 (1960) 566]; C.A.Baumgardner and T.O.Woodruff, Phys.Rev. 173 (1968) 746.

[13] M. A. C. S. Brown and E. G. S. Paige, Phys. Rev. Lett. 7 (1961) 84; M. A. C. S. Brown, E. G. S. Paige, and L. N. Simcox, Proc. Int. Conf. Semic. Phys. Exeter 1962 (A. C. Stickland, ed.), p. 111. London: The Institute of Physics and The Physical Society. 1962; A. C. Baynham and E. G. S. Paige, Phys. Lett. 6 (1963) 7; A. C. Baynham, Sol. State Comm. 3 (1965) 253.

[14] W. E. Pinson and R. Bray, Phys. Rev. Lett. 11 (1963) 268; Phys. Rev. 136 (1964) A1449; R. Bray, W. E. Pinson, D. M. Brown, and C. S. Kumar, Proc. Int. Conf. Semic. Phys. Paris 1964 (M. Hulin, ed.), p. 467. Paris: Dunod. 1964.

[15] O. Christensen, Phys. Rev. B 7 (1973) 763.

[16] W. Kaiser, R. J. Collins, and H. Y. Fan, Phys. Rev. 91 (1953) 1380.

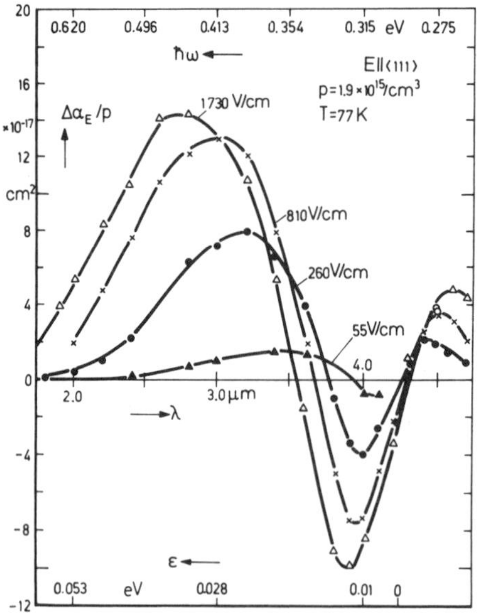

Fig. 11.62 Change of absorption coefficient (relative to the hole density) with electric field strength $\vec{E}$ in p-type germanium at 77 K. $\vec{E}$ is in the <111> direction. The peak at 3 μm corresponds to an energy of 0.412 eV for the 1→3 transition (after Pinson and Bray, ref. 14).

the consequence of an increase in conductivity [17]. The heavy-hole distribution function has been determined from the 1 → 3 transition at a low light intensity. In Fig. 11.62 the observed change in absorption cross section $\{a(E)-a(0)\}/p$ denoted by $\Delta a_E/p$ has been plotted vs wavelength λ for a given lattice temperature of 77 K and various electric field strengths $\vec{E} \parallel <111>$ and the photon wave vector $\vec{q}$. In Fig. 11.63 the product $aT^{3/2}$ has been plotted on a logarithmic scale vs $10^3/T$ for various photon energies in the range of this transition. From the slopes of the linear regions of the curves the heavy-hole energies, ϵ, have been determined as a function of $\hbar\omega$ by applying the relation

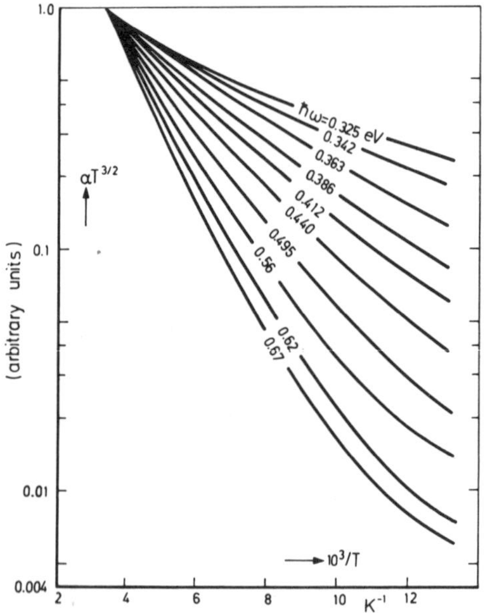

Fig. 11.63
The product of absorption coefficient a (normalized to unity at 300 K) and $T^{3/2}$ plotted vs the inverse temperature for photon energies near 0.412 eV corresponding to 1 → 3 transitions in p-type germanium (experimental, after Pinson and Bray, ref. 14).

[17] J. M. Feldman and K. M. Hergenrother, Appl. Phys. Lett. 9 (1966) 186.

$$a(0) \propto (p/N_v) g(\epsilon) \exp(-\epsilon/k_B T) \propto T^{-3/2} \epsilon^{1/2} \exp(-\epsilon/k_B T) \qquad (11j.48)$$

where $g(\epsilon) \propto \epsilon^{1/2}$ is the density of states in the heavy-hole band. Assuming for $a(E)$ the same factor of proportionality but a distribution function $f(\epsilon)$, this function can be determined from a comparison of both experiments in the range $0.015 \leqslant \epsilon \leqslant 0.06$ eV [14]. Outside this range one can easily extrapolate. In Fig. 11.64 the symmetrical part of the distribution function, $f_0(\epsilon)$, so obtained at 77 K for a field strength of 800 V/cm is shown together with a Maxwell-Boltzmann distribution for a carrier temperature of 169 K having the same average energy as $f_0(\epsilon)$. As mentioned in Chap. 8d the character of $f_0(\epsilon)$ is non-Maxwellian showing an impoverishment of heavy holes above the optical phonon energy of 0.037 eV and a containment below this energy. From the ratio of absorption coefficients for light polarization $\vec{a} \parallel \vec{E}$ and $\vec{a} \perp \vec{E}$, respectively, ($\vec{q} \perp \vec{E}$, of course) the anisotropy of the distribution, $f_2(\epsilon)/f_0(\epsilon)$, has been determined. This ratio

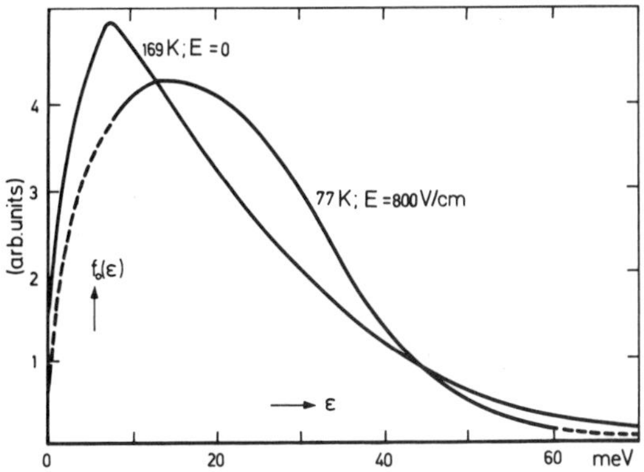

Fig. 11.64
Heavy-hole energy distribution (symmetrical part) in germanium at 77 K for a field strength of 800 V/cm along the ⟨100⟩ direction measured by optical absorption. For a comparison a Maxwell-Boltzmann distribution having the same average energy is shown (after Pinson and Bray, ref. 14).

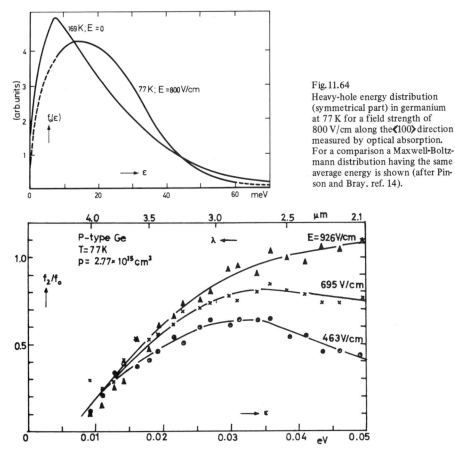

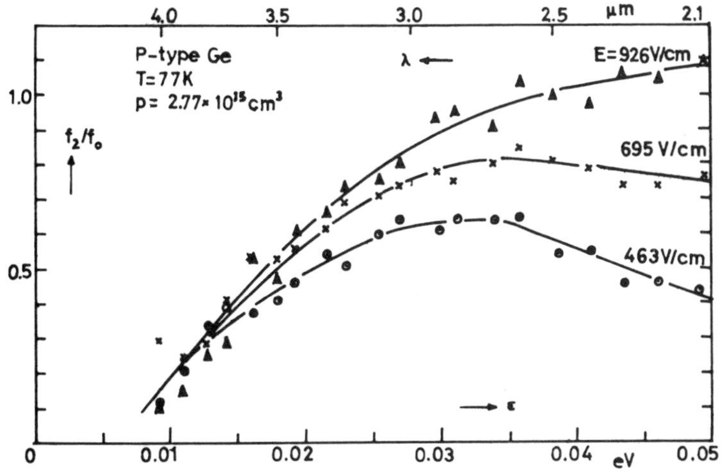

Fig. 11.65 The ratio of coefficients, f_2/f_0, of the expansion of the distribution function $f(\epsilon)$ in Legendre polynomials, obtained for heavy holes in p-type Ge at 77 K for various field strengths: (ref. 14).

has been plotted in Fig.11.65 as a function of ϵ for various electric field strengths. It is energy dependent and rises to a value ≈ 1. As mentioned in Chap.8d this indicates a "cyclical streaming effect" that an electric field has on heavy holes in $\vec{k}$-space. The functions $f_0(\epsilon)$ and $f_2(\epsilon)/f_0(\epsilon)$ have been obtained theoretically by Budd [18]. $f_0(\epsilon)$ may be approximated for $\epsilon < \hbar\omega_0$ by a Maxwell-Boltzmann distribution function with an electron temperature T_{e1} and for $\epsilon > \hbar\omega_0$ by a similar function with $T_{e2} < T_{e1}$ [15, 18]. The change in the distribution function upon application of a uniaxial stress has been observed by Christensen [15]. The average carrier energy increases with stress and almost saturates when the strain splitting of the two $p_{3/2}$ levels reaches the optical phonon energy, $\hbar\omega_0$. In some many-valley semiconductors a structure has been observed in the absorption spectrum which may be due to equivalent intervalley transitions. Fig.11.66 shows

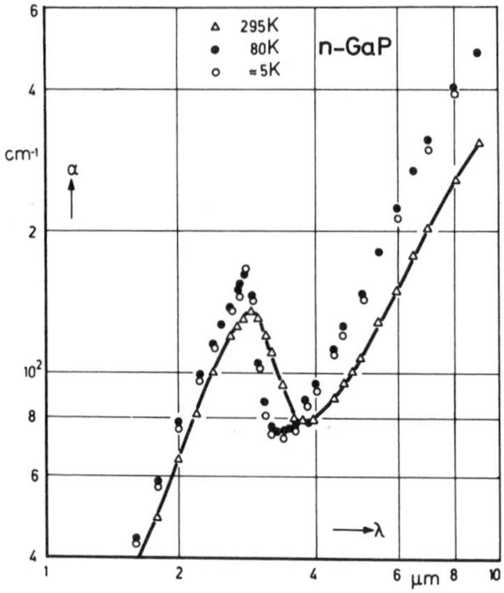

Fig.11.66
Free-carrier absorption spectrum for n-type gallium phosphide having a carrier density of $1 \times 10^{18}/cm^3$ (after Spitzer et al., ref. 19).

the absorption spectrum of n-type GaP where a maximum at 3 μm is observed [11]. In GaP_xAs_{1-x} its position is nearly independent of the composition x at $x \lesssim 1$. Similar maxima have been found in n-type GaAs [20, 21] and in n-type AlSb [22]. Free-hole absorption in GaAs shown in Fig.11.67 has also been explained by interband transitions [23].

[18] H. F. Budd, Phys. Rev. 158 (1967) 798.

[19] W. G. Spitzer, M. Gershenzon, C. J. Frosch, and D. F. Gibbs, J. Phys. Chem. Solids 11 (1959) 339.

[20] W. G. Spitzer and J. M. Whelan, Phys. Rev. 114 (1959) 59.

[21] I. Balslev, Phys. Rev. 173 (1968) 762: non-equivalent valley transitions.

[22] W. J. Turner and W. E. Reese, Phys. Rev. 117 (1960) 1003.

[23] R. Braunstein and E. O. Kane, J. Phys. Chem. Solids 23 (1962) 1423; I. Balslev, Phys. Rev. 177 (1969) 1173.

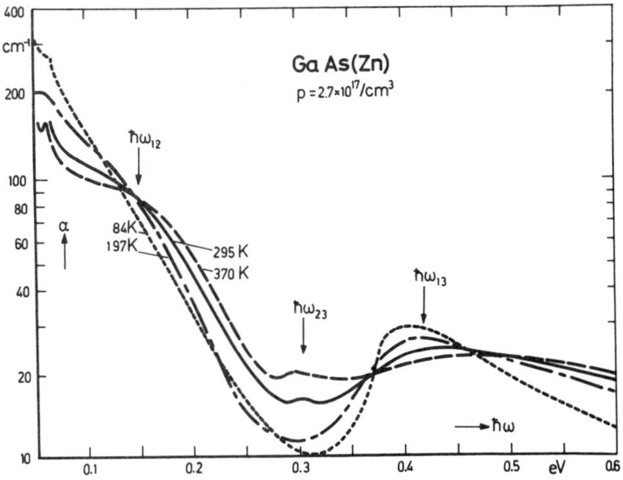

Fig. 11.67 Free-hole absorption in p-type GaAs at various temperatures, and explanation of peaks by the transitions indicated by the inset of Fig. 11.61 (after Braunstein and Kane, ref. 23).

Finally let us discuss the case of large carrier concentrations n where the refractive index $\sqrt{\kappa_{opt}}$ is no longer constant as has been assumed so far. From Eq. (11j.1) we obtain for the free-carrier contribution to the real part of the dielectric constant, κ_r,

$$\Delta\kappa_r = -\omega_p^2/(\omega^2+\gamma^2) \approx -ne^2/(m\kappa_0\omega^2) \qquad (11j.49)$$

The approximation is valid for small damping. Since the carrier gas is degenerate at the high densities considered here, the ratio n/m should be replaced by $(1/4\pi^3)$ $\langle v \rangle S/3\hbar$ where S is the area of the Fermi surface and $\langle v \rangle$ is the carrier velocity averaged over the Fermi surface [24]. For a small value of the extinction coefficient the reflectivity r_∞ becomes $(\sqrt{\kappa_r}-1)^2/(\sqrt{\kappa_r}+1)^2$ and may be nearly unity when the lattice contribution to κ_r is compensated by the negative carrier distribution to yield $\kappa_r \approx 0$. At a slightly different frequency the carrier contribution may be just sufficient for $\kappa_r = 1$ with the result that the reflectivity nearly vanishes. Observations shown in Fig. 11.68 are in agreement with these predictions [24]. The minimum near the "plasma frequency"

$$\overline{\omega}_p = \omega_p/\sqrt{\kappa_{opt}} = \sqrt{ne^2/(m\kappa_0\kappa_{opt})} \qquad (11j.50)$$

has been used for a determination of the effective mass m. Values of m increasing with carrier concentration from $0.023\ m_0$ to $0.041\ m_0$ have been determined in n-type InSb from Fig. 11.68. In n-type GaAs values between $0.078\ m_0$ at $n = 0.49\times10^{18}/cm^3$ and $0.089\ m_0$ at $5.4\times10^{18}/cm^3$ have been found [11]. In both cases the increase of the effective mass with the Fermi energy is due to the non-

[24] W. G. Spitzer and H. Y. Fan, Phys. Rev. 106 (1957) 882.

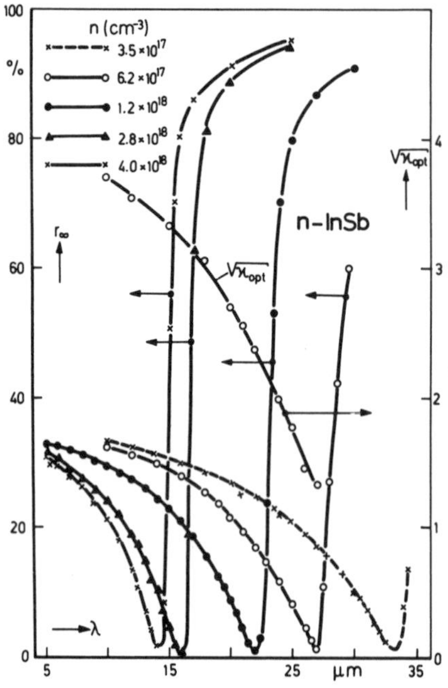

Fig. 11.68
Reflection spectra for five n-type InSb samples at 295 K; refractive index $\sqrt{\kappa_{opt}}$ valid for the sample with $6.2 \times 10^{17}/cm^3$ electrons (after Spitzer and Fan, ref. 24).

parabolicity of the conduction band.

The approximation of small damping, where $\omega\tau_m \gg 1$ (i.e. the free-space wavelength $\lambda \ll 185.9\ \mu m$ $\tau_m/10^{-13}$ sec), is valid for most semiconductors even in the far-infrared spectrum.

11k. Cyclotron Resonance

If a carrier moves in a dc magnetic field $\vec{B}$ at an oblique angle, its motion is a helix around the direction of the field with an angular frequency known as the "cyclotron frequency" given by*)

$$\omega_c = (e/m)B = 2\pi \frac{B}{m/m_0}\ 2.84\ \mathrm{GHz/kG} \tag{11k.1}$$

The corresponding wavelength in free space is

$$\lambda_c = 10.56\ \mathrm{cm}\ \frac{m/m_0}{B/kG} \tag{11k.2}$$

If the effective mass $m \approx m_0$ and B is a few kG, the wavelength is in the microwave range. Microwaves incident on the sample and polarized at an angle to B are absorbed; the absorption shows a resonance peak at $\omega = \omega_c$. This phenomenon is called "cyclotron resonance" [1]. From a measurement of the resonance frequency the value of the effective mass, m, is obtained (Eq.(11k.1). As in any

[1] For a review paper see e.g. B. Lax, Proc. Int. School of Physics (R. A. Smith, ed.), Vol. XXII, p. 240. New York: Acad. Press. 1963.

*) For numerical purposes, $1\ kG = 0.1\ Wb/m^2$. In the earth's magnetic field the free-electron cyclotron resonance frequency is of the order of magnitude of 1 MHz.

resonance observation a peak is not found for the case of strong damping, i.e. if most carriers make a collision before rotating at least through one radian. Since the number of collisions per unit time is $1/\tau_m$, the condition for a resonance peak is given by

$$\omega_c > 1/\tau_m \tag{11k.3}$$

This condition poses a severe limit on observations of microwave cyclotron resonance: τ_m should be larger than about 10^{-10} sec. This is true only at liquid helium temperatures and even then in only few semiconductors which can be purified highly enough so that at these low temperatures ionized and neutral impurity scattering are negligible.

Another method of observation uses infrared radiation instead of microwaves and strong magnetic fields which are either available in pulsed form (up to about 300 kG) or in superconducting magnets (up to about 120 kG). In this case even at room temperature the resonance condition may be fulfilled ($\tau_m = 10^{-13}$ sec corresponds to a wavelength of $\lambda_c = 185.9\ \mu$m which at $\leqslant 300$ kG requires $m/m_0 \leqslant 0.57$). Since due to induction losses it is difficult to work with pulsed magnetic fields, infrared cyclotron resonance at room temperature is limited to carriers with effective masses of less than about $0.2\ m_0$. At the large fields involved there is a spin splitting of the levels and a selection rule for the quantum number, $\Delta M = 0$, for linearly polarized fields in the "Faraday configuration" (Chap.11 l). For non-parabolic bands the effective mass m depends on B in large magnetic fields used in infrared cyclotron resonance. This poses another problem in this type of measurements.

It has been shown in Chap.9b that at temperatures T where $k_B T < \hbar\omega_c$, magnetic quantization is observed. For temperatures below 77 K and $\omega_c \approx 10^{13}$/sec, this condition is fulfilled. Cyclotron resonance may then be considered as a transition between successive Landau levels.

For a simplified treatment of cyclotron resonance let us consider the equation of motion of a carrier in an ac electric field and a static magnetic field B and not include an energy distribution of carriers:

$$d(m\vec{v}_d)/dt + m\vec{v}_d/\tau_m = e(\vec{E} + [\vec{v}_d\vec{B}]) \tag{11k.4}$$

We introduce a Cartesian coordinate system and choose for the z-axis the direction of $\vec{B}$. For eB/m we introduce the cyclotron frequency ω_c. With $\vec{E} \propto \exp(i\omega t)$ and $ne\vec{v}_d = \sigma\vec{E}$ we find for the non-zero components of the conductivity tensor σ (for $\omega = 0$ see Eqs.(4c.39) and (4c.42)):

$$\sigma_{xx} = \sigma_{yy} = \sigma_o \tau_m^{-1} \frac{\tau_m^{-1} + i\omega}{(\tau_m^{-1} + i\omega)^2 + \omega_c^2} \tag{11k.5}$$

$$\sigma_{xy} = -\sigma_{yx} = \sigma_o \tau_m^{-1} \frac{\omega_c}{(\tau_m^{-1} + i\omega)^2 + \omega_c^2} \tag{11k.6}$$

$$\sigma_{zz} = \sigma_o\, \tau_m^{-1}/(\tau_m^{-1} + i\omega) \tag{11k.7}$$

where the dc conductivity $\sigma_o = ne^2\tau_m/m$. We consider a right-hand circularly polarized field where $E_y = -iE_x$ and introduce σ_+ by

$$\sigma_+ = j_x/E_x = \sigma_{xx} + \sigma_{xy}E_y/E_x = \sigma_o\tau_m^{-1}\frac{\tau_m^{-1} + i\omega - i\omega_c}{(\tau_m^{-1} + i\omega)^2 + \omega_c^2} = \frac{\sigma_o\,\tau_m^{-1}}{\tau_m^{-1} + i(\omega + \omega_c)} \tag{11k.8}$$

Similarly, for a left-hand polarization we have

$$\sigma_- = \frac{\sigma_o\tau_m^{-1}}{\tau_m^{-1} + i(\omega - \omega_c)} \tag{11k.9}$$

We will consider here only the "Faraday configuration" where in a transverse electromagnetic wave $\vec{E}$ is perpendicular to the dc magnetic field $\vec{B}$. The power absorbed by the carriers is given by

$$P(\omega) = \frac{1}{2}\,\mathrm{Re}(jE^*) \tag{11k.10}$$

where E^* is the complex conjugate of E. For either polarization we have

$$P_\pm = \frac{1}{2}|\vec{E}|^2\,\mathrm{Re}\,\frac{\sigma_o\,\tau_m^{-1}}{\tau_m^{-1} + i(\omega \pm \omega_c)} = \frac{1}{2}|\vec{E}|^2\frac{\sigma_o\,\tau_m^{-2}}{\tau_m^{-2} + (\omega \pm \omega_c)^2} \tag{11k.11}$$

For a linear polarization which can be considered to be composed of two circular polarizations rotating in opposite directions

$$P(\omega) = P_+ + P_- = \frac{\sigma_o}{2}|\vec{E}|^2\left\{\frac{1}{1 + (\omega + \omega_c)^2\tau_m^2} + \frac{1}{1 + (\omega - \omega_c)^2\tau_m^2}\right\} = \sigma_o|\vec{E}|^2\frac{1 + (\omega^2 + \omega_c^2)\,\tau_m^2}{\{1 + (\omega^2 - \omega_c^2)\,\tau_m^2\}^2 + 4\omega_c^2\tau_m^2} \tag{11k.12}$$

At cyclotron resonance, $\omega = \omega_c$, assuming $\omega_c\tau_m \gg 1$, this becomes simply

$$P(\omega_c) = \frac{1}{2}\sigma_o|\vec{E}|^2 \tag{11k.13}$$

For this case the conductivities σ_+ for electrons ($\omega_c < 0$) and σ_- for holes ($\omega_c > 0$), given by Eqs.(11k.8) and (11k.9), are equal to the dc conductivity, σ_o, while outside the resonance they are smaller than σ_o. At low frequencies the power absorbed by the carriers is given by

$$P(0) = \sigma_o|\vec{E}|^2/\{1 + (\omega_c\tau_m)^2\} \tag{11k.14}$$

This equals, of course, $\sigma|\vec{E}|^2$ with σ given by Eq.(4c.44) neglecting in the present approximation the energy distribution of the carriers.

Fig.11.69 shows the ratio P/P_o plotted vs ω_c/ω where P_o stands for $\sigma_o|\vec{E}|^2$. Since usually in the experiment ω is kept constant and the magnetic field

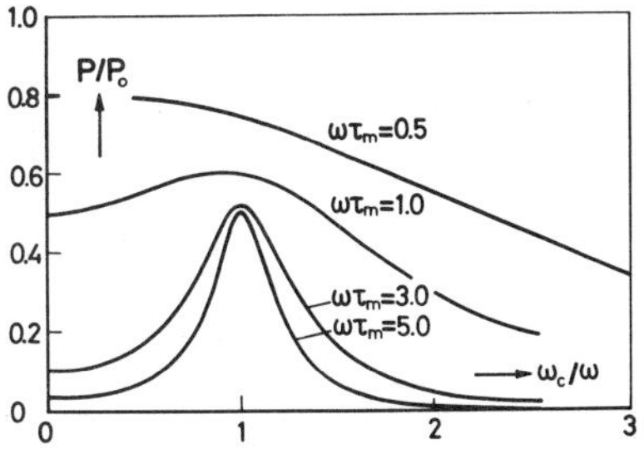

Fig. 11.69 Absorbed microwave power P as a function of $\omega_c/\omega \propto B$ where ω_c is the cyclotron frequency (after Lax et al., ref. 2).

strength is varied, it is quite natural to have ω_c/ω as the independent variable. Obviously the resonance is washed out for $\omega\tau_m < 1$.

Experimental results obtained on n-type silicon at 4.2 K are shown in Fig. 11.70 [2]. For a magnetic field in a <111> crystallographic direction there is a single resonance, while in <001> and <110> directions there are two resonances. This can be interpreted by the many-valley model of the conduction band treated in Chap. 7. Since in the <111> direction there is onlyone peak, the energy ellipsoids must be located along the <100> directions which are then equivalent relative to the magnetic field direction.

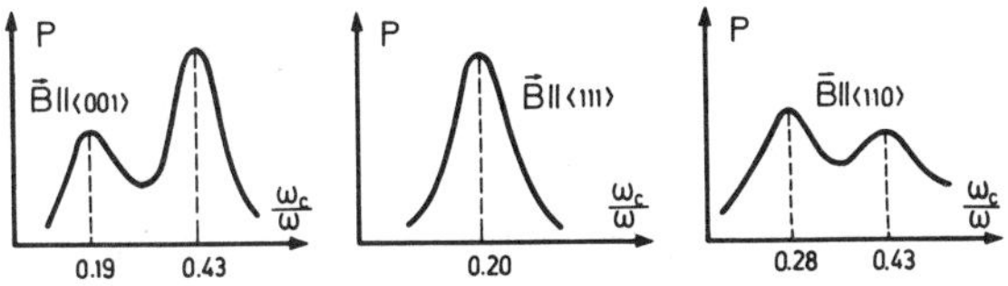

Fig. 11.70
Microwave absorption at 23 GHz in n-type silicon as a function of ω_c/ω for three different directions of the field (after Lax et al., ref. 2).

For an interpretation of Fig. 11.70 we solve Eq. (11k.4) with a mass tensor m in a coordinate system where the tensor is diagonal. For the resonance condition, $\omega = \omega_c$, where we may neglect $\vec{E}$ and τ_m^{-1} for simplicity, we have

$$\left.\begin{aligned} i\omega_c m_x v_x + e(v_z B_y - v_y B_z) &= 0 \\ i\omega_c m_y v_y + e(v_x B_z - v_z B_x) &= 0 \\ i\omega_c m_z v_z + e(v_y B_x - v_x B_y) &= 0 \end{aligned}\right\} \qquad (11k.15)$$

Let us denote the direction cosines of $\vec{B}$ with respect to the 3 coordinate axes by α, β, and γ, respectively. The secular equation

$$\begin{vmatrix} i\omega_c m_x & -eB\gamma & eB\beta \\ eB\gamma & i\omega_c m_y & -eB\alpha \\ -eB\beta & eB\alpha & i\omega_c m_z \end{vmatrix} = 0 \qquad (11k.16)$$

yields for ω_c:

$$\omega_c = eB\sqrt{\frac{\alpha^2 m_x + \beta^2 m_y + \gamma^2 m_z}{m_x m_y m_z}} \qquad (11k.17)$$

[2] B. Lax, H. J. Zeiger, and R. N. Dexter, Physica 20 (1954) 818.

For $\omega_c = (e/m)B$ the effective mass m becomes

$$m = \sqrt{\frac{m_x m_y m_z}{\alpha^2 m_x + \beta^2 m_y + \gamma^2 m_z}} \tag{11k.18}$$

If we take the valley in <001> direction ($m_x = m_y = m_t = m_l/K$; $m_z = m_l$) we obtain for $\vec{B}$ in a $[1\bar{1}0]$ plane with $\alpha = \beta = (\sin\theta)/\sqrt{2}$ and $\gamma = \cos\theta$:

$$m = m_l/\sqrt{K^2\cos^2\theta + K\sin^2\theta} \tag{11k.19}$$

The valleys in <100> and <010> directions have the same effective mass since they are symmetric to the $[1\bar{1}0]$ plane.

$$m = m_l/\sqrt{K\cos^2\theta + \tfrac{1}{2}K(K+1)\sin^2\theta} \tag{11k.20}$$

For the <111> direction ($\alpha = \beta = \gamma = 1/\sqrt{3}$) Eqs.(11k.19) and (11k.20) yield the same value

$$m = m_l/\sqrt{K(K+2)/3} \tag{11k.21}$$

In Fig.11.71 the curves calculated from Eqs.(11k.19) and (11k.20) have been fitted to the data points with effective masses of $m_l/m_o = 0.90 \pm 0.02$ and $m_t/m_o = m_l/(Km_o) = 0.192 \pm 0.001$. For any $\vec{B}$-direction which is not in the $[1\bar{1}0]$ plane there should be three resonant frequencies [3].

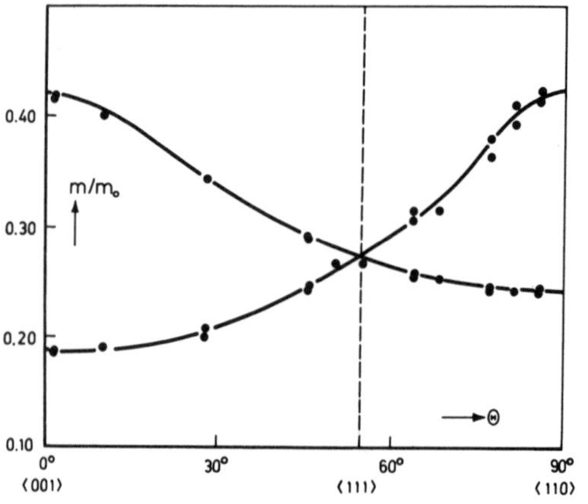

Fig.11.71 Effective cyclotron mass of electrons in silicon at 4 K as a function of the angle between the magnetic field and the <001> direction in the $[1\bar{1}0]$ plane (after Rauch et al., ref. 3).

In n-Ge having 4 ellipsoids on the <111> and equivalent axes there are in general 3 resonant frequencies if $\vec{B}$ is located in the $[1\bar{1}0]$ plane, otherwise there are 4. The analysis of the data yields [2]
$m_l/m_o = 1.64 \pm 0.03$ and
$m_t/m_o = (8.19 \pm 0.03)\times 10^{-2}$.

Observations of cyclotron resonance in p-Ge are shown in Fig.11.72 [4]. The two curves indicate two effective masses: The light-hole mass is isotropic in the $[1\bar{1}0]$ plane and has a value of 0.043 m_o, while the heavy-hole mass varies between 0.28 m_o and 0.38 m_o depending on the angle θ. In Chap.8 the various effective masses in the warped-sphere model have been treated in detail. The cyc-

[3] C. J. Rauch, J. J. Stickler, H. Zeiger, and G. Heller, Phys. Rev. Lett. 4 (1960) 64.
[4] G. Dresselhaus, A. F. Kip, and C. Kittel, Phys. Rev. 98 (1955) 368.

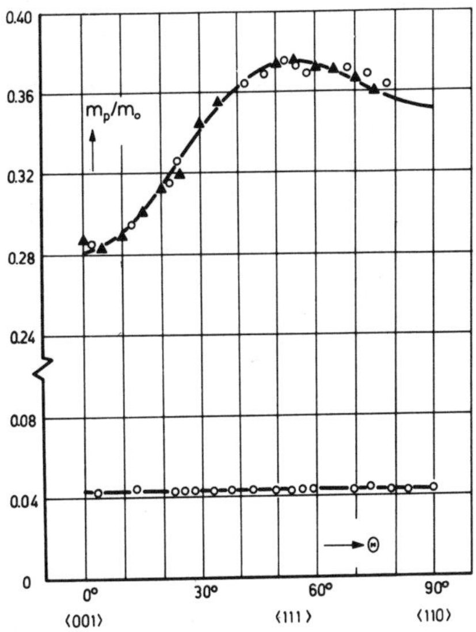

Fig. 11.72 Effective cyclotron mass of holes in p-type germanium at 4 K as a function of the angle between the magnetic field and the ⟨001⟩ direction in the [1$\bar{1}$0] plane (after Dresselhaus et al., ref. 4).

lotron effective mass can be calculated by a method proposed by Shockley [5]: The equation of motion (neglecting $\vec{E}$ and $1/\tau_m$ at resonance)

$$d(\hbar\vec{k})/dt = (e/\hbar)\,[\vec{\nabla}_k \epsilon \vec{B}] \qquad (11k.22)$$

is integrated in cylindrical coordinates where the z-axis is parallel to $\vec{B}$:

$$\hbar dk_\varphi/dt = \hbar k d\varphi/dt = (e/\hbar) B\, \partial\epsilon/\partial k \qquad (11k.23)$$

This yields

$$\frac{\hbar^2}{2\pi} \oint \frac{k d\varphi}{\partial\epsilon/\partial k} = m\omega_c \frac{1}{2\pi} \oint dt = m \qquad (11k.24)$$

The integration is around a constant-ϵ contour in $\vec{k}$-space. With $\epsilon(\vec{k})$ approximated by Eq.(8a.14) and $\vec{B}$ located in the [1$\bar{1}$0] plane we obtain

$$m_\pm/m_0 = (A_0 \pm B'_0)^{-1}\{1 + \frac{1}{32}(1 - 3\cos^2\theta)^2\Gamma + \cdots\} \qquad (11k.25)$$

where the subscript + refers to light holes and – to heavy holes as in Chap.8

$$B'_0 = \sqrt{B_0^2 + \tfrac{1}{4} C_0^2} \quad ; \quad \Gamma = \mp \frac{C_0^2}{2B'_0(A_0 \pm B'_0)} \qquad (11k.26)$$

The conductivity-heavy-hole mass in Ge obtained by fitting Eq.(11k.25) for the minus sign to the experimental data is 0.3 m_0. For the effective mass in the split-off valence band 0.075 m_0 is found.

Results of 27.4 μm infrared cyclotron resonance measurements in n-InSb at room temperature are shown in Fig.11.73 [6]. In the upper part of the figure the variation with $\vec{B}$ of the spin-split Landau levels is shown. To obtain a good fit of the data it was necessary to adjust ϵ_G to 0.20 eV while the room temperature "optical gap" is only 0.18 eV. At room temperature the effective mass is 0.013 m_0. At 77 K, m = 0.0145 m_0 and ϵ_G = 0.225 eV. The variation of the effective mass with temperature is due to the temperature dependence of the band gap ϵ_G and has been explained by Kane (see Chap.8). Infrared cylotron measure-

[5] W. Shockley, Phys. Rev. 79 (1950) 191.

[6] E. D. Palik, G. S. Picus, S. Teitler, and R. F. Wallis, Phys. Rev. 122 (1961) 475; see also ref. 7).

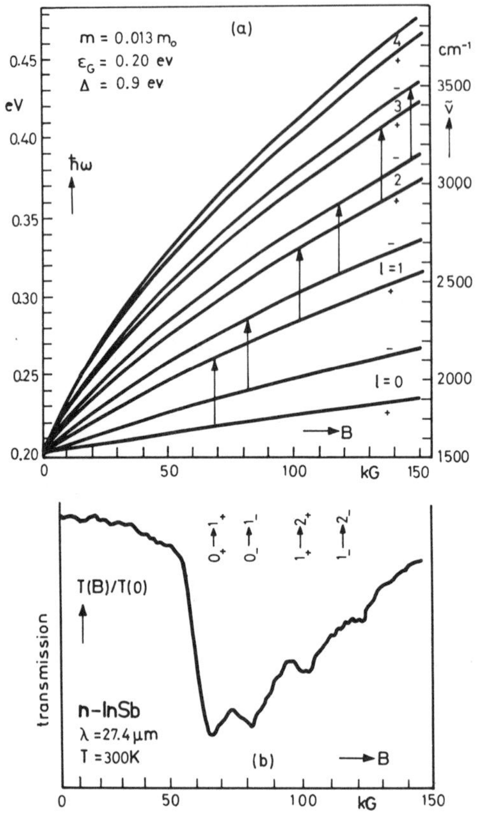

ments have also been made on n-InAs, n-InP, snf n-GaAs [7].

Microwave spin resonance of electrons in the conduction band is a magnetic dipole transition and therefore weak compared with the electric dipole transition at cyclotron resonance. Even so, the large negative g-value in n-InSb has been determined in this way [8] (Chap.11d).

Fig. 11.73
a Calculated conduction band Landau levels at $k_z = 0$ for InSb at room temperature, with allowed cyclotron resonance transitions.
b Observed cyclotron-resonance absorption in n-type InSb (after Palik et al., ref. 6).

11 l. Free-Carrier Magneto-Optical Effects

For a plane electromagnetic wave

$$\vec{E} = \vec{E}_1 \exp(i\omega t - i\vec{q}\vec{r}) \; ; \quad \vec{B} = \vec{B}_1 \exp(i\omega t - i\vec{q}\vec{r}) \tag{11 l.1}$$

In a non-magnetic conductor Maxwell's equations are :

$$[\vec{\nabla}\vec{E}] = -i[\vec{q}\vec{E}] = -\partial\vec{B}/\partial t = -i\omega\vec{B} \tag{11 l.2}$$

$$[\vec{\nabla}\vec{B}]/\mu_o = -i[\vec{q}\vec{B}]/\mu_o = \kappa\kappa_o \partial\vec{E}/\partial t + \sigma\vec{E} = i\omega\kappa\kappa_o\vec{E} + \sigma\vec{E} \tag{11 l.3}$$

where κ is the relative dielectric constant at the angular frequency ω (not the static dielectric constant). After eliminating $\vec{B}$ we obtain

[7] For a review see e.g. E. D. Palik and G. B. Wright: Semiconductors and Semimetals (R. K. Willardson and A. C. Beer, eds.), Vol. 3, p. 421. New York: Acad. Press. 1967.

[8] G. Bemski, Phys. Rev. Lett. 4 (1960) 62.

$$(q^2 - \kappa\omega^2/c^2 + i\mu_0\omega\sigma)\vec{E} = (\vec{q}\vec{E})\vec{q} \tag{11 l.4}$$

where c is the velocity of light in free space.

In the z-direction let us apply a static magnetic field for which we introduce the cyclotron resonance frequency, ω_c, given by Eq.(11k.1)*). The conductivity tensor is given by Eqs.(11k.5) - (11k.7). Since the z-components of this tensor do not depend on the static magnetic field, they shall not be of interest to us now; we therefore put $E_z = 0$.

Let us first consider the "Faraday configuration" where the wave propagates along the direction of the static magnetic field, i.e. $q_x = q_y = 0$ and therefore $(\vec{q}\vec{E}) = 0$. The right-hand side of Eq.(11 l.4) vanishes for this case. In components this equation is given by

$$(q^2 - \kappa\omega^2/c^2)E_x + i\mu_0\,\omega(\sigma_{xx}E_x + \sigma_{xy}E_y) = 0 \tag{11 l.5}$$

and

$$(q^2 - \kappa\omega^2/c^2)E_y + i\mu_0\,\omega(-\sigma_{xy}E_x + \sigma_{xx}E_y) = 0 \tag{11 l.6}$$

This is a homogeneous set of equations for E_x and E_y. It can be solved if the determinant vanishes:

$$(q^2 - \kappa\omega^2/c^2 + i\mu_0\omega\sigma_{xx})^2 + (i\mu_0\omega\sigma_{xy})^2 = 0 \tag{11 l.7}$$

There are two solutions for q^2:

$$q_{\mp}^2 = \kappa\omega^2/c^2 - i\mu_0\omega\sigma_{xx} \pm \mu_0\omega\sigma_{xy} \tag{11 l.8}$$

Introducing from Eqs.(11k.8) and (11k.9)

$$\sigma_{\pm} = \sigma_{xx} \mp i\sigma_{xy} \tag{11 l.9}$$

for the right- and left-hand circular polarization we obtain the "dispersion relation" for the Faraday configuration:

$$c^2q_{\mp}^2/\omega^2 = \kappa - i(\mu_0 c^2/\omega)\sigma_{\mp} = \kappa - i(\mu_0 c^2/\omega)\sigma_0\tau_m^{-1}/\{\tau_m^{-1} + i(\omega \mp \omega_c)\} \tag{11 l.10}$$

The left-hand side is the square of the refractive index; it depends on the direction of the polarization. If the refractive index vanishes, $n = cq/\omega = 0$, there is 100 % reflection (appendix D). This "plasma reflection" occurs at a frequency given by Eqs.(11 l.10), (11k.8), and (11k.9):

$$0 = \kappa - i(\mu_0 c^2/\omega)\sigma_0\tau_m^{-1}/\{\tau_m^{-1} + i(\omega \mp \omega_c)\} \tag{11 l.11}$$

We introduce the "plasma frequency" given by Eq.(11j.49)

$$\overline{\omega}_p = \omega_p/\sqrt{\kappa} = \sqrt{ne^2/(m\kappa\kappa_0)} = \sqrt{\mu_0 c^2\sigma_0/(\kappa\tau_m)} \tag{11 l.12}$$

and solve for ω assuming $\tau_m^{-1} \ll |\omega \mp \omega_c|$:

$$\omega = \overline{\omega}_p\,\{\sqrt{1 + (\omega_c/2\overline{\omega}_p)^2} \pm \omega_c/2\overline{\omega}_p\} \tag{11 l.13}$$

*) The static magnetic field, B, in Eq.(11k.1) should not be confused with the ac field $\vec{B}$ given by Eq.(11 l.1).

For weak magnetic fields $\sqrt{1 + (\omega_c/2\overline{\omega}_p)^2} \approx 1 + \frac{1}{2}(\omega_c/2\overline{\omega}_p)^2$ and $\Delta\omega = \omega - \overline{\omega}_p$ is given by

$$\Delta\omega = \pm \frac{1}{2}\,\omega_c + \frac{1}{8}\,\omega_c^2/\overline{\omega}_p \tag{11 l.14}$$

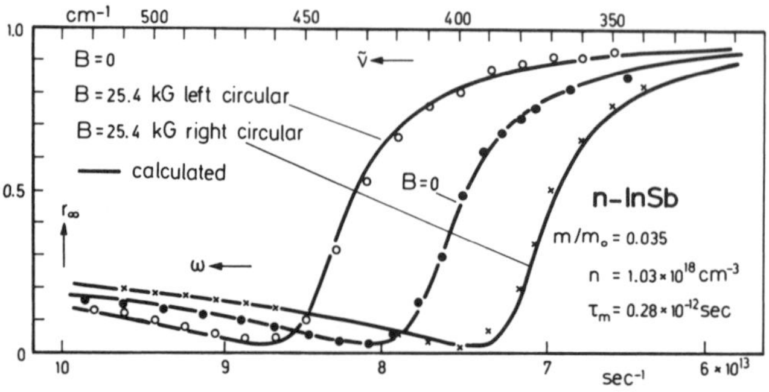

Fig. 11.74 Longitudinal magnetoplasma reflection in n-type InSb at room temperature (after Palik et al., ref. 1).

To a first approximation there is a linear shift of the plasma edge with magnetic field intensity. Fig. 11.74 shows the observed reflectivity of n-InSb at room temperature for left- and right-hand circular polarization [1]. The curve for B = 0 is similar to the curves in Fig. 11.68. The electron density is $1.03 \times 10^{18}/\mathrm{cm}^3$. A magnetic field of 25.4 kG causes a shift by $\Delta\omega = 0.65 \times 10^{13}/\mathrm{sec}$. In the low-field approximation the magneto-plasma shift is half the cyclotron frequency. The effective mass relative to the free electron mass is thus obtained from

$$m/m_o = (e/m_o)B/2\Delta\omega = 1.76 \times 10^{15} \times 2.54 \times 10^{-4}/1.3 \times 10^{13} = 0.035 \tag{11 l.15}$$

The observed frequency dependence may be compared with Eq.(11 l.10). From this comparison a value of 2.8×10^{-13} sec is obtained for the momentum relaxation time, τ_m. Fig. 11.74 shows for B = 0 that $\overline{\omega}_p$ is 7.5×10^{13}/sec and therefore the condition $(\overline{\omega}_p \tau_m)^2 \gg 1$ is satisfied ($\overline{\omega}_p \tau_m = 22.5$). This condition can be written in the form

$$\kappa\rho/\mathrm{ohm\ cm} \ll \tau_m/10^{-13}\mathrm{sec} \tag{11 l.16}$$

where κ is the relative dielectric constant and ρ is the resistivity. For strong magnetic fields where magnetoplasma waves occur a discussion will be given in Chap. 11.n.

Eq.(11 l.10) shows that in the presence of a magnetic field the right- and left-handed circularly polarized waves have different velocities of propagation, $\omega/q_{\mp}$. A plane-polarized wave can be thought to be composed of two circularly pola-

[1] E. D. Palik, S. Teitler, B. W. Henvis, and R. F. Wallis, Proc.Int.Conf.Phys. Semicond. Exeter 1962 (A. G. Stickland, ed.), p. 288. London: The Institute of Physics and the Physical Society. 1962.

rized components. After transmission through a sample of thickness d, by recomposition we obtain again a plane-polarized wave, with the plane of polarization being rotated by an angle θ_F relative to the initial location. This rotation is known as the "Faraday effect". Wave propagation is still along the static magnetic field ; this is the "Faraday configuration". The angle θ_F is the average of the rotations of the two circularly polarized waves, θ_+ and $-\theta_-$, shown in Fig.11.75:

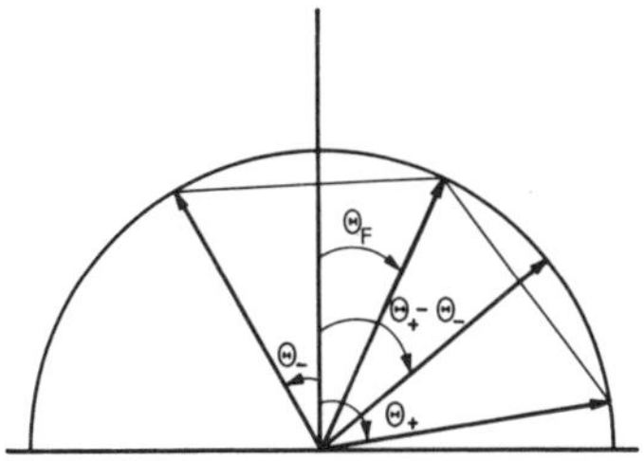

Fig.11.75 Construction of the Faraday angle θ_F as the average of rotations for right- and left-hand circular polarizations, θ_+ and $-\theta_-$, respectively.

$$\theta_F = \frac{1}{2}(\theta_+ - \theta_-) = \frac{1}{2}(q_+ - q_-)d \qquad (11\,l.17)$$

where d is the sample thickness.

From Eq.(11 l.10) we obtain for the Faraday rotation

$$\theta_F = \frac{\omega}{2c}\{\sqrt{\kappa - i(\mu_o c^2/\omega)\sigma_+} - \sqrt{\kappa - i(\mu_o c^2/\omega)\sigma_-}\}d \qquad (11\,l.18)$$

Introducing the plasma frequency given by Eq.(11 l.12) and assuming $\tau_m^{-1} \ll |\omega \pm \omega_c|$, we find

$$\sqrt{\kappa - i(\mu_o c^2/\omega)\sigma_\pm} = \sqrt{\kappa}\sqrt{1 - \overline{\omega}_p^2/\{\omega(\omega \pm \omega_c)\}} = \sqrt{\kappa}\sqrt{1 - \overline{\omega}_p^2(\omega \mp \omega_c)/\{\omega(\omega^2 - \omega_c^2)\}} \qquad (11\,l.19)$$

For the case $\overline{\omega}_p^2 \ll \omega(\omega \pm \omega_c)$ the square root can be expanded:

$$\sqrt{\kappa - i(\mu_o c^2/\omega)\sigma_\pm} = \sqrt{\kappa}\,\{1 - \frac{1}{2}\frac{\overline{\omega}_p^2}{\omega^2 - \omega_c^2} \pm \frac{1}{2}\frac{\overline{\omega}_p^2\,\omega_c}{\omega(\omega^2 - \omega_c^2)}\} \qquad (11\,l.20)$$

The difference of the square roots in Eq.(11 l.18) yields $\sqrt{\kappa}\,\overline{\omega}_p^2\omega_c/\{\omega(\omega^2 - \omega_c^2)\}$. The final result for θ_F given in degrees instead of radians, is thus

$$\theta_F = \frac{360^\circ}{2\pi} \cdot \frac{ne^3 Bd}{m^2\sqrt{\kappa_{opt}}\,\kappa_o 2c(\omega^2 - \omega_c^2)} \qquad (11\,l.21)$$

$$= \frac{1.64^\circ}{\sqrt{\kappa_{opt}}}\;\frac{(n/10^{18}\,cm^{-3})\,(\lambda/10\mu m)^2\,(B/kG)\,(d/cm)}{(m/m_o)^2\,(1 - \omega_c^2/\omega^2)}$$

where for the range of optical frequencies ω given by $\overline{\omega}_p^2 \ll \omega(\omega \pm \omega_c)$, the dielectric constant has been given a subscript "opt", and the free-space wavelength λ has been introduced. The sign of θ_F is different for electrons and holes and depends on whether the light propagates in the direction of $\vec{B}$ or in the opposite direction. θ_F/Bd is called the "Verdet constant".

In an analysis of data multiple reflections may have to be taken into account. The true rotation, θ_F, is then obtained from the observed rotation θ_F', by

$$\theta_F = \theta_F'/\{1 + 2r_\infty^2 \exp(-\alpha d)\cos(4\theta_F')\} \qquad (11\,l.22)$$

where α is the absorption coefficient [2] (Appendix D).

[2] H. Piller, J. Appl. Phys. 37 (1966) 763.

For the many-valley model the effective mass, m, is given by the "Hall effective mass" Eq.(7c.10). For this reason the Faraday effect may be considered as a "high-frequency Hall effect". Since the reciprocal relaxation time has been neglected, at small magnetic field intensities the Faraday effect provides a method for a determination of m which is not obscured by a Hall factor or by the anisotropy of τ_m, in contrast to the usual Hall effect.

For degenerate semiconductors with spherical energy surfaces the optical effective mass is given by [3]

$$\frac{1}{m} = \left(\frac{1}{\hbar^2 k} \cdot \frac{d\epsilon}{dk}\right)_\zeta \tag{11 l.23}$$

where the right-hand side is evaluated at the Fermi level ζ.

For heavy and light carriers of the same charge the factor n/m^2 in Eq.(11 l.21) is replaced by

$$n/m^2 \to n_h/m_h^2 + n_l/m_l^2 \tag{11 l.24}$$

where the subscripts h and l stand for "heavy" and "light".

The Faraday ellipticity, ϵ_F, is defined by

$$\epsilon_F = \frac{d}{2}\{\mathrm{Im}(q_+) - \mathrm{Im}(q_-)\} \tag{11 l.25}$$

where Im means the imaginary part. In an approximation for small values of τ_m^{-1} we find from Eq.(11 l.10)

$$\mathrm{Im}(q_\pm) = -\frac{\omega\sqrt{\kappa}\,\overline{\omega}_p^2(\omega \pm \omega_c)^{-2}\tau_m^{-1}}{2c\sqrt{1-\overline{\omega}_p^2/\{\omega(\omega \pm \omega_c)\}}} \tag{11 l.26}$$

For large values of ω we can neglect the square root in the denominator and obtain

$$\epsilon_F = \frac{\sqrt{\kappa_{opt}}\,\overline{\omega}_p^2\,\omega_c\,\omega d}{c(\omega^2-\omega_c^2)^2\tau_m} = \frac{ne^3Bd}{m^2\sqrt{\kappa_{opt}}\,\kappa_o c(1-\omega_c^2/\omega^2)\,\omega\tau_m} \tag{11 l.27}$$

For $\omega_c \ll \omega$ from Eqs.(11 l.21) and (11 l.27) a simple expression is found for the ratio ϵ_F/θ_F

$$\epsilon_F/\theta_F = 2/\omega\tau_m \tag{11 l.28}$$

where the unit of θ_F is the radian. A measurement of this ratio directly yields τ_m. However, since it has been assumed that $\tau_m^{-1} \ll \omega$, the ellipticity is small for the range of validity of Eq.(11 l.28) [4].

---

[3] H. Y. Fan: Semiconductors and Semimetals (R. K. Willardson and A. C. Beer, eds.), Vol. 3. New York: Acad. Press. 1967; the "optical effective mass" is the mass which appears in the expression for the free-carrier contribution to the dielectric constant.

[4] M. J. Stephen and A. B. Lidiard, J. Phys. Chem. Solids 9 (1959) 43; B. Donovan and J. Webster, Proc. Phys. Soc. (London) 79 (1962) 46 and 1081; ibid. 81 (1963) 90.

At microwave frequencies, where $\omega\tau_m \ll 1$, θ_F to a good approximation is obtained from Eq.(11 l.21) by replacing $1-\omega_c^2/\omega^2$ by $(\omega\tau_m)^{-2}$. The ratio ϵ_F/θ_F becomes $\omega\tau_m$ which is again small [5] The largest value of the ratio is obtained for $\omega\tau_m \approx 1$ and would be most useful for a determination of τ_m.

Experimental data on θ_F vs λ^2 obtained in n-type InSb for various electron densities at 77 and 296 K are shown in Fig.11.76 [6]. Effective masses obtained from these data increase both with temperature and electron density in agreement with the well-known non-parabolicity of the conduction band of InSb. In the range of λ = 10 to 20 μm, $\theta_F \propto \lambda^2$ in agreement with Eq.(11 l.21).

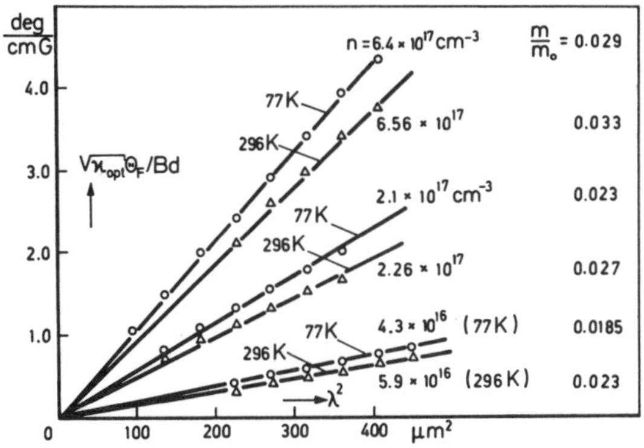

Fig.11.76 Faraday rotation in n-type InSb. A negative interband rotation (not shown here) causes deviation from linearity at short wavelengths (after Pidgeon, ref. 6).

Fig.11.77 shows data for θ_F vs B at 85 K for a sample with $n \approx 10^{15}\,cm^3$ [7]. Up to 20 kG, $\theta_F \propto B$; above 20 kG, however, there is cyclotron resonance where the assumption $\tau_m^{-1} \ll |\omega - \omega_c|$ made in the derivation of Eq.(11 l.21) is no longer valid since $\omega = \omega_c$. Two fits made with a more rigorous treatment are given in the figure. At the largest field strengths where $\omega_c \gg \omega$, θ_F is negative and $\propto 1/B$:

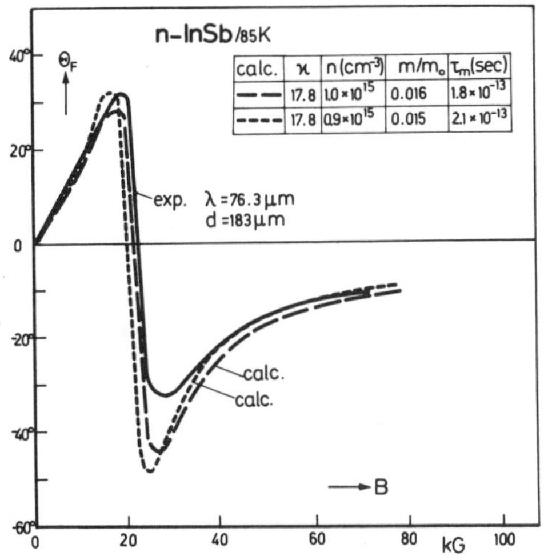

Fig.11.77 Observed and calculated Faraday rotation in n-type InSb in a range of magnetic fields which includes cyclotron resonance (after Palik, ref. 7).

[5] R. R. Rau and M. E. Caspari, Phys. Rev. 100(1955) 632; A. Bouwknegt, thesis. Univ. Utrecht 1965 (includes a review of experimental methods).
[6] C. R. Pidgeon, thesis. Univ. Reading 1962.
[7] E. D. Palik, Appl. Opt. 2 (1963) 527.

$$|\theta_F| = \frac{\mu_o c}{2\sqrt{\kappa_{opt}}} \frac{ne}{B} d = \frac{1.74^\circ}{\sqrt{\kappa_{opt}}} \frac{(n/10^{10}\,cm^{-3})\,(d/cm)}{B/kG} \qquad (11\,1.29)$$

This proportionality is in agreement with experimental data given in Fig.11.77.

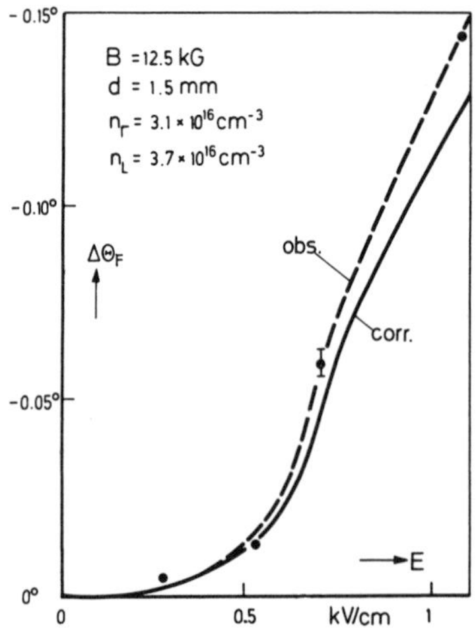

Fig.11.78 Observed decrease of the Faraday angle in n-type GaSb with electric field intensity (dashed) and values corrected for multiple reflections (full) (after Heinrich, ref. 8).

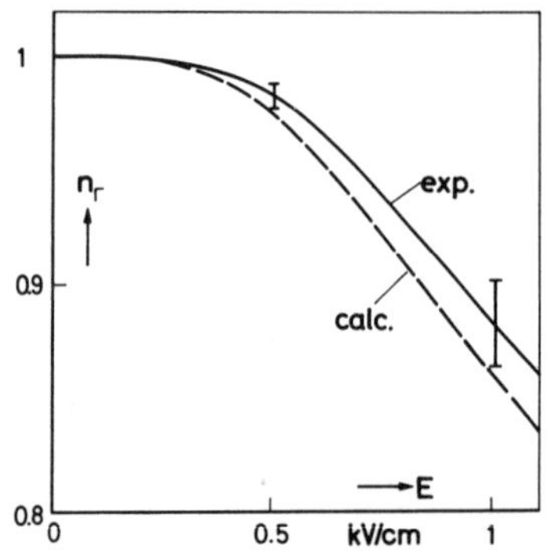

Fig.11.79 Decrease of the number of carriers in the central valley of the conduction band of GaSb with electric field E, normalized to unity for E = 0, as obtained from the experimental results of Faraday rotation given in Fig.11.78 (full curve). Dashed curve: Calculated (after Heinrich, ref. 8, and Heinrich and Jantsch, ref. 9).

In n-type GaSb heating of electrons by a static electric field, E, which causes electron transfer from the central valley to satelite valleys of the conduction band, has been observed by the infrared Faraday effect [8]. The (isotropic) effective mass in the central valley is only 0.047 m_o while in the <111> satellite valleys the transverse mass is 0.143 m_o and the longitudinal mass is even larger by a factor of $K_m = 8.6$. The zero-field mobilities are 3100 and 550 $cm^2/Vsec$, respectively, for the two valleys which in energy are $\Delta_L = 0.1$ eV apart. The experimental change of θ_F with E is plotted in Fig.11.78. The electron densities in the central valley, n_Γ, and the satellite valleys, n_L, are given by Eq.(3a.38) with T replaced by the electron temperature, T_e, which for simplicity has been assumed to be the same for both types of valleys. The central valley population, n_Γ, as a function of the electric field strength is shown in Fig.11.79. It decreases with carrier heating due to the transfer effect. The dashed line has been calculated from Eq.(3a.38) with T_e determined from Eqs. (4m.10), (4m.12), and (4m.13) where τ_ϵ was chosen for a best fit for the $\vec{j}$-$\vec{E}$-characteristics [9]. Instead of this semi-empirical approach, Ruch [10] calculated the effect taking non-equivalent intervalley scattering with $D_{iv} = 1.5 \times 10^9$ eV/cm into account and obtained good agreement with the experimental data.

We will now consider the "transverse" or "Voigt configuration" where $\vec{q} \perp \vec{B}$, e.g.

[8] H. Heinrich, Phys. Rev. B 3 (1971) 416.
[9] W. Jantsch and H. Heinrich, Phys. Rev. B 3 (1971) 420.
[10] J. G. Ruch, Appl. Phys. Lett. 20 (1972) 246.

$\vec{q} = (0, q, 0)$ for $\vec{E} = (E_x, E_y, 0)$. (An electromagnetic wave in a solid may be partly longitudinal.) As before we consider a linearly polarized wave. Eq.(11 1.4) in components is now given by Eq.(11 1.5) and by

$$(\kappa\omega^2/c^2 - i\omega\mu_o\sigma_{xx}) E_y + i\omega\mu_o\sigma_{xy} E_x = 0 \qquad (11\ 1.30)$$

The case $\tau_m^{-1} \ll \omega$ yields for the ratio of the field components

$$E_y/E_x = -i\overline{\omega}_p^2\omega_c/\{\omega(\overline{\omega}_p^2 + \omega_c^2 - \omega^2)\} \qquad (11\ 1.31)$$

Substituting the ratio E_y/E_x in Eq.(11 1.5) yields the "dispersion relation"

$$c^2q^2/(\omega^2\kappa) = 1 + \overline{\omega}_p^2/(\omega_c^2 - \omega^2) - \overline{\omega}_p^4\omega_c^2/\{\omega^2(\omega_c^2 - \omega^2)^2 + \omega^2\overline{\omega}_p^2(\omega_c^2 - \omega^2)\} \qquad (11\ 1.32)$$

For magnetic fields of small intensity where $\omega_c^2 \ll \omega^2$, a quadratic equation for ω^2 is obtained:

$$\omega^4 - (2\overline{\omega}_p^2 + c^2q^2/\kappa)\,\omega^2 + \overline{\omega}_p^2(\overline{\omega}_p^2 - \omega_c^2 + c^2q^2/\kappa) = 0 \qquad (11\ 1.33)$$

For 100% reflectivity, where $q = 0$, and for $\omega_c \ll \overline{\omega}_p$ we find for a solution

$$\omega_\pm = \overline{\omega}_p \pm \frac{1}{2}\omega_c \qquad (11\ 1.34)$$

while for a reflectivity minimum $c^2q^2 = \omega^2$ and therefore from Eq.(11 1.33)

$$\omega'_\pm = (\overline{\omega}_p/\sqrt{2(\kappa-1)})\sqrt{2\kappa - 1 \pm \sqrt{1 + 4\kappa(\kappa-1)\,\omega_c^2/\overline{\omega}_p^2}} \qquad (11\ 1.35)$$

is obtained which for $4\kappa(\kappa-1)\,\omega_c^2 \ll \overline{\omega}_p^2$ and $\kappa \gg 1$ can be approximated by

$$\omega'_\pm = \overline{\omega}_p\{1 \pm \kappa\omega_c^2/(2\overline{\omega}_p^2)\} = \overline{\omega}_p \pm \frac{1}{2}\omega_c(\kappa\omega_c/\overline{\omega}_p) \qquad (11\ 1.36)$$

Fig. 11.80 shows experimental data on reflectivity vs ω for n-InSb at room temperature [11]. The agreement with the theoretical curve is excellent if τ_m is taken into account. The experiment may serve for a determination of m, τ_m, and n.

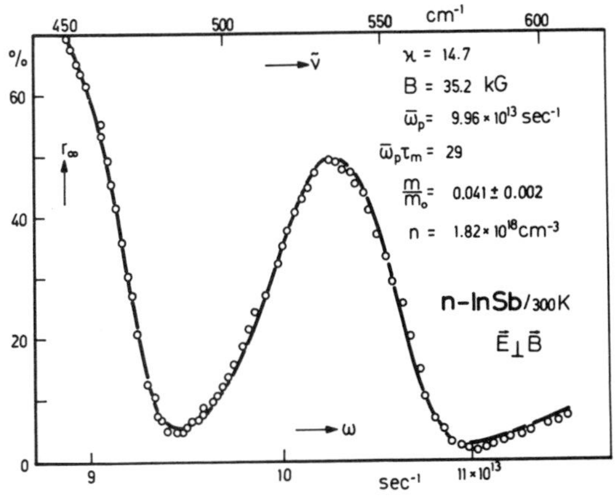

Fig. 11.80
Transverse magnetoplasma reflection in n-type InSb at room temperature) (after Wright and Lax, ref. 11).

---

[11] G. B. Wright and B. Lax, J. Appl. Phys. Suppl. 32 (1961) 2113.

As in the Faraday experiment the transmitted wave is thought to be decomposed into the two circularly polarized waves. The "Voigt phase angle" is defined as

$$\theta_V = (q_\perp - q_\parallel)\, d/2 \tag{11 l.37}$$

where the subcripts refer to the angle between the polarization and the static magnetic field. From the dispersion relation, Eq.(11 l.32), we obtain for q in the approximation $\omega^2 \gg \omega_c^2$ and $\overline{\omega}_p^2$:

$$q_\perp = (\omega/c)\sqrt{\kappa}\sqrt{1 - \overline{\omega}_p^2/\omega^2 - \overline{\omega}_p^2\,\omega_c^2/\omega^4} \approx (\omega/c)\sqrt{\kappa}(1 - \overline{\omega}_p^2/2\omega^2 - \overline{\omega}_p^2\,\omega_c^2/2\omega^4) \tag{11 l.38}$$

$q_\parallel$ is obtained from this relation by putting $\omega_c = 0$ since the electron motion parallel to the magnetic field is the same as without the field. Hence, we find for θ_V:

$$\theta_V = -\frac{d}{2}\,\omega\,\frac{\sqrt{\kappa}}{c}\,\frac{\overline{\omega}_p^2\,\omega_c^2}{2\omega^4} = -\frac{\mu_0 c}{4\sqrt{\kappa_{opt}}} \cdot \frac{ne^4 B^2 d}{m^3\omega^3}$$

$$= \frac{7.75^\circ}{\sqrt{\kappa_{opt}}}\;\frac{(n/10^{23}\,\text{cm}^{-3})\,(\lambda/10\,\mu\text{m})^3\,(B/\text{kG})^2\,(d/\text{cm})}{(m/m_0)^3} \tag{11 l.39}$$

valid for $\omega_c \ll \omega$. The Voigt phase angle is smaller than the Faraday angle by a factor of $\omega_c/2\omega$ and is therefore more difficult to measure.

In the many-valley model m is replaced by a "magneto-resistance mass", m_M, which may easily be defined by the right-hand side of Eq.(7d.5). For this reason the Voigt effect may be called the "high-frequency magneto-resistance effect". Since no τ_m-anisotropy is involved the Voigt anisotropy is directly related to the band structure, in contrast to the usual magnetoresistance.

For Voigt measurements the incident beam should be linearly polarized at an angle of 45° relative to the static magnetic field. Although there is no rotation in polarization upon transmission through the sample the wave becomes elliptically polarized, the ellipticity being [12]

$$\epsilon_V = \tan\theta_V \tag{11 l.40}$$

From measurements of the ellipticity the Voigt phase angle is thus determined.

Experimental data of θ_V vs λ^3 and B^2 obtained in n-InSb with $n = 1.6\text{x}10^{16}$ cm^{-3} at 85 K are shown in Figs.11.85 and 11.86, respectively [13]. A slight departure from the θ_V-vs-B^2 linear relationship indicates that the condition $\omega \gg \omega_c$ is not really valid here. At short wavelengths there is an interband transition to be discussed in Chap.11m. As in the case of the Faraday rotation in degenerate semiconductors the effective mass is the "optical mass" at the Fermi level.

Eq.(11 l.32) can be written in the form

---

[12] M. Born and E. Wolf: Principles of Optics, p.26. London: Pergamon. 1959.
[13] S. Teitler, E. D. Palik, and R. F. Wallis, Phys. Rev. 123 (1961) 1631.

$$c^2 q^2/(\omega^2 \kappa) = 1 + (\overline{\omega}_p/\omega)^2 \{1 + \omega_c^2/(\omega^2 - \omega_c^2 - \overline{\omega}_p^2)\} \tag{11 l.41}$$

This shows that in the Voigt configuration the cyclotron resonance ($q = \infty$) occurs at

$$\omega = \sqrt{\omega_c^2 + \overline{\omega}_p^2} \tag{11 l.42}$$

while in the usually applied configuration it is found at $\omega = \omega_c$.

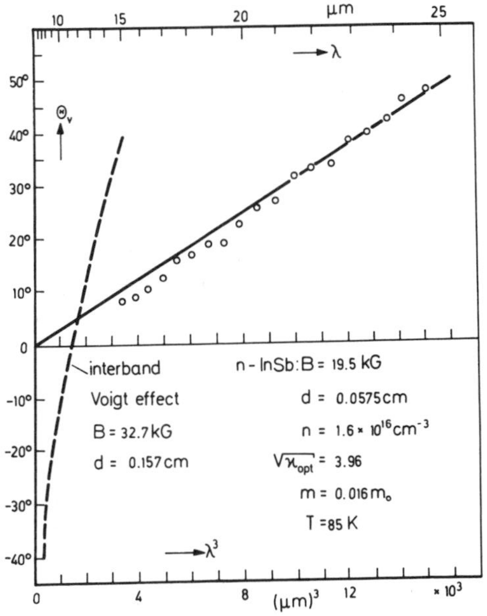

Fig. 11.81
Voigt effect in n-type InSb. An interband Voigt effect is shown at short wavelengths for somewhat different conditions (dashed curve). The straight line through the origin has been calculated (after Teitler et al., ref. 13, as quoted by Palik and Wright in ref. 7 of Chap. 11k).

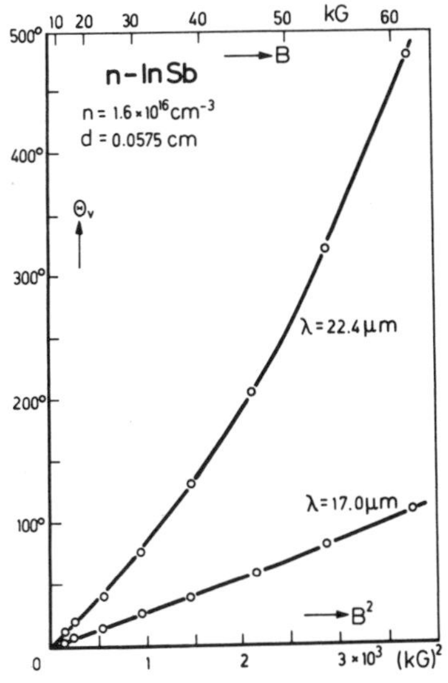

Fig. 11.82
Voigt angle in n-type InSb at 85 K, indicating departure from proportionality to B^2 (after Teitler et al., ref. 12, quoted by Palik and Wright in ref. 7 of Chap. 11k).

11m. Interband Magneto-Optical Effects

Magneto-optical effects for free carriers were calculated from the equation of motion, Eq.(11k.4). For bound electrons, however, we have to add a term $m\omega_e^2\vec{r}$ on the left-hand side and substitute $d\vec{r}/dt$ for $\vec{v}_d$:

$$md^2\vec{r}/dt^2 + m\omega_e^2\vec{r} = e(\vec{E} + [(d\vec{r}/dt)\vec{B}]) \tag{11m.1}$$

where the damping term $\propto \tau_m^{-1}$ has been neglected because the eigenfrequency of the harmonic oscillator $\omega_e \gg \tau_m^{-1}$. For the case of right- and left-hand circularly polarized radiation the conductivity becomes now instead of Eqs.(11k.8) and (11k.9)

$$\sigma_\pm = i\kappa_o\,\omega_p^2\omega/(\omega_e^2 - \omega^2 \pm \omega\,\omega_c) \tag{11m.2}$$

For the case of weak fields where $\omega_c \ll (\omega_e^2 - \omega^2)/\omega$, the Faraday rotation becomes

$$\theta_F = \frac{360^\circ}{2\pi}\,\frac{ne^3Bd}{2m^2\sqrt{\kappa_{opt}}\,\kappa_o^2 c\omega^2(\omega_e^2/\omega^2 - 1)^2} \tag{11m.3}$$

which is similar to Eq.(11 l.21) except that now the factor $(1 - \omega_c^2/\omega^2)$ in the denominator is replaced by $(\omega_e^2/\omega^2 - 1)^2$. The voigt angle is obtained by a similar calculation:

$$\theta_V = -\frac{\mu_o c}{4\sqrt{\kappa_{opt}}}\cdot\frac{ne^4B^2d}{m^3\omega^3(\omega_e^2/\omega^2 - 1)^3} \tag{11m.4}$$

Of course, at resonance, $\omega = \omega_e$, θ_F and θ_V are determined by τ_m^{-1} which has been neglected here.

Fig.11.83 shows the observed direct interband Faraday rotation in Ge at point Γ in the Brillouin zone as a function of the photon energy [1]. The reversal in sign cannot be accounted for by Eq.(11m.3). A quantum mechanical treatment with a consideration of the Zeeman splitting of the exciton levels [2] is for this case more adequate than the classical oscillator model. In the quantum mechanical treatment a phenomenological relaxation time τ of the exciton of the order of magnitude of 10^{-13}sec and a frequency γB are introduced, where $\gamma = g\mu_B/2\hbar$ and g and μ_B are the g-factor and the Bohr magneton. Sign reversals are then predicted for frequencies

$$\omega_\pm = \omega_e \pm \sqrt{\tau^{-2} + \gamma^2B^2} \tag{11m.5}$$

If there are several resonance frequencies ω_e, one resonance may give a contribution at the position of another one and thus prevent one of the two sign reversals

[1] Y. Nishina, J. Kolodziejczak, and B. Lax: Proc. Int. Conf. Phys. Semicond. Paris 1964 (J. Bok, ed.), p. 867. Paris: Dunod. 1964.

[2] J. Halpern, B. Lax and Y. Nishina, Phys. Rev. 134 (1964) A 140.

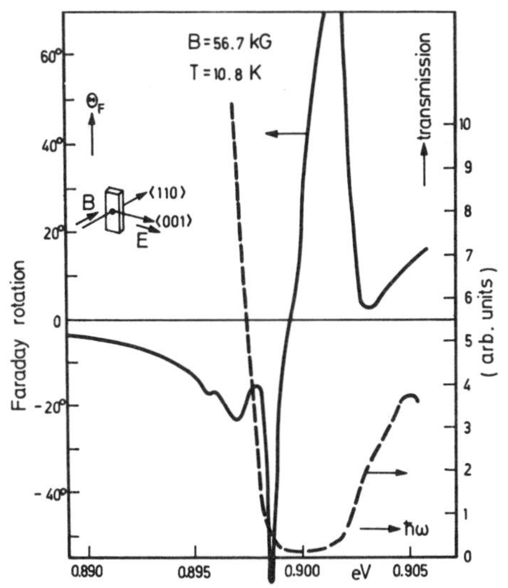

Fig. 11.83 Direct interband Faraday rotation in germanium. The dashed curve indicates the relative intensity of the transmitted radiation (after Nishina et al., ref. 1).

of Eq.(11m.5). This would be an explanation for having only one sign reversal in Fig. 11.83.

In the Voigt effect sign reversals may occur at frequencies

$$\omega_{\pm} = \omega_e \pm \sqrt{(\tau^{-2} + \gamma^2 B^2)/3} \tag{11m.6}$$

An oscillatory interband Faraday rotation has been observed in Ge at 8 K for magnetic fields of up to 103 kG [3]. The oscillatory effects are of the order of 2 % of the total rotation and have been correlated partly with exciton absorption and partly with Landau transitions.

11n. Magnetoplasma Waves

The dispersion relation for the Faraday configuration, Eq.(11 l.10) is now approximated for strong magnetic fields where $\omega_c \gg \omega$ [1]:

$$c^2 q_{\mp}^2/\omega^2 - \kappa = \omega_p^2/\{\omega(\mp\omega_c + i\tau_m^{-1})\} \tag{11n.1}$$

and the plasma frequency

$$\omega_p = \sqrt{ne^2/(m\kappa_o)} \tag{11n.2}$$

has been introduced for convenience.

For gases where the dielectric constant $\kappa = 1$, Eq.(11n.1) is known as the "Appleton Hartree equation". For the case of weak damping, $\tau_m^{-1} \ll \omega_c$, we expand Eq.(11n.1):

$$c^2 q_{\mp}^2/\omega^2 - \kappa = \pm\omega_p^2/\omega\omega_c - i\tau_m^{-1}\omega_p^2/\omega\omega_c^2 + \cdots \tag{11n.3}$$

[3] J. Halpern, J. Phys. Chem. Solids 27 (1966) 1505.

Chap. 11n

[1] For reviews see e.g. M. Glicksman: Solid State Physics (H. Ehrenreich, F. Seitz, and D. Turnbull, eds.), Vol. 26, p. 275. New York: Acad. Press. 1971; P. M. Platzman and P. A. Wolff: Solid State Physics (H. Ehrenreich, F. Seitz, and D. Turnbull, eds.), Suppl. 13, New York: Acad. Press. 1972.

In a typical plasma $\bar{\omega}_p = \omega_p / \sqrt{\kappa} \gg \sqrt{\omega\omega_c}$, and κ is negligible. For a first approximation we neglect also the damping term; the refractive index

$$cq_{\mp}/\omega = \sqrt{\pm\,\omega_p^2/\omega\omega_c} \tag{11n.4}$$

is real only for the positive sign of the radicand. Depending on the sign of ω_c, i.e. the type of conductivity, either the left- or right-handed circularly polarized wave is transmitted. Its phase velocity

$$\omega/q = c\sqrt{\omega\omega_c/\omega_p^2} = \sqrt{\omega B/ne\mu_0} = 1.77\ \text{cm/sec}\sqrt{\frac{(\nu/\text{Hz})\,(B/\text{kG})}{n/10^{22}\ \text{cm}^{-3}}} \tag{11n.5}$$

can be very low. E.g. for a semiconductor with 10^{18} carriers/cm^3 in a magnetic field of 1 kG at a frequency of 1 MHz the velocity is only $1.77\text{x}10^5$ cm/sec and the refractive index is about 10^5. Such a wave is called a "helicon". The penetration depth, δ, of the helicon wave is obtained from Eq.(11n.3):

$$\delta = \frac{\lambda}{2\pi}\cdot\frac{\omega\omega_c^2}{\tau_m^{-1}\omega_p^2}\sqrt{\frac{\omega_p^2}{\omega\omega_c}} = \frac{c\omega_c^{3/2}\tau_m}{\omega_p\,\omega^{1/2}} = \sqrt{\frac{1}{e\mu_0 c}}\,\frac{\mu}{\sqrt{n}}\,\frac{B^{3/2}\sqrt{\lambda}}{\sqrt{2\pi}}$$
$$= 1.628\ \text{cm}\ (\mu/\text{cm}^2\,\text{V}^{-1}\,\text{sec}^{-1})\sqrt{\lambda/\text{cm}}\ (B/\text{kG})^{3/2}/\sqrt{n/\text{cm}^{-3}} \tag{11n.6}$$

where μ is the carrier mobility and λ the free-space wavelength. E.g. for n-InSb at 77 K with $n = 1.2\text{x}10^{14}/\text{cm}^3$, $\mu = 3.5\text{x}10^5$ cm^2/Vsec at $B = 7.4$ kG, a 10 GHz wave ($\lambda \approx 3$ cm) has a range of 1.8 cm [2] which is larger by a factor of $\sqrt{2}/(\mu B)^{3/2}$ than the range of the order of 10^{-2} cm at $B = 0$. At optical wavelengths a magnetic field which is ten times stronger is required. In microwave experiments there are sometimes "dimensional resonances" which occur if the sample thickness is a multiple integer of half a wavelength in the sample. From measurements of the resonant wavelength Eq.(11n.5) yields the carrier density, n. Fig.11.84 shows resonances measured in n-InSb at a temperature of 77 K and a microwave frequency of 35.76 GHz [3].

In a plasma drifting with a velocity v_d there is a Doppler shift of the frequency. Because of the low phase velocity, ω/q, it is possible to have $v_d > \omega/q$, and energy would be transferred from the drifting carriers to the wave with the result of wave amplification. However, v_d should not be much larger than ω/q. In this case the refractive index would nearly vanish, and the phase velocity would again be high. In order to keep the refractive index high, the plasma is composed of two kinds of carriers, e.g. electrons and holes with different mobilities; since the hole drift velocity in general is different from the phase velocity of the wave, the refractive index is still very high even though $v_d \approx \omega/q$ for electrons. This is the idea of "two-stream instability" [4]. For practical applications, however, collision damping has been strong enough to prevent wave amplification.

[2] A. Libchaber and R. Veilex, Phys. Rev. 127 (1962) 774.

[3] N. Perrin, B. Perrin, and W. Mercouroff: Plasma Effects in Solids (J. Bok, ed.), p. 37. Paris: Dunod. 1964.

[4] J. Bok and P. Nozieres, J. Phys. Chem. Solids 24 (1963) 709.

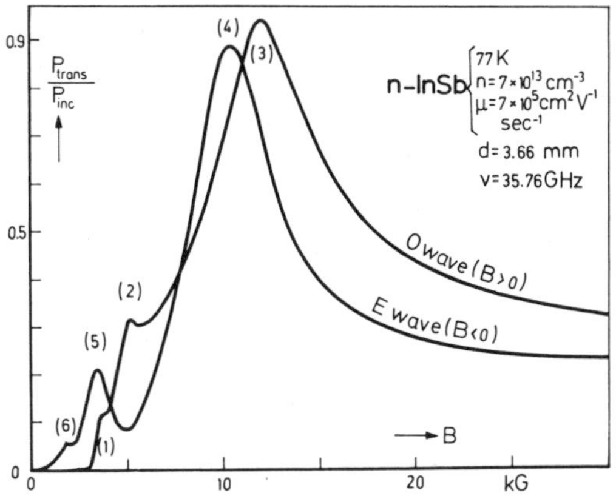

Fig. 11.84
Ratio of transmitted and incident microwave power for a magnetoplasma in n-type InSb at 77 K, as a function of the longitudinal magnetic induction for right- and left-hand circular polarization of the wave (after Perrin et al., ref. 3).

If in a polar crystal the phase velocity is equal to the velocity $\vec{u}_s$ of transverse sound waves ("shear waves") an interaction takes place; from Eq.(11n.5) we obtain for this case

$$|\vec{u}_s| = c\sqrt{\omega\omega_c/\omega_p^2} \tag{11n.7}$$

Solving for ω and taking $|\vec{u}_s| \approx 5\text{x}10^5$ cm/sec we find

$$\omega = (u_s/c)^2\,\omega_p^2/\omega_c \approx 5\,\frac{n/10^{11}\,\text{cm}^{-3}}{B/\text{kG}}\ \text{sec}^{-1} \tag{11n.8}$$

In an experiment [5] with Cd_3As_2 with $n = 10^{18}/\text{cm}^3$, shear waves were excited at a frequency, $\omega/2\pi$, of 525 kHz by a quartz transducer at one side of a disk-shaped sample. When a longitudinal field of 15.4 kG was applied, helicon waves produced by the coupling were detected by a coil of a few turns around the sample. Fig. 11.85 shows the variation of the coil voltage with the applied magnetic field. The experimental data are consistent with Eq.(11n.8).

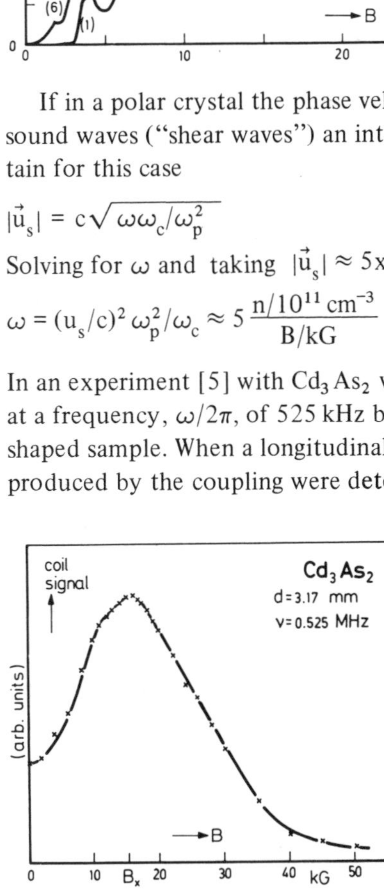

Fig. 11.85 Helicon phonon interaction: Coil signal vs magnetic induction (after Rosenman, ref. 5).

In intrinsic semiconductors and semimetals there are equal numbers of electrons and holes. If we can neglect τ_m^{-1} for <u>both</u> types of carriers*) the dispersion relation Eq.(11n.1) is given by

$$c^2q_\mp^2/\omega^2 = \sum_{i=1}^{2}\omega_{pi}^2/\{\omega(\mp\omega_{ci}-\omega)\}$$

$$\approx \sum_{i=1}^{2}\omega_{pi}^2/(\mp\omega_{ci}\omega) + \sum_{i=1}^{2}\omega_{pi}^2/\omega_{ci}^2 \tag{11n.9}$$

[5] I. Rosenman, Solid State Comm. <u>3</u> (1965) 405.

*) This is not the case e.g. in InSb at room temperature for a range of frequencies where for holes $\tau_m^{-1} > \omega$ while for electrons $\tau_m^{-1} < \omega$.

The approximation is valid for $\omega_c \gg \omega$. The first term on the right-hand side vanishes for equal numbers of electrons and holes. The second term yields a phase velocity

$$\omega/q = B/\sqrt{\mu_o n(m_n + m_p)} = 0.94\times10^{16}\,\text{cm/sec}\;\frac{B/\text{kG}}{\sqrt{n/\text{cm}^{-3}}\;\sqrt{(m_n + m_p)/m_o}} \quad (11n.10)$$

which is independent of the polarization and therefore valid also for a plane polarized wave. These waves are denoted as "Alfvén waves". E.g. in Bi at room temperature the refractive index cq/ω is 30 for a longitudinal magnetic field of 16 kG.

If the wave is incident at an oblique angle θ relative to the applied magnetic field there is an effect of birefringence: One wave which is polarized parallel to the plane of incidence is slower than the perpendicular one by a factor of $\cos\theta$. Alfvén wave measurements yield data on $m_n + m_p$ according to Eq.(11n.10) and therefore on the band structure.

11o. Nonlinear Optics

The intensity of light emitted by sources other than lasers is low enough to ensure a linear relationship between the polarization of a dielectric, $\vec{P}$, and the electric field strength, $\vec{E}$,

$$\vec{P} = \chi\vec{E} \quad (11o.1)$$

where the susceptibility

$$\chi = \kappa_o(\kappa - 1) \quad (11o.2)$$

and the dielectric constant, κ, in general are second-rank tensors and κ_o is the permittivity of free space. However, light emitted by a Q-switched ruby laser may have an intensity of up to $10^9\,\text{W/cm}^2$ corresponding to an ac field strength, $\vec{E}$, of about $10^6\,\text{V/cm}$ *). Just as in the case of hot carriers where the relationship between current density and $\vec{E}$ at such high electric field intensities is nonlinear, we have to take into account nonlinear terms in the $\vec{P}$-vs-$\vec{E}$ relationship [1];

$$P_i = \sum_k \chi_{ik} E_k + \sum_{kl} \chi_{ikl} E_k E_l + \sum_{klm} \chi_{iklm} E_k E_l E_m + \cdots \quad (11o.3)$$

---

[1] N. Bloembergen: Nonlinear Optics. New York: Benjamin. 1965; A. Yariv: Quantum Electronics, p. 340. New York: J. Wiley and Sons. 1967; for review articles see e.g. A. F. Gibson, Sci. Progr. 56 (1968) 479; J.A. Giordmaine, Physics Today 22 (1969) No. 1, p. 38; Sci. Am. 210 (1964) No. 4, p. 38.

*) The intensity of sun light on the earth at vertical incidence is about 2 cal/min = $0.135\,\text{W/cm}^2$ equivalent to $E \approx 10\,\text{V/cm}$. A 10^8 W laser beam focused to an area $(10\,\mu\text{m})^2$ yields even $10^8\,\text{V/cm}$ where, however, all solid material is destroyed.

Since in dielectrics the largest contribution to χ comes from bound electrons this nonlinearity is observed in insulators as well as in semiconductors. However, in semiconductors there is also a free-carrier contribution to the nonlinearity which will be discussed later in this chapter.

In a classical model the nonlinearity is due to the anharmonicity of the oscillations of the bound electrons in the ac field. In a one-dimensional model let us assume for the equation of motion of an oscillator

$$d^2 x/dt^2 + \omega_0^2 x - \epsilon x^3 = (e/m)E_1 \cos\omega \tag{11o.4}$$

where damping has been neglected for simplicity and the coefficient of the nonlinear term, ϵ, is assumed to be small. To a first approximation we take for a solution $x = x_1 \cos\omega t$, and by applying the mathematical identity $4\cos^3\omega t = \cos 3\omega t + 3\cos\omega t$, we find

$$\frac{3}{4}\epsilon x_1^3 + (\omega^2 - \omega_0^2)x_1 + (e/m)E_1 = 0 \tag{11o.5}$$

where a term $-\frac{1}{4}\epsilon x_1^3 \cos 3\omega t$ has been neglected in this approximation. The second approximation is obtained by integrating

$$d^2 x/dt^2 = -\omega^2 x_1 \cos\omega t + \frac{1}{4}\epsilon x_1^3 \cos 3\omega t \tag{11o.6}$$

which has been obtained from Eqs.(11o.4) and (11o.5) by eliminating E_1. The solution is nonlinear and given by

$$x = x_1 \cos\omega t - \epsilon(x_1^3/36\omega^2)\cos 3\omega t \tag{11o.7}$$

This yields a dipole moment $\propto \epsilon x$, with frequencies ω and 3ω. Since the polarization $\vec{P}$ is the dipole moment per unit volume, third harmonic generation is obtained which is adequately described by the third term on the right-hand side of Eq.(11o.3). Second harmonic generation can be obtained from a term proportional to x^2 in the equation of motion and hence from the second term in Eq.(11o.3). A restoring force proportional to x^2 implies a potential V proportional to x^3, i.e. $V(-x) \neq V(x)$: The dielectric does not possess inversion symmetry if a second harmonic is generated. Hence, it is the piezoelectric and ferroelectric materials which which are of primary interest here.

Eq.(11o.5) which yields the dependence of the oscillation amplitude, x_1, on frequency, ω, can be written in the form

$$\frac{3}{4}\frac{\epsilon}{\omega_0^2} x_1^3 = \{1 - (\omega/\omega_0)^2\}x_1 - (e/m\omega_0^2)E_1 \tag{11o.8}$$

When plotted vs x_1 the left-hand side is represented by a third-order parabola, while the right-hand side yields a straight line. For large values of ω there is a unique solution while for small values of ω the curves cross in 3 points yielding 3 possible values of the amplitude. Of course, in the latter case the smallest amplitude is the stable one. At a frequency ω_i where $d\omega/d(x_1^2)$ vanishes, there is an instability shown in Fig.11.86.

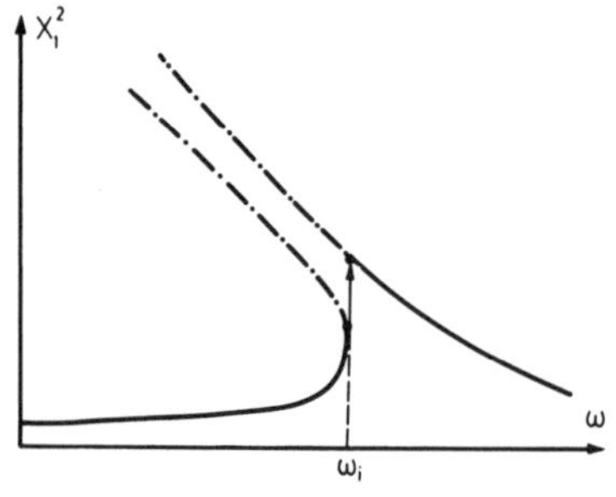

Fig.11.86 Instability of a classical anharmonic oscillator: the square of the amplitude plotted vs frequency shows instability at a frequency ω_i.

$$\omega_i = \omega_o \sqrt{1 - (\frac{3}{2})^{4/3} \epsilon^{1/3} (eE_1/m\omega_o^3)^{2/3}}$$

$$\simeq \omega_o \{1 - \frac{1}{2} (\frac{3}{2})^{4/3} \epsilon^{1/3} (eE_1/m\omega_o^3)^{2/3}\} \qquad (11o.9)$$

Since the series expansion in Eq.(11o.3) is valid only for $\epsilon E_1^2 \ll (m\omega^3/e)^2$ this shows that the frequency ω_i is smaller than the resonance frequency, ω_o, by only a small amount. Hence, in the vicinity of the resonance $\omega = \omega_o$ there is an enhanced multiple-harmonic generation.

The same applies to the generation of beat frequencies, $2\omega_1 - \omega_2$ and $2\omega_2 - \omega_1$, obtained by the application of a field composed of two frequencies, ω_1 and ω_2:

$$E = E_1 \exp\{i\omega_1 t - i(\vec{q}_1 \vec{r})\} + E_2 \exp\{i\omega_2 t - i(\vec{q}_2 \vec{r})\} \qquad (11o.10)$$

If the second frequency is twice the first frequency, $\omega_2 = 2\omega_1$, the beat frequencies are zero and $3\omega_1$. The former case yields "optical rectification" by "harmonic mixing".

Let us consider the case of long wavelengths where in Eq.(11.10) $(\vec{q}_1 \vec{r})$ and $(\vec{q}_2 \vec{r})$ can be neglected except for a phase shift φ, i.e. instead of propagating waves we simply have parallel ac fields:

$$E(t) = E_1 \cos(\omega t + \varphi) + E_2 \cos(2\omega t) \qquad (11o.11)$$

We will apply this field to a homogeneous isotropic semiconductor (i.e. no p-n structure) where for the range of warm carriers the $\vec{j}$-$\vec{E}$ characteristics are similar to the $\vec{P}$-vs-$\vec{E}$ relationship given by Eq.(11o.3):

$$j(t) = \sigma_o (E + \beta E^3) \qquad (11o.12)$$

A calculation of the time average of j yields a proportionality to $\beta \cos 2\varphi$. This is a dc current density, hence the notation "optical rectification". The lower part of Fig.11.87 shows E(t) for $E_1 = E_2$ and $\varphi = 0$. The time average of E vanishes which is true for any value of φ. The right-hand side of the figure shows j(t) obtained with a nonlinear symmetrical j(E) characteristic. The positive elongation is strongly reduced while the negative one is not. Hence, $\langle j \rangle \neq 0$. If in a calculation of $\langle j \rangle$ momentum relaxation and energy relaxation are taken into account, both the amplitude and the phase shift depend on τ_m and τ_ϵ:

$$\langle j \rangle = \frac{3}{4} \sigma_o \beta E_1^2 E_2 A \cos 2(\varphi + \psi) \qquad (11o.13)$$

where $A = A(\tau_m, \tau_\epsilon)$ and $\psi = \psi(\tau_m, \tau_\epsilon)$. For the simple case that $\omega\tau_m \ll 1$, the additional phase shift ψ is given by

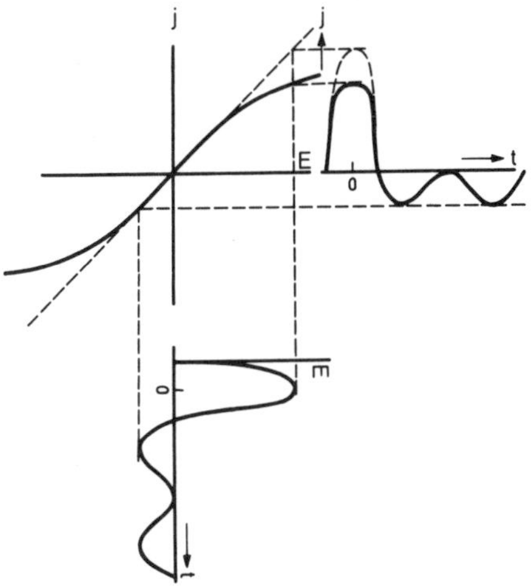

Fig. 11.87 Principle of optical rectification for ac electric fields.

$$\tan 2\psi = 2\omega^3\tau_\epsilon^3/(1 + 3\omega^2\tau_\epsilon^2) \quad (11o.14)$$

A measurement of ψ has served for a determination of τ_ϵ (Chap.4m, Figs. 6.21 and 6.28) [2].

Now considering again propagating waves, an important point for the generation of a frequency ω_3 from ω_1 and ω_2 in a dielectric is "phase matching". For the photons involved in this process, energy conservation yields the condition

$$\omega_3 = \omega_1 + \omega_2 \quad (11o.15)$$

while momentum conservation yields

$$\vec{q}_3 = \vec{q}_1 + \vec{q}_2 \quad (11o.16)$$

Assuming $\vec{q}_1$ and $\vec{q}_2$ to be parallel and introducing an index of refraction $n_i = c|q_i|/\omega_i$, both equations can be fulfilled simultaneously only for $n_1 = n_2 = n_3$. This, however, may not always be true. The power generated at the frequency ω_3 varies as [1]

$$P_3 \propto \frac{\sin\{\frac{1}{2}(q_1 + q_2 - q_3)x\}}{(q_1 + q_2 - q_3)^2} \quad (11o.17)$$

which is a maximum at

$$x = \pi/|q_1 + q_2 - q_3| = \frac{1}{2}|n_1/\lambda_1 + n_2/\lambda_2 - n_3/\lambda_3|^{-1} \quad (11o.18)$$

where the λ_i are the free-space wavelengths. E.g. for frequency doubling this is simply $(\lambda/4)|n_\omega - n_{2\omega}|^{-1}$ which is $\gg \lambda$ for $n_\omega \approx n_{2\omega}$ i.e. for a small dispersion. The drawback of having to use crystals of a length given by Eq.(11o.18) and thus not obtaining a 100% conversion can be overcome by taking birefringent crystals, usually uniaxial crystals, if the birefringence exceeds the dispersion. At a certain angle relative to the optical axis, the velocity of the ordinary wave of the frequency ω is the same as the velocity of the extraordinary wave at 2ω. The fundamental and the harmonic are polarized in different directions. Phases are matched over a "coherence length" of several cm, and conversion efficiencies of nearly 100% have been achieved in this way. A prerequisite is of course the coherence of the laser beam.

If $\omega_2 = 0$, i.e. a strong (pulsed) dc field is applied in addition to the electromagnetic field, frequency doubling is obtained by an oscillator obeying

[2] W. Schneider and K. Seeger, Appl. Phys. Lett. 8 (1966) 133; K. Hess and K. Seeger, Zeitschr. f. Physik 218 (1969) 431; ibid. 237 (1970) 252.

Eq.(11o.4). Large field amplitudes (of $\approx 10^6$ V/cm) may be obtained with small voltages by applying reverse-biased p-n junctions. But even with dc fields of only 10^2 V/cm applied to homogeneously doped n-type InAs at room temperature, the 10.6 μm radiation of a Q-switched CO_2 laser ($\approx 10^5$ W/cm^2) could be converted into 5.3 μm radiation. The experimental data displayed in Fig.11.88 show that the power $P_{2\omega}$ is proportional to the square of the dc current through the sample [3]. For electron densities between $1.5\text{x}10^{16}$/cm^3 and $3\text{x}10^{17}$/cm^3, $P_{2\omega}$ is proportional to the electron density which proves that this effect is due to the carrier plasma.*) A generation of beat frequencies from the $\lambda_1 = 10.6$ μm and $\lambda_2 = 9.6$ μm lines of the CO_2 laser, corresponding to frequencies $\omega_3 = 2\omega_1 - \omega_2$ and $2\omega_2 - \omega_1$ at 8.7 μm and 11.8 μm, has also been observed [5]. Fig.11.89 shows the relative output at 11.8 μm as a function of the electron density (curve a) and of the crystal length x (curve b). The slope of the latter is 2 which indicates a proportionality of the power P_3 to x^2. This dependence is obtained from Eq.(11o.13) for small values of the argument of the sine-function. In an extension of earlier papers, a treatment of optical mixing by mobile carriers in semiconductors has been given by Stenflo [6]. Wolff and Pearson [7] showed that band non-parabolicity strongly

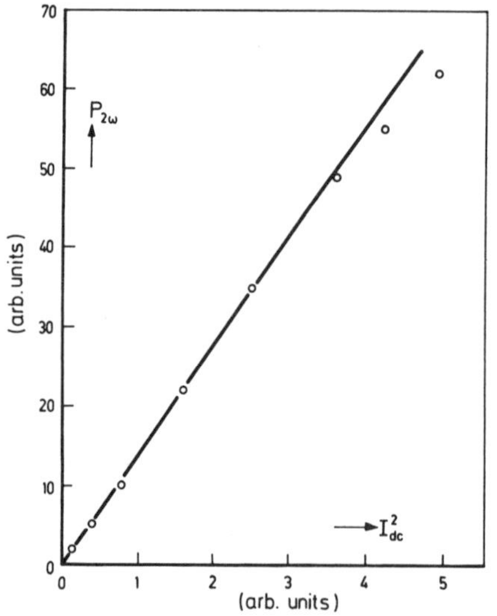

Fig.11.88 Second-harmonic power (5.3 μm wavelength) as a function of the square of the dc current through an n-type InAs sample ($4.4\text{x}10^{16}$/cm^3 carriers, conductivity 143 (ohm-cm)$^{-1}$). At a dc field of 100 V/cm, corresponding to roughly $I^2_{dc} = 1$ arbitrary units in the figure, the second-harmonic power is about 40 μW for 1 kW input at 10.6 μm (after McFee, ref.3).

[3] J. H. McFee, Appl. Phys. Lett. 11 (1967) 228.
[4] B. D. H. Tellegan, Nature 131 (1933) 840; V. A. Bailey and D. F. Martyn, Phil. Mag. 18 (1934) 369; ibid. 23 (Suppl.) (1937) 774.
[5] C. K. N. Patel, R. E. Slusher, and P. A. Fleury, Phys. Rev. Lett. 17 (1966) 1011.
[6] L. Stenflo, Phys. Rev. B 1 (1970) 2821.
[7] P. A. Wolff and G. A. Pearson, Phys. Rev. Lett. 17 (1966) 1015; J. Kolodziejczak, Proc. Int. Conf. Phys. Semicond. Moscow 1968, p. 233. Leningrad: Nauka. 1968.

*) The first discovery of frequency beating in (gaseous) plasma was the "Luxemburg effect": The program of Radio Luxemburg could be received at beat frequencies $2\omega_1 - \omega_2$ where ω_1 is the Luxemburg frequency and ω_2 that of other (weaker) radio stations [4].

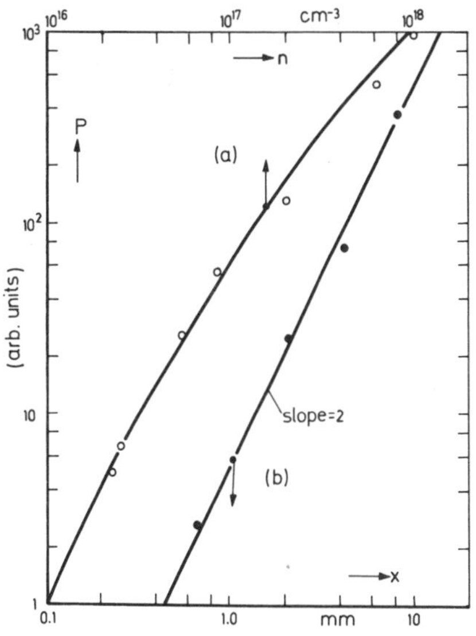

Fig.11.89 Total mixed signal (11.8 μm) output from n-type InAs as a function of carrier density (open circles: experimental; curve (a): calculated) and as a function of sample length (full circles: experimental, straight line (b): calculated) (after Patel et al., ref. 5).

enhances the nonlinearity. Lax et al. [8] suggested to increase the output at the beat frequency ω_3 by the application of a magnetic field of a magnitude such that ω_3 is near the cyclotron resonance frequency. As shown above, a resonance enhances the effect of a nonlinearity.

Without going into further details let us briefly consider "parametric amplification" [9]. The frequency ω_1 may be that of a signal which we want to amplify. In a nonlinear medium we obtain a beat frequency $\omega_3 = \omega_2 + \omega_1$ by mixing ω_1 with a "pump frequency" ω_2 of large amplitude. The beat frequency ω_3, which in this connection is also called the "idler frequency", is mixed with ω_2 as its amplitude grows to yield a second beat frequency $\omega_3 - \omega_2$ which turns out to be ω_1. Due to relations first given by Manley and Rowe [10] the generation at ω_3 is always accompanied by a growth in amplitude at the signal frequency, ω_1. Amplification at 17.88 μm using tellurium as the nonlinear crystal and a CO_2 laser for a pump has been observed by Patel [11]. With the principle of parametric amplification a c w parametric oscillator has been built [12] using $Ba_2NaNb_5O_{15}$ for a nonlinear crystal. Its main advantage over the usual combination of conventional light sources and monochromators is its large output at high resolution at a frequency which may be tuned continuously by changing the temperature, by applying an electric field or by turning the mixing crystal. Hence, the system acts as a tunable laser.

[8] B. Lax, W. Zawadski, and M. H. Weiler, Phys. Rev. Lett. 18 (1967) 462.
[9] M. Faraday, Phil. Trans. R. Soc. 121 (1831) 299; Lord Rayleigh, Phil. Mag. (Ser. 5) 24 (1887) 145.
[10] J. H. Manley and H. E. Rowe, Proc. IRE 44 (1956) 904.
[11] C. K. N. Patel, Appl. Phys. Lett. 9 (1966) 332.
[12] R. G. Smith, J. E. Gensic, H. J. Levinstein, J. J. Rubin, S. Singh, and L. G. Van Uitert, Appl. Phys. Lett. 12 (1968) 308.

11p. The Optoelectric Effect (Photon Drag)

An electromagnetic wave reflected by a mirror exerts a pressure on the mirror which is known as "radiation pressure". It is due to momentum transfer from photons to conduction electrons and can be explained in classical electrodynamics by the Lorentz force if the oscillation of an electron due to the ac electric field is considered in the presence of the ac magnetic field of the wave. Similarly, light propagating through a semiconductor acts upon mobile carriers with the effect of carrier transport in the direction of light propagation. Between a contact at the face of the sample where the light wave enters and another contact where it leaves the sample, a voltage is measured which is called the "optoelectric voltage" or "photon drag voltage".

Let us consider a transverse electromagnetic wave polarized in the x-direction and propagating in the z-direction:

$$E_x = E_1 \cos(\omega t - qz)\,; \quad B_y = B_1 \cos(\omega t - qz + \psi) \tag{11p.1}$$

where $q = N/\lambda\!\!\!^{-}$ is the absolute value of the wave vector in the z-direction. The equations of motion of a carrier are given by [1]

$$dv_x/dt + v_x/\tau_m = (e/m)\,(E_x - v_z B_y) \tag{11p.2}$$

$$dv_z/dt + v_z/\tau_m = (e/m)\,(E_z + v_x B_y) \tag{11p.3}$$

where E_z is the optoelectric dc field and diffusion and recombination shall be neglected for simplicity. The terms $v_z B_y$ and $v_x B_y$ introduce nonlinearities, and the optoelectric effect actually belongs to the field of nonlinear optics which has been treated in the previous chapter.

To a first approximation we neglect the nonlinear term in the first equation and obtain

$$v_x = \mu E_1 \frac{\cos(\omega t - qz) + \omega\tau_m \sin(\omega t - qz)}{1 + \omega^2 \tau_m^2} \tag{11p.4}$$

where the mobility $\mu = (e/m)\tau_m$ has been introduced. With this value of v_x we solve Eq.(11p.3):

$$v_z = \mu \left\{ E_z + \frac{\mu B_1 E_1}{2(1 + \omega^2 \tau_m^2)} \left(\cos\psi + \frac{\cos(2\omega t - 2qz + \psi) + 2\omega\tau_m \sin(2\omega t - 2qz + \psi)}{1 + 4\omega^2 \tau_m^2} \right) \right\} \tag{11p.5}$$

Neglecting terms in $2\omega t$ and, for $\omega \gg \tau_m^{-1}$, neglecting also the 1 in the denominator, we obtain for $v_z = 0$:

[1] The following treatment has been adapted from N. Bloembergen: Nonlinear Optics, p. 3-5. New York: Benjamin. 1965.

$$E_z = -\frac{1}{2}\mu E_1 B_1 \cos\psi/\omega^2\tau_m^2 \tag{11p.6}$$

The average light intensity is given by

$$I = \overline{[\vec{E}\vec{H}]} = \frac{1}{2\mu_o} E_1 B_1 \cos\psi \tag{11p.7}$$

From Eq.(11p.6) $|E_z|$ is finally obtained:

$$|E_z| = \mu\mu_o I/\omega^2\tau_m^2 \tag{11p.8}$$

Due to absorption of the light, I depends on z, and the voltage V_z is obtained by an integration over z:

$$V_z = \frac{\mu\mu_o}{\omega^2\tau_m^2}\frac{I_o}{a}\frac{(1-r_\infty)\{1-\exp(-ad)\}}{1+r_\infty\exp(-ad)} \tag{11p.9}$$

where I_o is the incident light intensity; corrections have been made for reflections at the front surface and for internal reflections. For an intensity I_o of about 10^5 W/cm^2 obtained from a Q-switched CO_2 laser at 10.6 μm, where $\omega^2\tau_m^2$ is typically of the order of magnitude of 10^3, for a mobility μ of about 10^3 cm^2/Vsec, an absorption coefficient a of about 1 cm^{-1}, a reflectivity of $r_\infty \approx 30\%$, and $d \gg 1/a$, we find $V_z \approx 10^{-3}$ V. Voltages of this order of magnitude have been measured in germanium, InSb, and GaAs at room temperature [2, 3].

In the treatment given above, the fact that the mobility μ is field-dependent and therefore a function of z, has been neglected. A more refined theory is given in ref. 2. Since the effect has a short response time it has been suggested to use it for a high-power laser radiation detector [3].

At microwave frequencies similar observations [4] have been explained as a Seebeck effect: Carriers are heated by the electromagnetic wave where it is incident on the sample while on the opposite sample face the wave amplitude and heating are smaller. Thermoelectricity by hot carriers has been treated by Conwell [5].

---

[2] P. M. Valov, A. A. Grinberg, A. M. Danishevsky, A. A. Kastalsky, S. M. Ryvkin, and I. D. Yaroshetsky, Proc. Int. Conf. Phys. Semicond., Cambridge, Mass. 1970 (S. P. Keller, J. C. Hensel, and F. Stern, eds.), p. 683. Oak Ridge/Tenn.: USAEC. 1970, and literature cited there.

[3] A. F. Gibson, M. F. Kimmitt, and A. C. Walker, Proc. Int. Conf. Phys. Semicond., Cambridge, Mass. 1970 (S. P. Keller, J. C. Hensel, and F. Stern, eds.), p. 690. Oak Ridge/Tenn.: USAEC. 1970.

[4] H. Kahlert and K. Seeger, Verhandl. DPG (VI) 3 (1968) 39.

[5] E. M. Conwell: Solid State Physics (F. Seitz, D. Turnbull, and H. Ehrenreich, eds.), Vol. 9, Suppl., p. 45-48 and 276-281. New York: Acad. Press. 1967.

12. Photoconductivity

In Chaps.5h and 5i we have considered diffusion of carriers which are generated by the absorption of light. In this chapter we will discuss in greater detail photoconduction with an emphasis on trapping processes.

12a. Photoconduction Dynamics

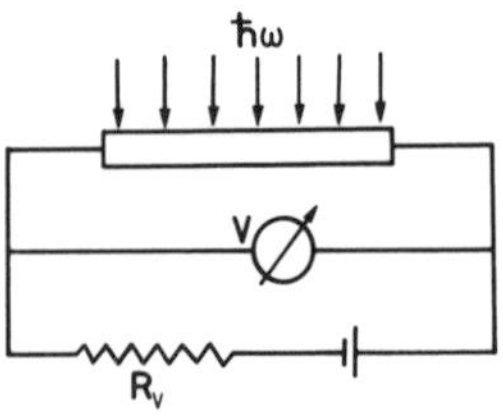

Fig.12.1 Schematic arrangement for photoconductivity measurements.

The experimental arrangement shown in Fig.12.1 allows the observation of photoconduction. Light incident on the semiconductor crystal is absorbed with the effect that additional carriers are produced. If the photon energy is smaller than the band gap, only one kind of carriers may be generated by e.g. impurity absorption; otherwise electrons and holes will be generated in pairs. The "dark conductivity"

$$\sigma_o = |e| \, (n_o \mu_n + p_o \mu_p) \tag{12a.1}$$

is thus increased by an amount $\Delta\sigma = |e|(\mu_n \Delta n + \mu_p \Delta p)$. This yields a relative increase of magnitude

$$\frac{\Delta\sigma}{\sigma_o} = \frac{\mu_n \Delta n + \mu_p \Delta p}{\mu_n n_o + \mu_p p_o} = \frac{b\Delta n + \Delta p}{bn_o + p_o} \tag{12a.2}$$

where the mobility ratio $b = \mu_n/\mu_p$ has been introduced. If in the experiment the sample current is kept constant by a large series resistor, the voltage V across the sample is reduced by an amount $\Delta V = V|\Delta\sigma/\sigma_o|$. For high sensitivity a light chopper and a phase sensitive detector are used. For investigations of transient behavior a flash tube or a Kerr cell chopper in combination with a cw light source and an oscilloscope are more appropriate.

If for simplicity we neglect diffusion processes the continuity equation, Eq. (5b.9), becomes

$$\frac{d\Delta n}{dt} = G - \frac{\Delta n}{\tau_n} \tag{12a.3}$$

where in case of uniform light absorption the "generation rate" G is given by

$$G = a\eta I/\hbar\omega \tag{12a.4}$$

I (in watt/cm²) is the light intensity, a is the absorption coefficient, and η the dimensionless quantum yield. In case of non-uniform light absorption the generation rate is a function of the distance x from the surface given by

$$G = (a\eta I/\hbar\omega)(1-r_\infty)\{\exp(-ax)+r_\infty \exp(-2ad+ax)\}/\{1-r_\infty^2 \exp(-2ad)\} \tag{12a.5}$$

where r_∞ is the reflectivity of an "infinitely thick" sample and d is the sample thickness.

For the case of photo-ionization of e.g. acceptor impurities the product $a\eta$ becomes

$$a\eta = \sigma_A N_A (1-f) \tag{12a.6}$$

where N_A is the total acceptor concentration, σ_A is the photo-ionization cross section, and f is the electron occupation probability of the acceptor.

For equilibrium Eq.(12a.3) yields $\Delta n = G\tau_n$. Assuming pair production, $\Delta n = \Delta p$, we obtain from Eq.(12a.2):

$$\frac{\Delta\sigma}{\sigma_0} = G\tau_n \frac{b+1}{n_0 b + p_0} \tag{12a.7}$$

This shows that it is natural for a sensitive photoconductor ($\Delta\sigma \gtrsim \sigma_0$) to have a long time constant τ_n.

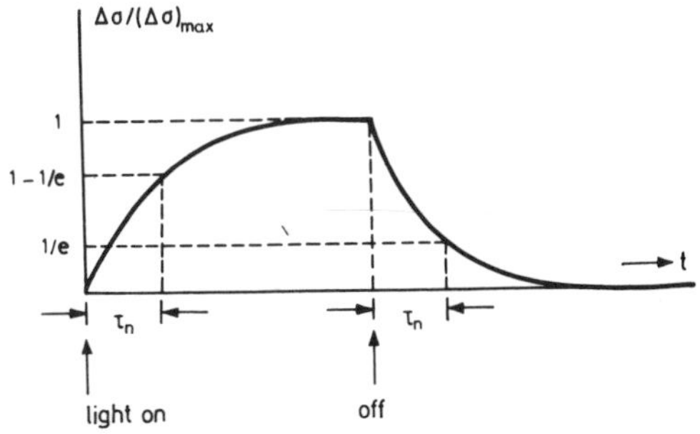

Fig.12.2 Dependence of photoconductivity on time t for a relaxation time τ_n.

The transient behavior which is observed if the light at low intensity is alternatively switched on and off, is shown in Fig.12.2. The time constant τ_n can easily be determined if the curves for the increase or decrease with time is exponential.

From measurements of τ_n and $\Delta\sigma/\sigma_0$ at equilibrium the quantum yield η may be determined. For an estimate of Δn let us assume typical values $\eta = 1$, $\tau_n = 10^{-4}$ sec, $I = 10^{-4}$ watt/cm², $a = 10^2$ cm^{-1} at a wavelength of 2 μm corresponding to $\hbar\omega = 10^{-19}$ wattsec. The results are $G = 10^{17}$ cm^{-3} sec^{-1} and $\Delta n = 10^{13}$ cm^{-3}.

If the energy of the incident photons is larger than the band gap, ϵ_G, radiative recombination of electron hole pairs by emission of photons $\hbar\omega = \epsilon_G$ may occur. The recombination rate is proportional to the product $np = (n_o + \Delta n) \cdot (p_o + \Delta p)$ and, hence, with a factor of proportionality, C,

$$\Delta n/\tau_n = C(np - n_o p_o) = C(n_o \Delta p + p_o \Delta n + \Delta n \Delta p) \tag{12a.8}$$

At equilibrium the left-hand side is equal to G. For the case $\Delta n = \Delta p$, this is a quadratic equation for Δn. Only at a low light intensity can the product $\Delta n \Delta p$ be neglected, and with $G \propto I$ we have $\Delta n \propto I$. At a high light intensity we can, on the other hand, neglect $n_o \Delta p + p_o \Delta n$ and find $\Delta n \propto \sqrt{I}$. This behavior is illustrated schematically in Fig.12.3.

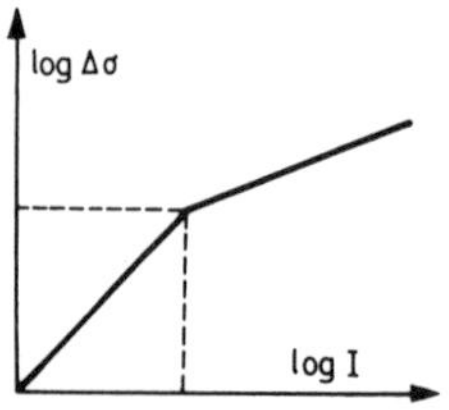

Fig.12.3 Photoconductivity for radiative recombination as a function of light intensity I, showing transition from a proportionality to I to a $\sqrt{I}$ behavior.

Besides radiative recombination with the emission of photons, $\hbar\omega = \epsilon_G$, there may also be recombination via "recombination centers" with the emission of photons $\hbar\omega < \epsilon_G$, or there may be no radiation at all. Recombination centers are either impurities or imperfections. Typical examples will be studied in Chap.12b.

The rate of electron capture by a single type of recombination center is given by

$$\Delta n/\tau_n = C_n n(1-f) \tag{12a.9}$$

where $f = N_r^x/N_r$ is the probability that the center is neutral (occupied), $C_n = N_r v\sigma_r$, N_r is the total concentration of recombination centers, and σ_r their capture cross section. The rate of thermal emission of electrons from recombination centers is given by

$$G = C'_n f \tag{12a.10}$$

where C'_n/N_r is the thermal ionization probability of a center. The condition for equilibrium in the dark, $\Delta n/\tau_n = G$, yields

$$C'_n/C_n = n(1-f)/f = g^{-1} N_c \exp(-\Delta\epsilon_r/k_B T) \tag{12a.11}$$

where the effective density of states, N_c is given by Eq.(3a.31) assuming nondegeneracy and $\Delta\epsilon_r = \epsilon_r - \epsilon_c$; the spin factor g depends on the kind of impurity. If we denote by n_1 the electron density in the conduction band for the case that the Fermi level coincides with the recombination level, the right-hand side of Eq.(12a.11) is given by $g_D^{-1} n_1$.

At non-equilibrium conditions which e.g. may be due to illumination, the net recombination rate of electrons is given by

$$\Delta n/\tau_n = C_n\{n(1-f) - g_D^{-1} n_1 f\} \tag{12a.12}$$

Similarly the net recombination rate of holes is obtained:

$$\Delta p/\tau_p = C_p\{pf - g_A p_1(1-f)\} \tag{12a.13}$$

Since electrons and holes recombine in pairs we find for a steady state illumination $\Delta n/\tau_n = \Delta p/\tau_p$ which we call simply $\Delta n/\tau$:

$$\Delta n/\tau = \frac{C_n C_p (np - n_i^2)}{C_n(n + g_D^{-1} n_1) + C_p(p + g_A p_1)} \tag{12a.14}$$

where the intrinsic concentration $n_i = \sqrt{n_1 p_1}$ has been introduced. Since a similar relation, $n_i = \sqrt{n_o p_o}$, holds for the carrier concentrations in the dark, n_o and p_o, we obtain from Eq.(12a.14) by introducing $n - n_o = \Delta n$; $p - p_o = \Delta p$:

$$\Delta n/\tau = \frac{n_o \Delta p + p_o \Delta n + \Delta p \Delta n}{C_p^{-1}(n_o + g_D^{-1} n_1 + \Delta n) + C_n^{-1}(p_o + g_A p_1 + \Delta p)} \tag{12a.15}$$

For shortness we introduce also $\tau_{po} = C_p^{-1}$; $\tau_{no} = C_n^{-1}$; $n_1' = g_D^{-1} n_1$; $p_1' = g_A p_1$ and obtain for the case $\Delta n = \Delta p \ll n_o$ and p_o:

$$\tau = \tau_{po}\frac{n_o + n_1'}{n_o + p_o} + \tau_{no}\frac{p_o + p_1'}{p_o + n_o} \tag{12a.16}$$

If $p_o \gg n_o$ and p_1', this yields $\tau = \tau_{no}$, while for $n_o \gg p_o$ and n_1', we obtain $\tau = \tau_{po}$, independent of p_o and n_o, respectively. It is noteworthy that this is in contrast to the behavior in the case of radiative recombination, given by Eq. (12a.8) where the lifetime does depend on the carrier concentration. For values of p_o near the intrinsic value, n_i, there is a maximum of the time constant. For the general case where $\Delta n = \Delta p$ and both are not small, Eq.(12a.15) is a second-order equation in Δn with the same kind of behavior as has been discussed in connection with Eq.(12a.8). For $\Delta n \neq \Delta p$ we have a change in f given by

$$\Delta f = (\Delta p - \Delta n)/N_r \tag{12a.17}$$

This change may be calculated from $\Delta n/\tau_n = \Delta p/\tau_p$ where both sides are given by Eqs.(12a.12) and (12a.13), respectively:

$$C_n\{\Delta n(1-f) - n\Delta f - n_1'\Delta f\} = C_p\{f\Delta p + p\Delta f + p_1'\Delta f\} \tag{12a.18}$$

Here for an approximation we replace f, n, and p by their equilibrium values $n_o/(n_o + n_1')$, n_o, and p_o, respectively. If we calculate the time constant as a function of n_o we find again a maximum which, however, in this case is generally not near the value of the intrinsic carrier concentration.

Recombination levels are usually "deep levels". "Shallow levels" on the other hand act as "traps". E.g. a shallow acceptor will trap a hole with a probability, which we denote as $1/\tau_1$, and it will keep it for an average time period of τ_2. The average concentration of trapped holes, ΔP, is given by the condition of charge neutrality,

$$\Delta P = \Delta n - \Delta p \tag{12a.19}$$

If $1/\tau_p$ is the probability for a hole to recombine with an electron, we have

$$\frac{d\Delta p}{dt} = G - \frac{\Delta p}{\tau_p} - \frac{\Delta p}{\tau_1} + \frac{\Delta P}{\tau_2} \tag{12a.20}$$

In equilibrium (indicated by a subscript o) the rate of trapping equals the rate of thermal re-excitation of holes into the valence band.

$$\Delta p_o/\tau_1 = \Delta P_o/\tau_2 \tag{12a.21}$$

Because phonons are involved in the process of re-excitation, the time constant τ_2 is strongly temperature dependent. The generation rate is then given by

$$G = \Delta p_o/\tau_p \tag{12a.22}$$

For Δn_o we obtain from Eqs.(12a.19), (12a.21), and (12a.22):

$$\Delta n_o = G\tau_p(1 + \tau_2/\tau_1) \tag{12a.23}$$

The relative conductivity change is now given by Eq.(12a.7) with τ_n being replaced by

$$\tau_p' = \tau_p \cdot \frac{1 + b\,(1 + \tau_2/\tau_1)}{1 + b} \tag{12a.24}$$

which for $\tau_2 \gg \tau_1$ may be approximated by

$$\tau_p' \approx \tau_p \{b/(1 + b)\} \cdot \tau_2/\tau_1 \tag{12a.25}$$

This is much larger than the time constant without trapping, τ_p. We find as a result that traps considerably increase the sensitivity of a photoconductor, but at the same time also its response time. The reason for this behavior is clear: as more holes are trapped, more electrons stay in the conduction band because of charge neutrality, Eq.(12a.19), and the larger is the conductivity. The number of trapped holes increases with the average time which a hole spends in a trap, τ_2. Of course, the same is true for electrons in a p-type semiconductor.

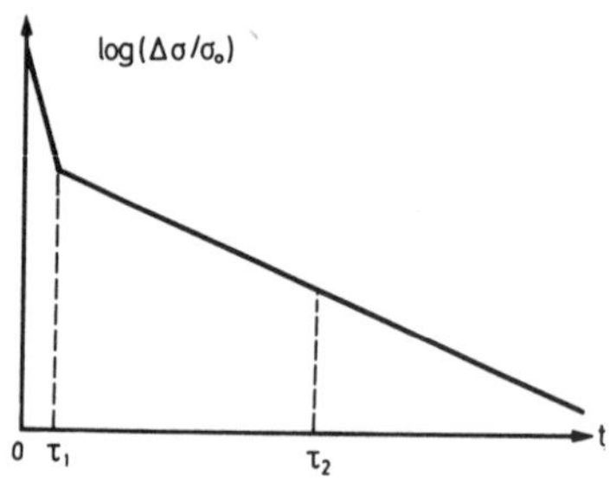

Fig.12.4 Relaxation of photoconductivity with two recovery time constants for the case of thermal release of carriers from traps.

The conductivity drop after switching off the illumination is illustrated in Fig. 12.4 valid for $\tau_2 \gg \tau_1$ and τ_p. For a short period, of the order of τ_1, the traps remain at equilibrium and the conductivity drops with the fast time constant τ_p. However, for longer time periods the time constant is given by τ_2. To a first approximation the sum of two exponentials with time constants τ_p and τ_2 with a smooth transition at $t = \tau_1$ adequately describes the photoconductivity transient behavior. Of course, with a slow apparatus not capable of measuring the fast time constant τ_p, only the slow drop of $\Delta\sigma$, which at low temperatures may even last hours, is observed.

Illumination of the sample with light of a long wavelength may empty the traps without generating electron-hole pairs, thus considerably decreasing the time constant τ_2 and "quenching" the high photo-sensitivity. If the exciting radiation is "white" and therefore contains a large amount of "quenching" infrared, the photo-sensitivity will be much lower than for monochromatic light.

Surface recombination has been defined by Eq.(5h.1). In a sample of thickness d a minority carrier has to drift for a distance of less than d/2 in order to arrive at the surface. Hence the rate of recombination is given by

$$d\Delta p/dt = -2s\Delta p/d \tag{12a.26}$$

which yields a time constant

$$\tau_p' = d/(2s) \tag{12a.27}$$

At an etched surface $s = 10^2$ cm/sec which e.g. for d/2 = 1mm yields $\tau_p' = 1$ msec. For a sandblasted surface ($s = 10^6$ cm/sec) we then have $\tau_p' = 100$ nsec. The combination of surface and bulk recombination yields a time constant given by

$$1/\tau_p' = 1/\tau_p + 2s/d \tag{12a.28}$$

Obviously the surface condition of a photoconductor considerably affects its overall performance.

Finally we mention that the signal-to-noise ratio of a photoconductor may be determined by generation-recombination noise. These processes (including trapping) are subject to statistics and occur at a random sequence. Therefore this kind of noise can only be reduced in magnitude by eliminating traps and recombination centers which, however, leads to a decrease in sensitivity as we have seen before. For practical applications see e.g. Sze [1].

12b. Deep Levels in Germanium

Recombination and trapping processes may conveniently be studied with deep levels in germanium which are either "double acceptors" or "double donors". We will see that the same deep-level impurity atom may act either as a recombination center or a trap depending on the type of conductivity which is determined by the shallow-impurity doping. Therefore germanium may serve as a model substance for this kind of investigation [1].

[1] S. M. Sze: Physics of Semiconductor Devices. New York: J. Wiley and Sons. 1969. For a review on deep impurities see e.g. H. J. Queisser: Festkörper-Probleme XI (O. Madelung, ed.). Braunschweig: Vieweg. 1971.

Chap.12b:

[1] R. Newman and W. W. Tyler: Solid State Physics (F. Seitz and D. Turnbull, eds.), Vol. 8, p. 49. New York: Acad. Press. 1957.

In Fig.3.7 multiple-acceptor levels in Ge are indicated, together with their "distribution coefficients" (ratios of impurity concentration in the solid phase to the concentration in the liquid phase which is in thermal equilibrium with the solid phase of germanium [2]). Let us e.g. consider Fe in p-type Ge. Thermal ionization yields a hole and a Fe^- ionized atom :

$$Fe + 0.35\,eV \rightleftharpoons e^+ + Fe^- \qquad (12b.1)$$

In n-type Ge, however, Fe at low temperatures is present in the form of a Fe^{--} ion which by thermal ionization yields an electron and also Fe^- :

$$Fe^{--} + 0.27\,eV \rightleftharpoons e^- + Fe^- \qquad (12b.2)$$

Therefore Fe is an acceptor and Fe^{--} is a donor, and usually there is no Fe in n-type Ge and no Fe^{--} in p-type Ge. The type of conductivity depends on the type of impurity, boron or phosphorus, which is usually present in iron at concentrations of a few ppm depending on the type of purification process. Since boron and phosphorus are so much more easily than iron introduced into a crystal when it grows from the melt, the content of iron is much smaller than that of the shallow impurities.

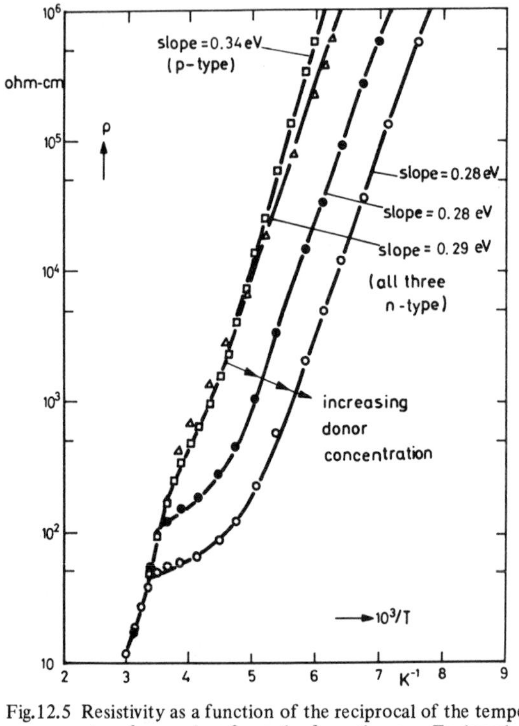

Fig.12.5 Resistivity as a function of the reciprocal of the temperature for a series of samples from the same Fe doped germanium crystal. With increasing donor concentration both the photosensitivity and the recovery time increase as shown in Fig.12.6 (after W. W. Tyler and H. H. Woodbury, Phys. Rev. 96 (1954) 874).

Fig.12.5 shows the temperature dependence of the dark resistivity of Fe-doped n- and p-type Ge. While in n-type samples the slope of the straight lines yields an energy of 0.28 eV, in p-type samples the slope yields 0.34 eV. After correcting for the temperature dependence of the mobility, the temperature dependence of the carrier density is obtained; this is determined by the values of 0.27 and 0.35 eV for the energy indicated in Eqs. (12b.1) and (12b.2). E.g. at a temperature of 150 K Fig. 12.5 shows a very distinct freeze-out of carriers. Illuminating the sample with

[2] C. D. Thurmond: Semiconductors (N. B. Hannay, ed.), p. 145. New York: Reinhold. 1960.

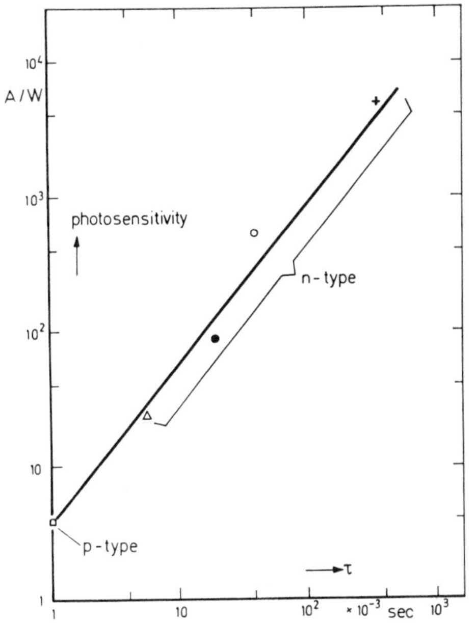

Fig. 12.6 Photosensitivity at a photon energy of 0.83 eV as a function of the recovery time for the samples of Fig. 12.5 and a sample marked + from another crystal.

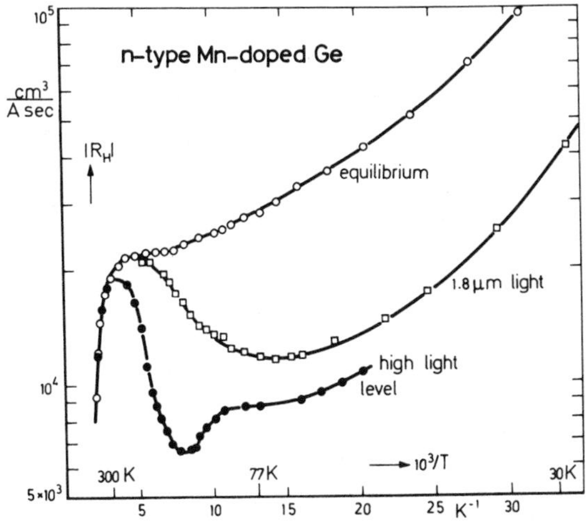

Fig. 12.7 Photo-Hall data for n-type Mn-doped germanium as a function of the inverse temperature at various light intensities. Data at high light levels were obtained using radiation from a tungsten lamp (3000 K) filtered through a germanium window 1 mm thick (after Newman and Tyler, ref. 1).

photons $\hbar\omega \gtrsim \epsilon_G$ yields electron-hole pairs. After turning off the illumination the resistivity increases non-exponentially with time. Let us define τ as the time period for an increase of ρ by 3 orders of magnitude. Fig. 12.6 shows the relation between the photoconductive sensitivity and τ for p- and n-type samples. Obviously the n-type samples at low temperatures contain traps, while the p-type samples do not.

Eq. (12b.2) shows that at low temperatures in n-type samples there are Fe^{--} ions. Minority carriers are positive holes, and these are attracted by the negative ions:

$$Fe^{--} + e^+ \rightleftharpoons Fe^- \qquad (12b.3)$$

The negatively charged electrons are repelled by negative ions, and this explains why Fe^{--} is a hole trap. The fact that an illuminated sample contains Fe^- ions which do not scatter electrons as strongly as the Fe^{--} ions in the dark, is illustrated by the temperature dependence of the mobility shown in Fig. 12.8 (Here the deep level is Mn which, however, acts similarly to Fe).

A strong illumination of n-type samples results in the reaction

$$Fe^- + e^+ \rightleftharpoons Fe \qquad (12b.4)$$

with the formation of neutral atoms. These no longer repel electrons and therefore act as recombination

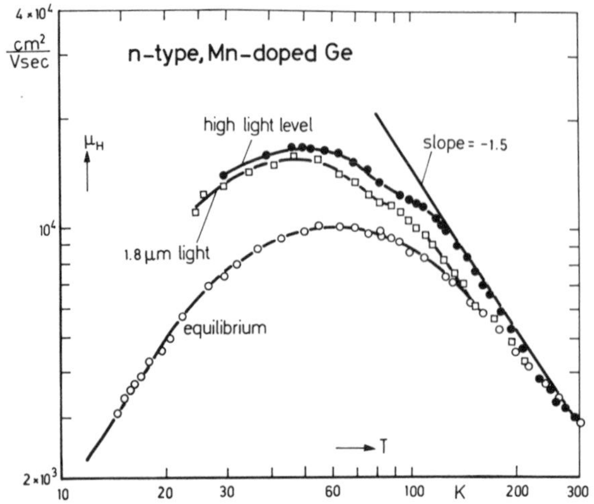

Fig. 12.8 Hall mobility data corresponding to the Hall coefficient data of Fig. 12.7 (after Newman and Tyler, ref. 1).

centers. The lifetime becomes very short and the sensitivity small. Fig. 12.7 shows the Hall coefficient as a function of temperature for various degrees of illumination.

In p-type samples we have recombination centers (neutral Fe atoms) also at low illumination levels (Eq. (12b.1)). The lifetime and the sensitivity are both small.

The dependence of the decay time τ on temperature for n- and p-type Ge doped with deep levels (Mn) is shown in Fig. 12.9. The resistivity of the samples had to be chosen low enough that the dielectric relaxation time, given by Eq. (5b.23), is much smaller than τ. The high-temperature slope of the curve for Mn doped n-Ge is determined by an energy of 0.31 eV. This energy is considered as an activation energy for either the capture of electrons over a repulsive potential barrier or the escape of a hole from a trap. By a "reverse current collection method" by Pell [3], it has been shown that in n-type Mn-doped samples, hole escape does take place.

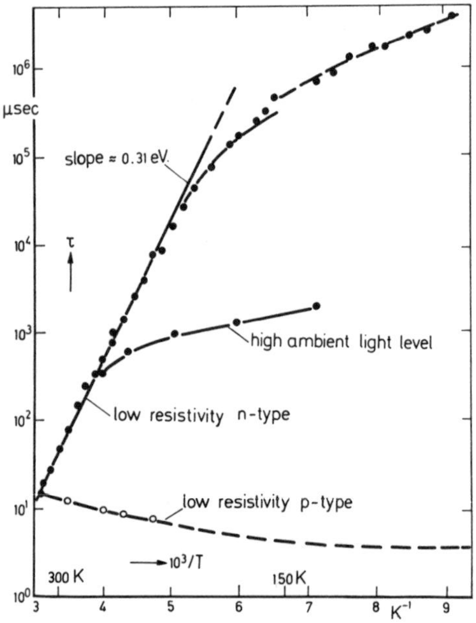

Fig. 12.9 Photoconductive response time as a function of the inverse temperature for Mn-doped germanium (after Newman and Tyler, ref. 1).

Tellurium in germanium is a double donor with levels 0.1 eV and 0.28 eV below the conduction band :

[3] E. M. Pell, J. Appl. Phys. 26 (1955) 658.

$$Te + 0.10\ eV \rightleftharpoons e^- + Te^+ \tag{12b.5}$$

$$Te^+ + 0.28\ eV \rightleftharpoons e^- + Te^{++} \tag{12b.6}$$

In p-type samples at low temperatures we have Te^{++} ions. Minority carriers (electrons) generated by illumination are trapped at these ionized atoms with the formation of Te^+. These do not attract positive holes and therefore act as traps. In n-type samples at low temperatures there are presumably Te atoms which for a short time trap minority carriers (positive holes):

$$Te + e^+ \rightleftharpoons Te^+ \tag{12b.7}$$

The Te^+ formed in this process, however, quickly attract negatively charged electrons:

$$Te^+ + e^- \rightleftharpoons Te \tag{12b.8}$$

and consequently act as recombination centers. Fig.12.10 shows the photoconductive decay time vs temperature for p-type germanium doped with Te.

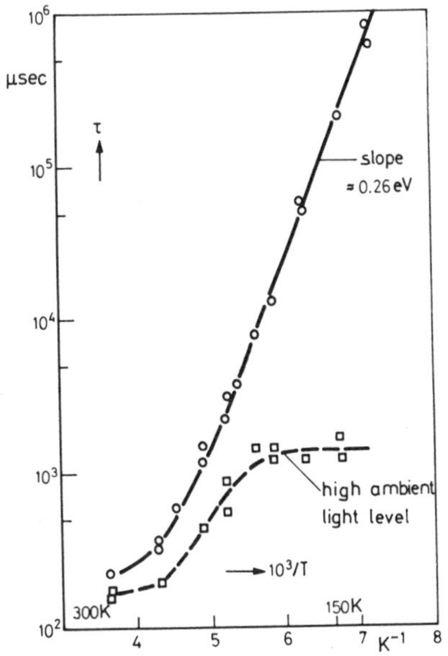

Fig.12.10 Photoconductive response time as a function of the inverse temperature for tellurium-doped p-type germanium (after Newman and Tyler, ref. 1).

The spectral sensitivity at 77 K of n- and p-type germanium doped with various transition metals is shown in Fig. 12.11a and b. The largest difference between n- and p-type samples is found in Mn-doped samples; this agrees with the position of levels indicated in Fig.3.7.

A rather exotic impurity in Ge is gold. Fig.3.7 shows that it may be either a donor 0.05 eV above the valence band or an acceptor with 3 different levels. As with the double-acceptors discussed above, the nature of the impurity level depends on the kind of shallow-level impurities which are present in addition to the deep-level impurity. From a statistical point of view on may say : It depends on the position of the Fermi level which is close to the conduction band in an n-type semiconductor and close to the valence band in a p-type semiconductor (Chap.3). Fig.12.12 shows on the left-hand side again the four gold levels in germanium and on the right-hand side the charge on the gold center depending on the position of the Fermi level. Since the lowest level is a donor, 0.05 eV above the valence band, and the next-highest level is an acceptor, 0.15 eV above the valence band, a transition between these two levels is possible if the Fermi level is located between these levels and the photon energy equals the energy difference.

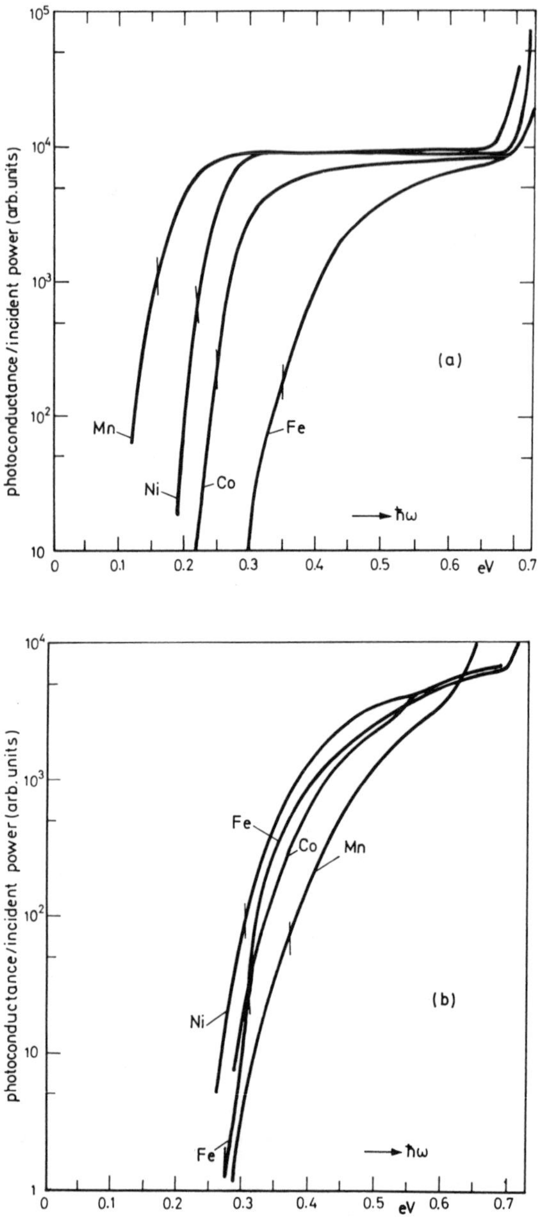

Fig.12.11 a Photoconductivity spectra at 77 K for p-type germanium with deep-level impurities. The vertical bars indicate the thermal ionization energies (after Newman and Tyler, ref. 1).

b Photoconductivity spectra for n-type germanium with deep-level impurities at 77 K except for Mn which is at 196 K. The vertical bars indicate the thermal ionization energies (after Newman and Tyler, ref. 1).

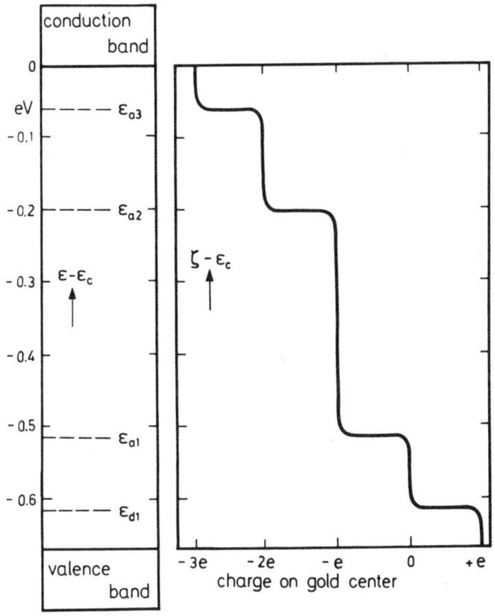

Fig.12.12
Gold levels in germanium and the dependence of the charge per gold atom on the Fermi level at low temperatures (after W. C. Dunlap: Progress in Semiconductors (A. F. Gibson, ed.), Vol. 2. London: Temple Press. 1957; and J. S. Blakemore: Semiconductor Statistics. London: Pergamon. 1962).

The following table 6 shows data for various Ge photodetectors.

Table 6: Ge photodetectors [1] (v.b. = valence band c.b. = conduction band)

Impurity	Level	Operating Temperature	Long-wavelength Limit
Zn	0.03 eV above v.b.	< 15 K	> 40 μm
Zn	0.09 eV above v.b.	60 K	18 μm
Te	0.10 eV below c.b.	60 K	15 μm
Au	0.05 eV below c.b.	20 K	12 μm
Au	0.05 eV above v.b.	20 K	25 μm
Au	0.15 eV above v.b.	77 K	12 μm
Cu	0.04 eV above v.b.	20 K	27 μm
Mn	0.16 eV above v.b.	65 K	8.5 μm

12c. Trapping Cross Section of an Acceptor

For a determination of the capture cross section σ_p of a negatively charged deep acceptor both the concentration of the deep level impurity and the concentration of shallow donors must be known. Instead of pulling a crystal from a melt doped with the double acceptor, the doping is done by diffusion of e.g. Ni into germanium crystals, having a predetermined concentration of e.g. As, at 1123 K for times long enough to saturate the concentration of Ni. The occupation fraction f of the Ni levels is calculated from the known concentrations of As and Ni: $f = N_{As}/N_{Ni}$. The first acceptor level of Ni is 0.22 eV above the valence band (Fig. 3.7). For $f < 1$ the samples are p-type. The product $f N_{Ni}$ is the concentration of negatively charged Ni acceptors. Denoting the thermal velocity of holes by v, the lifetime is given by

$$\tau_p = (\sigma_p v f N_{Ni})^{-1} \tag{12c.1}$$

Introducing the photo-ionization cross section σ_A of a neutral acceptor by

$$G = \sigma_A (1-f) N_{Ni} I/\hbar\omega \tag{12c.2}$$

we obtain from Eq.(12a.3) for equilibrium

$$\Delta p/I = (\sigma_A/\sigma_p)(1/\hbar\omega v)(1-f)/f \tag{12c.3}$$

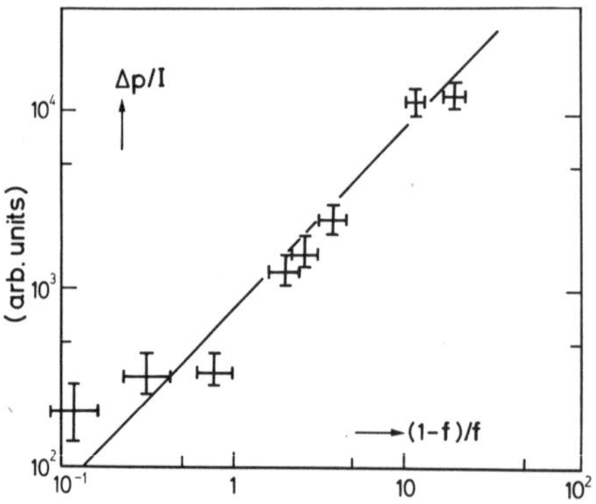

Fig.12.13 Photo-yield at a photon energy of 0.3 eV for Ni-doped p-type germanium at 77 K, as a function of $(1-f)/f$ (after Newman and Tyler, ref. 1).

The photo-yield which is proportional to $\Delta p/I$ is plotted vs $(1-f)/f$ in Fig.12.13 on a log-log scale [1]. The data points indicate the linear relationship given by Eq.(12c.3). From measurements of the absorption cross section, σ_A, ($\approx 10^{-16}$ cm^2 for Ni in p-Ge at 77 K and for a photon energy of 0.3 eV), and using the value of 10^7 cm/sec for the average thermal velocity of a hole, the cross section for capture of a hole by a singly-charged negative Ni center, σ_p, is found to be 10^{-13} cm^2 at 77 K. This corresponds to a radius of $\sqrt{\sigma_p/\pi} = 19$ Å.

[1] R. Newman and W. W. Tyler: Solid State Physics (F. Seitz and D. Turnbull, eds.), Vol. 8, p. 91. New York: Acad. Press. 1959.

13. Light Generation by Semiconductors

Electroluminescent devices emit incoherent visible or infrared light with typical linewidths of about 100 Å while the coherent radiation emitted by the semiconductor laser may have a linewidth as low as 0.1 Å. These devices together with photovoltaic diodes and solar cells (Chaps.5h, 5i, and 12) are called "optoelectronic devices". While the former convert electrical energy into optical radiation the latter do the inverse process. In this chapter we will consider light-emitting diodes ("LED") and diode lasers.

There is a variety of luminescence effects depending on the kind of excitation, such as photo-luminescence (excitation by optical radiation), cathodo-luminescence (by cathode rays, i.e. electron beams), radio-luminescence (by other fast particles including X- and γ-rays), and electro-luminescence (by dc or ac electric fields).If there are traps involved in the emission process the response times may be quite large, sometimes many seconds or even hours; the term "phosphorescence" is then used instead of "luminescence". By gradually raising the temperature the variation of the light output with temperature is recorded; this is called a "glow curve". Typical phosphors are the zinc and cadmium sulfides, selenides, and tellurides [1]; the phenomena in these substances are, however, so complex that we have to exclude these materials from the present considerations. We will also mention only the "Destriau effect" [2] which is the excitation of light in microcrystalline, nominally insulating solids inbedded in a dielectric medium, as the result of an applied ac electric field of a frequency of typically about 100 Hz. Electro-luminescence produced by the injection of minority carriers in a p-n junction at a forward bias and subsequent radiative recombination is sometimes called the "Lossev effect" [3].

---

[1] For details concerning II-VI compounds including luminescence see e.g. Physics and Chemistry of II-VI Compounds (M. Aven and J. S. Prener, eds.). Amsterdam: North-Holland Publ. Co. and New York: J. Wiley and Sons, 1967.

[2] G. Destriau, J. Chim. Phys. 33 (1936) 587.

[3] O. W. Lossev, Telegrafiya i Telefoniya 18 (1923) 61.

13a. The Luminescent Diode

Radiative electronic transitions following an excitation are possible (1) by interband transitions where $\hbar\omega \geqslant \epsilon_G$; (2) by transitions from the conduction band to an acceptor, or from a donor to the valence band, or from a donor to an acceptor; and (3) by intraband transitions involving hot carriers ("deceleration emission"). A typical nonradiative transition is the "Auger effect" where the transition energy is transferred to another electron [1] which is excited in this process either deep into a band or at an impurity and then loses its energy in a cascade of small photons or phonons.

For a calculation of the luminescence efficiency let us consider the following rate equations. The rate of change of electron density in the conduction band is given by

$$dn/dt = G - \sigma_t v_n nN_t(1-f_t) + AN_t f_t - \sigma_l v_n np_l \tag{13a.1}$$

where the first three terms on the right-hand side are essentially the same as in Eq.(12a.3) with Δn given by Eq.(12a.12), except that now they refer to traps (subscript t) while the last term describes the radiative recombination through capture of an electron by a luminescent center, p_l being the density of holes in the luminescent centers. σ_t and σ_l are the cross sections of traps and luminescence centers, respectively. A is a constant. The probability for an electron to stay in a trap, f_t, is given as the solution of the differential equation

$$df_t/dt = \sigma_t v_n n(1-f_t) - \sigma_r v_p pf_t - Af_t \tag{13a.2}$$

where the second term on the right-hand side describes nonradiative recombination through capture of a hole from the valence band; v_n and v_p are the electron and hole velocities, respectively. Under steady-state conditions the time derivatives in Eqs.(13.a1) and (13a.2) vanish. Eliminating the trap cross section, σ_t, from both equations and solving for G yields for equilibrium ($dn/dt = df_t/dt = 0$):

$$G = \sigma_l v_n np_l + \sigma_r v_p N_t f_t p \tag{13a.3}$$

The luminescence efficiency, η, is the ratio of the radiative recombination to the total recombination; it can be written in the form:

$$\eta = \frac{\sigma_l v_n np_l}{G} = \left(1 + \frac{\sigma_r v_p N_t f_t p}{\sigma_l v_n np_l}\right)^{-1} \tag{13a.4}$$

If the Fermi level is located between the electron trap level at energy ϵ_t and the luminescent level at energy ϵ_l, we find for the concentration of trapped electrons relative to the concentration of holes in luminescent centers:

$$N_t f_t / p_l = (N_t/N_l) \exp\{-(\epsilon_t - \epsilon_l)/k_B T\} \tag{13a.5}$$

[1] For a review see e.g. P. J. Dean, Trans. Metallurg. Soc. AIME 242 (1968) 384.

This finally yields for the luminescence efficiency

$$\eta = \left\{ 1 + \frac{N_t}{N_l} \cdot \frac{\sigma_r}{\sigma_l} \cdot \frac{v_p}{v_n} \cdot \frac{p}{n} \exp - \frac{\epsilon_t - \epsilon_l}{k_B T} \right\}^{-1} \tag{13a.6}$$

It is high if $N_l \gg N_t$ and $\epsilon_t - \epsilon_l \gg k_B T$, i.e. at low temperatures.

The highest efficiency reported is $\eta = 40\,\%$; it has been obtained in a GaAs diode at 20 K [2] (GaAs is a "direct" semiconductor, Chap. 11b). At room temperature η is not more than 7 %. The response time is typically 1 nsec or less. The emission spectra for 295 and 77 K are shown in Fig. 13.1. For a "direct semiconductor" the probability for a radiative transition from the conduction to the valence band is high. Even at low temperatures the emission is in the infrared part of the spectrum, however*).

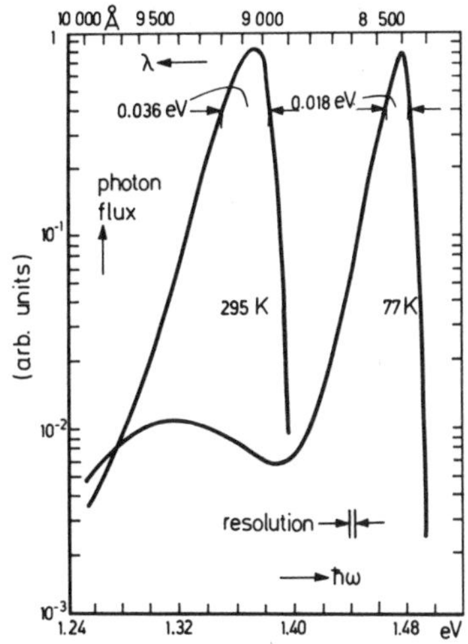

Fig. 13.1 Emission spectra observed on a forward biased gallium arsenide diode at 77 and 295 K (after Carr, ref. 2).

The diode is manufactured by the diffusion of Zn which is an acceptor; hence, it is a p^+n junction. Because of the band tail in the heavily doped p^+-side, which increases absorption, more light is emitted out of the n-side. A recent production method is "amphoteric doping": Silicon acts as an acceptor when it substitutes an As atom, and as a donor when it replaces a Ga atom. By a careful thermal treatment closely compensated structures are obtained which are particularly effective for emission of radiation.

Large bandgap semiconductors such as GaN, GaP, and SiC emit visible light. GaN is a direct semiconductor having an energy gap, ϵ_G, of 3.7 eV at room temperature. No p-n junction has been produced so far but Schottky barrier diodes emit blue and ultraviolet light at voltages which are, however, undesirably high. The emission spectra of GaP and SiC are shown in Figs. 13.2 and 13.3, respectively. Both substances are indirect semiconductors which explains their low efficiencies. The GaP p-n junction is doped with Cd and oxygen and contains a Cd-O complex. The emitted red light is due to a transition between the bound exciton level at this complex and a Cd acceptor level while the green light emission is due to a transition between an S donor level and the Cd acceptor level [3]. Excitons are also bound to the isoelectric impurity nitrogen (at phosphorus sites in GaP) and yield a green emission which

[2] W. N. Carr, IEEE Trans. Electron Devices ED-12 (1965) 531.

[3] M. Gershenzon, Bell Syst. Tech. J. 45 (1966) 1599.

*) Applied in combination with a silicon detector as a decoupling circuit.

appears as bright as the red emission since the human eye is more sensitive for green light [4].

The SiC emission spectrum depends also on the type of impurity (N, Al or B, Fig.13.3) [5]. The light output is proportional to the current which is of course determined by the forward bias. The response time is 0.5 μsec. The efficiency is at room temperature usually only about 10^{-3} %, but efficiencies up to 1 % have also been reported. The efficiency is, however, remarkably stable : diodes having an area of 1 mm^2 have been tested for 2 years at temperatures ranging from room temperature to 400 C and at forward currents of 50-200 mA with no measurable deterioration (other LED's degrade more quickly because of the high photon-induced stress). The brightness level is about 100 foot-lambert [6]. The main ap-

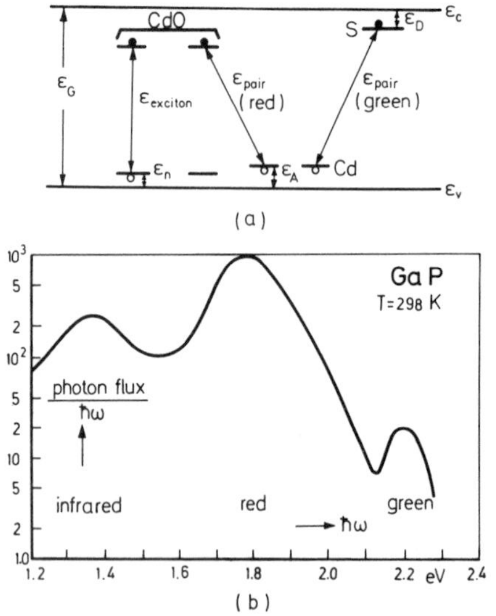

Fig.13.2a Energy band diagram for gallium phosphide containing Cd acceptor and S donor impurities and a Cd-O complex (after C. H. Henry, P. J. Dean, and J. D. Cuthbert, Phys. Rev. 166 (1968) 754).
b Emission from a gallium phosphide luminescent diode (after Gershenzon, ref. 3).

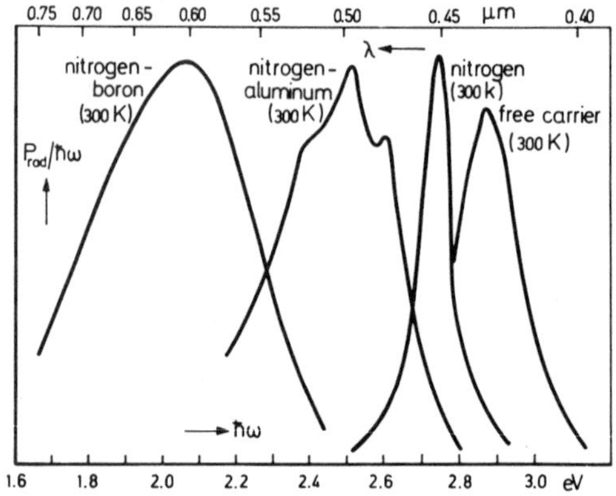

Fig.13.3
Observed emission spectra from silicon carbide luminescent diodes (crystal structure "6 H polytype") with n-regions containing the impurities given next to the curves (after Brander, ref. 5).

---

[4] For the relative eye response see D. E. Grey (ed.): Am. Inst. of Phys. Handbook, p. 6-139. New York: McGraw-Hill. 1963; W. D. Wright: The Measurement of Color. Princeton, N. J.: Van Nostrand. 1964.

[5] R. W. Brander, Proc. Int. Conf. on Silicon Carbide, University Park, Pa., 1968; Mat. Res. Bull. (H. K. Henisch and R. Roy, eds.), Vol. 4 , p. 187. London: Pergamon. 1969.

[6] Photometric terms are compared with radiometric terms in the review paper on LED's by A. A. Bergh and P. J. Dean, Proc. IEEE 60 (1972) 156.

plication of silicon carbide diodes are numeric and alphanumeric display devices. In this respect it may be of interest that crystals of more than 1 cm diameter have been grown. Depending on the crystallographic form (cubic or one of the polytypes) the gap varies between 2.2 and 3.1 eV. p-n junctions are produced by epitaxial growth from carbon saturated liquid silicon at 1500-1850 C on a p-type SiC substrate at a growth rate of 0.5 μm/min. A nitrogen atmosphere serves for doping with a donor. Contacts are made by a gold-tantalum alloying process. SiO_2 masks can be grown on SiC, and the photo-etch resist technique is applicable just as in silicon (Chap.5g).

At present the manufacture of signal lamps and displays is mostly based on $GaAs_xP_{1-x}$ which for $x \leqslant 0.55$ is "direct" with $\hbar\omega = 1.98$ eV (red light). For production an epitaxial process on a GaAs substrate is preferred.

A future application for electroluminescent diodes in the visible part of the spectrum may be as components for a flat TV screen. A still unsolved problem though is the spatial resolution of the screen: For a square with 500 points on a side (as in a cathode ray tube), 250 000 diodes would be required which poses problems in technology and marketing.

13b. The Semiconductor Laser

The acronym "laser" has been coined from the capitals of "Light Amplification by Stimulated Emission of Radiation". Before discussing how a semiconductor laser is different from a luminescent diode let us briefly consider the fundamentals of a two-level atomic or molecular laser in general [1].

Let ϵ_1 and ϵ_2 be two energy levels of an atom or molecule having electron concentrations of n_1 and n_2, respectively. A radiative transition involves the emission of a photon

$$h\nu = \epsilon_2 - \epsilon_1 \tag{13b.1}$$

The transition probability is given by n_2/τ_r where τ_r is the radiative lifetime of an electron in the upper level # 2. The transition is called "spontaneous". The absorption of radiation in a radiation field of energy density $\rho(\nu)\Delta\nu$ at frequencies between ν and $\nu + \Delta\nu$ is given by $B_{12}n_1\rho(\nu)\Delta\nu$ where $\Delta\nu$ is the linewidth and B_{12} is a coefficient to be determined later. According to Einstein [2] there occurs also "induced emission" $B_{21}n_2\rho(\nu)\Delta\nu$, which has its mechanical analogue in the

[1] For a comprehensive review on gaseous and solid state lasers see e.g. B. A. Lengyel: Introduction to Laser Physics. New York: J. Wiley and Sons. 1966; for semiconductor lasers see e.g. C. H. Gooch (ed.): Gallium Arsenide Lasers. New York: J. Wiley and Sons. 1969; M. H. Pilkuhn, phys. stat. sol. 25 (1968) 9.

[2] A. Einstein, Phys. Zeitschr. 18 (1917) 121; we have so far taken induced emission into account for free-carrier absorption, chap.11j.

forced oscillation of a classical oscillator at a "wrong" phase of the excitation: The oscillator is damped and transfers energy to the exciting system. The rate of change of the photon density, $\rho(\nu)$, is given by

$$d(\rho\Delta\nu)/dt = h\nu n_2/\tau_r + (B_{21}n_2 - B_{12}n_1)\rho\Delta\nu \tag{13b.2}$$

and vanishes at equilibrium where the electron concentrations

$$n_i \propto g_i \exp(-\epsilon_i/k_B T)\,; \quad i = 1; 2 \tag{13b.3}$$

and the factor g_i is the spin degeneracy of level # i. The photon density at equilibrium is the black-body radiation density given by

$$\rho(\nu)\Delta\nu = \frac{8\pi\nu^2\Delta\nu}{c^3} \cdot \frac{h\nu}{\exp(h\nu/k_B T) - 1} \tag{13b.4}$$

For equilibrium (zero left-hand side) we divide Eq.(13b.2) by ρ and consider the case for $T \to \infty$ where according to Eq.(13b.4) the radiation density $\rho \to \infty$, and we find

$$B_{21}n_2 = B_{12}n_1 \tag{13b.5}$$

For $T \to \infty$, Eq.(13b.3) yields $n_2/n_1 = g_2/g_1$ and therefore we have

$$B_{21}g_2 = B_{12}g_1 \tag{13b.6}$$

According to Einstein this relation is also valid at finite temperatures. From Eqs. (13b1)-(13b.4) and (13b.6) we thus obtain

$$B_{21} = \frac{c^3}{8\pi\nu^2\tau_r\Delta\nu} \tag{13b.7}$$

The time dependent photon distribution ρ obtained as a solution of Eq.(13b.2) is a linear function of $\exp(-\alpha L)$ where for the time t the ratio L/c has been substituted and L is the optical path length. In a crystal we have to take the optical dielectric constant, κ_{opt}, into account, if we consider L to be the crystal length, and obtain for the absorption coefficient α from Eq.(13b.2)

$$\alpha = -2\sqrt{\frac{\ln 2}{\pi}} \cdot \frac{c^2}{8\pi\nu^2\tau_r\Delta\nu\kappa_{opt}}\left(n_2 - \frac{g_2}{g_1}n_1\right) \tag{13b.8}$$

Here a factor $2\sqrt{(\ln 2)/\pi} \approx 0.94$ has been introduced to correct for a Gaussian line shape [3].

A negative value of α would indicate amplification of the wave. For $g_2 = g_1$ the gain coefficient $-\alpha$ which is usually denoted by g is simply proportional to $n_2 - n_1$. Since at thermal equilibrium $n_2 < n_1$, a "population inversion" must be maintained for amplification.

The main difference between the output of a luminescent diode and of a laser is that the latter emits coherent radiation. Hence, a Fabry-Perot resonator is an

[3] A. L. Schawlow, Proc. Int. School of Physics (P. A. Miles, ed.), Vol. XXXI, p. 1. New York: Acad. Press. 1964.

essential part of a laser. In a crystal two cleaved parallel faces (see Fig.13.4) serve for such a resonator. If we denote by r_∞ the reflectivity of a cleaved surface assuming $L \gg \lambda$, the wavelength, the feedback condition is

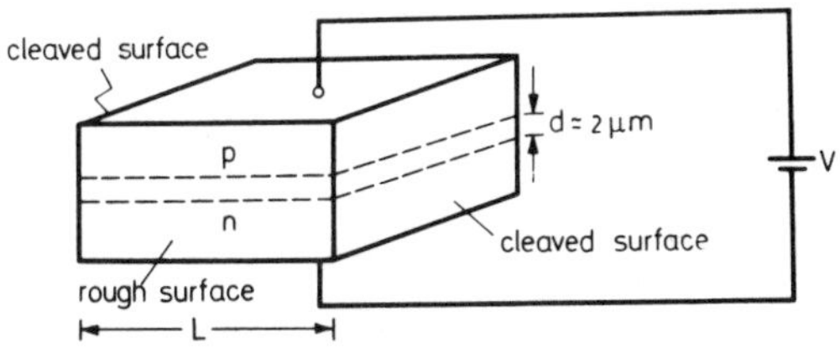

Fig.13.4
p-n junction laser structure (schematic).

$$r_\infty \exp \int_0^L (g - a')\, dz = 1 \tag{13b.9}$$

where a' is the absorption coefficient due to non-radiative recombination; the gain coefficient g depends on the photon flux which is a function of position, z. If for simplicity we neglect this dependence, the integration yields

$$g = \frac{1}{L} \ln (1/r_\infty) + a' \tag{13b.10}$$

With an assumption $n_1 = 0$ the threshold for oscillation is determined by Eqs. (13b.8) and (13b.10):

$$\frac{n_2}{\tau_r} = \frac{\kappa_{opt} 4\pi\nu^2 \Delta\nu}{c^2 \sqrt{(\ln 2)/\pi}} \left\{ \frac{1}{L} \ln \left(\frac{1}{r_\infty}\right) + a' \right\} \tag{13b.11}$$

This is the density of electrons which must be supplied per unit time to the energy level at ϵ_2 if the level at ϵ_1 should remain empty as has been assumed.

We now discuss the potential distribution in the p-n junction. Fig.9.2a shows a simplified schematic diagram with the Fermi level ζ being in the conduction band on the n-side and in the valence band on the p-side. If we now apply a forward bias the Fermi level is split into quasi-Fermi levels ζ_n^* and ζ_p^* which have been introduced in Chap.5d. Fig.9.2b shows the potential distribution for this case. An important point to consider is that there is an overlap of states in the valence band filled with holes and of states in the conduction band filled with electrons. The frequency ν of the recombination radiation for band-to-band transitions is given by $\epsilon_G \approx h\nu < (\zeta_n^* - \zeta_p^*)$. The voltage applied to the junction is usually about ϵ_G/e. The current density through the junction, j, is given by Eq. (5c.28) where the saturation current density, j_s, is given by Eq.(5c.40) except that n_i is replaced by n in the degenerate case with a hole mobility much less than the electron mobility:

$$j_s = |e|\, nd/\tau \tag{13b.12}$$

d is the junction width of about 2 μm and τ is the total lifetime which is written equal to $\eta\tau_r$ and η is the internal quantum efficiency for generating a photon from an electron-hole pair. Here we have an example for the case assumed for the derivation of Eq.(13b.11): The upper level # 2 is the conduction band where the states near the band edge are completely filled: $n_2 = n$, while the energy level # 1 is the valence band edge where the states are empty. (They are completely filled with holes which, of course, has the same meaning.) Since Eq.(13b.11) determines the threshold for laser operation we denote the current threshold by j_t and obtain from Eqs.(13b.11) and (13b.12) [4]:

$$j_t = \frac{1}{\beta}\left\{\frac{1}{L}\ln\left(\frac{1}{r_\infty}\right) + a'\right\} \tag{13b.13}$$

where a "gain factor" has been introduced:

$$\beta = \frac{\eta}{|e|\, d} \cdot \frac{c^2\sqrt{(\ln 2)/\pi}}{\kappa_{opt}\, 4\pi\nu^2\Delta\nu} \tag{13b.14}$$

a' is the "loss factor". For long crystals $j_t \approx a'/\beta$. The total laser power emitted from both ends of the device is

$$P = \frac{\eta\epsilon_G\, LW(j-j_t)\ln(1/r_\infty)}{|e|\{\ln(1/r_\infty) + La'\}} \tag{13b.15}$$

where W is the width of the device [5]. The linear dependence of the threshold current density, j_t, on the inverse crystal length, $1/L$, has been observed in a GaAs laser at various temperatures between 4.2 K and room temperature. Exper-

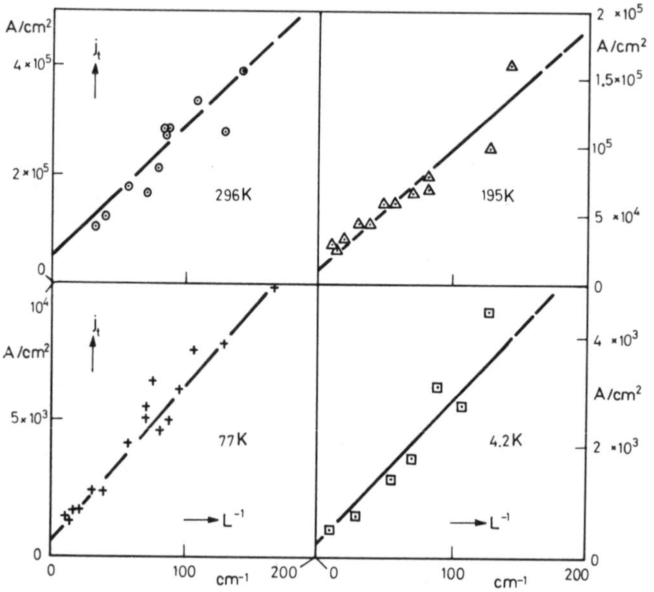

Fig.13.5 Laser threshold current density as a function of reciprocal GaAs diode length for various temperatures (after Pilkuhn et al., ref. 6).

[4] G. Lasher and F. Stern, Phys. Rev. 133(1964) A 553.

[5] H. S. Sommers, Jr., Solid State Electronics 11 (1968) 909; J. R. Biard, W. N. Carr, and B. S. Reed, Trans. AIME 230 (1964) 286.

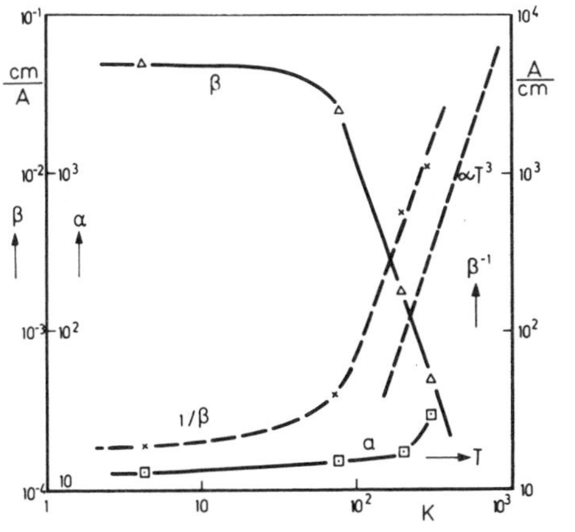

Fig. 13.6 Temperature dependence of loss and gain factor of a GaAs laser diode (after Pilkuhn and Rupprecht, ref. 7).

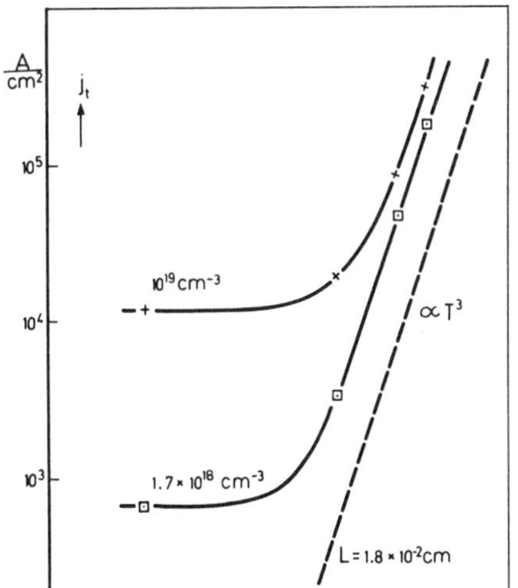

Fig. 13.7 Threshold current density of two GaAs lasers as a function of temperature; the substrate doping concentrations are written next to the curves (after Pilkuhn and Rupprecht, ref. 7).

imental data are shown in Fig. 13.5 [6]. Values of the loss and gain factors as a function of temperature are plotted in Fig. 13.6 [7]. There is a rapid rise of $1/\beta$ with temperature, approximately as T^3, for $T > 100$ K. Such a temperature dependence resembles that of the threshold current density which is shown in Fig. 13.7 for two different substrate doping levels [7] (substrate n-type, doped with Sn; p-side produced by a 3-hr Zn diffusion at 850 C and subsequent plating with a gold-nickel film).

The spectrum below and above the threshold current is shown in Fig. 13.8 [8]. The line narrows from about 125 Å to a few Å at 77 K and less than 1 Å at 4.2 K.

A magnetic field causes a shift of the emission line as indicated by Figs. 13.9 and 13.10 for GaAs and InAs, respectively [8]. In GaAs the shift is proportional to B^2 and can be correlated with the quadratic Zeeman effect of the ground state of a shallow donor or exciton as the initial state. An evaluation of the electron effective mass on this basis yields $m/m_0 = 0.074$ from Fig. 13.9a and 0.071 from Fig. 13.9b which is in close agreement with cyclotron resonance data. In InAs diodes the

---

[6] M. Pilkuhn, H. Rupprecht, and S. Blum, Solid State Electronics 7 (1964) 905.
[7] M. H. Pilkuhn and H. Rupprecht: Radiative Recombination in Semiconductors (C. Benoit a la Guillaume, ed.), p. 195. Paris: Dunod. 1964.
[8] B. Lax: Proc. Int. School of Physics (P. A. Miles, ed.), Vol. XXXI, p. 17. New York: Acad. Press. 1964.

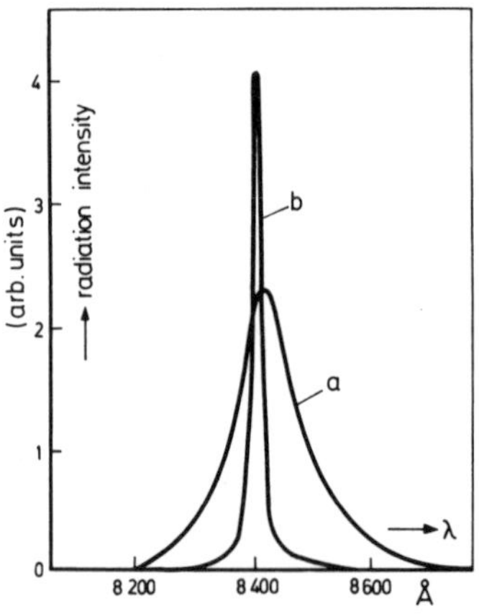

Fig.13.8 Emission spectra of a GaAs laser at 77 K (a) below the threshold for lasing operation and (b) above threshold (after B. Lax, ref.8).

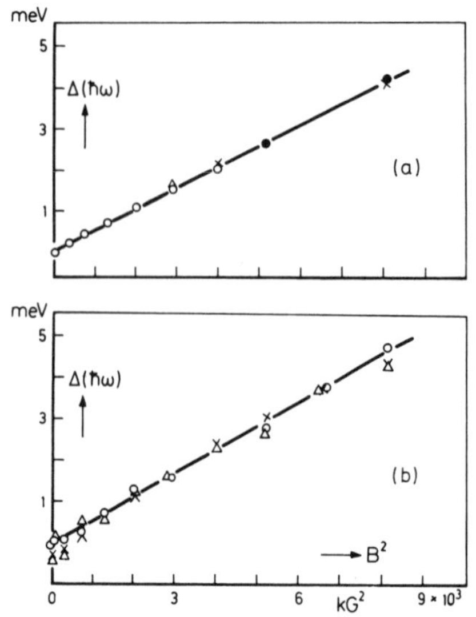

Fig.13.9 Zeeman effect on emission at 4.2 K from (a) GaAs laser (b) GaAs luminescen t diode (after F. L. Galeener, G. B. Wright, W. E. Krag, T. M. Quist, and H. J. Zeiger, Phys. Rev. Lett. 10 (1962) 472).

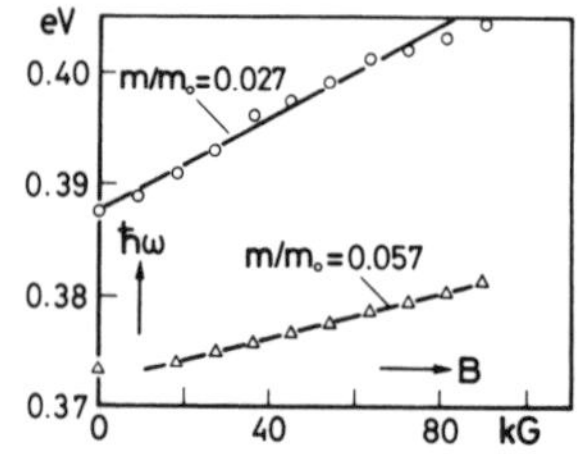

Fig.13.10 Zeeman effect on emission from an InAs laser at 77 K for electron concentrations of $2x10^{16}/cm^3$ (circles) and about $10^{19}/cm^3$ (triangles) (after I. Melngailis, Bull. Am. Phys. Soc. 8 (1963) 202, quoted by B. Lax, ref. 8).

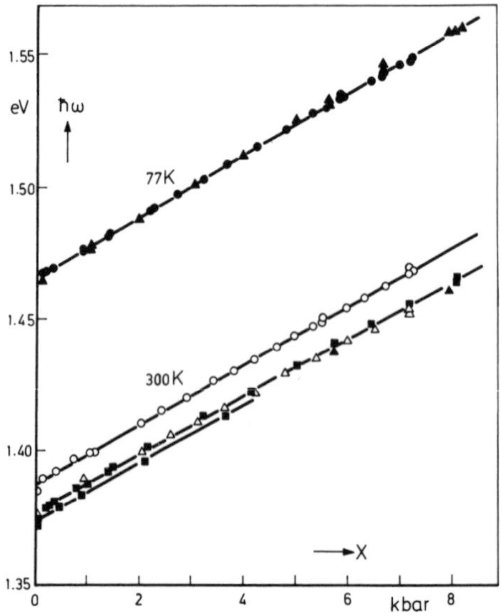

Fig.13.11 Dependence of the emission frequency of a GaAs laser on hydrostatic pressure at 77 and 300 K for various samples (after J. Feinleib, S. Groves, W. Paul, and R. Zallen, Phys. Rev. 131 (1963) 2070).

linear shift is consistent with a band-to-band transition.

A shift of the emission line of a GaAs laser with applied hydrostatic pressure is shown in Fig.13.11 [8]. From the shift the deformation potential constant has been determined.

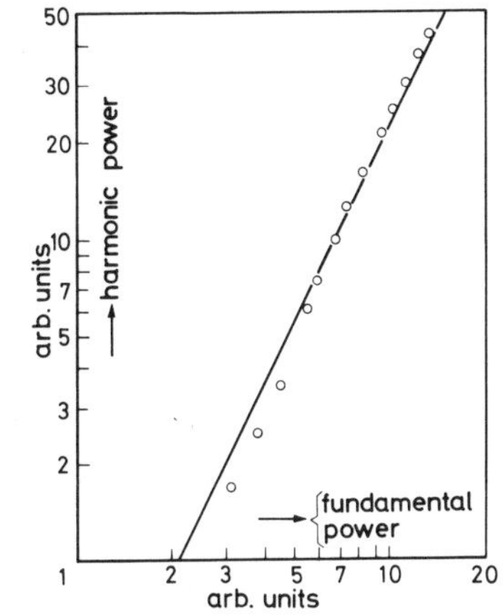

Fig.13.12 Second-harmonic emission from a GaAs laser (after J. A. Armstrong, M. I. Nathan, and A. W. Smith, Appl. Phys. Lett. 3 (1963) 68).

When an injection laser emits a power of e.g. 100 W distributed over an active region 1 μm thick and 100 μm wide, the field amplitude is there of the order of magnitude 10^5 V/cm. It has been shown in Chap.11o that at such high amplitudes a second harmonic may be generated. Fig.13.12 shows the second harmonic emission $P_{2\omega}$ from a GaAs laser as a function of the power output P_ω at the fundamental frequency [9]. As one would expect from the theory of non-linear optics, $P_{2\omega} \propto P_\omega^2$.

In comparison with other types of lasers available commercially, injection lasers are still rarely used. A typical GaAs laser for pulsed operation at room temperature (current 150 A) yields an output of 800 W at 0.916 μm [10]; a GaAs laser for cw operation at liquid nitrogen temperature (current 2.2 A) yields 0.1 W at 0.84 μm [11]. Typical threshold current densities are 10^2 A/cm^2 at temperatures below 77 K and $5\text{x}10^4$ A/cm^2 at room temperature.

CW operation at room temperature (output of more than 20 mW) has been achieved by a slight modification of the GaAs laser: A "heterojunction" GaAs/p-$Ga_{1-x}Al_xAs$ with a band gap of about 2 eV at the mixed-crystal side (GaAs: 1.4 eV) has been formed such that the difference in band gap produces a potential barrier essentially in the conduction band with the consequence of injected-electron accumulation in a thin p-GaAs transition region (Fig.13.13). Due to this electron accumulation the transition rate for stimulated emission is high even at room temperature. The index of refraction is somewhat higher on the GaAs side of the junction; this yields a light pipe effect with a further reduction in losses. The n-type side has been formed by a second heterojunction n-$Ga_{1-x}Al_xAs$-GaAs. There are no interface electronic states present because

[9] J. A. Armstrong, M. I. Nathan, and A. W. Smith, Appl. Phys. Lett. 3 (1963) 68.

[10] Sperry LA 0800-1, Sperry Gyroscope Division, electro-optics group, Sperry-Rand Corp., Great Neck, N. Y., USA.

[11] Compagnie Générale de Télégraphie Sans Fil, Boite Postale 10-91, Orsay, France.

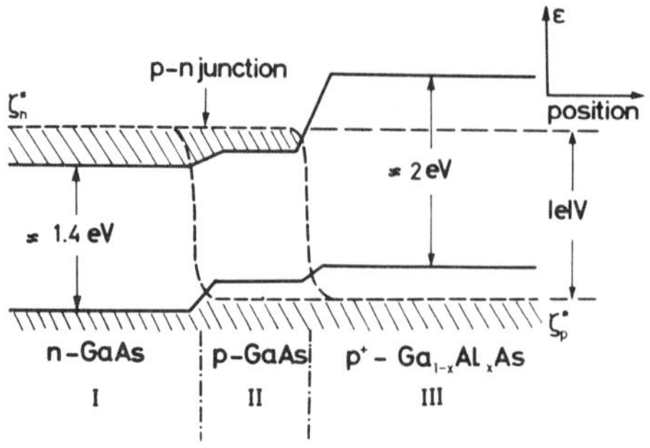

Fig.13.13 Energy band diagram of a GaAs-$Ga_{1-x}Al_xAs$ heterojunction with a limited active volume (II) (after Hayashi and Panish, ref. 13).

of the nearly equal lattice constants in GaAs and AlAs, and therefore no additional nonradiative recombination. However, the narrow active region ($\approx 0.5\,\mu m$, breadth $100\,\mu m$) yields light diffraction like a slit. The double heterojunction has been produced by liquid-phase epitaxy [12]. For a heat sink a tin-coated diamond on a copper substrate has been used. The threshold current density was $10^3\,A/cm^2$ at room temperature, the output power about 10^{-4} W at an applied voltage of about 2 V. The total efficiency was about 10% [13].

With an $ZnSe_xTe_{1-x}$ ($0.4 \leqslant x \leqslant 0.5$) diode laser operated at 77 K visible light ($\lambda = 0.627\,\mu m$) has been produced with an efficiency of 18% [14]. A metal semiconductor rectifying contact, which would have many advantages for use as a laser, contains too many recombination centers for lasing action to take place. So far it has not been possible to make large-band-gap semiconductor p-n junctions which are degenerate on both sides [15].

Narrow-gap semiconductor diode lasers $Pb_{1-x}Sn_xTe$ and $Pb_{1-x}Sn_xSe$ have been produced [16], the former also by proton bombardment [17]. Operation is possible only at low temperatures. Because of the narrow gap the radiation

[12] J. M. Woodall, H. Rupprecht, and W. Reuter, J. Electrochem. Soc. 116 (1969) 899; M. B. Panish, S. Sumski, and I. Hayashi, Metallurgical Transactions 2 (1971) 795; the same type of heterojunction has also been useful for solar cells (Chap.5i).

[13] I. Hayashi and M. B. Panish, J. Appl. Phys. 41 (1970) 150 and 3195; Sci. Am. 225 (July 1971) 32; I. Hayashi, M. B. Panish, P. W. Foy, and S. Sumski, Appl. Phys. Lett. 17 (1970) 109; IEEE-Trans. Quantum Electronics QE-6 (1970) 4.

[14] M. Aven, Appl. Phys. Lett. 7 (1965) 146.

[15] F. F. Morehead, Jr., Sci.Am. 216 (1967) No.5, p. 109; an interesting aspect is ion implantation: In SiC diodes have been produced in this way by H. L. Dunlap and P. J. Marsh, Appl. Phys. Lett. 15 (1969) 311.

[16] T. C. Harman: The Physics of Semimetals and Narrow-Gap Semiconductors (D. L. Carter and R. T. Bate, eds.), p. 363. Oxford: Pergamon. 1971.

[17] J. P. Donnelly, A. R. Calawa, T. C. Harman, A. G. Foyt, and W. T. Lindley, Solid State Electronics 15 (1972) 403.

wavelength is very long: From the telluride system at $x = 0.315$ a wavelength of 31.8 μm has been obtained at 40 K, the threshold being at 640 A/cm^2. The single-mode, cw diode laser can be compositionally tailored to emit any desired wavelength in the range 6.5 to 31.8 μm [16]. In Fig.2.21 the variation of the energy band gap with composition is shown. It is similar in the selenide system where the zero-gap composition is at $x = 0.15$. At $\lambda \approx 80\,\mu$m ($h\nu = 16$ meV) optical phonon-photon interaction should become quite interesting but has not been investigated so far. A serious problem in very far-infrared semiconductor lasers is self-absorption of the radiation in the crystal by free carriers (Chap.11j).

A future application of injection lasers in combination with glass fiber cables will be communication by pulse-code modulation techniques, an advantage being the enormous band width. A survey of LED and diode laser preparation has been given by Casey and Trumbore [18]. Laser modes have been discussed by Lax [19]and laser instabilities by Ripper [20].

An interesting modification of the semiconductor laser is the tunable "spin-flip Raman laser". In contrast to the injection laser, it requires another laser for a pump, such as CO_2 gas laser ($\lambda = 10.6\,\mu$m; $P \approx 3$ kW). The pump radiation is incident on a sample of n-type InSb ($n = 3 \times 10^{16}$ cm^{-3}) at 25-30 K in a magnetic field of 30-100 kG. The Landau levels in the conduction band of InSb are spin-split, according to Eq.(9b.13), the spacing between the spin sublevels, $g\mu_B B$, being proportional to the magnetic field strength. The angular frequency of the scattered light, ω_s, is different from that of the incident light, ω_0, by just this amount:

$$\omega_s = \omega_0 - g\mu_B B/\hbar \tag{13b.16}$$

Since the inelastic scattering process is similar to Raman scattering it is also called a "Raman" process although it involves no phonons. In contrast to the usual Raman process, the frequency of the scattered light can be tuned by a variation of the magnetic field strength. Lasing action is obtained if the pump power exceeds a certain threshold. Fig.13.14 shows the scattered light above and below threshold (upper and lower curve, respectively) [21]. Under stimulated-emission conditions the line is narrow (width estimated $\leqslant 0.03$ cm^{-1}). In Fig.13.15 the stimulated Raman scattering wavelength is plotted vs the magnetic field strength [22]. Tuning has been achieved in the range from 10.9 to 13.0 μm. For a peak input power of 1.5 kW the peak output was 30-100 W, the pulse length ≈ 100 nsec (about one third of the pump pulse) and the repetition rate 120/sec. The

[18] H. C. Casey, Jr., and F. A. Trumbore, Mater. Sci. Eng. 6 (1970) 69.
[19] B. Lax, ref. 8, p. 39.
[20] J. E. Ripper, IEEE-Trans. Quantum Electronics QE-5 (1969) 391; J. E. Ripper and J. C. Dyment, IEEE-Trans. Quantum Electronics QE-5 (1969) 396.
[21] C. K. N. Patel and E. D. Shaw, Phys. Rev. Lett. 24 (1970) 451.
[22] C. K. N. Patel, E. D. Shaw, and R. J. Kerl, Phys. Rev. Lett. 25 (1970) 8.

effect has been predicted by Yafet [23] in an extension of earlier work by Wolff [24] on a Raman laser based on Landau transitions.

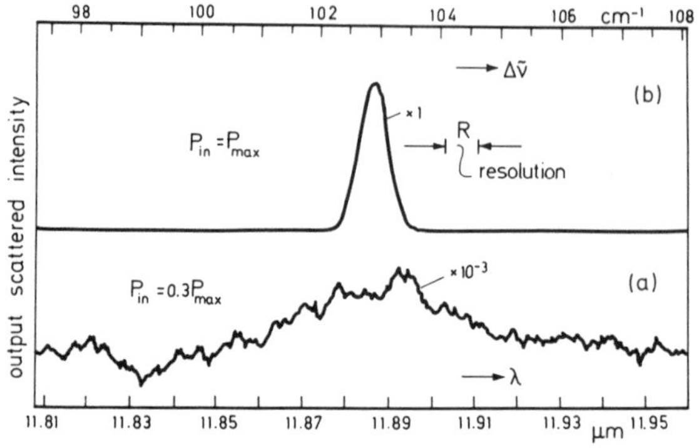

Fig.13.14 Spin-flip scattered light intensity as a function of wavelength above the stimulated emission threshold (upper curve, R = spectrometer resolution) and below threshold (lower curve, 10^3 times magnified) (after Patel and Shaw, ref. 21).

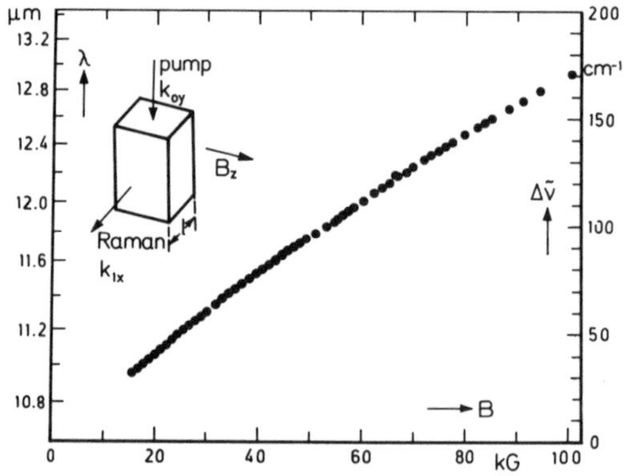

Fig.13.15 Stimulated spin-flip scattering wavelength vs magnetic field. Laser geometry shown by the inset (after Patel et al., ref. 22).

---

[23] Y. Yafet, Phys. Rev. 152 (1966) 858.
[24] P. A. Wolff, Phys. Rev. Lett. 16 (1966) 225; J. Quantum Electronics 2 (1966) 659.

14. Properties of the Surface

So far we have dealt with the bulk properties of semiconductors and tacitly assumed that the crystal is extended infinitely. We will now briefly discuss the influence of the crystal surface on the transport properties of semiconductors.

14a. Surface States

The success of the Kronig-Penney energy band model (Chap.2a) in explaining the bulk properties of crystals inspired I. Tamm [1] to investigate the energy levels of an atomic chain of finite length. Fig.14.1 shows the electric potential

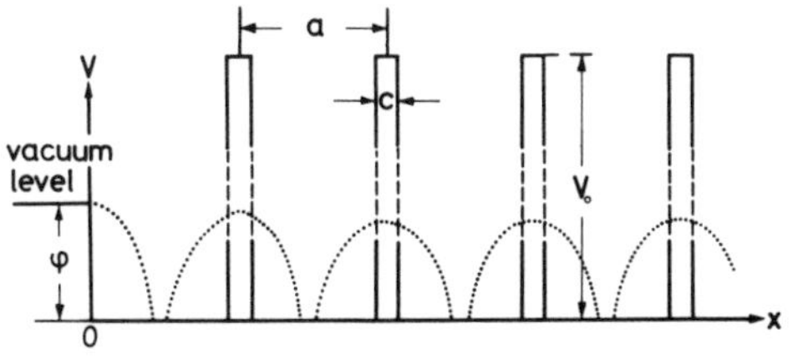

Fig.14.1
Simplified potential distribution in a one-dimensional crystal having a surface at x = 0 (after Tamm, ref. 1). A more realistic distribution is shown by the dotted curve.

[1] I. Tamm, Zeitschr. f. Physik 76 (1932) 849; Phys. Z. Sowjetunion 1 (1932) 733; Zh. Eksp. Teor. Fiz. 3 (1933) 34; for surface state calculations see S. G. Davison and J. D. Levine: Solid State Physics (H. Ehrenreich, F. Seitz, and D. Turnbull, eds.), Vol. 25 , p. 1. New York: Acad. Press. 1970 ; for a general review on semiconductor surfaces see e.g. D. R. Frankl: Electrical Properties of Semiconductor Surfaces. Oxford: Pergamon. 1967; A. Many, Y. Goldstein, and N. B. Grover: Semiconductor Surfaces. Amsterdam: North-Holland Publ. 1965; I. M. Lifschitz and S. I. Pekar, Fortschr. Phys. 4 (1956) 81; J. Koutecký, J. Phys. Chem. Solids 14 (1960) 233; G. Heiland, Fortschr. Phys. 9 (1961) 393.

of the Tamm model at one end of the chain. In the vacuum, $x < 0$, the potential is taken as constant, $V = \varphi$, where the constant may be considered as the work function of the solid, and $x = 0$ corresponds to the surface of the crystal.

For a solution of the Schrödinger equation (2a.3) where for β^2 we substitute

$$\gamma^2 = 2m\hbar^{-2}(\varphi - \epsilon) \tag{14a.1}$$

and $\epsilon < \varphi$ is assumed, we obtain

$$\psi = A' e^{\gamma x} \text{ for } x < 0 \tag{14a.2}$$

which satisfies the boundary condition $\psi \to 0$ for $x \to -\infty$. For $x > 0$ we take Eq.(2a.21) for a solution and solve Eqs.(2a.23) and (2a.24) for the constant B of Eq.(2a.21) :

$$B = -A \frac{\beta e^{i(\alpha - k)b} + \{i\alpha \sinh(\beta c) - \beta \cosh(\beta c)\} e^{ikc}}{\beta e^{-i(\alpha + k)b} - \{i\alpha \sinh(\beta c) + \beta \cosh(\beta c)\} e^{ikc}} \tag{14a.3}$$

In a simplified version of the Kronig-Penney model we have $c \to 0$ and $V_0 \to \infty$ such that the product $V_0 c$ remains finite. Since $b \to a$ for $c \to 0$, Eq.(14a.3) in this model is simply

$$B = -A \frac{e^{i(\alpha - k)a} - 1}{e^{-i(\alpha + k)a} - 1} \tag{14a.4}$$

The condition of continuity for ψ and $d\psi/dx$ at $x = 0$ yields

$$\left.\begin{aligned} A + B &= A' \\ i\alpha(A - B) &= \gamma A' \end{aligned}\right\} \tag{14a.5}$$

since ψ in the first potential trough of the crystal, i.e. for $0 < x < a$, is $Ae^{i\alpha x} + Be^{-i\alpha x}$. After substituting for B its value given by Eq.(14a.4) this is a homogeneous set of equations for A and A' which has solutions if the determinant vanishes. This condition can be written

$$e^{ika} = (\gamma/\alpha)\sin(\alpha a) + \cos(\alpha a) \tag{14a.6}$$

In the approximation $c \to 0$, $V_0 \to \infty$ where $m\hbar^{-2}aV_0c$, denoted now as p, is finite, Eq.(2a.27) yields in this approximation

$$\frac{1}{2}(e^{ika} + e^{-ika}) = \cos(ka) = (p/\alpha a)\sin(\alpha a) + \cos(\alpha a) \tag{14a.7}$$

If we eliminate e^{ika} from the last two equations, take Eqs.(2a.1) and (14a.1) into account, and introduce for shortness

$$q = (a/\hbar)\sqrt{2m\varphi} = \alpha a\sqrt{1 + (\gamma/\alpha)^2} \tag{14a.8}$$

we find after some algebra

$$(2p/q^2)(\alpha a)\cot(\alpha a) = 1 - (2p/q^2)\cdot\sqrt{q^2 - (\alpha a)^2} \tag{14a.9}$$

where $\alpha a = (a/\hbar)\sqrt{2m\epsilon}$ and the constant $2p/q^2 = V_0 c/(\varphi a)$. In the range $\epsilon < \varphi$, i.e. $\alpha a < q$, this equation has solutions calculated by applying the Kronig-Penney model, Eq.(14a.7). This is shown in Fig.14.2 for p = 7.2 and q = 6 and 12.

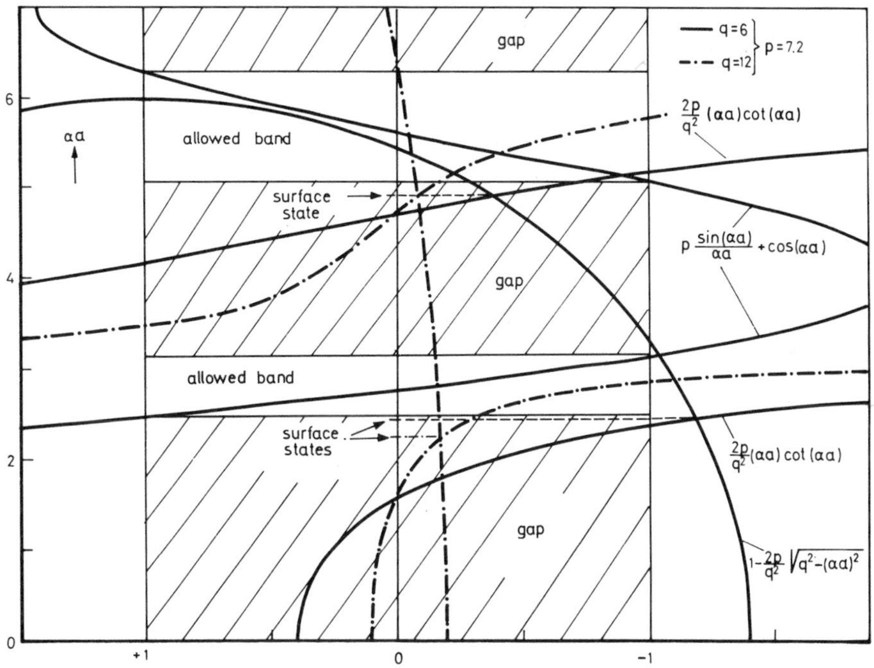

Fig.14.2 Kronig-Penney plot (similar to Fig.2.4 except that the ordinate scale $\propto \epsilon^{1/2}$) extended to include Tamm surface states for p = 7.2 and q = 6 (full curves) and q = 12 (dash-dotted). The curves represent the right- and left-hand sides of Eq.(14a.9) and the right-hand side of Eq.(14a.7). The ordinate $\alpha a = (a/\hbar)\sqrt{2m\epsilon}$. Surface states are obtained only in gaps where $\alpha a < q$.

For a comparison with Fig.2.4, the constants p and q have been chosen equal to $2m\hbar^{-2}V_0(b/2)^2(2c/b)$ and $2\sqrt{2m\hbar^{-2}V_0(b/2)^2\varphi/V_0}$, respectively. For the values given in the text after Eq.(2a.27) we find p = 7.2 and for $\varphi/V_0 = 1/4$, q = 6.

These energy levels in the band gaps are called "surface states". One can show that the wave functions are localized at the surface. Inside the crystal the wave function shows damped oscillations with increasing distance from the surface while outside the crystal it decreases exponentially. Within the surface the states are not localized, however, but form a band. Conductivity in this band in all experiments has been too small to be detected besides the conductivity in the space charge next to the surface inside the crystal. Impurity atoms adsorbed at the surface can introduce localized states, just as impurities in the bulk.

Shockley [2] calculated the energy levels of a linear chain of 8 atoms as a function of the lattice parameter as shown in Fig.14.3.

---

[2] W. Shockley, Phys. Rev. 56 (1939) 317.

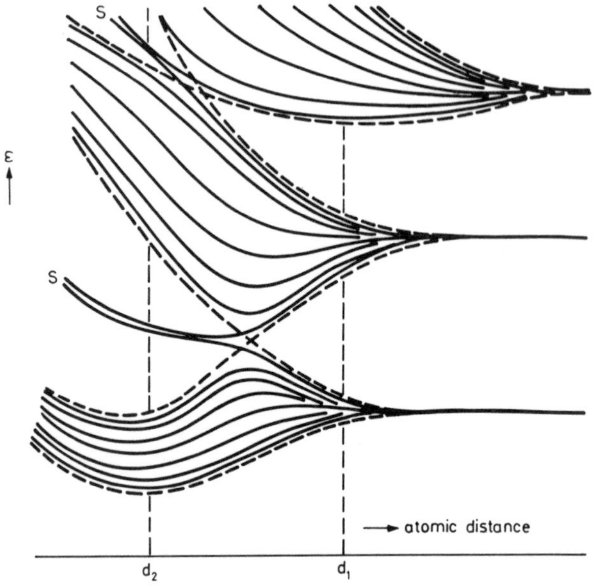

Fig.14.3 Calculated energy bands and surface states (labeled "S") of a one-dimensional crystal containing eight atoms, as a function of the atomic distance (after Shockley, ref. 2).

For an interatomic spacing less than a critical value there are both band gaps and two states (one per surface atom) within each gap; the states arise as a consequence of the existence of a boundary. One volume state simultaneously disappears from each of the adjacent bands.

In a 3-dimensional crystal one would expect one surface state per surface atom. Surface states can also be recognized as being due to the dangling bonds of the surface atoms. The density of surface atoms has an order of magnitude of $10^{15}/cm^2$. There are steps at the surface, and between 1 and 20% of the surface atoms are located at these steps. Careful annealing reduces the number of steps. Usually the surface is covered by impurity atoms except when special precautions are taken. A "clean" surface may be obtained by cleavage of a crystal in ultrahigh vacuum as e.g. 3×10^{-10} torr, although pressure bursts of 10^{-7} torr upon cleavage have been observed. After a time of several hours a monoatomic layer has been formed on a clean surface in ultrahigh vacuum which consists mainly of oxygen atoms. What is considered a "clean surface" depends very much on the type of measurement. For a given observation a surface is often considered clean if the experimental results are not changed by further purification of the surface.

If the surface under consideration is formed by the tip of a needle of radius

10^{-4} mm, a field intensity of about 10^8 V/cm can be produced. Such a field is sufficient to tear off atoms including impurity atoms by the electrostatic force when the needle is positively charged. This method is used in the field electron microscope [3]. Other methods of purification are argon bombardment and subsequent annealing in ultrahigh vacuum. A sensitive method for the detection of surface impurities and dislocations is low-energy electron diffraction ("LEED") [4].

Impurity atoms on a surface of a semiconductor may be ionized. A monoatomic layer would yield about 10^{15} elementary charges per cm^2 assuming that each impurity atom would be ionized. However, it is not possible to have such a tremendous charge at the surface. Even with only 5×10^{13} carriers/cm^2 the electrostatic field energy is already equal to the surface energy of the crystal lattice. Observed surface charges usually vary between 10^{11} and 10^{13} carriers/cm^2.

Experimental evidence of surface states first came from the rectification properties of silicon-metal contacts. Meyerhof [5] observed that these were practi-

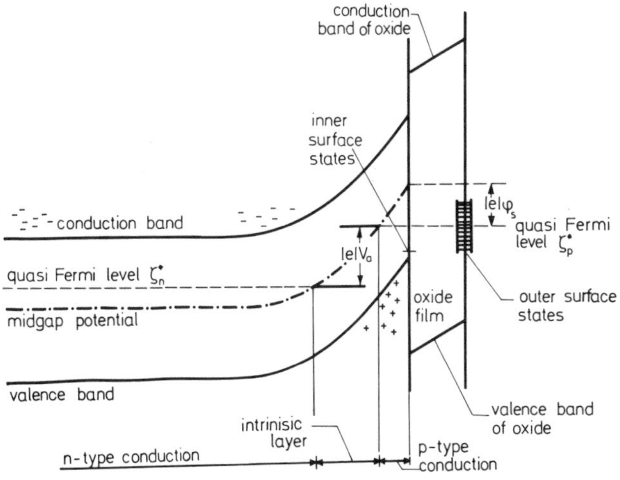

Fig. 14.4 Energy band and surface state structure at semiconductor surfaces (after H. Statz, G. A. de Mars, L. Davis, Jr., and A. Adams, Jr.: Semiconductor Surface Physics (R. H. Kingston, ed.), p. 139. Philadelphia: Univ. of Pennsylvania Press. 1957).

---

[3] E. W. Müller: Ergebn. d. exakt. Naturwiss. Vol. XXVII, Kap. III, 8, p. 290. Berlin-Göttingen-Heidelberg: Springer. 1953.

[4] R. E. Schlier and H. E. Farnsworth: Semiconductor Surface Science (R. H. Kingston, ed.), p. 3. Philadelphia: Univ. of Pennsylvania Press. 1957; for a review see Proc. Symp. Structure of Surfaces, Surface Sci. 8 (1967) 1; a brief description of surface characterization by LEED is given by e.g. S. G. Davison and J. D. Levine, ref. 1, p. 110.

[5] W. E. Meyerhof, Phys. Rev. 71 (1947) 727.

cally independent of the difference in work function between the metal and the silicon. Bardeen [6] explained this observation by assuming surface states which are due to impurities on the interface between the metal and the semiconductor. As a rule of thumb, such interface states are more pronounced in smaller-gap materials. The interface is often formed by an oxide film, especially if the semiconductor has been etched with an oxidizing agent (e.g. HF + HNO_3) before making the metallic contact. An energy band diagram for this case is shown in Fig.14.4. There are inner surface states called "fast states" and outer surface states called "slow states" according to their different response times to the

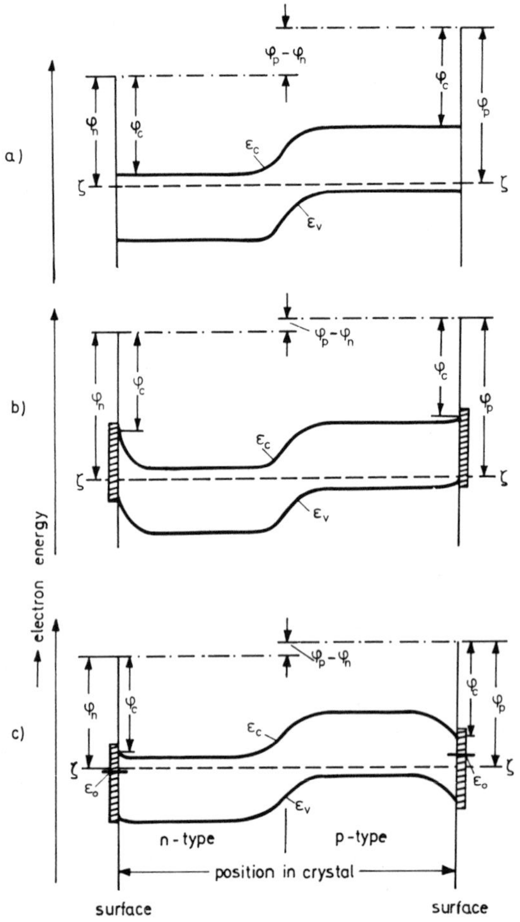

Fig.14.5 Energy band diagram for a p-n junction (a) without surface states, (b) with surface acceptor levels (hatched) distributed over the band gap, and (c) with an additional discrete level ϵ_0 (after G. Heiland, Fortschr. d. Physik 9 (1961) 393).

[6] J. Bardeen, Phys. Rev. 71 (1947) 717.

application of a strong electric field perpendicular to the surface. The surface in Fig. 14.4 is negatively charged while the bulk is n-type. The conduction electrons are repelled from the negative surface charge with the formation of either a depletion layer or, as shown here, of a p-type inversion layer if the valence band is bent above the bulk quasi Fermi level. For positively charged surface states an accumulation layer is formed. Fig. 14.5 shows how band bending due to charged surface states yields about the same work function on both sides of a germanium p-n junction. In spite of a built-in potential of 0.34 V a difference in work function of only 0.002 ± 0.004 eV has been measured [7] by the Kelvin method [8] which can be explained by band bending (Fig. 14.5).

The main problem in all semiconductor surface investigations is the determination of the energy distribution of surface states. In spite of a great number of experiments there is still controversy about the distribution, essentially because (1) it has not been possible to vary the doping and therefore the Fermi level within a large range and (2) it is difficult to get reproducible surfaces independent of processing. The following effects have been applied for surface investigations: surface conductance, field effect, photoelectric methods, surface recombination, field emission, photo-surface conductance, optical absorption and reflection. Of these we will consider here only the transport effects.

14b. Surface Transport Effects

The surface resistance is measured by the standard four probe methods discussed in Appendix C. Since the thickness d of the charge accumulation layer or inversion layer at the surface is not known, the conductance $d/\rho = \sigma_\square$ is obtained in units of "mho/square" (which is sometimes written ohm^{-1}), rather than the conductivity σ.

Fig. 14.6 shows the surface conductance, Hall coefficient, and the surface carrier density as functions of oxygen coverage (actually oxygen pressure) after the surface has been cleaned by ionic bombardment and subsequent heat treatment [1]. The Hall mobility μ_H has been calculated from the Hall coefficient R_H and the surface conductance $\sigma_\square$ according to

$$\mu_H = \{R_H \sigma_\square^2 - (R_H \sigma_\square^2)_o\}/(\sigma_\square - \sigma_{\square o}) \qquad (14b.1)$$

and the excess carrier density at the surface Δp according to

[7] A. B. Fowler, J. Appl. Phys. 30 (1959) 556.

[8] see e.g. A. Eberhagen, Fortschr. Phys. 8 (1960) 245.

Chap. 14. b

[1] R. Missman and P. Handler, J. Phys. Chem. Solids 8 (1959) 109; there is a suspicion that during the outgassing procedure of the glass tube boron has been transferred from the glass to the Ge.

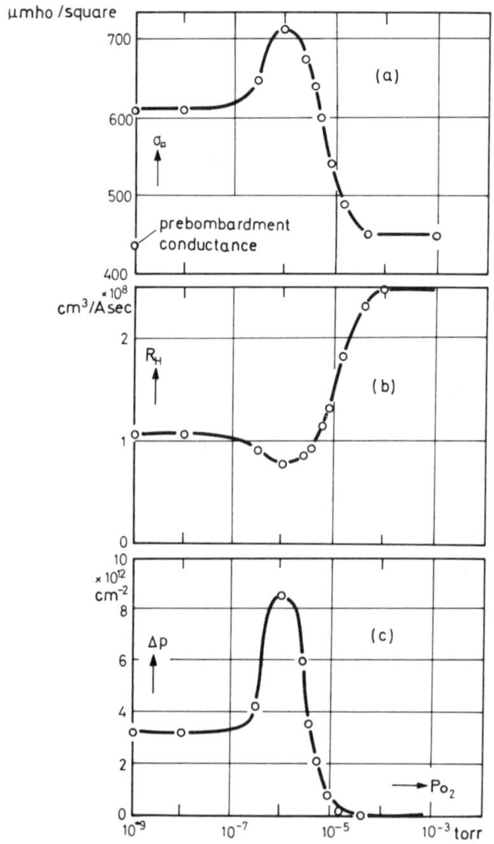

Fig. 14.6 (a) Conductance, (b) Hall coefficient, and (c) surface carrier density of Ge with an argon bombarded and annealed surface as a function of oxygen coverage (after Missman and Handler, ref. 1).

$$\Delta p = (\sigma_{\square} - \sigma_{\square o})\, d/2e\mu_H \qquad (14b.2)$$

where the subscript o indicates the flat-band condition. Experimentally the surface conductance can only be determined if the bulk conductivity is negligibly small. The bulk was n-type Ge with a resistivity of 20 ohm-cm. A clean surface should be p-type according to the idea of dangling bonds. Therefore the n-type bulk is expected to be covered by an inversion layer. It is not changed by an oxygen pressure up to 10^{-7} torr. At higher pressures oxygen is absorbed by the surface with the effect of an increase in surface charge. This has a maximum at 10^{-6} torr. Above 10^{-4} torr the surface charge is negligible ("flat-band condition"). At the maximum the energy band bending is equivalent to a change in surface potential equal to about 0.25 eV.

A field effect experiment is a means of varying the surface potential without changing the ambient: By a capacitively applied electric field normal to the surface the surface conductance of the sample is varied [2]. This is essentially the idea of the field effect transistor, Fig.5.16. For an electric field E the total induced charge on the semiconductor per unit area is $q = \kappa\kappa_o E$. It consists of the charge in the space-charge region, q_{sc}, and the charge in the surface states, q_{ss}; the latter can be assumed to be mobile, and the conductance change is almost entirely due to a change in space charge. The field effect mobility is defined by

$$\mu_{FE} = \delta\Delta\sigma_\square/\delta q = \delta\Delta\sigma_\square/(\delta q_{sc} + \delta q_{ss}) \approx \delta\Delta\sigma_\square/\delta q_{sc} \qquad (14b.3)$$

where $\Delta\sigma_\square = \sigma_\square - \sigma_{\square o}$ is the excess surface conductance above the flat-band condition. Fig.14.7 shows an experimental tube for measurements of μ_{FE} and of $\Delta\sigma_\square$ [3]. In Fig.14.8 the observed field effect mobility is shown as a function of time

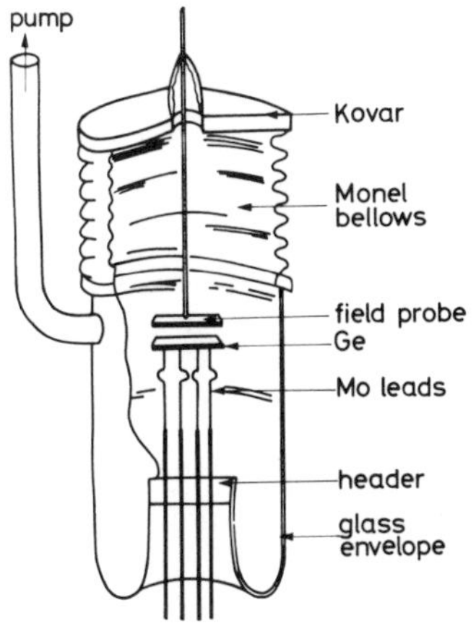

Fig.14.7 Schematic diagram of an experimental tube used for measurements of surface conductance and field effect mobility (after Handler, ref. 3).

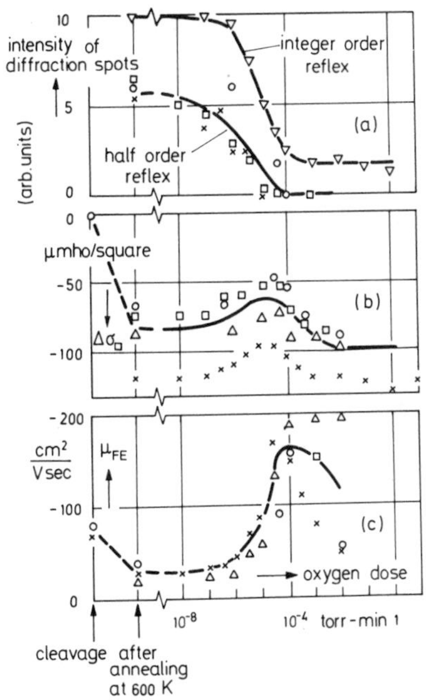

Fig.14.8 (a) LEED spot intensity, (b) surface conductance, and (c) field effect mobility simultaneously measured as a function of oxygen coverage on a clean germanium [111] surface. At the right-hand side of the figure there is about a monolayer (after Henzler, ref. 4).

[2] W. Shockley and G. L. Pearson, Phys. Rev. 74 (1948) 232.

[3] P. Handler: Semiconductor Surface Physics (R. H. Kingston, ed.), p. 23. Philadelphia: Univ. of Pennsylvania Press. 1957.

after introduction of oxygen into the tube at a pressure of 10^{-6} - 10^{-7} torr [4]. The field effect mobility is negative and of the other magnitude of 100 $cm^2/Vsec$; it bears no relation to the carrier mobility in the bulk of the crystal.

There exists a unique relationship between the space charge and the surface potential which is derived from Poisson's equation. Fig.14.9 shows the excess conductance vs space charge (calculated) and vs total induced charge (measured),

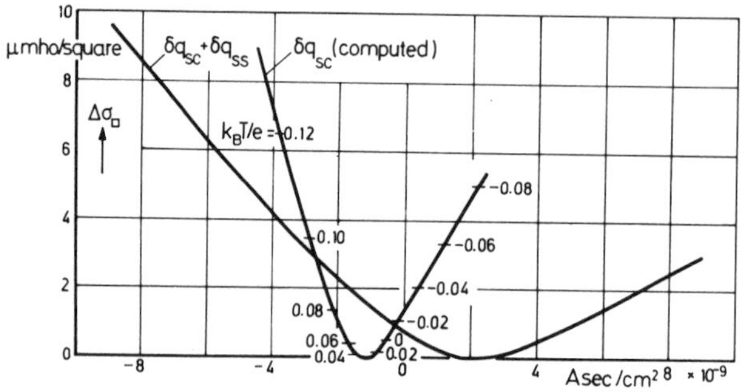

Fig.14.9 Observed change in surface conductance as a function of changes in total induced charge, δq, and calculated change in surface conductance as a function of changes in space charge, δq_{sc} (after Brown et al., ref. 5).

with values of the surface potential (in units of $k_BT/|e|$) given along the calculated curve [5]. At the right-hand side of the diagram the surface is p-type and a positive increment in space charge yields an increase in conductance; at the left-hand side the surface is n-type and a positive increment in space charge results in a decrease of the number of negatively charged electrons and consequently in a decrease in conductance. Because of the different mobilities of electrons and holes the minimum is not at zero space charge. From an observation of the conductance minimum the surface potential can be determined for all values of the total induced charge.

In a field effect experiment an ac field of frequency e.g. 30 Hz has been applied in addition to a dc field and the ac conductance is plotted vs the ac voltage for various ambients as shown in Fig.14.10 [5]. In fact, the experimental curve in Fig.14.9 has been composed of the portions shown in Fig.14.10. Oxygen produces an n-type surface. Water vapor makes the surface p-type which may be a consequence of the dipole nature of the water molecule.

Field effect measurements have also been made on the surface recombination velocity [6] s and calculations of s as a function of the surface charge have been

[4] M. Henzler: Festkörper-Probleme (O. Madelung, ed.), Vol. XI, p. 187. Braunschweig: Vieweg, Oxford: Pergamon. 1971.

[5] W. L. Brown, W. H. Brattain, C. G. B. Garrett, and H. C. Montgomery: Semiconductor Surface Physics (R. H. Kingston, ed.), p. 111. Philadelphia: Univ. of Pennsylvania Press. 1957.

[6] G. C. Dousmanis, Phys. Rev. 112 (1958) 369.

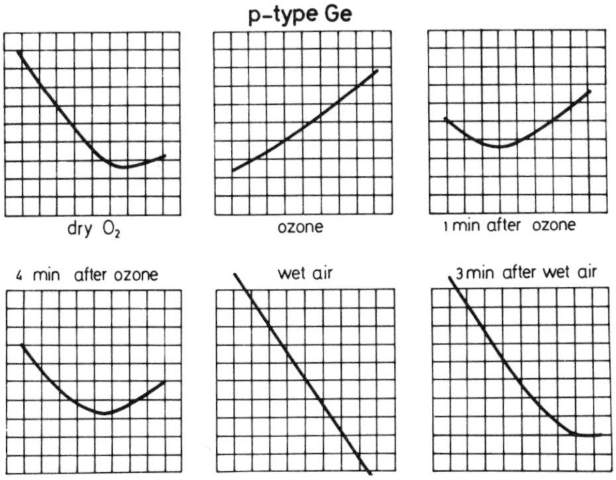

Fig. 14.10 Oscilloscope patterns of conductance change with induced charge for a series of gaseous ambients (after Brown et al., ref. 5).

done by Many et al. [7]. The surface mobility as a function of barrier height has been calculated by Schrieffer [8]. In other experiments the surface potential has been varied by illumination [9]. Surface noise is also interesting from a technical point of view [10]. In photoemission measurements one of the main problems is to distinguish the surface state contribution from the various bulk contributions [11]. An interesting problem which is still in a controversial state is the delayed emission of electrons from semiconductor surfaces [12]. After an excitation by either electron bombardment ($\approx$ 1 keV) or ultraviolet irradiation germanium emits electrons. The emission decreases with time elapsed after the excitation. If the temperature is raised "glow curves" of the emission can be recorded. There is a close resemblance of this phenomenon to phosphorescence. The emission is thought to arise from filled surface states close to the vacuum level. Since the emission current is so small that every single emitted electron is detected by an electron multiplier in combination with pulse counting techniques, a correlation of the results with other surface phenomena has not been possible.

[7] A. Many and D. Gerlich, Phys. Rev. 107 (1957) 404.

[8] J. R. Schrieffer, Phys. Rev. 97 (1955) 641.

[9] C. G. B. Garrett, W. H. Brattain, W. L. Brown, and H. C. Montgomery: Semiconductor Surface Physics (R. H. Kingston, ed.), p. 126. Philadelphia: Univ. of Pennsylvania Press. 1957.

[10] R. H. Kingston and A. L. McWhorter, Phys. Rev. 103 (1956) 534.

[11] S. G. Davison and J. D. Levine, ref. 1, p. 106 and literature cited there; for a correlation of photoemission originating from the bulk with band structure see e.g. U. Gebhardt, Festkörper-Probleme (O. Madelung, ed.), Vol. X, p. 175. Braunschweig: Vieweg, Oxford: Pergamon. 1970.

[12] K. Seeger, Zeitschr. f. Physik 149 (1957) 453.

15. Miscellaneous Semiconductors

15a. Superconducting Semiconductors

Superconductivity, which is normally observed in metals at temperatures below a transition temperature $T_c \lesssim 21$ K may also occur in degenerate semiconductors where T_c is below 0.5 K. The effect has been predicted by Gurevich, Larkin, and Firsov [1] in 1962 and by Cohen [2] in 1964 and verified experimentally in the semiconductors GeTe [3], SnTe [4], and $SrTiO_3$ [5] in 1964.

From the theory of superconductivity by Bardeen, Cooper, and Schrieffer [6] a transition temperature

$$T_c = 1.14\,\Theta \exp\{-1/N(0)V\} \qquad (15a.1)$$

is obtained where Θ is the Debye temperature, N(0) is the density of states at the Fermi level, and V is the average matrix element for the interaction of two electrons [7]. For a degenerate electron gas, $N(0) \propto mn^{1/3}$ where n is the carrier

[1] V. L. Gurevich, A. I. Larkin, and Yu. A. Firsov, Fiz.Tverd.Tela 4 (1962) 185 [Engl.: Sov. Phys.-Solid State 4 (1962) 131].

[2] M. L. Cohen, Phys. Rev. 134 (1964) A 511; M. L. Cohen and C. S. Koonce: Proc. Int. Conf. Phys. Semic. Kyoto 1966, p. 633. Tokyo: Phys. Soc. Japan. 1966.

[3] R. A. Hein, J. W. Gibson, R. Mazelsky, R.C. Miller, and J. K. Hulm, Phys. Rev. Lett. 12 (1964) 320.

[4] R. A. Hein, J. W. Gibson, R. S. Allgaier, B. B. Houston, Jr., R. L. Mazelsky, and R. C. Miller: Proc. 9th Int. Conf. Low-Temp. Phys. (J. A. Daunt, ed.), p. 604. New York: Plenum 1965.

[5] J. F. Schooley, W. R. Hosler, and M. L. Cohen, Phys. Rev. Lett. 12 (1964) 474.

[6] J. Bardeen, L. N. Cooper, and J. R. Schrieffer, Phys. Rev. 108 (1957) 1175.

[7] See e.g. P. G. de Gennes: Superconductivity of Metals and Alloys. New York: Benjamin Inc. 1966.

density and m the effective mass. For a high transition temperature, the product N(0)V should be large. Many-valley semiconductors have an advantage over the single-valley type in that they have a larger density of states and a larger effective mass. The semiconductors mentioned above are believed to be of the many-valley type (e.g. twelve <110> valleys in GeTe and SnTe).

GeTe at normal temperatures is p-type with a mobility of about 100 $cm^2/Vsec$ and an effective mass which is about equal to the free electron mass. Because of natural deviations from stochiometry the carrier density is about $0.9 \times 10^{21}/cm^3$; because of this large carrier density the band gap can only be estimated as being between 0.5 and 1 eV. Fig. 15.1 shows the observed changes in the magnetic susceptibility of samples of various compositions as the temperature is lowered and Fig. 15.2 the transition temperature as a function of the hole density. The hole density of the sample with the lowest transition temperature is consistent with the assumption that there are two holes per germanium vacancy. The samples were pressed and sintered. Apparently the minimum carrier density for superconductive behavior to occur is $8.5 \times 10^{20}/cm^3$ in this material (Fig. 15.2). Superconducting semiconductors may be classified as "type II superconductors".

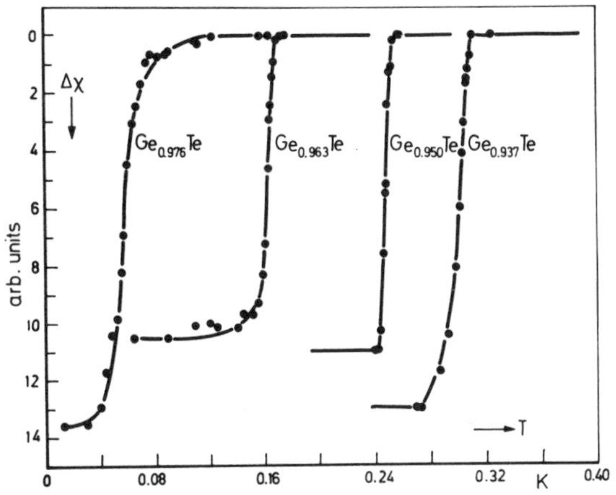

Fig. 15.1
Changes in the initial susceptibility of germanium telluride samples as a function of temperature. The large differences in the values of the susceptibilities in the superconducting state are due to differences in the mutual inductance circuits employed (after Hein et al., ref. 3).

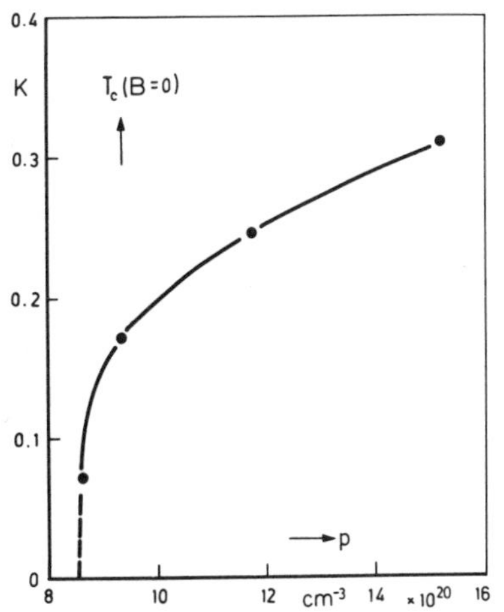

Fig. 15.2 Zero-field transition temperatures as a function of the carrier concentration p for the samples of Fig. 15.1 (p deduced from Hall data) (after Hein et al., ref. 3).

Superconducting transitions of a sample of $SrTiO_3$ with 2.5×10^{19} electrons/cm^3

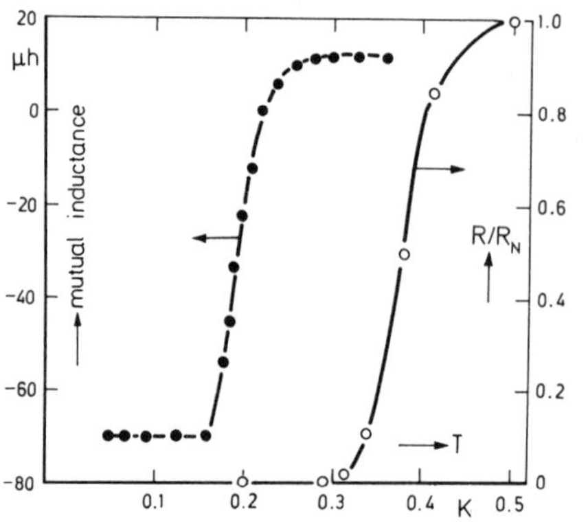

Fig.15.3 Superconducting transition of reduced strontium titanate, $SrTiO_3$, for a carrier concentration of $2.5 \times 10^{19}/cm^3$. Full circles are data reflecting the magnetic susceptibility (left ordinate scale); open circles show the ratio of electrical resistance relative to the normal value (the latter is constant up through the liquid helium temperature range) (after Schooley et al., ref. 8).

are shown in Fig.15.3 [8]. While the full circles indicate data reflecting the magnetic susceptibility, the open circles represent the electrical resistance normalized to unity at 0.5 K. Compared with metallic superconductors the transitions are rather broad which may possibly be due to strain set up during cooling. Fig.15.4 shows the experimental values of T_c as a function of the electron concentration. There is a maximum of T_c at $10^{20}/cm^3$. The curves have been calculated [1, 9] for two values of a fitting parameter; the lower curve seems to fit the data quite well above $10^{20}/cm^3$. The decrease of T_c with n is characteristic for polar semiconductors like $SrTiO_3$. This material is not ferroelectric, in contrast to $BaTiO_3$, but its phonon spectrum has also a low optical branch which has been claimed to induce the superconductive behavior [9] ("soft mode").

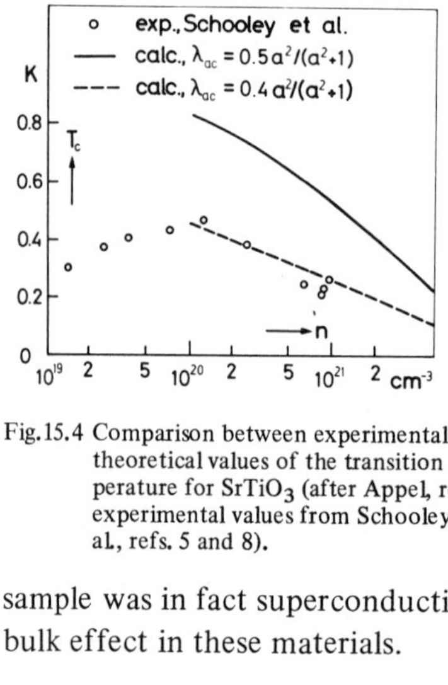

Fig.15.4 Comparison between experimental and theoretical values of the transition temperature for $SrTiO_3$ (after Appel, ref. 9; experimental values from Schooley et al., refs. 5 and 8).

One might suppose that the superconductive behavior were confined to unrepresentative regions of the sample. Fig. 15.5 shows the observed heat capacity divided by T of GeTe plotted vs T^2 [10]. For the case of no superconductivity (e.g. in a strong magnetic field) this plot yields a straight line. The maximum is apparently due to the fact that the major part of the sample was in fact superconducting, i.e. superconductivity appears to be a true bulk effect in these materials.

[8] J. F. Schooley, W. R. Hosler, E. Ambler, J. H. Becker, M. L. Cohen, and C. S. Koonce, Phys. Rev. Lett. 14 (1965) 305.
[9] J. Appel, Phys. Rev. Lett. 17 (1966) 1045.
[10] L. Finegold, Phys. Rev. Lett. 13 (1964) 233.

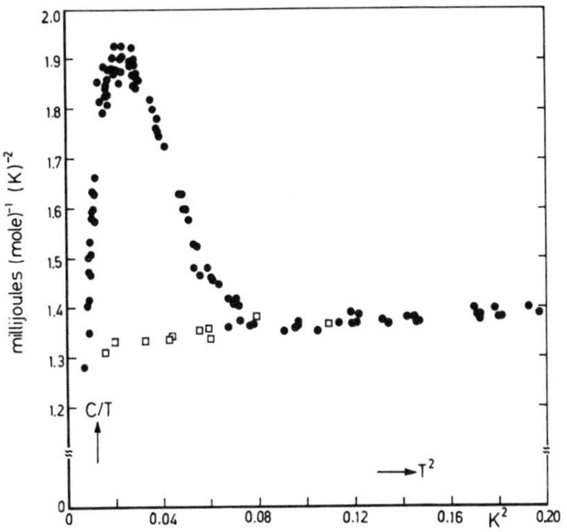

Fig.15.5
Heat capacity of germanium telluride as a function of the square of the temperature, with no magnetic field (circles) and with 500 G applied (squares) (after Finegold, ref. 10).

15b. Liquid, Vitreous, and Amorphous Semiconductors

Upon melting, semiconductors either retain their semiconducting properties (Se, Te, ZnTe, PbTe) or become metallic (Ge, Si, AlSb, GaSb, InAs, InSb) where the metallic character is that of e.g. liquid mercury [1]. Melting is accompanied by abrupt changes in the mass density ρ, the conductivity σ, the thermoelectric power $d\Theta/dT$, and the magnetic susceptibility χ. The changes in mass density are a few % and negative (as in Se, Te, and the tellurides) or positive (as in Ge, Si, and the II-V compounds). In the first group of materials, melting is accompanied by the formation of molecular or chain-like structures, while in the second group a destruction of the covalent bonds in a structure with a closer packing prevails. X-ray diffraction of molten germanium shows that even at 70 K above the melting point there are traces of crystalline structure which disappear when the temperature is increased by about 300 K.

Figs.15.6a-d show ρ, σ, $d\Theta/dT$, and χ as a function of temperature for Ge, and Figs.15.7a and b show ρ and σ for Te. In Ge the metallic behavior above the melting point is characterized by a decrease of the conductivity with temperature, a very small thermoelectric power and a lower value of the total magnetic susceptibility; the latter may be attributed to an increased spin paramagnetism of conduction electrons caused by the increased number of the carriers. Hall effect measurements show that in Ge the electron mobility is decreased upon melting by nearly three order of magnitude to a value of 0.5 $cm^2/Vsec$, while the car-

[1] V. M. Glazov, S. N. Chizhevskaya, and N. N. Glagoleva: Liquid Semiconductors. New York: Plenum. 1969.

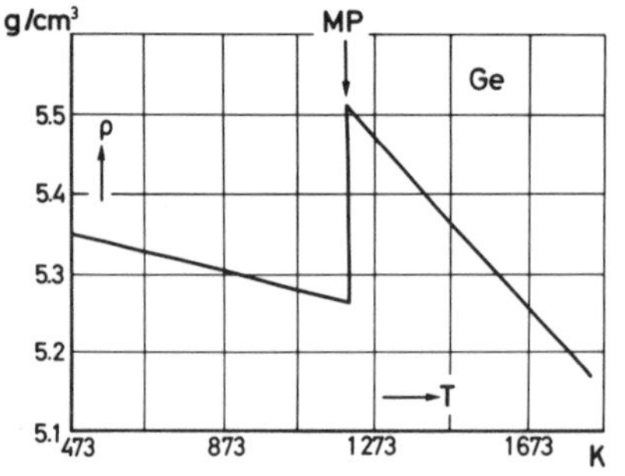

Fig.15.6a Mass density of germanium near the melting point (MP), as a function of temperature.

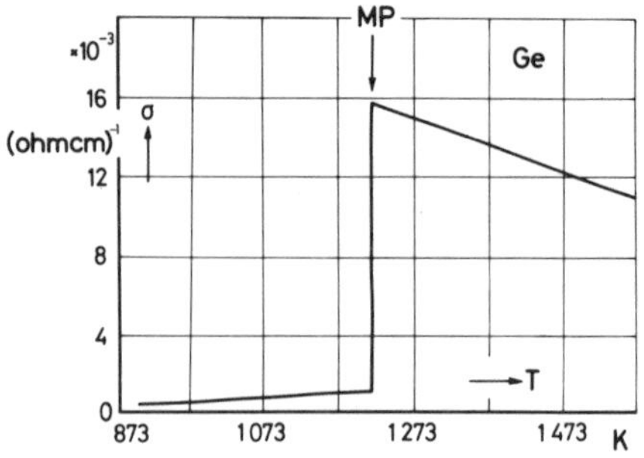

Fig.15.6b Conductivity of germanium near MP.

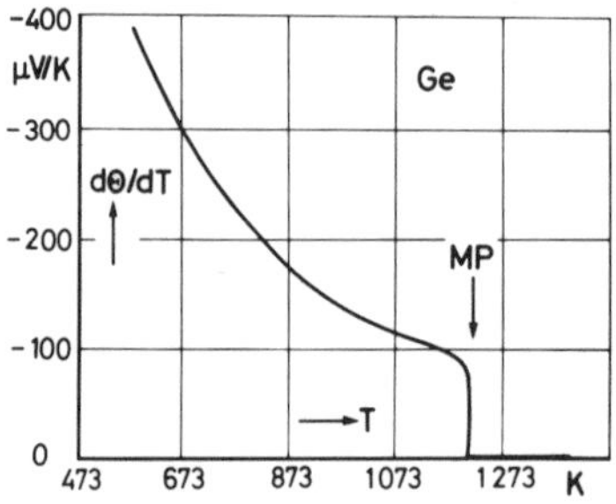

Fig.15.6c Thermoelectric power of germanium near MP.

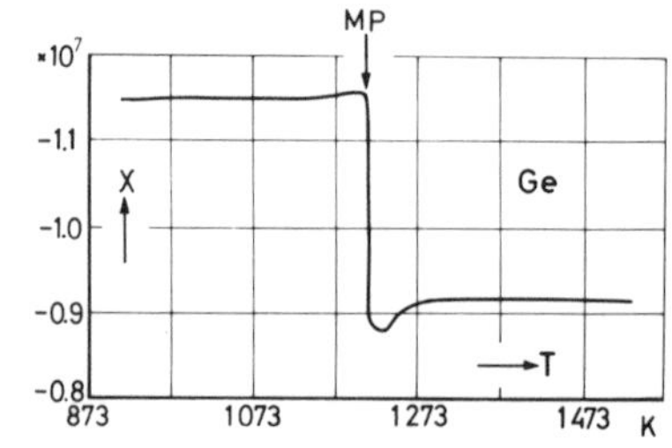

Fig.15.6d Magnetic susceptibility of germanium near MP (after Glazov et al., ref. 1).

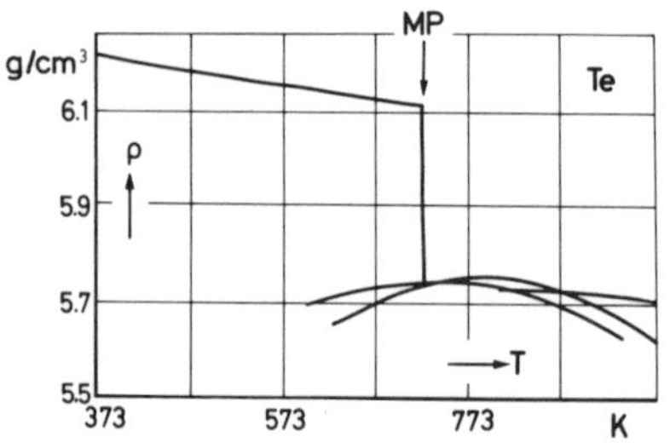

Fig.15.7a Mass density of tellurium near MP.

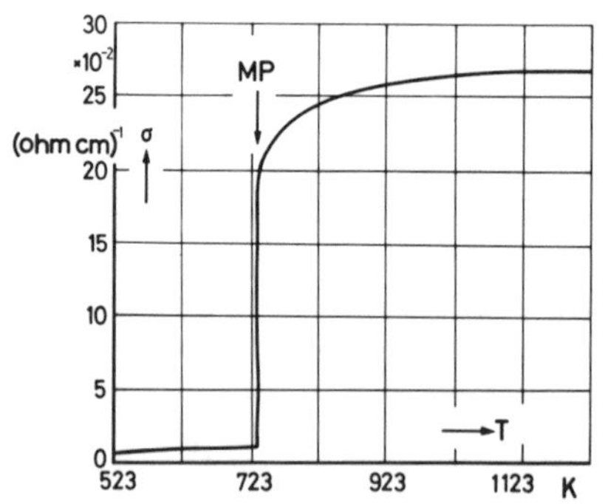

Fig.15.7b Conductivity of tellurium near MP (after Glazov et al., ref. 1).

rier concentration is increased by nearly four orders of magnitude up to a value of $1.2 \times 10^{23}/cm^3$, corresponding to about 3 electrons per atom. The semiconducting behavior of liquid tellurium is indicated by its positive temperature coefficient of conductivity. Hall effect measurements show that the hole concentration is increased by about a factor of 10 upon melting, and since the conductivity increases by about the same amount the mobility is not changed appreciably (30-40 cm^2/Vsec). According to optical measurements the concentration of free carriers in liquid tellurium is only about 0.17 per atom [2].

Selenium, which as tellurium is found in the sixth group of the periodic table and remains semiconducting when liquified, can be obtained in a vitreous (glassy) form when supercooled. Vitreous selenium contains Se_8 rings. In the gradual transition from the vitreous to the liquid state the conductivity remains practically unchanged, and it has been assumed that also the liquid contains such ring-shaped structures.

Vitreous semiconductors gained much interest recently because of their potential technical applications as switches and memory devices. In Chap. 10a the breakdown properties of compensated Ge have been discussed. A characteristic similar to the one shown in Fig. 10.9 has been observed at room temperature in some vitreous semiconductors such as the chalcogenide glasses As_2Se_3, $Ge_{16}As_{35}$, $Te_{28}S_{21}$, $Si_6Te_{24}As_{15}Ge_5$ (known as "STAG") etc. by Ovshinsky [3] and others.

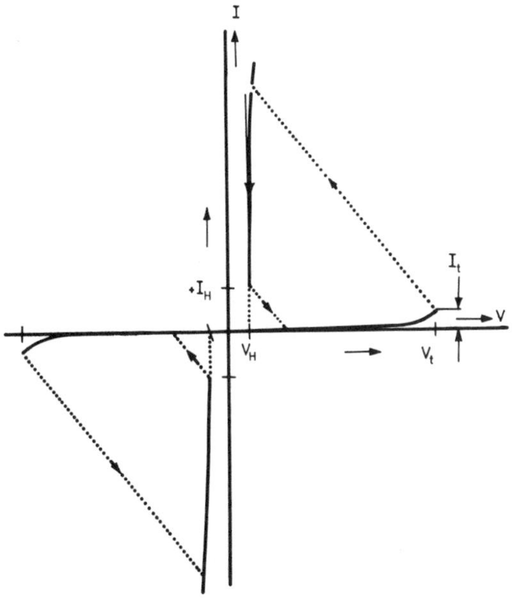

Fig. 15.8
Schematic current-voltage characteristic for a vitreous chalcogenide semiconductor: "threshold switch" (after Ovshinsky, ref. 3, and A. Csillag and H. Jäger, J. Non-Cryst. Solids 2 (1970) 133).

Fig. 15.8 shows a schematic I-V diagram. Initially at voltages below V_t the current is very small. At V_t breakdown occurs within a time of less than a nanosecond along the dotted line which is determined by the ohmic resistor in series with the sample under investigation. The breakdown which is non-destructive is associated with the formation of a current filament as has been discussed in Chap. 10a. The holding voltage V_H is between 1 and 10 V independent of the sample length which suggests that the voltage drop occurs near one or both of

[2] J. N. Hodgson, Phil. Mag. 8 (1963) 735.
[3] S. R. Ovshinsky, Phys. Rev. Lett. 21 (1968) 1450.

the electrodes. When the current is lowered below a value which is denoted by I_H the system switches back into the original state of high resistivity (Fig.15.8, lower dotted line). The characteristic is independent of the polarity of the applied voltage. A device of this kind is called a "threshold switch".

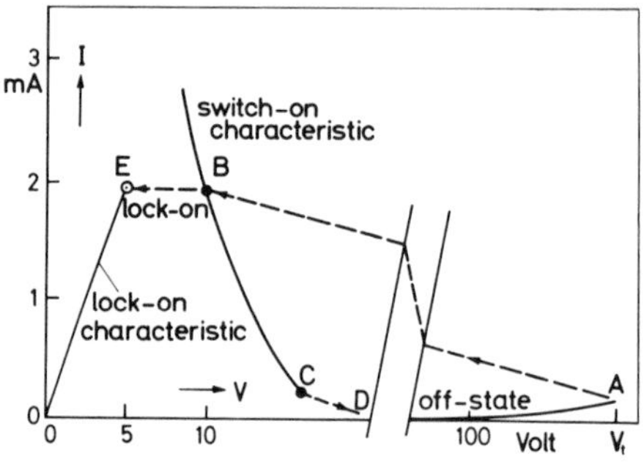

Fig.15.9 "Lock-on" process in amorphous semiconductors: "memory switch" (after Kikuchi and Iizima, ref. 4).

If in certain vitreous compounds the high current is sustained for times of between 10^{-6} and 10^{-1} sec the voltage decreases further from point B to E as shown in Fig. 15.9. During this process, which is called "lock-on" [4], a crystalline filament of 15 μm thickness grows from the anode towards the cathode at a speed of between 10^{-2} and 10^{2} cm/sec depending on the material [5, 6] as shown schematically in Fig.15.10. This phenomenon may be used for a memory device. If at the lock-on state a short current pulse is applied, which is high enough to bring by Joule heat the crystallized channel quickly above the melting temperature, this region is transformed into the original vitreous state of high resistivity as a consequence of the quick cooling effect by its environment; this is the process of "erasing" the memory.

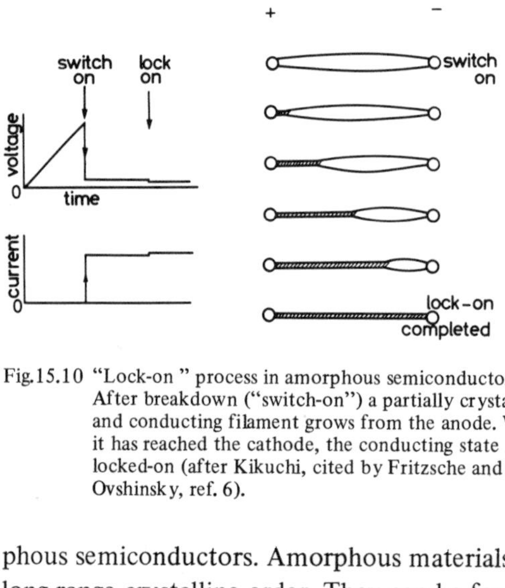

Fig.15.10 "Lock-on" process in amorphous semiconductors: After breakdown ("switch-on") a partially crystallized and conducting filament grows from the anode. When it has reached the cathode, the conducting state is locked-on (after Kikuchi, cited by Fritzsche and Ovshinsky, ref. 6).

The potential technical applications have stimulated research in vitreous and amorphous semiconductors. Amorphous materials just as glasses or liquids have no long-range crystalline order. They can be formed by evaporation, sputtering, electrolytic deposition or chemical vapor deposition. They are different from glasses because upon cooling their melt would crystallize. Both the ideal glassy state and more so the amorphous state are metastable: If the material is held

[4] M. Kikuchi and S. Iizima, Appl. Phys. Lett. 15 (1969) 323.
[5] R. Uttrecht, H. Stevenson, C. H. Sie, J. D. Griener, and K. S. Raghavan, J. Non-Cryst. Solids 2 (1970) 358.
[6] H. Fritzsche and S. R. Ovshinsky, J. Non-Cryst. Solids 4 (1970) 464.

for a while at a high enough temperature, well below the melting point, it will transform into a polycrystalline form, in the case of films with the formation of clusters.

The question is still open, what causes the very fast electric breakdown, and at present one can only speculate about the answer. Various band models for glassy and amorphous semiconductors have been proposed. In view of a short-range crystalline order observed in these materials the band model developed for the ideal crystal has only been slightly modified. In a modification of the Kronig-Penney model N. F. Mott [7] suggested random well depths in a periodic well structure while A. I. Gubanov [8] put forward the idea of a random spacing of wells of a uniform depth. Finally Cohen, Fritzsche, and Ovshinsky [9] develop their model from the band structure of a heavily doped crystalline semiconductor where with increased doping impurity bands in the forbidden gap develop and finally merge with the allowed bands, thus forming "band tails" by the high local electric fields (Chap.9a). The tails of localized states developing from both the conduction band and the valence band may even overlap, giving localized states throughout the "forbidden" gap. The valence band tail states above the Fermi level operate as electron traps while the conduction band states below act as hole traps. The mobility in the band tails is determined by hopping processes (Chap.6n) and therefore much lower than in the bands themselves. The steep mobility changes at the former band edges justifies the substitution of a "mobility gap" for the density-of-states gap in pure crystalline semiconductors.

This picture is corroborated by the temperature dependence of the conductivity and of the thermoelectric power shown in Fig.15.11 for liquid and amorphous As_2Se_3 [10]. The conductivity is represented by a function

$$\sigma = \sigma_0 \exp(-\Delta\epsilon/2k_B T) \qquad (15b.1)$$

where σ_0 and $\Delta\epsilon$ are constants. This law resembles Eq.(4b.59) with n_i given by Eq.(3b.20) valid for intrinsic conduction in a crystalline semiconductor. In contrast to the latter where at low temperatures there is an extrinsic range with a rise of σ with increasing $1/T$, Eq.(15b.1) is valid down to conductivity values where the material must be classified as "insulating". The thermoelectric power

[7] N. F. Mott and E. A. Davis: Electronic Processes in Non-Crystalline Materials. Oxford: Clarendon. 1971.

[8] A. I. Gubanov: Quantum Electron Theory of Amorphous Conductors. New York: Consultants Bureau. 1965.

[9] M. H. Cohen, H. Fritzsche, and S. R. Ovshinsky, Phys. Rev. Lett. 22 (1969) 1065.

[10] R. E. Drews, R. Zallen, R. C. Keezer, Bull. Am. Phys. Soc. 13 (1968) 454; J. T. Edmond, Brit. J. Appl. Phys. 17 (1966) 979; B. T. Kolomiets, T. F. Mazarova, V. P. Shylo: Proc. Int. Conf. Phys. Semic., Exeter 1962, p. 159; data compiled by J. Stuke, J. Non-Cryst. Solids 4 (1970) 1.

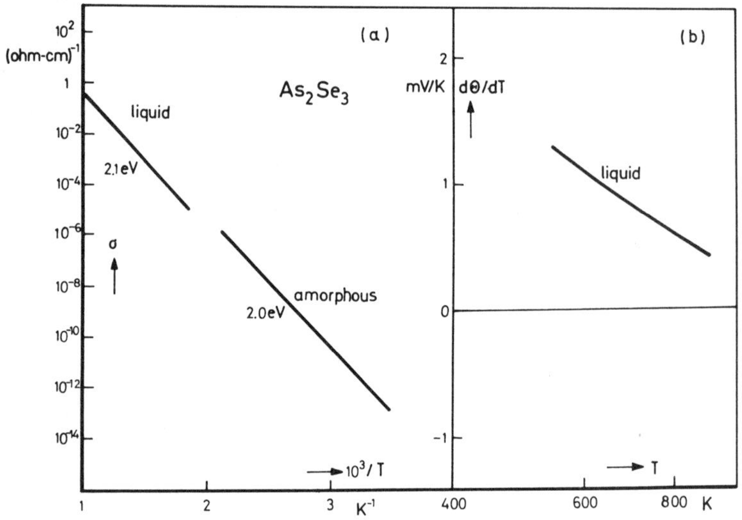

Fig.15.11 a Temperature dependence of the electrical conductivity for liquid and amorphous As_2Se_3.
b Temperature dependence of the thermopower for liquid As_2Se_3 (after Drews et al., ref.10).

of the liquid resembles that valid for intrinsic conduction, with a larger mobility of holes than of electrons since its sign is positive (similar to Fig.4.19). This behavior is rather independent of an addition of impurities to the material and one has no trouble in making non-rectifying contacts, just as would be expected for a material having a high density of localized states, 10^{18} - $10^{19}/cm^3$. Hall effect measurements if at all possible yield a very small Hall mobility. It iz puzzling to find a negative Hall coefficient, even when the thermopower shows that conductivity is due to holes.

An extrapolation of σ to $1/T = 0$ for various semiconductors in the amorphous state is shown in Fig.15.12. All curves lead to approximately the same value of σ_0 which is between 10^3 and 10^5 (ohm-cm)$^{-1}$.

At low temperatures there are deviations from Eq.(15b.1). Fig.15.13 shows the conductivity of amorphous Ge plotted vs $T^{-1/4}$ where a straight line is obtained [11]:

$$\sigma = \sigma_0' \exp\{-(\Delta/T)^{1/4}\} \quad (15b.2)$$

where σ_0' and Δ are constants. This law has been explained in terms of a 3-dimensional hopping model [12]. A 2-dimensional hopping model would yield

$$\sigma = \sigma_0' \exp\{-(\Delta/T)^{1/3}\} \quad (15b.3)$$

as has been observed in some organic semiconductors (Chap.15c).

[11] A. H. Clark, Phys. Rev. 154 (1967) 750.
[12] N. F. Mott, Phil. Mag. 19 (1969) 835.

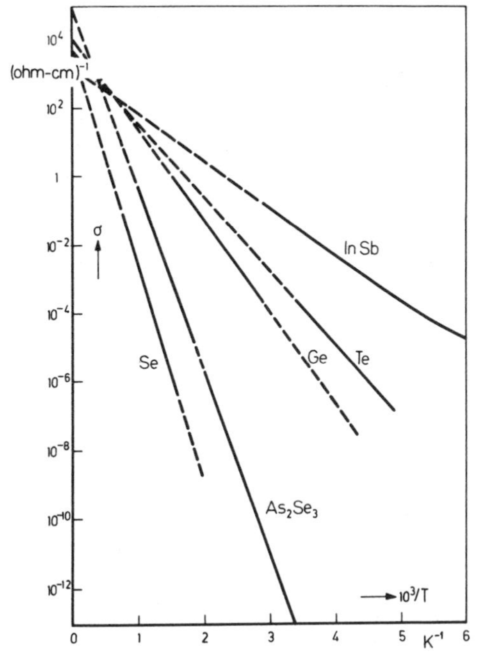

Fig.15.12 Temperature dependence of the electrical conductivity for various amorphous materials and extrapolation to 1/T=0 (after J. Stuke, J. Non-Cryst. Solids 4 (1970) 1).

The frequency dependence of the ac conductivity of amorphous As_2Se_3 at room temperature up to 10 GHz is shown in Fig.15.14. The increase of conductivity with frequency has been considered strong evidence for hopping processes. For hopping processes an increase with frequency $\propto \omega^{0.8}$ has been calculated (Chap.6n) while for a pure crystalline semiconductor a decrease with frequency is found (Chap.4n). An extension of the hopping model to the case of amorphous semiconductors leads to a $\omega\,\{\ln(\nu_p/\omega)\}^4$-dependence instead of the $\omega^{0.8}$dependence mentioned above where ν_p is a phonon frequency [13].

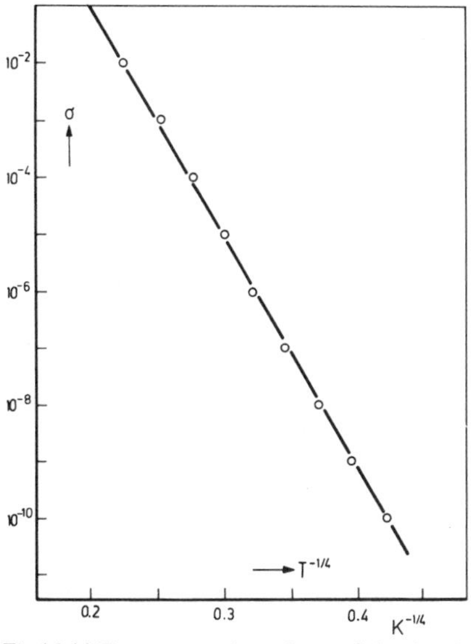

Fig.15.13 Temperature dependence of the electrical conductivity for amorphous germanium at low temperatures (after Clark, ref. 11).

One of the main methods of investigating a band structure is the observation of fundamental absorption. Fig.15.15a shows a comparison of the reflectivities for amorphous and crystalline germanium while in Fig.15.15b the imaginary parts of the dielectric constant are compared. The sharp reflectivity maximum at 4.5 eV is completely washed out in the amorphous state while the 2 eV peak is considerably broadened. Since the 4.5 eV peak is due to a transition on the <100> axis in the Brillouin zone while the nearest neighbor in the diamond lattice is found in the <111> direction, it is concluded that the long-range order is strongly disturbed (<100> and neighboring directions) while

[13] I. G. Austin and N. F. Mott, Advan. Phys. 18 (1969) 14.

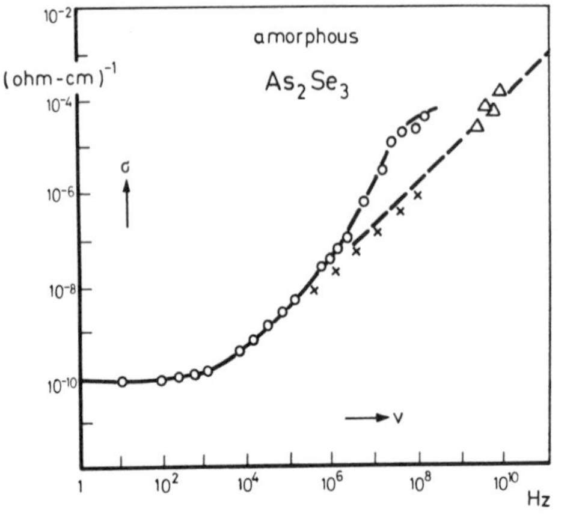

Fig.15.14 Frequency dependence of the ac electrical conductivity for amorphous As_2Se_3 (after A. E. Owen and J. M. Robertson, J. Non-Cryst. Solids 2 (1970) 40; E. B. Ivkin and B. T. Kolomiets, J. Non-Cryst. Solids 3 (1971) 41; and P. C. Taylor, S. C. Bishop, and D. L. Mitchell, Solid State Comm. 8 (1970) 1783).

the short-range order is only weakly disturbed (<111> direction).

An "optical band gap", ϵ_0, can be determined from the fundamental absorption edge. Fig. 15.16 shows for amorphous As_2Te_3 a plot of $(\alpha\hbar\omega)^{1/2}$ vs $\hbar\omega$ which according to Eqs.(11b.12) and (11b.2) yields ϵ_0 [14]. For a parameter the crystal temperature T has been chosen. In the inset ϵ_0 is plotted vs T for a determination of ϵ_0 extrapolated to absolute zero. A value of 0.98 eV is obtained. From the samples a mobility gap, $\Delta\epsilon$, of 0.80 ± 0.04 eV has been obtained which, although somewhat smaller than the extrapolated optical gap, indicates that the band model for crystalline solids requires only slight modifications for an application to amorphous semiconductors.

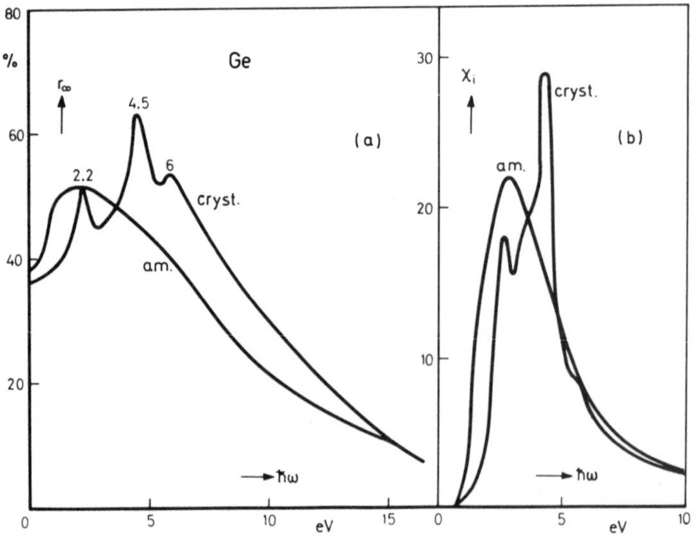

Fig.15.15 (a) Reflectivity and (b) imaginary part of the dielectric constant, $\kappa_i = 2nk$, of crystalline and amorphous Ge (after J. Tauc, A. Abraham, L. Pajasova, R. Grigorovici, and A. Vancu: Proc. Conf. Physics of Non-Crystalline Solids, Delft 1964 (J. A. Prins, ed.), p. 606. Amsterdam: North Holland Publ. Co. 1965).

[14] K. Weiser and M. H. Brodsky, Phys. Rev. B 1 (1970) 791.

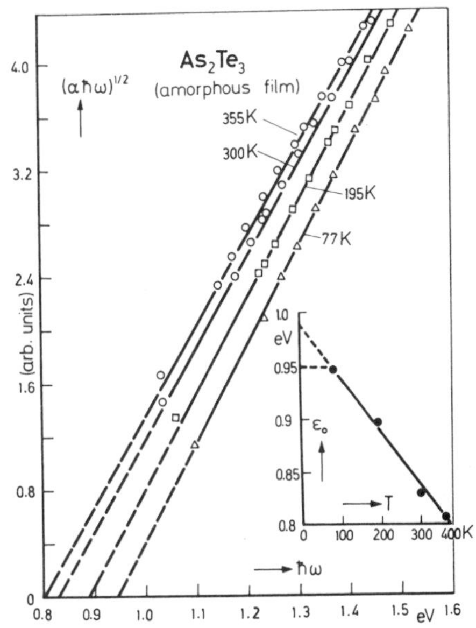

Fig. 15.16 Optical absorption edge for amorphous As_2Te_3 for various temperatures; the inset shows the optical gap so determined as a function of temperature (after K. Weiser and M. H. Brodsky, ref. 14).

A puzzle still to be solved is that the optical absorption is insensitive to the larger number of localized states in the band gap. Fritzsche [15] suggested a model (Fig. 15.17) where irregular potential fluctuations, which are synchronous in the valence and conduction band, do not affect the optical absorption edge; this corresponds to the observations that a p-n junction in a crystal is not noticeable optically, except for secondary effects due to the built-in potential which are extremely small. The band edges are defined here as local averages over regions of the order of the coherence lengths of the material waves of the carriers which take part in optical transitions. The band tails extend through the middle of the gap as shown on the left-hand side of Fig. 15.17. The hatched regions are then "localized states" in a sense that e.g. an electron in a potential well only after supplementing an "activation energy" will be able to contribute to charge trans-

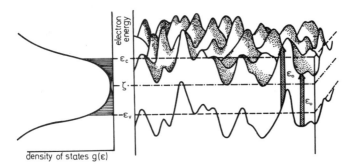

Fig. 15.17 Potential fluctuations of the initial and final electron states assumed for optical transitions, corresponding to an "optical gap" ϵ_0. The left-hand side shows the density of states. The region of localized states lies between the band edges, ϵ_c and ϵ_v. The short range potential wells which give rise to many of the localized states are not shown here. Only that part of the long wave-length potential fluctuations is shown which causes a parallel shift of the valence and conduction band states. The part which causes a spatial variation of ϵ_0 is omitted for clarity (after Fritzsche, ref. 15).

[15] H. Fritzsche, J. Non-Cryst. Solids 6 (1971) 49.

port in a conduction experiment, as is the case for a trapped electron in a crystalline semiconductor. It is in this sense that one may call $\Delta\epsilon$ of Eq.(15b.1) a "mobility gap" which does not have to be identical with the "optical gap" ϵ_o. Obviously, strong electric fields should exist between electrons and holes trapped in adjacent potential extrema. These fields should cause tails in the optical absorption edge due to the Franz-Keldysh effect (Chap.11e). However, in Fig.15.16 there is no indication for band tails in the optical absorption.

An interesting point concerning the electrical properties of amorphous semiconductors has been made by van Roosbroeck and coworkers [16]. Due to the large number of recombination centers the lifetime of injected carriers, τ, is very small. On the other hand, the resistivity is very large, with the consequence of an unusually long dielectric relaxation time, $\tau_d \gg \tau$. This is called the "relaxation case", in contrast to the usual "lifetime case" of low resistivity crystalline semiconductors. By considering Poisson's equation, the continuity equations, and the usual definitions of current densities (including diffusion terms) and recombination rate, the unusual properties of a "relaxation case semiconductor" have been investigated. Injection causes majority carrier depletion and therefore maximum resistivity [17, 18]. Independent of doping, the material appears to be intrinsic at all temperatures. The effect has been clearly demonstrated in single-crystal GaAs, doped with a deep-level impurity (oxygen) and therefore having an electron concentration of only $3 \times 10^7/cm^3$ at room temperature, although the mobility was quite high, $4500\ cm^2/Vsec$. By Zn diffusion at one end a p^+-n

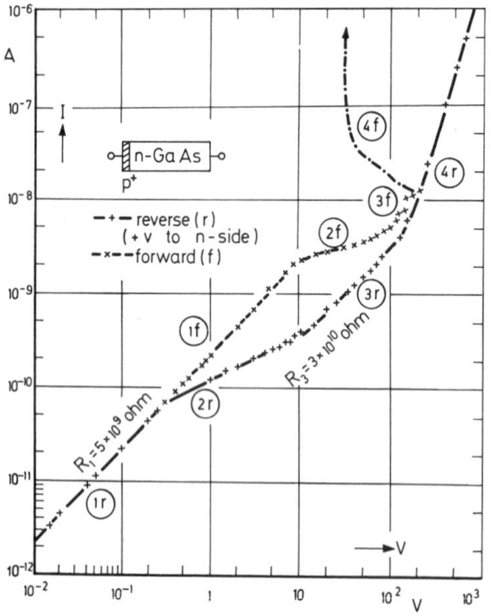

Fig.15.18 Current-voltage characteristics (forward and reverse) of a relaxation-case semiconductor: GaAs p^+-n junction at room temperature in darkness (after Queisser et al., ref. 18).

[16] For a review see H. J. Queisser, European Solid State Device Res. Conf., Lancaster 1972 (P. N. Robson, ed.). London: The Institute of Physics and the Physical Society. 1973.

[17] W. van Roosbroeck, Phys. Rev. 123 (1961) 474; W. van Roosbroeck and H. C. Casey, Jr., Phys. Rev. B 5 (1972) 2154.

[18] H. J. Queisser, H. C. Casey, Jr., and W. van Roosbroeck, Phys. Rev. Lett. 26 (1971) 551.

junction was produced. The dielectric relaxation time was 10^{-4} sec which is large compared with a lifetime of less than 10^{-8} sec. Forward and reverse characteristics are shown in Fig.15.18.

The reverse bias region "1r" is linear, the resistance being about 3 times larger than expected from the n-region alone. The main resistance contribution comes from the space charge region (≈ 70 μm) next to the p-n junction. In this region the resistivity is a maximum

$$\rho_{max} = (\mu_n/\mu_p)^{1/2}/2en_i\mu_n \qquad (15b.4)$$

where $n_i = 9\times10^5/cm^3$ is the intrinsic carrier concentration and the mobility ratio $\mu_n/\mu_p \approx 20$. At a higher bias the space charge region widens through a majority carrier depletion, the current being proportional to the square root of the bias. When the entire sample is depleted this region ends and the resistance of the total device is determined by ρ_{max}. At a still higher bias a space-charge-limited current yields a superlinear behavior.

The forward resistance equals the low-field reverse resistance up to a region where injected holes deplete the n region of the more mobile electrons by recombination after having traversed the space charge region. Finally (region 4f) double injection occurs with the formation of a negative differential resistance region.

This experiment clearly shows that at moderate field strengths the resistance of a relaxation case semiconductor is dominated by the intrinsic carrier density in a space charge region which may be extended through the whole sample, independent of doping. The temperature dependence of the resistance is then exponential and given by the band gap while the optical behavior is not influenced by the space charge near the fundamental edge.

The typical behavior of a relaxation case semiconductor is not only shown by amorphous semiconductors but also by almost all organic semiconductors to be treated in the following chapter.

15c. Organic Semiconductors

Interest in electronic conduction in organic materials arose both from the sensitization of photographic emulsions by organic dyestuffs and from the suggestion that it might play an important role in the fundamental physical processes of living organisms [1].

Aromatic hydrocarbons such as naphthalene, anthracene, and tetracene (phenacene) are the best-known organic semiconductors. They are derivatives of benzene. The structural diagrams are shown in Fig.15.19. The "double bonds" indi-

---

[1] A. Szent-Györgyi, Nature 157 (1946) 875; for a review of early work on organic semiconductors see C. G. B. Garrett: Semiconductors (N. B. Hannay, ed.), chap. 15. New York: Reinhold Publ. Co. 1959.

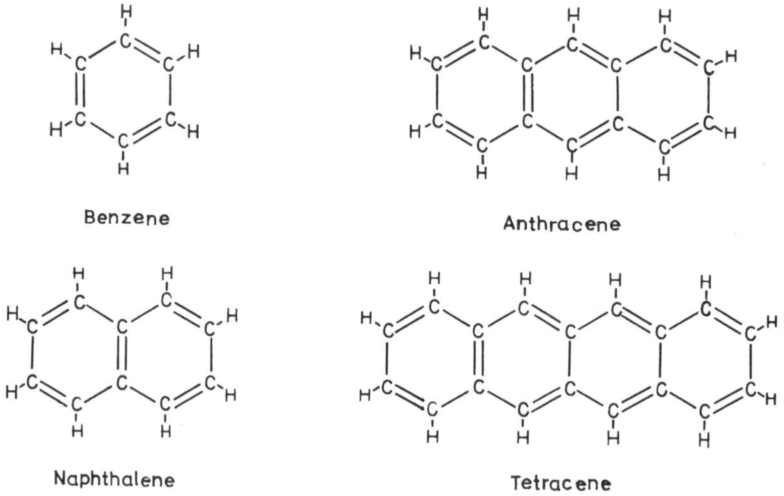

Fig.15.19 Molecular diagrams of four conjugated aromatic hydrocarbons. A single bond represents a pair of electrons, a double bond two pairs.

cated in the diagrams in the usual manner are actually single bonds; an equivalent number of "π electrons" is present in cloud-like regions above and below the central plane of the molecule. The conduction of an electric current from one molecule to the next is assumed to involve only π electrons; this is considered the "slow step" in the conduction process.

In single crystals of naphthalene, anthracene, and many others, semiconductivity has been investigated. These crystals are monoclinic (space group C_{2h}^5). For a typical example, the structure of anthracene is shown in Fig.15.20 [2]. The two axes a and b are at right angles to each other, but the third axis c is at an oblique angle. In the ab plane which is the cleavage plane the conductivity in anthracene is isotropic, while the conductivity perpendicular to this plane is lower by at least an order of magnitude.

In many respects there is a striking similarity between amorphous inorganic and crystalline organic semiconductors. First of all the temperature dependence of the conductivity of the latter is also represented by Eq.(15b.1). The "mobility gap" $\Delta\epsilon$ ranges from 0.2 eV (in powdered cyananthrone [3]) up to more than 3 eV (in anthracene single crystals [4]). Ohm's law is valid up to fields of about 10^3 V/cm. At higher fields there is a superlinear region; at about 10^6 V/cm a non-destructive reversible breakdown occurs. At frequencies between 10^5 and 10^7 Hz an ac conductivity has been observed which is of the same order of mag-

[2] M. Pope, Sci. Am. 216 (1967) No.1, p. 86.
[3] H. Inokuchi, Bull. Chem. Soc. Japan 25 (1952) 28.
[4] H. Inokuchi, Bull. Chem. Soc. Japan 29 (1956) 131.

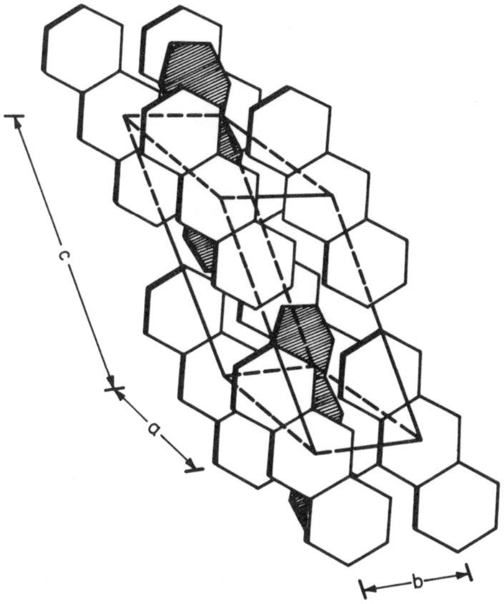

Fig.15.20 Crystal structure of anthracene. The monoclinic unit cell is indicated (after Pope, ref. 2).

nitude as the dc conductivity and whose temperature dependence is similar to that of the dc conductivity, although the dielectric relaxation time may be as high as 100 sec (e.g. in phthalocyanine). Hall effect measurements present problems; one may assume that the Hall mobility is very small. A very small Seebeck coefficient, 5×10^{-5} V/K, has been reported for phthalocyanine [4]; the sign is in agreement with a conduction by holes. In the same material point-contact rectification was observed in a direction which indicates that holes are the majority carriers. Otherwise, as in amorphous semiconductors, there are no contact problems of the very small conductivities that were observed: Metal foils fused onto the surface, silver paste, or evaporated metal electrodes have been used. Sometimes guard rings were applied in order to distinguish between bulk and surface conduction. The role of impurities (of concentration 0.1 %) seems to be only to introduce a second activation energy $\Delta\epsilon'$, which is characteristic for the impurity, such that

$$\sigma = \sigma_o \exp(-\Delta\epsilon/2k_BT) + \sigma_o' \exp(-\Delta\epsilon'/2k_BT) \qquad (15c.1)$$

However, $\Delta\epsilon'$ is only about 3/4 of its value for the impurity material in a pure state.

The optical absorption edge and the threshold for photoconduction which are generally not very different do not seem to be related to the energy $\Delta\epsilon$ which determines the conductivity. Photoconduction depends both on surface preparation and on the ambient. Energy transport by excitons seems to be important in photoconduction experiments. Although the lifetime of a singlet exciton is only about 10^{-8} sec, it moves fast enough to pass 10^4 molecules during this time. For anthracene and tetracene, singlet exciton energies of 3.15 and 2.4 eV, respectively, have been reported [5], while the triplet exciton energies are only 1.8 and 1.4 eV, respectively.

---

[5] P. E. Fielding and F. Gutman, J. Chem. Phys. 26 (1957) 411.

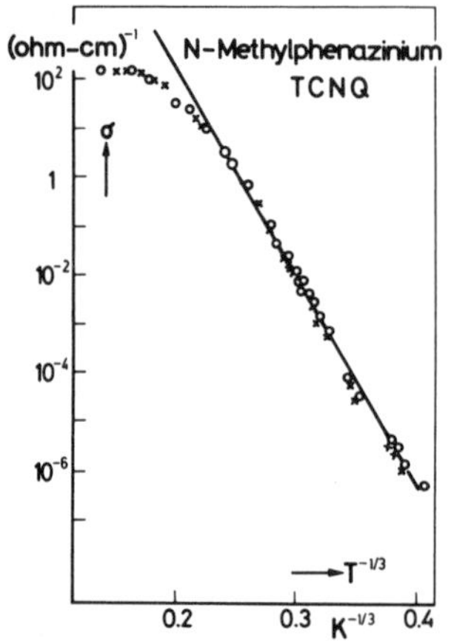

Fig. 15.21 Electrical conductivity for NMP-TCNQ plotted vs $1/T^{1/3}$ (after Brenig et al., ref. 6).

Just as in amorphous semiconductors, hopping processes have been discussed in attempts to explain organic semiconduction. In contrast to amorphous materials, however, many organic semiconductors have a layered structure and hopping in two dimensions as described by Eq. (15b.3) may be expected. Fig. 15.21 shows the conductivity of a single crystal of N-Methylphenazinium Tetracyanoquinodimethan ("NMP-TCNQ") plotted on a log scale vs $T^{-1/3}$ [6]. Over most of the region a straight line is followed indicating the validity of Eq. (15b.3). One of the TCNQ salts has been chosen because these organic materials are known for their high electronic conductivities.

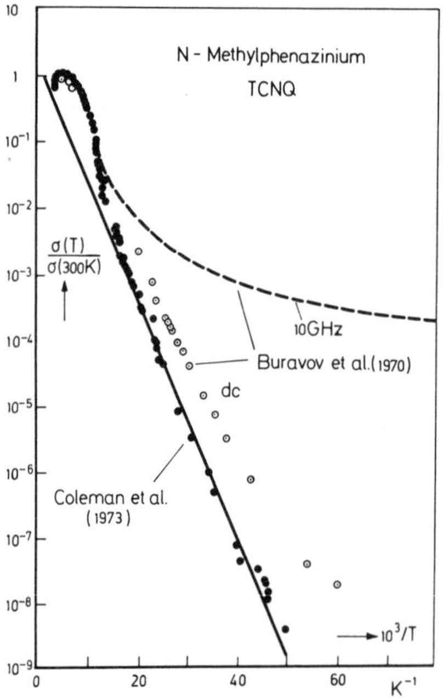

Fig. 15.22 Temperature dependence of the electrical conductivity for NMP-TCNQ, normalized to its value at room temperature, 380 (ohm-cm)$^{-1}$ (after Coleman et al., ref. 7); the dashed curve: data at 10 GHz (after Burovov et al., ref. 8).

As for amorphous semiconductors an increase of the ac conductivity with frequency is a strong argument for hopping processes. In Fig. 15.22 the dc conductivity [7] is compared with the microwave conductivity at 10 GHz [8]. At room temperature the two conductivities practically coincide. At the lowest temperature (≈ 10 K) the microwave conductivity is higher by several orders of magnitude. An interesting observation is the high dielectric constant of 350 for the present complex and of 800 for Acridinium TCNQ at 4.2 K and 10 GHz. As in ferroelectric materials the existence of a soft mode in the phonon spectrum is possibly correlated with the high dielectric constant.

In another one of the TCNQ complexes called TTF-TCNQ (Dimethyltetrathiofulvalen-TCNQ) metallic

[6] W. Brenig, G. H. Döhler, and H. Heyszenau, Phys. Letters 39A (1972) 175.
[7] L. Coleman, J. A. Cohen, A. F. Garito, and A. J. Heeger, Phys. Rev. B7 (1970) 2122.
[8] L. I. Buravov, M. L. Khidekel', I. F. Shchegolev, and E. B. Yagubskii, JETP Pisma Red. 12 (1970) 142 [Engl.: JETP Letters 12 (1970) 99].

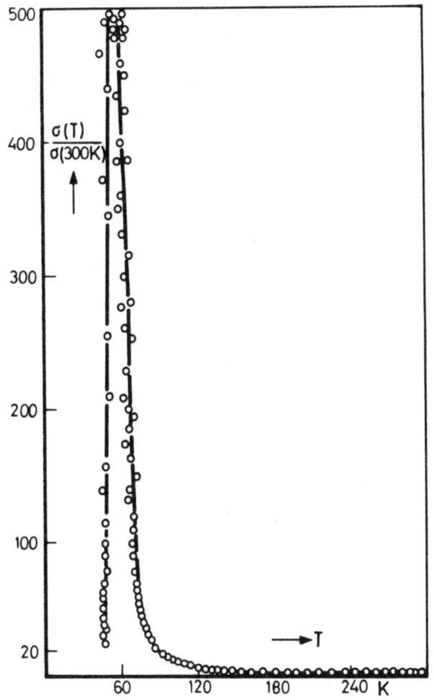

Fig.15.23 Temperature dependence of the electrical conductivity for TTF-TCNQ (after Heeger et al., ref. 9).

conductivity ($\sigma > 10^6\,(\text{ohm-cm})^{-1}$) has been observed at 58 K. Because of experimental problems only a lower limit for σ has been found so far. The room temperature conductivity, σ (300 K), is already extremely high for an organic material, 1837 $(\text{ohm-cm})^{-1}$. In a narrow range of temperatures assumed to be associated with a phase transition, there is a steep increase in conductivity shown in Fig.15.23 [9]. The data have been interpreted by superconducting fluctuations associated with a tendency towards high-temperature superconductivity. As pointed out by J. Bardeen, H. Fröhlich suggested in 1954 that a lattice instability in a one-dimensional crystalline system might show superconducting behavior [10], and there were theoretical considerations for ferroelectric phase transitions showing a "soft phonon mode" in correlation with an observed superconductivity (Chap.15a).

[9] L. B. Coleman, M. J. Cohen, D. J. Sandman, F. G. Yamagishi, A. F. Garito, and A. J. Heeger, Solid State Comm. 12 (1973) 1125.

[10] H. Fröhlich, Proc. Roy. Soc. London, Ser. A 223 (1954) 296.
A brief account of this work has been given by J. Bardeen: Handbuch der Physik (S. Flügge, ed.), Vol. 15, p. 366/7. Berlin-Göttingen-Heidelberg: Springer. 1956.

Appendix A

The Tensors of Electrical and Thermal Conductivities

For the case of the simple model of band structure the carrier contributions to the electrical and thermal current densities, $\vec{j}$ and $\vec{w}$, in an electric field of intensity $\vec{E}$, a magnetic induction of intensity $\vec{B} = (B_x, B_y, B_z)$ and a temperature gradient $\vec{\nabla}_r T$ are given by

$$\vec{j} = \begin{pmatrix} \sigma - \beta B_x^2 & \gamma B_z - \beta B_x B_y & -\gamma B_y - \beta B_x B_z \\ -\gamma B_z - \beta B_x B_y & \sigma - \beta B_y^2 & \gamma B_x - \beta B_y B_z \\ \gamma B_y - \beta B_x B_z & -\gamma B_x - \beta B_y B_z & \sigma - \beta B_z^2 \end{pmatrix} (E + \vec{\nabla}\zeta/|e|)$$

$$+ \begin{pmatrix} \sigma' - \beta' B_x^2 & \gamma' B_z - \beta' B_x B_y & -\gamma' B_y - \beta' B_x B_z \\ -\gamma' B_z - \beta' B_x B_y & \sigma' - \beta' B_y^2 & \gamma' B_x - \beta' B_y B_z \\ \gamma' B_y - \beta' B_x B_z & -\gamma' B_x - \beta' B_y B_z & \sigma' - \beta' B_z^2 \end{pmatrix} (\vec{\nabla} T)/T \qquad \text{(A.1)}$$

and similarly one has the same expression for $-\vec{w}$ except that σ, β, and γ are replaced by σ', β', and γ' while σ', β', and γ' are replaced by σ'', β'', and γ'' :

$$-\vec{w} = \left(\sigma' - \beta' B_x^2 \cdots\right) (\vec{E} + \vec{\nabla}\zeta/|e|) + \left(\sigma'' - \beta'' B_x^2 \cdots\right) (\vec{\nabla} T)/T \qquad \text{(A.2)}$$

The coefficients σ, β, γ, σ', . . . have the following meanings :

$$\sigma = e^2 \{(n/m_n) \langle Z\tau_m \rangle_n + (p/m_p) \langle Z\tau_m \rangle_p\} \qquad \text{(A.3)}$$

$$\sigma' = -|e| \{(n/m_n) \langle Z\tau_m (\epsilon - \zeta) \rangle_n - (p/m_p) \langle Z\tau_m (\epsilon - \zeta) \rangle_p\} \qquad \text{(A.4)}$$

$$\sigma'' = (n/m_n) \langle Z\tau_m (\epsilon - \zeta)^2 \rangle_n + (p/m_p) \langle Z\tau_m (\epsilon - \zeta)^2 \rangle_p \qquad \text{(A.5)}$$

$$\gamma = -|e|^3 \{(n/m_n^2) \langle Z\tau_m^2 \rangle_n - (p/m_p^2) \langle Z\tau_m^2 \rangle_p\} \qquad \text{(A.6)}$$

$$\gamma' = e^2 \{(n/m_n^2) \langle Z\tau_m^2 (\epsilon - \zeta) \rangle_n + (p/m_p^2) \langle Z\tau_m^2 (\epsilon - \zeta) \rangle_p\} \qquad \text{(A.7)}$$

$$\gamma'' = -|e| \{(n/m_n^2) \langle Z\tau_m^2 (\epsilon - \zeta)^2 \rangle_n - (p/m_p^2) \langle Z\tau_m^2 (\epsilon - \zeta)^2 \rangle_p\} \qquad \text{(A.8)}$$

$$\beta = -e^4 \{(n/m_n^3) \langle Z\tau_m^3 \rangle_n + (p/m_p^3) \langle Z\tau_m^3 \rangle_p\} \qquad \text{(A.9)}$$

$$\beta' = |e|^3 \{(n/m_n^3) <Z\tau_m^3 (\epsilon - \zeta)>_n - (p/m_p^3) <Z\tau_m^3 (\epsilon - \zeta)>_p\} \quad (A.10)$$

$$\beta'' = -e^2 \{(n/m_n^3) <Z\tau_m^3 (\epsilon - \zeta)^2>_n + (p/m_p^3) <Z\tau_m^3 (\epsilon - \zeta)^2>_p\} \quad (A.11)$$

where Z is given by Eq.(4e.3) :

$$Z = (1 + \tau_m^2 e^2 B^2 / m^2)^{-1} \quad (A.12)$$

One should remember that the momentum relaxation time τ_m may depend on the carrier energy ϵ. The average of an arbitrary function $h(\epsilon)$ is given by

$$<h(\epsilon)> = \frac{-\frac{2}{3}\int_0^\infty (\epsilon/k_BT)^{3/2} h(\epsilon) \partial f_0 / \partial(\epsilon/k_BT)\, d(\epsilon/k_BT)}{\int_0^\infty (\epsilon/k_BT)^{1/2} f_0\, d(\epsilon/k_BT)} \quad (A.13)$$

where f_0 is the equilibrium distribution function of the carriers. In the following approximations we assume a Maxwellian distribution function. $<\tau_m>$ is then given by Eq.(4c.41). We also assume $\tau_m(\epsilon) \propto \epsilon^r$ where r is an arbitrary constant. Besides the conductivities for electrons and holes,

$$\sigma_n = (ne^2/m_n) <\tau_m>_n ; \quad \sigma_p = (pe^2/m_p) <\tau_m>_p \quad (A.14)$$

we introduce the well-known Hall mobilities

$$\mu_{Hn} = (|e|/m_n) (<\tau_m^2> / <\tau_m>)_n ; \quad \mu_{Hp} = (|e|/m_p) (<\tau_m^2> / <\tau_m>)_p \quad (A.15)$$

and a new kind of mobility occurring in magnetoresistance phenomena :

$$\mu_{Mn}^2 = (e/m_n)^2 (<\tau_m^3> / <\tau_m>)_n ; \quad \mu_{Mp}^2 = (e/m_p)^2 (<\tau_m^3> / <\tau_m>)_p \quad (A.16)$$

In the approximation valid for a weak magnetic field where we neglect in $\vec{j}$ and $\vec{w}$ powers of $\vec{B}$ higher than the second, we obtain for the coefficients $\sigma, \beta, \gamma, \ldots$

$$\sigma = \sigma_n + \sigma_p - (\sigma_n \mu_{Mn}^2 + \sigma_p \mu_{Mp}^2) B^2 \quad (A.17)$$

$$\sigma' = -\frac{1}{|e|} \{(\sigma_n - \sigma_p) k_BT(r + 5/2) - (\sigma_n \zeta_n - \sigma_p \zeta_p) - (\sigma_n \mu_{Mn}^2 - \sigma_p \mu_{Mp}^2) k_BT (3r + 5/2) B^2 + (\sigma_n \mu_{Mn}^2 \zeta_n - \sigma_p \mu_{Mp}^2 \zeta_p) B^2\} \quad (A.18)$$

$$\sigma'' = \frac{1}{e^2} \{(\sigma_n + \sigma_p)(k_BT)^2 (r + 7/2)(r + 5/2) - (\sigma_n \zeta_n + \sigma_p \zeta_p) k_BT (2r + 5) + \sigma_n \zeta_n^2 + \sigma_p \zeta_p^2 - (\sigma_n \mu_{Mn}^2 + \sigma_p \mu_{Mp}^2)(k_BT)^2 (3r + 7/2)(3r + 5/2) B^2 + (\sigma_n \mu_{Mn}^2 \zeta_n + \sigma_p \mu_{Mp}^2 \zeta_p) k_BT (6r + 5) B^2 - (\sigma_n \mu_{Mn}^2 \zeta_n^2 + \sigma_p \mu_{Mp}^2 \zeta_p^2) B^2\} \quad (A.19)$$

$$\gamma = -\sigma_n \mu_{Hn} + \sigma_p \mu_{Hp} \quad (A.20)$$

$$\gamma' = \frac{1}{|e|} \{(\sigma_n \mu_{Hn} + \sigma_p \mu_{Hp}) k_BT (2r + 5/2) - (\sigma_n \mu_{Hn} \zeta_n + \sigma_p \mu_{Hp} \zeta_p)\} \quad (A.21)$$

$$\gamma'' = -\frac{1}{e^2} \{(\sigma_n \mu_{Hn} - \sigma_p \mu_{Hp})(k_BT)^2 (2r + 7/2)(2r + 5/2) - (\sigma_n \mu_{Hn} \zeta_n - \sigma_p \mu_{Hp} \zeta_p) k_BT \ (4r + 5) + \sigma_n \mu_{Hn} \zeta_n^2 - \sigma_p \mu_{Hp} \zeta_p^2\} \quad (A.22)$$

$$\beta = -(\sigma_n \mu_{Mn}^2 + \sigma_p \mu_{Mp}^2) \quad (A.23)$$

$$\beta' = \frac{1}{|e|} \{(\sigma_n \mu_{Mn}^2 - \sigma_p \mu_{Mp}^2) k_B T (3r + 5/2) - (\sigma_n \mu_{Mn}^2 \zeta_n - \sigma_p \mu_{Mp}^2 \zeta_p)\} \quad (A.24)$$

$$\beta'' = -\frac{1}{e^2} \{(\sigma_n \mu_{Mn}^2 + \sigma_p \mu_{Mp}^2)(k_B T)^2 (3r + 7/2)(3r + 5/2) - (\sigma_n \mu_{Mn}^2 \zeta_n + \sigma_p \mu_{Mp}^2 \zeta_p) k_B T (6r + 5) + \sigma_n \mu_{Mn}^2 \zeta_n^2 + \sigma_p \mu_{Mp}^2 \zeta_p^2\} \quad (A.25)$$

Quite often the Fermi energies, ζ_n and ζ_p, can be combined to give the negative band gap energy, $-\epsilon_G$:

$$\zeta_n + \zeta_p = -\epsilon_G \quad (A.26)$$

For $\vec{B} = 0$ the terms concerning electrons in σ, β, and γ, are denoted as σ_o, β_o, and γ_o, respectively, in Eqs.(4c.40), (4c.51), and (4c.43). In the approximation valid for a strong magnetic field, the coefficients are given by

$$\sigma = B^{-2} (\sigma_n / \mu_{Hn}^2 + \sigma_p / \mu_{Hp}^2) \{(2r + 3/2)!\}^2 (3/2\text{-}r)! / \{(3/2 + r)!\}^3 \quad (A.27)$$

$$\sigma' = -\frac{B^{-2}}{|e|} \{(\sigma_n / \mu_{Hn}^2 - \sigma_p / \mu_{Hp}^2) k_B T (5/2 - r)! - (\sigma_n \zeta_n / \mu_{Hn}^2 - \sigma_p \zeta_p / \mu_{Hp}^2)(3/2 - r)!\} \cdot \{(3/2 + 2r)!\}^2 / \{(3/2 + r)!\}^3 \quad (A.28)$$

$$\sigma'' = \frac{B^{-2}}{e^2} \{(\sigma_n / \mu_{Hn}^2 + \sigma_p / \mu_{Hp}^2)(k_B T)^2 (7/2 - r)! - 2(\sigma_n \zeta_n / \mu_{Hn}^2 + \sigma_p \zeta_p / \mu_{Hp}^2) k_B T (5/2 - r)!\} \cdot \{(3/2 + 2r)!\}^2 / \{(3/2 + r)!\}^3 \quad (A.29)$$

$$\gamma = -|e| B^{-2} (n - p) \quad (A.30)$$

$$\gamma' = B^{-2} \{(n + p) k_B T \cdot 5/2 - (n\zeta_n + p\zeta_p)\} \quad (A.31)$$

$$\gamma'' = -\frac{B^{-2}}{|e|} \{(n - p)(k_B T)^2 \, 35/4 - (n\zeta_n - p\zeta_p) \, 5 k_B T + n\zeta_n^2 - p\zeta_p^2\} \quad (A.32)$$

$$\beta = -B^{-2} (\sigma_n + \sigma_p) \quad (A.33)$$

$$\beta' = \frac{B^{-2}}{|e|} \{(\sigma_n - \sigma_p) k_B T (r + 5/2) - (\sigma_n \zeta_n - \sigma_p \zeta_p)\} \quad (A.34)$$

$$\beta'' = -\frac{B^{-2}}{e^2} \{(\sigma_n + \sigma_p)(k_B T)^2 (r + 7/2)(r + 5/2) - (\sigma_n \zeta_n + \sigma_p \zeta_p) k_B T (2r + 5) + \sigma_n \zeta_n^2 + \sigma_p \zeta_p^2\} \quad (A.35)$$

Appendix B

Modified Bessel Functions

In calculations of averages over the Maxwell-Boltzmann distribution function the result frequently is one of the "modified Bessel functions" $K_n(t)$, n = 0, 1, or 2 [1, 2]. These are solutions, vanishing for $t \to \infty$, of the following differential equation:

$$t^2 d^2 K_n/dt^2 + t dK_n/dt - (t^2 + n^2) K_n = 0 \,; \quad n = 0, 1, 2, \cdots \tag{B.1}$$

For a negative value of the argument t, we take $K_n(|t|)$ throughout this book.

The recursion formula

$$K_{n+1}(t) = (2n/t) K_n(t) + K_{n-1}(t) \,; \quad t > 0 \tag{B.2}$$

is easily verified. Hence, we might limit our considerations to $K_0(t)$ and $K_1(t)$. However, $K_2(t)$ is also included because of its frequent occurrence.

The following integral representations of the functions are important (ref. 1, p. 181.) valid for $t > 0$:

$$K_0(t) = \int_1^\infty \frac{e^{-t\xi}}{\sqrt{\xi^2 - 1}} d\xi = e^{-t} \int_0^\infty \frac{e^{-\xi} d\xi}{\sqrt{\xi(\xi + 2t)}} = \frac{1}{2} \int_0^\infty \frac{1}{\xi} \exp\{-\frac{1}{2}(\xi + t^2/\xi)\} d\xi \tag{B.3}$$

$$K_1(t) = t \int_1^\infty \sqrt{\xi^2 - 1}\, e^{-t\xi} d\xi = \frac{e^{-t}}{t} \int_0^\infty \sqrt{\xi(\xi + 2t)}\, e^{-\xi} d\xi = \frac{1}{2t} \int_0^\infty \exp\{-\frac{1}{2}(\xi + t^2/\xi)\} d\xi \tag{B.4}$$

$$K_2(t) = \frac{t^2}{3} \int_1^\infty (\xi^2 - 1)^{3/2} e^{-t\xi} d\xi = \frac{1}{2t^2} \int_0^\infty \xi \exp\{-\frac{1}{2}(\xi + t^2/\xi)\} d\xi \tag{B.5}$$

$$K_0(t) + K_1(t) = \int_1^\infty \sqrt{\frac{\xi + 1}{\xi - 1}}\, e^{-t\xi} d\xi = \frac{e^{-t}}{t} \int_0^\infty \sqrt{1 + 2t/\xi}\, e^{-\xi} d\xi \tag{B.6}$$

$$K_2(t) - K_1(t) = t \int_1^\infty (\xi - 1)^{3/2} (\xi + 1)^{1/2} e^{-t\xi} d\xi = \frac{e^{-t}}{t^2} \int_0^\infty \xi^{3/2} (\xi + 2t)^{1/2} e^{-\xi} d\xi \tag{B.7}$$

[1] G. N. Watson: A Treatise on the Theory of Bessel Functions. London: Cambridge Univ. Press. 1944.

[2] The functions are tabulated e.g. in M. Abramowitz and I. A. Stegun: Handbook of Mathematical Functions. New York: Dover Publ. Co. 1968.

Eq.(B.2) yields for n = 1 :

$$K_2(t) = \frac{2}{t} K_1(t) + K_o(t) \tag{B.8}$$

The derivatives of the functions are given by

$$K_o'(t) = -K_1(t) \tag{B.9}$$

$$K_1'(t) = -K_o(t) - \frac{1}{t} K_1(t) = \frac{1}{t} K_1(t) - K_2(t) \tag{B.10}$$

$$K_2'(t) = -K_1(t) - \frac{2}{t} K_2(t) \tag{B.11}$$

Approximations for large values of the argument, $t \gg 1$, are given by

$$K_o(t) = \sqrt{\frac{\pi}{2t}}\, e^{-t}(1 - \frac{1}{8t} + - \cdots) \tag{B.12}$$

$$K_1(t) = \sqrt{\frac{\pi}{2t}}\, e^{-t}(1 + \frac{3}{8t} + - \cdots) \tag{B.13}$$

$$K_2(t) = \sqrt{\frac{\pi}{2t}}\, e^{-t}(1 + \frac{15}{8t} + - \cdots) \tag{B.14}$$

Approximations for small values of the argument, $t \ll 1$, are

$$K_o(t) = \{\ln \frac{2}{t} - 0.5772\cdots\}(1 + \frac{t^2}{4} + - \cdots) + \frac{t^2}{4} + - \cdots \tag{B.15}$$

$$K_1(t) = \frac{1}{t} - (\frac{t}{2} \ln \frac{2}{t})\,(1 + \frac{t^2}{8} + - \cdots) + \frac{t}{2}\, 0.072\cdots + - \cdots \tag{B.16}$$

$$K_2(t) = \frac{2}{t^2} - \frac{1}{2} + - \cdots \tag{B.17}$$

The functions $K_o(t)$, $K_1(t)$, and $K_2(t)$ are shown in Fig.B.1.
For computer calculations the following representations of K_o and K_1 are given where $x = 2/t$:

For $0 < t \leqslant 2$ (i.e. $x \geqslant 1$) :

$$K_o(t) = -k_{oo} + \ln x + \sum_{n=1}^{6} (k_{on} + i_{on} \ln x)/x^{2n} \tag{B.18}$$

$$K_1(t) = x/2 - \sum_{n=1}^{6} (k_{1n} + i_{1n} \ln x)/x^{2n-1} \tag{B.19}$$

For $2 \leqslant t < \infty$ (i.e. $0 \leqslant x \leqslant 1$)

$$K_o(t) = \frac{e^{-2/x}}{\sqrt{2/x}} \sum_{n=0}^{6} K_{on} x^n \tag{B.20}$$

$$K_1(t) = \frac{e^{-2/x}}{\sqrt{2/x}} \sum_{n=0}^{6} K_{1n} x^n \tag{B.21}$$

where the constants $k_{oo} \cdots K_{ln}$ are given by the following table [ref. 2, p. 379] :

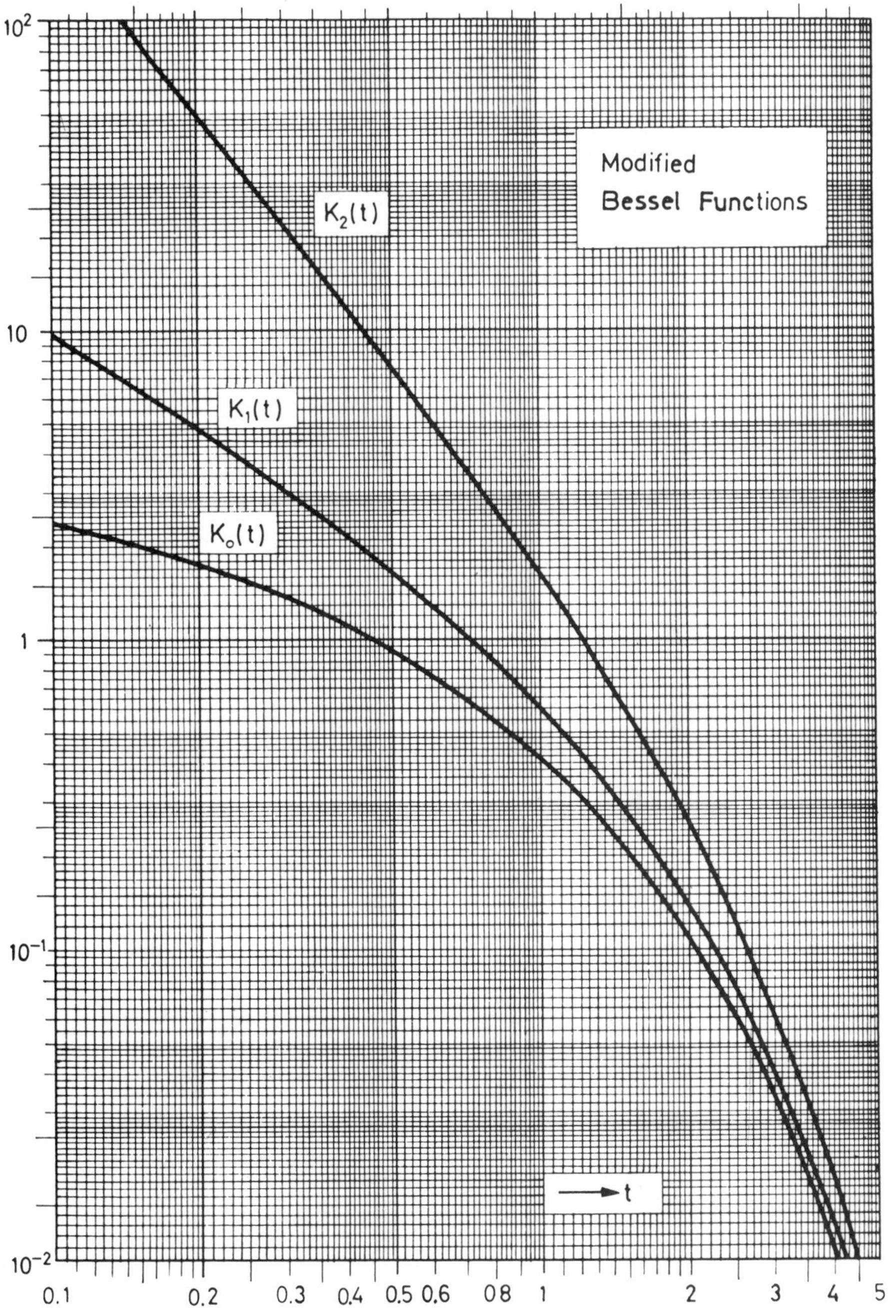

Fig.B1: Modified Bessel functions.

Table of constants.

n	k_{on}	i_{on}	k_{1n}	i_{1n}	K_{on}	K_{1n}
0	0.5772 1566	—	—	—	+1.2533 1414	+1.2533 1414
1	0.4227 8420	0.9999 9941	−7.7215 72x10^{-2}	1.0000 0000	−0.0783 2358	+0.2349 8619
2	0.2306 9756	2.5000 3043x10^{-1}	+3.3639 29x10^{-1}	4.9999 9824x10^{-1}	+0.0218 9568	−0.0365 5620
3	0.0348 8590	2.7772 1427x10^{-2}	+9.0784 485x10^{-2}	8.3334 0711x10^{-2}	−0.0106 2446	+0.0150 4268
4	0.0026 2698	1.7411 1598x10^{-3}	+0.9597 01x10^{-2}	6.9432 9756x10^{-3}	+0.0058 7872	−0.0078 0353
5	0.0001 0750	6.7176 1776x10^{-5}	+5.5202 x10^{-4}	3.4809 2403x10^{-4}	−0.0025 1540	+0.0032 5614
6	0.0000 0740	2.4264 6185x10^{-6}	+2.343 x10^{-5}	1.1229 2483x10^{-5}	+0.0005 3208	−0.0006 8245

The accuracy is

Eq.(B.18) : $|\epsilon| < 1\text{x}10^{-8} + 1.6\text{x}10^{-7} \ln(2/t)$

Eq.(B.19) : $|\epsilon| < 8\text{x}10^{-9} \{1 + t^2 \ln(2/t)\}$

Eq.(B.20) : $|\epsilon| < 1.9\text{x}10^{-7}\, t^{-1/2} e^{-t}$

Eq.(B.21) : $|\epsilon| < 2.2\text{x}10^{-7}\, t^{-1/2} e^{-t}$

Appendix C

Methods of Resistivity and Hall Effect Measurements

The voltage drop across a semiconductor sample carrying an electric current always contains unknown metal-semiconductor contact potentials unless the voltage drop is measured between potential probes as shown in Fig.C.1. For routine determinations of low resistivities a simplified four-point method may be applied: 4 metal pins at equal distance D are pressed by springs against the semiconductor sample. If the outer pins carry a current of intensity I a voltage drop is measured between the inner probes of magnitude V. Assuming the semiconductor sample to be much thicker than the pin distance D, the resistivity ρ is given by [1].

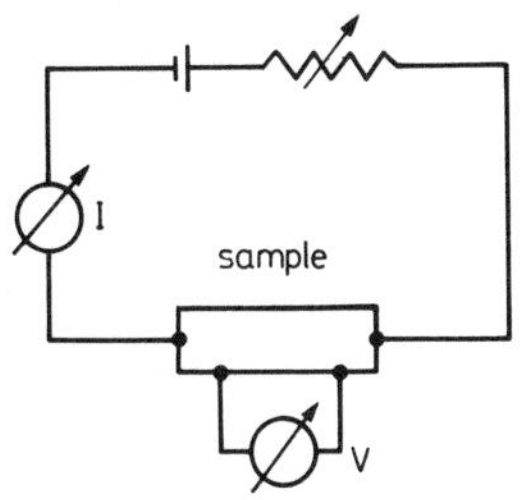

Fig.C.1 Resistivity measurement with potential probes.

$$\rho = 2\pi DV/I \tag{C.1}$$

Let us prove this relation by considering the fact that a current I carried by the lower half-space completely filled by the semiconductor material causes the same voltage drop as a current 2I carried by the full-space also assumed to be completely filled. Therefore, in a distance r from the current carrying electrode, the current density $j = 2I/4\pi r^2$ and the electric field intensity

$$E(r) = \rho j = \rho\, 2I/4\pi r^2 \tag{C.2}$$

Hence, the voltage drop between the inner probes in Fig.C.2 is given by

$$V' = \int_{2D}^{D} E dr = \rho I/4\pi D \tag{C.3}$$

Since there is a current drain at the electrode farthest to the right as shown in Fig.C.2, this causes a second voltage drop

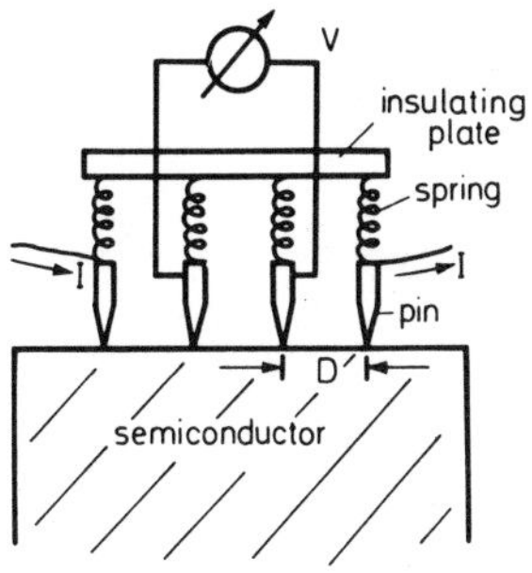

Fig.C.2 4-point probe resistivity measurement.

[1] L. Valdes, Proc. IRE 42 (1954) 420.

$$V'' = \rho(-I)/4\pi(-D) \tag{C.4}$$

which turns out to be equal to the first voltage drop, V'. The total voltage V = V' + V'' yields ρ as given by Eq.(C.1).

In this calculation the current is thought of first going to infinity from the first electrode causing V' and then going from there to the second electrode, causing V''. In potential theory it is proved that this is equivalent to the real current flow [2].

In case that the sample is a plate of thickness d much smaller than the probe distance D, the current density is only $j = I/2\pi rd$ which yields

$$V' = -\int_{2D}^{D} E dr = -\rho I/2\pi d \int_{2D}^{D} dr/r = \rho I \cdot \ln 2/2\pi d = V'' = V/2 \tag{C.5}$$

and

$$\rho = \pi d V/(I \cdot \ln 2) \tag{C.6}$$

For convenience more often a square arrengement of the four probes is used rather than the linear arrangement. The diagonal distance between two probes is $D\sqrt{2}$ which yields for the thick sample

$$V = -2 \int_{D\sqrt{2}}^{D} E dr = (\rho I/\pi)\left(\frac{1}{D} - \frac{1}{D\sqrt{2}}\right) = (\rho I/\pi D)(1 - 1/\sqrt{2}) \tag{C.7}$$

and

$$\rho = \pi D V/\{I(1 - 1/\sqrt{2})\} \tag{C.8}$$

and for the thin sample

$$V = 2\{\rho I/2\pi d\} \ln(D\sqrt{2}/D) = \rho I \cdot \ln 2/2\pi d \tag{C.9}$$

and

$$\rho = 2\pi d V/(I \cdot \ln 2) \tag{C.10}$$

The influence of the distance between probes and sample edge has been considered by Keywell [3]. An extension of these equations to the case of anisotropic semiconductors has been given by Wasscher [4].

Of great practical importance is a method by van der Pauw [5]. At the circumference of a plane parallel disk of thickness d, 4 point contacts are attached which in Fig.C.3 are labeled A-B-C-D. From A to B there is a current flow of magnitude I_{AB}. Between C and D there is a voltage drop V_{CD}. Let us define a "resistance"

$$R_{AB,CD} = |V_{CD}|/I_{AB} \tag{C.11}$$

[2] See e.g. J. A. Stratton: Electromagnetic Theory. New York: McGraw-Hill. 1941.
[3] F. Keywell, Rev. Sci. Instr. 31 (1960) 833.
[4] J. D. Wasscher, Philips Res. Repts. 16 (1961) 301.
[5] L. J. van der Pauw, Philips Techn. Rdsch. 20 (1958/9) 230.

In a second experiment we have a current I_{BC} from B to C and a voltage drop V_{AD} between A and D. We define another "resistance"

$$R_{BC,DA} = |V_{DA}|/I_{BC} \tag{C.12}$$

Van der Pauw proved that the resistivity ρ of the sample is given by

$$\exp(-\pi \frac{d}{\rho} R_{AB,CD}) + \exp(-\pi \frac{d}{\rho} R_{BC,DA}) = 1 \tag{C.13}$$

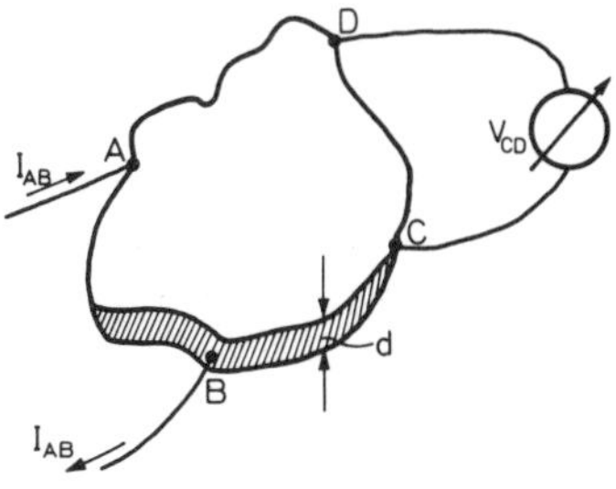

Fig.C.3 Van der Pauw arrangement.

This equation cannot be solved for ρ analytically. However, if we define a factor f by the relation

$$\rho = \frac{\pi}{\ln 2}\, d\, \frac{R_{AB,CD} + R_{BC,DA}}{2}\, f \tag{C.14}$$

we obtain from Eq.(C.13)

$$\cosh\left\{\frac{\ln 2}{f} \cdot \frac{R_{AB,CD}/R_{BC,DA} - 1}{R_{AB,CD}/R_{BC,DA} + 1}\right\} = \frac{1}{2}\exp\left(\frac{\ln 2}{f}\right) \tag{C.15}$$

which can be solved for the ratio $R_{AB,CD}/R_{BC,DA}$. If we give f arbitrarily values between 0 and 1 we obtain values of this ratio which are listed in Table 7.

Table 7. Van der Pauw factor.

$R_{AB,CD}/R_{BC,DA}$	1	2	5	10	20	50	100	200	500	1000
f	1	0.95	0.81	0.69	0.59	0.46	0.40	0.34	0.29	0.25

Eq.(C.14) thus allows the determination of ρ from two resistance measurements at the same sample. The prerequisite of point contacts can be somewhat diminished by a sample shape shown in Fig.C.4. Such a sample is called "clover-shaped". Because of the 4-fold symmetry the resistance ratio $R_{AB,CD}/R_{BC,DA}$ should be unity. If the experiment yields a ratio > 2 it is common practice to renew the contacts. If this is of no help there is some reason to believe that the sample is not homogeneously doped and may even contain local p-n junctions.

Fig.C.4 Clover shaped sample for van der Pauw measurements.

Now let us prove Eq.(C.13). At first we assume a sample shape as shown in Fig.C.5: The plane-parallel plate of thickness d is

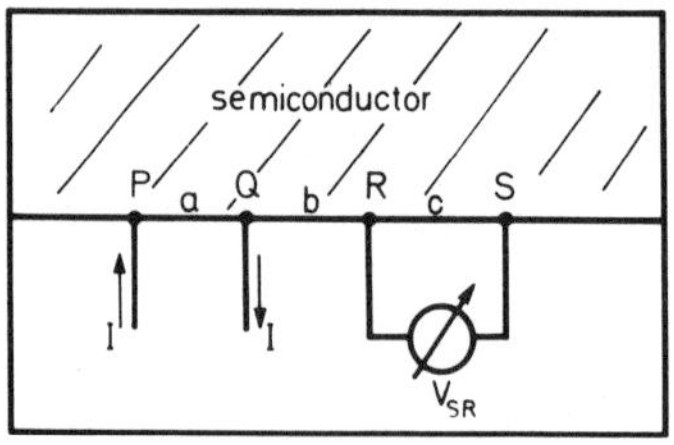

Fig.C.5 Sample shape assumed for a proof of Eq.(C.13).

assumed to completely fill the upper semi-infinite half-plane. The current carrying point contacts are denoted by P and Q. Between potential probes at R and S a voltage drop V_{SR} is observed. Current lines and potentials remain unaffected if supplementing the upper half-plane by the lower one and at the same time doubling the current. The current density at a distance r from P is therefore $j = 2I/2\pi rd$ and the electric field is $E = \rho j$. This yields a voltage drop between R and S of magnitude

$$V'_{SR} = \int_S^R E dr = \frac{\rho I}{\pi d} \int_S^R \frac{dr}{r} = \frac{\rho I}{\pi d} \ln \frac{\overline{PR}}{\overline{PS}} = \frac{\rho I}{\pi d} \ln \frac{a+b}{a+b+c} \tag{C.16}$$

a, b, and c are shown in Fig.C.5. Since the current 2I leaves the conductor at Q this causes a second voltage drop

$$V''_{SR} = -\frac{\rho I}{\pi d} \ln \frac{\overline{QR}}{\overline{QS}} = -\frac{\rho I}{\pi d} \ln \frac{b}{b+c} \tag{C.17}$$

The total voltage drop $V_{SR} = V'_{SR} = V''_{SR}$ is thus given by

$$V_{SR} = -\frac{\rho I}{\pi d} \ln \frac{(a+b+c)b}{(a+b)\cdot(b+c)} \tag{C.18}$$

We divide both sides by I and obtain $R_{PQ,RS}$. Solving for the argument of the logarithm yields

$$\exp\{-\frac{\pi d}{\rho} R_{PQ,RS}\} = \frac{(a+b+c)b}{(a+b)\cdot(b+c)} \tag{C.19}$$

In a second expriment Q and R are current leads and S and P potential probes. We obtain by cyclic permutation where e.g. $\overline{PQ} = a$ is replaced by $\overline{QR} = b$ and $\overline{QS} = b+c$ by $\overline{RP} = \overline{PR} = a+b$,

$$\exp\{-\frac{\pi d}{\rho} R_{QR,SP}\} = \frac{a\cdot c}{(b+c)\cdot(a+b)} \tag{C.20}$$

The right-hand sides of the last two equations add to unity which proves van der Pauw's equation for the sample shape given by Fig.C.5. For an arbitrary sample shape we introduce a complex function

$$w = f(x+iy) = u + iv; \quad u = u(x, y); \quad v = v(x, y) \tag{C.21}$$

such that u(x, y) is the potential field in the upper half-plane. The electric field intensity

$$\vec{E} = (-\frac{\partial u}{\partial x}, -\frac{\partial u}{\partial y}) \tag{C.22}$$

is determined by u(x, y). The current I which crosses the connecting line between two arbitrary points T_1 and T_2 is given by

$$I = \frac{d}{\rho} \int_{T_1}^{T_2} (\vec{E}\, d\vec{n}) = \frac{d}{\rho} \int_{T_1}^{T_2} (-\frac{\partial u}{\partial y}\, dx + \frac{\partial u}{\partial x}\, dy) \qquad (C.23)$$

where $d\vec{n} = (-dy, dx)$ is normal to the path differential $d\vec{s} = (dx, dy)$. We can easily solve this integral by means of the Cauchy-Riemann relations

$$\frac{\partial u}{\partial x} = \frac{\partial v}{\partial y} \;;\; \frac{\partial u}{\partial y} = -\frac{\partial v}{\partial x} \qquad (C.24)$$

and obtain for I

$$I = \frac{d}{\rho} \int_{T_1}^{T_2} (\frac{\partial v}{\partial x}\, dx + \frac{\partial v}{\partial y}\, dy) = \frac{d}{\rho}\, \{v(T_2) - v(T_1)\} \qquad (C.25)$$

The product of ρ and the current density $\vec{j}$ has components

$$\rho\vec{j} = (-\frac{\partial v}{\partial y}, \frac{\partial v}{\partial x}) \qquad (C.26)$$

v remains constant if in Fig.C.5 we go along the real x-axis from $-\infty$ to $+\infty$ which marks the boundary of the sample, until we pass the point P where the current enters the sample. Here, according to Eq.(C.25), v is increased by the amount $I\rho/d$. When passing the second current lead Q, v is decreased by the same amount.

The upper half-plane can be mapped on the sample shown in Fig.C.4 by conformal mapping introducing an analytical function t(z):

$$w = f(z) = f(z(t)) = h(t) = l + im \qquad (C.27)$$

where l and m are real and imaginary parts, respectively, of h(t). Now m remains constant along the sample edge except when passing the currents leads at A and B just as did v in Fig.C5. At point A, m is increased by $I\rho/d$, at B it is reduced by the same amount. Therefore l in the t-plane has to be interpreted as a potential just as u in the z-plane. $V_{DC} = V_{SR}$. From the theory of conformal mapping it follows that $(d/\rho)R_{AB,CD}$ is invariant [5]. The same is true for $(d/\rho)R_{BC,DA}$. This proves Eq.(C.13) for an arbitrary sample shape.

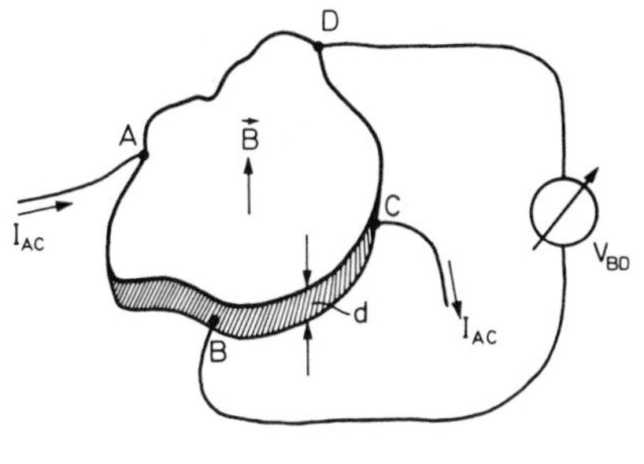

Fig.C.6 Hall effect arrangement according to van der Pauw.

For a Hall measurement the arrangement shown in Fig.C3 is slightly modified (Fig.C6). At C rather than at B the current leaves the sample and the voltage between B and D, V_{BD}, is measured. The resistance given by

$$R_{AC,BD} = V_{BD}/I_{AC} \qquad (C.28)$$

is changed by an amount $\Delta R_{AC,BD}$ when a homogeneous magnetic field is applied perpendicular to the plate-shaped sample. The Hall coefficient

R_H is obtained from

$$R_H = \frac{d}{|\vec{B}|} \Delta R_{AC,BD} \tag{C.29}$$

This at first sight somewhat surprising relationship can be proved quite easily. Due to the deflection of carriers in the magnetic field opposite sample edges are charged in such a way that the current lines are the same as they were before the application of the magnetic field. The potentials at the points B and D are changed due to the Hall field $\vec{E}_t$ perpendicular to j, while the current lines remain unchanged. In Fig.C7 the points B and D are on the dashed potential line and the points B' and B are on the same current line which goes along the sample edge. For a path element $d\vec{s}$ along the current line where $\vec{E}_t$ is perpendicular to $d\vec{s}$, the integral

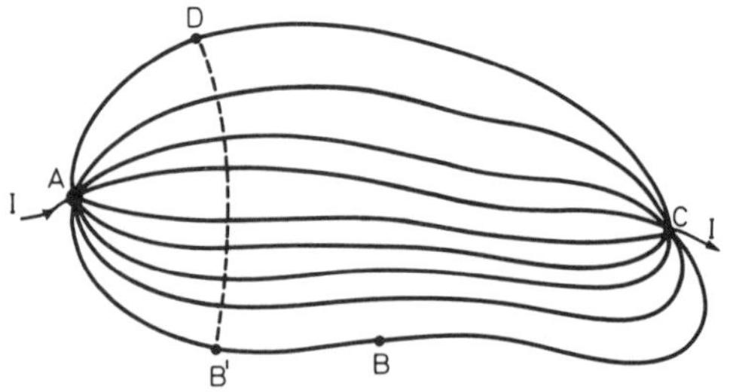

Fig.C.7 Current lines in a disk-shaped sample under the influence of a transverse magnetic field.

$$\int_{B'}^{B} (\vec{E}_t d\vec{s}) = 0 \tag{C.30}$$

which makes the integral from D to B equal to that from D to B'. Since the Hall voltage between opposite points D and B' is given by $R_H \cdot I \cdot |\vec{B}| / d$, the resistance change is given by

$$\Delta R_{AC,BD} = \frac{\Delta V_{BD}}{I_{AC}} = R_H |\vec{B}|/d \tag{C.31}$$

where the current I_{AC} is determined by

$$I_{AC} = d \int_{D}^{B'} |\vec{j}| \, ds \tag{C.32}$$

Solving Eq.(C.31) for R_H yields Eq.(C.29).

An extension of Eqs.(C.13) and (C.29) for anisotropic conductors has been given by van der Pauw [6]. We denote by x_1, x_2, and x_3 the main axes of the resistivity tensor ρ such that it is a diagonal tensor with elements ρ_1, ρ_2, and ρ_3. If the sample thickness in x_i-direction is d and ρ is replaced by $\sqrt{\rho_j \rho_k}$ with (i, j, k) = (1, 2, 3) cyclically permuted, Eq.(C.13) is still valid. From 3 measurements with d in the 3 x_i-directions, using 3 differently oriented samples from the same single crystal, the resistivities ρ_1, ρ_2, and ρ_3 can be determined. With the same samples, Hall measurements with 3 different directions of the magnetic field in each case, totalling to 9 measurements altogether, yield a complete determination of the Hall tensor.

[6] L. J. van der Pauw, Philips Res. Repts. 16 (1961) 187; J. Hornstra and L. J. van der Pauw, J. Electronics and Control 7 (1959) 169.

Appendix D

Optical Transmittivity and Reflectivity; Kramers Kronig Relations

We consider a plane electromagnetic wave propagating in a semiconductor assumed to be non-magnetic and optically isotropic [1]. Let us take the z-axis of a Cartesian coordinate system in the direction of propagation. The electric and magnetic field vectors in complex notation depend on position z and time t according to

$\vec{E}$ and $\vec{H} \propto e^{i(\omega t - Nz/\bar{\lambda})}$

where ω is the angular frequency, $\bar{\lambda} = \lambda/2\pi$, λ is the free-space wavelength, $N = n - ik$, n is the refractive index and k is the extinction coefficient. The normal component of the wave incident at an angle φ on a surface has an $\vec{E}$-vector perpendicular to the plane of incidence. Its reflectivity (ratio of reflected energy flux to incident energy flux) for an "infinitely thick" plate is given by

$$r_n = \frac{(\bar{a}^2 + \bar{\beta}^2) + \cos^2\varphi - 2\bar{a}\cos\varphi}{(\bar{a}^2 + \bar{\beta}^2) + \cos^2\varphi + 2\bar{a}\cos\varphi} \tag{D.1}$$

where

$$\bar{a}^2 + \bar{\beta}^2 = \sqrt{(n^2 - k^2 - \sin^2\varphi)^2 + (2nk)^2} \tag{D.2}$$

and

$$\bar{a} = \sqrt{\frac{1}{2}\{n^2 - k^2 - \sin^2\varphi + (\bar{a}^2 + \bar{\beta}^2)\}} \tag{D.3}$$

The reflectivity of the transverse component ($\vec{E}$ parallel to the plane of incidence) is given by

$$r_p = r_n \cdot \frac{(\bar{a}^2 + \bar{\beta}^2) + \sin^2\varphi\tan^2\varphi - 2\bar{a}\sin\varphi\tan\varphi}{(\bar{a}^2 + \bar{\beta}^2) + \sin^2\varphi\tan^2\varphi + 2\bar{a}\sin\varphi\tan\varphi} \tag{D.4}$$

At $\varphi = 45°$, $r_p = r_n^2$ and a measurement with unpolarized light yields for the reflectivity $r = \frac{1}{2}(r_n + r_n^2)$.

---

[1] For a derivation of the following relations see e.g. M. Born: Optik. Berlin: Springer. 1933; M. Born and E. Wolf: Principles of Optics, 2nd ed. Oxford: Pergamon. 1959; T. S. Moss: Optical Properties of Semiconductors. London: Butterworth. 1959.

There is a minimum of the function r_p vs φ at the "Brewster angle" φ_B. For $k = 0$ the minimum is zero and the Brewster angle is given by $\tan \varphi_B = n$. In this case the reflected beam is perpendicular to the refracted beam. Fig.D.1 shows plots of r_n, r_p, and r_p/r_n vs φ for the case $n = 3$ and $k = 1$ which has been arbitrarily chosen. r_n can be approximated by

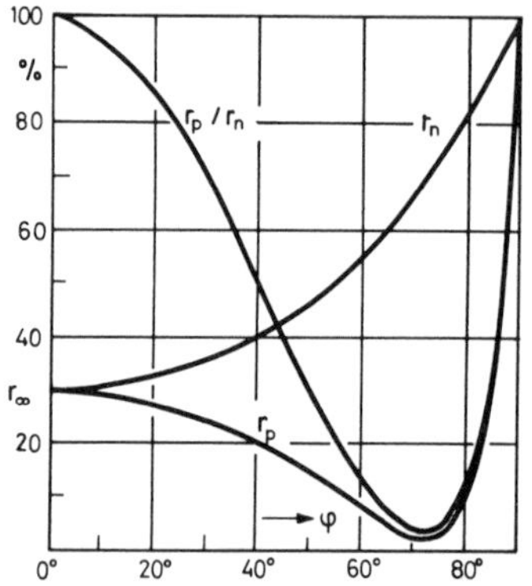

Fig.D.1 Theoretical reflectance as a function of the angle of incidence for n = 3; k = 1 (after T. S. Moss: Optical Properties of Semiconductors. London: Butterworth. 1959).

$$r_n = \frac{(n - \cos \varphi)^2 + k^2}{(n + \cos \varphi)^2 + k^2} \tag{D.5}$$

For vertical incidence, $\varphi = 0$, there is no distinction between normal and parallel components; in this case we denote the reflectivity of the infinitely thick plate by r_∞. Eq. (D.1) yields rigorously for arbitrary values of n and k

$$r_\infty = \frac{(n - 1)^2 + k^2}{(n + 1)^2 + k^2} \tag{D.6}$$

For a plate of finite thickness d, we introduce the absorption coefficient*) $a = 2k/\lambda\!\!\!^{-} = 4\pi k/\lambda$, the phase shift $\delta = nd/\lambda\!\!\!^{-}$ and the quantities

$$\gamma = \ln \sqrt{1/r_\infty} \tag{D.7}$$

and

$$\tan \psi = \frac{2k}{n^2 + k^2 - 1} \tag{D.8}$$

We consider only the case of vertical incidence. The reflectivity is given by

$$R = \frac{\sinh^2(\frac{1}{2} ad) + \sin^2 \delta}{\sinh^2(\frac{1}{2} ad + \gamma) + \sin^2(\delta + \psi)} \tag{D.9}$$

and the transmittivity by

$$T = \frac{\sinh^2 \gamma + \sin^2 \psi}{\sinh^2(\frac{1}{2} ad + \gamma) + \sin^2(\delta + \psi)} \tag{D.10}$$

If there is no absorption, $k = 0$, the sum $R + T = 1$. The dependence of R and T on λ yields interference effects. For negligible absorption

$$T_{max}/T_{min} = (n^2 + 1)^2/4n^2 \tag{D.11}$$

This relation is useful for a determination of the index of refraction, n.

For non-polished surfaces or polychromatic light the average reflectivity

*) For metals $2/a$ is known as the "skin depth".

$$<R> = r_\infty \{1 + <T> \exp(-ad)\} \tag{D.12}$$

and the average transmittivity

$$<T> = 2\,\frac{\sinh^2\gamma + \sin^2\psi}{\sinh(ad + 2\gamma)} \tag{D.13}$$

may be used for a determination of the absorption coefficient a. For $k \ll n$ one may neglect ψ and obtain a from

$$2\sinh(ad) = \frac{(1-<R>)^2}{<T>} - <T> \tag{D.14}$$

Fig.D.2 is a plot of $<T>$ vs $<R>$ with r_∞ and $\exp(-ad)$ as parameters [2]. From a determination of r_∞ and a one can obtain the refractive index, n, and the extinction coefficient k.

Neglecting ψ in Eq.(D.13) this can be written

$$<T> = \frac{(1-r_\infty)^2\, e^{-ad}}{1-r_\infty^2\, e^{-2ad}}\,. \tag{D.15}$$

which for $(r_\infty e^{-ad})^2 \ll 1$ further simplifies to

$$<T> = (1-r_\infty)^2\, e^{-ad} \tag{D.16}$$

In this approximation, $<R> = r_\infty$.

For the case of strong absorption, $k \approx n$, in a thin plate where $d \ll \lambda/(2\pi n)$, the "Woltersdorff thickness" [3], there is no dispersion in reflection and transmission.

The Kramers-Kronig relations [4] have been used in Chap.11a for the evaluation of absorption spectra over a wide frequency range. It may be of interest to note that they are also used in e.g. the theory of electrical circuit analysis and are of a very general nature [5]. The problem has been treated e.g. by Stern [6].

Maxwell's equations relate the complex index of refraction, $N = n-ik$, to the relative dielectric constant, κ. For the following purposes it is useful to introduce a complex frequency-dependent dielectric constant, $\bar{\kappa}(\omega)$, which in terms of the index of refraction, n, and the extinction coefficient, k, is given by

$$\bar{\kappa}(\omega) = \kappa_r - i\kappa_i \tag{D.17}$$

---

[2] W. Nazarewicz, P. Rolland, E. da Silva, and M. Balkanski, Appl. Optics 1 (1962) 369.

[3] W. Woltersdorff, Zeitschr. f. Physik 91 (1934) 230.

[4] R. de L. Kronig, J. Opt. Soc. Am. 12 (1926) 547.

[5] H. W. Bode: Network Analysis and Feedback Amplifier Design. New York: Van Nostrand. 1945.

[6] F. Stern: Solid State Physics (F. Seitz and D. Turnbull, eds.), Vol. 15, p. 328. New York: Acad. Press. 1963.

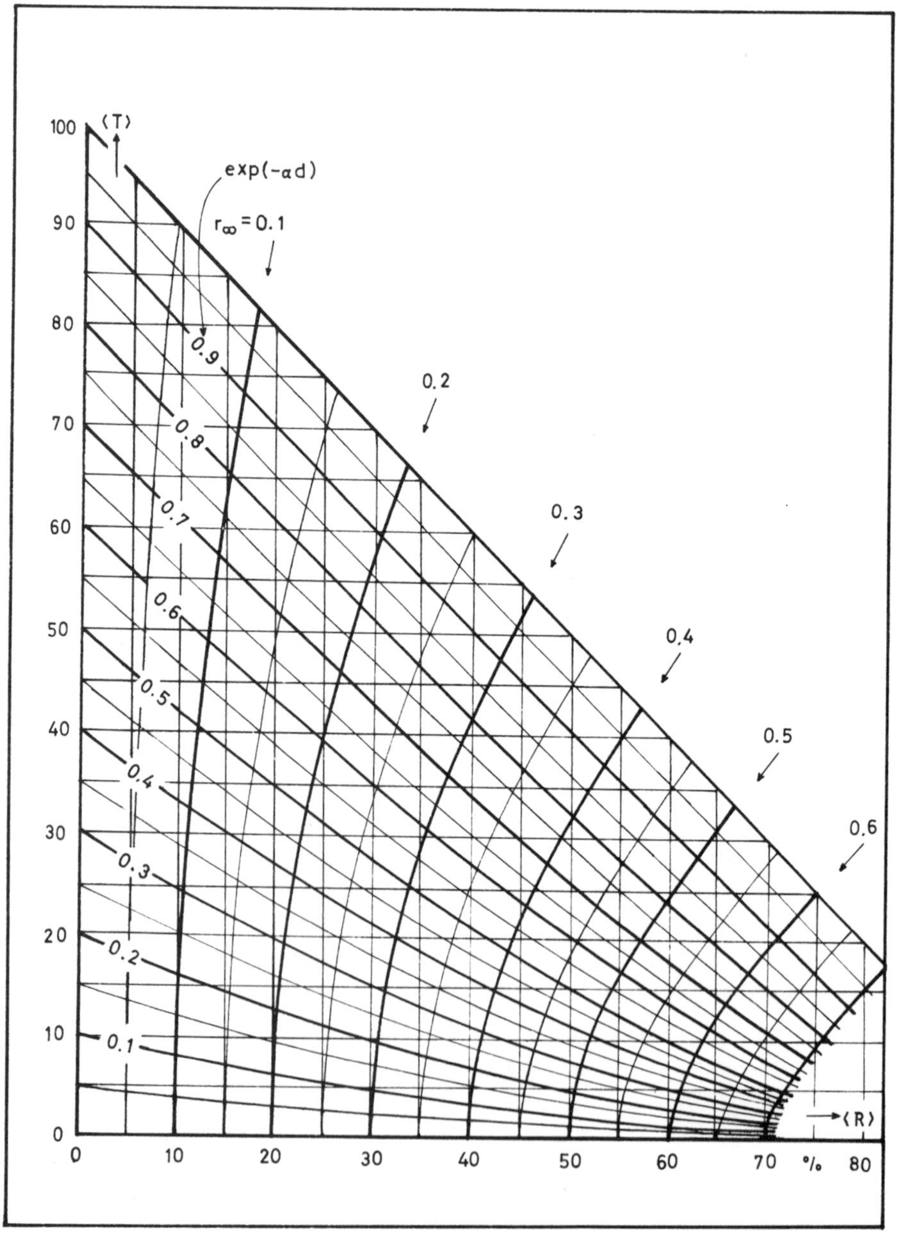

Fig.D.2 Abac chart for a determination of a and r_{∞} from the observed average values of transmittivity $\langle T \rangle$ and reflectivity $\langle R \rangle$; e.g. $\langle T \rangle = 31\,\%$ and $\langle R \rangle = 35.6\,\%$ yields $\exp(-\alpha d) = 0.6$ and $r_{\infty} = 0.3$ (after Nazarewicz et al., ref. 2).

where

$$\kappa_r = n^2 - k^2 ; \quad \kappa_i = 2nk \tag{D.18}$$

The dielectric displacement vector, $\vec{D} = \bar{\kappa}\kappa_0\vec{E}$, is thus also complex and frequency dependent, $\vec{D} = \vec{D}(\omega)$. A Fourier transformation

$$D(t) = \int_{-\infty}^{\infty} D(\omega)\, e^{-i\omega t} d\omega = \int_{0}^{\infty} \{D(\omega)e^{-i\omega t} + D(-\omega)\, e^{i\omega t}\} d\omega \tag{D.19}$$

gives D as a function of time, t. Since D(t) should be real, we find $D(-\omega)$ as the complex conjugate of $D(\omega)$:

$$D(-\omega) = D^*(\omega) \tag{D.20}$$

and consequently

$$\bar{\kappa}(-\omega) = \bar{\kappa}^*(\omega) \tag{D.21}$$

which for the real and imaginary parts of $\bar{\kappa}$ results in

$$\kappa_r(-\omega) = \kappa_r(\omega) ; \quad \kappa_i(-\omega) = -\kappa_i(\omega) \tag{D.22}$$

One can show that the Kramers-Kronig relations between κ_r and κ_i hold, namely:

$$\kappa_r(\omega) - 1 = \frac{2}{\pi} \int_0^{\infty} \frac{\omega' \kappa_i(\omega') - \omega \kappa_i(\omega)}{\omega'^2 - \omega^2} d\omega' \tag{D.23}$$

and

$$\kappa_i(\omega) = -\frac{2}{\pi}\, \omega \int_0^{\infty} \frac{\kappa_r(\omega') - \kappa_r(\omega)}{\omega'^2 - \omega^2} d\omega' \tag{D.24}$$

Since κ_r and κ_i are given by Eq.(D.18) as functions of n and k the Kramers-Kronig relations yield the surprising result that the index of refraction and the extinction coefficient depend on one another. A special case is a dielectric substance which at low frequencies is an insulator (k = 0) with an index of refraction, $n(0) = n_0$, given by

$$n_0^2 - 1 = \frac{2}{\pi} \int_0^{\infty} \frac{2nk}{\omega'} d\omega' = \frac{2}{\pi} \int_0^{\infty} \frac{2nk}{\lambda} d\lambda \tag{D.25}$$

The static dielectric constant $\kappa = n_0^2$ is > 1 only if there is absorption of light somewhere in the electromagnetic spectrum.

The Kramers-Kronig relations (D.23) and (D.24) are strictly valid only for non-conducting media. Toll [7]has shown that for non-vanishing conductivity the dispersion relations have to be modified.

A relation even more useful for practical applications is given by

---

[7]J. S. Toll, Phys. Rev. 104 (1956) 1760.

$$\psi(\omega) = \frac{1}{\pi} \int_0^\infty \ln \left| \frac{\omega' - \omega}{\omega' + \omega} \right| \frac{d\gamma(\omega')}{d\omega'} \, d\omega' \tag{D.26}$$

where ψ and γ are given by Eqs.(D.8) and (D.7), respectively. Since for small absorption $\psi \approx 2k/(n^2 - 1)$ and due to the logarithmic factor only a small range of frequencies at ω contributes significantly to the integral, one can estimate the extinction coefficient from an observed frequency dependence of the reflectivity, $r_\infty = e^{-2\gamma}$, of a "thick" plate. In the approximation $r_\infty \approx (n-1)^2/(n+1)^2$ this also yields a value for the denominator in ψ, $(n^2 - 1)$, at the frequency ω.

The extension of the Kramers-Kronig analysis to oblique incidence has been given by Roessler [5]. For methods of practical application see e.g. Greenaway and Harbeke [6].

[8] D. M. Roessler, Brit. J. Appl. Phys. 16 (1965) 1359.

[9] D. L. Greenaway and G. Harbeke: Optical Properties and Band Structure of Semiconductors, p. 12/13. Oxford: Pergamon. 1968.

Appendix E

Measurement of Typical Semiconductor Properties

Conductivity type : Sign of the Hall effect, Seebeck effect or Righi-Leduc effect. Forward direction of Schottky barrier or p-n junction. Intraband Faraday effect. (Problems of interpretation arise at near-intrinsic conduction if electrons and holes have very different mobilities.)
Band gap : Conductivity or Hall effect as a function of temperature at intrinsic conduction. Edge of fundamental optical absorption or photoconduction. Interband transition in a magnetic field.
Band structure : Interband optical absorption. Near band edge : Magnetoresistance. Voigt effect. Hot-electron galvanomagnetic effects.
Carrier concentration : Hall effect and conductivity at extrinsic conduction (For intrinsic conduction the mobility ratio must be known; it can occasionally be extrapolated from the extrinsic range). Intraband Faraday and Voigt effect.
Fermi level : Seebeck effect and effects listed under "Carrier concentration".
Carrier mobility : Hall effect and conductivity. (See also "Carrier diffusion " in connection with the Einstein relation).
Carrier diffusion : Drift experiment (Haynes Shockley) at extrinsic conduction.
Effective mass : Cyclotron resonance. Intraband Faraday effect. Plasma reflection. Shubnikov-de Haas effect. Magnetophonon effect. Interband transitions in a magnetic field. Burstein shift. Magneto-Kerr effect.
Momentum relaxation time : Microwave and far-infrared absorption. Cyclotron resonance line width. Microwave Faraday effect.
Energy relaxation time : Microwave and far-infrared harmonic mixing (nonlinear optics). Frequency conversion. Microwave "photoconduction" vs frequency.
Lifetime of minority carriers and diffusion length : Haynes Shockley experiment. Photoconductivity and PEM effect. Photovoltaic effect.
Surface recombination velocity : Photovoltaic effect. Dember effect.
Impurities : Influence of ionized and neutral impurity scattering on the temperature dependence of the mobility or on the cyclotron resonance line width. Low-temperature electrical breakdown. Low-temperature microwave phonon attenuation. Optical absorption.
Dislocation density : Count of etch pits; X-ray topography.

Display of a p-n junction : Galvanic metal plating or electrophoresis (barium titanate powder in CCl_4) at reverse bias. Etching.
Dielectric constant : Microwave reflectivity. Helicon propagation in extremely high magnetic fields.
Crystallographic orientation : Laue diagram. Light figure method.

Numerical Values of Important Quantities

$h = (4.135708 \pm 14) \times 10^{-15}$ eVsec

$\hbar = 1.05459193\ (\pm\ 123) \times 10^{-34}$ Wattsec^2 = $6.58218292 \times 10^{-16}$ eVsec
(Planck's constant divided by 2π)

$k_B = 8.61573 \times 10^{-5}$ eV/K = 1.38046×10^{-23} Wsec/K
(Boltzmann's constant)

$|e| = 1.60219177\ (\pm\ 44) \times 10^{-19}$ Asec (electron charge)

$m_0 = 9.10955854\ (\pm\ 600) \times 10^{-35}$ Wattsec^3/cm^2
(free electron mass)

$|e|/m_0 = 1.758834 \times 10^{15}$ cm^2/Vsec^2 = 17.58834 GHz/kG
(electron charge per unit mass)

$N_{Av} = 6.0247 \times 10^{23}$/mole (Avogadro's number, physical scale)

$c = 2.997925010\ (\pm\ 330) \times 10^{10}$ cm/sec (velocity of light)

$4\pi\kappa_0\hbar c/e^2 = 137.0360221\ (\pm\ 15)$ (reciprocal fine structure constant)

$\kappa_0 = 8.859 \times 10^{-14}$ Asec/Vcm = $1/\mu_0 c^2$
(permittivity of free space)

$\mu_0 = 1.25602 \times 10^{-8}$ Vsec/Acm (permeability of free space)

$1/\kappa_0 c = \mu_0 c = 376.732$ ohm (impedance of free space)

$a_B = 4\pi\kappa_0\hbar^2/m_0 e^2 = 0.5291$ Å (Bohr radius)

$\mu_B = e\hbar/2m_0 = 5.788 \times 10^{-6}$ eV/kG (Bohr magneton)

$Ry = m_0 e^4/2(4\pi\kappa_0\hbar)^2 = 13.607$ eV (Rydberg energy)

(Errors are in the last digits)

Author Index

C

D

E

I

J

K

O

P

Q

R

Subject Index

Substance Index